QUANTITATIVE ECOTOXICOLOGY

Second Edition

QUANTITATIVE ECOTOXICOLOGY

Second Edition

Michael C. Newman

CRC Press
Taylor & Francis Group
Boca Raton London New York

CRC Press is an imprint of the
Taylor & Francis Group, an **informa** business

CRC Press
Taylor & Francis Group
6000 Broken Sound Parkway NW, Suite 300
Boca Raton, FL 33487-2742

Printed in the United States of America on acid-free paper
Version Date: 20120615

International Standard Book Number: 978-1-4398-3564-7 (Hardback)

Library of Congress Cataloging-in-Publication Data

Newman, Michael C.
Quantitative ecotoxicology / Michael C. Newman. -- 2nd ed.
p. cm.
Summary: "This book provides a quantitative treatment of the science of ecotoxicology. The first chapters consider fundamental concepts and definitions essential to understanding the fate and effects of toxicants at various levels of ecological organization as covered in the remaining chapters. Scientific ecotoxicology and associated topics are defined. The historical perspective, rationale, and characteristics are outlined for the strong inferential and quantitative approach advocated in this book. The general measurement process is discussed, and methodologies for defining and controlling variance, which could otherwise exclude valid conclusions regarding ecotoxicological endeavors, are considered. Ecotoxicological concepts at increasing levels of ecological organization are discussed in the second part of the book. Quantitative methods used to measure toxicant effects are outlined in this section. The final chapter summarizes the book with a brief discussion of ecotoxicological assessment. Numerous figures and tables accompany text, with many statistical tables found in the appendix for quick reference. Although the book primarily focuses on aquatic systems, with appropriate modification the concepts and methods can be applied to terrestrial systems"-- Provided by publisher.
Includes bibliographical references and index.
ISBN 978-1-4398-3564-7 (hardback)
1. Pollution--Environmental aspects--Measurement. 2. Environmental toxicology--Mathematics. I. Title.

QH545.A1N493 2012
363.70072--dc23 2012021096

Visit the Taylor & Francis Web site at
http://www.taylorandfrancis.com

and the CRC Press Web site at
http://www.crcpress.com

To Peg, Ben, and Ian

... what we hope ever to do with ease we may learn first to do with diligence.

—Samuel Johnson (from Green 1984*)

* Greene, D., ed., *Samuel Johnson*, Oxford, Oxford University Press, 1984.

Contents

Preface

IMPETUS FOR THIS BOOK

> It is, therefore, urged without reason, as a discouragement to writers, that there are already books sufficient in the world; that all topics of persuasion have been discussed, and every important question clearly stated and justly decided; ... [However], whatever be the present extent of human knowledge, it is not only finite, and therefore in its own nature capable of increase; but so narrow, that almost every understanding may, by a diligent application of its powers, hope to enlarge it. It is, however, not necessary, that a man forbear to write, till he has discovered some truth unknown before; he may be sufficiently useful, by only diversifying the surface of knowledge, and luring the mind by a new appearance to a second view of those beauties which it had passed over unattentively before.
>
> **—Samuel Johnson (from Murphy 1836)**

While browsing the preface to Moriarty's 1983 book, *Ecotoxicology: The Study of Pollutants in Ecosystems*, in preparation for writing the preface for the first edition of this book, I was struck by the similarity of our intentions. He suggests, and I still concur as this second edition is being prepared, that an abundance of ecotoxicological data exists, but much of it is insignificant. Considering myself an unwitting, albeit minor, contributor to this dilute database and dismayed at the prospect of mediocrity as an inescapable theme in my professional career, I dedicated considerable thought to factors contributing to this situation. Certainly, ecotoxicology is not trivial, and as so is not characterized by practitioners lacking sufficient acumen or funding. For the most part, the professionals involved in the field are well trained, well intended, and well funded. Further, the prosaic argument that ecosystems are too complex to understand to any practical extent is inconsistent with the contrastingly rapid progress made in disciplines such as molecular genetics, immunology, computer/information sciences, and physics, which deal with equally complex and less tangible subjects. Further, substantial quantitative advancement in ecology has occurred in the last 70 years, but these advances have failed *en bloc* to permeate the ecotoxicology literature. Statistics demonstrating that the majority of publications in many sciences go practically uncited (Hamilton 1990) and are seldom read by scientists outside the specific field in which they appeared provided me with a broader appreciation of the problem but no conceptual tools to improve a condition seemingly more severe in ecotoxicology than in many other fields.

I had resigned myself to the fact that all fields go through a presynthetic or descriptive phase prior to maturation. I assigned ecotoxicology (and its predecessors such as aquatic and wildlife toxicology) to this unsatisfying status until I stumbled upon an article entitled "Strong Inference" (Platt 1964). Platt's arguments regarding qualities affecting the relative rates of advancement of various disciplines suggested that solidification of ecotoxicological principles and paradigms could be greatly accelerated by adapting a stronger inferential approach. In the process, we would also become much better environmental stewards. One goal of this book is to encourage a more rigorous inferential approach. This goal is shared with the companion books, *Fundamentals of Ecotoxicology* and *Ecotoxicology: A Comprehensive Approach*, in which Platt's strong inference approach was enlarged to a strongest possible inference approach by incorporating quantitative Bayesian inference.

Each chapter of this volume treats ecotoxicology as a scientific endeavor. As is the case with all sciences, the focus is "the organization and classification of knowledge on the basis of explanatory principles" (Nagel 1961). This book emphasizes the strongest possible inferential and quantitative themes. Consequently, many aspects relevant to regulatory activities, e.g., standard methods and environmental legislation, will not be presented in balance with their importance in addressing ecotoxicological problems facing society today. By no means should this omission be considered a mute assignment of the regulatory aspects of ecotoxicology to a status inferior to those scientific. These critical topics are covered clearly and thoroughly in other volumes, such as Rand's 1995 *Fundamentals of Aquatic Toxicology*, and numerous publications by the U.S. Environmental Protection Agency (EPA), the Organization for Economic Cooperation and Development (OECD), and the United Nations (UN).

The topics covered in this book are arranged in order of increasing biological organization and scale. This being the case, early chapters may contain information that some workers may not consider sufficiently high in biological organization to be considered pertinent to ecotoxicology, i.e., not community or higher-level topics. My objection to this artificial limitation has been expressed often elsewhere. "Few ecologists would disagree that progress in ecology would have been slowed by exclusion of all but community and system level research. It seems illogical to assume that growth in this new field of ecology would not be similarly compromised by such a restriction" (Newman and McIntosh 1991). Regardless of the level examined, the intent in all chapters is to better understand the fate and effects of toxicants in the biosphere, and to frame this understanding in quantitative terms.

ORGANIZATION OF THIS BOOK

This book explores quantitative features of the science of ecotoxicology. It is revised to more neatly complement *Fundamentals of Ecotoxicology* (Newman 2010) and *Ecotoxicology: A Comprehensive Treatment* (Newman and Clements 2008). The first is an introductory textbook that provides a general description of ecotoxicology useful in graduate or upper-level undergraduate courses. The second is intended as a textbook for a more intensive or advanced course, and also as a reference book for professionals. Each is organized into chapters that move from biological themes emerging at lower to higher levels of organization. The overarching goal is to produce three interconnected books with complementary discussions of key concepts and approaches. Topics and organization of this second edition of *Quantitative Ecotoxicology** overlap coverage of the aforementioned two textbooks but from the vantage of quantitative concepts and methods. The central role played by quantification in modern science mandates this third treatment of ecotoxicology.

> The most distinct and beautiful statement of any truth must take at last the mathematical form.
>
> **—Thoreau (from Walls 1999)**

> A precise statement can be more easily refuted than a vague one, and it can be better tested. This consideration also allows us to explain the demand that qualitative statements should if possible be replaced by quantitative ones….
>
> **—Popper (1972)**

> As we enter the twenty-first century, the statistical revolution in science stands triumphant. It has vanquished determinism from all but a few obscure corners of science.
>
> **—Salsburg (2001)**

The first chapters of *Quantitative Ecotoxicology* explore fundamental concepts and definitions essential to understanding the fate and effects of toxicants at various levels of biological organization

* The first edition had the longer title *Quantitative Methods in Aquatic Ecotoxicology*, reflecting an initial bias toward freshwater ecotoxicology.

as covered in the remaining chapters. Scientific ecotoxicology and associated topics are defined in Chapter 1. The historical perspective, rationale, and characteristics are outlined there for the quantitative approach taken later in this book. The second chapter discusses the general measurement process. It considers methodologies for defining and controlling variance, which could otherwise exclude valid conclusions from ecotoxicological endeavors. Ecotoxicological concepts at increasing levels of ecological organization are discussed in Chapters 3 through 7. Quantitative methods used to measure toxicant accumulation and effects are outlined in each of these chapters. The importance of establishing type II, in addition to type I, error rates, as emphasized in the revised Chapter 5, necessitated more discussion of design issues, especially sample size and power estimation (Appendix 27). The final chapter summarizes the book with a brief discussion of ecotoxicology from a nonregulatory vantage.

Michael C. Newman
A. Marshall Acuff Jr. Professor of Marine Science
College of William and Mary
Virginia Institute of Marine Science
Gloucester Point, Virginia

REFERENCES

Hamilton, D.P. Publishing by—and for?—the numbers. *Science* 250:1331–1332 (1990).
Moriarty, F. *Ecotoxicology: The study of pollutants in ecosystems.* Orlando, FL: Academic Press, 1983.
Murphy, A., ed. *The works of Samuel Johnson, LL.D.* Vol. 1. New York: George Dearborn Publishers, 1836.
Nagel, E. *The structure of science. Problems in the logic of scientific explanation.* New York: Harcourt, Brace and World, 1961.
Newman, M.C. *Fundamentals of ecotoxicology.* 3rd ed. Boca Raton, FL: Taylor & Francis/CRC Press, 2010.
Newman, M.C., and W.H. Clements. *Ecotoxicology: A comprehensive treatment.* Boca Raton, FL: Taylor & Francis/CRC Press, 2008.
Newman, M.C., and A.W. McIntosh. Preface. In *Metal ecotoxicology: Concepts and applications*, ed. M.C. Newman and A.W. McIntosh. Chelsea, MI: Lewis Publishers, 1991.
Platt, J.R. Strong inference. *Science* 146:347–353 (1964).
Popper, K.R. *Objective knowledge. An evolutionary approach.* Oxford: Clarendon Press, 1972.
Rand, G.M., ed. *Fundamentals of aquatic toxicology: Effects, environmental fate, and risk assessment.* Washington, DC: Taylor & Francis, 1995.
Salsburg, D. *The lady tasting tea: How statistics revolutionized science in the twentieth century.* New York: Henry Holt and Company, 2001.
Walls, L.D. *Material faith: Thoreau on science.* Boston: Houghton Mifflin Company, 1999.

Acknowledgments

If I have seen further it is by standing on the shoulders of giants.*

The sentiment of the above quote needs modification to accurately reflect my thoughts—perhaps to "If I have seen more, it is by standing on the tips of my toes to peer over the shoulders of some very remarkable people." I claim none of the concepts and methods in this book as mine. Most came out of decades of insightful and diligent work by others who are acknowledged clearly throughout the book. The author is extremely grateful to the many scientists, statisticians, and mathematicians who fashioned the fascinating concepts and valuable tools gathered together here. The intent is not to present my concepts and tools but simply to be "sufficiently useful, by only diversifying the surface of knowledge, and luring the mind by a new appearance to a second view of those beauties which it had passed over unattentively before" (see Samuel Johnson quote opening the preface). Of course, any errors based on misreading or misperception while peering over the shoulders of my predecessors I do claim as mine alone.

Along a more specific vein, Drs. James Oris and Mark Sandheinrich provided insightful guidance at the onset of this revision. Drs. Michael Hooper, Margaret Mulvey, and Paul Story provided valuable criticism of draft chapters. Mr. Jincheng Wang reviewed meticulously several equations and computer code applied in several illustrations of model behavior.

* Sir Isaac Newton penned this now hackneyed phrase in a 1676 letter to Robert Hooke. In keeping with the theme of this acknowledgment, it is instructive to note that Newton borrowed the phrase from an 1159 treatise by John of Salisbury, who in turn had borrowed it from Bernard of Chartres (Troy, S.D., ed., *Medieval Rhetoric: A Casebook*, Routledge Medieval Casebooks, vol. 36. New York: Routledge, 2004).

About the Author

Michael C. Newman is currently the A. Marshall Acuff Jr. Professor of Marine Science at the College of William and Mary's Virginia Institute of Marine Science, where he also served as Dean of Graduate Studies for the School of Marine Sciences from 1999 to 2002. Previously, he was a faculty member at the University of Georgia's Savannah River Ecology Laboratory. His research interests include quantitative ecotoxicology, environmental statistics, risk assessment, population effects of contaminants, metal chemistry and effects, and bioaccumulation and biomagnification modeling. In addition to more than 125 articles, he has authored 5 books and edited another 6 on these topics. The English edition and Mandarin and Turkish translations of *Fundamentals of Ecotoxicology* have been adopted widely as the textbook for introductory ecotoxicology courses. He has taught at universities throughout the world, including the College of William and Mary, University of California–San Diego, University of Georgia, University of South Carolina, Jagiellonian University (Poland), University of Antwerp (Belgium), University of Hong Kong, University of Joensuu (Finland), University of Koblenz–Landau (Germany), University of Technology–Sydney (Australia), Royal Holloway University of London (UK), Central China Normal University, and Xiamen University (China). He has served numerous international, national, and regional organizations, including the OECD, U.S. EPA Science Advisory Board, Hong Kong Areas of Excellence Committee, and the U.S. National Academy of Science NRC. In 2004, the Society of Environmental Toxicology and Chemistry awarded him its Founder's Award, "the highest SETAC award, given to a person with an outstanding career who has made a clearly identifiable contribution in the environmental sciences."

CHAPTER 1

Introduction

Science is built up of facts, as a house is built of stones; but an accumulation of facts is no more a science than a heap of stones is a house.

—Poincaré (1952)

We speak piously of taking measurements and making small studies that will 'add another brick to the temple of science.' Most such bricks just lie around the brickyard.

—Platt (1964)

1.1 ECOTOXICOLOGY AS A SCIENTIFIC DISCIPLINE

The [ecotoxicology] literature is both enormous and, in large part, trivial.

—Moriarty (1983)

Truhaut (1977) is credited as the first to use the term *ecotoxicology* to define the "natural extension of toxicology, the science of the effects of poisons on individual organisms, to the ecological effects of pollutants" (Moriarty 1983). Cairns and Mount (1990) similarly defined *ecotoxicology* as "the study of the fate and effect of toxic agents in ecosystems," a phrasing that adds the study of pollutant fate but removes the word *science*. It is puzzling to read these definitions if one accepts, as I do, that the aspirations of environmental scientists during the last half century have always been ecotoxicological in nature. What could be so lacking in the body of knowledge that had accumulated up to 1977 as to necessitate the implied reformation around a new field of ecotoxicology?

The impetus for defining this "new" approach seems to grow out of frustration with our continued inability to predict or, in many cases, clearly document effects at any but the lowest levels of ecological organization. Effective prediction remains an elusive goal, despite decades of sincere effort with ample funding. The frustration grows acute as the need for accurate description and prediction becomes more and more pressing (Clemmitt 1992). Discomfort is invoked periodically by statements such as Lederman's (1991), then president-elect of the American Association for the Advancement of Science, that "understanding ... ecological and environmental issues and providing guidance to policymakers" is one of the major tasks facing U.S. scientists today. Or Al Gore's more recent statement, "... we are colliding with the planet's ecological system, and its most vulnerable components are crumbling as a result" (Gore 2006). Predictably, the ever-present banter about the relative virtues of applied versus basic science, standard versus nonstandard methods, field versus mesocosm versus laboratory studies, and reductionist versus holistic approaches took on Babelian proportions. As the din increased, an attempt emerged in the late 1970s to regroup under the new standard, ecotoxicology.

Such reformation provides the opportunity for great advances and equally great mistakes. This is particularly true in a melding of synthetic disciplines like ecology and toxicology (Maciorowski 1988). A series of insightful and timely papers that provide much needed perspective (e.g., Cairns 1989, 1991; Cairns and Mount 1990) was published several decades ago. A contextual framework also emerged for the application of ecotoxicological methods to environmental regulation and remediation (e.g., Connell 1987; Duffus 1986; Adams 1990). What still remains distinctly absent is a focused effort to provide an effective consilient scientific framework for this field. This absence of a consistent scientific framework, in my opinion, has slowed progress during the last 30 years. Basic principles are left to be pondered as afterthoughts as legitimate and immediate needs for standardization or for information on the next of a seemingly endless number of new toxicants are satisfied. This opinion seems to have been shared by Moriarty (1983), who stated in the last paragraph of his book, "I have tried to relate the problems of ecotoxicology to their ecological context. Failure so to do has led to much muddled thinking and to unreliable conclusions." Schwetz (in Clemmitt 1992) expressed a similar opinion that toxicology is "sometimes too much of an applied science. So most ideas come from other sciences."

The goal of this second edition is to further contribute to the development of "an organization and classification of knowledge on the basis of explanatory principles" (Nagel 1961) for the science of ecotoxicology. The emphasis will be on detailing quantitative methods because they lend themselves most readily to explicit formulation of conceptual models (hypotheses), falsification, and estimation of likelihood or statistical confidence for competing explanations. However, it should not be forgotten while reading this volume that "the mathematical box is a beautiful way of wrapping up a problem, but it will not hold the phenomena unless they have been caught in a logical box to begin with" (Platt 1964). General concepts of scientific inference will be discussed to aid in avoiding such recurrent logical errors. They will be extended to quantitative expressions where warranted. Further, explicit definitions fundamental to the discipline will be formulated at the onset to avoid confusion and ambiguity.

Another definition is required here to distinguish the science of ecotoxicology from the impressive body of information fulfilling essential regulatory or monitoring needs. Ecotoxicology is the science that attempts to organize knowledge about the fate and effects (including those to humans) of toxic agents in the biosphere based on explanatory principles (Newman and Clements 2008; Newman 2010). This definition is very similar to that of Jørgensen (1990) ("the science of toxic substances in the environment and their impact on the living organisms") but emphasizes a higher ecological level focus and several important qualities of scientific knowledge. The remainder of this chapter will be used to define and clarify the basic components of this definition: biosphere and ecosystem, toxicant effect, toxicant fate, and the organization of knowledge based on explanatory principles.

1.2 TOXICANTS AND BIOSPHERE

> Any ecosystem under study has to be delimited by arbitrary decision, but one has to remember always that the imposed boundaries are open.
>
> **—Margalef (1968)**

> The relationship between structure and functioning is a fundamental one in ecosystems science. Ecosystems, and indeed the global biosphere, are prototypical examples of complex adaptive systems, in which macroscopic system properties such as trophic structure, diversity-productivity relationships, and patterns of nutrient flux emerge from interactions among components, and feed back to influence the subsequent development of these interactions.
>
> **—Levin (1998)**

The functional unit of the biosphere in classical ecology is the ecosystem.* Many ecotoxicologists discuss only the biotic community residing in a defined area when dealing with ecosystem effects. However, an ecosystem includes the biotic community and its abiotic environment functioning together as a unit to direct the flow of energy and cycling of materials. The ecosystem approach embraces the concept of components functioning to maintain the system through a complex of feedback loops. Margalef (1968) offers the context that ecosystems are systems in which "individuals or whole organisms may be considered elements of interaction, either among themselves, or with a loosely organized environmental matrix."

It is important to keep in mind that the ecosystem concept is an artificial construct used to frame concepts and hypotheses. It is not a concept without limitations (Margalef 1968; Gutmann 1976). The ecosystem model should not be applied as a perfect depiction of reality despite its enormous usefulness. How closely the qualities of an operationally defined ecosystem conform to those of the abstract one depends on many factors, including spatial and temporal scale, distinctness of system boundaries, and the specific qualities under study. For this reason, comparison of the qualities of a speculatively impacted ecosystem to those of an idealized ecosystem may be a worthwhile mental exercise, but it could not be used to conclude definitively that an adverse effect occurred. The most appropriate comparisons would be to properties of a reference ecosystem or to the same ecosystem before contamination. Obviously, only temporal comparison seems possible for the global ecosystem, that is, the biosphere.

1.3 TOXICANT EFFECTS IN ECOSYSTEMS

1.3.1 Classification Based on the Stress Concept

> Everybody knows what stress is and nobody knows what it is.
>
> —**Selye (1973)**

1.3.1.1 Stress

Effects of toxicants on ecosystem components are often measured along a spectrum ranging from the molecular (e.g., induced detoxification proteins), to the whole ecosystem level (e.g., a shift in system respiration:production or nutrient cycling), to the entire biosphere (e.g., global warming or ozone thinning). Most often, a measurement of stress is implied for any significant shift in some quality regardless of the ecological level examined. The measured quality may be a primary (e.g., modified level of a hormone directly influenced by the toxicant) or higher-order (e.g., increased intensity of parasitic infection as a consequence of the debilitating effects of a toxicant) indicator of change.

But what is stress? Stress in the common vernacular is something that causes a tension that can alter a normal state, such as a heavy load might stress one's back. Unfortunately, this general definition is often confused with the more explicit biological definition of stress to individual organisms. Hans Selye (1956, 1973) developed the concept of stress and the associated general adaptation syndrome (GAS) as applied to individuals. "Stress is the state manifested by a specific syndrome which consists of all the nonspecifically induced changes within a biological system"

* From the vantage of a physicist or chemist, the phrasing "classical ecology" must seem an oxymoron. Ecology is a new and rapidly evolving science itself. "The word ecology was coined not much more than 100 years ago, and the oldest professional society, the British Ecological Society, is less than a century old" (May and McLean 2007). For this reason, some concepts in the first edition of this book needed modification to conform to revisions in modern ecology. This was especially needed as the spatial scale of ecological issues broadened and the associated tools became more convenient.

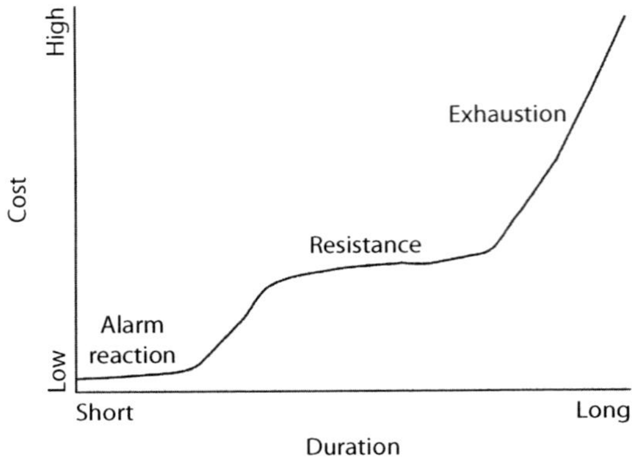

Figure 1.1 Cost associated with the three phases of Selyean stress.

(Selye 1956). He stated quite clearly that it is a *specific syndrome* in response to an external agent. It is a state achieved during any of a variety of activities, including many nondetrimental activities such as intense exertion. Further, it is a *distinctive or specific suite of responses that are beneficial or compensatory*. The stress response fulfills the purpose of turning the individual back toward the state of homeostasis. This stress response-achieved state approaching homeostasis is mediated by hypothalamic-pituitary-adrenal axis changes and was called heterostasis by Hans Selye. The associated general adaptation syndrome has three phases (Figure 1.1): the alarm reaction, adaptation (or resistance), and exhaustion components (Selye 1973). The alarm component involves immediate reactions, such as increased pulse rate and blood pressure. If the stressor continues to exert an effect on the organism, responses that stimulate tissue defense, such as enlargement of the adrenal cortex and shrinkage of the thymus, occur. After a period of exposure to the stressor, the individual enters a characteristic exhaustion phase, indicating that the organism's finite amount of adaptive energy has been exhausted. With continued stress, the individual will be unable to maintain itself and will die. The concerted expression of specific mechanisms acts to regain or resist deviation from homeostasis.

But what is stress in the context of ecotoxicology? The explicit definition originally given by Selye for individuals exposed to stressors and maintained in the medical literature has not been retained in studies of higher levels of ecological organization. The precise definition of stress depends on the level at which an effect is measured. Consequently, it is important to understand the various ecotoxicological meanings given to this concept in the literature.

Definitions vary, even at the individual level. Adams (1990) compiled the following definitions of stress, which focus on the level of the individual:

1. "The sum of all physiological responses that occur when animals attempt to establish or maintain homeostasis, the stressor being an environmental alteration and stress the organism's response."
2. "Adaptive physiological changes resulting from a variety of environmental stressors."
3. "A diversion of metabolic energy from an animal's normal activities."
4. "The sum of all the physiological responses by which an organism tries to maintain or reestablish normal metabolism in the face of chemical or physical changes."
5. "Alteration of one or more physiological variables to the point that long-term survival may be impaired."
6. "The effect of any environmental alteration that extends homeostatic or stabilizing processes beyond their normal limits."

Examining population level stress, including long-term, genetic consequences, the following definitions have been forwarded:

7. "An environmental change that results in reduction of net energy balance (i.e. growth and reproduction).... Any reduction in production (somatic growth, reproduction or both) in response to an environmental change signifies reduced Darwinian fitness, and therefore represents a result of environmental stress" (Koehn and Bayne 1989).
8. "An environmental condition that, when first applied, reduces Darwinian fitness; for example, reduces survivorship (S) and/or fecundity (m) and/or increases time (t) between life-cycle events" (Sibly and Calow 1989).
9. "Anything which reduces growth or performance, it follows that, in a situation where a particular stress operates, there must be a reduction in fitness.... [If genotypes vary in fitness and stress is occurring consistently] evolutionary changes are to be expected" (Bradshaw and Hardwick 1989).
10. "A recent anthropogenic change in the environment affecting a population's reproductive reserve or reducing its environmentally controlled abundance limit" (Shuter 1990).

At the community or ecosystem levels, the following definitions have been advanced:

11. "A detrimental or disorganizing influence ... negative responses to unusual external disturbances, or stressors of low probability to which a community of organisms is not preadapted" (Odum 1985).
12. "An external force or factor, or stimulus that causes changes in the ecosystem, or causes the ecosystem to respond, or entrains ecosystematic dysfunctions that may exhibit symptoms" (Rapport et al. 1985).
13. "A stressor is any condition or situation that causes a system to mobilize its resources and increase its energy expenditure. Stress is the response of the system to the stressor. Responses to stressors may include adaptation or functional disorder" (Lugo 1978).
14. "A perturbation (stressor) applied to a system (a) which is foreign to that system or (b) which is natural to that system but applied at an excessive amount" (Barrett et al. 1976).

Esch and Hazen (1978) provided the following definition that attempts to cover all levels of ecological organization.

15. "The effect of any force which tends to extend any homeostatic or stabilizing process beyond its normal limit, at any level of biological organization."

Hoffman and Parsons (1991), although focusing on population genetics, gave a similar definition that covers all levels of organization.

16. "The term 'stress' [is used] to represent an environmental factor causing change in a biological system which is potentially injurious."

Careful review of these definitions suggests that stress is used to identify either: (1) a response, (2) a characteristic or specific response, (3) an effect, or (4) an external factor causing a response or effect. In this book, the external factor is referred to as the stressor. The response or effect is stress. Inclusion of an effect that does not also constitute a response is contrary to a central theme of Selye's original concept. However, repeated omission of this theme in definitions necessitates the inclusion of nonresponse effects. It also necessitates establishment of classes of stress that clarify its meaning when used in ecotoxicology.

Four qualities are present in the above definitions regardless of the level of ecological organization. First, stress is a response to or effect of an external factor that is detrimental or disorganizing. Unlike the original concept of stress as advanced by Selye, stress does not include a response to or effect of a nondetrimental factor. Selye would have classified the body's response to extreme physical exertion as stress, although the exertion caused no detriment. Second, the detrimental or disorganizing factor is atypical or present at atypical levels. Implied here is the idea that the system has not adjusted itself previously to the specific stressor in such a way as to mediate its effects during predictable or highly probable exposures at a future time. Third, the system responds by or is characterized by a modification of energy flow or system structure. In the case of a response, the shift

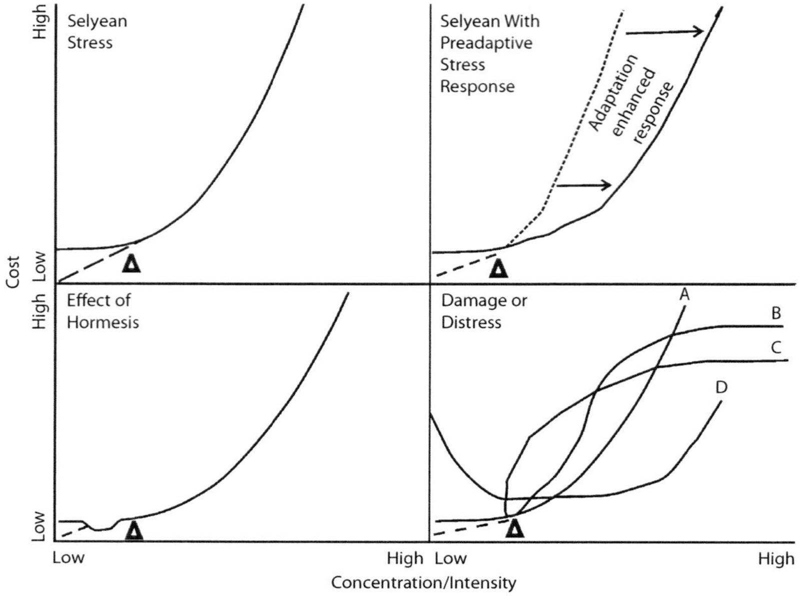

Figure 1.2 Examples of toxicant effects. The triangle denotes a possible threshold concentration. Below the threshold, the cost may be constant (solid line). Alternatively, it could increase with concentration (broken line). A simple Selyean and a Selyean effect with a superimposed preadaptation are illustrated in the top panels. An effect with hormesis is shown in the bottom left panel. Damage or distress effects can display a variety of trajectories. Examples shown here include the following: (A) zinc-induced gill damage (Hodson 1974), (B) DDT-induced fish mortality (King 1962), (C) oxygen consumption by gill tissue exposed to cadmium (Dawson et al. 1977), and (D) energy cost as influenced by salinity (Koehn and Bayne 1989). The curve shapes are products of the range of concentrations and units of effect as much as they are influenced by the type of effect.

acts to establish a condition similar to some norm or homeostasis, or to reestablish such a condition. Steady state is not an essential component of the response, although it is implied in several definitions. Fourth, temporal qualities are central to the concept of stress. Stress is a response to a recent stressor. In contrast to these four qualities common to the above definitions, the specific syndrome quality critical to Selye's stress concept has not been retained as an essential quality of stress at higher levels of organization.

With these common qualities identified, a clear, general definition of stress can be offered in the context of ecotoxicology: stress at any level of ecological organization is a response to or effect of a recent, disorganizing, or detrimental factor. A stress response represents an effort to maintain or reestablish a state approximating homeostasis, i.e., normal energy flow, material cycling, or system structure.

The following qualifiers are forwarded to differentiate between the various types of stress that can occur. The three categories of stress and three nonstress effects (Figure 1.2) defined here employ Selye's original stress concept as a yardstick for comparison. Selye's theory is a sensible and common yardstick bolstered by an enormous literature and refinement of ideas. This lends considerable justification to using Selye's concept as a touchstone.

1. Selyean stress is a specific or characteristic response to a recent, disorganizing, or detrimental factor. Its purpose is to maintain or reestablish a state similar to homeostasis (i.e., energy flow, material cycling, or system structure) within a defined norm. It is not characterized by any previous adaptation to the specific stressor. The increase in pulse rate or blood pressure associated with physical exertion is a typical example of this category of stress. If the stressor continues to elicit a response

over a longer period of time, other characteristics, such as those described above for individuals exposed to a stressor, may be expressed. Rapport et al. (1985) describe details of an ecosystem general adaption syndrome analogous to Selye's GAS. However, many examples cited suggest ecosystem effects of a type described in the next category.

2. Preadaptive stress is similar to Selyean stress, but it is characterized by a previous adaptation to the stressor, i.e., the system has specific information with which to mediate the effect of the stressor. The adaptation may be recent and transient (e.g., acclimation), or long term and relatively permanent (e.g., genetic adaptation). These responses will tend to be specific to the stressor. According to Rapport et al. (1985), this definition contains aspects of Selye's concept of eustress, a response to events that organisms or systems expect or anticipate. Induction of the cytochrome P-450 monooxygenase system by polycyclic aromatic hydrocarbons is a response to a specific class of toxicants resulting from genetic adaptation.* At the ecosystem level, preadaptive stress might be the type of response elicited by a regular or predictable stressor such as that associated with tides or seasonal fluctuations in soil moisture. To avoid confusion, it must be noted that this concept is not associated with that of genetic preadaptation.
3. Damage or distress is the adverse effect(s) of a stressor that is not a consequence of a system response. This category of effect is often defined as stress and, reluctantly, will be discussed as stress here. At lower levels of organization, the term is most commonly used to denote cell, tissue, or organ damage. Rapport et al. (1985) use the term *distress syndrome* when discussing this type of effect on ecosystems. A toxicant modifies the normal function or structure of a system without involving an active response by the system. For example, an effect might be measured if sufficient numbers of cells are damaged in a target organ. Similarly, signs of ecosystem distress would include a reduction in species diversity, increased nutrient loss, or a shift in the balance between productivity and biomass (Rapport et al. 1985; Levin 1998). In both cases, the effect is not an active response by the system to the toxicant: it is an adverse consequence of intoxication.

1.3.1.2 Nonstress Effects

What are some other effects of toxicants in ecosystems? Several types of effects are outside the concept of stress as defined above:

1. A hormetic effect (Figure 1.2) is a stimulatory effect exhibited upon exposure to low levels of some toxicants or physical agents. It is not normally characterized by a toxicant-specific system response (Stebbing 1982). Although seemingly counterintuitive, the effect of a toxic substance at a certain level of exposure can appear as beneficial. This general phenomenon is called hormesis. Southam and Ehrlich (1943) defined it as "a stimulatory effect of subinhibitory concentrations of any toxic substance on any organism." More recently, Calabrese (2008) defined it as "a biphasic dose-response phenomenon characterized by a low-dose stimulation and a high-dose inhibition." Recent treatments of this topic include effects of cadmium on growth of wood ducks (Brisbin et al. 1987) and radiation effects (Sagan 1987). Reviews of chemically induced hormesis were developed two decades ago (Calabrese et al. 1987) and more recently by Calabrese (2008).
2. A neutral effect is a measurable change that has no apparent impact (adverse or beneficial) on the system's overall qualities or probability of persistence. Although most measurable effects are likely to be positive or negative, it is illogical to reject the possibility of a neutral effect. Definition of this effect category can be particularly useful in formulation of null hypotheses for statistical or logical assessments of effect.
3. An ambiguous effect is a measured effect of undefined qualities relative to the degree of detriment/benefit, passivity, or preadaption. The present state of our understanding in ecotoxicology necessitates this category. Often effects measured at higher levels of organization fall into this category.

* In the unifying scheme of Selye, cytochrome P-450 monooxygenase induction is an example of a response controlled by catatoxic hormones, that is, by hormones designed to enhance stressor destruction. In contrast, syntoxic hormones, such as those orchestrating the Selyean stress response, facilitate the organism's ability to coexist with a stressor during exposure. Additional details can be found in Selye (1956) and Newman and Clements (2008, pp. 138–140).

Table 1.1 Categories of Ecotoxicological Effect

Category	Beneficial Response Deleterious Consequence Neutral Effect	Specific Characteristic to Response?	Response Involves Preadaptation to Stressor	Examples
Selyean stress	B	Yes	No	Increased pulse rate and blood pressure with exertion
Preadaptive stress	B	Yes	Yes	Metallothionein induction
Hormetic response	B	No	No	Stimulation of algal population growth
Damage/distress	D	NA	NA	Thinning of eggshell due to DDT exposure; decreased hunting efficiency of a predator due to intoxication
Neutral effect	N	NA	NA	Increase in toxicant concentration in a species with no adverse consequences
Ambiguous effect	?	?	?	

1.3.1.2.1 Balance between Beneficial and Adverse Effects

The discussion above has been focused on a restricted portion of the range of toxicant concentrations. For many toxicants or physical agents, the response curve can assume a shape similar to D in the bottom right-hand corner of Figure 1.2. Below certain concentrations, the effect becomes increasingly detrimental. Familiar examples include those associated with concepts such as Liebig's law of the minimum (e.g., phosphorus effects on crop yield) or Shelford's law of tolerance (e.g., salinity or temperature effects on marine species). An essential element can exert such a pattern of effect on individuals. Odum's push-pull model of stress suggests that disordering effects within a certain degree can be beneficial at the ecosystem level also. Odum's discussion of pulse stability in ecosystems (Odum 1969) presents such a beneficial effect in the context of a preadaptive stress. Lugo (1978) used Odum's push-pull or positive-negative effects model to describe numerous ecosystem level effects of stressors, including toxicant-associated effects.

1.3.1.3 Summary

The characteristics of effects based on the concept of stress are summarized in Table 1.1. Hypothetical diagrams of costs to a system with change in toxicant concentration are shown for each type in Figure 1.2. The responses are not necessarily exclusive of one another. For example, metallothionein induction may minimize cost at a low concentration of copper but, at a point of metallothionein saturation, "spillover" of significant amounts of metal to other cellular fractions occurs (see Klaverkamp et al. 1991). At that point, the Selyean stress response may become increasingly important. Prior to metallothionein induction, a hormetic response could have occurred. Damage to kidney tissues could have occurred during the exposure.

1.3.2 Classification of Effects Based on Other Criteria

Other classification systems of effects have been derived at various levels of biological organization. For example, a toxicant may be carcinogenic to an individual. At the population level, a toxicant may act to increase the risk of local population extinction. Community diversity might be decreased by a contaminant. A brief discussion of some common systems follows, and each will be discussed in detail in later chapters.

1.3.2.1 Temporal Context

Effects associated with individuals are frequently categorized in a temporal context, e.g., as acute or chronic. Often these definitions are used in discussions of toxicity testing methodologies. For example, an acute effect occurs immediately as a result of an intense exposure event. Casarett and Doull (1975) defined acute effects as "those that occur or develop rapidly after a single administration of a substance." They defined chronic effects as "those that are manifest after an elapse of time." Three decades later, Eaton and Gilbert (2008) defined acute in the latest version of the same classic toxicology textbook as "acute exposure is defined as exposure to a chemical for less than 24 hours ... acute exposure usually refers to a single administration, [but] repeat exposures may be given." According to Rand and Petrocelli (1985), a chronic effect "may occur when the chemical produces a deleterious effect as a result of a single exposure, but more often they are a consequence of repeated or long-term exposures." The difference between these two types of effects is a matter of degree. To illustrate this point, Casarett explained that an acute exposure to a toxicant such as beryllium can produce an effect that will take some time to manifest itself. Finally, Suter et al. (1987) briefly discussed interpretations of the term *chronic effect* to mean those arising from exposure over "greater than 10% of the organism's lifespan." They suggested that all life stages and processes must be exposed to detect chronic effects. Although seemingly more applicable to nonhuman exposure situations than the above definitions, the proposed conventions of Suter et al. (1987) are no less arbitrary and do not clarify the intended distinction in the literature.

1.3.2.2 Lethality

Making the distinction between lethal and sublethal effects is also difficult (Moriarty 1983). Rand and Petrocelli (1985) discussed death or failure to produce viable offspring in the context of lethal effects. Sublethal effects include deleterious behavioral, anatomical, or physiological changes. Unfortunately, it is often impossible to say whether a sublethal effect (e.g., diminution of predator avoidance behavior) would or would not result in death (lethality) of an individual within a natural setting.

1.3.2.3 Site of Action

Most toxicological treatments (e.g., Casarett and Doull 1975; Rand and Petrocelli 1985) also distinguish between effects in the context of their sites of action. A systemic effect involves action on systems such as the central nervous, immune, or cardiovascular system. A local effect occurs at the primary site of damage,* such as a gill lesion caused by direct contact with the toxicant. Toxicants might also be classified according to their target organ. As examples, substances might be classified as hepatotoxicants, immunotoxicants, nephrotoxicants, neurotoxicants, or perhaps endocrine modifiers.

1.3.3 Summary of Toxicant Effects

As described in this section, toxicant effects, including system responses, can be discussed relative to the concept of stress. Six classes of effect (Selyean stress, preadaptive stress, hormetic response, damage/distress, neutral effect, and ambiguous effect) were described. Other classification schemes important in regulatory activities or traditional toxicology are based on time frame,

* Relative to the Selyean framework, stress-related changes can also be local or systemic (Selye 1956). In contrast to the GAS-related responses described above, local adaptation syndrome (LAS)-linked responses exist and include such responses as the change in immunological response in a local tissue experiencing inflammation.

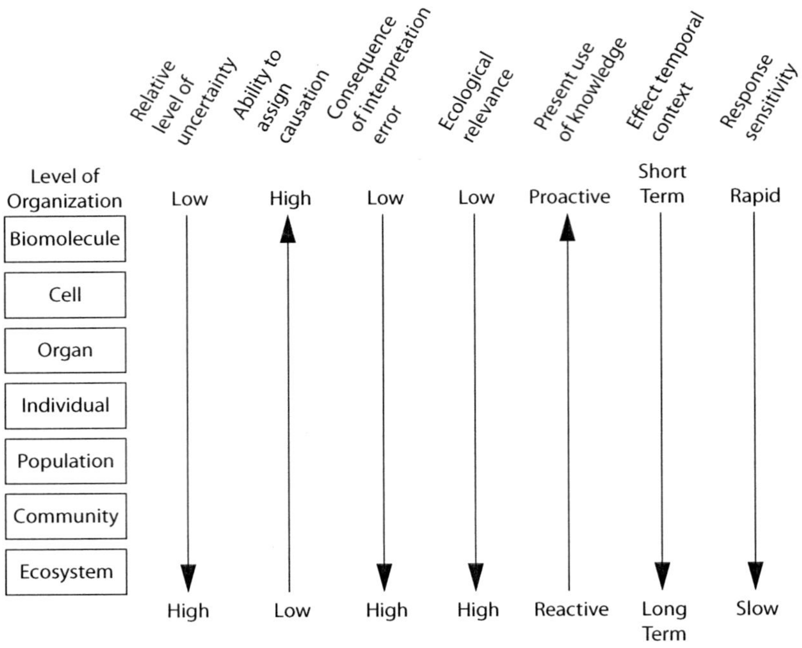

Figure 1.3 Features of ecotoxicological effects based on level of ecological organization. This figure is a composite derived from those of Haux and Forlin (1988, Figure 1), Adams et al. (1989, Figure 1), Chapman (1991, Figure 4), and Burton (1991, Figure 2).

lethality, or site of action criteria. Further, effects can be classified based on whether they exacerbate a preexisting condition, involve synergism with another agent, or represent an atypical reaction of a hypersensitive individual. Precise classifications of effects in the context of regulatory testing are detailed in sources such as Sprague (1969, 1971), Buikema et al. (1982), Suter et al. (1987), and Weber et al. (1989). Those classifications that focus on lower levels of organization will be discussed in later chapters.

What is the present state of knowledge relative to ecotoxicological effects? Figure 1.3 summarizes the present perception of our state of knowledge of effects along the ecological spectrum of organization. Our abilities to understand and assign causal relationships are best at the lower levels of organization and become poorer as the level of organization increases. The lower-level effects are generally believed to be more sensitive (i.e., manifested at lower toxicant concentrations) than effects at higher levels of organization. They often respond more rapidly to a toxicant. There is an associated advantage in that biochemical or physiological indicators can be used proactively (i.e., prior to irreversible or major ecological harm). When an ecosystem level effect is noted, it is most often used to document a degraded state. Such reactive documentation has little predictive value because the degradation has already occurred by the time an effect is seen. However, beyond documentation, this effect can be useful to establish a baseline for comparison as remedial actions are implemented. The responsiveness (rate at which the effect manifests itself after a toxic exposure occurs) and the temporal context (duration of time that the effect will be significant after removal of the toxic agent) of an effect will be shorter and more rapid at lower levels of organization. Chapman (1991) suggested that effects at lower levels tend to be more reversible than effects at higher levels of organization. At first glance, all of these qualities seem to indicate that effects at lower levels of organization are superior to higher-level effects as tools for managing ecosystem health. They respond quickly to change, are more readily understood, are more easily assigned causation, and are more effectively used prior to permanent or significant ecological degradation. Further, "the immediate effects of pollutants are on individual organisms, by either direct toxicity or altering the environment" (Moriarty 1983).

Despite the virtues of lower-level responses, they have no overall superiority to higher-level responses. Moriarty (1983) continues the above quote by stating that although the immediate effects are at the individual level, "the ecological significance, or lack of it, resides in the indirect impact on the populations of species," that is, on the interacting populations in the ecological community. The goal of ecotoxicological stewardship is the protection of ecological systems, not biochemical moieties, physiological homeostasis, or even individuals, in most instances. The probability of falsely assigning an adverse ecotoxicological effect is increased when higher-level effects are neglected in favor of lower-level responses (Chapman 1991). Since our ability to relate lower-level responses to ecosystem degradation is limited, it follows that the ecological relevance of a lower-level response (e.g., a 50% decrease of total metal bound to metallothionein) is much more ambiguous than that of a higher-level effect (e.g., a 50% drop in species richness).

1.4 TOXICANT FATE IN ECOSYSTEMS

Rand and Petrocelli (1985) succinctly stated that toxicant fate is the "disposition of material in various environmental compartments (e.g., soil or sediment, water, air, biota) as a result of transport, transformation, and degradation." Concentration, distribution, speciation, and phase association of the toxicant in the various ecosystem components are considered, as well as toxicant sources and sinks (Figure 1.4). The toxicant will accumulate in both the biotic or abiotic components of the ecosystem under study.

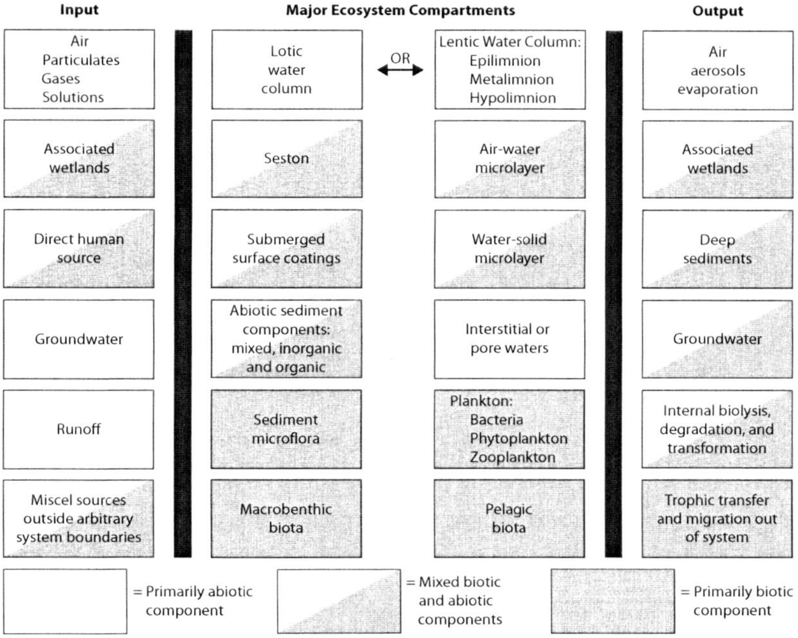

Figure 1.4 Compartments with significant ecotoxicological relevance to the fate of toxicants in aquatic ecosystems.

1.4.1 Fate in Biotic Components

Bioaccumulation is defined here as the accumulation of a toxicant in (or on, in certain situations) an individual. Results of bioaccumulation studies are very often extrapolated to make population or food web level implications. Studies may focus on taxonomic groups that accumulate relatively high concentrations of toxicants (e.g., metals in algae) or on groups more sensitive than others (e.g., DDT in nesting waterfowl). Alternatively, focus may be on assemblages associated physically with or linked functionally to components containing high concentrations of toxicant (e.g., infaunal species inhabiting contaminated sediments in the first case; scraper species grazing on metal-contaminated periphyton in the second case).

Accumulation can include direct uptake from air through lungs and associated pulmonary structures, from water through gills or epithelial tissues, and input from food via the gut. Usually, dietary sources include soil, detritus, living plant material, or animal tissue. For this discussion, the concept of bioaccumulation also explicitly includes accumulation of toxicants as a consequence of parasite-host, parent-egg/embryo/juvenile, or general symbiotic interactions. Some of these exchanges may not be direct trophic interactions as commonly envisioned. Further, the accumulation of toxicants from items processed coincidentally with food items is also included as a component of bioaccumulation. As an important example, pica (intentional or unintentional ingestion of soil) can be a substantial source for some terrestrial herbivores. Bioaccumulation will be discussed in more detail in Chapter 3.

As noted above, the transfer of toxicant from one individual to another during trophic interactions involves bioaccumulation. Biomagnification is said to occur if the toxicant concentration increases during successive trophic transfers. This trophic transfer of toxicants will be discussed in Chapter 7.

1.4.2 Fate in Abiotic Components

The intimate association of biotic and abiotic components of ecosystems is emphasized in Figure 1.4 using examples of significant inputs, outputs, and locations of toxicants in ecosystems. Indeed, any discussion of toxicant fate within ecosystems that ignores either abiotic or biotic components would be incomplete. Not only are these components in close physical contact, but they modify one another in such ways as to facilitate or inhibit toxicant exchange or transformation. For example, bioturbation may enhance a lipophilic toxicant movement between sediments and overlying waters, and chemical speciation of a metal dissolved in water will influence bioaccumulation.

Physical and chemical mechanisms influence toxicant fate in and exchange between ecosystem components. Equilibrium partitioning may determine the distribution of organic pollutants. Redox reactions and dissolution/precipitation or coprecipitation phenomena can have strong influences on metal or radionuclide movement within ecosystems. Complexation, photolysis, and adsorption/desorption also play key roles. Examples of physical processes include soil weathering and erosion, sedimentation, atmospheric fallout, bulk water movement, and bottom scouring.

1.5 ORGANIZATION OF KNOWLEDGE BASED ON EXPLANATORY PRINCIPLES

> While ecologists have sought to elucidate the nature of these "parallel effects" of [human-induced] stress on ecosystems, they have perhaps more often followed what Dyson identified as the predominant methodology of the biological sciences: a preoccupation with description of the diversity of phenomena, by and large to the exclusion of consideration of unifying themes.
>
> **—Rapport et al. (1985)**

> It has also been argued that ecology is largely ideographic (explains particulars) rather than nomothetic (derives universal laws). For a certain class of questions about ecosystems it is true that only particular explanations are needed (e.g., grass dominates this system because of frequent fires, a historical event) but the same is true in physics when known laws are applied to a given situation. This does not preclude the existence of laws.
>
> **—Loehle (1988)**

> Well, there are two kinds of biologists, those who are looking to see if there is one thing that can be understood, and those who keep saying it is very complicated and that nothing can be understood.... You must study the simplest system you think has the properties you are interested in.
>
> **—Cy Levinthal (quoted in Platt 1964)**

1.5.1 Introduction

The ecotoxicological literature is replete with statements that a working knowledge of ecosystems is impossible due to ecosystem complexity. Yet, such statements are inconsistent with the rapid growth of knowledge in equally complex disciplines. A casual glance at the intricate charts of metabolic pathways or the human genome should bring such statements immediately to question. The complex of metabolic networks functioning within each individual is staggering, but taken *en masse*, our working knowledge of such systems is impressive. A similar impression can be brought away after browsing the molecular genetics literature: entire genomes have been mapped for several species, including humans.

What then is the reason for our limited knowledge base in ecotoxicology? The belief is forwarded here that a significant impediment to the growth of ecotoxicological knowledge lies in the approach by which such knowledge is sought, not the complexity of the system under study. Although some approaches are adopted of necessity, others are selected out of habit. Some were learned from a parent discipline. The predisposition for description in ecology (see Loehle and Rapport et al. quotes above) and the propensity for single-species endpoint toxicity tests transplanted from traditional toxicology are two significant pieces of evidence supporting this argument. Other habits are more pervasive in life sciences (see Levinthal quote above). Regardless of their origins, all habits are not equally fruitful and, consequently, should be subject to review, modification, replacement, or rejection. Based on this premise, the remainder of this chapter will examine scientific habits worth considering as ecotoxicology matures as a science.

1.5.2 The Structure of Scientific Knowledge

1.5.2.1 Historical Perspective

Chamberlin (1897) suggested that the format of scientific inquiry has changed throughout history. Initially, knowledge was so limited that it appeared to be within the abilities of learned individuals to develop ruling theories that explained all phenomena. No sooner was a phenomenon presented than a ruling theory was used to explain it. This process built a large knowledge structure by repeated application alone. Chamberlin referred to this habit as precipitate explanation, the immediate and sufficient application of a theory to explain an observation. The ruling theory approach slowly was replaced by the working hypothesis approach. "Under the ruling theory, the stimulus is directed to the finding of facts for the support of the theory. Under the working hypothesis, the facts are sought for the purpose of ultimate induction and demonstration, the hypothesis being but a means for the more ready development of facts and their relations" (Chamberlin 1897). The working hypothesis approach questions theory but still retains a propensity toward precipitate explanation. When a theory or hypothesis is put forward, it tends to be given favored status in testing. It was Chamberlin's thesis that, although its roots are no longer considered valid, the habit of precipitate explanation continues into modern scientific inquiry. He suggested the method of multiple hypotheses to minimize such bias. The method of multiple working hypotheses considers all potential hypotheses simultaneously. Equal amounts of effort are spent on the hypotheses. In this manner, the tendency to unintentionally favor one hypothesis is lessened.

Karl Popper (1965, 1968) argued that scientific inquiry should test hypotheses by a formal process of falsification. No theory can be proven true, but it can be shown to be false through observation or experimental challenge. Repeated survival of a theory or hypothesis through a rigorous falsification process confers a favored status to it. Its strength is enhanced if it continues to be "corroborated by past experience." Regardless, it is never deemed true. Acknowledging that no scientist is totally objective, Popper referred to the testability of a hypothesis as "subject to intersubjective testing." A good theory is one that can be tested by scientists, each of whom approaches it with his or her own prejudices. A process that Popper likened to natural selection occurs: the fittest theory survives.

> The empirical basis of objective science has thus nothing "absolute" about it. Science does not rest upon solid bedrock. The bold structure of its theory rises, as it were, above a swamp. It is like a building erected on piles. The piles are driven down from above into the swamp, but not down to any natural or "given" base; and if we stop driving the piles deeper, it is not because we have reached firm ground. We simply stop when we are satisfied that the piles are firm enough to carry structure, at least for the time being.
>
> **—Popper (1968)**

Survival of intersubjective testing alone does not constitute corroboration and consequent enhanced status for a theory. The testing must be rigorous. Some tests have higher powers to falsify based on logic alone. A classic illustration of this point involves Einstein's theory of relativity. The explanation of eccentricities in planetary orbits was less powerful in supporting Einstein's theory than the observation that light was attracted by gravity because it had more alternative explanations than the latter observation (see Popper 1965, pp. 35–36, for more details of this example). The theory of relativity was at higher risk of rejection during the second test. Rousseau (1992) suggests that avoidance of high-risk testing is one of three symptoms of pathological science, that is, science practiced without objectivity. A tradition of low-risk testing of theories in any field has an associated danger because "our habit of believing in laws is a product of frequent repetition" (Popper 1965). The ability to separate dogma from paradigm becomes impaired if low-risk testing is common in a discipline.

Regardless of the logical power of a test, a test with insufficient measurement precision or accuracy is valueless in the process of falsification (see discussion of condensation bounds in Popper 1968, pp. 123–127). It may even slow progress due to the confusing or ambiguous nature of associated conclusions. For this reason, Popper (1968) advanced the opinion of "superiority of methods that employ measurements over purely qualitative methods." This is the reason why emphasis is placed on quantitative methods in this book.

This very brief discussion of the history of scientific inquiry will end here. Certainly, cessation does not suggest that Popper provides the capstone for scientific logic. Bayesian theory extends many associated topics pertinent to our discipline. The interested reader is referred to Howson and Urbach (1989) for an excellent presentation of the Bayesian approach. A Bayesian expansion of this approach will be sketched below and revisited throughout later chapters. Regardless, this discussion has served to identify three habits to be avoided in ecotoxicology: precipitate explanation, low-risk hypothesis testing, and imprecise or inaccurate measurement.

1.5.2.2 Strong Inference

> Strong inference is just the simple and old-fashioned method of inductive inference that goes back to Francis Bacon.... The difference comes in their systematic application.
>
> **—Platt (1964)**

As mentioned in the preface, Platt (1964) observed that scientific disciplines progress at very different rates as a consequence of the general approach taken to extract and organize knowledge. Some lack a tradition of strong inference, that is, rigorous hypothesis formulation and falsification procedures. Strong inference includes the well-known steps of hypothesis formulation, execution of experiments designed to falsify, alternate hypothesis generation, and continued testing until one hypothesis remains corroborated. However, Platt suggested that the distinction between disciplines lies in the value placed on rigorous and consistent application of such techniques.

Some fields focus too much on "surveys, taxonomic, design of equipment, systematic measurements and tables, theoretical calculations—all [of which] have their proper and honored place, provided they are parts of a chain of precise induction of how nature works" (Platt 1964). He argued that such preoccupation is taught by example to students: it is not inherent to the field.

Some disciplines are characterized by a focused effort on rigorous testing of hypotheses. Others show a meandering tendency toward precipitate explanation. Experiments are inattentively designed to support favored hypotheses. Platt advocated the use of formal experimental inference methods (the scientific method), coupled with the method of multiple hypotheses, to minimize this problem. He recommended the following practices to foster strong inference in a discipline.

1. Apply methods of inductive inference consistently and systematically.
2. Formulate hypotheses such that they are amenable to falsification. Use the "logic of exclusion" when possible.
3. State all reasonable alternate explanations of observations when presenting results.
4. Use the method of multiple hypotheses.
5. "Be explicit and formal and regular about it, to devote a half hour or an hour to analytical thinking every day, writing out a logical tree and the alternatives and crucial experiments explicitly in a permanent notebook."
6. After hearing a scientific explanation, ask two questions. Is there an experiment which could disprove it? What explanation does the present explanation exclude?

1.5.2.3 Selection of Hypotheses

Hypothesis or theory selection is a subjective process. The Bayesian treatment presented by Howson and Urbach (1989) supports this statement using the familiar "all ravens are black" argument. They explain that, if approached objectively, the possible number of theories to be tested is equal to the number of ravens in existence (n) raised to the number of possible color patterns (m) or n^m. Unless some level of subjective experience is used to select profitable hypotheses for testing, the inferential process would become impossibly cumbersome.

How then are hypotheses selected? A variety of criteria have been advanced. Popper would favor hypotheses that are easily falsifiable (practice 2 above). Loehle (1990) suggested that an optimum region (Medawar zone) exists relative to the tractability of the hypothesis and payoff for solving the problem. Loehle (1988) also recommended that ecologists should be concerned with theory reduction during the selection process. Theory reduction strives to explain one theory on the basis of another. This enhances parsimony and gives "two levels of explanation for the same phenomena." Linked also to parsimony is the application of Occam's razor to theory or hypothesis selection. Popper (1968) favors simple hypotheses as "they tell us more; because their empirical content is greater and because they are better testable." A strong, quantitative argument favoring parsimonious hypotheses can also be developed from the Bayesian vantage (Jeffreys and Berger 1992).

Chamberlin (1897) advocated a process in which a series of plausible hypotheses are considered equally and simultaneously. He observed that the movement from ruling theory to working hypotheses still remains biased toward a central hypothesis. The multiple working hypotheses approach remains subjective but lessens the tendency for a single hypothesis to become the controlling idea. "In developing the multiple hypotheses, the effort is to bring up into view every rational explanation for the phenomenon in hand and to develop every tenable hypothesis relative to its nature, cause or origin, and give all of these as impartially as possible a working form and a due place in the investigation. The investigator thus becomes the parent of a family of hypotheses; and by his parental relations to all is morally forbidden to fasten his affections unduly upon any one." Admittedly, phrases such as "rational explanation" and "tenable hypothesis" are permeated with subjectivity. Regardless, bias is lessened in the process of falsification, and a habit of thoroughly considering all hypotheses is fostered. The multiple working hypotheses approach has one additional advantage. It lessens the tendency to stop inquiry when a single "cause" is found. It increases the probability of detecting multiple or complex causes by evenly distributing effort between a set of hypothetical causes. The approach fosters thoroughness as well as lessening bias. As powerful as the multiple hypothesis approach is, its formal application in a falsification process is logically compromised in many situations that scientists confront. However, the approach about to be described extends the approach described so far in a way that minimizes this difficulty.

1.6 BAYESIAN INFERENCE

> It is occasionally possible to encapsulate a method of science as a recipe. The most satisfying is that based on multiple competing hypotheses, also known as strong inference. It works only on relatively simple processes under restricted circumstances and particularly in physics and chemistry, where context and history are unlikely to affect the outcome.
>
> **—E.O. Wilson (1998)**

Although Wilson confuses the strong inference and multiple hypotheses concepts in the above quote, he is correct about how difficult it can be to apply Platt's approach to situations with several plausible explanations. This difficulty is periodically invoked as justification for resorting to weak inference. However, there is a solution to the difficulty that combines the strong inference approach with quantitative Bayesian methods, that is, inference favoring explanations or hypotheses that are most probable based on evidence in hand. Newman (Newman and Clements 2008) refers to this straightforward extension of Platt's strongest inference as the strongest possible inference approach. The strongest possible inference approach assumes four conditions: (1) Chamberlin's multiple working hypothesis context is the best available, (2) two or more working hypotheses have been identified to explain a set of data or evidence, (3) the most probable hypothesis based on the existing evidence is the most credible, that is, the most likely of the candidate hypotheses to be true, and (4) any favored causal hypothesis is a working hypothesis subject to reevaluation as new information emerges. Formally, that hypothesis with the highest (posterior) probability is said to be the most credible one, i.e., most worthy of belief based on the available data (Woodworth 2004). Evidence-based, conditional levels of credibility for the candidate explanations are assigned based on probabilities. If additional evidence emerges in the future, the probabilities are recalculated and levels of belief/credibility in the candidate multiple working hypotheses adjusted accordingly.

Platt's strong inference approach is an ideal special case in the strongest possible inference approach. Platt emphasized rigorous hypothesis formulation and strong testing based on dichotomous falsification decisions, that is, a logic tree with branching associated with reject/do not reject decisions. Every effort should be expended to achieve the testing design he describes, but realistically, there are some stages in the process of inquiry and some situations in which Platt's strong inference is not possible. Also, as illustrated in Example 1.1, type I and II errors complicate application of even this simplest of approaches. The strongest possible inference approach expands the strong inference approach to include less discriminating testing situations that are inevitably part of most scientific inquiry schemes. Less discriminating testing situations are especially common at early stages of investigation in a research program. As Wilson states above, many situations do not permit a strong, discerning single test of the plausibility of two competing explanatory hypotheses. Although Platt's strong inference is the undeniable ideal to strive toward,* the broader framework provided by the strongest possible inference approach is required.

The strongest possible inference is essentially quantitative abductive inference, that is, inference that favors the most probable explanation or hypothesis.† This syllogism from Josephson and Josephson (1996) presents abductive inference in more formal terms:

D is a collection of data about a phenomenon,
H explains the data collection, D,
No alternate hypothesis (H_A) explains D as effectively as H does,
$\therefore$ H is probably true.

Bayesian equations facilitate abductive inference quantification by more explicitly defining the qualifiers, "as effectively as" and "probably true," in the above syllogism. Using the assessment of a single hypothesis (H) as a simple example, the Bayesian context to estimate the amount of support, credulity, belief, or plausibility warranted by data (D) is the following (Howson and Urbach 1989):

D provides support for H if $P(H|D) > P(H)$.
D draws support away from H if $P(H|D) < P(H)$.
D provides neither undermining nor supportive information if $P(H|D) = P(H)$

where $P(H)$ = probability of H being true before any consideration of the data, and $P(H|D)$ = probability of H being true given D. At this point, the task becomes estimating the probabilities in this scheme. Evidence (D) is combined with a prior probability of H being true to produce a statement of (posterior) probability given the evidence—a new probability of an explanation being true is established. If more evidence (D_{NEW}) was then collected during an inquiry, the newly established probability can be used as the new prior probability‡ and combined with (D_{NEW}) to calculate a new post probability reflecting the plausibility of H, given H and D_{NEW}. Bayes's theorem (Equation 1.1) can be used to estimate $P(H|D)$ in this case,

$$P(H \mid D) = \frac{P(H)P(D \mid H)}{P(D)} \tag{1.1}$$

where $P(D|H)$ = the (prior) probability of getting D if the hypothesis were true, and $P(D)$ = the probability of getting D whether or not H is true. The resulting $P(H|D)$ can become the new prior probability ($P(H_{\mathrm{NEW}})$) with the collection of additional data, D_{NEW}.

* This is true only if false positive and negative error rates are handled properly, as is often not the case during the associated statistical testing of hypotheses (Newman 2008).

† This explanation of the strongest possible inference comes directly from Chapter 36 in Newman and Clements (2008).

‡ The probability is a "prior probability" relative to the collection of the new data.

$$P(H \mid D_{NEW}) = \frac{P(H_{NEW})P(D_{NEW} \mid H)}{P(D_{NEW})} \tag{1.2}$$

This process can be repeated with the addition of data until the associated probability is sufficient to make an evidence-based judgment about hypothesis plausibility. The $P(H|D_{NEW})$ can be recalculated as more data are collected.

This same process can also be applied to judging any hypothesis against its negation ($\sim H$), a single alternate (H_A), or several alternate hypotheses (e.g., H_{A1}, H_{A2}, H_{A3} …). Equation (1.3) illustrates how the posterior odds for H versus H_A being true ($P(H|D)/P(H_A/D)$) can be calculated from the prior odds ($P(H)/P(H_A)$) and likelihood ratio ($P(D|H)/P(D|H_A)$):

$$\frac{P(H \mid D)}{P(H_A \mid D)} = \frac{P(H)}{P(H_A)} \frac{P(D \mid H)}{P(D \mid H_A)} \tag{1.3}$$

Equation (1.4) estimates the probability of a hypothesis from its prior ($P(H)$), the prior of its negation ($P(\sim H)$), $P(D|H)$, and $P(D|\sim H)$ (e.g., the null hypothesis is true versus the null hypothesis is untrue) (Hacking 2001):

$$P(H \mid D) = \frac{P(H)P(D \mid H)}{P(H)P(D \mid H) + P(\sim H)P(D \mid \sim H)} \tag{1.4}$$

Equation (1.5) is a generalization in which the probability of the ith hypothesis or explanation (H_i) of n hypotheses/explanations is true given the information (D),

$$P(H_i \mid D) = \frac{P(D \mid H_i)P(H_i)}{\sum_{i=1}^{n} P(D \mid H_i)P(H_i)} \tag{1.5}$$

So, the relative credulities for a set of alternate explanatory hypotheses can be judged quantitatively using these evidence-based probabilities. Lane et al. (1987), Lane (1989), and Hutchinson and Lane (1989) provide convincing demonstrations in which this approach greatly improved causal assessment in medical diagnostics. Equally important, estimates of belief warranted by evidence can be recalculated as evidence accumulates through time. These calculations can easily accommodate the most discriminating accept/reject context described by Popper (1959) and advocated by Platt (1964) in his strong inference approach. They can also be applied in more complex contexts characterized by higher uncertainty, as referred to by Wilson (1998) and encountered often at different stages of ecotoxicological investigations.

Example 1.1

It ain't what you don't know that gets you in trouble. It's what you know for sure that just ain't so.

—Attributed to both Artemis Ward and Mark Twain

Rizak and Hurdey (2006) posed a hypothetical, but very realistic, question to a large number of environmental professionals in an attempt to understand how accurate one should expect his or her responses to be during an actual situation requiring a sound decision. The question was the following:

Monitoring evidence for a(n) [Australian] city has indicated that in treated drinking water, a pesticide, say "atrazine," is truly present above the recognized standard methods detection limit once every 1,000 water samples from consumers' taps. The analytical test for the pesticide has the following characteristics:

- *95% of tests will be positive for detection when the contaminant is truly present above the detection limit, and*
- *98% of the tests will be negative for detection when the contaminant is truly not present above the detection limit.*

With these characteristics, given a positive result (detection) on the analytical test for the pesticide in the drinking water system, how likely do you think this positive result is true?

Most of the 352 respondents were highly experienced with these kinds of environmental issues. They were split into two general categories, water professionals and environmental engineering/science professors, although both categories produced similar responses. Of the water professionals, 32% believed the pesticide was almost certainly (0.95–1.00 probability) present and another 36% believed the pesticide was very likely (0.80–0.95 probability) present. At the other extreme, only 7% believed it was extremely unlikely (0.00–0.05 probability) or very unlikely (0.05–0.20 probability) to be present. The overall judgment of the professionals was a 0.80 probability that the pesticide was truly present given a positive test. (This book's author obtained very similar results after conducting a survey with this question while teaching two large classes in southern China and another in Vietnam, a seminar in Virginia, and giving keynote talks in Seville, Spain, and Kerala, India.)

The information provided by Rizak and Hrudey (2006) can be incorporated into Equation (1.4) to estimate the probability for the hypothesis of the pesticide being truly present given a positive test ($^{+}Test$), that is, $P(Pesticide|^{+}Test_1)$.

$$P(Pesticide \mid {}^{+}Test_1) = \frac{P(Pesticide)P({}^{+}Test_1 \mid Pesticide)}{P(Pesticide)P({}^{+}Test_1 \mid Pesticide) + P({}^{-}Pesticide)P({}^{+}Test_1 \mid {}^{-}Pesticide)}$$

$$P(Pesticide \mid {}^{+}Test_1) = \frac{(0.001)(0.95)}{(0.001)(0.95) + (0.999)(0.02)} = \frac{0.00095}{0.00095 + 0.01998} = 0.0454$$

The probability that the pesticide is truly present given a positive test is 0.045, not 0.80. Despite the opinions of most polled professionals, the plausibility is very low for the hypothesis that the pesticide was present. The basic error made often in these kinds of judgments is paying too little attention to the base rate (1 in 1,000 in this case).

Revision of such a posterior probability ($p = 0.0454$) based on more evidence can be illustrated by extending this fictitious example. Let us assume that a positive test for a tap water sample resulted automatically in it being subjected to further testing with a more sensitive, accurate, and precise analytical method. The probability of a positive test if the pesticide is truly present is 0.99, and the probability of a negative test when the pesticide is truly not present is 0.98. Assume that a positive test results from this second analysis and a new posterior probability is estimated,

$$P(Pesticide \mid {}^{+}Test_2) = \frac{(0.0454)(0.99)}{(0.0454)(0.99) + (0.9546)(0.02)} = 0.702$$

Positive results for the first and the follow-up test result in a new posterior probability of 0.702 for the hypothesis that the pesticide is truly present. Now the plausibility of the pesticide being present in the tap water, as calculated correctly from the evidence in hand, is high enough to warrant concern and consequent action. Perhaps this moderately high plausibility is sufficient to then prompt a second sampling of the tap from which the original positive sample was taken, and this second sample also produces a positive test. The new posterior probability can be calculated as follows:

$$P(Pesticide \mid {}^{+}Repeat\ Sample\ /\ Test_1) = \frac{(0.702)(0.95)}{(0.702)(0.95)+(0.298)(0.02)} = 0.991$$

A positive test from the first sample, a positive confirmation test for that same first sample, and a positive test for a second follow-up sample result in an estimated 0.991 probability that the pesticide is present in the water from that tap. Most decision makers would assume correctly from this information that the pesticide was present in this water source and take appropriate action.

Formal application of these equations is not always needed to apply quantitative abductive inference. In fact, it is often easier to communicate insights about such situations by using diagrams of natural frequencies (Gigerenzer 2002). This approach is illustrated in Figure 1.5 for the first two steps of the above abductive inference exercise.

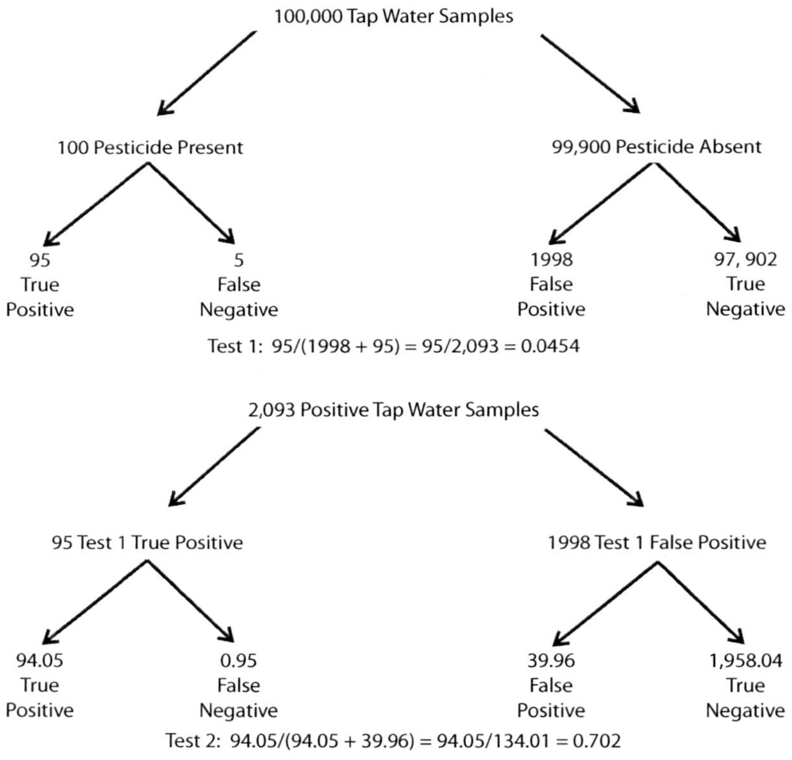

Figure 1.5 Diagrams of natural frequencies, instead of Bayesian equations, are used here to estimate the probability that a pesticide is present in a sample for which a positive test was obtained for an initial test (test 1, top of figure) and then a second, follow-up test (test 2, bottom of figure). In this illustration, 100,000 tap water samples were chosen as the initial number of samples analyzed for the pesticide. Frequencies of true positive and false positive samples were estimated using the given 1 in 1,000 prior frequency of the pesticide being present, and test 1 false negative and false positive rates of 0.05 and 0.02, respectively. Only 95 of the total number of positive tests (2,093) were true positive tests, so the posterior probability of the pesticide being present given test 1 produces a positive result is 95/2,093, or 0.0454. For 100,000 tap water samples, 2,093 will produce a positive test 1 result and be tested further with test 2. The frequency-based method of estimating the posterior probability of the pesticide being present if test 2 also shows a positive result is shown at the bottom of the figure. The posterior probability of the pesticide being present after positive results from the two tests are obtained for a tap water sample is now 94.05/134.01, or 0.702.

1.7 TOWARD STRONGEST POSSIBLE INFERENCE AND CLEAR ECOLOGICAL RELEVANCE

Emphasis is placed in this chapter on the development of ecotoxicology as a science. The discipline is characterized by a strong need for prediction at all levels of biological organization, but it is generally lacking in sufficient knowledge for making such predictions, especially at higher levels of organization. This condition is unfortunate because ecological relevance is highest for effects at the higher levels.

Increased complexity was rejected as the explanation for our lack of understanding at higher levels. Instead, the opinion is forwarded that the approaches employed in ecotoxicology, especially at higher levels of organization, do not foster rapid growth of knowledge (Figure 1.6). A strongest possible inferential approach, the best case of which is Platt's strong inference scheme, is advocated to alleviate some of this difficulty. Strongest possible inferential methods are applicable to all levels of organization, although powerful techniques with more explicit probabilities such as random assignment experiments or even quasi-experiments are logistically easier at lower levels. Criticisms based on the relative values of field versus mesocosm versus laboratory, holistic versus reductionist,

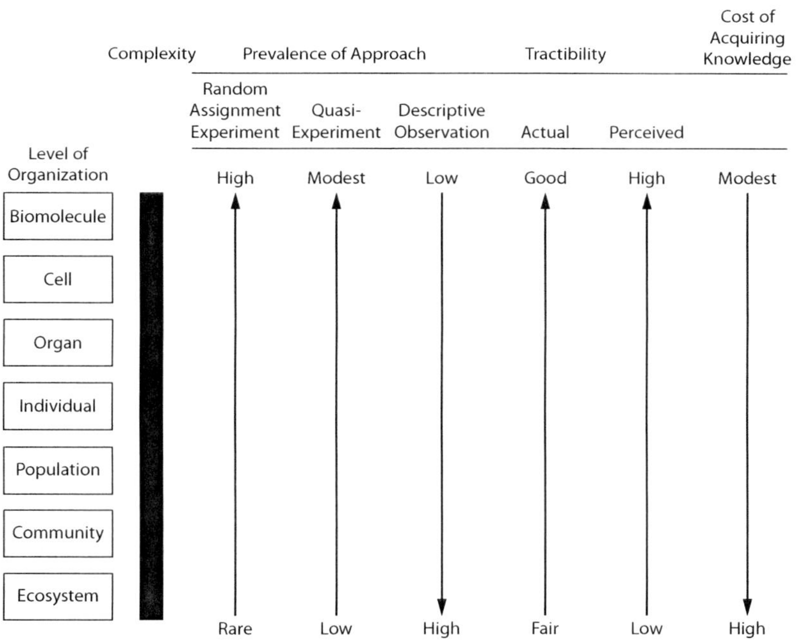

Figure 1.6 Elements of ecotoxicological knowledge and predominant approaches used to increase knowledge at each level of organization. This figure, used with Figure 1.3, qualitatively defines our present state of ecotoxicological inquiry. General approaches listed here include experiments with random assignment of experimental units to treatments, quasi-experiments (treatments; some measure of outcome and experimental units are present, but there is no random assignment of experimental units to treatments), and descriptive observation studies (see Cook and Campbell 1979 for more details). All are valuable approaches, but the values given to them relative to implying causal relationships are random assignment experiments > quasi-experiments > descriptive observation. The decrease in tractability (ability to extract knowledge) with increasing level of organization is not a consequence of system complexity. It is a consequence of the predominant approaches customarily taken and differences in cost of inquiry at the various levels of organization.

or standard versus nonstandard approaches become unfounded if strongest possible inference is an integral theme in all approaches. The decreased tractability of study at higher levels of organization or in the field must be counterbalanced by increased efforts and intensified inferential structure. Some difficulties can be offset to a degree by using the microcosm/mesocosm or large research team approaches, such as those taken by Likens and coworkers (1977). Contrary to common belief, stronger inferential techniques are most valuable at higher levels of organization where ecological relevance is high but costs can be prohibitive. An analogy would be the intense effort invested in planning crucial and decisive experiments employing limited beam time on particle accelerators.

Strongest possible inference is advocated as the most effective means of obtaining useful knowledge in the emerging and socially obligated field of ecotoxicology. It is characterized by the systematic and consistent application of standard methods of inference, especially Bayesian inference techniques. It includes the concept of multiple hypotheses as implemented by abductive methods as a means of minimizing subjectivity in selection and comparison of hypotheses. Methods for selecting hypotheses should be influenced by amenability to falsification/abductive computations and quantitative formulation, and parsimony, including theory reduction. To be most effective, the measurement process must be demonstrably precise and unbiased. Hypotheses should be readily amenable to abductive analysis such that statistical rejection criteria or evidence-based plausibilities are defined in terms of probability.

Based on the materials discussed above, 12 habits are suggested to enhance the rate at which ecotoxicological knowledge is acquired.

1. Be aware of and avoid the habits of precipitate explanation and informal abduction.
2. While recognizing the value of such approaches, avoid an excessive preoccupation with "surveys, taxonomic, design of equipment, systematic measurements and tables, [and] theoretical calculation."
3. Systematically and consistently apply inductive/abductive inference techniques.
4. Favor experiments or observations with a high risk of logical falsification or producing evidence with the most influence on estimated posterior probabilities.
5. Favor quantitative methods under rigorous precision/accuracy control.
6. Preferentially formulate hypotheses amenable to falsification or the strongest testing possible.
7. Apply the principle of multiple hypotheses.
8. Favor hypotheses that are easily falsifiable or subject to the strongest possible testing.
9. Favor hypotheses that are tractable and have good probability of solving the problem.
10. Favor parsimonious hypotheses.
11. Favor hypotheses that enhance theory reduction.
12. Recognize negative results as a critically important component of the process, not a consequence of the worker's failure to pick "the right question."

REFERENCES

Adams, S.M., ed. Biological indicators of stress in fish. *Am. Fisheries Soc. Symp.* 8:1–8 (1990).

Adams, S.M., K.L. Shepard, M.S. Greeley Jr., B.D. Jimenez, M.G. Ryon, L.R. Shugart, and J.F. McCarthy. The use of bioindicators for assessing the effects of pollutant stress on fish. *Mar. Environ. Res.* 28:459–464 (1989).

Barrett, G.W., G.M. Van Dyne, and E.P. Odum. Stress ecology. *Bioscience* 26(3):192–194 (1976).

Bradshaw, A.D., and K. Hardwick. Evolution and stress—Genotypic and phenotypic component. *Biol. J. Linn. Soc.* 37:137–155 (1989).

Brisbin, I.L., Jr., K.W. McLeod, and G.C. White. Sigmoid growth and the assessment of hormesis: A case for caution. *Health Phys.* 52(5):553–559 (1987).

Buikema, A.L., Jr., B.R. Niederlehner, and J. Cairns Jr. Biological monitoring part IV—Toxicity testing. *Water Res.* 16:239–262 (1982).

Burton, G.A., Jr. Assessing the toxicity of freshwater sediments. *Environ. Toxicol. Chem.* 10:1585–1627 (1991).

Cairns, J., Jr. Will the real ecotoxicologist please stand up? *Environ. Toxicol. Chem.* 8:843–844 (1989).

Cairns, J., Jr. Restoration ecology: A major opportunity for ecotoxicologists. *Environ. Toxicol. Chem.* 10:429–432 (1991).

Cairns, J., Jr., and D.I. Mount. Aquatic toxicology. Part 2 of a four-part series. *Environ. Sci. Technol.* 24(2):154–161 (1990).

Calabrese, E.J. Hormesis. Why is it important to toxicology and toxicologists? *Environ. Toxicol. Chem.* 27:1451–1474 (2008).

Calabrese, E.J., M.E. McCarthy, and E. Kenyon. The occurrence of chemically induced hormesis. *Health Phys.* 52(5):531–541 (1987).

Casarett, L.J., and J. Doull. *Toxicology: The basic science of poisons*. New York: Macmillan Publishing Co., 1975.

Chamberlin, T.C. The method of multiple working hypotheses. *J. Geol.* 5:837–848 (1897).

Chapman, P.M. Environmental quality criteria. What type should we be developing? *Environ. Sci. Technol.* 25(8):1353–1359 (1991).

Clemmitt, M. Public, private health concerns spur rapid progress in toxicology. *Scientist* 6(4):1–7 (1992).

Connell, D.W. Ecotoxicology—A framework for investigations of hazardous chemicals in the environment. *Ambio* 16(1):47–50 (1987).

Cook, T.D., and D.T. Campbell. *Quasi-experimentation design and analysis issues for field settings*. Boston: Houghton Mifflin Company, 1979.

Dawson, M.A., E. Gould, F.P. Thurberg, and A. Calabrese. Physiological response of juvenile striped bass, *Morone saxatilis*, to low levels of cadmium and mercury. *Chesapeake Sci.* 18(4):353–359 (1977).

Duffus, J.H. 1986. Interpretation of ecotoxicity. In *Toxic hazard assessment of chemicals*, ed. M. Richardson. London: Royal Society of Chemistry, Burlington House, pp. 90–116.

Eaton, D.L., and S.G. Gilbert. 2008. Principles of toxicology. In *Casarett and Doull's toxicology. The basic science of poisons*, ed. C.D. Klaassen. 7th ed. New York: McGraw-Hill Co., chap. 2.

Esch, G.W., and T.C. Hazen. Thermal ecology and stress: A case history for red-sore disease in largemouth bass. In *Energy and environmental stress in aquatic systems,* ed. J.H. Thorp and J.W. Gibbons. CONF-771114. Springfield, VA: National Technical Information Service, 1978, pp. 331–363.

Gigerenzer, G. *Calculated risks. How to know when numbers deceive you*. New York: Simon & Schuster, 2002.

Gore, A. *An inconvenient truth. The planetary emergency of global warming and what we can do about it*. New York: Rodale, 2006.

Guttman, B.S. Is "levels of organization" a useful biological concept? *Bioscience* 26(2):112–113 (1976).

Hacking, I. *An introduction to probability and inductive logic*. Cambridge: Cambridge University Press, 2001.

Haux, C., and L. Förlin. Biochemical methods for detecting effects of contaminants on fish. *Ambio* 17(6):376–380 (1988).

Hodson, P.V. *The effect of temperature on the toxicity of zinc to fish of the genus Salmo*. PhD thesis, University of Guelph, Ontario, 1974.

Hoffman, A.A., and P.A. Parsons. *Evolutionary genetics and environmental stress*. Oxford: Oxford University Press, 1991.

Howson, C., and P. Urbach. *Scientific reasoning. The Bayesian approach*. La Salle, IL: Open Court Publishing Company, 1989.

Hutchinson, T.A., and D.A. Lane. Assessing methods for causality assessment of suspected adverse drug reactions. *J. Clin. Epidemiol.* 42:5–16 (1989).

Jeffreys, W.H., and J.O. Berger. Ockham's razor and Bayesian analysis. *Am. Sci.* 80:64–72 (1992).

Jørgensen, S.E. 1990. *Modelling in ecotoxicology*. New York: Elsevier.

Josephson, J.R., and S.G. Josephson. *Abductive Inference. Computation, philosophy, technology*. Cambridge: Cambridge University Press, 1996.

King, S.F. *Some effects of DDT on the guppy and the brown trout*. U.S. Fish and Wildlife Service Scientific Report 399. Washington, DC: U.S. Fish and Wildlife Service, 1962.

Klaverkamp, J.F., M.D. Dutton, H.S. Majewski, R.V. Hunt, and L.J. Wesson. Evaluating the effectiveness of metal pollution controls in a smelter by using metallothionein and other biochemical responses in fish. In *Metal ecotoxicology, concepts and applications*, ed. M.C. Newman and A.W. McIntosh. Chelsea, MI: Lewis Publishers, 1991, pp. 33–64.

Koehn, R.K., and B.L. Bayne. Towards a physiological and genetical understanding of the energetics of the stress response. *Biol. J. Linn. Soc.* 37:157–171 (1989).

Lane, D.A. Subjective probability and causality assessment. *Appl. Stoch. Model Data Anal.* 5:53–76 (1989).

Lane, D.A., M.S. Kramer, T.A. Hutchinson, J.K. Jones, and C. Naranjo. The causality assessment of adverse drug reactions using a Bayesian approach. *Pharmaceut. Med.* 2:265–283 (1987).

Lederman, L.M. *Science: The end of the frontier?* A supplement to *Science*. Washington, DC: AAAS, January 1991, p. 20.

Levin, S.A. Ecosystems and the biosphere as complex adaptive systems. *Ecosystems* 1:431–436 (1998).

Likens, G.E., F.H. Bormann, R.S. Pierce, J.S. Eaton, and N.M. Johnson. *Biogeochemistry of a forested ecosystem*. New York: Springer-Verlag, 1977.

Loehle, C. Philosophical tools: Potential contributions to ecology. *Oikos* 51:97–104 (1988).

Loehle, C. A guide to increased creativity in research—Inspiration or perspiration? *Bioscience* 40:123–129 (1990).

Lugo, A.E. Stress and ecosystems. In *Energy and environmental stress in aquatic systems*, ed. J.H. Thorp and J.W. Gibbons. Springfield, VA: National Technical Information Service CONF-771114, 1978, pp. 62–101.

Maciorowski, A.F. Populations and communities: Linking toxicology and ecology in a new synthesis. *Environ. Toxicol. Chem.* 7:677–678 (1988).

Margalef, R. *Perspectives in ecological theory*. Chicago: University of Chicago Press, 1968, p. 13.

May, R.M., and A.R. McLean. 2007. *Theoretical ecology. Principles and applications*. Oxford: Oxford University Press.

Moriarty, F. *Ecotoxicology: The study of pollutants in ecosystems*. Orlando, FL: Academic Press, 1983.

Nagel, E. *The structure of science. Problems in the logic of scientific explanation*. New York: Harcourt, Brace and World, 1961.

Newman, M.C. "What exactly are you inferring?" A closer look at hypothesis testing. *Environ. Toxicol. Chem.* 27:1013–1019 (2008).

Newman, M.C. *Fundamentals of ecotoxicology*. 3rd ed. Boca Raton, FL: CRC Press, 2010.

Newman, M.C., and W.H. Clements. *Ecotoxicology: A comprehensive treatment*. Boca Raton, FL: CRC Press, 2008.

Odum, E.P. The strategy of ecosystem development. *Science* 162:262–270 (1969).

Odum, E.P. Trends expected in stressed ecosystems. *BioScience* 35:419–422 (1985).

Platt, J.R. Strong inference. *Science* 146:347–353 (1964).

Poincaré, H. *Science and hypothesis*. New York: Dover Publishing, 1952. English translation of 1905 *La Science Et l'Hypothèse* by G.B. Halsted.

Popper, K.R. *The logic of scientific discovery*. London: Routledge, 1959.

Popper, K.R. *Conjectures and refutations. The growth of scientific knowledge*. New York: Harper and Row, 1965.

Popper, K.R. *The logic of scientific discovery*. London: Hutchinson and Company, 1968.

Rand, G.M., and S.R. Petrocelli. *Fundamentals of aquatic toxicology*. New York: Hemisphere Publishing Corporation, 1985.

Rapport, D.J., H.A. Regier, and T.C. Hutchinson. Ecosystem behavior under stress. *Am. Nat.* 125(5):617–640 (1985).

Rizak, S.N., and S.E. Hrudey. Misinterpretation of drinking water quality monitoring data with implications for risk management. *Environ. Sci. Technol.* 49(17):5244–5250 (2006).

Rousseau, D.L. Case studies in pathological science. *Am. Sci.* 80:54–63 (1992).

Sagan, L.A. What is hormesis and why haven't we heard about it before? *Health Phys.* 52(5):521–525 (1987).

Selye, H. *The stress of life*. New York: McGraw-Hill Book Company, 1956.

Selye, H. The evolution of the stress concept. *Am. Sci.* 61:692–699 (1973).

Sibly, R.M., and P. Calow. A life-cycle theory of responses to stress. *Biol. J. Linn. Soc.* 37:101–116 (1989).

Shuter, B.J. Population-level indicators of stress. *Am. Fisheries Soc. Symp.* 8:145–166 (1990).

Southam, C.M., and J. Ehrlich. Effects of extract of western red cedar heartwood on certain wood-decaying fungi in culture. *Phytopathology* 33:517–524 (1943).

Sprague, J.B. Measurement of pollutant toxicity to fish. I. Bioassay methods for acute toxicity. *Water Res.* 3:793–821 (1969).

Sprague, J.B. Measurement of pollutant toxicity to fish. III. Sublethal effects and "safe" concentrations. *Water Res.* 5:245–266 (1971).

Stebbing, A.R.D. Hormesis—The stimulation of growth by low levels of inhibitors. *Sci. Total Environ.* 22:213–234 (1982).

Suter, G.W., A.E. Rosen, E. Linder, and D.F. Parkhurst. Endpoints for responses of fish to chronic toxic exposures. *Environ. Toxicol. Chem.* 6:793–809 (1987).

Truhaut, R. Ecotoxicology: Objectives, principles and perspectives. *Ecotoxicol. Environ. Saf.* 1:151–173 (1977).

Weber, C.I., W.H. Peltier, T.J. Norberg-King, W.B. Horning II, F.A. Kessler, J.R. Menkedick, T.W. Neiheisel, P.A. Lewis, D.J. Klemm, Q.H. Pickering, E.L. Robinson, J.M. Lazorchak, L.J. Wymer, and R.W. Freyberg. *Short-term methods for estimating the chronic toxicity of effluents and receiving waters to freshwater organisms.* 2nd ed., EPA/600/4-89/001. Cincinnati, OH: EMSL U.S. Environmental Protection Agency, 1989.

Wilson, E.O. *Consilience. The unity of knowledge*. New York: Random House, 1998.

Woodworth, G.G. *Biostatistics. A Bayesian introduction*. Hoboken, NJ: John Wiley & Sons, 2004.

CHAPTER 2

The Measurement Process

Boswell: "Sir Alexander Dick tells me, that he remembers having a thousand people a year to dine at his house; that is reckoning each person as one, each time he dines there." Johnson: "That is about three a day." Boswell: "How your statement lessens the idea." Johnson: "That, Sir, is the good of counting. It brings everything to a certainty, which before floated in the mind indefinitely." Boswell: "But Omne ignotum pro magnifico est:* one is sorry to have this diminished." Johnson: "Sir, you should not allow yourself to be delighted with error." Boswell and Glover

—Samuel Johnson (1750)

2.1 GENERAL

2.1.1 Overview

Rousseau (1992) defines three symptoms of pathological science (the excessive loss of scientific objectivity). The first is an aversion to crucial experiments that could disprove a favored theory. The second is a disregard for prevailing ideas and theories. Traditional theories are given inadequate consideration as the researcher becomes more and more enamored with a new discovery. These first two symptoms should seem familiar to the reader as they were discussed in Chapter 1 as low-risk testing and precipitate explanation. The third symptom was mentioned only very briefly in that chapter. Rousseau suggests that the last symptom of pathological science often begins with an effect that is "at the limits of detectability or has very low statistical significance.... Once the investigator [is] convinced that something new and important has been discovered, the fact that all of the parameters involved ... are not under control is viewed as having little consequence" (Rousseau 1992). The improperly controlled or poorly understood measurement process acts as the seed from which increasingly biased behavior grows. Fortunately, the means are available for minimizing such misinterpretations of the results of most measurement processes. Several methods for controlling measurement difficulties in or defining the limitations of the measurement process are outlined here at the beginning of this book because, in their absence, later methods would be useless.

Although sometimes believed to be pertinent primarily to chemical analyses, the techniques described in this chapter for assessing such qualities as accuracy and precision are amenable to and necessary for any measurement process. For example, if a plankton net is towed, the measured numbers of individuals of each species from that tow have associated limits of detection. (Indeed, this point will be discussed again in Chapter 7 regarding species abundance.) Further, questions regarding precision and accuracy of measurements must be answered prior to any meaningful data analysis. In this example, precision may be quantified with replicate tows of the same net or one tow of

* "The unknown always passes for the marvelous."

two identical nets coupled by a common yoke. Precision of the enumeration process might involve having several skilled individuals count the same plankton sample in the laboratory. Accuracy during enumeration could be estimated with species "spikes" to a portion or aliquot of a sample.

2.1.2 The Necessity of Controlled Measurement

It is difficult to find a sound reason why this crucial aspect of ecotoxicology was given such low priority for many years. Although formal methods for implementing quality control have been available since the 1920s (Grant 1964; Shewart 1986), quality control was followed informally with varying degrees of commitment until environmental regulations mandated otherwise. Inexplicably, it remains underemphasized in college course work outside of statistics, applied chemistry, and engineering. Indeed, programs implemented to ensure controlled measurement can still elicit extreme responses. "Some of the scientists seem to consider quality assurance an insult to their professional integrity, an obstruction to 'real work' and an inference that scientists will cheat or falsify data or results" (Zimmerman 1990). Such an immoderate attitude may be a remnant of concepts abandoned early in the development of science (see Chapter 1 or Chamberlin 1897). Regardless, intersubjective testing as practiced today requires clear documentation of measurement conditions (Ayala et al. 1989). Failure to do so inhibits progress and improvement of skills (Taylor 1987; Rayl 1991), decreases the effectiveness of the decision-making process (Keith et al. 1983; Palca 1991), fosters self-delusion (Rousseau 1992), and inhibits our ability to detect the infrequent occasion of fraud (Koshland 1987; Culliton 1988).

2.2 REGIONS OF QUANTITATION

2.2.1 Overview

Keith et al. (1983) and Taylor (1987) provide lucid explanations of quantitation regions; consequently, their presentations are condensed into this overview with only minor modification. The interested reader is urged to examine the original materials for enriching details.

The certainty of a measured value can be gauged relative to the standard deviation (s_o) for samples with concentrations near 0, i.e., signals near the baseline noise of the measurement process. The relative uncertainty is most commonly used for this purpose (Taylor 1987).

$$\text{Relative Uncertainty (\%)} = 100\frac{z\sqrt{2}}{N} \tag{2.1}$$

where $z = z$ statistic at a confidence level of $100(1 - \alpha)\%$, and N = measured value of an observation expressed as a multiple of s_o.

For example, the relative uncertainty of a value three times larger than s_o at a 95% confidence level is $100[(1.960\sqrt{2})/3]$, or ±92%. Such an observation has a measurement uncertainty nearly as large as its mean value. A more acceptable measurement uncertainty might be that associated with a $10s_o$ value, i.e., $100[(1.960\sqrt{2})/10]$, or ±28%.

The information needed to calculate the relative uncertainty (mean value and s_o) can be generated in several ways. The sample signal (S_t) may be used to estimate the mean value if there is no significant blank signal (S_b). If the blank signal is measurable then the difference between the sample and blank signals ($S_t - S_b$) may be used. Next, one of two methods can be used to estimate the standard deviation of the measurement process (s_o). As recommended by Taylor (1987), s_o can be

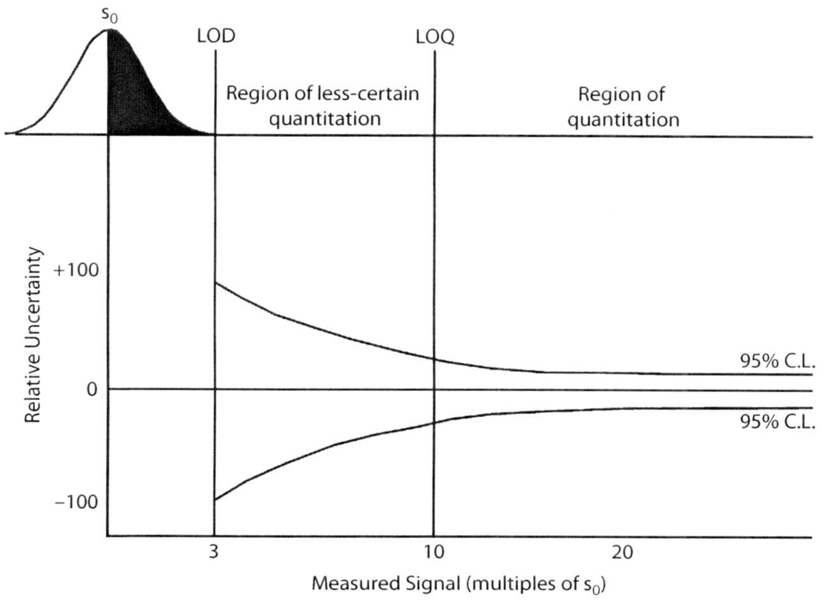

Figure 2.1 Regions of quantitation. (Modified from Keith, L.H. et al., *Anal. Chem.*, 55, 2210–2218, 1983, and Taylor, J.K., *Quality Assurance of Chemical Measurements*, Lewis Publishers, Chelsea, MI, 1987.)

estimated by plotting the standard deviation of signal versus concentration for a series of replicate standards (or samples), including one set close to a concentration of 0. The Y-intercept of the plot is then used to estimate s_o. A minimum of three concentration levels with a total of at least seven measured signals is recommended for estimating s_o. Although this linear extrapolation method is preferred by Taylor (1987), s_o can also be estimated with signals from replicates at one concentration near the limit of quantitation. Taylor (1987) recommends using a concentration close to approximately $20s_o$ in this single concentration approach, although no explanation is given for choosing this concentration. The author uses replicates of samples at or near that of the background.

Using the information generated above, each measurement can now be assigned to a region of quantitation (Figure 2.1). Again, the measured values are expressed as multiples of s_o. At $3s_o$ above the baseline signal (the limit of detection (LOD)), the measured value is estimated to be within nearly ±100% of the true value with a 95% confidence level. Values below the LOD are often reported as below the detection limit (<DL or <LOD) or not detected (ND). Above the limit of detection but below $10s_o$ is the region of less certain quantitation (also called the region of qualitative analysis by Currie (1968)). In this region, a measurement is detectable but associated with a sufficiently large uncertainty (greater than approximately ±30%) to render it semiquantitative. Above $10s_o$ (the limit of quantitation (LOQ)) but below the point at which linearity of the standard curve ends (limit of linearity (LOL)) is the region of quantitation.

A major goal in design of quantitative samplings and procedures should be to generate a data set with all observation values within the region of quantitation. Indeed, recommendations have been forwarded that only data in the region of quantitation should be used to make quantitative decisions (Keith et al. 1983). Unfortunately, constraints such as available instrumentation, regulation-mandated methodologies, temporal variation in concentrations, and incorporation of control treatments or uncontaminated sites often produce data sets with observations in all or several measurement regions.

2.2.2 Data Sets with Below Detection Limit Observations

2.2.2.1 Definitions

Extending the above discussion, a straightforward definition of the limit of detection can be stated. The limit of detection is defined as the signal level $3s_o$ above the baseline signal expressed in the units of interest, e.g., concentration units. Below the limit of detection, the measurement uncertainty is approximately equal to or greater than the value itself. Note that specific applications may necessitate modification of this general definition, e.g., radiological methods assuming a Poisson distribution for signal variability. Estimation for nonlinear calibration methods (Schwartz 1983), comparison of instruments (Arellano et al. 1985), estimation using chromatographic techniques (Synovec and Yeung 1985), measurement of radioactivity (Currie 1968; Donn and Wolke 1977), and measurement of mixture analytes such as Aroclor 1242 (Alford-Stevens 1987) or dioxins (Helsel 2010) are a few important examples that might require refinement of the above definition.

Further, several categories of limits of detection can be defined depending on the intended use of the resulting statistic. An instrument detection limit (IDL) measures the signal-to-noise ratio associated with the measurement equipment. The method detection limit (MDL) includes variation in all measurement steps leading to estimation of s_o. For example, the mass of sample used, extraction consistency, and other procedural steps contributing to signal variation may be incorporated during estimation of a MDL.

2.2.2.2 Reporting

The limit of reporting (LR) is "a limit above which data values are reported without qualification by an analytical laboratory" (Helsel 2005). The LR may be the LOD or even the LOQ (Keith et al. 1983; Helsel and Gilliom 1986). Most often, data sets are reported with all observations assigned numerical values except those below the LOD. Observations with values less than the LOD are noted as below the LOD (BDL, <DL, <LR, or ND). The laudable intent of this practice is to report only values that are above the noise of the measurement process by a statistically defined amount. The resulting data set that contains a subset of observations with no assigned numerical value below a certain point is said to be censored. The data can be further defined as left censored because censoring at the LOD involves low values from the left portion of the sample distribution. Such an approach can cause interpretation problems later because it gives a false impression to unwary data end users that the quality of the information associated with reported values from the region of less certain quantitation is equivalent to that of values from above the limit of quantitation.

Left censoring of data sets creates many problems. Censoring precludes use of valuable information by the decision maker. Fortunately, arguments against universal censoring are increasing in frequency (Gilbert and Kinnison 1981; Gilliom et al. 1984; Gilbert 1987; Porter et al. 1988; Newman et al. 1989). According to these authors, it could be more effective in some situations to report values for all observations along with the associated measurement uncertainty. For example, the relative uncertainty could be estimated using Equation (2.1) if s_o was reported. Then the end user could examine and manipulate the data set as appropriate for his or her particular need. For example, a rank order test may extract valuable information more effectively from a data set with a moderate proportion of observations below the LOD or LOQ. Such an approach would require considerable—perhaps unrealistic—levels of diligence during information extraction from data sets in order to avoid equally serious compromises as those imposed by the current practice of censoring.

2.2.2.3 *Estimating Mean and Standard Deviation for Censored Data*

2.2.2.3.1 Deletion and Substitution Methods

Regardless of arguments against censoring, left censoring of data sets remains a widespread practice. Commonly used, informal methods of coping with censored data sets are deletion or substitution techniques (Newman et al. 1989). In the deletion procedure, values below the reporting limit are not used in estimation of mean and standard deviation. In the substitution procedure, some value such as 0, ½LOD, or the LOD is substituted for all values below the reporting limit when the mean and standard deviation are to be calculated. Although substitution of ½LOD for a single or few LOD observations in a large data set may not cause major problems, the general application of these statistically unsubstantiated techniques biases estimates of mean and standard deviation. For example, deletion and substitution methods will produce biased estimates of the mean. In Figure 2.2A, two distributions with identical means but dissimilar standard deviations are used to illustrate this bias. The estimate of the mean for the distribution with the wider standard deviation will be biased upward more than that of the distribution with the narrower standard deviation if observations below the LOD are deleted, or if the LODs are substituted for <DL observations. If 0 or ½LOD is substituted for the censored values, the estimated mean for the broader distribution will be smaller than that of the narrow distribution.

Similarly, standard deviations estimated with deletion or substitution methods are biased (Figure 2.2B). For example, deletion of or substitution of the LOD for censored values will bias the

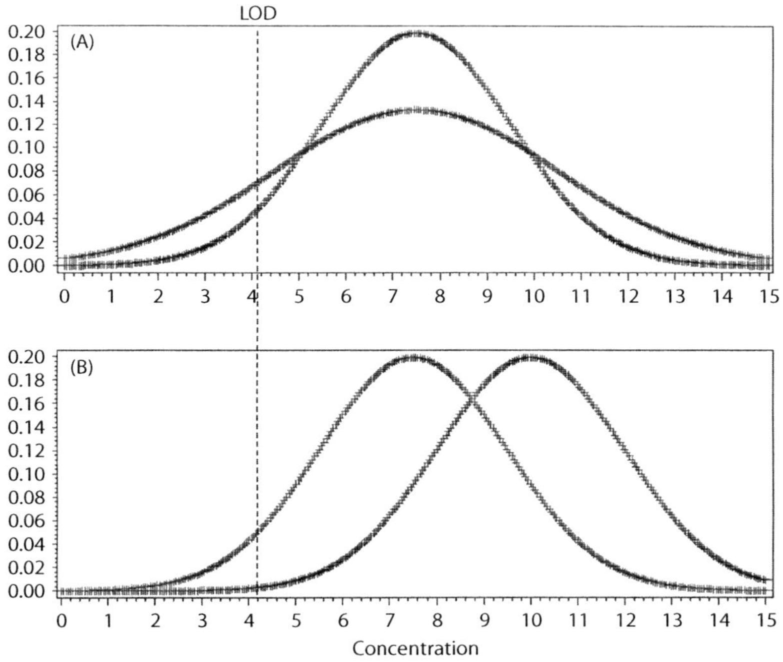

Figure 2.2 Problems associated with deletion or substitution techniques used to estimate means and standard deviations in left-censored data sets. Panel A shows two normal distributions with identical means but different standard deviations. Panel B shows two normal distributions with different means but identical standard deviations. The points censored (below the LOD) are those in the curve areas to the left of the vertical dashed line. (Modified from Newman, M.C., and P.M. Dixon, *Am. Environ. Lab.*, 4/90, 27–30, 1990.)

standard deviation for the distribution with the smaller mean more than that with the larger mean. Substitution of 0 will have the opposite effect.

2.2.2.3.2 *Winsorized Mean and Standard Deviation*

Winsorized estimates of mean and standard deviation can be used for censored data sets. The assumption is made that the data are distributed symmetrically. The <DL values are replaced by the smallest observation value above the LOD; however, the same number of largest values are also replaced by the value of the next smallest observation. The mean and standard deviation are calculated for this modified data set. The simple, arithmetic mean for the modified data (Winsorized mean) is unbiased (Equation 2.2). However, Equation (2.3) is needed to provide an unbiased estimate of the Winsorized standard deviation.

$$\bar{x}_w = \frac{\sum_{i=1}^{n} x_i}{n} \tag{2.2}$$

$$s_w = \frac{s(n-1)}{v-1} \tag{2.3}$$

where s = standard deviation of the modified data set, n = the total number of observations, and v = the number of observations not modified.

Example 2.1

The following sulfate concentrations (mg/L) were measured during a routine water quality survey of the Savannah River (South Carolina) (n = 21, mean = 5.1, standard deviation = 2.1):

1.3 ("<2.5")	3.5	5.2	6.5
2.3 ("<2.5")	3.6	5.6	6.9
2.6	4.0	5.7	7.1
3.3	4.1	6.1	7.7
3.5	4.5	6.2	7.9
			9.9

For this illustration, let's assume that the lowest two observations were censored; i.e., we have an embarrassing detection limit of only 2.5 mg/L. The smallest two observations are replaced by the next largest value (<2.5 and <2.5 become 2.6 and 2.6). Next, the two largest values are replaced by the value for the next smallest observation (7.9 and 9.9 become 7.7 and 7.7). The modified data set is now the following:

2.6	3.5	5.2	6.5
2.6	3.6	5.6	6.9
2.6	4.0	5.7	7.1
3.3	4.1	6.1	7.7
3.5	4.5	6.2	7.7
			7.7

The Winsorized mean ($\bar{x}_w$) is 5.08. The standard deviation of the modified data set (s) is 1.79. The number of observations (n) is 21 and the number of unmodified observations (v) is 17 (or n

– 4). The Winsorized standard deviation (s_w) is estimated with Equation (2.3) to be [1.79(21 – 1)]/(17 – 1), or 2.24. Rounding off, the Winsorized mean (5.1) for the censored data set is the same as that estimated for the uncensored data. The Winsorized standard deviation is slightly higher (2.2) than that estimated for the uncensored data (2.1).

The SAS code below performs the Winsorization just described and also the similar approach of trimming data from the two distribution tails. A 95% confidence interval for the mean (4.04 to 6.12) is estimated. In this case, a trimming of 0.952 from each tail results by intent in two observations being removed from each tail.

```
DATA ALL;
INPUT SO4 @@;
DATALINES;
1.3 2.3 2.6 3.3 3.5 3.5 3.6 4.0 4.1 4.5 5.2 5.6 5.7 6.1 6.2 6.5 6.9 7.1
7.7 7.9 9.9
;
RUN; /* WINSORIZATION OF TWO LOWEST AND TWO HIGHEST VALUES */
PROC CAPABILITY WINSOR=2(TYPE=TWOSIDED ALPHA=.05);
VAR SO4;
RUN; /* TRIMMING OF TWO LOWEST AND TWO HIGHEST VALUES */
PROC CAPABILITY TRIMMED=.0952;
VAR SO4;
RUN;
```

Winsorization or trimming is useful if the distribution is symmetrical. In Example 2.1, the assumption of a normal distribution is made after careful examination of the data set. The technique can also be used for some skewed data if a transformation is performed that produces a symmetrical distribution. For example, a symmetrical (normal) distribution is produced if a set of lognormal observations is log transformed. However, a backtransformation bias would then have to be estimated for log-transformed data, as will be described below. Further discussion of Winsorized estimates can be found in Dixon and Massey (1969), Sokal and Rohlf (1981), Gilbert (1987), and Berthouex and Hinton (1991).

A Winsorized t can be calculated that allows a confidence interval for the Winsorized mean to be generated (Dixon and Massey 1969),

$$t = \frac{v-1}{n-1} t_w \tag{2.4}$$

where t_w = a t from a conventional table or software function using a specified probability (p) and df of v – 1. The Excel® function TINV(1 – confidence level,df) can be used and, in this example, produces TINV(0.05,16) = 2.12. The confidence interval is then constructed,

$$\bar{x}_w \pm \frac{n-1}{v-1} t_{0.975(v-1)} \frac{s_w}{\sqrt{n}} \tag{2.5}$$

In this example, the 95% confidence interval is 5.08 ± 1.03, that is, 4.05 to 6.11. Notice in the PROC CAPABILITY statement given in Example 2.1 that a two-sided 95% confidence interval was requested during Winsorization. The estimated confidence interval from SAS was 4.04 to 6.12.

2.2.2.3.3 *Probability Plotting*

Perhaps the most straightforward method for estimating the mean and standard deviation of censored data sets is probability plotting. The sulfate data from Example 2.1 are used in Figure 2.3

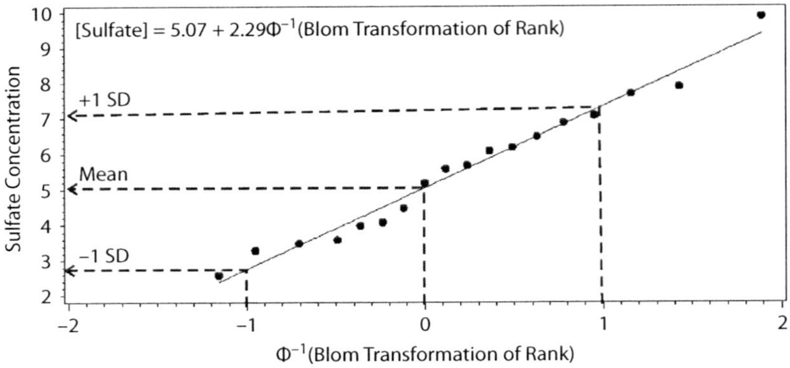

Figure 2.3 An example of estimating mean and standard deviation by probability plotting. The two lowest of the 21 sulfate concentrations are treated as less than the LOD. It is clear that, with the points above the censored values, a reasonable estimate of the mean and standard deviation can be obtained using concentrations corresponding to the regression line intercept and slope.

to illustrate this approach. Again, the lowest two observations are assumed to be below the LOD. The observations above the LOD are ranked from smallest to largest. The ranks for the 19 observations begin at 3 and progress to 21 because the censored observations occupied the first two lowest ranks of 1 and 2. The ranks can then be converted to proportions using one of several approximations, the van der Waerden (Equation 2.6), Tukey (Equation 2.7), or Blom (Equation 2.8) scaling transformations. Then these ranks are converted to their corresponding inverse normal scores (Φ^{-1}(proportion)), which express the proportions in units of standard deviations from the mean. The inverse normal score can be obtained readily with functions of numerous software packages such as Excel NORMINV(0.XX,0,1), where 0.XX is the estimated proportion and 0,1 specify a unit normal curve with a mean of 0 and standard deviation of 1.

$$\Phi^{-1}\left(\frac{Rank}{n+1}\right) \tag{2.6}$$

$$\Phi^{-1}\left(\frac{Rank-\frac{1}{3}}{n+\frac{1}{3}}\right) \tag{2.7}$$

$$\Phi^{-1}\left(\frac{Rank-\frac{3}{8}}{n+\frac{1}{3}}\right) \tag{2.8}$$

The Φ^{-1}(rank transformation) values are then plotted against their corresponding concentrations for all observations above the LOD (Figure 2.3). A linear regression model might be performed to produce a line intercept and slope that are estimates of the mean and standard deviation, respectively. The following code was used to generate Figure 2.3 and the regression estimates given in that figure. The scaling transformation preferred by this author (NORMAL = BLOM) was specified but the van der Waerden (NORMAL = VW) or Tukey (NORMAL = TUKEY) transformation might have been chosen instead.

```
GOPTIONS RESET=ALL BORDER;
DATA ALL;
INPUT SO4 @@;
IF SO4<2.5 THEN SO4=2.4; RANKSO4=SO4;
DATALINES;
1.3 2.3 2.6 3.3 3.5 3.5 3.6 4.0 4.1 4.5
5.2 5.6 5.7 6.1 6.2 6.5 6.9 7.1 7.7 7.9 9.9
;
RUN;
PROC RANK TIES=MEAN NORMAL=BLOM OUT=NEW;
VAR RANKSO4;
RUN;
DATA NPLOT;
SET NEW;
IF SO4 NE 2.4;
RUN;
PROC GLM;
MODEL SO4=RANKSO4/CLPARM;
OUTPUT OUT=PRED PREDICTED=PSO4;
RUN;
SYMBOL1 INTERPOL=JOIN VALUE=NONE COLOR=BLACK;
SYMBOL2 VALUE=DOT COLOR=BLACK;
PROC GPLOT DATA=PRED;
PLOT (PSO4 SO4)*RANKSO4/OVERLAY;
RUN;
```

The mean and standard deviation as estimated from the intercept and slope, respectively, were 5.07 and 2.29. The 95% confidence limits are also readily generated for the intercept and slope by including CLPARM in the GLM MODEL statement. The estimates are similar to the mean and standard deviation for the uncensored data set (5.1 and 2.1). This approach is referred to by Helsel (2005) as the fully parametric regression on order statistics approach, and is judged less useful than the robust regression on order statistics described below. It is worth noting that Antweiler and Taylor (2008) agree with Helsel's assessment. With the robust method (robust regression on order statistics) (Helsel 2005), the regression model is used to generate predicted concentrations for the censored observations and the predicted values for these observations are added to the uncensored data. The mean and standard deviation are then estimated with this filled-in data set. In this way, only the fill-in observation values require any assumption of a particular distribution to produce and the original values are used for the above LOD observations. Helsel often advocates for the assumption of a lognormal distribution (that is, log transformation of observations prior to method application) when generating the fill-in observation values because it is difficult to determine distribution type for many environmental data sets and his simulations have shown that the lognormal distribution is the most robust assumption to make. Helsel provides a SAS macro on his webpage (http://www.practicalstats.com/nada/) that implements this procedure. His macro was applied to the sulfate data with the following SAS code:

```
%MACRO cros(f,rem,value);
PROC PRINT;
RUN;
DATA SO4;
INPUT value rem $ @@;
DATALINES;
```

```
2.5 < 2.5 < 2.6 X 3.3 X 3.5 X 3.5 X 3.6 X 4.0 X 4.1 X 4.5 X 5.2 X
5.6 X 5.7 X 6.1 X 6.2 X 6.5 X 6.9 X 7.1 X 7.7 X 7.9 X 9.9 X
;
RUN;
%cros(SO4, rem, value);
```

Although not expanded here to show all of the steps in the macro, the cros macro is embedded in the code at the "%MACRO cros(f,rem,value);" line and the sulfate data entered as the actual observation values or the LOD associated with the <LOD observations followed by either a < to indicate a censored observation or an X to indicate a noncensored observation. Then the cros macro is called with the "%cros(SO4, rem, value);" statement. The resulting estimates of the mean (5.16) and standard deviation (2.07) are close to those of the uncensored data (5.1 and 2.1).

Probability plotting or the robust regression on order statistics approach might be attempted for other distributions if linearizing transformations for the rank-derived proportions can be produced. The software package SAS has many such transformations available as cumulative density function (CDF). For lognormal distributions, the uncensored measurements might be log transformed and the resulting normally distributed log-transformed observations treated as just described for normally distributed data. The resulting slope and intercept might then be backtransformed from logarithmic to arithmetic units, but a backtransformation bias is introduced that should be corrected as described below. Helsel (2005) points out that the fill-in approach avoids the need for this backtransformation bias. Regardless, an approach of general utility is described below for correcting the bias in various backtransformed estimates.

2.2.2.3.4 Distributional Methods

Although difficult with a complete data set, censoring makes assignment of an underlying distribution to a data set even more difficult. Indeed, this is one of the major objections to censoring. However, several methods are applicable for estimating mean and standard deviation if the type of underlying distribution can be identified with confidence.

Maximum likelihood estimators (MLEs) were developed for a variety of distributions, including the normal distribution (Cohen 1950, 1957, 1959, 1961). With transformation and backtransformation bias correction as discussed below, methods for normal distributions can be applied to data sets from lognormal distributions. Cohen also developed MLEs for the Poisson (Cohen 1954), Weibull (Cohen 1975), and three-parameter gamma (Cohen and Norgaard 1977) distributions. The restricted MLE (Cohen 1957, 1961) as applied to normal or lognormal data sets will be presented here. Cases with small and large data sets will also be considered as the MLEs generated with few observations should be adjusted for a bias that becomes increasingly significant as sample size decreases (Schneider 1986).

As an example of one MLE approach, the restricted MLEs for the mean and standard deviation are the following:

$$\bar{X}_{MLE} = \bar{X}_c - \lambda(\bar{X}_c - LOD) \tag{2.9}$$

$$s_{MLE} = \sqrt{s_c^2 + \lambda(\bar{X}_c - LOD)^2} \tag{2.10}$$

where $\bar{X}_c$ = the mean of observations above the LOD, LOD = limit of detection, s^2_c = variance of observations above the LOD, and λ = a statistic from Table 2 of Cohen (1961). Two values (h and γ) are needed to extract λ from Table 2 (Cohen 1961). The h is the number of observations below the LOD divided by the total number of observations. The γ is $s_c^2 / (\bar{X}_c - LOD)^2$.

Example 2.2

The sulfate data from Example 2.1 will be used to generate MLEs. Again, assume that the lowest two observations are below the LOD.

$$\text{LOD} = 2.5 \qquad \bar{X}_c = 5.47 \qquad s^2_c = 3.70 \qquad h = 2/21 = 0.095$$

$$\gamma = 3.70/(5.47 - 2.5)^2 = 3.70/8.82 = 0.42$$

Linearly interpolating from values given in Table 2 of Cohen (1961), λ is estimated to be approximately 0.130162. The $\bar{X}_{MLE}$ is estimated with Equation (2.9) to be approximately 5.47 – 0.130162(5.47 – 2.5), or 5.08 mg/L. Applying Equation (2.10), the s_{mle} is calculated to be approximately the square root of 3.70 + 0.130162(5.47 – 2.5)^2, or 2.20.

Because MLEs become increasingly biased as the number of observations in the data set decreases, estimates of this bias provided by Schneider (1986) and Schneider and Weissfeld (1986) may have to be applied to MLE estimates such as those from Equations (2.9) and (2.10). These bias corrections are recommended if the number of observations is 20 or less (Schneider and Weissfeld 1986).

$$\bar{X}_{bc} = \bar{X}_{MLE} - \left[\frac{s_{MLE}}{n+1}\right] B_X \tag{2.11}$$

where $\bar{X}_{bc}$ = the bias-corrected MLE for the mean, and

$$B_X = -e^{2.692 - 5.439[(n-k)/(n-2k+1)]} \tag{2.12}$$

$$s_{bc} = s_{MLE} - \left[\frac{s_{MLE}}{n+1}\right] B_s \tag{2.13}$$

where s_{bc} = the bias-corrected MLE for the standard deviation, and

$$B_s = -\left[0.3121 + 0.859\frac{n-k}{n+1}\right]^{-2} \tag{2.14}$$

where n = the total number of observations, and k = the number of observations below the LOD.

Other techniques that require the assumption of a specific underlying distribution include conventional order statistic (Gupta 1952; Sarhan and Greenberg 1956, 1958, 1962) and "fill-in" with expected values (Gleit 1985) methods. The order statistic methods are unbiased regardless of sample size but have higher mean square errors than the maximum likelihood methods (Newman and Pinder 1987). It should be noted if using order statistics techniques that pertinent tables such as those in Sarhan and Greenberg's publications (1956, 1958, 1962) are designed for right-censored data sets. They must be converted to cope with left censoring. Alternatively, the original data set can be multiplied by –1 to convert it to a right-censored data set amenable to use with these tables. After estimations are made, the sign of the mean is changed back to a positive number.

If the distribution is lognormal, the above procedures can be used to estimate the mean and standard deviation but they are applied to the log-transformed observation values. When doing this, it is important to realize that the backtransformed values are biased estimates of the arithmetic mean and standard deviation. Aitchison and Brown (1957) provide the following equations to correct for the backtransformation bias.

$$\bar{X}_{lc} = e^{\bar{X}_l \Psi(s_l^2/2)} \quad (2.15)$$

$$s_{lc}^2 = e^{2\bar{X}_l [\Psi(2s_l^2) - \Psi[((n-2)/(n-1))s_l^2]]} \quad (2.16)$$

where s_l = standard deviation of the transformed values, $\bar{X}_l$ = mean of the transformed values, n = the number of observations, and $\Psi_n(t)$ = a value with argument t extracted from Table A2 of Aitchison and Brown (1957).

To avoid using a table, $\Psi_n(t)$ can be estimated with Equation (2.17) (Finney 1941).

$$\Psi n(t) = 1 + \frac{(n-1)t}{n} + \frac{(n-1)^3 t^2}{2!n^2(n+1)} + \frac{(n-1)^5 t^3}{3!n^3(n+1)(n+3)} + \frac{(n-1)^7 t^4}{4!n^4(n+1)(n+3)(n+5)} + \cdots \quad (2.17)$$

The equation can be expanded beyond the five terms shown above: the author applies the equation conveniently with Excel using seven terms.

2.2.2.3.5 Robust Methods

Again, some methods work better than others if the underlying distribution from which the observations are taken is unknown (Gilliom and Helsel 1986; Helsel and Gilliom 1986; Helsel 2005; Antweiler and Taylor 2008). Helsel (1990) recommends the robust regression on order statistics method described and illustrated with the cros SAS macro above. A regression line is constructed for the log-transformed observation values above the LOD versus their corresponding transformed proportions. Next, the regression is used to predict fill-in values for the below LOD values. All values including the predicted fill-in values are then backtransformed to arithmetic units. The mean and standard deviation are estimated using the data set that now includes fill-in values for the censored observations. Helsel (2005) concludes with simulations to back his arguments that this method is most robust because it only requires that a specific distribution be assumed to generate the fill-in observation values.

2.2.2.3.6 Modified Kaplan–Meier (Product Limit) Approach

A completely nonparametric approach can be taken to estimating central tendency and dispersion for censored data (Blackwood 1989; She 1997; Helsel 2005). As will be described in much more detail in Chapter 4, one such nonparametric approach, the Kaplan–Meier approach, was initially developed to analyze survival data sets in which some individuals survived beyond the end of the trial or survey. Times to death were known for the individuals dying during the trial or survey period, but the times to death for the survivors were only known to be greater than the duration of the trial or survey, that is, greater than some time at the right-hand side of the distribution of times to death for the study population. Although more involved methods now exist for applying these methods to left-censored data as well as conventional right-censored survival data sets, a simple trick can be applied that allows basic Kaplan–Meier methods to be applied with many software packages to left-censored environmental data sets containing <LOD observations. The approach described as flipping in the recent literature involves simply subtracting some number larger than the highest observation in the data set from each observation and then taking the absolute value of

each resulting flipped observation. The original left-censored data set is now right censored after this flipping and can be used in basic Kaplan–Meier methods. Figure 2.4 is a screen capture of the results for the sulfate data inserted into Helsel's KMStats v. 1.4 Excel spreadsheet (obtainable from www.practicalstats.com/nada) that implements Kaplan–Meier methods to estimate univariate statistics for censored data.

Example 2.3

The sulfate data used in previous examples are used again to generate Kaplan–Meier estimates for the median and its 95% confidence interval with the SAS PROC LIFETEST. To produce "flipped" observations, each sulfate observation has 8 subtracted from it before its absolute value is taken.

```
DATA SULFATE;          /* A FLAG OF 1 DENOTES AN UNCENSORED OBSERVATION */
INPUT SO4 FLAG @@;     /* A FLAG OF 0 DENOTES A <DL OBSERVATION
*/
FLIP = ABS(SO4-8);     /* CONVERSION OF LEFT TO RIGHT CENSORED DATA      */
DATALINES;
7.9 1 7.7 1 7.1 1 6.9 1 6.5 1 6.2 1 6.1 1 5.7 1 5.6 1 5.2 1 4.5 1 4.1 1
4.0 1 3.6 1 3.5 1 3.5 1 3.3 1 2.6 1 2.5 0 2.5 0
 ;
PROC LIFETEST CONFTYPE=LINEAR; /* A NONPARAMETRIC METHOD USUALLY USED */
TIME FLIP*FLAG(0);             /* FOR SURVIVAL/FAILURE DATA IS NOW TO */
RUN;                           /* BE USED FOR THE FLIPPED OBSERVATIONS*/
/* REMEMBER THAT THE FLIPPED VALUES FOR MEDIAN AND OTHER STATISTICS    */
/* WILL NEED TO BE CONVERTED BACK TO THE ORIGINAL UNITS, I.E., BY      */
/* X=ABS(SO4-8). FOR THE ESTIMATED MEDIAN FLIP VALUE OF 3.15, THE      */
/* MEDIAN SO4 CONCENTRATION WOULD BE 8-3.15 OR 4.85.                   */
```

The SAS output indicates a median of 3.15 and a 95% confidence interval of 1.8 to 4.5. Converting flip transforms back to sulfate concentration yields a median of 4.8 and a 95% confidence interval of 3.5 to 6.2. (The LINEAR option was selected here for confidence interval estimation instead of the LOGLOG default.)

2.2.2.3.7 Programs Available for Mean and Standard Deviation Estimation

The SAS software and similar programs can be applied to estimate univariate statistics for censored data sets. Helsel's software and spreadsheets are also very convenient and can be downloaded from the web address mentioned above.

2.2.2.4 Hypothesis Testing with Censored Data

Methods are also available for hypothesis tests to compare censored data sets. Some, such as the conventional nonparametric (Wilcoxon) tests illustrated below, allow testing of two data sets sharing a common, single LOD. Others, such as the Gehan–Wilcoxon test and other modified survival analysis methods, can test two data sets with observations censored at several LOD. The sulfate example will be used again to illustrate several of these methods. The Gehan–Wilcoxon test will be performed manually below using an example of cadmium in tissues of fish from two locations (from Helsel 2005) and then again with the SAS software package after data flipping.

Before progressing, it is important to note that fundamental, overlooked mistakes in the normal conduct of hypothesis (or significance) testing are emerging as a central concern of modern statisticians. The reader should keep in mind the issues described in Chapter 5 when conducting

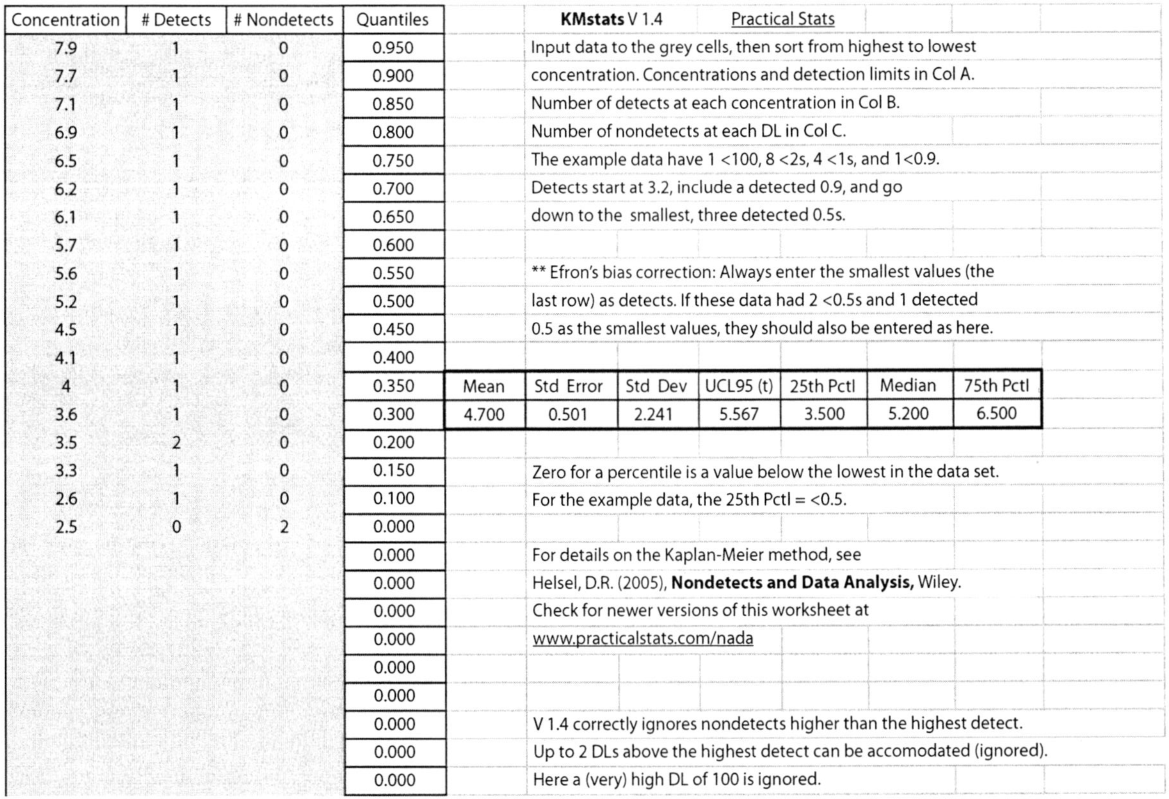

Concentration	# Detects	# Nondetects	Quantiles
7.9	1	0	0.950
7.7	1	0	0.900
7.1	1	0	0.850
6.9	1	0	0.800
6.5	1	0	0.750
6.2	1	0	0.700
6.1	1	0	0.650
5.7	1	0	0.600
5.6	1	0	0.550
5.2	1	0	0.500
4.5	1	0	0.450
4.1	1	0	0.400
4	1	0	0.350
3.6	1	0	0.300
3.5	2	0	0.200
3.3	1	0	0.150
2.6	1	0	0.100
2.5	0	2	0.000
			0.000
			0.000
			0.000
			0.000
			0.000
			0.000
			0.000
			0.000
			0.000

KMstats V 1.4 Practical Stats

Input data to the grey cells, then sort from highest to lowest concentration. Concentrations and detection limits in Col A.

Number of detects at each concentration in Col B.

Number of nondetects at each DL in Col C.

The example data have 1 <100, 8 <2s, 4 <1s, and 1<0.9.

Detects start at 3.2, include a detected 0.9, and go down to the smallest, three detected 0.5s.

** Efron's bias correction: Always enter the smallest values (the last row) as detects. If these data had 2 <0.5s and 1 detected 0.5 as the smallest values, they should also be entered as here.

Mean	Std Error	Std Dev	UCL95 (t)	25th Pctl	Median	75th Pctl
4.700	0.501	2.241	5.567	3.500	5.200	6.500

Zero for a percentile is a value below the lowest in the data set.

For the example data, the 25th Pctl = <0.5.

For details on the Kaplan-Meier method, see

Helsel, D.R. (2005), **Nondetects and Data Analysis,** Wiley.

Check for newer versions of this worksheet at

www.practicalstats.com/nada

V 1.4 correctly ignores nondetects higher than the highest detect.

Up to 2 DLs above the highest detect can be accomodated (ignored).

Here a (very) high DL of 100 is ignored.

Figure 2.4 Screen shot of Helsel's KMStats v1.4 Excel spreadsheet that applies Kaplan–Meier methods to left-censored data to generate univariate statisitics. Here, the sulfate data set has been entered into the spreadsheet with two observations included as <2.5.

hypothesis testing. Application of confidence limits instead to statistical inference avoids some issues, so confidence interval estimation is integrated into examples in this chapter.

2.2.2.4.1 Two Data Sets with One LOD

With thoughtfulness, rank sum-based nonparametric methods can be applied to compare two sets of observations censored at a single LOQ. The decrease in statistical power associated with using nonparametric methods is an acceptable sacrifice to avoid misapplication of the more powerful parametric methods. (Approximately 5% more observations are needed from normal distributed populations for a rank sum test to match the power of a *t*-test for differences in means (Dixon and Massey 1969).)

The Wilcoxon (Mann–Whitney–Wilcoxon) rank sum test can be applied by considering all observations censored at a single LOQ as tied observations. The specifics of the approach about to be shown are appropriate if the number of observations taken from each population is 10 or greater. It is assumed that n_A and n_B observations are taken from the two populations to be compared (*A* and *B*) and then ranked from smallest to largest. If some observations have the same value, the ranks for these tied observations are the average rank of the tied observations. The resulting ranks (Wilcoxon scores) are summed for each population (ΣR_A and ΣR_B). For a two-sided test, the lowest rank sum (T') is then compared to that expected under the assumption that the observations were drawn from the same population. For a one-sided test, the rank sum is chosen for the observation set suspected to have the lower distribution of values (e.g., T' from ΣR_B). The following equations can be used to estimate mean and standard deviation for T' assuming more than approximately 10 observations per sample and a normal distribution for T':

$$\mu_{T'} = \frac{n_A(n_A + n_B + 1)}{2} \tag{2.18}$$

$$\sigma_{T'} = \sqrt{\frac{n_A n_B(n_A + n_B + 1)}{12}} \tag{2.19}$$

The T' and estimates from Equations (2.18) and (2.19) are used to produce a z,

$$z = \frac{T' - \mu_{T'}}{\sigma_{T'}} \quad \text{or} \quad \frac{T' \pm 0.5 - \mu_{T'}}{\sigma_{T'}} \tag{2.20}$$

Sokal and Rohlf (1981) use the conventional $T' - \mu_{T'}$ in the numerator of Equation (2.20); however, because discontinuous ranks are being considered, many others (e.g., Dixon and Massey 1969; Noether 1971; Sheskin 2004) use $T' + 0.5 - \mu_{T'}$ if $T' - \mu_{T'} > 0$ or $T' - 0.5 - \mu_{T'}$ if $T' - \mu_{T'} < 0$ in the numerator, i.e., either adding or subtracting 0.5 to make the z smaller. A p value can be generated for a z with a cumulative normal distribution function, such as the Excel function, NORMDIST(z).

Example 2.4

The sulfate concentration data set used in previous examples (A) is expanded by adding a fictitious treatment to reduce sulfate concentrations. A posttreatment set (B) of another 21 observations is generated: <2.5, <2.5, <2.5, <2.5, <2.5, <2.5, <2.5, 2.6, 2.7, 3.2, 3.3, 3.7, 3.7, 3.8, 4.0, 4.2, 4.4, 4.6, 4.9, 5.1, and 6.3. A one-sided test can be applied to compare these data sets, asking, "Has treatment reduced sulfate concentration?" The number of observations taken pretreatment (n_A) and posttreatment (n_B) are intentionally equal to make hand calculations easy; however, this general approach can be taken if the numbers of observations are unequal. The two observation sets are tabulated below with their ranks shown in the right-hand column.

Observation	Observation Set	Rank	Observation	Observation Set	Rank
<2.5	A	5	4.0	A	22.5
<2.5	A	5	4.0	B	22.5
<2.5	B	5	4.1	A	24
<2.5	B	5	4.2	B	25
<2.5	B	5	4.4	B	26
<2.5	B	5	4.5	A	27
<2.5	B	5	4.6	B	28
<2.5	B	5	4.9	B	29
<2.5	B	5	5.1	B	30
2.6	A	10.5	5.2	A	31
2.6	B	10.5	5.6	A	32
2.7	B	12	5.7	A	33
3.2	B	13	6.1	A	34
3.3	A	14.5	6.2	A	35
3.3	B	14.5	6.3	B	36
3.5	A	16.5	6.5	A	37
3.5	A	16.5	6.9	A	38
3.6	A	18	7.1	A	39
3.7	B	19.5	7.7	A	40
3.7	B	19.5	7.9	A	41
3.8	B	21	9.9	A	42

The T' is the sum of the ranks for the B observations, i.e, 341.5. Next, $\mu_{T'}$ and $\sigma_{T'}$ can be estimated using Equations (2.18) and (2.19), and finally, Equation (2.20) solved with the results.

$$\mu_{T'} = \frac{21(21+21+1)}{2} = 451.5 \qquad \sigma_{T'} = \sqrt{\frac{21*21*(21+21+1)}{12}} = 39.75$$

$$z = \frac{341.5 - 451.5}{39.75} = -2.77$$

The Excel function NORMDIST(–2.77) produces a p value of 0.0028, which would be interpreted based on current statistical testing convention as a statistically significant result (α = 0.05). (Including the continuity factor of –0.5 yields a z of –2.78 and p value of 0.0027.) The following SAS code performs these calculations:

```
DATA SULFATE;
INPUT SO4 SITE $ @@;
DATALINES;
 2.5 A 2.5 A 2.6 A 3.3 A 3.5 A 3.5 A 3.6 A 4.0 A 4.1 A 4.5 A
5.2 A 5.6 A 5.7 A 6.1 A 6.2 A 6.5 A 6.9 A 7.1 A 7.7 A 7.9 A 9.9 A
2.5 B 4.6 B 3.7 B 4.2 B 5.1 B 2.5 B 2.5 B 4.0 B 3.7 B 2.5 B
2.5 B 6.3 B 4.9 B 3.8 B 2.5 B 3.3 B 2.6 B 2.7 B 2.5 B 3.2 B 4.4 B
;
RUN;
PROC SORT;
BY SITE;
```

```
RUN;
PROC NPAR1WAY WILCOXON; /* ADDING CORRECT=NO TO THIS LINE WILL RESULT IN */
VAR SO4;                /* NO CONTINUITY CORRECTION BEING APPLIED        */
CLASS SITE;
RUN;
```

SAS subtracts 0.5 if the $T' - \mu_{T'} < 0$ or adds 0.5 if $T' - \mu_{T'} > 0$ to compute z.

The above SAS code produces the following output:

Mean Score	SITE	N	Sum of Scores	Expected Under H0	Std Dev Under H0
26.738095	A	21	561.50	451.50	39.550523
16.261905	B	21	341.50	451.50	39.550523

```
Average scores were used for ties.

      Wilcoxon Two-Sample Test
Statistic                   561.5000

      Normal Approximation
Z                             2.7686
One-Sided Pr > Z              0.0028
Two-Sided Pr > |Z|            0.0056

      t Approximation
One-Sided Pr > Z              0.0042
Two-Sided Pr > |Z|            0.0084
```

2.2.2.4.2 Two Data Sets with Several LOD

A generalized Wilcoxon test called the Gehan or Gehan–Wilcoxon test (for details see Sprent 1993; Gehan 1965) applies score sums for observations from two groups to test for statistically significant differences between groups. Formally, the null hypothesis of this test is that the cumulative distributions for the two populations are the same, $F_Y = F_X$ (Gehan 1965). It was originally developed by Gehan with the aid of D.R. Cox to compare groups of children treated differently to prolong a state of remission from acute leukemia, i.e., right-censored time-to-event information (Gehan 1979). Children entered into the study at various times, resulting in several durations of censoring for those who maintained good health until the end of the trial. Because of the manner in which scoring is done, it can also be applied to data sets with several LODs, that is, several points of left censoring. The number of observations in each of the two data sets should be approximately 10 or more.

The handy fish tissue cadmium data from Helsel (2005) that has 9 and 10 observations for the two locations is ideal for illustrating the approach (Figure 2.5). The 10 samples from one location (Y) include one observation of <0.2 mg/kg and the 9 samples from the other location (X) include three censored observations of <0.6, <0.4, and <0.3 mg/kg, perhaps because different amounts of tissue were digested for analyses. Scores are produced from all 19 observations and used to estimate a z. Some table or software statistical function is then used to produce a corresponding p value, such as Excel NORMDIST(z).

The observations (y_j and x_i) for the two locations are ranked from smallest to largest, creating a table like that shown at the bottom left of Figure 2.5. The table is filled with U_{ij} scores (top left of Figure 2.5) using the following rules:

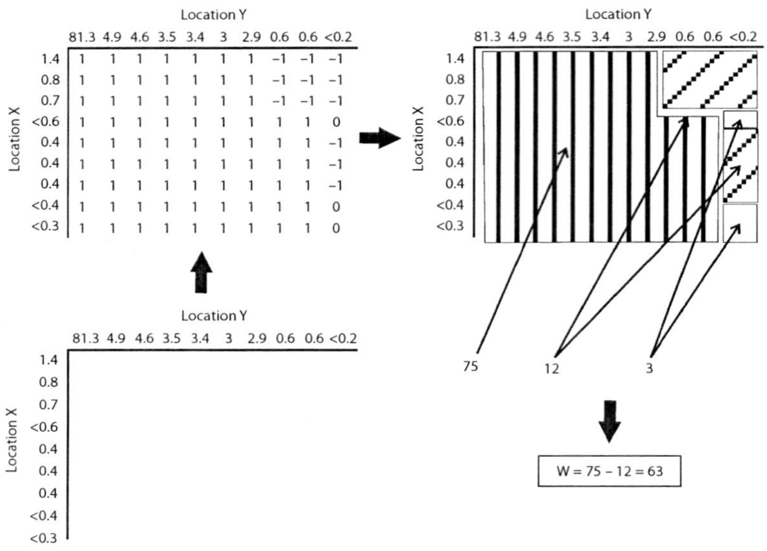

Figure 2.5 Depiction of computing *W* for the Gehan method using the fish tissue cadmium concentration data of Helsel (2005). The values for the two locations (*Y* and *X*) are ranked along the horizontal and vertical edges of a table (bottom left), and then U_{ij} values filled into the table (top left) to produce clusters of U_{ij} of 1 (vertical bars), –1 (diagonal bars), or 0 (white) (top right). The U_{ij} are summed to get an estimate of 63 for *W* (bottom right).

$U_{ij} = -1$ if $x_i > y_j$ (y_j can be a <LOD observation)
$U_{ij} = 1$ if $x_i < y_j$ (x_i can be a <LOD observation)
$U_{ij} = 0$ if $x_i = y_j$ or the comparison is indeterminant

The only point needing some clarification involves indeterminant comparisons. In Figure 2.5, the first instance of an indeterminant comparison involves a y_j of <0.2 paired with a x_i of <0.6. There is no means of deciding which is the larger observation so a 0 is inserted.

The U_{ij} scores in the table are summed to produce *W*, a statistic related to the Wilcoxon score already discussed (see Gehan 1965). The *W* is $75 \cdot (+1) + 3 \cdot (0) + 12 \cdot (-1) = 63$ in this example. The *W* would be 0 if $F_Y = F_X$. This *W* needs to be standardized to its standard deviation as follows to produce the *z*,

$$z = \frac{W}{Std\ Dev\ (W)} = \frac{W}{\sqrt{\dfrac{(n_Y n_X \Sigma h^2)}{[(n_Y + n_X)(n_Y + n_X - 1)]}}} \tag{2.21}$$

where n_Y and n_X = the number of observations in the samples taken from *Y* and *X*, respectively, and Σh^2 = the sum of the h^2 about to be described. The *h* for an observation is estimated as the difference between the number of observations greater (*G*) and less (*L*) than it when all observations have been pooled and ranked from smallest to largest. The h^2 are summed over all pooled observations to generate the Σh^2 that happens to be 2,220 in this example.

$$z = \frac{63}{\sqrt{\dfrac{(10 \cdot 9 \cdot 2220)}{[(10+9)(10+9-1)]}}} = \frac{63}{24.17} = 2.61$$

The p value generated by Excel with the estimated z of 2.61 is the following, 1 – NORMSDIST(2.61) = 0.0045. The cumulative distribution of tissue cadmium concentrations is inferred to be significantly higher in population Y than population X ($\alpha = 0.05$).

Example 2.5

The SAS PROC LIFETEST can be used to execute the Gehan after data are flipped, as already described. Helsel's fish tissue cadmium data set is used here to illustrate the approach. It needs to be noted at the onset of this example that the Wilcoxon is one of the tests applied as specified by the STRATA statement line. This generalized Wilcoxon test is the Gehan test.

```
/* IN THIS PROGRAM, A CONVENTIONAL NONPARAMETRIC METHOD USUALLY       */
/* USED FOR RIGHT-CENSORED SURVIVAL TIME DATA IS "TRICKED" INTO       */
/* DOING A HYPOTHESIS TEST FOR A LEFT-CENSORED <DL DATA SET.          */
/* SAS DOES A WILCOXON RANK SUM TEST THAT CAN INCLUDE CENSORED        */
/* DATA. IN ORDER FOR THE LEFT CENSORED DATA TO BE USED IN THE SAS    */
/* PROCEDURE, WE NEED TO DO A SIMPLE "FLIP" IN WHICH THE LEFT         */
/* CENSORED DATA ARE FLIPPED SO THAT THEY ARE RIGHT-CENSORED. THIS    */
/* IS EASILY DONE BY SELECTING A NUMBER GREATER THAN THE LARGEST      */
/* NUMBER IN THE DATA (I.E., 100 HERE). THE ABSOLUTE VALUE OF THE     */
/* CD CONCENTRATION - THE LARGEST NUMBER IS THE "FLIPPED" DATA        */
/* POINT. THE FLIPPED DATA CAN NOW BE ANALYZED AS CENSORED DATA       */
/* WITH THE SAS PROCEDURE LIFETEST.                                   */
DATA CADMIUM;       /* EXAMPLE AND METHOD WERE TAKEN FROM D.R. HELSEL */
                    /* NONDETECTS AND DATA ANALYSIS JOHN WILEY& SONS  */
                    /* 2005, TABLE 9.4, PAGE 147                      */
                    /* A FLAG OF 0 DENOTES A <DL OBSERVATION          */
INPUT CD SITE $ FLAG @@;
FLIP = ABS(CD-100);       /* CONVERSION OF LEFT TO RIGHT CENSORED DATA
*/
DATALINES;

81.3 A 1 4.9 A 1 4.6 A 1 3.5 A 1 3.4 A 1 3.0 A 1 2.9 A 1
 1.4 B 1 0.8 B 1 0.7 B 1 0.6 A 1 0.6 A 1 0.6 B 0 0.4 B 1
 0.4 B 1 0.4 B 1 0.4 B 0 0.3 B 0 0.2 A 0
 ;
PROC SORT;
 BY SITE FLIP;
RUN;
PROC LIFETEST PLOT=(S); /* A WILCOXON (=GEHAN) TEST OFTEN USED        */
TIME FLIP*FLAG(0);      /* FOR SURVIVAL/FAILURE DATA IS NOW TO        */
STRATA SITE;            /* BE USED TO TEST THE NULL HYPOTHESIS        */
RUN;                    /* THAT THE DISTRIBUTIONS ARE THE SAME        */
                        /* FOR THE TWO SITES.                         */
```

The following output is generated as a result.

Rank Statistics

SITE	Log-Rank	Wilcoxon
A	4.0114	63.000
B	-4.0114	-63.000

Test of Equality over Strata

Test	Chi-Square	DF	Pr > Chi-Square
Log-Rank	5.5260	1	0.0187
Wilcoxon	7.1707	1	0.0074

The formulations applied by SAS PROC LIFETEST are slightly different from the straightforward ones used in the hand-calculated illustration above, and a χ^2 statistic is used to generate an associated p value for the hypothesis of equality for the two populations from which the samples were taken. If requested in the STRATA statement, other useful tests such as the Peto-Peto test or tests for trends can also be done. Using the flip technique, many tests for censored data can be made using methods described in survival analysis sources, such as Cox and Oakes (1984), Marubini and Valsecchi (1995), Der and Everitt (2006), and Selvin (2008).

2.2.2.5 *Summary*

Censoring creates difficulties during analysis of environmental data. When censoring does occur, deletion or substitution methods should not be used for estimating mean and standard deviation. A variety of techniques are available to estimate these univariate statistics. They each have their advantages and disadvantages. When uncertain, it seems best to follow Helsel's recommendation that the robust techniques be used. Regardless of the technique selected, the original descriptions of the technique should be reviewed prior to application to familiarize oneself with the assumptions being made.

2.3 BLANK CORRECTION

2.3.1 Overview

Blank control is a critical part of many measurement processes. Failure to properly control or define blanks can lead to wasted resources and misappropriation of regulatory effort. For example, Settle and Patterson (1980) indicate that pervasive lead contamination in the United States went essentially undetected due to poorly controlled blanks. A decade later, Windom et al. (1991) argued that the U.S. Geological Survey (USGS) NASQAN dissolved metals data used to assess trends in U.S. waters were suspect because analytical procedures were inadequate for blank control. Although Boatman (1991) counters Windom and coworkers' specific conclusions about the USGS data, he also expresses the opinion that most dissolved metals data are unreliable as a consequence of pervasive sample contamination.

Blank control can involve inexpensive (Jay 1985) or very expensive (Moody 1982) procedures. Consequently, careful consideration should be given to blank control at various steps in the measurement procedure in order to avoid easily eliminated problems that compromise data integrity or, at the other extreme, to avoid the accrual of unnecessary costs that limit the number of samples that can be analyzed.

A sequence of blanks may be used in the sampling and measurement process. An empty sample container may be filled in the field with water containing no analyte and then serve as a travel or sampling blank. A similar sampling blank may be generated during laboratory experiments. A solution containing no detectable analyte may be processed and analyzed along with a sample set to generate a procedural blank. A reagent blank may be used in calibration. Ideally, such a reagent blank is a solution containing everything that the sample contains (reagents and similar matrix) except the analyte being measured. Each of these blanks has its use in tracing sources of unacceptable contamination and quantifying acceptable contamination.

2.3.2 Estimating the Effects of the Blank on the Measurement Process

General methods for calculating confidence intervals for differences between means were used by Taylor (1984, 1987) to estimate the 95% confidence interval for the true blank-corrected mean.

$$C_s = (C_m - C_b) \pm z\sqrt{\frac{s_m^2}{m} + \frac{s_b^2}{b}} \qquad (2.22)$$

where C_s, C_m, and C_b = the mean concentrations estimated in the sample after blank correction, in the original sample, and in the blank, respectively; s_m^2 and s_b^2 = the variances associated with the samples and blanks, respectively; m and b = the number of determinations of the sample and blank, respectively; and z = z statistic for $100(1 - \alpha/2)\%$ confidence interval (1.96 for a 95% confidence level, e.g., Excel NORMSINV(0.975)).

Several factors influencing the effectiveness of blank correction can be illustrated with Equation (2.22) (Figure 2.6). First, the correctness of the estimated mean blank determination (C_b) is as important as that of the estimated mean sample value (Figure 2.6A). Indeed, as the mean blank concentration might be subtracted from many samples, an argument could be made that its accurate estimation is more important than that of any single sample value. Also, the relative magnitudes of the sample and blank values are important (Figure 2.6A versus 6B). As the sample value approaches that of the blank, its confidence interval overlaps more and more with that of the blank. Second, the variation around the blank can be as important as the magnitude of the mean blank value. Taylor argues convincingly from Equation (2.22) that as much effort should be made to estimate the blank value as is made to measure sample values in the trace range (ppm range) (Figure 2.6C). As the sample concentration approaches the blank (as analyses extend down toward the LOD with $s_m^2 \rightarrow s_b^2$), an equal number of blanks and samples are advocated. Regardless, Taylor (1987) suggests the rule of thumb that blank correction should not exceed 10 times the acceptable limit of error in trace analyses. He feels that 10% error in the blank is acceptable.

2.3.3 Blank Control Charts

It is recommended that the mean and standard deviation of the blanks associated with analyses be monitored formally using the control chart methods described in the next section of this chapter. These methods use mean values and acceptable variation for in-control analytical conditions as a basis for judging the quality of subsequent measurements. Relative to blanks, control charts with only upper limits should suffice. This should not be taken to imply that a significant decrease in the measured blank value is not valuable information. A consistent drop in the blank below a previously defined level could indicate an unsuspected way of improving the measurement process. A reevaluation of procedures would then be made to find the reason. It could also suggest an inconsistency in calculation methods such as those that may arise during the training of new personnel.

Example 2.6

A fictitious series of lead digestion blanks can be used to illustrate the construction of a blank control chart. In this example, individual, not property, charts are produced, as will be explained later in this chapter. Ten blanks obtained when contamination was in control were used to generate warning and control limits. Then the blanks from a series of future digestion/analytical sessions were generated and plotted on the control charts.

```
TITLE "PB DIGESTION BLK IR CHARTS WITH INITIAL LIMITS";
DATA QCBLKS;
INPUT CONC REP @@;
DATALINES;
.001 1 .007 2 .002 3 .010 4 .002 5 .002 6 .004 7 .001 8 .006 9 .012 10
;
PROC SHEWART DATA=QCBLKS;
IRCHART CONC*REP/LIMITN=2 OUTLIMITS=BLKLIMIT NOCHART;
```

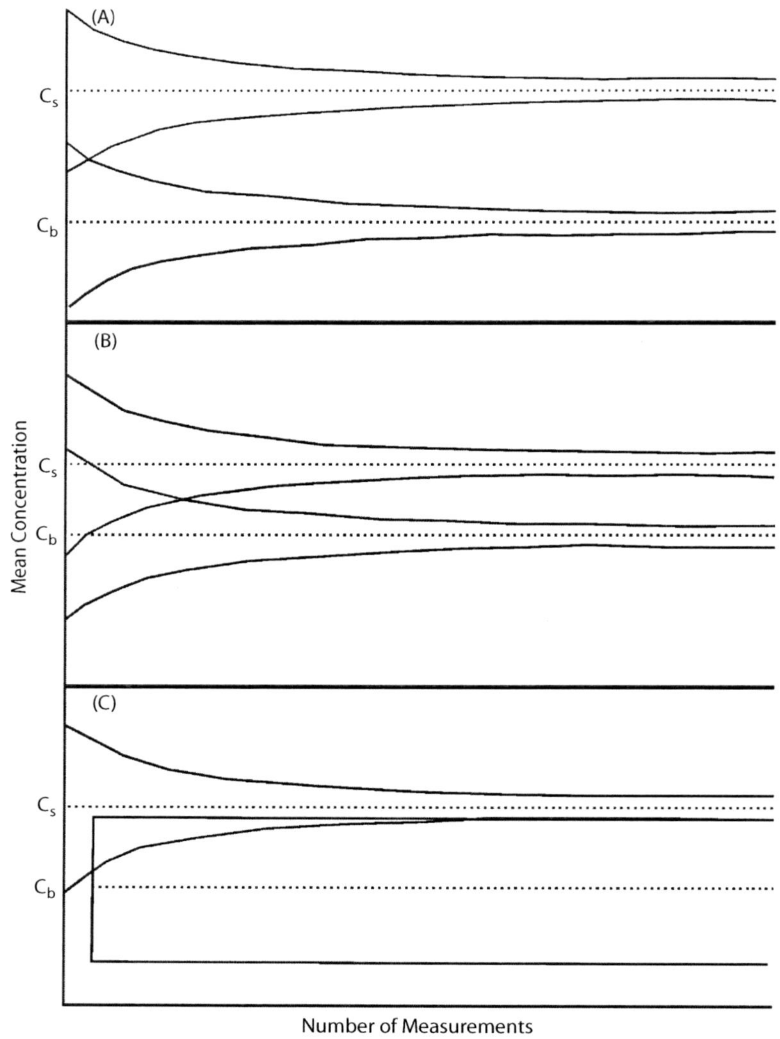

Figure 2.6 An example illustrating the influence of the relative magnitude of the blank and sample concentrations, and number of blank measurements. In panel A, the 95% confidence intervals for the sample and blank separate as the number of measurements increases. Panel B shows that the number of measurements needed to get a similar separation increases as the sample mean value comes closer to that of the blank. Panel C demonstrates the difficulty encountered if the number of blank determinations remains constant at a small number. In panel C, the confidence intervals do not separate as in panel B because the number of blank measurements is inadequate.

```
RUN;
DATA LEADBLK;
INPUT CONC REP @@;
DATALINES;
0.001 1 0.001 2 0.005 3 0.003 4 0.005 5
0.002 6 0.003 7 0.002 8 0.020 9 0.023 10
0.015 11 0.010 12 0.002 13 0.002 14 0.002 15
0.006 16 0.002 17 0.002 18 0.004 19 0.006 20
0.005 21 0.003 22 0.007 23 0.008 24
;
PROC SORT;
BY REP;
RUN;
SYMBOL V=STAR C=BLUE;
PROC SHEWHART GRAPHICS DATA=LEADBLK LIMITS=BLKLIMIT GRAPHICS;
     IRCHART CONC*REP /LIMITN=2 SIGMAS=2 COUTFILL=RED CCONNECT=BLACK
     COUT=RED;
     IRCHART CONC*REP /LIMITN=2 SIGMAS=3 COUTFILL=RED CCONNECT=BLACK
     COUT=RED;
RUN;
```

In this example, individual range charts are produced with the preestablished control limits from 10 blanks using the first PROC SHEWHART in the above code. The LIMITN = 2 option requests a moving range be plotted with 2 being the number of consecutive measurements used to calculate the moving range. The charts produced indicate unacceptable blank results for values (REP) 9 and 10.

2.4 ACCURACY AND PRECISION

2.4.1 Accuracy

2.4.1.1 Overview

Accuracy is the closeness of a measured value to the true value. The true value may be established by adding a known spike to a sample or using an analyte-containing material with a matrix similar to that of the sample. Spiking allows the analyst to produce a standard material with a concentration similar to that of the sample to be analyzed. This has its advantages as accuracy estimates can be concentration dependent. However, it is extremely difficult to develop a spiking methodology that incorporates the added analyte into the sample matrix so that it truly represents the analyte condition in the sample. For example, the digestion and analysis of a solid sample spiked with a metal solution would give an analyst very little information about the accuracy associated with the sample digestion. Alternatively, a standard material may be used that has the analyte incorporated into a matrix similar to that of the sample being studied. The standard material is supplied with a certified value and an acceptance interval for any analysis of the material. Both approaches have their place in improving and documenting accuracy.

2.4.1.2 Estimation of Accuracy

Accuracy is often expressed as percent recovery. If the sample has no background concentration to consider then the percent recovery is estimated with the simple formula

$$P = 100\frac{MSC}{CSC} \tag{2.23}$$

where MSC = measured spike concentration, and CSC = calculated spike concentration.

In this case, estimates of the mean and standard deviation of percent recovery data (P) are made with the familiar equations

$$\bar{P} = \frac{\sum_{i=1}^{n} P_i}{n} \tag{2.24}$$

$$S_p = \sqrt{\frac{\sum_{i=1}^{n} (P_i - \bar{P})^2}{n-1}} \tag{2.25}$$

where $\bar{P}$ = the mean P, P_i = the ith measure of P, and n = the number of measured P values.

If the spiked material has a background concentration, percent recovery is estimated with the following formula.

$$P = 100\frac{MSC - UC}{CSC} \tag{2.26}$$

where UC = unspiked concentration.

An unbiased estimate of the mean percent recovery would be the same as given in Equation (2.24) for percent recovery without background concentration. However, an unbiased estimate of the variance of percent recovery data would be the following (Provost and Elder 1983):

$$S_p^2 = \left[\frac{\left(\bar{P}\frac{CV}{100}\right)^2}{n}\right]\left[\left(1+\frac{1}{\frac{CSC}{UC}}\right)^2 + \left(\frac{n}{m\left(\frac{CSC}{UC}\right)^2}\right)\right] \tag{2.27}$$

where m = the number of measurements used to estimate UC, and CV = the coefficient of variation for the measurement process, i.e., the standard deviation/mean. By taking the square root of this variance estimate, the standard deviation for P can be obtained when there is a significant background concentration.

As demonstrated by Provost and Elder (1983) and reiterated above, it is important to realize that the variance of percent recoveries is not the same for samples with and without background concentrations. The variance associated with measurement of the background concentration should also be taken into consideration during calculations. When a significant background concentration is involved, the variance will be influenced by the ratio of the spike concentration to background concentration (CSC/UC). As the ratio decreases, the variance associated with the estimated mean percent recovery increases.

2.4.2 Precision

2.4.2.1 Overview

Precision is the effectiveness of a measurement process in producing similar results upon repeated application. For example, precision of the measurements 1.0, 2.5, 0.7 is inferior to that of

1.1, 0.9, 1.1. By analogy, precision would be the closeness of bullet holes in a target to one another, while accuracy would be the closeness of a hole to the bull's-eye of the target. A measurement process can be precise but inaccurate, e.g., a tight cluster of bullet holes out at the very edge of the target. Although a process can also be imprecise but accurate (a widely spread cluster with a very minimal average distance from the bull's-eye), imprecision very often spawns inaccuracy.

2.4.2.2 Estimation of Precision

Precision is estimated using replicate analyses. The standard deviation may be estimated when several replicates are analyzed. Another estimator of precision, range may be calculated from replicate analyses. Most often the range is used for duplicate analyses. The range is the absolute value of the difference between the smallest and largest measurement in a set. The mean of a set of ranges is computed with Equation (2.28).

$$\bar{R} = \frac{\sum_{i=1}^{n} R_i}{n} \tag{2.28}$$

where R_i = the ith estimate of range, and n = the number of ranges estimated.

The relationship between the range and the standard deviation depends on the distribution of the data. If the data can be assumed to be normally distributed, the standard deviation can also be estimated from the range (Grant and Leavenworth 1996).

$$S = \frac{\bar{R}}{d_2\sqrt{n}} \tag{2.29}$$

where $\bar{R}$ = the mean range, and d_2 = a value obtained from Appendix 1.

2.4.3 Control Charts

2.4.3.1 Overview

Control charts have traditionally been used to monitor precision, accuracy, and blank acceptability of measurement processes. Average values and acceptable variation for these qualities might be established when the measurement process is functioning properly, that is, in control. Checks against these in-control standards of quality are then made through time or analytical sessions. There are many variations on two basic control chart approaches, Shewhart and cusum methods. Both will be discussed here in general terms. The interested reader is referred to Grant and Leavenworth (1996) for a more in-depth discussion of Shewhart charts and to Van Dobben De Bruyn (1968) for a similar treatment of cusum charts. General discussions of these methods relative to environmental sciences may also be found in U.S. EPA (1972a, 1972b).

In general, an expected average value is estimated for the measurement process when it is in control. An estimate of the variation to be expected is also calculated. Critical limits are then set based on the probability of exceeding certain thresholds. Limits may include upper limits only or both upper and lower limits. For example, a control chart used to monitor a blank may not have a lower limit because a blank significantly lower than normal is not viewed as a problem. Similarly, precision better than usual is not a problem. In contrast, the measurement accuracy as estimated by percent recovery can be either too low or too high.

In all of these methods as described below, the number of values used to calculate the mean and limits must be explicitly defined. Also, the number of observation values used to compare an analytical session's results against the chart should also be the same.

2.4.3.2 Accuracy

2.4.3.2.1 Shewhart Charts

Taylor (1987) describes two approaches to Shewhart charts. The first (property chart) uses individual property values, e.g., individual percent recovery values. The second ($\bar{X}$ chart) involves the mean of individual values, e.g., the mean of 6% recovery values generated during an analytical session, for example. The $\bar{X}$ chart method will be discussed here, as it tends to be less sensitive to spurious observations.

Accuracy may be estimated using mean percent recovery as mentioned above or using the mean value of a standard material measured in replicate along with the samples. In Figure 2.7A, a control chart with warning ($\pm 2S_p / \sqrt{n}$) and control limits ($\pm 3S_p / \sqrt{n}$) is constructed for percent recovery data. Roughly 95% (95.5%) of observed mean percent recoveries are expected to fall within the warning limits if the measurement process is properly controlled. More than 99% (99.7%) of all mean values generated by the in-control measurement process should be within the control limits. Rules for rejection of data may be generated such as those below, as outlined by Taylor (1987) and Van Dobben De Bruyn (1968).

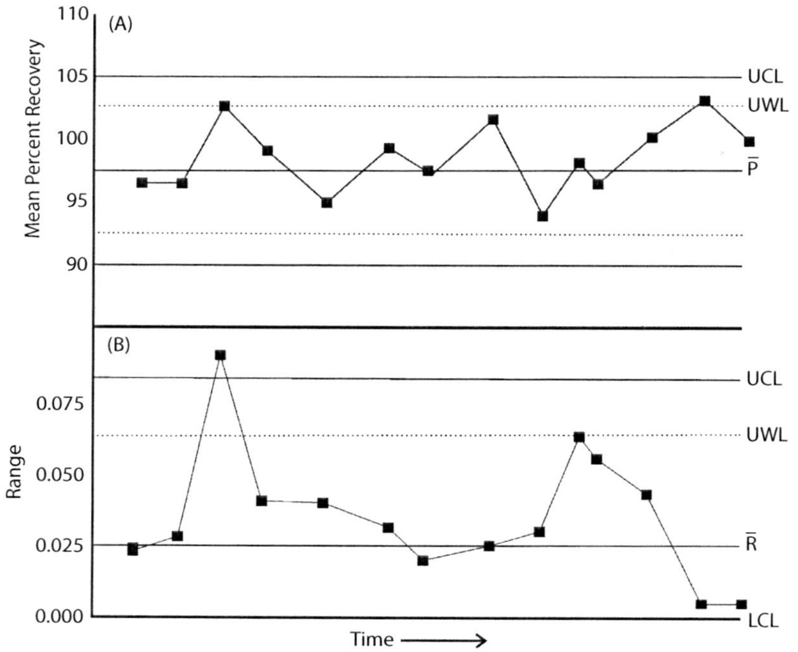

Figure 2.7 An example of Shewhart chart techniques. Panel A allows the analyst to monitor accuracy (see Example 2.7). Panel B allows precision to be estimated (see Example 2.9). Taken together, such accuracy and precision charts are called $\bar{x}$ -*R* charts.

1. If a value is outside of the upper or lower control limits, the measurement process is judged out of control.
2. If two successive points are outside the upper or lower warning limits, the measurement process is judged out of control. Note that the two points must be outside the same (upper or lower) warning limit.
3. If there is an obvious trend, the system is out of control. An example of such a trend would be 7 consecutive points on one side of the mean.

The associated measurements are invalid if the process is judged out of control. The system should be examined and the problem resolved. Quality control samples should be analyzed more frequently until it is certain that the problem has been resolved.

Control charts may also be used to suggest reexamination of the measurement process before it becomes out of control. For example, a mean value beyond the warning limit or 4 points in a row on one side of the mean may prompt close examination of the process to catch any possible emerging problem.

Example 2.7

Twenty measurements were made of a standard material with a certified lead concentration of 0.25 μg/g. Percent recovery (P_i) is used to estimate accuracy.

Measurement	$Pi(\%)$	$(P_i - \bar{P})^2$
0.27	108	112.36
0.22	88	88.36
0.21	84	179.56
0.31	124	707.56
0.23	92	29.16
0.23	92	29.16
0.25	100	6.76
0.24	96	1.96
0.25	100	6.76
0.23	92	29.16
0.27	108	112.36
0.28	112	213.16
0.20	80	302.76
0.25	100	6.76
0.22	88	88.36
0.26	104	43.56
0.22	88	88.36
0.23	92	29.16
0.24	96	1.96
0.26	104	43.56
Σ	**1948**	**2,120.80**

The mean percent recovery $(\bar{P})$ is estimated to be $\Sigma P_i/n$ = 1,948/20 = 97.4%.

According to Equation (2.25), the standard deviation for percent recovery values is estimated to be $s_p = \sqrt{2120.80/(20-1)} = 10.6$. The standard deviation for the mean percent recovery is $s_p/\sqrt{n}$, or $10.6/\sqrt{20} = 2.4\%$.

Shewhart chart limits can now be constructed using these estimates (and their estimated standard deviations).

Upper control limit = + $3(s_p/\sqrt{n}) = 97.4 + 3(10.6/\sqrt{20}) = 104.5\%$
Upper warning limit = + $2(s_p/\sqrt{n}) = 97.4 + 2(10.6/\sqrt{20}) = 102.1\%$
Lower warning limit = $-2(s_p/\sqrt{n}) = 97.4 - 2(10.6/\sqrt{20}) = 92.7\%$
Lower control limit = $-3(s_p/\sqrt{n}) = 97.4 - 3(10.6/\sqrt{20}) = 90.3\%$

These results can now be used to construct an accuracy control chart for future lead analyses conducted under identical conditions as those used to generate the above initializing data. Figure 2.7A shows the resulting chart. Estimates of mean percent recovery are plotted during any future analyses and assessed relative to the limits established above.

2.4.3.2.2 Cusum Charts

Cusum (cumulative sum) charts as illustrated here use all information generated before the present measurement session to evaluate quality. Deviations are summed through all previous measurement sessions, whereas with the Shewhart charts described herein, initializing data are used to establish quality criteria. The change in slope of the cusum chart is used to monitor the quality of analyses. Cusum charts are more sensitive to consistent small deviations than Shewhart charts (van Dobben De Bruyn 1968).

In the U.S. EPA's *Handbook for Analytical Quality Control in Water and Wastewater Laboratories* (U.S. EPA 1972a) and *Region VI Laboratory Quality Control Manual* (U.S. EPA 1972b), cusum charts for accuracy are developed using differences between observed and known values for a spiked or standard sample. That approach will be used here; however, alternate procedures are described in Van Dobben De Bruyn (1968).

Cusum charts require defined α (probability of falsely judging the measurement process as out of control) and β (probability of falsely judging the measurement process as in control) values. Values between 0.05 and 0.15 are recommended in the U.S. EPA manuals noted above. These values along with an estimate of the standard deviation for the difference between the measured and known values are used to estimate the cusum chart limits.

First, the variance of the differences (s_d^2) is estimated using Equation (2.30).

$$s_d^2 = \frac{\sum_{i=1}^{n} d_i^2 - \frac{\left[\sum_{i=1}^{n} d_i\right]^2}{n}}{n-1} \tag{2.30}$$

where d_i = the ith difference between the true and measured value, and n = the number of differences used in the estimation.

With this estimate of the variance for the differences, the minimum (s_o^2) and maximum (s_l^2) acceptable variations are estimated.

$$s_0^2 = 0.64\, s_d^2 \tag{2.31}$$

$$s_l^2 = 1.44\, s_d^2 \tag{2.32}$$

The upper ($UPPER(M)$) and lower ($LOWER(M)$) control limits for M sets of differences to be plotted are the following:

$$UPPER(M) = \frac{2\ln\left[\frac{1-\beta}{\alpha}\right]}{\frac{1}{s_0^2} - \frac{1}{s_l^2}} + M\frac{\ln\left[\frac{s_l^2}{s_0^2}\right]}{\frac{1}{s_0^2} - \frac{1}{s_l^2}} \tag{2.33}$$

$$LOWER(M) = \frac{2\ln\left[\frac{\beta}{1-\alpha}\right]}{\frac{1}{s_0^2}-\frac{1}{s_l^2}} + M\frac{\ln\left[\frac{s_l^2}{s_0^2}\right]}{\frac{1}{s_0^2}-\frac{1}{s_l^2}} \tag{2.34}$$

Example 2.8

Again, the 20 measurements of the standard material with a certified concentration of 0.25 µg Pb/g are used. The numbers here are identical to those used in Example 2.7.

Measurement	Measured – Certified Concentration (*d*)	d^2
0.27	0.02	0.0004
0.22	–0.03	0.0009
0.21	–0.04	0.0016
0.31	0.06	0.0036
0.23	–0.02	0.0004
0.23	–0.02	0.0004
0.25	0.00	0.0000
0.24	–0.01	0.0001
0.25	0.00	0.0000
0.23	–0.02	0.0004
0.27	0.02	0.0004
0.28	0.03	0.0009
0.20	–0.05	0.0025
0.25	0.00	0.0000
0.22	–0.03	0.0009
0.26	0.01	0.0001
0.22	–0.03	0.0009
0.23	–0.02	0.0004
0.24	–0.01	0.0001
0.26	0.01	0.0001
Σ	**–0.13**	**0.0141**

Using Equations (2.30) to (2.32):

$$s_d^2 = [\Sigma d_i^2 - (\Sigma d_i)^2/(n)]/(n-1)$$

$$= [0.0141 - (-0.13)^2/(20)]/19$$

$$= 0.000698$$

$$s_o^2 = 0.64sd^2 = 0.64(0.000698) = 0.000446$$

$$s_l^2 = 1.44sd^2 = 1.44(0.000698) = 0.001005$$

With s_o^2 and s_l^2, the upper and lower limits can be established for the cusum chart. For initial construction of the chart, they will be estimated for a range of M from 0 to 30 using Equations (2.33) and (2.34). Values for α and β are set at 0.15.

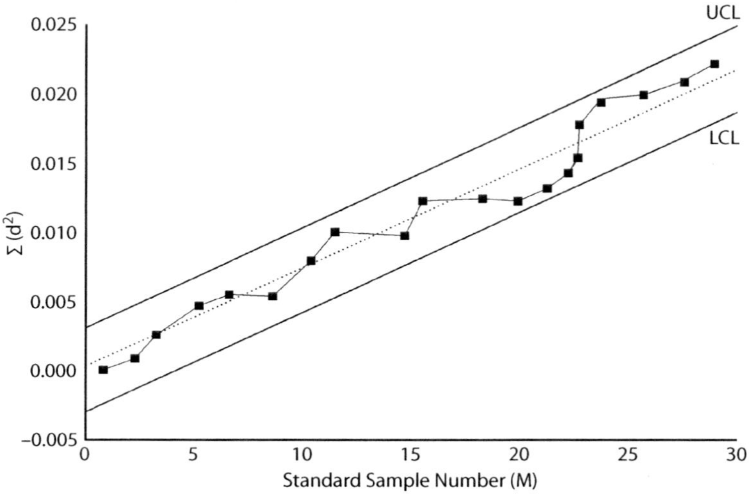

Figure 2.8 An example of a cusum chart for accuracy of a series of lead analyses (see Example 2.8).

$$\begin{aligned}
\text{Upper}(0) &= (2\ln((0.85/0.15))/(1/0.000446 - 1/0.001005) \\
&\quad + 0(\ln(0.001005/0.000446)/(1/0.000446 - 1/0.001005)) \\
&= 0.00278 \\
\text{Upper}(30) &= (2\ln((0.85/0.15))/(1/0.000446 - 1/0.001005) \\
&\quad + 30(\ln(0.001005/0.000446)/(1/0.000446 - 1/0.001005)) \\
&= 0.02232 \\
\text{Lower}(0) &= (2\ln((0.015/0.85))/(1/0.000446 - 1/0.001005) \\
&\quad + 0(\ln(0.001005/0.000446)/(1/0.000446 - 1/0.001005)) \\
&= -0.00278 \\
\text{Lower}(30) &= (2\ln((0.15/0.85))/(1/0.000446 - 1/0.001005) \\
&\quad + 30(\ln(0.001005/0.000446)/(1/0.000446 - 1/0.001005)) \\
&= 0.0168
\end{aligned}$$

Figure 2.8 is the cusum chart resulting from these calculations. The cumulative sum of the differences squared (Σd_i^2) is plotted against M after each analytical session, as shown. If the Σd_i^2 value goes above the UPPER(M) then the analysis is out of control and the associated measurements are invalid. The problem must be resolved before proceeding with further analyses.

2.4.3.3 Precision

2.4.3.3.1 Shewhart Charts

Precision charts (Figure 2.7B, for example) are often generated from duplicate analyses. The mean range and standard deviation about the mean range can be estimated using equations such as

Equation (2.28), (2.29), and (2.30). However, the upper warning limit (UWL) and upper (UCL) and lower (LCL) control limits for the $\bar{R}$ can be fabricated as follows:

$$UCL = D_4\bar{R} \tag{2.35}$$

$$UWL = \bar{R} + \frac{2}{3}(D_4\bar{R} - \bar{R}) \tag{2.36}$$

$$LCL = D_3\bar{R} \tag{2.37}$$

The values for D_4 and D_3 depend on the number of measurements in the subgroup from which the range was derived. For ranges estimated from duplicate measurements (n = 2), D_4 = 3.27 and D_3 = 0. Values can be taken from Appendix 1 for range estimates from subgroup sizes of up to 20 replicates. Note in Appendix 1 that D_3 becomes greater than 0 when the replicate number increases to 6. The control and warning limits for $\bar{R}$ charts are used like those for the $\bar{x}$ charts. However, there is no lower warning limit.

Frequently the $\bar{X}$ chart for accuracy and $\bar{R}$ chart for precision are combined into a composite graph such as Figure 2.7. Such a chart is called an $\bar{X}$-R chart (X bar–R chart).

Example 2.9

Twenty duplicates of lead analyses for a set of similar samples are analyzed to give the following results.

Duplicate 1	Duplicate 2	Range
0.25	0.28	0.03
0.25	0.22	0.03
0.25	0.22	0.03
0.27	0.28	0.01
0.21	0.26	0.05
0.31	0.28	0.03
0.23	0.25	0.02
0.23	0.22	0.01
0.18	0.28	0.10
0.17	0.16	0.01
0.29	0.26	0.03
0.37	0.35	0.02
0.24	0.26	0.02
0.29	0.31	0.02
0.30	0.30	0.00
0.28	0.31	0.03
0.17	0.19	0.02
0.30	0.27	0.03
0.28	0.29	0.01
0.29	0.29	0.00
Σ		**0.50**

The mean range ($\bar{R}$) is 0.50/20 = 0.025.

Using Equations (2.35), (2.36), and (2.37), the limits for the precision control chart can be estimated to be the following.

$$\text{Upper Control Limit} = D_4\bar{R} = 3.27(0.025) = 0.082$$

$$\text{Upper Warning Limit} = +2/3(D_4\bar{R} - \bar{R})$$

$$= 0.025 + 2/3[3.27(0.025) - 0.025]$$

$$= 0.063$$

$$\text{Lower Control Limit} = D_3\bar{R} = 0(0.025) = 0.000$$

Figure 2.7B is the control chart resulting from these calculations. The mean range for each analytical session is plotted as shown and compared to the limits. For example, the third mean range indicates an out-of-control condition for analytical precision. The results of that session's work are invalid. This SAS code will generate a $\bar{X} - R$ chart like Figure 2.7, except percent recovery is used instead of the actual standard material concentration. An in-control data set of standard material lead concentrations is produced and used in the first PROC SHEWHART below to produce limits (PBLIMITS) used later (second PROC SHEWHART) to track analytical quality during a series of analytical sessions.

```
TITLE "SHEWHART CHART - XRCHARTS WITH PRE-ESTABLISHED LIMITS";
/* PROPERTIES OF TRIPLICATE ANALYSES BEING USED */
/* TRUE VALUE OF STANDARD MATERIALS IS 0.25   */
DATA CONTROL;
INPUT CONC DAY REP;
PERCENT=(CONC/.25)*100;
DATALINES;
0.25 1 1
0.21 1 2
0.23 1 3
0.28 2 4
0.21 2 5
0.25 2 6
0.24 3 7
0.25 3 8
0.26 3 9
;
RUN;
PROC SHEWHART DATA=CONTROL;
XRCHART PERCENT*DAY/OUTLIMITS=PBLIMITS NOCHART;
RUN;
DATA LEAD;
INPUT CONC DAY REP @@;
PERCENT=(CONC/.25)*100;
DATALINES;
0.27 1 1  0.22 1 2  0.21 1 3  0.31 2 4  0.23 2 5
0.23 2 6  0.25 3 7  0.24 3 8  0.25 3 9  0.17 4 10
0.15 4 11 0.18 4 12 0.20 5 13 0.25 5 14 0.22 5 15
0.26 6 16 0.22 6 17 0.23 6 18 0.24 7 19 0.26 7 20
0.25 7 21 0.23 8 22 0.27 8 23 0.28 8 24
;
RUN;
PROC SORT;
BY DAY;
```

```
RUN;
PROC PRINT;
RUN;
SYMBOL V=STAR C=BLUE;
PROC SHEWHART GRAPHICS DATA=LEAD LIMITS=PBLIMITS GRAPHICS;
     XRCHART PERCENT*DAY/SIGMAS=2 COUTFILL=RED CCONNECT=BLACK COUT=RED;
     XRCHART PERCENT*DAY/SIGMAS=3 COUTFILL=RED CCONNECT=BLACK COUT=RED;
RUN;
```

2.4.3.3.2 Cusum Charts

When ranges for duplicate analyses are used to estimate precision, the procedures described above for accuracy cusum charts can be used. The approach would be the same but the difference between the duplicate measurements would be used instead of the difference between the observed and true value of the samples.

2.5 VARIANCE STRUCTURE

During most measurement processes, variation is contributed to the final measured signal at several steps. Variation is introduced from field sampling, to sample preparation, to sample analysis. The variance structure for the measurement process can be written as $s^2_{total} = s^2_{sampling} + s^2_{preparation} + s^2_{analysis}$ assuming independence of variance components. Variance associated with individual steps in sampling (sampling strata, e.g., among sites or treatments), sample preparation (aliquot selection, extraction, digestion), or analyses (replicate analyses) can be similarly defined.

Prior to any use of a data set, the variance structure of the measurement process should be understood. With proper experimental design, analysis of variance (ANOVA) methods can be used to estimate these variance components. As demonstrated in Examples 2.10 and 2.11, failure to define such structure can result in suboptimal data interpretation.

Example 2.10

In a survey of mercury in muscle of the deep-water great lanternshark, *Etmopterus princeps*, Newman et al. (2010) examined the variance structure of their sampling methodologies. Muscle samples from lanternshark from three Atlantic Ocean locations were taken from two parts of each fish (near large and small dorsal spine) and duplicate samples at these two parts. The variance structure included variation associated with three geographical locations, many individuals per location, two parts of each fish, and duplicate analyses of each sample. The question addressed by variance component analysis was whether or not it mattered if only one sample from one part of a shark was taken. The variability among locations and individual sharks was the feature of most interest, so it was desirable that the variability among parts of the shark and duplicate analyses of a sample be small relative to that associated with those of the locations and individual sharks. The SAS code below defined the variance associated with location (LOC), individual sharks (FISH), part sampled (SPINE), and replicate sample analyses (SPNELOC). The table below was produced as part of the output.

```
PROC NESTED; /* NESTED ANOVA EXAMINING MERCURY SAMPLE STRUCTURE  */
CLASS LOC FISH SPINE SPNELOC;
VAR HGDW;
RUN;
```

Variance Source	DF	Sum of Squares	Mean Square	Variance Component	Percent of Total
Total	159	1210.098274	7.610681	12.932834	100.0000
LOC	1	409.504006	409.504006	4.879884	37.7325
FISH	34	609.957869	17.939937	3.738720	28.9087
SPINE	36	49.989200	1.388589	0.161130	1.2459
SPNELOC	72	74.197600	1.030522	-2.810320	0.0000
Error	16	66.449600	4.153100	4.153100	32.1128

HGDW Mean	8.46681250
Standard Error of HGDW Mean	1.59981250

The variance structure indicates that 38% of the total variation in the data set was assigned to differences among the three locations and another 29% was associated with differences among individual sharks. The differences in duplicate samples taken from one part of a shark or samples taken from two different parts of a shark were minor, that is, approximately 1% and <1%, respectively. Whether one takes a muscle sample from the vicinity of the front or back dorsal fin, or whether or not one analyzes two samples from one part of a shark did not seem to matter relative to the goal of the survey. It would be more important to focus on gathering samples from different locations and from more sharks at each location.

Example 2.11

In anticipation of a planned discharge containing small amounts of mercury into a southeastern U.S. stream, a biomonitoring survey was conducted of mercury concentrations in an endemic fish, the dusky shiner (*Notropis cummingsae*). Although this shiner occurs in distinct schools, the fidelity of individuals to a specific school and the extent of movement of a school within the watershed were unknown. Consequently, an effective sampling protocol could not be developed for monitoring mercury concentrations in fish from the stream above, at, and below the proposed point of discharge. After a preliminary survey indicating that a pooled sample of six individuals was sufficient to minimize variance between samples, the following survey was conducted.

Three regions (upstream, midstream at the site of the proposed future discharge, and downstream) were sampled in the watershed during two seasons (October and immediately prior to spawning in April). Duplicate pooled samples from two schools at each of two sites within each region of the watershed were used to examine the variance structure. As fish size was suspected to influence mercury concentration, a mixed ANOVA model was used to examine variance structure for average fish wet weight as well as mercury concentration in the pooled samples. The components of variance estimated during this predischarge survey are tabulated below as percentages of the total variance for three variables.

Variable	Season	Watershed Region	Site	School	Error
Wet weight	<1%	28%	<1%	66%	6%
Mercury	20%	6%	2%	5%	67%
Size-normalized mercury	24%	22%	11%	1%	42%

There is a clear tendency for shiners of similar size to school together. This is important to note because mercury concentration is influenced by fish size. Sixty-six percent of the total variance in average shiner size was associated with among school differences. The strong influence of watershed region on variance in average wet weight could easily be attributed to adult spawning movement upstream prior to spawning and juvenile movement downstream after hatching in feeder streams in upper reaches of the watershed. When the mercury concentrations are normalized to an average fish wet weight, the unexplained variance drops by 25% (from 67% to 42%).

Once the contribution of average wet weight on mercury concentrations in pooled samples was taken into consideration, there appeared to be little motivation for sampling more individual schools as only 1% of the total variance was attributed to among school variation. Instead, the focus of the monitoring became the variation between regions, with samples being taken at different sites within each region. Sampling was limited to one season when the least amount of size-dependent migration was occurring (October).

2.6 SAMPLE SIZE

2.6.1 Overview

If too few individuals or too little material is taken for measurement, the confidence in the resulting measurements may be unacceptable for the intended purpose of the study. On the other hand, sampling too many individuals or too much mass can waste resources that might otherwise be available to address additional questions. For example, if the mean concentration of DDT in fish from a certain river were to be determined, it would be wise to ascertain the minimum number of individual fish to be sampled in order to obtain a reasonable estimate. If too few fish were sampled, the ability to estimate the mean concentration with a reasonable degree of precision could be lost. As a result, subsequent statistical analyses would be compromised. On the other extreme, if too many fish were sampled, resources may be wasted and sampling of additional locations on the river could be precluded. Further, if too small a piece of muscle tissue were taken from each fish, the resulting measurement may not accurately represent the concentration in fish muscle. Even if an adequate number of individuals were sampled, the nonrepresentative nature of the mass of muscle tissue used would compromise future statistical analyses.

Methods for estimating minimum sample size are described in this section. Gilbert (1987) and Taylor (1987) discuss determination of minimum number of individual samples or replicate analyses in more detail. The discussion below is essentially that of Gilbert (1987). Only techniques based on a prespecified relative error (|measured mean – actual mean|/actual mean) will be outlined, although Gilbert (1987) also details methods applicable if variance about the estimated mean or margin of error (|measured mean – actual mean|) is prespecified. Taylor (1987) discusses means of incorporating monetary cost into decisions regarding the number of measurements to make at various steps in the process. Visman (1969), Ingamells (1969), Ingamells and Switzer (1973), Taylor (1987), and Wallace and Kratochvil (1987) discuss methods for determining adequate sample weight in heterogeneous samples. This discussion draws heavily on these five publications.

2.6.2 Number of Individuals to Sample

2.6.2.1 Overview

Stuart (1976) illustrates the influence of sample number on estimation of population characteristics in the following way. Since the variance of the sample mean is the product of the population variance and $[1/n][(N-n)/(N-1)]$, it follows that the variance of the sample mean approaches 0 as n (the number of observations in the sample) increases toward N (the total number of individuals in the population). The more samples taken, the better the estimate will be. If n is very small relative to N, e.g., $(N-n)/(N-1) \rightarrow 1$, the term above relating sample variance to that of the population approaches $1/n$. Two facts should be clear from this observation. First, the effectiveness of sample variance in estimating the population variance improves as sample number increases. Second, the improvement is minimal beyond a certain sample number. It follows that estimation of the minimum sample size aids in accurate estimation and effective use of analytical resources.

2.6.2.2 Minimum Number of Individuals if Analytical Error Is Negligible

In this approach, we assume that the magnitude of the analytical error is insignificant relative to that of the sampling error. However, at the end of this section, the situation where both are significant will be discussed.

Gilbert (1987) describes a two-step approach to estimating the necessary number of samples. In his approach, the population variance (σ^2) need not be known with assurance. The first step uses a z statistic to generate an initial estimate of the minimum sample number. In the second step, t statistics with $n - 1$ degrees of freedom are used iteratively to get the final estimate of sample number. If the number of degrees of freedom ($n - 1$) is large then the second step may not be necessary as the z statistic will approximate the t statistic.

The method described here assumes a previously specified, acceptable relative error ($d_r = |\bar{x} - \mu|/\mu$). The estimate of the coefficient of variation required for this approach may be obtained during a preliminary survey or with an educated guess. (By a simple substitution of d or $|\bar{x} - \mu|$ for d_r and standard deviation for CV in Equations (2.38), (2.39), (2.40), and (2.41), the sample number based on a specified margin of error (d) can be estimated. In that case, the standard deviation is estimated during a preliminary survey or by best guess.)

If the size of the population (N) sampled is small relative to the variance of the population (σ^2), the initial estimate of the required sample size is the following:

$$n = \frac{\dfrac{[z_{1-\alpha/2}CV]^2}{d_r}}{1 + \dfrac{\left[\dfrac{z_{1-\alpha/2}CV}{d_r}\right]^2}{N}} \qquad (2.38)$$

However, if the size of the population is large relative to the variance of the population, then Equation (2.38) reduces to the following:

$$n = \left[\frac{z_{1-\alpha/2}CV}{d_r}\right]^2 \qquad (2.39)$$

The initial estimate of minimum sample number (n) is made with previously defined α and d_r, estimated CV, and Equation (2.38) or (2.39). Using this initial n to define the associated degrees of freedom ($n - 1$) for the t statistic, the sample number is again estimated using either Equation (2.40) (if N is small relative to σ^2) or Equation (2.41).

$$n = \frac{\left[\dfrac{t_{1-\alpha/2,n-1}CV}{d_r}\right]^2}{1 + \dfrac{\left[\dfrac{t_{1-\alpha/2,n-1}CV}{d_r}\right]^2}{N}} \qquad (2.40)$$

$$n = \left[\frac{t_{1-\alpha/2,n-1}CV}{d_r}\right]^2 \qquad (2.41)$$

The new estimate of n is then used again in either Equation (2.40) or (2.41) to get yet another estimate. This process is iterated until no large changes occur in n between iterations.

It has been noted that the traditional approach described above often underestimates the minimum sample size (Kupper and Hafner 1989). Blackwood (1991), in commenting on implications in environmental sampling, attributed the bias to the neglect of assurance levels (the probability that the specified confidence interval (CI), e.g., 95% CI, will have the specified interval width). Kupper and Hafner (1989) provide formulas and tables for generating better estimates (n^*_m) given an initial estimate (n_m) (from Equation 2.39) and assurance level ($1 - \gamma$). Appendix 2 is taken from Kupper and Hafner (1989) to facilitate estimation here.

Example 2.12

Lead concentrations in a large population of green sunfish will be estimated for a roadside pond. During a preliminary sampling, the analytical variance was determined to be insignificant relative to that between individual fish from this site. A coefficient of variation of approximately 0.47 was estimated during this initial survey. The minimum number of fish to be sampled for measurement of lead concentration in this population can be estimated assuming an acceptable relative error of 0.25 and an α of 0.10 ($z_{1-\alpha/2} = 1.645$). (The z can be obtained with Excel NORMSINV(0.95).) An initial estimate of n from Equation (2.39) is the following:

$$n = [(1.645 \times 0.47)/0.25]^2 = 9.56 \text{ or } 10 \text{ individuals}$$

Now, with an α of 0.10 and $n - 1$ or 9 degrees of freedom, a t statistic can be used in Equation (2.41) to refine the estimated sample number. (The t can also be obtained using Excel TINV(0.1,9).)

$$n = [(1.833 \times 0.47)/0.25]^2 = 11.88 \text{ or } 12 \text{ individuals}$$

Using this new estimate of n (12), the calculation using Equation (2.41) is repeated. Each new estimate of n is used in further iterations until the estimates of n stabilize.

$$n = [(1.796 \times 0.47)/0.25]^2 = 11.40 \text{ or } 12 \text{ individuals}$$
$$n = [(1.796 \times 0.47)/0.25]^2 = 11.40 \text{ or } 12 \text{ individuals}$$

The minimum sample size of 12 individuals is selected for the survey.
If the initial estimate from Equation (2.39) is used in Appendix 2 to estimate the minimum sample size ($\alpha = 0.10$, $1 - \gamma = 0.90$), the estimate is 17. Note that the assurance level for the initial estimate ($1 - \gamma'$) was only 0.39.

A wide array of software, including shareware, is available for easy implementation of these methods (see Appendix 27). For example, the SAS Release 9.2 PROC POWER will generate sizes needed for samples such as those described here or those used for other purposes.

2.6.2.3 Minimum Number of Replicate Analyses

Now, in the above discussion, it was assumed that the variances associated with other measurement components (tissue subsampling, analyses) were minimal. This is not an assumption to be made lightly. During preliminary design, it is helpful to estimate the number of replicate measurements needed to obtain an acceptable relative error. The process is identical to that described above for estimating minimum number of individuals to sample from a population. The CV for the measurement process (not including sampling) is used along with a predetermined α and an acceptable relative error.

Example 2.13

Let's assume that, in Example 2.12, portions of a muscle sample were analyzed for lead. The resulting CV was 0.07. An α of 0.05 and relative error of 0.05 were designated as acceptable for the analytical process. The estimated number of replicate analyses was determined in the following steps. The initial estimate using z statistic was

$$n = [(1.96 \cdot 0.07)/0.05]^2 = 7.53 \text{ or } 8 \text{ replicate analyses}$$

Iterations using t statistics were

$$n = [(2.365 \cdot 0.07)/0.05]^2 = 10.96 \text{ or } 11$$
$$n = [(2.228 \cdot 0.07)/0.05]^2 = 9.73 \text{ or } 10$$
$$n = [(2.262 \cdot 0.07)/0.05]^2 = 10.03 \text{ or } 11$$

Eleven replicate analyses would be needed to meet the tight relative error at an α of 0.05 for this survey. Kupper and Hafner's (1989) corrections for underestimation may be applied here as well to obtain a better estimate.

2.6.2.4 Sample and Replicate Numbers under Other Variance Structures

As illustrated very briefly in Section 2.5, components of variance can be significant at several steps in the sampling and postsampling stages of the measurement process. Consequently, estimation of minimum sample number or replicate number can be more involved than demonstrated above. Reference to methods associated with the most common situations is provided here.

Taylor (1987) outlines methods for estimating sample and replicate sizes when both sampling and analytical variance are significant. He also presents methods of incorporating measurement cost into determinations of sample numbers as well as methods applicable when samples are taken from several design strata. Cochran (1977) also discusses incorporation of costs in estimations and estimation of sample number for stratified data. He presents more detail on the techniques described above for determining sample number, including estimation during studies of proportions in populations. Gilbert (1987) outlines methods for sample number estimation when measurements are correlated. (In the above methods, measurements are assumed to be uncorrelated.) Correlated measurements may arise when measurements are taken over time at sites near one another. Currently, shareware (Visual Sampling Plan (VSP)) implementing many of the methods described by Gilbert (1987) can be downloaded from http://vsp.pnl.gov/. As detailed in Chapter 5 and Appendix 27, sample size is central to significance testing power analysis. Cohen (1988) provides excellent discussion of sample size relative to power analysis.

2.6.3 Weight (or Volume) of Sample

2.6.3.1 Overview

Many populations sampled in ecotoxicology are not composed of discrete units such as individual organisms. For example, subsamples of weights or volumes are taken routinely for materials such as soil, sediment, water, or tissue. The distribution of the analyte of interest in such a sample may be homogeneous or heterogeneous. If it is heterogeneous, the discrete particles within it can be

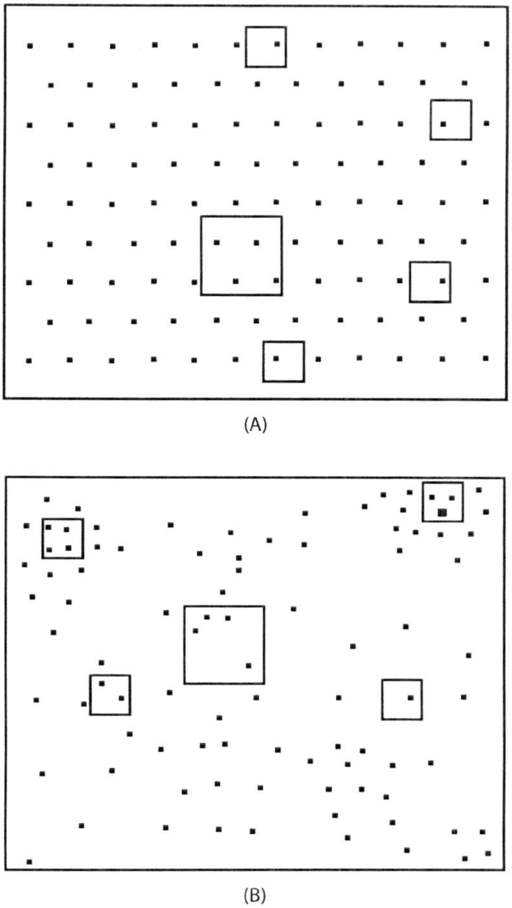

Figure 2.9 An illustration of sampling of heterogeneous samples. The sample may be well mixed (A) or poorly mixed (B). Means of estimating sample increments for such materials are described in the text.

well mixed (Figure 2.9A) or poorly mixed (Figure 2.9B). For a heterogeneous sample, the variability among sample increments (aliquots or units of sample analyzed) decreases as the weight analyzed increases (Figure 2.9). Therefore, it is important to analyze increments of sufficient weight to obtain an acceptable level of variability. But how large must the sample be in order to bring the variance down to that acceptable level?

First, to answer this question, the variance associated with increment sampling might be estimated. To this end, the standard deviation of the overall measurement process is represented with the following equation:

$$s_{ov}^2 = s_s^2 + s_{ss}^2 + s_a^2 \tag{2.42}$$

where s_{ov}^2 = overall variance, s_s^2 = variance associated with increment sampling, s_{ss}^2 = variance associated with replicate measurements, and s_a^2 = variance associated with a measurement.

The s_s may then be estimated as described by Wallace and Kratochvil (1987). This is accomplished by taking the difference after the other variance components in Equation (2.42) are determined. The s^2_{ss} might be estimated by analyzing a homogeneous standard material to derive s_a^2 and subsamples of the material in question used to derive s_{ss+a}^2. The s_a^2 is assumed to be similar for the reference material and the sample.

$$s_{ss}^2 = s_{ss+a}^2 - s_a^2 \quad (2.43)$$

If there is no suitable reference material, the combined s^2_{ss+a} can be used in place of $s_{ss}^2 + s_a^2$. Alternatively, s^2_{ss+a} and s_s^2 can be estimated using nested ANOVA methods as described in Examples 2.10 and 2.11 above.

The sampling variance is then estimated by analyzing a series of increments of sample to derive s_{ov} and solving for s_s^2 in Equation (2.42). A large s_s^2 suggests substantial variance between increments that must be controlled by estimating an acceptable minimum sample weight. A sample weight is estimated that is associated with a predetermined maximum sampling uncertainty.

2.6.3.2 *Increment Weight (or Volume) for Well-Mixed Materials*

A sampling constant (K_s or Ingamells' constant) is defined by convention as "the weight of a single increment that must be withdrawn from a well-mixed material to hold the relative sampling uncertainty, RSD, to 1% at the 68% level of confidence" (Wallace and Kratochvil 1987).

$$K_s = wRSD^2 \quad (2.44)$$

where w = weight, and RSD = relative standard deviation associated with analysis of samples of weight, w. In terms of the coefficient of variation as used previously, $RSD = 100*CV$.

This K_s is used primarily to characterize the material, not to determine sampling design (Ingamells 1974). Assuming that the sample is well mixed, the sample weight required to obtain a specified relative standard deviation can be estimated with K_s.

Example 2.14

The weight of a sample required to get a relative standard deviation of 5% for a well-mixed material with a K_s of 3.5 g is estimated.

$$K_s = wRSD^2$$
$$w = K_s/RSD^2$$
$$w = 3.5/5^2$$
$$w = 0.14 \text{ g}$$

If the sample is well mixed then K_s will be constant across different weight increments. If K_s is not constant then the material is not well mixed.

To determine if K_s is constant over a range of weights, estimates of s_S^2 are obtained for large (s_L^2) and small (s_S^2) weight increments. Wallace and Kratochivil (1987) recommend that the ratio w_L/w_S be at least 10. The w_S should be kept as small as reasonable. If the sample is well mixed, then the following relationship should be true.

$$s_L^2 w_L^2 = s_S^2 w_S^2 \quad (2.45)$$

where w_S = weight of the small increments, and w_L = weight of the large increments.

Wallace and Kratochvil (1987) suggest that a large number of small and large increments be used to estimate s_L^2 and s_S^2. They recommend that the sample not be considered well mixed if the null hypothesis that $\sigma_L > \sigma_s\sqrt{w_S/w_L}$ is rejected. A one-tailed F ratio test with an α of 0.10 or lower is recommended by convention.

2.6.3.3 *Increment Weight (or Volume) for Poorly Mixed Materials*

If the sample is not well mixed, the s_L and s_S are used to estimate a Visman's constant (degree of segregation). Wallace and Kratochvil (1987) recommend 20 to 30 increments of each weight. Calculations begin with estimation of random- and segregation-associated variance in s_s^2.

$$s_S^2 = \frac{A}{wn} + \frac{B}{n} \tag{2.46}$$

where A = the random variance component, B = the segregation component, and n = the number of weight increments.

Notice that the random variance component is lowered by increasing the number of increments analyzed or the weight of each increment. However, that associated with segregation in the sample is influenced only by the number of increments analyzed. This assumption of no correlation between increment size and variance due to segregation has been questioned in some instances. Wallace and Kratochvil (1987) present a detailed treatment of estimation when there is a correlation.

Using w_L, w_S, s_L^2, and s_S^2 estimated previously, A and B can be determined.

$$A = \frac{w_L w_S}{w_L - w_S}[s_S^2 - s_L^2] \tag{2.47}$$

$$B = s_L^2 - \frac{A}{w_L} = s_S^2 - \frac{A}{w_S} \tag{2.48}$$

An estimation of the degree of segregation (Z_s) can be determined with the above information (Visman 1969). As will be shown, Z_s is very useful in determining sample weights during design of sampling schemes.

$$Z_S = \sqrt{\frac{B}{A}} \tag{2.49}$$

The Z_s increases as the degree of segregation in the sample increases. A thoroughly segregated sample has a Z_s of 1. Taylor (1987) suggests that it is invalid to use a single K_s if Z_s exceeds 0.05.

The optimum increment weight to be taken from the material (w_{opt}) was defined by Ingamells (1974) with Equation (2.50). The optimum increment weight is the sample weight giving the best precision when a total weight of W_T is sampled from the field in n increments. Wallace and Kratochvil (1987) state that "when increment weights equal to A/B … are collected …, 50% of the sampling variance will arise from segregation regardless of the absolute values of A and B." Means of estimating weights for contributions of A to the overall variance other than 50% are discussed by Wallace and Kratochvil (1987).

$$w_{opt} = \frac{A}{B} = \frac{W_T}{n} \tag{2.50}$$

where W_T = the total weight of sample, and n = the number of increments analyzed.

Example 2.15

The following example is a modified version of one for molybdenum concentrations measured in cores of various lengths from an ore bed. To remain pertinent to ecotoxicology, measurements become molybdenum concentrations in contaminated sediments and core lengths become sample weights. Further details for this specific example are provided in Ingamells (1974).

Sixty-one samples are taken for weight increments of 0.3333 g (w_S) and 10.0000 g (w_L). The mean concentration of molybdenum was estimated to be 29.0000 mg/g. The s_s for the 61 w_S samples was 0.8380 and that (s_L) for the 61 w_L samples was 0.2450.

A. Using the small increment samples, K_s is estimated.

$$K_s = wRSD^2 = 0.3333(2.89) = 0.96 \text{ g}$$

B. Is the sediment well mixed?

$$\begin{aligned} F_s &= \sigma_L^2 / (w_S / w_L)\sigma_S^2] \text{ as estimated by } s_L^2 / [(w_S / w_L)s_S^2] \\ &= 0.2450^2 / [(0.3333/10.0000)0.8380^2] \\ &= 2.56 \end{aligned}$$

To extract $F\alpha_{(v1,v2)}$ from a table of F statistics or from some software function, the degrees of freedom must be estimated for the denominator and numerator. Since both s^2_S and s^2_L were generated using 61 sample increments each, the degrees of freedom for both are equal to $n - 1$, or 60, and an $F_{0.05(60,60)}$ statistic is used. The Excel function FINV(0.05,60,60) will generate the required F statistic of 1.53. Since $F_s > F_{0.05(60,60)}$, we conclude that the material is not well mixed.

C. What is the degree of segregation for the sediments?

To answer this question, A, B, and Z_s are estimated with Equations (2.47), (2.48), and (2.49).

$$A = 0.2210$$
$$B = 0.0379$$
$$Z_s = 0.41$$

D. What is the optimum increment weight for this sediment?

$$W_{opt} = A/B = 0.2210/0.0379 = 5.85 \text{ g per sample aliquot}$$

2.7 OUTLIERS

2.7.1 Overview

Robert A. Millikan's work, including that estimating the charge of an electron, earned him worldwide recognition and, in 1923, a Nobel Prize. More recently, his work has earned an additional distinction. Although his conclusions and character remain unquestioned, his tendency to omit "bad" data pushed him to the middle of the debate about acceptable scientific behavior (Ayala et al. 1989; Sigma Xi 1991; Goodstein 1992). In his paper involving electron charge, he stated that data generated from all oil droplets were examined, but later review of his notebooks revealed that data from 49 of 140 droplets were excluded (Sigma Xi 1991). His criteria for rejection of all but the "good" data remain the focus of criticism (Ayala et al. 1989).

Advocating the other equally questionable extreme during the 1930s, Bessel published geodetic research in which no observations were rejected for any reason (Anscombe 1960). Clearly, protocol grounded in sound reasoning and statistics must be followed to avoid the obvious problems that could arise during the use of either of these two extreme approaches.

It is unfortunate that such extreme behaviors must be linked to two otherwise remarkable scientific careers. It was not due to a lack of available methods. Anscombe (1960) indicates that formulation of statistical criteria for the rejection of outliers began in the 1850s. He also states that outlier rejection methodologies were common by 1925. Regardless, such behaviors can be avoided by application of methods such as those described below.

Grubbs (1969) suggests that the following points be considered prior to using a statistical method to reject or retain any suspect observations. First, if the individual making the measurement is skilled and is aware of a gross departure from the normal measurement process, then the associated observation should be rejected. It should be rejected regardless of its degree of conformity to expectations. Second, prior to using a statistical method, one must be certain that the apparent extreme deviation from the expected distribution of observations does not result from the assumption of an incorrect model. Perhaps the usual assumption of a normal distribution is invalid and extreme values at one tail should be expected as in the case of a lognormal distribution. Last, when a reason cannot be found for the anomalous observation, it should be reported and a clear statement of how the observation was or was not used in interpreting results should be presented. If no reason is found for rejection but the observation is rejected using a statistical test, the data set should be considered censored. Unlike our previous discussion of censoring relative to <DL observations, there is no specified threshold of censoring. Methods for coping with this type of censoring are similar to those discussed previously and are discussed in detail in the associated references.

There are general rules for highlighting potential outliers within a data set (Barnett and Lewis 1994). Woodworth (2004) articulated a commonly applied nonparametric approach based on an observation's distance from the first (Q1) and third (Q3) quartiles.* Outliers are defined according to this scheme as observations more than 1½ interquartile ranges above Q3 or below Q1. Extreme outliers are those observations more than three interquartile ranges above Q3 or below Q1. Box plots are often convenient tools to identify such observations. Note that this scheme is not intended to identify "wrong" or "bad" observations: it simply is one common way to initially identify extraordinary observations in a data set.

2.7.2 Single Suspect Observation

Several techniques are available for coping with outliers. The Winsorizing or trimming methods described for coping with left-censored data sets could be used to avoid the effects of outliers on estimation of mean and standard deviation. Dixon's test (see Dixon and Massey 1969; Grubbs 1969) is another frequently used technique for single outliers. Sokal and Rohlf (1981) give a detailed description of Dixon's test along with an example. However, Grubbs (1969) suggests the following method is better than Dixon's test for samples with single suspected outliers. Test criteria for high values and low values are defined by Equations (2.51) and (2.52), respectively.

* Recollect that if 100 observations were ranked from smallest to largest, the Q1 would be the 25th observation (25th percentile) and Q3 would be the 75th observation (75th percentile). The interquartile range is Q3 − Q1, which contains the central 50% of the observations.

$$T = \frac{outlier - \bar{x}}{s} \tag{2.51}$$

$$T = \frac{\bar{x}_i - outlier}{s} \tag{2.52}$$

where $\bar{x}$ = mean of all observations, *outlier* = value of the suspected outlier, and s = standard deviation of all observations.

The T value estimated from Equation (2.51) or (2.52) is then compared to critical values in Appendix 3, which are for a one-sided test.

Example 2.16

Mercury concentrations are measured in a series of fish samples. The following concentrations (µg/g wet weight) are generated: 0.052, 0.058, 0.084, 0.088, 0.092, 0.093, 0.124, and 1.719 (mean = 0.289, standard deviation = 0.578).

The obvious question is then asked, "Is the highest value an outlier under the assumption of a normal distribution?" Equation (2.51) is used to estimate T since the suspected outlier is the largest observation.

$$T = (1.719 - 0.289)/0.578 = 2.474$$

From Appendix 3, a critical value for T ($\alpha = 0.05$, one-tailed test) for a sample size of 8 is found to be 2.03. Since T exceeds this critical value, the suspected outlier is judged not to come from the same normal distribution as the other observations.

2.7.3 Several Suspect Observations

If several observations in a set of data are outliers, the ability to detect them can be compromised in the test described above because multiple outliers tend to mask each other in such tests. Rosner (1983) developed a generalized extreme studentized deviate (ESD) procedure for examining outliers when many might be present. This generalized ESD many-outlier method can be used for data sets with single or multiple outliers when the sample size is between 25 and 500.

Rosner's general approach involves the following steps. A maximum number of possible outliers (k) is specified. Then k possible outliers are identified in the data using the following procedure. First, the mean ($\bar{x}_i$) and standard deviation (s_i) are estimated using all observations in the data set. Next, all observations (x_i values) are ranked by their R_i (Equation 2.53). An R_1 or the MAX($|x_i - \bar{x}|/s_i$) is identified. This observation with the largest $|x_i - \bar{x}_i|/s_i$ is the first potential outlier. The suspect observation is now omitted prior to proceeding to the next step. The mean (Equation 2.54) and standard deviation (Equation 2.55) are estimated for the $n - 1$ remaining observations. They are used to estimate $|x_i - \bar{x}_i|/s_i$ for each of the remaining observations. A MAX($|x_i - \bar{x}_i|/s_i$) is identified and designated as R_2. The process is continued and generates values for R for $n, n - 1, \ldots, n - k + 1$ numbers of observations. The k observations associated with these R values are the suspect observations.

$$R_{i+1} = \frac{|x_i - \bar{x}_i|}{s_i} \tag{2.53}$$

where i = the number of observations omitted from the calculations, e.g., $i = 0$ for estimation of R_1, $i = 1$ for estimation of R_2, etc.

$$\bar{x}_i = \frac{\sum_{p=1}^{n-i} x_p}{n-i} \tag{2.54}$$

where n = the number of observations in the original data set (including suspected outliers), and x_p = the value of observation p from 1 to $n-i$.

$$s_i = \sqrt{\frac{\sum_{p=1}^{n-i} [x_p - \bar{x}_i]^2}{n-i-1}} \tag{2.55}$$

Each of the R values corresponding to the k suspect observations is compared to tabulated critical values (Appendix 4). Each critical value (λ_i) is extracted from this table using the associated total number of observations in the original sample (n), α, and $\ell + 1$ (the number of outliers that have been removed plus 1). There are no outliers if all R values are less than or equal to the corresponding λ_i values. If some R_i values are greater than the corresponding λ_i values then there are ℓ outliers.

An example of Rosner's test is given below. Further discussion with another example can be found in many statistics books, including Gilbert (1987) and Barnett and Lewis (1994).

Example 2.17

A sample of fish are analyzed for mercury concentration (μg/g wet weight). Several outliers are suspected. A maximum number of outliers is set at 5 and Rosner's techniques are applied.

	Concentration				
(x_i)	**R(n)**	**R($n-1$)**	**R($n-2$)**	**R($n-3$)**	**R($n-4$)**
0.052	0.479	0.476	0.591	1.261	1.317
0.052	0.479	0.476	0.591	1.261	1.317
0.058	0.464	0.456	0.550	1.130	1.171
0.065	0.448	0.432	0.503	0.978	1.000
0.077	0.419	0.391	0.423	0.717	0.707
0.084	0.402	0.367	0.376	0.565	0.537
0.088	0.393	0.354	0.349	0.478	0.439
0.092	0.383	0.340	0.322	0.391	0.341
0.093	0.381	0.337	0.315	0.370	0.317
0.093	0.381	0.337	0.315	0.370	0.317
0.094	0.379	0.333	0.309	0.348	0.293
0.097	0.371	0.323	0.289	0.283	0.220
0.104	0.355	0.299	0.242	0.130	0.049
0.112	0.336	0.272	0.188	0.043	0.146
0.114	0.331	0.265	0.174	0.087	0.195
0.124	0.307	0.231	0.107	0.304	0.439
0.129	0.295	0.214	0.074	0.413	0.561
0.137	0.276	0.187	0.020	0.587	0.756
0.148	0.250	0.150	0.054	0.826	1.024

	Concentration				
(x_i)	**R(n)**	**R(n – 1)**	**R(n – 2)**	**R(n – 3)**	**R(n – 4)**
0.199	0.129	0.024	0.396	1.935	2.268
0.205	0.114	0.044	0.436	2.065	**2.415(R$_5$)**
0.210	0.102	0.061	0.470	**2.174(R$_4$)**	
0.793	1.286	2.044	**4.383(R$_3$)**		
1.390	2.707	**4.075(R$_2$)**			
1.719	**3.490(R$_1$)**				

$$R_1 = \text{MAX}(|x_i - \bar{x}_i|/s_i) \text{ for } n = 3.490;\ \bar{x}_i = 0.253,\ s_i = 0.420$$
$$R_2 = \text{MAX}(|x_i - \bar{x}_i|/s_i) \text{ for } n - 1 = 4.075;\ \bar{x}_i = 0.192,\ s_i = 0.294$$
$$R_3 = \text{MAX}(|x_i - \bar{x}_i|/s_i) \text{ for } n - 2 = 4.383;\ \bar{x}_i = 0.140,\ s_i = 0.149$$
$$R_4 = \text{MAX}(|x_i - \bar{x}_i|/s_i) \text{ for } n - 3 = 2.174;\ \bar{x}_i = 0.110,\ s_i = 0.046$$
$$R_5 = \text{MAX}(|x_i - \bar{x}_i|/s_i) \text{ for } n - 4 = 2.415;\ \bar{x}_i = 0.106,\ s_i = 0.041$$

The estimated R's for the complete data set are given in the second column (from the left). The R_1 is typed in bold at the bottom of that column. The corresponding observation value (1.719) is removed from the data set and the process is repeated. The results are tabulated in column three. R_2 is identified (bold) and the corresponding, suspect observation value (1.390) is removed before proceeding. The process is repeated again and again (columns 4, 5, and 6) until k (= 5) suspect observations are identified.

The five observations associated with R_1, R_2, R_3, R_4, and R_5 are tested as potential outliers. If $R_{i+1} > \lambda_{\ell+1}$ then the corresponding observation is an outlier. (Note that ℓ is the number of outliers with the maximum value for R being k.)

i	x_i	$n - i$	$\bar{x}_i$	s_i	R_{i+1}	$\lambda_{\ell+1}$
0	1.719	25	0.253	0.420	3.490	2.82
1	1.390	24	0.192	0.294	4.075	2.80
2	0.793	23	0.140	0.149	4.383	2.78
3	0.210	22	0.110	0.046	2.174	2.76
4	0.205	21	0.106	0.1041	2.415	2.73

The $\lambda_{\ell+1}$ values are taken from Rosner's table (Appendix 4 of this book) for values of n (total number of sample observations), α, and ℓ+1 (actual number of outliers plus 1). They are tabulated above in the extreme right column. There are three outliers (1.719, 1.390, and 0.793), as three observations have $R_{i+1} > \lambda_{\ell+1}$.

Linear interpolation may be used in Rosner's table (Appendix 4). The tables in Rosner (1983) do not have values for samples with less than 25 observations. The details of filling in these values are provided at the end of the appendix table. Rosner suggests that when the sample has less than 25 observations, the very similar methods in Hawkins (1980, Appendix 4) be employed. Gilbert (1987) suggests that Dixon's test can be used but outlier masking may occur.

2.8 SUMMARY

In this chapter, quantitative techniques are presented for addressing the most common questions to arise during the measurement process. By no means should these techniques be viewed as sufficient to avoid illogical or invalid interpretation of quantitative data. The reader is directed to the references for more complete treatment of these and related topics. The reader is also referred to publications dealing with calibration curves and associated statistical methods (Van Arendonk et al. 1981; Meyer 1982; Christian and Tucker 1984; Helland 1987), bias in regression methods employing log transformation of the dependent variable (Beauchamp and Olson 1973; Duan 1983; Koch and Smillie 1986; Newman 1991; Newman and Heagler 1991), closure problems in data sets with variables summing to a constant (for example, percentage distribution of a metal in various sediment phases) (Johansson et al. 1984), concentrations as ratios and general treatment of ratios (Doctor et al. 1980), and completeness of water quality analyses (Johnsson and Lord 1987; Peden et al. 1986; U.S. EPA 1988).

In these first two chapters, the concept of scientific ecotoxicology was developed and basic quantitative methods pertinent to ecotoxicology were outlined. The intent of these chapters was to provide a conceptual and quantitative foundation upon which the remaining chapters are built. In the remaining chapters, quantitative methods will be discussed at increasing levels of ecological organization. In the final chapter, these methods will be summarized with a brief discussion of scientific and practical applications of the described concepts and methods.

REFERENCES

Aitchison, J., and J.A.C. Brown. *The lognormal distribution with special reference to its use in economics.* New York: Cambridge University Press, 1957.

Alford-Stevens, A.L. Mixture analytes. Procedures for determining detection limits should be defined. *Environ. Sci. Technol.* 21(2):137–139 (1987).

Anscombe, F.J. Rejection of outliers. *Technometrics* 2(2):123–147 (1960).

Antweiler, R.C., and H.E. Taylor. Evaluation of statistical treatments of left-censored environmental data using coincident uncensored data sets. I. Summary statistics. *Environ. Sci. Technol.* 42:3732–3738 (2008).

Arellano, S.D., M.W. Routh, and P.D. Dalager. Criteria for evaluation of ICP-AES performance. *Am. Lab.*, 1985:20–32 (1985).

Ayala, F., R.M. Adams, M.-D. Chilton, G. Holton, D. Hull, K. Patel, F. Press, M. Ruse, and P. Sharp. *On being a scientist.* Washington, DC: National Academy Press, 1989.

Barnett, V., and T. Lewis. *Outliers in statistical data.* 3rd ed. New York: John Wiley & Sons, 1994.

Beauchamp, J.J., and J.S. Olson. Correction of bias in regression estimates after logarithmic transformation. *Ecology* 54:1403–1407 (1973).

Berthouex, P.M., and S.W. Hinton. *Estimating the mean of data sets that include measurements below the limit of detection.* NCASI Technical Bulletin 621. New York: National Council of the Paper Industry for Air and Stream Improvement, 1991.

Blackwood, L.G. Analyzing censored environmental data using survival analysis: Single sample techniques. *Environ. Monit. Assess.* 18:25–40 (1989).

Blackwood, L.G. Assurance levels of standard sample size formulas. *Environ. Sci. Technol.* 25(8):1366–1367 (1991).

Boatman, C.D. Comment on "Inadequacy of NASQAN data for assessing metal trends in the nation's rivers." *Environ. Sci. Technol.* 25(11):1940–1941 (1991).

Chamberlin, T.C. The method of multiple working hypotheses. *J. Geol.* 5:837–848 (1897).

Christian, S.D., and E.E. Tucker. Least squares analysis with the microcomputer. Part five: General least squares with variable weighting. *Am. Lab.* 84:18–32 (1984).

Cochran, W.G. *Sampling techniques.* 3rd ed. New York: John Wiley & Sons, 1977.

Cohen, A.C., Jr. Estimating the mean and variance of normal populations from singly truncated and doubly truncated samples. *Ann. Math. Stat.* 21:557–569 (1950).

Cohen, A.C., Jr. Estimation of the Poisson parameter from truncated samples and from censored samples. *J. Am. Stat. Assoc.* 3:158–168 (1954).

Cohen, A.C., Jr. On the solution of estimation equations for truncated and censored samples from normal populations. *Biometrika* 44:225–236 (1957).

Cohen, A.C., Jr. Simplified estimators for the normal distribution when samples are singly censored or truncated. *Technometrics* 1(3):217–237 (1959).

Cohen, A.C., Jr. Tables for maximum likelihood estimates: Singly truncated and singly censored samples. *Technometrics* 3(4):535–541 (1961).

Cohen, A.C., Jr. Multicensored sampling in the three parameter Weibull distribution. *Technometrics* 17(3):347–351 (1975).

Cohen, A.C., Jr., and Norgaard, N.J. Progressively censored sampling in the three-parameter gamma distribution. *Technometrics* 19(3):333–340 (1977).

Cohen, J. *Statistical power analysis for the behavioral sciences.* 2nd ed. Hillsdale, NJ: Lawrence Erlbaum Association, 1988.

Cox, D.R., and D. Oakes. *Analysis of survival data.* London: Chapman & Hall, 1984.

Culliton, B.J. Random audits of papers proposed. *Science* 242:657–658 (1988).

Currie, L.A. Limits of qualitative detection and quantitative determination. *Anal. Chem.* 40(3):586–593 (1968).

Der, G., and B.S. Everitt. *Statistical analysis of medical data using SAS.* Boca Raton, FL: Chapman & Hall/CRC, 2006.

Dixon, W.J., and F.J. Massey Jr. *Introduction to statistical analysis.* New York: McGraw-Hill Book Company, 1969, pp. 330–332.

Doctor, P.G., R.O. Gilbert, and J.E. Pinder III. An evaluation of the use of ratios in environmental transuranic studies. *J. Environ. Qual.* 9(4):539–546 (1980).

Donn, J.J., and R.L. Wolke. The statistical interpretation of counting data from measurements of low-level radioactivity. *Health Physics* 32:1–14 (1977).

Duan, N. Smearing estimate: A nonparametric retransformation method. *J. Am. Stat. Assoc.* 78(383):605–610 (1983).

Finney, D.J. On the distribution of a variate whose logarithm is normally distributed. *J. R. Stat. Soc. Suppl.* 7:155–161 (1941).

Gehan, E.A. A generalized Wilcoxon test for comparing arbitrarily singly censored samples. *Biometrika* 52:203–213 (1965).

Gehan, E.A. Citation classic. A generalized Wilcoxon test for comparing arbitrarily singly censored samples. *Current Content* 39:255 (1979).

Gilbert, R.O. *Statistical methods for environmental pollution monitoring.* New York: Van Nostrand Reinhold Co., 1987.

Gilbert, R.O., and R.R. Kinnison. Statistical methods for estimating the mean and variance from radionuclide data sets containing negative, unreported or less-than values. *Health Physics* 40:377–390 (1981).

Gilliom, R.J., and D.R. Helsel. Estimation of distributional parameters for censored trace level water quality data. 1. Estimation techniques. *Water Resources Res.* 22(2):135–146 (1986).

Gilliom, R.J., R.M. Hirsch, and E.J. Gilroy. Effect of censoring trace-level water-quality data on trend-detection capability. *Environ. Sci. Technol.* 18:530–539 (1984).

Gleit, A. Estimation of small normal data sets with detection limits. *Environ. Sci. Technol.* 19:1201–1206 (1985).

Goodstein, D. What do we mean when we use the term 'science fraud'? *Scientist* 3/2:11–12 (1992).

Grant, E.L. *Statistical quality control.* 3rd ed. New York: McGraw-Hill Book Company, 1964.

Grant, E.L., and R.S. Leavenworth. *Statistical quality control.* 7th ed. New York: McGraw-Hill Book Company, 1996.

Grubbs, F.E. Procedures for detecting outlying observations in samples. *Technometrics* 11(1):1–21 (1969).

Gupta, A.K. Estimation of the mean and standard deviation of a normal population from a censored sample. *Biometrika* 39:260–273 (1952).

Hawkins, D.M. *Identification of outliers.* New York: Chapman & Hall, 1980.

Helland, I.S. On the interpretation and use of R^2 in regression analysis. *Biometrics* 43:61–69 (1987).

Helsel, D.R. Less than obvious. Statistical treatment of data below the detection limit. *Environ. Sci. Technol.* 24(12):1766–1774 (1990).

Helsel, D.R. *Nondetects and data analysis. Statistics for censored environmental data.* Hoboken, NJ: John Wiley & Sons, 2005.

Helsel, D.R. Summing nondetects: Incorporating low-level contaminants in risk assessment. *Integr. Environ. Assess. Manag.* 6:1–6 (2010).

Helsel, D.R., and R.J. Gilliom. Estimation of distributional parameters for censored trace level water quality data. 2. Verification and applications. *Water Resour. Res.* 22(2):147–155 (1986).

Ingamells, C.O. New approaches to geochemical analysis and sampling. *Talanta* 21:141–155 (1974).

Ingamells, C.O., and P. Switzer. A proposed sampling constant for use in geochemical analysis. *Talanta* 20:547–568 (1973).

Jay, P.C. Anion contamination of environmental water samples introduced by filter media. *Anal. Chem.* 57:780–782 (1985).

Johansson, E., S. Wold, and K. Sjodin. Minimizing effects of closure on analytical data. *Anal. Chem.* 56:1685–1688 (1984).

Johnson, S. *The complete works of Samuel Johnson*. Troy, NY: Princeton Pafraets Book Company, 1750; reprinted, 1903.

Johnsson, P.A., and D.G. Lord. *A computer program for geochemical analysis of acid-rain and other low-ionic-strength, acidic waters*. USGS Water-Resources Investigations Report 87-4095. Denver, CO: USGS Books and Open-File Reports, 1987.

Keith, L.H., W. Crummett, J. Deegan Jr., R.A. Libby, J.K. Taylor, and G. Wentler. Principles of environmental analysis. *Anal. Chem.* 55:2210–2218 (1983).

Koch, R.W., and G.M. Smillie. Bias in hydrologic prediction using log-transformed regression models. *Water Resour. Bull.* 22(5):717–723 (1986).

Koshland, D.L., Jr. Fraud in science. *Science* 235:141 (1987).

Kupper, L.L., and K.B. Hafner. How appropriate are popular sample size formulas? *Am. Stat.* 43(2):101–105 (1989).

Marubini, E., and M.G. Valsecchi. *Analyzing survival data from clinical trials and observational studies*. Chichester, UK: John Wiley & Sons, 1995.

Meyer, E.F. Comments on curve-fitting methods. *Anal. Chem.* 54:1878–1879 (1982).

Moody, J.R. NBS clean laboratories for trace element analysis. *Anal. Chem.* 54(13):1358A–1376A (1982).

Newman, M.C. A statistical bias in the derivation of hardness-dependent metals criteria. *Environ. Toxicol. Chem.* 10:1295–1297 (1991).

Newman, M.C., and P.M. Dixon. UNCENSOR: A program to estimate means and standard deviations for data sets with below detection limit observations. *Am. Environ. Lab.* 4/90:27–30 (1990).

Newman, M.C., P.M. Dixon, B.B. Looney, and J.E. Pinder III. Estimating mean and variance for environmental samples with below detection limit observations. *Water Resour. Bull.* 25(4):905–916 (1989).

Newman, M.C., and M.G. Heagler. Allometry of metal bioaccumulation and toxicity. In *Metal ecotoxicology*, ed. M.C. Newman and A.W. McIntosh. Chelsea, MI: Lewis Publishers, 1991, pp. 91–130.

Newman, M.C., and J.E. Pinder III. Coping with uncertainty: Limits of detection, limits of quantitation and nested sources of error. In *Proceedings of AWWA Water Quality Technology Conference*. Denver: AWWA Water Quality Association, 1987, pp. 509–532.

Newman, M.C., X. Xu, C.F. Cotton, and K.R. Tom. High mercury concentrations reflect trophic ecology of three deep-water Chondrichthyans. *Arch Environ Contam Toxicol* DOI: 10.1007/s00244-010-9584-4 (2010).

Noether, G.E. *Introduction to statistics*. New York: Houghton Mifflin, 1971.

Palca, J. Get-the-lead-out guru challenged. *Science* 253:842–844 (1991).

Peden, M.E., S.R. Bachman, C.J. Brennan, B. Demir, K.O. James, B.W. Kaiser, J.M. Lockard, J.E. Rothert, J. Sauer, L.M. Skowron, and M.J. Slater. *Development of standard methods for the collection and analysis of precipitation*. EPA/600/4-86/024. Cincinnati, OH: Environmental Monitoring and Support Laboratory, 1986.

Porter, P.S., R.C. Ward, and H.F. Bell. The detection limit. Water quality monitoring data are plagued with levels of chemicals that are too low to be measured precisely. *Environ. Sci. Technol.* 22:856–861 (1988).

Provost, L.P., and R.S. Elder. Interpretation of percent recovery data. *Am. Lab.* December 1983:57–63 (1983).

Rayl, A.J.S. Misconduct case stresses importance of good notekeeping. *The Scientist* November 11:18–19 (1991).

Rosner, B. Percentage points for a generalized ESD many-outlier procedure. *Technometrics* 25(2):165–172 (1983).

Rousseau, D.L. Case studies of pathological science. *Am. Sci.* 80:54–63 (1992).

Sarhan, A.E., and B.G. Greenberg. Estimation of location and scale parameters by order statistics from singly and doubly censored samples. Part I. The normal distribution up to samples of size 10. *Ann. Math. Stat.* 27:427–451 (1956).

Sarhan, A.E., and B.G. Greenberg. Estimation of location and scale parameters by order statistics from singly and doubly censored samples. Part II. Tables for the normal distribution for samples of size $11 \leq n \leq 15$. *Ann. Math. Stat.* 29:79–105 (1958).

Sarhan, A.E., and B.G. Greenberg. *Contributions to order statistics*. New York: John Wiley & Sons, 1962.

Schneider, H. *Truncated and censored samples from normal populations*. New York: Marcel Dekker, 1986.

Schneider, H., and L. Weissfeld. Inference based on type II censored samples. *Biometrics* 42:531–536 (1986).

Schwartz, L.M. Lower limit of reliable assay measurement with nonlinear calibration. *Anal. Chem.* 55:1424–1426 (1983).

Selvin, S. *Survival analysis for epidemiologic and medical research: A practical guide*. Cambridge: Cambridge University Press, 2008.

Settle, D.M., and C.C. Patterson. Lead in albacore: Guide to lead pollution in Americans. *Science* 207:1167–1176 (1980).

She, N. Analyzing censored water quality data using a non-parametric approach. *J. Am. Water Resour. Assoc.* 33(3):615–624 (1997).

Sheskin, D.J. *Handbook of parametric and nonparametric statistical procedures*. 3rd ed. Boca Raton, FL: Chapman & Hall/CRC, 2004.

Shewhart, W.A. *Statistical method: From the viewpoint of quality control*. New York: Dover Publ., 1986.

Sigma Xi. *Honor in science*. Research Triangle Park, NC: Sigma Xi, Scientific Research Society, 1991.

Sokal, R.R., and F.J. Rohlf. *Biometry*. 2nd ed. New York: W.H. Freeman and Company, 1981.

Sprent, P. *Applied nonparametric statistical methods*. 2nd ed. London: Chapman & Hall, 1993.

Stuart, A. *Basic ideas of scientific sampling*. New York: Hafner Press, 1976.

Synovec, R.E., and E.S. Yeung. Improvement of the limit of detection in chromatography by an integration method. *Anal. Chem.* 57:2162–2167 (1985).

Taylor, J.K. Guidelines for evaluating the blank correction. *J. Testing Eval.* 12(1):54–55 (1984).

Taylor, J.K. *Quality assurance of chemical measurements*. Chelsea, MI: Lewis Publishers, 1987.

U.S. EPA. *Handbook for analytical quality control in water and wastewater laboratories*. Cincinnati, OH: Analytical Quality Control Laboratory, 1972a.

U.S. EPA. *Region VI laboratory quality control manual*. 2nd ed. Ada, OK: U.S. EPA Analytical Quality Control Program, 1972b.

U.S. EPA. *National stream survey phase I: Quality assurance report*. EPA/600/4–88/018. Las Vegas, NV: U.S. EPA, Environmental Monitoring Systems Laboratory, 1988.

Van Arendonk, M.D., R.K. Skogerboe, and C.L. Grant. Correlation coefficients for evaluation of analytical calibration curves. *Anal. Chem.* 53:2349–2350 (1981).

Van Dobben De Bruyn, C.S. *Cumulative sum tests in theory and practice*. New York: Hafner Publishing Company, 1968, p. 82.

Visman, J. A general sampling theory. *Mater. Res. Stand.* 9:9–64 (1969).

Wallace, D., and B. Kratochvil. Visman equations in the design of sampling plans for chemical analysis of segregated bulk materials. *Anal. Chem.* 59:226–232 (1987).

Windom, H.L., J.T. Byrd, R.G. Smith Jr., and F. Huan. Inadequacy of NASQAN data for assessing metal trends in the nation's rivers. *Environ. Sci. Technol.* 25(6):1137–1142 (1991).

Woodworth, G.G. *Biostatistics. A Bayesian introduction*. Hoboken, NJ: John Wiley & Sons, 2004.

Zimmerman, S.W. Quality assurance and science—Oil and water? *Energy Update*, September 3–4 (1990).

CHAPTER 3

Bioaccumulation

Models are, for the most part, caricatures of reality, but if they are good, then, like good caricatures, they portray, though perhaps in distorted manner, some of the features of the real world.

—Kac (1969)

3.1 GENERAL

Toxicant bioaccumulation became a topic of public and scientific concern early in the 1950s. Contributing to this awakening were the tragic consequences of metal bioaccumulation in food species so apparent after outbreaks of Minamata and Itai-itai disease. Accumulation of fission products in food species and humans from nuclear weapons open-air testing became an issue of global concern. In the early 1960s, Rachel Carson's *Silent Spring* thrust the consequences of pesticide accumulation by wildlife species into the public's awareness.

Fortunately, much has been accomplished in the area of bioaccumulation during the last six decades. After World War II, considerable talent was focused on quantitative prediction of radionuclide uptake and elimination by human and nonhuman species. An abundance of quantitative techniques grew from these efforts (as evidence see Gibaldi and Perrier 1982; Whicker and Schultz 1982; Connell 1990; Mackay 1991; Bacci 1994). Sentinel species were also developed to monitor pollutants in marine and freshwater systems. Pharmacokinetic models developed in the late 1930s and synthesized in such sources as Atkins (1969) and Wagner (1979) were readily adopted by ecologists predisposed to compartment modeling.

Today, an impressive body of knowledge has been developed for toxicant bioaccumulation. Although this body of knowledge provides a firm foundation upon which to build, there remains much more to be learned. Indeed, basic issues remain unresolved despite this body of knowledge. For example, ambiguity still exists regarding the major factors underlying variation in toxicant concentrations among individuals from the same population (Lobel et al. 1991). As pointed out by Barron et al. (1990a), many advances made in pharmacokinetic or toxicokinetic modeling have yet to be incorporated into ecotoxicological studies, and as clarified by Landrum et al. (1992), incorrect units are sometimes still specified for associated model parameters. The purpose of this chapter is to explore techniques presently in use and those potentially useful in ecotoxicology.

3.2 MODELING BIOACCUMULATION: GENERAL APPROACH

Models of mathematically distinct compartments are the rule in ecotoxicology. Although these compartments are sometimes linked to specific organs (excretion by the kidneys) or tissue pools (loss from fatty tissues), more often than not they are treated as simply kinetically distinct (fast or slow elimination) compartments. Some speculation might be made to link the mathematical compartment to a physical compartment. In other studies, clear linkage to physical compartments is made. A good example of such effective linkage is the work of Lyon et al. (1984) with metal elimination from the hemolymph of crayfish. The elimination of several metals from a fast compartment was correlated with protein ligand binding and conformed to predictions of the Irving–Williams series. Unfortunately, speculation can be based as much on preconceptions as on the direct results of the study. As a consequence, it becomes critical for the reader to keep the distinction between physical and kinetic compartments clearly in mind when interpreting reports of bioaccumulation modeling.

3.2.1 Elimination

3.2.1.1 General

Distinctions have been emphasized by Barron et al. (1990a) for the terms *depuration*, *clearance*, and *elimination*, although they are used frequently as synonyms in ecotoxicology. Depuration is the loss of a toxicant from the organism after it has been placed into an environment devoid of toxicant. This term is associated with a particular experimental design involving the transfer of organisms that have accumulated elevated levels of toxicant to an environment with no significant sources of the toxicant. In the design, much effort is made to ensure that toxicant lost to the environment is not available to be taken up again during the course of depuration. Clearance is the rate of substance transferred from a compartment, normalized to a concentration. It has units of flow, such as ml/h for a water compartment. It can describe transfer of materials between compartments within the organism or between the organism and an external compartment. Landrum et al. (1992) provide the following, more explicit definition relative to a model involving an organism and an environmental compartment: "Clearance is defined as the volume or mass of a compartment scavenged of the contaminant per mass of organism per [unit] time and is contrasted to the rate constants that describe the fractional change in a compartment concentration [or mass] per [unit] time." It will be discussed later in the specific context of clearance volume-based models. In this text, elimination is the summation of metabolic, excretory, physiochemical, and under some circumstances, radioactive decay processes resulting in the decrease of toxicant in the organism. (This definition adds physiochemical and radioactive decay to processes that Barron et al. (1990a) list as contributing to elimination.) Biochemical breakdown of an organic toxicant would contribute to elimination under this definition as well as renal excretion. Radioactive decay might be included as part of elimination only if the loss of a radioactive contaminant was being modeled. However, if the radionuclide is simply being used as a tracer in the experimental design, radioactive decay would not be incorporated as a component of elimination. Growth dilution, the apparent decrease in toxicant concentration resulting from organism growth, would not be considered a component of elimination. With growth dilution, the total amount of toxicant in the organism would not decrease as a consequence of growth. The concentration would decrease because the amount of tissue in which the toxicant was distributed increased.

Elimination occurs through an array of physical, chemical, and biological mechanisms. As mentioned briefly above, phase I oxidation or hydrolysis reactions linked perhaps to an additional phase II biotransformation (Greenblatt and Shader 1985; Newman and Clements 2008) can render a nonpolar organic toxicant to a water-soluble form amenable to excretion. (Note that increased water solubility does not always lead to enhanced elimination.) Other general processes, including transport across gills, exhalation, secretion via the gallbladder or hepatopancreas, molting, excretion, or egg deposition, have been noted in reviews by Whicker and Schultz (1982), Spacie and Hamelink (1985), and Newman and Clements (2008). Many such biological processes change with time, e.g., developmental changes influence biliary excretion of methylmercury (Ballatori and Clarkson 1982). Physical mechanisms can also be important. Radioactive decay can be significant in modeling elimination of many short-lived radionuclides. Also, physical desorption of surface-bound toxicants (Crist et al. 1988) can contribute to elimination. Toxicants can be lost from plants by leaching, herbivore grazing, and leaf fall (Whicker and Schultz 1982).

Despite this diversity of elimination mechanisms, most quantitative treatments of elimination draw on a small number of mathematical models. Perhaps this is primarily a consequence of some fundamental underlying principle, such as the predominance of apparent first-order kinetics. This is at least partially true and has greatly enhanced progress. Regardless, it is profitable to assume that habit or tradition also plays a role in the uniformity of approaches taken. Under this assumption, less conventional approaches are presented in this chapter alongside commonly applied methods. They are not presented as superior or inferior to common methods: they are presented as alternatives to be critically evaluated during model selection and development.

3.2.1.2 Reaction Order

During this discussion, assumptions will be made regarding the apparent reaction order of elimination kinetics. Consequently, a brief discussion of reaction order is warranted.

In the case of a reaction involving only one reactant, reaction order refers to the power to which the reactant concentration is raised in the differential equation describing the reaction rate, $-dC/dt = kC^n$. Most reactions germane to this chapter are described by low order (zero, first, or occasionally, second; $n = 0$, 1, or 2) or saturation kinetics. Figure 3.1 summarizes these four reaction kinetics models. With zero-order kinetics, reaction rate is independent of concentration, i.e., proportional to concentration raised to the 0 power ($C^0 = 1$). The rate is determined by C^1 with first-order kinetics. First-order kinetics is so pervasive in bioaccumulation modeling that it is often assumed unless specified otherwise (Barron et al. 1990a; Newman and Clements 2008). Second-order kinetics might involve a single reactant or two different reactants. In the first case, the reaction rate is determined by C^2. When two different reactants are involved (C_1, C_2), the order of the reaction is determined by the sum of their exponents. For example, if $-dC_1/dt = kC_1^1C_2^1$ then the reaction order is 1 + 1 or 2.

With many enzyme-catalyzed reactions or receptor molecule-facilitated transport, saturation kinetics (Michaelis–Menten kinetics) can be relevant. With Michaelis–Menten kinetics, the reactant (substrate, S) combines with the enzyme (E) to form a complex (ES). Next, the substrate in the complex is converted to product (P). The assumption is made that the complex formation step is reversible and the product formation step proceeds as a first-order reaction. However, the overall rate of reaction for $E + S \leftrightarrow ES \rightarrow E + P$ will shift from first order to zero order if the concentration of the reactant (S) increases beyond a certain point. With a finite amount of enzyme present, a concentration of substrate can be reached above which essentially all of the available enzyme molecules are saturated. Beyond this concentration, the reaction will proceed at a rate independent of substrate concentration. This is also true for receptor molecule-based transport across membranes.

3.2.1.3 Linear Elimination Model

In the context of zero-order elimination kinetics, the rate at which the toxicant is eliminated from the organism (or compartment) would be independent of the concentration in the organism (Figures 3.2 and 3.3, linear). Such behavior could be modeled using Equation (3.1):

$$C_t = C_0 - k_e t \tag{3.1}$$

where C_t = concentration at time, t, C_0 = concentration at $t = 0$, and k_e = a constant with units of concentration/h.

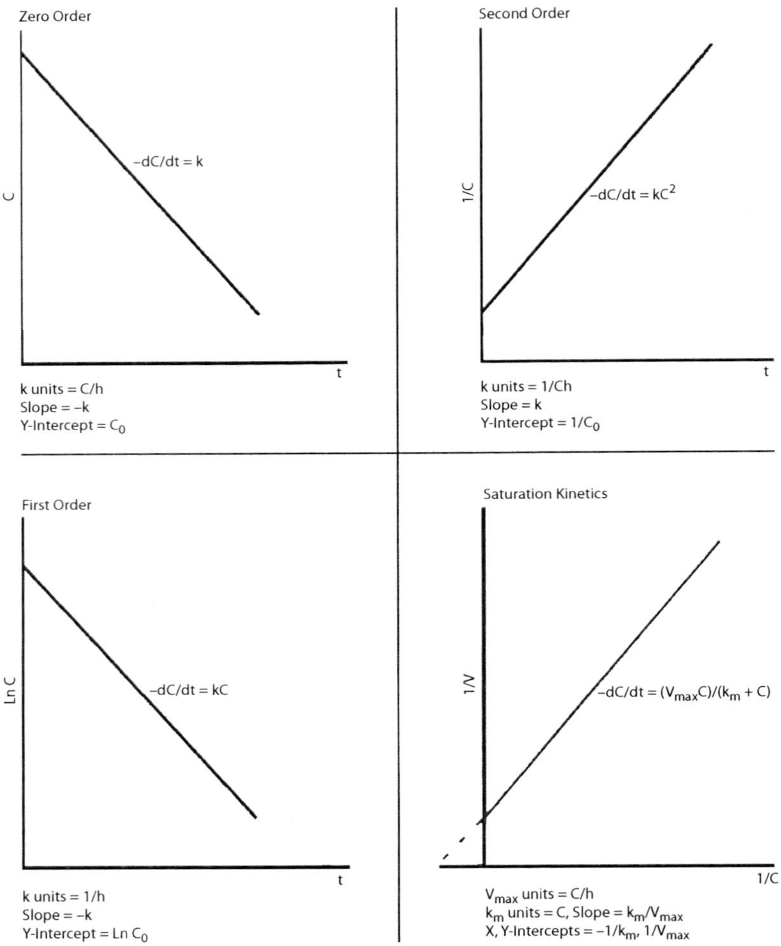

Figure 3.1 Zero-order, first-order, second-order, and saturation kinetics. Slopes and intercepts from linear plots such as those shown here are often used to estimate parameters describing the reaction kinetics.

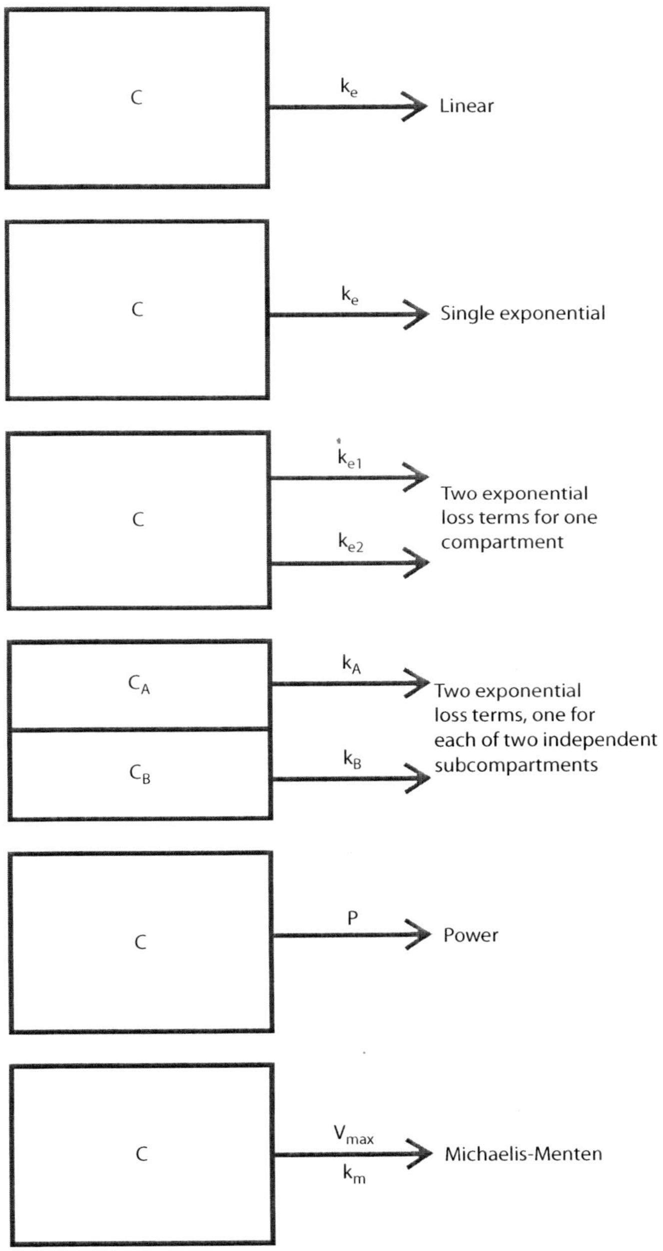

Figure 3.2 Six examples of common, simple models for toxicant elimination. See text for further detail.

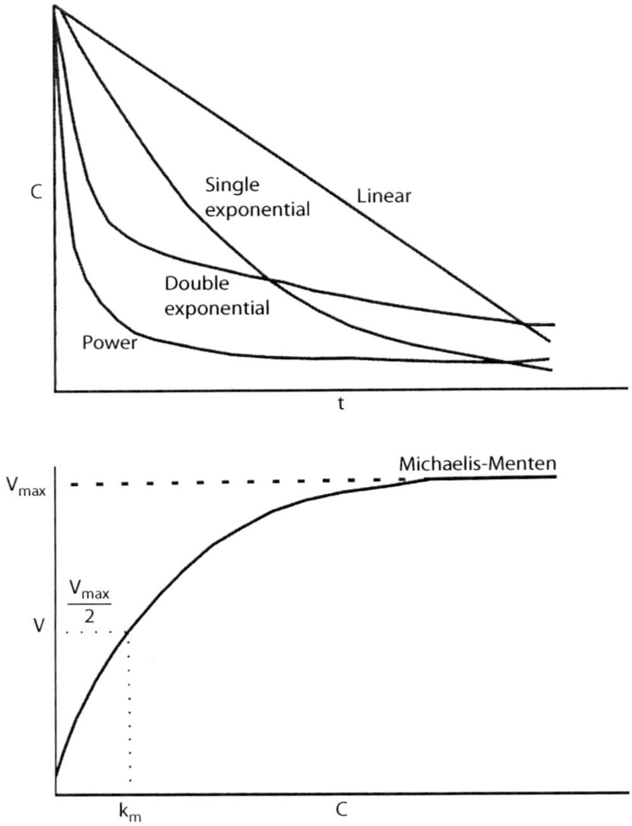

Figure 3.3 Illustrations of elimination kinetics associated with the models shown in Figure 3.2 and described in the text.

Extremely slow release of lead from bone can approximate zero-order kinetics if measured over a short time span. Schulz-Baldes (1974) provide a detailed treatment of apparent linear kinetics for lead elimination from the blue mussel, *Mytilus edulis*. As mentioned above, an enzyme- or transport molecule-mediated mechanism under saturation conditions is another situation where kinetics will display zero-order-like behavior.

3.2.1.4 Single (Mono-) Exponential Elimination Model

The most common model for elimination is based on first-order kinetics. In this case, the rate of elimination would be directly proportional to the concentration of toxicant in the organism at any time. A simple, first-order model with one pool of toxicant is described in Equation (3.2) and illustrated in Figures 3.2 and 3.3 (single exponential).

$$C_t = C_0 e^{-k_e t} \tag{3.2}$$

where k_e = elimination rate constant with units of 1/h.

The k_e in this equation describes the fractional change in concentration per unit of time. There are many examples of toxicant elimination displaying such monoexponential elimination. This model also describes simple radioactive decay kinetics; however, in that case, the decay rate constant is normally designated as λ, not k_e.

The time needed for the concentration to drop by a factor of 2 (one-half that present at the beginning of the time interval) is the elimination half-life. For the simple model described by Equation (3.2), it can be estimated by Equation (3.3).

$$t_{1/2} = \frac{Ln\left(\frac{1}{0.5}\right)}{k_e} = \frac{Ln\,2}{k_e} \tag{3.3}$$

Half-life estimates are most often derived by measuring loss at several intervals over time. However, design limitations can necessitate estimation with two measurement times only, e.g., Steele et al. (1986) or Scott et al. (1986). It is important to realize that two-measurement methods are more sensitive to error than methods using several measurement intervals. Simulations by Phillips (1989) suggest that selection of large time intervals and reduction of analytical error to a minimum can reduce the associated difficulties. Regardless, prior to use of two-measurement methods, the reader is urged to examine the work of Phillips (1989).

The mean lifetime of a particle in the compartment (τ) as defined by Equation (3.4) is $1.44t_{1/2}$ (Whicker and Schultz 1982). This term is also called the mean residence time or turnover time (Cutshall 1974).

$$k_e = \frac{1}{\tau} \tag{3.4}$$

Example 3.1

Anthracene elimination is measured in fish removed from a contaminated environment and placed into a laboratory tank receiving a continuous flow of uncontaminated water. The (fictitious) data set for elimination over 4 h is given in the leftmost two columns of the table below. (Ln C = natural logarithm of nmoles of anthracene/g; ε = regression residual.)

Time (h)	Anthracene (nmol/g)	Ln C	Predicted Ln C	ε	ε^2
0	100	4.605	4.597	0.008	0.000064
1	53	3.970	3.917	0.053	0.002809
2	24	3.178	3.237	–0.059	0.003481
3	12	2.485	2.557	–0.072	0.005184
4	7	1.946	1.877	0.069	0.004761
					$\Sigma\varepsilon^2$ = 0.016299

The natural logarithm of anthracene concentration (Ln C) and time (t) were fit using least-squares, linear regression to produce the following model: Ln C = $-0.680t + 4.597$. The associated correlation coefficient ($r^2 = 0.996$) indicated that more than 99% of the variation in Ln C could be accounted for by the model. The elimination rate constant is estimated to be –slope (see Figure 3.1, first order) or 0.680/h. The $t_{1/2}$ = Ln $2/ke$ = 0.693/0.680 = 1.02 h.

According to currently accepted practices in the field, the linear equation above can be backtransformed to arithmetic units to yield the following model, C (nmol/g) = $99.19e^{-0.680t}$. Predictions of concentrations at any time are then made. However, unless there is no error in the model or the predicted median value is desired, predictions from such a backtransformed model will be biased. The bias arises from the fact that, although the arithmetic mean of Ln Y is predicted accurately in the linear model, the arithmetic mean of Y is not accurately predicted in the backtransformed model. The median of Y (maximum likelihood estimate) is predicted in the backtransformed model. (See Beauchamp and Olson 1973, Newman and Heagler 1991, Newman 1991, or Sprugel 1983 for discussion of this bias.) The magnitude of this bias will increase as the error variance of the model increases. The prediction bias can be estimated from the mean square error (MSE) if the residuals (ε) appear to be normally distributed as in the present example. The MSE is the sum of the ε_i^2 values divided by $n - 2$, where n is the number of data pairs. The MSE is 0.016299/3 or 0.005433 in this example. The bias can be estimated using $e^{\mathrm{MSE}/2}$ to be $e^{0.005433/2}$ or 1.0027. Any predicted value can be multiplied by 1.0027 to generate an unbiased prediction from the backtransformed model. For example, the initial anthracene concentration is predicted to have been the following.

Biased estimate: $C = 99.19$ nmol/g
Unbiased estimate: $C_{unbiased} = 1.0027 \cdot 99.19$
$= 99.46$ nmol/g

The bias is insignificant in this example, but in other cases, it may be important to consider.

If the residuals were not normally distributed, the bias could have been estimated with a smearing estimate (Koch and Smillie 1986). The smearing estimate is $1/n(\Sigma e^{\varepsilon_i})$. Each individual residual is used as the exponent for e and then summed. This sum is then multiplied by $1/n$ to get an estimate of the transformation bias.

In this example, the bias was insignificant. Often this is not the case. The bias should be estimated if models are fit using logarithmic transforms of the dependent variable and the results are used for predictive purposes. If the model is generated for descriptive rather than predictive purposes, the MSE should still be included in model presentation. Inclusion of the MSE will allow evaluation of the bias prior to future implementation for predictive purposes by other researchers. Finally, if the predicted median Y is acceptable or preferred, this bias correction is unnecessary.

3.2.1.5 Elimination Model with Two Loss Terms for One Compartment

If there are two components of elimination removing toxicant from the compartment (organism), Equation (3.2) can be modified to describe the resulting kinetics (Figure 3.2, two exponential loss terms for one compartment). The resulting time course of elimination would be similar to that of the single exponential in Figure 3.3.

$$C_t = C_0 e^{-(k_{e1}+k_{e2})t} \tag{3.5}$$

where k_{e1} and k_{e2} = two distinct elimination rate constants.

Equation (3.5) is useful for study of elimination of a radioisotope with significant radioactive decay. In that case, the decay rate constant (λ) is used in place of one of the elimination rate constants.

The elimination half-life of a compartment with more than one elimination component is defined as an effective half-life by Whicker and Schultz (1982) and estimated by Equation (3.6).

$$t_{eff} = \frac{Ln\,2}{k_{eff}} \tag{3.6}$$

where

$$k_{eff} = \sum_{i=1}^{n} k_{ei} \tag{3.7}$$

with k_{ei} being the ith k_e. Each elimination component has an elimination half-life defined by Equation (3.3) above.

Example 3.2

Cutshall (1974) estimated depuration of ^{65}Zn from oysters removed downstream from the Department of Energy's Hanford Facility nuclear production reactors. The objective of the analyses was to define the kinetics of radioactive contaminant loss from a food species. Visually extracted data from Cutshall's Figure 1a were used in this illustration of modeling elimination kinetics. (It should be noted here and elsewhere that when visually extracted data are used in examples, the data will be slightly different from the original data, and consequently, results are not directly comparable to those in the original paper.)

What is the k_e for ^{65}Zn elimination?

The following SAS (SAS 1988) program fits these data to Equation (3.5) using nonlinear least-squares regression methods with and without weighting. It also fits the transformed data (Ln ^{65}Zn activity) to the model (Equation 3.5) using a linearized form, Ln ^{65}Zn activity = $b \cdot$ day + a (see Figure 3.1; first-order kinetics). It uses a decay rate constant (λ) of 0.00283 for ^{65}Zn.

```
DATA OYSTERZN;
INPUT DAY ZINC @@;
WNLIN2=DAY**2;
LZINC=LOG(ZINC);
DATALINES;
001 700 001 695 001 675 001 630 001 606 001 540 001 505 001 470
001 460 001 455 001 395 001 355 001 300 001 275 001 270 001 260
010 340 010 500 020 280 020 570 020 525 025 535 025 345 030 302
030 345 035 320 040 275 045 275 060 375 060 260 060 245 075 155
080 190 080 350 090 410 090 360 095 370 100 195 115 400 125 175
125 095 130 320 130 325 140 325 160 145 160 140 170 150 175 290
185 125 200 135 210 180 220 055 255 100 260 140 260 110 280 090
295 160 315 055 325 075 325 045 355 085 365 075 385 065 410 045
420 070 420 030 450 060 465 065 505 025 515 025 550 015 555 020
625 020
;
/* USING NONLINEAR REGRESSION WITH PARTIAL DERIVATIVES GIVEN.     */
/* INITIAL ESTIMATES AND BOUNDS ARE SELECTED BY CURSORY DATA      */
/* EXAMINATION. KE = THE ELIMINATION RATE CONSTANT, INITACT =     */
/* THE ACTVIVITY OF 65-ZN IN THE OYSTER AT TIME = 0 DAYS, AND     */
/* 0.00283 IS THE DECAY RATE CONSTANT FOR 65-ZN                   */
/*                                                                */
/* FIRST - NONLINEAR REGRESSION WITH NO WEIGHTING                 */
PROC NLIN;
   PARMS KE=0.004, INITACT=500;
   BOUNDS KE>0,
         1000>INITACT>0;
   MODEL ZINC=INITACT*EXP((-(KE+0.00283))*DAY);
      DER.INITACT=EXP((-(KE+0.00283))*DAY);
      DER.KE=-INITACT*DAY*EXP((-(KE+0.00283))*DAY);
      OUTPUT OUT=PRED PREDICTED=PPCI RESIDUAL=RPCI;
RUN;
/* PLOTTING THE PREDICTED AND ORIGINAL ACTIVITY DATA VERSUS DAY   */
```

```
/* AND THEN THE REGRESSION RESIDUALS VERSUS DAY                    */
PROC PLOT;
PLOT ZINC*DAY PPCI*DAY="*"/OVERLAY;
PLOT RPCI*DAY="*"/VREF=0;
RUN;
/* SECOND - NONLINEAR REGRESSION WITH DAY-SQUARED WEIGHTING        */
PROC NLIN;
  PARMS KE=0.004, INITACT=500;
  BOUNDS KE>0,
      1000>INITACT>0;
  MODEL ZINC=INITACT*EXP((-(KE+0.00283))*DAY);
    _WEIGHT_=WNLIN2;
    DER.INITACT=EXP((-(KE+0.00283))*DAY);
    DER.KE=-INITACT*DAY*EXP((-(KE+0.00283))*DAY);
    OUTPUT OUT=PRED PREDICTED=W2PPCI RESIDUAL=W2RPCI;
RUN;
/* PLOTTING THE PREDICTED AND ORIGINAL ACTIVITY DATA VERSUS DAY  */
/* AND THEN THE REGRESSION RESIDUALS VERSUS DAY.                  */
PROC PLOT;
PLOT ZINC*DAY W2PPCI*DAY="*"/OVERLAY;
PLOT W2RPCI*DAY="*"/VREF=0;
RUN;
/* NOW LEAST-SQUARES LINEAR REGRESSION OF LN 65-ZN VERSUS DAY    */
/* WILL BE USED TO ESTIMATE KE. NOTE THAT, IN THIS MODEL, THE    */
/* DECAY RATE CONSTANT ISN'T INCLUDED. THEREFORE, THE ELIMINA-   */
/* TION RATE CONSTANT IS THE SUM OF KE AND THE DECAY RATE CON-   */
/* STANT FOR 65-ZN. TO DERIVE KE, THE DECAY RATE CONSTANT IS     */
/* SUBTRACTED FROM THE ESTIMATE FROM THE MODEL.                  */
PROC GLM;
MODEL LZINC=DAY;
OUTPUT OUT=LINEAR PREDICTED=PPCILIN RESIDUAL=RPCILIN;
RUN;
/* PLOTTING THE PREDICTED AND ORIGINAL ACTIVITY DATA VERSUS DAY  */
/* AND THEN THE REGRESSION RESIDUALS VERSUS DAY.                  */
PROC PLOT;
PLOT LZINC*DAY PPCILIN*DAY="*"/OVERLAY;
PLOT RPCILIN*DAY="*"/VREF=0;
RUN;
```

The results of these calculations are illustrated in Figures 3.4 and 3.5. The unweighted, nonlinear regression model fit these data adequately (Figure 3.4A). The resulting estimates (±asymptotic standard error) of *ke* and the initial ^{65}Zn activity were 0.00268 ± 0.00067/h and 465 ± 20 pCi/g, respectively. However, examination of the regression residuals (Figure 3.4B) indicates that the assumption of independence between the variance of ^{65}Zn activity and day is incorrect. The variance decreases rapidly as duration of depuration increases.

Weighting by some function of the independent variable (day) is one means of coping with such unequal variance or heteroskedasticity. For example, if the error variance increased linearly with an increase in time, a weighting of 1/day could be used. An observation toward the end of the time course would have less weight in the fitting than one at the beginning of the depuration process. In the present case, the error variance seemed to decrease rapidly as day increased and a weighting of day^2 was used. The commonly used weighting (Boxenbaum et al. 1974) of Y^2, or in this case pCi^2, was examined but was less effective than a weighting of day^2. (For a further discussion of weighting, please see Neter et al. 1990, pp. 420–423, or Christian and Tucker 1984). Weighted, nonlinear regression results (k_e = 0.00244 ± 0.00035/h; initial ^{65}Zn activity = 454± 3 8 pCi/g) were only slightly lower than those from the nonweighted regression.

Another means of fitting these data to Equation (3.5) would involve transformation of the dependent variable prior to linear regression. The Ln ^{65}Zn activity was regressed against day to produce the following model: Ln ^{65}Zn = –0.0053 day + 6.07.

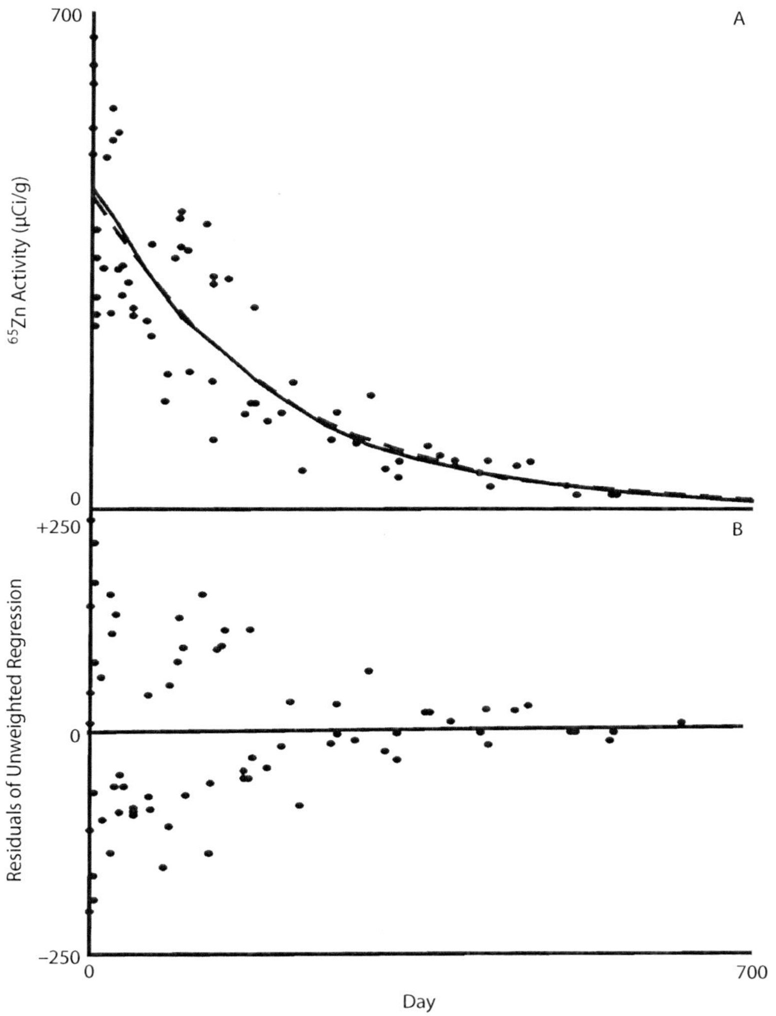

Figure 3.4 Elimination of ^{65}Zn from oysters. (From Cutshall, N., *Health Phys.*, 26, 327–331, 1974.) Panel A shows the original data extracted from the above-mentioned citation and the results of nonlinear regression analyses. Both unweighted (solid line) and weighted (dashed line) regression results are shown. The clear decrease in the regression residuals from the unweighted regression is shown in panel B.

The results of the linear regression are given in Figure 3.5. The linear model fits these data adequately as indicated by the uniform distribution of the data points about the predicted line (Figure 3.5A) and the lack of pattern in a plot of regression residuals against time (Figure 3.5B). The absence of any obvious heteroscedasticity indicates the effectiveness of transformation as another means of coping with unequal variance with time.

As shown in Figure 3.1 (first order), the slope is used to estimate $-k$. The Y-intercept is the mean predicted estimate of Ln ^{65}Zn activity at time = 0. Remember that the k includes the radioactive decay rate constant ($\lambda = 0.00283$) in this case. The ke could be estimated by $k - \lambda$. The estimated k (±standard error) is 0.00531 ± 0.00024. The k_e is $0.00531 - 0.00283$, or 0.00248. This estimate is very close to that obtained with the weighted, nonlinear regression methods.

The Y-intercept (6.07 ± 0.06) is the estimate of the mean predicted value of Ln ^{65}Zn activity at time = 0. Normally, the value $e^{6.07}$ or 433 pCi/g dry wt would be used as the estimate of

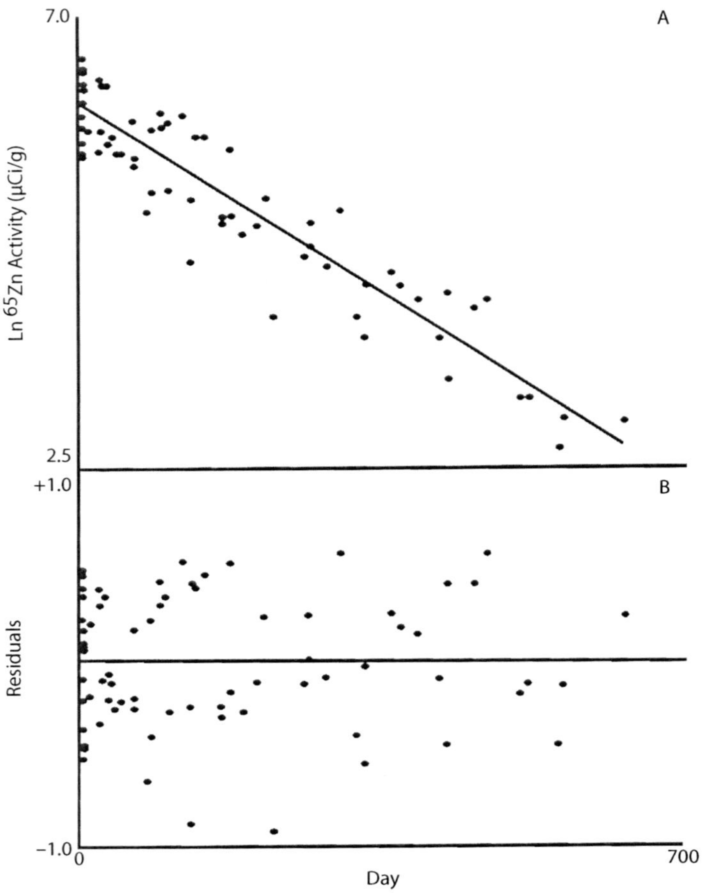

Figure 3.5 Elimination of ^{65}Zn from oysters. (From Cutshall, N., *Health Phys.*, 26, 327–331, 1974.) Linear regression was used to analyze these data. The linear regression model results fit the transformed data nicely. Panel B shows the lack of any strong time dependence of the regression residuals.

^{65}Zn activity at time = 0. Again, this is not an estimate of the mean predicted value in arithmetic units. It was derived by backtransformation of the predicted mean of values expressed as logarithms. Upon backtransformation, the *Y*-intercept (predicted mean value at time = 0) becomes the predicted median value on an arithmetic scale, not the predicted arithmetic mean value. Consequently, the usual approach of predicting the mean initial concentration by backtransformation is biased. If the regression residuals are normally distributed as in this example, the bias can be approximated using the mean square error (the SAS program generated a MSE of 0.125 for this model). The bias is estimated to be $e^{\mathrm{MSE}/2}$ or 1.06. The unbiased estimate of the mean ^{65}Zn activity at time = 0 is 1.06 · 433 or 460 pCi/g.

What are the effective and biological half-lives for ^{65}Zn in the oysters during depuration? The results of the linear regression procedures can be used in these estimations.

$$t_{eff} = \mathrm{Ln}\ 2/k_{eff} = 0.693/0.00531 = 130.5 \text{ days}$$
$$t_{1/2} = \mathrm{Ln}\ 2/k_e = 0.693/0.00248 = 279.4 \text{ days}$$

What is the mean lifetime of a particle as determined by k_e?

$$\tau = 1.44 t_{1/2} = 1.44 \cdot 279.4 = 402.4 \text{ days}$$

3.2.1.6 *Elimination Model with Two Loss Terms for Two Subcompartments (Biexponential)*

Other permutations on first-order elimination models are useful. If one is considering the loss from the whole body and elimination is taking place from two distinct subcompartments in the body, the following model is applicable (Figure 3.2, two exponential loss terms, one for each of two independent subcompartments; Figure 3.2, double exponential; Figure 3.3). As discussed in Newman and Clements (2008), a model with two subcompartments with exchange between subcompartments but elimination from only one will also display the same elimination kinetics as described here. An example of such a model would be a system with a storage subcompartment. In that case, the k_A and k_B have very different interpretations (see Atkins 1969, pp. 44–49; Newman and Clements 2008, chap. 8; Whicker and Schultz 1982, pp. 53–54).

$$C_t = C_A e^{-k_A t} + C_B e^{-k_B t} \tag{3.8}$$

where C_A and C_B = concentrations in subcompartments *A* and *B*, respectively, and k_A and k_B = elimination rate constants (1/h) for subcompartments *A* and *B*, respectively.

Example 3.3

The elimination of methylmercury ($^{203}HgCH_3$) is measured in sunfish over 94 days. The ^{203}Hg activities (pCi/g) measured during the time course are tabulated below in the two left-hand columns. A plot of Ln activity versus day showed a steeply sloped line initially, then a distinct break to a shallower slope after 22 days of elimination. Two compartments within the fish were tentatively identified based on this plot. Backprojection or backstripping methods (see Newman and Clements 2008, pp. 119–120, Figure 8.1) were used to get initial estimates of the rate constants and initial concentrations for these two subcompartments.

Time (d)	Activity (pCi/g)	Predicted Activity	ε (Predicted Activity – Observed Activity)
3	27,400	12,038	15,362
6	17,601	11,598	6,003
10	12,950	11,037	1,913
14	11,157	10,503	654
22	9,410	9,511	−101
31	**9,150**	8,507	
38	**6,060**	7,799	
45	**6,830**	7,151	
52	**7,270**	6,556	
59	**6,280**	6,011	
66	**5,950**	5,512	
73	**5,480**	5,053	
80	**5,310**	4,633	
87	**3,940**	4,248	
94	**3,310**	3,895	

In the first step of backstripping, data from the time period during which the slow compartment dominated the kinetics (days 31 to 94, as printed in boldface) were used to estimate the associated C_B and k_B. The natural logarithm of activity was regressed against day for all data collected after day 22 to yield the following relationship $C = 12{,}469e^{-0.0124\text{day}}$. This equation and the minor bias correction factor (1.002, $MSE = 0.018$) were used to generate predicted activities in this compartment through the entire elimination time course (column 3, "Predicted Activity"). The residuals (measured activities, as shown in column 2, minus the predicted activities in the slow compartment (B), as shown in column 3) were estimated for days 3 through 22 and placed in column 4. These residuals are estimates of the activity associated with the fast compartment during this period. The natural logarithms of these residuals were then regressed against day to yield $C = 34{,}773e^{-0.2863\text{day}}$. The bias correction was very small for this compartment also (1.001, $MSE = 0.0029$). The random distribution of residuals from this second (fast component) regression model suggested that no further backstripping was warranted.

As estimates from backstripping can be quite misleading (Van Liew 1962), they are used here only as initial estimates for iterative, nonlinear regression analysis as implemented by SAS. (It is important to have reasonable initial estimates, as iterative, nonlinear regression methods can converge on suboptimal local solutions. Inaccurate estimators and poor model fit result from such solutions.) Initial estimates of C_A, C_B, k_A, and k_B from this backstripping technique were 34,807, 12,494, 0.2863, and 0.0124, respectively. The estimates of C_A, C_B, k_A, and k_B (± asymptotic standard error) from the nonlinear regression model were 37,908 ± 4,824, 12,245 ± 858, 0.297 ± 0.044, and 0.012 ± 0.001, respectively. Roughly 76% of the activity was associated with the compartment displaying rapid elimination (37,908/(37,908 + 12,245)) and 24% was associated with the second compartment (12,245/(37,908 + 12,245)).

Remember that the k_A and k_B estimates include radioactive decay ($\lambda = 0.0145$ for ^{203}Hg). Consequently, 0.0145 must be subtracted from these estimates to determine the rate constants for biological elimination. It becomes immediately apparent upon doing so that the slow compartment actually had no detectable biological elimination. The loss was dominated by radioactive decay. In contrast, the first compartment had a large biological elimination rate constant of approximately 0.282.

Considering the negligible elimination from the second compartment, this relationship for methylmercury elimination (Equation 3.8) can be reduced to $C_t = C_A e^{-kAt} + C_B$. The methylmercury in C_B was unavailable for elimination. With this new expression of the model, the limiting value of C_t would be C_B, not 0.

The SAS code used in these calculations was the following. Predicted results and associated residuals were also produced by this code.

```
DATA MERCURY;
   INPUT DAY PCI @@;
   LPCI=LOG(PCI);
   IF DAY>22;
DATALINES;
03 27400 06 17601 10 12950 14 11157 22 9410 31 9150 38 6060
45 6830 52 7270 59 6280 66 5950 73 5480 80 5310 87 3940
94 3310
;
/* RUNNING LINEAR MODEL FOR SLOW COMPARTMENT                        */
PROC GLM;
      MODEL LPCI=DAY;
      OUTPUT OUT=EXPO PREDICTED=PEXPO RESIDUAL=REXPO;
RUN;
/* PLOTTING PREDICTED AND ORIGINAL ACTIVITIES VERSUS DAY            */
/* AND REGRESSION RESIDUALS VERSUS DAY                              */
PROC PLOT;
   PLOT LPCI*DAY PEXPO*DAY/OVERLAY;
   PLOT REXPO*DAY="*"/VREF=0;
RUN;
PROC PRINT;
   VAR DAY PCI LPCI PEXPO REXPO;
RUN;
/* NOW FITTING "FAST" COMPONENT                                     */
```

```
DATA FAST;
   INPUT DAY CPI @@;
   LPCI=LOG(PCI);
DATALINES;
03 15362 06 6003 10 1913 14 654 22 -101
;
/* RUNNING LINEAR MODEL FOR FAST COMPARTMENT                    */
PROC GLM DATA=FAST;
   MODEL LPCI=DAY;
   OUTPUT OUT=FEXPO PREDICTED=PFEXPO RESIDUAL=RFEXPO;
RUN;
/* PLOTTING ORIGINAL AND PREDICTED ACTIVITIES VERSUS DAY        */
/* AND REGRESSION RESIDUALS VERSUS DAY                          */
PROC PLOT;
   PLOT LPCI*DAY PFEXPO*DAY="*"/OVERLAY;
PLOT RFEXPO*DAY="*"/VREF=0;
RUN;
DATA MERCURY;
   INPUT DAY PCI @@;
   LPCI=LOG(PCI);
DATALINES;
03 27400 06 17601 10 12950 14 11157 22 9410 31 9150 38 6060
45 6830 52 7270 59 6280 66 5950 73 5480 80 5310 87 3940
94 3310
;
/* NOW USING ESTIMATES FROM BACKSTRIPPING ABOVE TO FIT THESE    */
/* DATA TO A NONLINEAR, DOUBLE EXPONENTIAL MODEL                */
PROC NLIN DATA=MERCURY;
   PARMS KCB=0.0124 KCA=0.2863 CB=12494 CA=34807;
   BOUNDS 0<KCB<1, 0<KCA<1, 5000<CB<20000, 20000<CA<45000;
   MODEL PCI=CA*EXP(-KCA*DAY)+CB*EXP(-KCB*DAY);
   OUTPUT OUT=DOUBLE P=PDOUB R=RDOUB;
RUN;
/* PLOTTING THE RESULTS OF THE NONLINEAR REGRESSION MODEL       */
PROC PLOT;
   PLOT PCI*DAY PDOUB*DAY="*"/OVERLAY;
   PLOT RDOUB*DAY="*"/VREF=0;
RUN;
```

But how many components should be stripped from a data set? There are several methods of estimating the appropriate number in addition to that used too frequently—the researcher's temperament and sense of optimism. Myhill's simulations (1967) of two component elimination models suggest the following rules of thumb for data acceptability for stripping two components. If the ratio of the two k_e values is roughly 2, then acceptable convergence can be expected using iterative regression methods if the data error doesn't exceed 1%. It would require a very high-quality data set to extract these two close components. For ratios of 4, 6, and 10, the data error should not exceed 5, 10, or 10%, respectively.

But what information is available for making the judgment between models of increasing numbers of components? Certainly, the MSE, plots of the original data's fit to the predicted model, and plots of regression residuals give some indication of the goodness of model fit as the number of exponential compartments increases. However, these do not provide an objective means of comparing models derived from data with a constant number of observation pairs but different numbers of estimated parameters. Barron et al. (1990a) suggest the methods of Boxenbaum et al. (1974) or Akaike's information criterion (Yamaoka et al. 1978) for this purpose. Akaike's information criterion (AIC) can be calculated from the following general equation:

$$AIC = n\left[\ln\left(\sum_{i=1}^{n} w_i(Y_i - \hat{Y}_i)^2\right)\right] + 2p \tag{3.9}$$

or, as used later in this book, $AIC = -2\ \mathrm{Log}(Maximum\ Likelihood) + 2p$, where n = the number of data pairs fit to the model, Yi = the Y value of the ith observation, $\hat{Y}_i$ = the predicted Y value of the ith observation, w_i = weighting for the ith observation (1 for unweighted regression), and p = the number of parameters estimated by the model.

The AIC estimates the information content within the various candidate models. It assumes that the more complex model will contain more information and then estimates the consequent increase in the information. For example, if two models with the same residual sums of squares are compared, the one with the least number of estimated parameters will be favored. The process of selecting the models with the lowest AIC is called minimum AIC estimation or MAICE. Other appropriate criteria, such as the equally widely applied Bayesian information criterion (BIC), might be used in similar fashion. The interested reader is directed to Wang and Liu (2006), who compare the AIC and BIC for such use, or to Bozdogan (2000). Another widely applied approach involves computation of Mallows's C_p statistics (Equation 3.10) for models of increasing complexity (Mallows 1973, 1995; Hocking 1976; Der and Everitt 2006).

$$C_p = \frac{\sum_{i=1}^{n}(Y_i - \hat{Y}_i)^2}{s_r^2} + 2p - n \tag{3.10}$$

where s_r^2 = the residual (error) mean square for the model, including all explanatory variables and an intercept. All candidate variables are selected for model inclusion and C_p statistics computed for all $2^p - 1$ possible models. The C_p is used to select the best one, two, three, and more variable models; i.e., the best models are those with the lowest C_p statistic. The SAS PROC GLMSELECT implements these selection methods.

As a specialized example of applying information criteria to toxicokinetics, Micromath Scientific Software (1989) points out that the AIC is dependent on the magnitude of the data points and formulates a modified AIC to avoid this shortcoming in toxicokinetics modeling. An alternative, model selection criterion (MSC) based on the AIC is put forward as a substitute for the AIC. Its value is independent of the magnitude of the data and, consequently, provides a "normalized" estimate of acceptability. They suggest that the following general judgments of model fit can be assigned to MSCs: <2 (unacceptable), 3 (marginal), 4 (typical of reasonably well-fit model), 5 (very good fit), >6 (exceptional and perhaps suspect). The MSC is estimated with Equation (3.11).

$$MSC = Ln\left[\frac{\sum_{i=1}^{n} w_i(Y_i - \bar{Y})^2}{\sum_{i=1}^{n} w_i(Y_i - \hat{Y}_i)^2}\right] - \frac{2p}{n} \tag{3.11}$$

Example 3.4

Using the program RSTRIP (Micromath Scientific Software 1989) to analyze the mercury data in Example 3.3, MSC values of 1.24, 4.09, 2.32, −0.49, and −0.78 were generated for one-, two-, three-, four-, and five-exponential components models, respectively. The MSC values supported our previous decision to extract two components from these data because the highest MSC was associated with the two-exponential component model. The MSC for the two-exponential model was within the range of "typical of a reasonably well-fit model." The results of the two-exponential model are used here to illustrate these calculations (assuming unweighted regression or weighting = $1/(Y_i^0) = 1$).

Y	$Y-\overline{Y}$	$(Y-\overline{Y})^2$	$\overline{Y}$	($Y-\overline{Y}$ or e)	$Wt\varepsilon^2$
27,400	18,193.5	331,003,442.3	27,357.0	43.0	1,849.0
17,601	8,394.5	70,467,630.3	17,769.0	–168.0	28,224.0
12,950	3,743.5	14,013,792.3	12,808.0	142.0	20,164.0
11,157	1,950.5	3,804,450.3	10,952.0	205.0	42,025.0
9,410	203.5	41,412.3	9,472.8	–62.8	3,943.8
9,150	–56.5	3,192.3	8,462.4	687.6	472,793.8
6,060	–3,146.5	9,900,462.3	7,781.0	–1,721.0	2,961,841.0
6,830	–2,376.5	5,647,752.3	7,156.8	–326.8	106,798.2
7,270	–1,936.5	3,750,032.3	6,582.9	687.1	472,106.4
6,280	–2,926.5	8,564,402.3	6,055.2	224.8	50,535.0
5,950	–3,256.5	10,604,792.3	5,569.7	380.3	144,628.1
5,480	–3,726.5	13,886,802.3	5,123.1	356.9	127,377.6
5,310	–3,896.5	15,182,712.3	4,712.4	597.6	357,125.8
3,940	–5,266.5	27,736,022.3	4,334.6	–394.6	155,709.2
3,310	–5,896.5	34,768,712.3	3,987.1	–677.1	458,464.4
Σ		**549,375,610.5**			**5,403,585.3**

$$
\begin{aligned}
MSC &= \text{Ln}\,(549{,}375{,}610.5/5{,}403{,}585.3) - 2(4)/15 \\
&= \text{Ln}(101.67) - 8/15 \\
&= 4.62 - 0.53 \\
&= 4.09
\end{aligned}
$$

3.2.1.7 Power Elimination Model

Occasionally, a power function (see Figures 3.2 and 3.3, power) is used to best describe elimination, e.g., Whicker and Schultz (1982) or Newman and McIntosh (1983a).

$$C_t = C_1 t^{-P} \tag{3.12}$$

where C_1 = concentration at day 1, and P = a constant.

In such cases, a double logarithm plot (Ln C versus Ln t) results in a straight line. Initially, such a model would seem to have no direct link to our previous discussion of reaction order. However, kinetics described with a power model could result from several simultaneous yet inseparable elimination components such as those associated with first-order kinetics. Whicker and Schultz (1982) give several examples of processes producing apparent power models. The simultaneous decay of a mixture of nuclear fission products can fit a power model, although when taken separately, each radionuclide involved displays exponential decay. Also, the release of radionuclides from bone can be described with a power model. Whicker and Schultz (1982) explain this observation by describing bone as a very heterogeneous "compartment" with an initial rapid loss of superficially bound radionuclide and then increasingly slower clearance from pools deeper within the bone.

Example 3.5

Newman and McIntosh (1983a) placed freshwater snails (*Physa integra*) from a contaminated reservoir into clean water under controlled laboratory conditions. Over a 22-day period, snails were removed and analyzed for lead concentration. Visual inspection of the resulting depuration curve suggested a power model. The following code was used to fit these data.

```
DATA LEAD;
   INPUT DAY LEAD @@;
   LLEAD=LOG(LEAD);
   LDAY=LOG(DAY);
DATALINES;
0.16 41.0 0.16 31.0 0.16 25.3 1.00 30.5 1.00 22.7
1.00 22.0 2.00 13.5 2.00 15.0 2.00 16.5 4.00 10.0
4.00 12.0 4.00 15.0 8.00 08.7 8.00 13.0 8.00 15.0
12.0 08.0 12.0 09.0 12.0 12.0 16.0 08.0 16.0 10.5
16.0 12.0 22.0 12.5 22.0 10.6 22.0 05.5
;
/* USING LINEAR REGRESSION OF LN TRANSFORMED X AND Y  */
/* VARIABLES FOR POWER CURVE FITTING OF ELIMINATION   */
PROC GLM;
   MODEL LLEAD=LDAY;
   OUTPUT OUT=POWER PREDICTED=PPOWER RESIDUAL=RPOWER;
RUN;
/* PLOTTING PREDICTED AND ORIGINAL LN CONCENTRATION  */
/* VERSUS LN DAY AND REGRESSION RESIDUALS VERSUS LN DAY */
PROC PLOT;
   PLOT LLEAD*LDAY PPOWER*LDAY="*"/OVERLAY;
   PLOT RPOWER*LDAY="*"/VREF=0;
RUN;
```

The resulting model (Ln Pb = −0.272Ln day + 3.001) showed no pattern to its residuals and had a significant r^2 of 0.77 (α = 0.05). The backtransformed model was Pb = 20.1day$^{-0.272}$ with a prediction bias correction factor of 1.03. The slope and intercept had small standard errors of 0.031 and 0.064, respectively.

When a double exponential model was imposed on these data as in Example 3.3, an adequate fit also resulted. The slow and fast compartment sizes were estimated to be 11.6 (± 3.1) and 24.1 (± 3.6), respectively. The k_e for the fast component was adequately estimated (0.737 ± 0.259/h). That for the slow compartment was near 0 and had a relatively large standard error (0.009 ± 0.018/h), suggesting negligible clearance from this compartment. Nearly 32% of the lead was essentially unavailable for elimination.

3.2.1.8 Michaelis–Menten Elimination Model

Finally, if Michaelis–Menten kinetics are warranted (Figures 3.1 to 3.3), the following equation defines the kinetics. As discussed in considerable detail in Piszkiewicz (1977), such kinetics will conform to first order when concentrations are below saturation and zero order when an excess of substrate is present. Spacie and Hamelink (1985) discuss apparent Michaelis–Menten kinetics in data generated by Mayer (1976) for DEHP elimination by fathead minnow.

$$C_0 - C_t + k_m Ln\left[\frac{C_0}{C_t}\right] = V_{max}t \tag{3.13}$$

where V_{max} = the maximum velocity of substrate conversion, and k_m = the substrate concentration at which the velocity of conversion is $V_{max}/2$.

The rate of substrate conversion is defined by the equation

$$v = \frac{V_{max}C}{C + k_m} \tag{3.14}$$

where v = the velocity of substrate conversion, and C = substrate concentration.

One of several linearizing plots is shown in Figure 3.1 (double reciprocal or Lineweaver–Burk plot; Equation 3.15). The Lineweaver–Burk transformation (Equation 3.15) is obtained by taking the reciprocal of Equation (3.14). (Derivations of these transformations are presented in Piszkiewicz 1977.) Although the Lineweaver–Burk plot is the most commonly used transformation for deriving *km* and *Vmax*, it is statistically the worst-behaved transformation (Raaijmakers 1987). Other common linear plots based on other transformations include Eadie–Hofstee (v versus v/C, Equation (3.16)), Scatchard (v/C versus v, Equation 3.17), and the least used but preferrable Woolf (C/v versus C, Equation 3.18) plots (Raaijmakers 1987).

$$\frac{1}{v} = \frac{k_m + C}{V_{max}C} = \frac{1}{V_{max}} + \frac{k_m}{V_{max}C} \tag{3.15}$$

$$v = V_{max} - \frac{k_m v}{C} \tag{3.16}$$

$$\frac{v}{C} = \frac{V_{max}}{k_m} - \frac{v}{k_m} \tag{3.17}$$

$$\frac{C}{v} = \frac{k_m}{V_{max}} + \frac{C}{V_{max}} \tag{3.18}$$

Relative to regression analysis, the Eadie–Hofstee (v versus v/C) and Scatchard (v/C versus v) plots have the complication that v (the measured rate of substrate conversion) is present in both the dependent and independent variables. Consequently, the assumption of trivial error in the independent variable becomes a concern. Raaijmakers (1987) discusses additional problems associated with such plots. Many of the same transformations employed are also used in modeling adsorption (see below). They are subject to the same biases discussed in detail by Kinniburgh (1986).

To incorporate Michaelis–Menten kinetics into elimination, zero-order kinetics may be used during that period of elimination when saturation is occurring and then first-order kinetics thereafter. The time at which the transition from pseudolinear kinetics to first-order kinetics (t^*) occurs is given by Wagner (1979) as

$$t^* = \left[\frac{1 - \frac{1}{e}}{V_{max}}\right] C_0 + \frac{k_m}{V_{max}} \tag{3.19}$$

where e = base e (approximately 2.71828).

The slope of the pseudolinear phase of elimination (k_o^*) is also estimated by Wagner (1979).

$$k_0^* = V_{max} - \frac{k_m}{t^*} \tag{3.20}$$

Note that, because t^* depends on C_0, the slope will also be dependent on initial concentration.

3.2.1.9 Comment on Interpreting Two Compartment Kinetics

As discussed in Section 3.2.1.6., several compartment models can result in biexponential elimination kinetics. The one shown in Figure 3.2 (two exponential loss terms, one for each of two independent subcompartments) involves two independent subcompartments. (In reality, each of these subcompartments is actually a mathematical compartment. They were considered subcompartments here only in the context of whole body elimination kinetics.) That model was explored initially only because it was the easiest to define quantitatively, not because it was the most commonly modeled.

Another more generally useful compartment model that displays identical elimination kinetics is shown in Figure 3.6. This model consists of two compartments (central A and peripheral B) with exchange between them. Dose (D) is introduced into the central compartment and elimination occurs only from the central compartment. Such a model might be appropriate for modeling elimination of a toxicant from the blood that is also subject to storage in the body tissues. If elimination from the two compartments were monitored, very different curves would result for the two compartments (Figure 3.6). Although toxicant elimination from both compartments in the first model (top panel) remained independent of each other through time, elimination becomes more

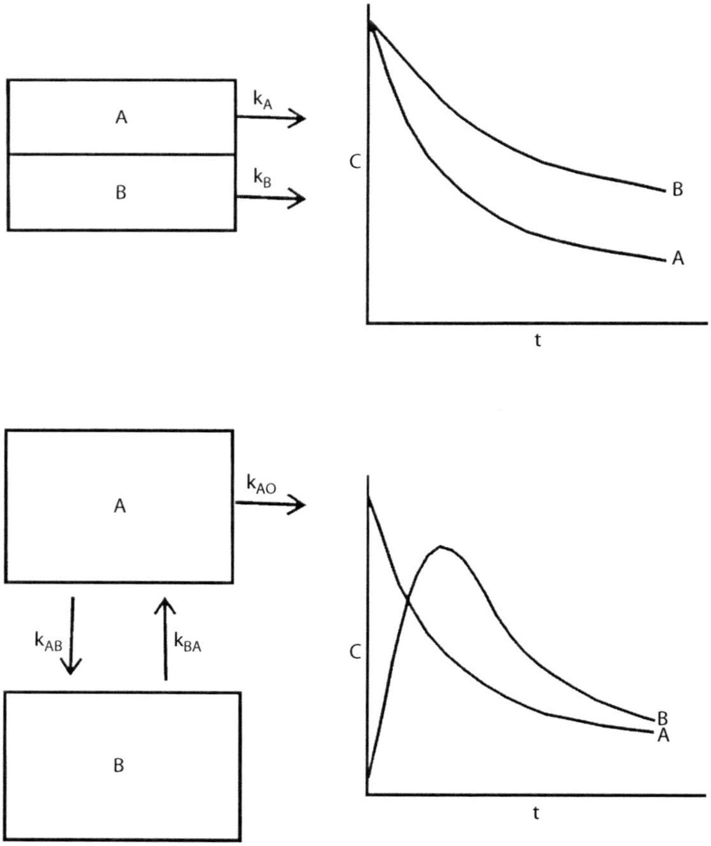

Figure 3.6 Examples of two-compartment models (with and without exchange between compartments) producing very different elimination kinetics from each compartment.

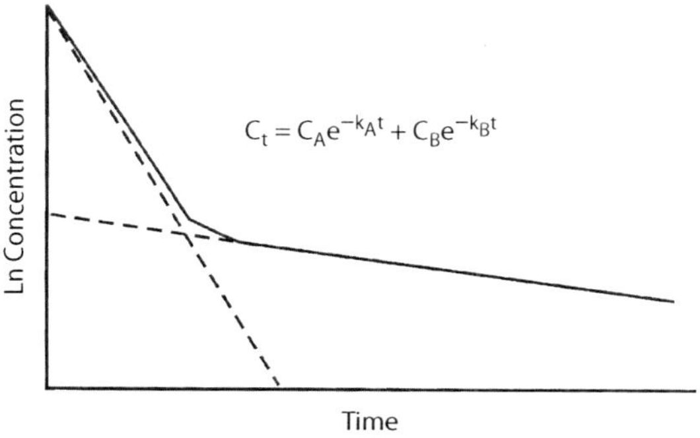

Figure 3.7 A graphical illustration of the backstripping approach to determine the constants of an apparent two-component, first-order model. The constants might be macroconstants that are used to derive microconstants for a model, such as one with exchange between a central and a peripheral compartment. The concentration versus time data might be generated from the central compartment (e.g., blood) for this purpose.

complex when the toxicant is introduced into compartment A in this second model (bottom panel). A curve such as that shown in the bottom panel of Figure 3.6 (labeled A) would be seen if one measured the concentration in the central compartment A (e.g., blood) through time. Fortunately, the macroconstants, C_A, C_B, k_A and k_B, estimated with Equation (3.8) for the central compartment allow calculation of the microconstants, k_{AB}, k_{AO}, and k_{AB}, for the two-exchanging-compartment model (Gibaldi and Perrier 1982; Renwick 2008). C_A and C_B are analogous to mathematical initial concentrations in the two compartments or the *Y*-intercepts of the two phases from a logarithm-transformed plot such as Figure 3.7. The k_A and k_B are derived from the slopes of the lines of the transformed data plots.

$$k_{BA} = \frac{C_A k_B + C_B k_A}{C_A + C_B} \tag{3.21}$$

$$k_{A0} = \frac{k_A k_B}{k_{BA}} \tag{3.22}$$

$$k_{AB} = k_A + k_B - k_{BA} - k_{A0} \tag{3.23}$$

Because the C_0 is a consequence of the initial distribution of the administered dose (*D*) to the central compartment (*V*) and the C_0 is equal to *CA* + *CB* of Equation (3.8), the volume of the central compartment can be estimated from *D*, C_A, and C_B, i.e., $V = D/(C_A + C_B)$ (Renwick 2008). The half-life in this two-compartment model can also be estimated as $t_{1/2}$ = Ln $2/k_B$. Although most applications of this model involve drug pharmacokinetics, Holleman, Luick, and Whicker's work with ^{134}cesium elimination from reindeer (Holleman et al. 1971) is an early example of an ecotoxicological application.

Another model very similar to that described by Equations (3.21) to (3.23) could also generate biexponential elimination kinetics (Equation 3.8). This model involves two interconnected compartments, but elimination is from the peripheral second compartment (rate constant = k_{B0}) instead of the central compartment (rate constant = k_{A0}). In this case, the toxicant is introduced into the first compartment (A), exchanges with the second compartment (B), and is eliminated from this second compartment. The equations relating the biexponential parameters to constants for this model are the following:

$$k_{AB} = \frac{C_A k_A + C_B k_B}{C_A + C_B} \quad (3.24)$$

$$k_{B0} = \frac{k_A k_B}{k_{AB}} \quad (3.25)$$

$$k_{BA} = k_A + k_B - k_{AB} - k_{B0} \quad (3.26)$$

A final two-compartment situation that can be easily modeled is one in which the toxicant is introduced into the first compartment (A) and passes (rate constant = k_{AB}) into the second compartment (B), from which it is eliminated (rate constant = k_{BO}). The k_A and k_B estimated from the biexponential elimination curve are estimates of k_{AB} and k_{B0}, respectively.

Derivations similar to Equations (3.21) to (3.26) can be had for more complex models. For example, parameters estimated from an apparent triexponential elimination curve (three extracted elimination components) can be used to estimate constants for associated parameters for a model with storage in two compartments and elimination from a third, central compartment (Gibaldi and Perrier 1982). However, as the models become more complex, the importance of high-quality elimination data becomes more critical during parameter derivation.

3.2.2 Adsorption

3.2.2.1 General

Toxicant adsorption on surfaces such as those of algae (Crist et al. 1988; Fujita and Hashizume 1975; Stary and Kratzer 1982; Les and Walker 1984; Wang and Wood 1984), periphytic microflora (Newman and McIntosh 1989), zooplankton (Ellgehausen et al. 1980), fish (McKone et al. 1971), and fish gill (Pagenkopf 1983) plays a critical role in the initial stages of uptake. Adsorption, "the process through which a net accumulation occurs at the common boundary of two contiguous phases" (Sposito 1984), may involve weak, reversible bonding such as that associated with van der Waals forces (physical adsorption) or strong, irreversible bonding (chemical adsorption). Desorption can be significant for substances that undergo physical adsorption but only minimal after chemical adsorption. The terms *adsorption* and *physical adsorption* are often used interchangeably as in this treatment. Although often visualized as simple ion exchange at specific binding sites, adsorption can include more complex processes. For example, adsorption of some ionic organic toxicants may not be driven by specific sites on the surface but, rather, by the lower affinity of nonpolar groups for the aqueous phase relative to the solid phase (Stumm and Morgan 1970).

Adsorption of a toxicant might result from a direct toxicant-surface interaction. However, adsorption may also be indirect, e.g., involve initial complexation with a ligand. For example, Sposito (1984) lists the following metal-ligand-surface interactions that could occur: "1. The ligand has a high affinity for the metal and forms a soluble complex with it, and this complex has a high affinity for the adsorbent. 2. The ligand has a high affinity for the adsorbent and is adsorbed, and the adsorbed ligand has a high affinity for the metal. 3. The ligand has a high affinity for the metal and forms a soluble complex with it, and this complex has a low affinity for the adsorbent. 4. The ligand has a high affinity for the adsorbent and is adsorbed, and the adsorbed ligand has a low affinity for the metal." Such interactions may give rise to shifts in adsorption such as those associated with bacterial uptake of ^{241}Am in the presence of dissolved organic matter (Giesy and Paine 1977), chromate accumulation by algae in the presence of various inorganic ligands (Stary and Kratzer

1982), and algal uptake of nickel in the presence of humic acids (Wang and Wood 1984). Complex ligand-toxicant interactions can also be important in estimating adsorption in tissues during pharmacokinetic modeling (Wagner 1979).

To add further complications, other compounds or ions may compete for adsorption sites. As examples, group Ia metals, e.g., Li, Na, K, Rb, and Cs, or oxyanions, e.g., AsO_4^{-3} and PO_4^{-3}, may compete for sites. Sposito (1984) and Stumm and Morgan (1970) discuss adsorption of several metals varying in affinity for a solid adsorbent. Pertinent examples involving competition include pH effects on algal uptake of metals (Crist et al. 1988; Stary and Kratzer 1982; Les and Walker 1984), intermetal effects on algal uptake (Crist et al. 1988), or hardness ion effects on metal toxicity (Pagenkopf 1983).

3.2.2.2 Common Forms of Freundlich and Langmuir Equations

Several methods have been developed to quantify adsorption based on the assumptions of a fixed number of binding sites, reversible adsorption, a homogeneous surface, and fixed adsorbate binding (that is, the steric configuration of the bound toxicant does not change). The two most commonly used are the Freundlich (Equation 3.27) and Langmuir (Equation 3.28) isotherm equations.

$$\frac{X}{M} = KC^{1/n} \tag{3.27}$$

where X/M = the amount adsorbed/unit weight of adsorbent, C = the equilibrium concentration in the solution after adsorption has taken place, and K,n = derived constants.

$$\frac{X}{M} = \frac{abC}{1+bC} \tag{3.28}$$

where a,b = derived constants.

The K and n constants for the Freundlich equation are often derived with the following linear transformation. The slope and intercept are used to estimate n and K, respectively.

$$\text{Log } \frac{X}{M} = \text{Log } K + \frac{1}{n}\text{Log } C \tag{3.29}$$

The a and b constants for the Langmuir equation are similarly derived with the following linear equation.

$$\frac{C}{X/M} = \frac{1}{ab} + \frac{1}{a}C \tag{3.30}$$

First, the slope is used to estimate a. Next, b is derived with that a and the intercept ($1/ab$). Unlike the Freundlich isotherm equation, which is wholly an empirical model, the Langmuir isotherm equation is derived from principles that allow direct physical meaning to be assigned to the associated constants. Equation (3.28) can be rewritten to conform to this theoretical context for the convenience of any reader who may wish to compare this discussion to the well-established literature on adsorption theory.

$$n = \frac{KCM}{1 + KC} \tag{3.31}$$

where n = amount adsorbed (mmol) per unit mass (g), C = equilibrium concentration in the aqueous phase (mmol/ml), M = the adsorption maximum (mmol), and K = the affinity parameter, which is a measure of bond strength.

Figure 3.8 illustrates the shape of this model and its linear transform. The maximum number of binding sites can be estimated with Equation (3.31). The affinity parameters for various chemicals can be compared under different conditions. For example, Crist et al. (1988) used this approach to estimate the relative affinity of alkali, alkali earth, and transition metals for algal cells. They also quantified the influence of pH on metal adsorption to these cells.

The traditional means of fitting adsorption data to the Langmuir isotherm model have been reexamined by Kinniburgh (1986). Table 3.1 summarizes the transformations examined.

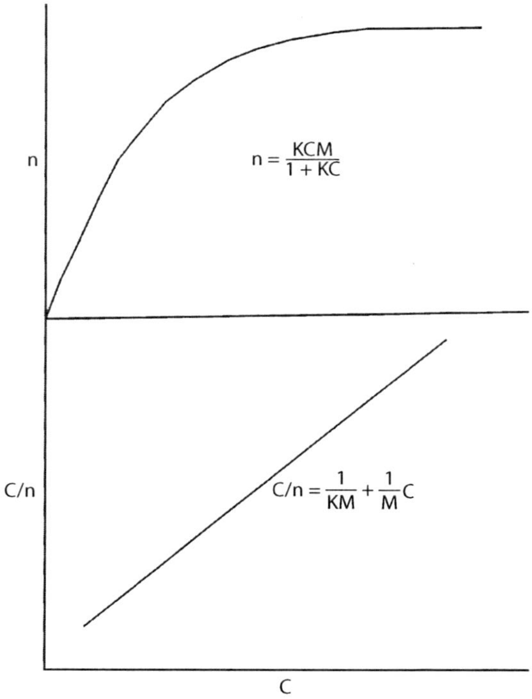

Figure 3.8 An adsorption model and its linear transformation. See text for details.

Table 3.1 Linear Transformations for the Langmuir Isotherm Model

Transformation	Y vs. X	K, M Estimates
Lineweaver–Burk	$1/n$ vs. $1/C$	Intercept/slope, 1/intercept
Reciprocal	C/n vs. C	Slope/intercept, 1/slope
Eadie–Hofstee	n vs. n/C	−1/slope, intercept
Scatchard	n/C vs. n	−Slope, −intercept/slope

Source: Kinniburgh, D.G., *Environ. Sci. Technol.*, 20, 895–904, 1986, Table II.

He found that unweighted, least-squares regression of these transformed data weighted low n observations more than high n observations. Further, the Eadie–Hofstee and Scatchard transformations had the experimentally measured n incorporated in the independent variable, leaving open the possibility of significant error in the independent variable. He also noted that, as mentioned previously in our discussion of Michaelis–Menten kinetics, the frequently used Lineweaver–Burk transformation behaves poorly during regression analysis. However, the Langmuir reciprocal behaved acceptably. (It is reassuring to note that the Langmuir reciprocal transformation is the same as the Woolf transformation found by Raaijmakers (1987) to be the best for fitting Michaelis–Menten kinetics.) Kinniburgh suggests that the Lineweaver–Burk, Langmuir reciprocal, Eadie–Hofstee, and Scatchard transformations fit data better (single-pass, least-squares regression) when the following weightings were used: n_i^4, n_i^4/C_i^2, n_i^2, and n_i^2, respectively. With these weightings, the regression results were closer to those derived by nonlinear regression methods. Example 3.6 is based partially on the data analysis of Kinniburgh (1986) to illustrate his recommended procedures. Only the surface experiencing adsorption is changed to remain relevant to bioaccumulation.

Example 3.6

(Derived from Table III of Kinniburgh 1986)

Adsorption of zinc from solution is measured for a freshwater, filamentous alga and the following data are generated. What are the K and M parameters for zinc adsorption?

n (mmol Zn adsorbed/g)	Equilibrium Zn Concentration (mmol/L)
0.075	0.030
1.40	0.069
1.95	0.118
2.51	0.166
3.03	0.217
3.53	0.270
4.02	0.325
4.41	0.388
4.79	0.453
5.22	0.512

Kinniburgh (1986) outlined several linear transformations of the Langmuir equation and examined their abilities to fit this type of data. The recommended Langmuir reciprocal transformation with and without weighting will be used here as well as nonlinear regression. The following SAS code performs these calculations.

```
DATA ZINC;
   INPUT N C @@;
   WGT = (N**4)/(C**2); /* RECOMMENDED WEIGHTING            */
   Y=C/N;
DATALINES;
0.75 0.030 1.40 0.069 1.95 0.118 2.51 0.166 3.03 0.217
3.53 0.270 4.02 0.325 4.41 0.388 4.79 0.453 5.22 0.512
;
/* USING NONLINEAR REGRESSION PROVIDING NO DERIVATIVES      */
/* AND NO WEIGHTING                                         */
PROC NLIN;
   PARMS K=3.00, M=9.00;
   BOUNDS K>0, M>0;
   MODEL N = (K*C*M)/(1+K*C);
   OUTPUT OUT=PRED PREDICTED=PN RESIDUAL=RN;
```

```
RUN;
/* PLOTTING NONLINEAR REGRESSION DATA AND PREDICTIONS     */
/* VERSUS CONCENTRATION, AND REGRESSION RESIDUALS VERSUS  */
/* CONCENTRATION                                          */
PROC GPLOT;
  PLOT N*C PN*C="*"/OVERLAY;
  PLOT RN*C="*"/VREF=0;
RUN;
/* USING UNWEIGHTED LEAST-SQUARES LINEAR REGRESSION ON    */
/* LANGMUIR RECIPROCAL PLOT OF ADSORPTION DATA            */
PROC GLM;
  MODEL Y=C;
  OUTPUT OUT=LINEAR PREDICTED=PLIN RESIDUAL=RLIN;
RUN;
PROC GPLOT;
  PLOT Y*C PLIN*C="*"/OVERLAY;
  PLOT RLIN*C="*"/VREF=0;
RUN;
/* USING WEIGHTED LEAST-SQUARES, LINEAR REGRESSION ON     */
/* LANGMUIR RECIPROCAL PLOT OF ADSORPTION DATA            */
PROC GLM;
  MODEL Y=C;
  WEIGHT WGT;
  OUTPUT OUT=WLINEAR PREDICTED=PWLIN RESIDUAL=RWLIN;
RUN;
PROC GPLOT;
  PLOT Y*C PWLIN*C="*"/OVERLAY;
  PLOT RWLIN*C="*"/VREF=0;
RUN;
```

The nonlinear regression fit these data with very little pattern in the regression residuals. The estimates (± asymptotic standard errors) of K and M were 2.10 ± 0.19 and 9.89 ± 0.52 mmol, respectively. (These are essentially the same estimates as given in Kinniburgh's Table IV.)

Although both linear regressions had high r^2 values (unweighted = 0.966, weighted = 0.976), residuals from the lowest C data points were large. A $\cap$-shaped pattern to the residuals was apparent when plotted against C. K and M were calculated from the unweighted regression results as follows (see Table 3.1):

K = Slope/Intercept = 0.1140/0.044 = 2.59
M =1/Slope = 1/0.1140 = 8.77 mmol

Similarly, the estimates from the weighted regression were the following:

K = 0.1037/0.047 = 2.21
M = 1/0.1037 = 9.64 mmol

These estimates are the same as those given in Kinniburgh (1986, Table IV). Consistent with the conclusion of Kinniburgh from these data, the weighted regression produced estimates closer to those of the nonlinear regression model than the unweighted regression.

3.2.2.3 Other Useful Adsorption Equations

Sposito (1984) describes four classes of adsorption isotherms: S-, H-, L-, and C-curves for soils. It is reasonable to assume that adsorption onto biological surfaces may also deviate from the one class of adsorption isotherm models described above (L-curve). The reader is directed to Sposito (1984) for a general discussion of other classes of adsorption isotherms and to the excellent article by Kinniburgh (1986) for means of fitting data to more complex models.

3.2.3 Accumulation Models: One Compartment

3.2.3.1 General

Uptake (the movement of a toxicant into, or in some special cases onto, the organism) can occur directly from water, food, soil, or sediments. Movement may be across the general integument, gills, pulmonary surfaces, or the gut. Initial phases of uptake often involve adsorption as described above. Uptake can vary tremendously depending on the specific species and environmental conditions. Indeed, uptake can vary even for individuals held under constant conditions. Changes may be a consequence of such processes as general physiological changes, modified calcium flux associated with insect molting, or fluctuating nonpolar organic uptake associated with seasonal lipid dynamics of mollusks.

According to Gordon (1972), there are six major mechanisms for movement of solutes across membranes: passive diffusion, active transport, facilitated diffusion or transport, exchange diffusion, pinocytosis, and solvent drag. Passive diffusion occurs as a movement of solute down a chemical, electrical, or activity gradient. Diffusion may involve exchange diffusion, which is the simple exchange of ions of the same type across the membrane in both directions. Exchange diffusion involves a carrier molecule. Facilitated diffusion or transport also occurs down a chemical gradient but at a rate greater than predicted by passive diffusion alone. It involves a carrier molecule. Active transport requires energy to facilitate movement up a chemical gradient. Regulation of potassium and sodium (potassium higher in the cell than outside and sodium lower in the cell than outside) by a coupled ion pump (membrane-bound ATPase) is a good example of active transport. Such a system is capable of saturation kinetics and requires energy as demonstrated by its failure upon metabolic inhibition. Elements such as cesium, which chemically mimics potassium, can be actively transported by such a mechanism. Another similar example is active calcium transport at low environmental concentrations (Hunn 1985). Movement via energy-requiring pinocytosis involves enclosing material from outside the cell in a vacuole that then migrates across the cell to discharge its contents inside the organism. Occasionally relevant, as in the case of a toxicant-damaged gill, solvent drag involves the movement of the solute in the direction of the solvent movement. The relative importance of each mechanism will vary depending on many factors, including route of introduction (food versus water), the toxicant being studied, the external conditions, and the organism's state. Several mechanisms can work simultaneously or one can dominate uptake. Associated uptake dynamics can be influenced by energy requirements (active transport, facilitated diffusion, exchange diffusion, pinocytosis) or the possibility of saturation kinetics (active transport, facilitated diffusion, exchange diffusion).

Passive diffusion down an activity gradient is often the focus of model formulation for nonpolar organic toxicant uptake, e.g., Barber et al. (1988) and Erickson and McKim (1990). It may also be significant in metal uptake under a variety of conditions (Simkiss 1983). Passive diffusion is described by Fick's law (Equation 3.32).

$$\frac{dS}{dt} = -DA\frac{dC}{dX} \tag{3.32}$$

where dS/dt = the rate of movement across the membrane, D = the diffusion coefficient, A = the area across which diffusion is occurring, and dC/dX = the change in concentration across the membrane.

This model (Equation 3.32) constitutes the core of models designed to quantify specific cases of contaminant uptake such as the uptake of polychlorinated biphenyls by protozoa (Kujawinski et al. 2000). Although the focus during modeling is frequently on the concentration gradient (dC/dX) between the organism and the bulk external media, other aspects of this relationship are also critical. For example, the gill area available for uptake increases disproportionately with increases in body volume during growth (Hughes 1984). (Such allometric relationships are normally described by power functions, as will be discussed later in this chapter.) Area available for uptake from food can also change because relative intestine length can change disproportionately with size (Ribble and Smith 1983). Thus, the change in A must be considered for uptake over periods of time during which growth can be significant. In the case of uptake via gills, the diffusion gradient is influenced by many factors, including water flow across the gill and blood flow through the gills (Erickson and McKim 1990a, 1990b). Such factors are often at the heart of physiologically based pharmacokinetic (PBPK) models such as those described later.

Uptake from food may be defined with the direct application of processes such as those described above if one considers a very simple system such as a gutless endoparasite exposed to a relatively uniform environment, e.g., ^{137}Cs bioaccumulation in liver flukes within white-tailed deer inhabiting a contaminated site. However, in most cases, uptake from food is more complex. For example, lead uptake by a snail scrapping and ingesting a heterogeneous mixture of microflora and associated abiotic debris can be extremely difficult to predict relative to uptake of dissolved lead from soft water (Newman and McIntosh 1983b; Wang and Wood 1984). Assessment of bioavailability, the degree to which a contaminant in a potential source is free for uptake, becomes very critical to understand in such cases. Often assimilation efficiency will be experimentally estimated based on the overall concentration in the food or some defined fraction of the food. Normalization of concentration can involve a major fraction suspected as being particularly available or unavailable for uptake.

Models of bioaccumulation range from very simple to highly complex. The discussion below provides a framework for such models beginning with simple compartment models. By far the most common models in use are those discussed initially as they often provide adequate prediction in many instances with minimum information on the system being modeled. However, complex models such as physiologically based pharmacokinetic models can provide in-depth understanding of the most significant processes controlling accumulation kinetics, estimation of target organ exposures, and enhanced extrapolation among species or toxicants.

3.2.3.2 Compartment Models

3.2.3.2.1 Rate Constant-Based Models

The elimination model formulations described to this point were based on rate constants. This type of model has been the mainstay of bioaccumulation modeling in ecotoxicology. Rate constant-based models will be expanded in this section to include uptake. However, the short shrift given to compartment volumes in our discussions must be addressed before proceeding.

Although our discussion so far has dealt solely with changes in concentrations, rate constant models can describe changes in mass of toxicant also. We have focused on concentrations because one is often concerned with the concentration in the organism relative to concentrations in its environment. Further, effects are more readily related to toxicant concentration within the organism or target organ than to amounts or mass of toxicant. Indeed, this is the reasoning behind the extensive use of concentrations in pharmacokinetics (see Wagner 1979; Greenblatt and Shader 1985; Gibaldi 1991; Renwick 2008). However, certain aspects of modeling concentrations must now be clarified to avoid confusion later.

Barron et al. (1990), Newman and Clements (2008), and Landrum (2010) indicate that, if concentration units are used, they should be considered as amounts normalized to unit mass or volume. In fact, the rate constant model is more accurately described as a mixed model, that is, one including rate constant and clearance components. This is an important point to be made. Potential misinterpretation that may result from use of concentrations can be easily illustrated with one of our previous examples, elimination of methylmercury from two compartments (Example 3.3). Note that the C_A and C_B "concentrations" estimated for these two compartments would only be reasonable if the two compartments in question were of equal size (mass or volume assuming equal densities between compartments). However, this restriction is not pertinent if C_A and C_B are envisioned as the amounts in each compartment. For this and related reasons, it is often easier to use mass in these equations if complex models with compartments of differing volumes are anticipated. Early books by Atkins (1969) and Whicker and Schultz (1982) provide extensive model formulations based on amounts. Detailed examples of one-compartment model formulations involving transfer of amounts of materials are given in Willis and Jones (1977) and Giesy et al. (1980). Analogous examples of formulations involving two compartments are found in Goldstein and Elwood (1971), Holleman et al. (1971), and Konemann and Van Leeuwen (1980).

When dealing with concentration models involving multiple compartments, the volumes of each compartment must be estimated. The volume when multiplied by the concentration in the compartment gives an estimate of the amount of substance in the compartment. For example (Wagner 1979), the total amount or dose (D_t) of a substance injected intravenously will be the sum of the amounts in all of the compartments in the organism.

$$D_t = C_b V_b + \sum_{i=1}^{n} C_i V_i \tag{3.33}$$

where C_b = the concentration in the blood, V_b = the estimated blood volume, C_i = the concentration in the ith compartment, and V_i = the volume of the ith compartment.

It follows that the total volume in which the dose is distributed in the organism (V_t) can be defined.

$$V_t = V_b + \sum_{i=1}^{n} V_i \tag{3.34}$$

One need only know the volumes and concentrations in the various compartments to define the distribution of mass within the organism.

A justifiable questioning of the above argument can be levied based on the previously stated fact that measured mathematical compartments are not per se linked to physical entities such as an organ or tissue type having an easily measured volume. If the compartments are not necessarily physical entities, then how can compartment volumes be determined?

The apparent volume of distribution concept as used in pharmacokinetics can be adapted for this purpose (see Greenblatt and Shader 1985; Gibaldi 1991). The apparent (or effective) volume of distribution (V_d) is the mathematically determined volume of a compartment in which the total amount of material is associated. The apparent volume is expressed in units of volume of a reference compartment. For example, the following procedure may be used to estimate the V_d for a material in the notional blood compartment. A known amount of drug or toxicant is injected intravenously and allowed to distribute itself throughout the circulatory system. The blood concentration is then measured. The apparent volume of distribution of the blood (L) would be the total amount of the dose (mg) injected divided by the concentration (mg/L) measured in the blood after the material has been allowed to distribute itself in the circulatory system, i.e., V_d = dose/concentration. When several mathematical compartments are involved, each will have a V_d. These V_d values will be the apparent volumes expressed in units of volume of a reference compartment—blood in this case.

With this background, the models described in Figure 3.6 can now be further characterized by the addition of two V_d values, one for each of the two compartments. In the case of the two compartments with no exchange (Figure 3.6, top model), the V_d values can be estimated to be the following.

$$V_{dA} = \frac{D}{C_A} \quad (3.35)$$

where V_{dA} = the apparent volume of compartment A, C_A = the concentration in compartment A, and D = the amount of substance introduced into each compartment of the organism.

$$V_{dB} = \frac{D}{C_B} \quad (3.36)$$

where V_{dB} = the apparent volume of compartment B, C_B = the concentration in compartment B, and D = as defined above.

In the second model shown in Figure 3.6 (dose administered to a central compartment, distributed to a peripheral compartment, and eliminated from the central compartment), estimation of apparent volumes can be done with the following equations.

$$V_{dA} = \frac{D}{C_A + C_B} \quad (3.37)$$

$$V_{dB} = V_{dA}\left[\frac{k_{AB}}{k_{BA}}\right] \quad (3.38)$$

With constant exposure to the toxicant, the apparent V_d for the total organism at equilibrium ($V_{ss} = V_A + V_B$) is described by Barron et al. (1990) as relating "the amount of chemical in the [organism] to its concentration in the [source] compartment at steady state.... [It] expresses the affinity or capacity of the aquatic animal for a particular chemical in terms of the equivalent volume of exposure water holding the same quantity of chemical. [It] is viewed as the steady-state partitioning of compound between animal and water." It is expressed in units of volume/volume.

With the above refinements added to our discussion, we can now move to more involved models. Five of the most common single-compartment models of bioaccumulation are shown in Figure 3.9. The exponential model is identical to that described for single exponential elimination except a constant source (C_1) has been added with an uptake rate constant of k_u. The k_u has units of ml/(g·h). (Some authors will give units of 1/h for k_u under the assumption of equal densities for the source and tissue.)

Normally, uptake can be defined as conforming to first-order kinetics. However, as recommended by Barron et al. (1990a), this assumption should be tested by exposing the organism to a range of concentrations. Barron et al. (1990a) recommended concentrations differing by fivefold at least. The following transformation of Equation (3.42) is recommended by Spacie and Hamelink (1985) for this purpose.

$$C_t = k_u C_1\left[\frac{1 - e^{-k_e t}}{k_e}\right] \quad (3.39)$$

where C_t = the concentration (amount normalized to mass in the compartment) in the compartment at time t, k_u = the uptake rate constant (ml/(g·h)), and C_1 = the concentration (constant) in the (assumed infinite) source.

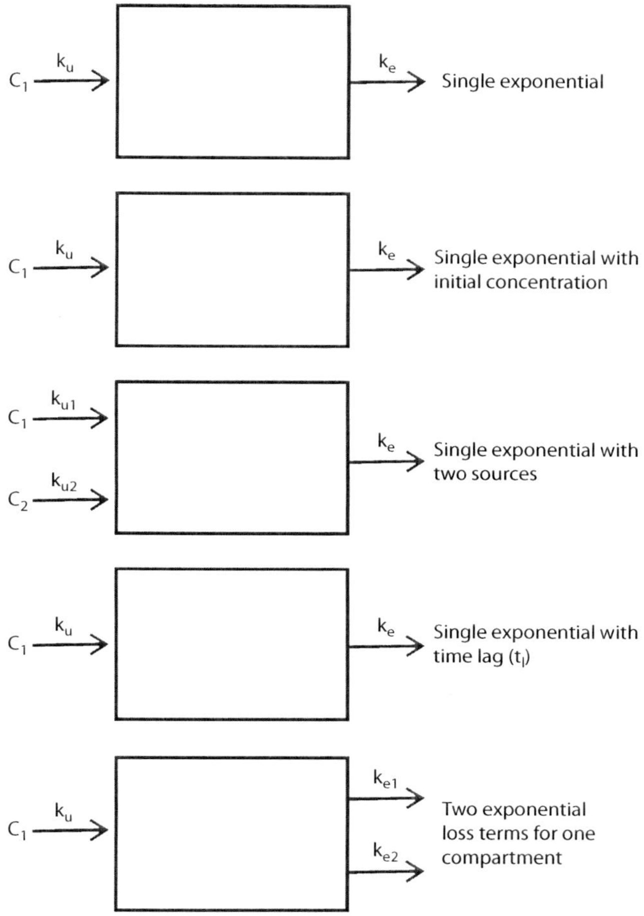

Figure 3.9 Five common single-compartment models for accumulation.

The bracketed term in Equation (3.39) is plotted against the various C_t values. A linear plot will result if C_1 and *ku* are constant. (To verify first-order elimination kinetics, Ln Ct/C_0 versus time for all elimination time courses, regardless of C_0, should yield a single, straight line.)

In this case of first-order kinetics, the single exponential elimination equation describing the change in concentration with time (Equation 3.40) becomes Equation (3.41).

$$\frac{dC}{dt} = -k_e C \tag{3.40}$$

$$\frac{dC}{dt} = k_u C_1 - k_e C \tag{3.41}$$

The integrated form of Equation (3.40) is Equation (3.2). That of Equation (3.41) is the following:

$$C_t = \frac{k_u}{k_e} C_1 (1 - e^{-k_e t}) \tag{3.42}$$

This model describes accumulation through time that eventually approaches steady-state equilibrium with the source. The concentration in the compartment at steady-state equilibrium (C_s) can be estimated.

$$C_s = \frac{k_u}{k_e} C_1 \tag{3.43}$$

Upon division of both sides of Equation (3.43) by C_1, it becomes obvious that k_u/k_e is a measure of the distribution of toxicant between the source (C_1) and organism (C_s) at steady-state equilibrium. As such, it has come to be referred to as the bioconcentration factor (BCF) when water is used as the source.

If, prior to the exposure being studied, the compartment has an initial concentration of toxicant (C_0), then the following modification of Equation (3.42) can be used.

$$C_t = \frac{k_u}{k_e} C_1[1 - e^{-k_e t}] + C_0 e^{-k_e t} \tag{3.44}$$

If there are two sources, Equation (3.42) can be modified to Equation (3.45).

$$C_t = \frac{k_{u1}C_1 + k_{u2}C_2}{k_e}[1 - e^{-k_e t}] \tag{3.45}$$

Equation (3.46) is pertinent if there is a time lag (t_l) between the beginning of exposure and initiation of uptake, e.g., induction of a transport mechanism or a delay in commencement of feeding.

$$C_t = \frac{k_u}{k_e} C_1[1 - e^{-k_e(t - t_l)}] \tag{3.46}$$

Finally, as an obvious extension of our discussion of elimination via two components from one compartment, accumulation under these conditions can be modeled with Equation (3.47).

$$C_t = \frac{k_u}{k_{e1} + k_{e2}} C_1[1 - e^{-(k_{e1}+k_{e2})t}] \tag{3.47}$$

It is important to remember that although a common assumption, uptake need not be considered a first-order process. For example, Schulz-Baldes (1974) describe linear accumulation kinetics ($C_t = k_t + C_0$) for lead in blue mussels (*Mytilus edulis*). Kolehmainen (1972) provides an example of cyclic changes associated with seasonal changes in ^{137}cesium kinetics in bluegill sunfish (*Lepomis macrochirus*). Further, saturation of uptake mechanisms can be anticipated in many cases. Whicker and Schultz (1982, pp. 74–78) provide an in-detail discussion of uptake models exhibiting exponential, linear, power, and cyclic time dependence.

Example 3.7

Ionic mercury accumulation in mosquitofish (*Gambusia holbrooki*) was measured over 6 days by Newman and Doubet (1989). The ambient concentration in water was 0.24 ng Hg/ml. Data for one of the fish studied were fit to Equation (3.42) using the following SAS code:

```
DATA MERCURY;
   INPUT DAY HG @@;
DATALINES;
0 000 1 380 2 540 3 570 4 670 6 780
;
/* FITTING A SINGLE EXPONENTIAL ACCUMULATION MODEL      */
/* NOTE THAT THE HG CONC IN THE WATER WAS 0.24 NG/ML    */
PROC NLIN;
   PARMS KU=1000 KE=0.5;
   BOUNDS 0<KU, 0<KE<1;
   MODEL HG = 0.24*(KU/KE)*(1-EXP(-KE*DAY));
   OUTPUT OUT=MERCURY P=PHG R=RHG;
RUN;
```

The estimates of k_e and k_u (± asymptotic standard error) were 0.59 ± 0.11/day and 1,867 ± 242 ml/(g · day), respectively. The BCF is estimated to be 1,867/0.59, or 3,164. The predicted concentration in the fish at equilibrium is (1,867/0.59) · 0.24, or 759 ng/g.

A more involved method of extracting accumulation parameters includes an exposure period during which organisms accumulate toxicant (elimination and uptake occurring together) followed by a depuration period during which the toxicant is eliminated in a toxicant-free environment (elimination only). Spacie and Hamelink (1985) suggest that the accumulation portion of this exercise continue until the concentration in the organism approaches practical equilibrium. Practical equilibrium can be defined as the duration necessary to achieve concentrations 95% of those at final equilibrium ($t_{95\%} = -(\text{Ln } 0.05/k_e)$). The elimination phase should continue for approximately three half-lives. However, there are cases with very slow elimination or very rapid uptake where accumulation to near equilibrium may not be practical or desirable (Hayton, personal communication, 1992).

Example 3.8

Bioaccumulation of bromophos (O-(4-bromo-2,5-dichlorophenyl) O,O-dimethyl ester) from water by the guppy (*Poecilia reticulata*) was measured by De Bruijn and Hermens (1991). Initially, fish were allowed to accumulate bromophos for 264 h (C_1 approximately 10.5 ng/L, although it was 5.3 for the first several hours). The fish were then allowed to eliminate it in a bromophos-free environment. Concentrations (ng/g extractable fat) in the fish were assayed throughout the accumulation and elimination phases of the experiment. Data were visually extracted from De Bruijn and Hermens' Figure 1 to illustrate fitting of a single exponential bioaccumulation model with one source. Data are replotted in Figure 3.10 along with the results of the following regression analyses. The uptake and elimination rate constants were estimated simultaneously and then sequentially (k_e estimated from depuration phase data and then used in estimating k_u from the accumulation phase data) with these data. Although both nonlinear (weighted and unweighted) and linear regression on transformed data were used to estimate k_e for the sequential estimation method, only the linear regression estimate is presented here. The residuals from the linear regression model showed no obvious trends, while those of the nonlinear regressions did.

```
DATA ACCUM;
   INPUT HOUR BRPHOS @@;
DATALINES;
0.5 1900 001 3000 002 5200 004 6900 008 24000 024 50000
072 200000 144 400000 240 500000 264 500000
;
/* FITTING A SINGLE EXPONENTIAL ACCUMULATION MODEL WITH SIM-       */
/* ULTANEOUS ESTIMATION OF KE AND KU. THE BROMOPHOS CONCEN-        */
/* TRATION WAS 10.5 NG/ML FOR ALL BUT THE INITIAL HOURS.           */
/* FISH CONC. WERE NG/G OF EXTRACTABLE FAT.                        */
```

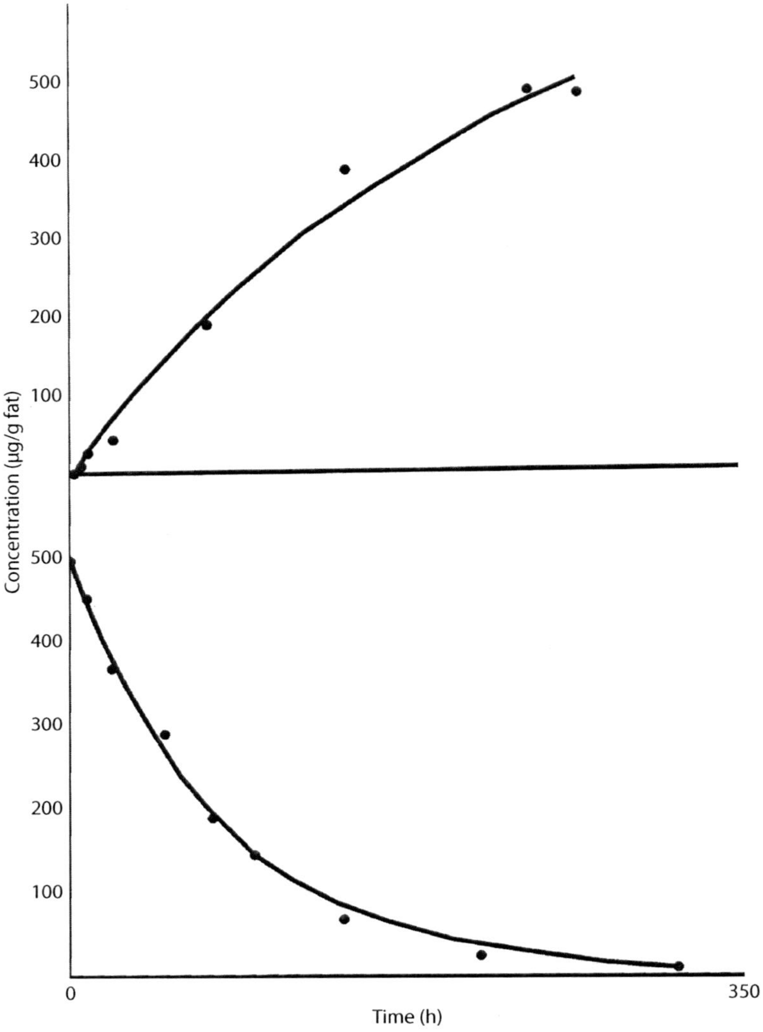

Figure 3.10 Fit of regression results to data from De Bruijn, J., and J. Hermens, *Environ. Toxicol. Chem.*, 10, 791–804, 1991. Accumulation and elimination of bromophos from water in the guppy were fit in this example. See text for more details.

```
PROC NLIN;
   PARMS KU=0.001 KE=0.01;
   BOUNDS 0<KU, 0<KE<1;
   MODEL BRPHOS = 10.5*(KU/KE)*(1-EXP(-KE*HOUR));
   OUTPUT OUT=BROMO P=PBRPHOS R=RBRPHOS;
RUN;
DATA ELIMIN;
   INPUT HOUR BRPHOS @@;
   LBROMO=LOG(BRPHOS);
DATALINES;
000 500000 012 450000 024 370000 048 290000 072 190000 096 150000
144 70000 216 21000 319 5000
;
/* NOW ESTIMATING KE FROM THE ELIMINATION DATA FOR LATER SUB-    */
```

```
/* STITUTION INTO THE NONLINEAR REGRESSION (AS KE) FOR ESTI-     */
/* MATION OF KU.                                                 */
PROC GLM;
   MODEL LBROMO=HOUR;
   OUTPUT OUT=BROMO2 P=P2BRPHOS R=R2BRPHOS;
RUN;
/* NOW USING THE ESTIMATED KE (0.01469 FROM THE PROC GLM ABOVE   */
/* IN THE ACCUMULATION MODEL AND SOLVING FOR KU.                 */
PROC NLIN DATA=ACCUM;
   PARMS KU=0.001;
   BOUNDS 0<KU;
   MODEL BRPHOS=10.5*(KU/0.01469)*(1-EXP(-0.01469*HOUR));
   OUTPUT OUT=BROMO3 P=P3BRPHOS R=R3BRPHOS;
RUN;
```

The k_u and k_e (± asymptotic standard error) derived simultaneously from the accumulation data were 345 ± 32 ml/(g fat · h) and 0.0052 ± 0.0010/h, respectively. Plots of predicted and original data, and plots of the regression residuals show good fit of the model. However, the k_e estimated from the elimination data (0.0147 ± 0.0002/h) was quite different from this estimate. The elimination data fit to the model was excellent ($r^2 = 0.998$) with no pattern to the residuals. When this estimate was used in the accumulation model, the k_u was estimated to be 644 ± 40 ml/(g fat · h). The residuals were lower than expected in the initial phases of accumulation but increased to above predicted values later.

There exists a certain degree of covariance between k_e and k_u estimates when they are derived simultaneously. Such covariance can be avoided by the sequential approach described here. However, the simultaneous and sequential estimation methods should give similar results if nothing significant had changed between the accumulation and elimination phases of the experiment. The results shown above suggest that something had changed. The authors describe an initial (0 and 8 h analyses) concentration of bromophos of 5.3 ng/ml that increased to 10.5 ng/ml thereafter. Concentrations in the water shown in their Figure 1 suggest that, for times less than 72 h, concentrations were below this average value. The low concentrations initially in the water could have produced the lower than predicted pattern in the data residuals from the first portion of the accumulation model. Under this assumption, it would be best to use the estimate of k_e from the elimination phase and then estimate k_u using the sequential approach (k_e supplied to the non-linear, regression model to derive k_u). The final estimates of *ke* and *ku* using this approach were 0.0147 ± 0.0002/h and 644 ± 40 ml/(g fat·h).

Growth dilution can be incorporated into Equation (3.42) in a manner similar to the incorporation of radioactive decay into Equation (3.5). If it is justified to model growth as a first-order process, a growth rate constant (g) can be included to produce the following relationship:

$$C_t = \frac{k_u}{k_e + g} C_1 (1 - e^{-(k_e + g)t}) \tag{3.48}$$

in which the units for g are g/(g per unit time). First-order growth should be assumed only after careful thought: in some cases, the motivation behind this assumption is mathematical convenience.

The rate constant-based modeling exploration to this point has not considered the dynamics within a finite system with a source that is depleted during uptake. The clear description of rate constant-based models for uptake from a limited volume of water is provided by Landrum et al. (1992). They begin their description by defining the amount of a toxicant in the entire system (A, mg) as the sum of the amounts in the organism (Q_O, mg) and the water source (Q_W, mg). A system-dependent uptake rate constant (k_{um}) is defined that will vary among systems with different volumes of water or masses of organisms. The mass of toxicant in the organism at time t can be defined by Equation (3.49):

$$Q_{Ot} = \frac{(k_{um}A)(1 - e^{-(k_{um}+k_e)t})}{k_{um} + k_e} \tag{3.49}$$

The system dependence of the model can be removed to generate a system-independent *ku* by considering the volume of water (V_W, ML) and total mass of organism (M_O, g), and the identity $k_u = k_{um}(V_W/M_O)$,

$$Q_{Ot} = \frac{\left(\left[\frac{k_u}{\left(\frac{V_W}{M_O}\right)}\right]A\right)(1 - e^{-([k_u/(V_W/M_O)]+k_e)t})}{\left[\frac{k_u}{\left(\frac{V_W}{M_O}\right)}\right] + k_e} \tag{3.50}$$

3.2.3.2.2 *Clearance Volume-Based Models*

Pharmacokinetics models often are formulated using the concept of clearance (Barron et al. 1990a; Greenblatt and Shader 1985; Landrum and Lydy 1991; Landrum et al. 1992). As defined above, clearance can be thought of as the volume (relative to that of a reference compartment) cleared per unit time (ml/h). If normalization to mass of tissue is employed in the model, then the units are ml/(g·h). For example, if toxicant uptake from water is being considered, then clearance would be interpreted as the volume of the toxicant source (water) cleared of toxicant per gram of organism every hour. If one is considering loss of a drug from an organ after intravenous injection, then clearance may be seen as the volume cleared from the organ (in terms of volume of a reference compartment such as the blood) during some unit of time. (As noted in Barron et al. (1990) and Landrum (2010), uptake rate constants such as that described above are actually clearances, not proportional rate constants like elimination rate constants.)

Clearance (*Cl*) is defined in terms of apparent volume of distribution and rate constant as the following.

$$Cl = kV_d \tag{3.51}$$

For example, the clearance associated with elimination from one compartment would be $k_e \cdot V_d$. By simple substitution into Equation (3.3),

$$t_{1/2} = \frac{Ln\ 2 * V_d}{Cl} \tag{3.52}$$

The total clearance from the system being modeled is the sum of all the individual clearances.

$$Cl_t = \sum_{i=1}^{n} Cl_i \tag{3.53}$$

where Cl_i = clearance from the *i*th compartment.

The units of the Cl_i can be expressed relative to the external environment, e.g., a water compartment, or they may be expressed relative to an internal compartment. The internal compartment is often the blood or plasma. In the case of blood, the toxicant bound to ligands in the blood may be treated as yet another separate compartment. In that case, the reference compartment may become the free toxicant in the blood.

By substitution, the models above can now be expressed in terms of clearance. Remember that V_d expresses the distribution of the toxicant between the reference compartment and the compartment in question. Landrum and Lydy (1991) point out that, this being the case, the V_d is an estimation of the BCF. Since the BCF is estimated in the rate constant model as k_u/k_e, then V_d estimates *ku/ke*. With this information and rearranging Equation (3.51), Equation (3.42) can be modified to a clearance volume-based model of accumulation,

$$C_t = V_d C_1 (1 - e^{-(C_1/V_d)t}) \tag{3.54}$$

As a relevant example of clearance volume-based modeling, Peters et al. (1999) applied a two-compartment, clearance volume model to compare the toxicokinetics of two alkali elements, cesium and rubidium, introduced into channel catfish. The clearances and V_d for these two metals were compared to provide insight about their behaviors as potassium analogs. The results were also used to understand better the distribution of these metals, including radioactive cesium ^{137}Cs in tissues of fishes.

At this point, basic rate constant- and clearance volume-based models have been described. A third, equally valuable approach to modeling bioaccumulation is fugacity-based modeling. Fugacity-based models are exceptionally useful if multiple compartments are being modeled. For this reason alone, fugacity model discussion will be delayed until later in this chapter during discussions of more complicated models.

3.2.4 Accumulation Models: Several Compartments or Sources

3.2.4.1 General

Accumulation compartment models can be expanded beyond the treatment described above to include multiple sources or several internal compartments. With enriched description comes the difficulty of deriving many parameters and ascertaining the sensitivity of the outcome to errors in each parameter estimate. Regardless, the richness of detail within such models can greatly enhance understanding of the system in question. Some of the simpler models are described here.

3.2.4.2 One Compartment with Uptake from Food and Water

Significant uptake from both food and water can occur for a wide range of toxicants (Schulz-Baldes 1974; Preston 1971; Luoma et al. 1992). The consequent accumulation can be accommodated in the model structure developed so far. Spacie and Hamelink (1985) used Thomann's model (1981) for accumulation of toxicant from two sources (food (C_f) and water (C_w)) to illustrate incorporation of a variety of factors into bioaccumulation models. To incorporate ingestion into the model, two new terms were introduced, assimilation efficiency and weight-specific ration. The assimilation efficiency (α) is the amount of toxicant absorbed per amount of toxicant in the food.* The weight-

* Although assimilation efficiency is used in a general sense here, Penry (1998) provides a relevant and discerning discussion of digestive, absorption, and assimilation efficiencies. Newman and Clements (2008, pp. 127–131) also discuss these terms. Digestion efficiency is determined as the difference in the amount of substance fed to the organism and the amount of that substance remaining in the egested feces. Absorption efficiency is the efficiency of uptake of the products of digestion into the organism's cells. Assimilation efficiency is a measure of the factional amount of absorbed substance that is actually assimilated into the organism's tissues.

specific ration (R) is the amount of food provided per amount of organism present, e.g., 1 g food/100 g fish. Obviously, the feeding rate (f, g food/g fish) for free-ranging organisms can be used instead of R, e.g., Harrison and Klaverkamp (1989).

$$C_t = \frac{k_u C_w + \alpha R C_f}{k_e}[1 - e^{-k_e t}] + C_0 e^{-k_e t} \tag{3.55}$$

Thomann's model also incorporates an initial concentration of toxicant (see Equation 3.44). Although not included in Equation (3.55), the model was also expanded to include growth dilution (see Equation 3.48). A growth rate constant was used based on the assumption of exponential growth during the period of exposure.

Each set of parameters associated with elimination, uptake from food, uptake from water, and growth would be estimated through a series of experiments such as those described above. The resulting parameters would then be incorporated into Equation (3.55). Assimilation efficiency from food is estimated relative to the entire animal here. If the model has several compartments, then assimilation efficiency may be relative to the reference compartment, e.g., blood compartment in pharmacokinetics modeling of oral drug administration. Estimation of assimilation efficiency (or fractional absorption) in this context will be described in Section 3.3.1.

3.2.4.3 Multiple Sources and Elimination Components

Generalizing to multiple uptake and elimination components from one compartment that has an initial concentration of toxicant, Whicker and Schultz (1982) provided the model

$$C_t = \frac{\sum_{j=1}^{m} C_j k_{uj}}{\sum_{i=1}^{n} k_{ei}}[1 - e^{-(\Sigma_{i=1}^{n} k_{ei})t}] + C_0 e^{-(\Sigma_{i=1}^{n} k_{ei})t} \tag{3.56}$$

3.2.4.4 Complex Multiple Compartment Models

More complex models range from those involving two compartments (Blaylock et al. 1982), to three compartments (Waiwood et al. 1987), to many compartments (Giblin and Massaro 1973; Skrable et al. 1980). In many, the focus of uptake and elimination kinetics can move from mathematical compartments to specific organs or tissues. For example, Waiwood et al. (1987) modeled zinc kinetics in lobster using the gills, hemolymph, hepatopancreas, and tail muscle as the major compartments. Such linkage of compartments to physical structures has the virtue of allowing direct physical interpretation of the associated parameters as well as leaving open the opportunity for incorporating physiologic and allometric relations into model development. Barron et al. (1990b) and Nichols et al. (1990, 1991) provide good examples in which such models became physiologically based pharmacokinetic (or toxicokinetic) models (PBPK models). This genre of models will be discussed in more detail below.

Important in many models is the internal degradation or conversion of the parent toxicant to metabolites. Such metabolism can involve first-order or saturation kinetics. If the parent toxicant concentration is high, metabolism may be best described by zero-order kinetics. Metabolites or chemical species can also be incorporated. Leversee et al. (1982) and Foster and Crosby (1986) provide good examples of metabolism of organic toxicants by aquatic biota. Walker (1978, 1987) relates such degradation to elimination kinetics. Bioaccumulation of the herbicide, 3,5,6-trichloro-2-pyridinyloxy acetic acid as modeled by Barron et al. (1990b) is one good example of a clearance volume-based model incorporating metabolism.

Internal conversion may be important in modeling accumulation of inorganic toxicants as well as organic toxicants. A quick review of the complex, biologically mediated transformations of arsenic should make this obvious, e.g., Lunde (1972), Unlu and Fowler (1979), Edmonds and Francesconi (1981), and Zingaro (1983).

3.2.5 Physiologically Based Pharmacokinetic Models

3.2.5.1 General

Compartment models derived from multiple exponential components are often limited to two or three compartments by the associated data requirements (Nichols et al. 1990). For example, we have already discussed the high quality of data estimated necessary by Myhill (1967) to strip two components from an elimination curve, that is, data error less than 1% for a ratio of *ke* values of 2. Further, extrapolation from the results is quite limited because compartments are specific to the species and conditions of the experiment. Beyond two or three components, simple rate constant-based models are often abandoned in favor of physiologically based models.

Physiologically based pharmacokinetic (PBPK) or physiologically based toxicokinetic (PBTK) models "define an organism in terms of its anatomy, physiology, and biochemistry" (Nichols et al. 1990). This class of models, developed first in the medical sciences, is not based on mathematically identified compartments. Rather, each compartment has an assigned physical meaning. Because of this linkage, physiological and allometric relationships can be incorporated. For example, accumulation can be linked to such processes as gill oxygen exchange (Landrum and Stubblefield 1991), bivalve filtration rates (Watkins and Simkiss 1988), or temperature-dependent metabolism (Rose et al. 1989). This being the case, extrapolation between similar species and conditions is more readily accomplished. Further, concentrations in target organs are more clearly assessed. Normally, simulation software such as SIMUSOLV (e.g., Neely et al. 1987), Advanced Continuous Simulation Language (e.g., Nichols et al. 1990), and NONLIN (e.g., Schultz et al. 1996; Schultz and Newman 1997) are used to develop such models.

Action of a tissue compartment on the toxicant can involve binding (storage) as described by adsorption in Section 3.2.2 or elimination as described in Section 3.2.1. Elimination can be physical removal from the system, such as urinary excretion, or biochemical conversion, such as hepatic breakdown of organic toxicants. Kinetics can include saturation kinetics. Barron et al. (1990) give the following equations as descriptions of tissues displaying only storage (muscle, Equation 3.57) and tissues displaying metabolism also (liver, Equation 3.58).

$$\frac{dC_m}{dt} = \frac{Q_m\left[C_i - \dfrac{C_m}{R_m}\right]}{V_m} \tag{3.57}$$

$$\frac{dC_h}{dt} = \frac{Q_h\left[\left(C_i - \dfrac{C_h}{R_h}\right) - Cl_h C_i\right]}{V_h} \tag{3.58}$$

where V_m,V_h = weight of muscle and hepatic compartments, R_m,R_h = blood/tissue partition coefficient, C_i = concentration entering the tissue, C_m/R_m,C_h/R_h = blood concentrations exiting the tissues, Q_m,Q_h = blood flow to the tissues, and Cl_h = clearance constant for hepatic compartment.

3.2.5.2 Uptake from Water

Although some have involved metals, e.g., Part and Svanberg (1981), Van Der Putte and Part (1982), most PBPK models of toxicant accumulation have been developed for organic contaminants. Regardless of the class of toxicants being modeled, most models of accumulation via the gills begin with Fick's law (Equation 3.32). For example, Hayton and Barron (1990) define the rate of transfer across the gill with the following relationship.

$$\frac{dX}{dt} = D_m A K_m \left[\frac{C_0 - C_i}{d} \right] \tag{3.59}$$

where D_m = the diffusion coefficient (length^2/time), A = area of gill epithelium available for exchange, K_m = epithelium/water distribution coefficient, C_i,C_o = concentration inside and outside of the gill, and d = the thickness of the epithelium.

Basic physiological and anatomical relationships are incorporated into the process of gill uptake to produce the relationship

$$P = \left[\frac{d}{D_m A K_m} + \frac{h}{D_0 A} + \frac{1}{K_b V_b} + \frac{1}{V_m} \right]^{-1} \tag{3.60}$$

where P = the uptake clearance (volume/time), D_a = the diffusion coefficient in water (length^2/time), h = thickness of stagnant water layer on gill surface, K_b = blood/water distribution coefficient, V_b = effective blood flow through the gills, and V_w = effective water flow across the gills.

In this model, some parameters (d, h, A, V_b, V_w) are related to the specific organism while others (D_m, D_a, K_m, K_b) are linked to properties of the organic chemical in question. By defining these parameters for various organisms and organic chemicals, our ability to predict bioaccumulation for untested species and chemicals is enhanced. Further, our ability to identify critical parameters is also enhanced.

This general approach is actively being developed for organic toxicants by such workers as Erickson and McKim (1990a, 1990b), Nichols et al. (1990, 1991, 1998, 2004), and Yang et al. (2000). Although a seemingly powerful approach, it has yet to be applied in a focused manner for inorganic toxicants. Further, linkage of concentrations in various compartments, including target organs to effect, has lagged behind PBPK models used in mammalian toxicology. Such models that incorporate effect compartments in toxicological research are termed pharmacokinetic-pharmacodynamic models (Gibaldi and Perrier 1982).

Example 3.9

Because of the complexity of such models and the specialized software applied to model formulation, a detailed example would be difficult to provide. However, a general discussion of the model developed by Nichols et al. (1990a) could provide an appreciation for this approach. A more detailed description is given in the appendix of Nichols et al. (1990a).

Figure 3.11 is a visualization of this model. The organic toxicant enters the fish via the gills. This entry is controlled by a variety of physiological and anatomical features such as those described above (Equation 3.60). The toxicant is distributed between several compartments, including fat tissue, kidney, liver, richly perfused tissue, and poorly perfused tissue. Elimination occurs via saturation or first-order kinetics in the liver.

A large amount of information was needed to parameterize the model. Information on cardiac and respiratory function, metabolic and excretory rates, compartment volumes, and

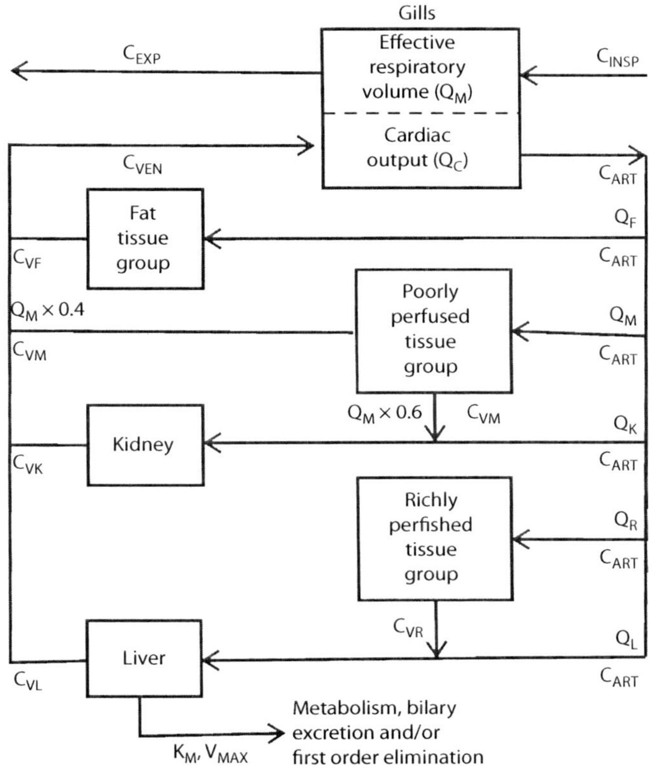

Figure 3.11 An example of a physiologically based pharmacokinetic model from Nichols et al. (1990). An organic toxicant enters via the gills as determined by *Qw* and *Qc*. It is then distributed between several internal compartments and eliminated from the liver according to Michaelis–Menton kinetics. (From Nichols, J.W., J.M. McKim, M.E. Andersen, M.L. Gargas, H.J. Clewell, III, and R.J. Erickson, *Toxicol. Appl. Pharmacol.*, 106, 433–447, 1990. Reprinted with permission from Elsevier.)

blood flow rates was estimated or derived from laboratory experiments with rainbow trout (*Oncorhynchus mykiss*).

3.2.5.3 Uptake from Food

General uptake from food has already been described in Section 3.2.4.2 and Equation (3.55). However, few complex PBPK models such as those derived for organic chemical bioaccumulation from water have been derived for uptake from food, e.g., Nichols et al. (2004). Bioenergetic considerations have been added to the general model (Equation 3.55) by Pentreath (1976). The ingestion of methylmercury was also related to food intake for maintenance and growth. Growth dilution was also incorporated into the model. Intake of methylmercury was defined by Equation (3.61).

$$I_f = \Phi \xi W^j + \phi \frac{1}{\varepsilon}\frac{dW}{dt} \tag{3.61}$$

where ϕ = the concentration of methylmercury in the food, ξ = the maintenance food coefficient (size dependent), W = the weight of the fish, j = allometric coefficient relating ξ to weight, and ε = efficiency of food utilization for growth.

The excretion efficiency (K) was also linked to fish weight.

$$K = a_2 W^{-b_2} \tag{3.62}$$

where a_2 and b_2 are derived parameters.

From these relationships, Equation (3.63) was derived to describe the change in methylmercury concentration with time as a function of feeding bioenergetics.

$$\frac{dC_t}{dt} = \phi \xi W_t^{j-1} + \phi \frac{1}{\epsilon} \frac{dW_t}{dt} \frac{1}{W_t} - \left[\frac{dW_t}{dt} \frac{1}{W_t} + \frac{a_2}{W_t^{b_2}} \right] C_t \tag{3.63}$$

3.2.5.4 Growth and Allometric Considerations

In Equation (3.63), any of several growth models can be added in place of W_t. Pentreath (1976) selected the von Bertelanffy growth equation. In Equation (3.55), exponential growth was assumed. Other models such as the Richards growth equation (McCallum and Dixon 1990) could be incorporated.

Allometric considerations (the effects of changes in size of the whole organism or an anatomical feature) are also critical to PBPK models. For example, the area available for absorption of a contaminant in food may change disproportionately with the volume or mass of an organism. As an example, Ribble and Smith (1983) noted such a size-related shift in relative intestine length in fish. Size-specific metabolic rate shifts may also be critical in toxicant elimination and uptake (Newman and Heagler 1991; Reichle 1967; Fagerström 1977).

3.2.6 Fugacity-Based Models

Fugacity modeling can be directly related to PBPK models described to this point. Gobas and McKay (1987) give an excellent demonstration of this in their derivation of a bioaccumulation model for uptake via the gills. Fugacity-based models can enhance modeling because they allow expression of all amounts of toxicant in many potentially heterogeneous compartments in the same units. This facilitates easy linkage to compartments outside of the organism (MacKay and Paterson 1982; Mackay and Fraser 2000; Arnot and Mackay 2010). Models dealing with gas, solid, and liquid phases express all phase activities in terms of the gas phase fugacity (Jorgensen 1990; Pankow 1991; Paasivirta 1991). Mackay (1991) discusses this approach in detail in his excellent book *Multimedia Environmental Models: The Fugacity Approach*. Excellent examples of fugacity modeling can be found in the special issue of *Environmental Toxicology and Chemistry* (October 2004) that honored the fugacity research of Don Mackay.

Fugacity is the escaping tendency of a substance from a phase and is expressed in units of pressure (Pa). This tendency for a substance to partition itself between phases is based on chemical potential or activity (MacKay and Paterson 1982). It (f) is related linearly to concentration (C) by the fugacity capacity (Z) (Gobas and MacKay 1987).

$$C = fZ \tag{3.64}$$

where C = concentration (mol/m^3), f = fugacity (Pa), and Z = the fugacity capacity (mol/(m^{3}Pa)).

When two phases are in equilibrium, the partition coefficient (P_{12}) will equal C_1/C_2 or Z_1/Z_2. Transport between phases is expressed with the transport constant, D. The rate of transport (mol/h) between two phases can be defined,

$$N = D(f_1 - f_2) \tag{3.65}$$

where D = the transport constant (mol/(h·Pa)), f_1 = the fugacity of phase 1, and f_2 = the fugacity of phase 2.

In the context of the single compartment bioaccumulation model (Equation (3.41)),

$$k_e = \frac{D_0}{V_0 Z_0} \tag{3.66}$$

$$k_u = D_0 V_0 Z_w \tag{3.67}$$

where D_o = the transport constant for the organism, V_o = the volume of the organism, Z_o = the fugacity capacity of the organism, and Z_w = the fugacity capacity of the water.

Gobas and MacKay (1987) give the following fugacity-based model analogous to the rate constant-based (Equation 3.42) and clearance volume-based (Equation 3.54) models described previously.

$$C_t = C_w \left(\frac{Z_0}{Z_w} \right) (1 - e^{-(D_0/V_0 Z_0)t}) \tag{3.68}$$

This model can be expressed in fugacities terms as well by simple substitution (Mackay 1987),

$$f_o = f_w [1 - e^{-(D_0/(V_0 Z_0))t}] \tag{3.69}$$

In Mackay's book (1991), a more detailed, fugacity-based bioaccumulation model that incorporates uptake from food and water, fish metabolism, and growth is described.

Example 3.10

As in Example 3.9, it would be impractical to provide a detailed derivation of a fugacity-based model here. Instead, a fugacity-based model derived by Mackay and Paterson (1982) for bioaccumulation of styrene by ingestion and inhalation is depicted in Figure 3.12. It has the same general form as the PBPK model shown in Figure 3.11, but fugacity (f_i) and transfer parameters (D_i) are used to define the flow of contaminant within the organism.

3.3 MODELING BIOACCUMULATION: ALTERNATIVE APPROACHES

3.3.1 Statistical Moments Approach

Except for a very brief discussion of mean residence time estimation from half-life, models have been discussed in a solely deterministic context to this point. However, alternative models based on the distribution of contaminant molecules can also be formulated. In this statistical approach, the residence times within compartments can be estimated using only concentration versus time curves. Not only does this allow estimation without imposition of a specific model, but it also allows estimation of statistics that are comparable among model types (mean residence time). Such an approach has been used very successfully in pharmacokinetics but infrequently in ecotoxicology.

Yamaoka et al. (1978) brought statistical moments methods to pharmacokinetics. In their development of this approach, the first three moments (zero, first normal, and second central moments)

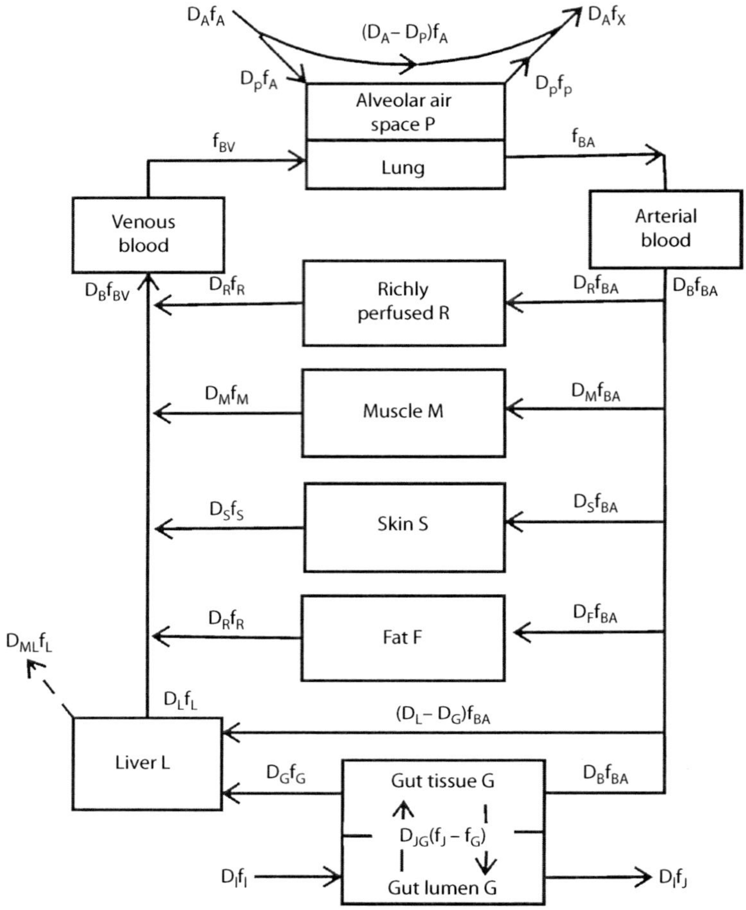

Figure 3.12 An example of a fugacity-based model developed by Mackay and Paterson (1982) for styrene bioaccumulation. Styrene is taken up by inhalation and ingestion. Note that this model is very similar in form to that in Figure 3.10. However, the units are changed to accommodate fugacity and transfer parameters. (From Mackay, D., and S. Paterson, *Environ. Sci. Technol.*, 116(12), 654A–660A, 1982. Reprinted with permission from *Environmental Toxicology and Chemistry*. Copyright © 1982 American Chemical Society.)

of the plasma concentration versus time curve were used to estimate the mean and variance of drug residence time. These moments were defined by Equations (3.70) to (3.72):

$$AUC = \int_0^\infty C_p dt \tag{3.70}$$

$$MRT = \frac{\int_0^\infty tC_p\, dt}{\int_0^\infty C_p\, dt} \tag{3.71}$$

$$VRT = \frac{\int_0^\infty (t - MRT)^2 C_p \, dt}{\int_0^\infty C_p \, dt} \tag{3.72}$$

where AUC = the area under the curve, MRT = the mean residence time, VRT = the variance of residence time, C_p = the plasma drug concentration, and t = time.

Because the C_p is normally followed for some finite maximum time (T), the area under the right tail of the curve ($T \rightarrow \infty$) must be approximated. This is usually done assuming a terminal exponential component from T to ∞, e.g., $C_{po}e^{-kt}$. The following unnormalized moments are used for this purpose.

$$S_0 = \int_0^T C_p dt + \frac{C_{p0}}{k} e^{-kT} \tag{3.73}$$

$$S_1 = \int_0^T t C_p dt + \left[\frac{C_{p0}}{k^2} + \frac{C_{p0}T}{k} \right] e^{-kT} \tag{3.74}$$

$$S_2 = \int_0^T t^2 C_p dt + \left[2\frac{C_{p0}}{k^3} + 2\frac{C_{p0}}{k^2} + \frac{C_{p0}T^2}{k} \right] e^{-kT} \tag{3.75}$$

The integrated portions of Equations (3.73) to (3.75) can be estimated using the trapezoidal method and the plasma concentrations (C_{pi}) measured at n time intervals (t_i) up to T.

$$\int_{t_0}^{t_n} C_{pi} dt \approx \sum_{i=0}^{n-1} (t_{i+1} - t_i) \frac{C_{pi} + C_{pi+1}}{2} \tag{3.76}$$

$$\int_{t_0}^{t_n} t C_{pi} dt \approx \sum_{i=0}^{n-1} (t_{i+1} - t_i) \frac{t_i C_{pi} + t_{i+1} C_{pi+1}}{2} \tag{3.77}$$

$$\int_{t_0}^{t_n} t^2 C_{pi} dt \approx \sum_{i=0}^{n-1} (t_{i+1} - t_i) \frac{t_i^2 C_{pi} + t_{i+1}^2 C_{pi+1}}{2} \tag{3.78}$$

Using Equations (3.73) to (3.75) with the substitutions given in Equations (3.76) to (3.78) for these integrated portions, the S_0, S_1, and S_2 are estimated. The area under the concentration-time curve (AUC) is estimated simply by S_0. The mean and variance of the residence time are estimated with Equations (3.79) and (3.80).

$$MRT = \frac{S_1}{S_0} \tag{3.79}$$

$$VRT = \frac{S_2}{S_0} - \left[\frac{S_1}{S_0}\right]^2 \tag{3.80}$$

If a specific model is defined to describe the data set and an iterative regression method is used to fit a multiexponential model, these statistics can also be estimated from the results.

$$S_0 = \sum_{i=1}^{n} \frac{C_{p0i}}{k_i} \tag{3.81}$$

$$S_1 = \sum_{i=1}^{n} \frac{C_{poi}}{k_i^2} \tag{3.82}$$

$$S_2 = \sum_{i=1}^{n} \frac{2C_{p01}}{k_i^3} \tag{3.83}$$

where C_{p0i} = the concentration in the ith exponential component, k_i = the elimination rate constant for the ith component, and n = the total number of components in the model.

Example 3.11

The elimination of anthracene from fish as presented previously in Example 3.1 (single exponential elimination model) will be used to illustrate the statistical moments approach. In the previous example, one elimination component was obvious. The estimated initial concentration was 99.46 nmol/g and the elimination rate constant was 0.680/h. According to Equation (3.4) and associated assumptions, the mean residence time should be approximately 1.47 h. Now let's use the method of statistical moments to describe elimination.

i	Time (h)	Anthracene (nmol/g)	$t_{i+1} - t_i$	$(C_i + C_{i+1})/2$
0	0	100	1	76.5
1	1	53	1	38.5
2	2	24	1	18.0
3	3	12	1	9.5
4	4	7	—	—

Assuming no specific model except to estimate the area beyond the last sampling time (4 h), the calculations are the following.

From Equations (3.73) and (3.76):

$$S_0 = 1(76.5) + 1(38.5) + 1(18.0) + 1(9.5) + (99.46/0.680)e^{-0.680(4)}$$
$$= 142.5 + 9.6 = 152.1$$

From Equations (3.74) and (3.77):

$$S_1 = 142.5 + (99.46/0.680^2 + 99.46(4)/0.680)e^{-0.680(4)}$$
$$= 151.0 + 52.7 = 203.7$$

From Equations (3.75) and (3.78):

$$S_2 = 313 + (2(99.6)/0.680^3 + 2(99.6)(4)/0.680^2 + 99.6(4^2)/0.680)e^{0.680(4)}$$
$$= 313 + 309.6 = 622.3$$

Using Equations (3.79) and (3.80):

$$MRT = S_1/S_0 = 203.7/152.1 = 1.34 \text{ h}$$
$$VRT = S_2/S_0 - (S_1/S_0)^2 = 622.3/152.1 - (1.34)^2 = 2.3 \text{ h}^2$$

If the original, single exponential model is assumed, then the calculations can be done using Equations (3.81) to (3.83) instead of Equations (3.73) to (3.75).

$$S_0 = 99.6/0.680 = 146.5$$
$$S_1 = 99.6/0.680^2 = 215.4$$
$$S_2 = 2(99.6)/0.680^3 = 633.5$$
$$MRT = S_1/S_0 = 215.4/146.5 = 1.47 \text{ h}$$
$$VRT = S_2/S_0 - (S_1/S_0)^2 = 633.5/146.5 - (215.4/146.5)^3 = 2.2 \text{ h}^2$$

The errors associated with the estimates of AUC, MRT, and VRT are dependent on t. The longer the period during which the concentrations are measured, the better the estimates will be. Yamaoka et al. (1978) give results of simulations (oral administration) indicating that measurement of concentrations until values drop to 5% of the maximum resulted in relative errors of 5, 10, and 40% for AUC, MRT, and VRT, respectively. When concentrations are measured to 1% of the maximum, relative errors become 1, 2, and 10% for AUC, MRT, and VRT, respectively.

With these estimates of AUC and MRT, clearance from the plasma (or blood, Cl_b) and volume of distribution at steady state (V_{ss}) can be determined (Barron et al. 1990a) as shown in Equations (3.84) and (3.85), respectively. If a first-order elimination process is assumed, the MRT can be related to k_e as shown in Equation (3.86).

$$Cl_b = \frac{D}{AUC} \tag{3.84}$$

$$V_{SS} = Cl_b MRT \tag{3.85}$$

$$k_e = \frac{1}{MRT} \tag{3.86}$$

If a dose is administered separately orally and also intravenously, the median absorption time for the substance across the gut (MAT) can be defined. The MAT and $VRT_{gastric}$ can be estimated by the differences between the oral and intravenous estimates, i.e., $MRT_{oral} - MRT_{iv}$ and $VRT_{oral} - VRT_{iv}$ (Yamaoka et al. 1978):

$$MAT = MRT_{oral} - MRT_{iv} \tag{3.87}$$

$$VRT_{gastric} = VRT_{oral} - VRT_{iv} \tag{3.88}$$

and the first-order rate constant for such absorption (k_a) is the following (Gilbaldi 1991):

$$k_a = \frac{1}{MAT} \tag{3.89}$$

Another potentially useful, yet generally overlooked, application of this approach is the estimation of metrics of bioavailability (e.g., Figure 3.13). Yamaoka et al. (1978) and Greenblatt and Shader (1985) discuss the use of plasma AUC to estimate the bioavailability of orally administered drugs. The general approach involves estimation of AUC after intravenous injection (AUC_{iv}) and after oral administration (AUC_{oral}). The fractional absorption (f), a measure of bioavailability of orally administered chemicals, is estimated.

$$f = \frac{AUC_{oral}}{AUC_{iv}} \tag{3.90}$$

The f estimates the percentage of an orally administered drug absorbed and entering the blood relative to that in the blood after direct injection. Equation (3.90) assumes that the same dose was given by both routes. It may be changed to accommodate different doses,

$$f = \frac{D_{iv}AUC_{oral}}{D_{oral}AUC_{iv}} \tag{3.91}$$

where D_{iv} and D_{oral} = intravenous and oral doses.

The AUCs for various routes may be estimated using urinary excretion. The method is very similar to that described above. However, flow estimates are also incorporated into the calculations.

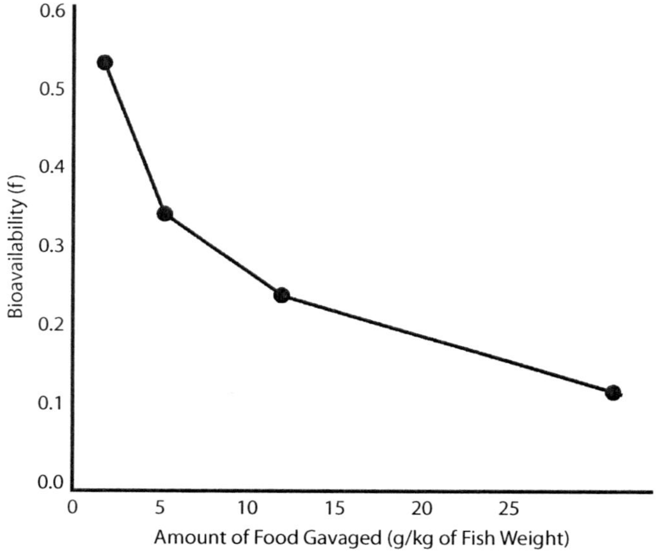

Figure 3.13 An illustration of using Equation (3.91) to calculate fractional absorption, one measure of bioavailability. The statistical moments approach was used to estimate *f* for channel catfish receiving different amounts of food (McCloskey et al. 1998). Note that the fraction of the dosed methylmercury that passes through the gut wall into the blood increased as the amount of food in which it was contained decreased. Although not shown here, McCloskey et al. (1998) also estimated the fractional absorptions by fitting the data to a specific model. Plots of both sets of estimates are provided in Figure 8.4 of Newman and Clements (2008).

The assumption is made that urinary excretion reflects the amount of a substance in the plasma. Yamaoka et al. (1978) and Greenblatt and Shader (1985) provide details of this approach.

This approach can also be used for other routes of toxicant introduction and comparison of various toxicants entering by the same route. A common example in pharmocokinetics involves intramuscular versus intravenous administration. The approach is identical to that described above for intravenous versus oral administration. Obviously, this approach has much to offer to estimation of bioavailability of toxicants in various environmental components. Unfortunately, it remains underutilized.

3.3.2 Stochastic Models

All of the compartment models described in previous sections are based on the assumption that a deterministic system was being modeled. Constants and initial compartment concentrations were assumed invariant. All molecules within a compartment were assumed to have identical probabilities of leaving the compartment during any time interval. In reality, such assumptions are rarely, if ever, valid. For example, the elimination rate is a consequence of a multitude of processes including metabolic rate, gill ventilation, rate of urinary excretion, gut clearance, bile secretion, etc. Even for an individual, some distribution of elimination rate constants with an "average" value should be expected. If a group of individuals were used in a depuration experiment and subsets of the group were taken at time intervals for measurement of concentrations, a distribution of initial concentrations between individuals must be expected except in the most fortuitous experiment. Further, all molecules of a contaminant are unlikely to have the same probability of leaving a compartment over the course of depuration. Regardless, the bioaccumulation models used in ecotoxicology today are deterministic models based on these unlikely assumptions. For the reasons given above, Matis and coworkers (Matis and Wehrly 1979; Matis and Tolley 1980; Matis et al. 1983, 1991) advocated the wider use of stochastic models.

"A stochastic system has been defined as one whose output is uncertain and a stochastic model as one that is structured to account for at least part of the uncertainty [in stochastic systems]" (Matis and Tolley 1980). Stochastic models allow description of systems with heterogeneous compartments and stochastic flows between compartments. Whicker and Schultz's (1982) description given earlier of radionuclide elimination from bone can be viewed as an example of a heterogeneous compartment. As mentioned, a series of exponential compartments could have been used instead of a power model except that extraction of too many elimination rate constants would be required. Alternatively, a stochastic model based on a gamma distribution instead of multiple exponentials could describe this system with fewer parameter estimates. The gamma model would do so by incorporating a series of exponentials within its parameter estimates. It would also maintain a higher fidelity to the underlying mechanisms than a power model.

Let's use the explanation of Matis and Tolley (1980) to illustrate a simple stochastic model. In this approach, the amount of a toxicant in a compartment experiencing elimination is modeled. The familiar form of the deterministic, monoexponential model is Equation (3.2). This model can be restated in the following form under the assumption that the mean value (μ_t) was being measured.

$$\mu_t = X_0 e^{-kt} \tag{3.92}$$

The model is now stochastic in its form. Although similar to that in Equation (3.2), this form predicts the mean concentration over time. It leaves open the possibility of random changes in the model (Matis et al. 1991). Functions such as exponential, Weibull, gamma, Poisson, or negative binomial can be used to describe such uncertainty. For example, k may be treated as a random variable. Matis and Tolley (1980) provide details on a variety of such models with different distributions. Further, the probability of a molecule leaving the compartment within a time interval (the hazard rate) can be varied easily within the stochastic model format. A detailed example of modeling mercury accumulation in fish using stochastic models is also given in Matis et al. (1991). A software package (KINETICA) for constructing stochastic models is also described. Perhaps with the increased availability of such software, stochastic models might become more common in ecotoxicology. It seems more likely that inclusion of stochasticity with Monte Carlo simulation, which has become mainstream in other areas of ecotoxiocology modeling, will fill this niche.

3.4 INTRINSIC FACTORS AFFECTING BIOACCUMULATION

3.4.1 General

Factors intrinsic to the individual and extrinsic factors modify bioaccumulation. Intrinisic factors will be discussed briefly in the remainder of this chapter.

Intrinsic factors include the ability of individuals to transform, modify, or metabolize the toxicant. An obvious example is the internal transformation of organic toxicants such as benzo(a) pyrene (Leversee et al. 1982) or p-nitrophenol (Foster and Crosby 1986). However, transformations of inorganic toxicants may also be significant. For example, arsenic metabolism (Edmonds and Francesconi 1981; Zingaro 1983), especially relative to the role of arsenate as an analog in phosphate biochemistry (Benson and Summons 1981; Morris et al. 1984), plays a critical role in bioaccumulation. Radionulcide bioaccumulation is strongly influenced by the biochemistry of chemical analogs, e.g., Morgan (1964), Blaylock and Frank (1979), Blaylock (1982), Rodgers (1986), and Hinton et al. (1992). Less direct effects include such biochemical processes as calcium and phosphate effects on zinc uptake as controlled by calcium pyrophosphate granule formation (Jones et al. 1984) within the organism. Another intrinsic factor contributing to bioaccumulation is age (Bache et al. 1971; Williamson 1980) and associated developmental changes (Ballatori and Clarkson 1982). Sexual differences also exist (Hirayama and Yasutake 1986; Nicoletto and Hendricks 1988). Two other factors, individual size and lipid content, are often expressed in quantitative terms and, as a result, will be discussed in detail here.

3.4.2 Lipid Content

Bioaccumulation of organic compounds is influenced by the lipid pool within individuals. Neely et al. (1974) compared the lipophilic tendencies of eight organic compounds to the equilibrium bioconcentration factor (BCF, or concentration in the organism/concentration in the water source) reported in fish. The partition coefficient between water and n-octanol (K_{ow}) was used to estimate the lipophilic nature, and k_u/k_e (see Equation 3.43) was used to estimate the BCF for each compound. The linear regression model describing this relationship was log BCF = 0.542 log K_{ow} + 0.124. They stated that this relationship could be used to predict BCF for compounds with K_{ow} values within the tested range (log K_{ow} from 2.88 to 7.62). (Note that reanalysis of these data (Newman 1993) indicates a prediction bias of $10^{0.1173/2}$, or 1.14 (MSE = 0.1173) associated with these data.) Veith et al. (1979) and Mackay (1982) produced similar linear relationships using more organic compounds with log K_{ow} values in the range of 1.0 to 7.05. Geyer et al. (1982) found a strong relationship between log BCF and log K_{ow} for the marine mussel, *Mytilus edulis*. Chiou (1985) compared the K_{ow} to the partition coefficient of triolein (glyceryl trioleate, a lipid similar to triglycerides in animals) and verified the K_{ow} as a viable surrogate measure of organic contaminant lipophilicity. However, Figure 2 in Chiou's paper suggests that this relationship does deviate from linearity above a log K_{ow} of approximately 6. As suspected by Mackay (1982), the linearity of relationship between log BCF and log K_{ow} did not persist with high K_{ow} compounds.

Connell and Hawker (1988) discuss possible mechanisms associated with deviations from linearity in this relationship. They use the following models in their explanation. Equation (3.93) is a direct modification of Equation (3.42) using BCF instead of k_u/k_e in the equation. The BCF is linearly related to k_u and $1/k_e$. Equation (3.94) was proposed by Mackay and Hughes (1984) based on the fugacity approach. Note that the inverse of k_e is linearly related to K_{ow} in this equation.

$$BCF = \frac{k_u}{k_e}[1 - e^{-k_e t}] \tag{3.93}$$

$$\frac{1}{k_e} = \frac{V_l}{k_e} K_{OW} + \frac{V_l}{Q_0} \tag{3.94}$$

where V_l = the lipid phase volume in the animal, Q_w = the effective flow rate of water, and Q_o = the effective flow rate of octanol or organic matter between the water and lipid phase.

Gobas et al. (1986) suggested that uptake and elimination for organics with relatively low lipophilicity are controlled by membrane permeation. For those with higher lipophilicity (log K_{ow} > 3–4) permeation of membranes was rapid, and therefore diffusion likely controls their uptake and elimination. Between log K_{ow} values of approximately 3 and 6, the diffusion control produces a linear relationship between log K_{ow} and log k_u or log k_e. (Note that, relative to internal compartments, cell permeability has little effect on compartmental modeling for very lipophilic compounds also (Gobas et al. 1986).) However, as lipophilicity increases (and K_{ow} increases), the k_u and k_e begin to decrease and the relationships become nonlinear (Figure 3.14A). This nonlinearity is thought to result from the decreased diffusion associated with the large molecular size of the highly lipophilic compounds. The result of these influences on k_e and k_u are shown in Figure 3.14B. The relationship between log BCF and log K_{ow} is linear between log K_{ow} of 3 and approximately 6.5, where diffusion controls uptake and elimination. The relationships deviate from linearity below approximately 3, as they are dominated by membrane permeation, not diffusion. Above a log K_{ow} of about 6.5, the log BCF decreases due to molecular size effects on diffusion rates. (Chiou's observations (1985) mentioned above that K_{ow} is less effective as a surrogate measure of fish lipid/water partitioning at high log K_{ow} values could also contribute to this trend.) Consequently, Connell and Hawker (1988) fit polynomial equations to log BCF/log K_{ow} data for organic compounds with a very wide range of lipophilicity.

3.4.3 Size

The allometry (the study of size and its consequences) of bioaccumulation has been given less attention than warranted. This is surprising as size variation among individuals or samples from populations is one of the most predictable of characteristics influencing bioaccumulation. Allometric or scaling effects are also important in interspecies comparisons if species vary significantly in size.

The effect of size on toxic or therapeutic drug action has been clear to pharmacologists and physiologists since the 1920s, e.g., Campbell (1926). However, incorporation of scaling into environmental sciences did not begin with any intensity until the end of World War II. At that time, studies conducted with the emergence of nuclear weapons testing suggested that animal size was very important in determining body concentrations (or activities) of fission products. By the mid-1960s, a large body of literature on metal biomonitoring emerged to reinforce these conclusions. By the mid-1970s, Boyden (1974, 1977) had adopted the power models used in physiological and anatomical allometry to quantitatively describe scaling effects on contaminant bioaccumulation. The scaling of body burden (amount of a substance/individual) for metals in shellfish was described with the equation

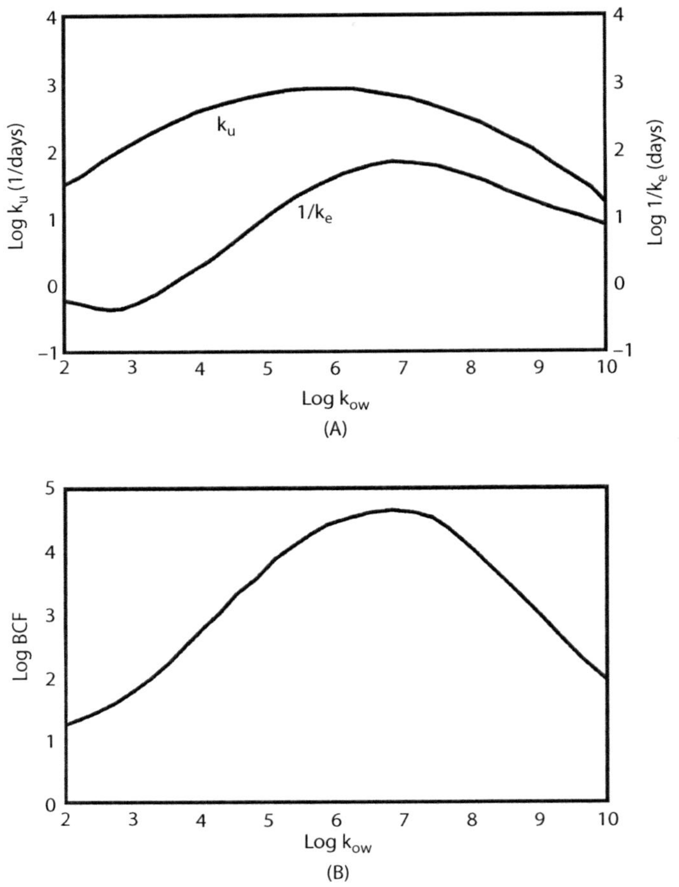

Figure 3.14 The effect of *Kow* on the rate constants and BCF for organic toxicants. Panel A shows the nonlinear effect of log K_{ow} on both log k_u and log k_e, and panel B shows the consequent effect of log K_{ow} on log BCF. (Modified from Connell, D.W., and D.W. Hawker, *Ecotoxicol. Environ. Saf.*, 16, 242–257, 1988.)

$$B = aW^b \tag{3.95}$$

where B = body burden (mg/individual), W = weight of the individual, and a and b = constants.

The relationship is most often transformed to facilitate least-squares, linear regression.

$$Log\ B = \text{Log}\ a + b\ \text{Log}\ W \tag{3.96}$$

To express this relationship in terms of concentration, Boyden gave the following simple conversion:

$$B' = \frac{B}{W} = \frac{aW^b}{W} = aW^{b-1} \tag{3.97}$$

where B' = concentration (mg/g of tissue).

Equation (3.97) can also be transformed as was Equation (3.95) to yield a linear relationship. However, because of the potential presence of W in both the independent (W) and dependent (B/W)

variables, it is advisable to use Equation (3.96) instead of the analogous transformation of Equation (3.97) to fit this type of data.

Boyden (1974, 1977) described three classes of relationships based on estimated b values (Figure 3.15). The two most common classes had b values of approximately 1 or lower. The first ($b \approx 0.75$) was characterized by increasing concentrations with decreasing animal size (weight). Boyden suggested that bioaccumulation was dominated by some process directly linked to metabolism because a b of approximately 0.75 was similar to the b often found for size-specific metabolic rate. The second class ($b \approx 1.00$) had concentrations independent of animal size. Boyden speculated that such a relationship may be directly linked to the number of binding sites available in the tissues.

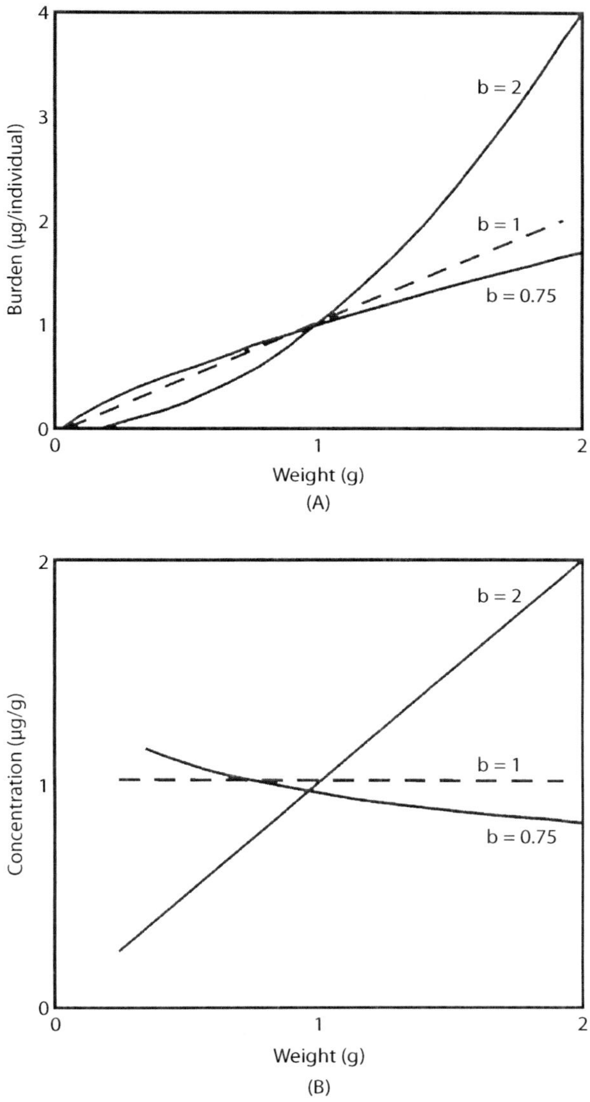

Figure 3.15 The effect of animal size (weight) on body burden (panel A, amount/individual) and concentration (panel B, amount/g of tissue) of a toxicant. Three general curves with b values of 0.75, 1.00, and 2.00 are shown here to illustrate the various power models possible. See the text for more details.

As animal size increases, the number of binding sites increases but the amount of element bound per g of tissue remains constant. In the final and least common class ($b \approx 2.00$), the metal concentration increased with increasing size. Speculating, Boyden suggested that sequestration after avid binding in specific tissues or organs could result in such a relationship.

There are several statistical complications associated with the techniques described above. First, the potential prediction bias detailed earlier in this chapter must be kept in mind. Second, using b may not be valid as discussed by Boyden (1977) and Newman and Heagler (1991). The use of predictive regression methods assumes no significant measurement error or inherent variability in the independent variable (weight): predictive regression techniques minimize variation about the dependent variable only. In many cases, this is a questionable approach as estimates of weight or size are often subject to significant error. For example, an average fish weight may be used when very small individuals of similar size are pooled to obtain sufficient tissue for analysis. In such a case, functional regressions that minimize the sum of the products of deviations of the independent and dependent variables from the regression line are more appropriate (for more details see Newman and Heagler 1991; Ricker 1973). Fortunately, if b and the correlation coefficient (r) are available from predictive regression results, the slope of the functional regression line (v) can be estimated by $v = \pm\, b/r$. The sign of v is the same as that of b and r.

Several conceptual complications have also been discussed regarding Boyden's speculation as described above. Fagerström (1977) argues that it is a mistake to compare b values from body burden or concentration to those from metabolic rates. The crux of his reasoning is that b values for states (burden or concentration) cannot be directly compared to that for a flux (metabolic rate). He uses the accumulation of an indeterminant element* to demonstrate this point. For an indeterminant element, the following relationships are pertinent.

$$t_{1/2} \propto W^{1-b} \tag{3.98}$$

$$B' = W^{0} \tag{3.99}$$

$$\theta \propto W^{1-b} \tag{3.100}$$

$$B \propto W^{1} \tag{3.101}$$

where Θ = metabolic turnover rate.

The rate of elemental turnover (Θ) will be directly proportional to the animal's metabolic rate. Contrary to Boyden's interpretations, Fagerström argues that a b value of 1, not 0.75, suggests direct linkage to metabolic rate (Equation 3.101).

Newman and Heagler (1991) reanalyzed Boyden's data supplemented with more current data to reassess the general classification of relationships based on b (or v) values. They found no evidence supporting the three different classes of b values described by Boyden. Instead, continuous, skewed distributions of b (median = 0.80) and v (median = 0.85) values were evident. They also suggested that linear models may be more appropriate than power models for narrow allometry (description of relationships over a narrow range of sizes such as those often associated with populations). Power models may be more useful in the context of broad allometry, e.g., inter- or intraspecies comparisons with very wide ranges of animal sizes.

* An element is indeterminant if its concentration within the organism is directly proportional to that in the environment (source). It is determinant if its concentration in the organism remains constant over a range of environmental concentrations; i.e., it is biologically controlled over a range of concentrations.

Several studies consider scaling effects on bioaccumulation kinetics, not observed body burdens or concentrations. It is clear that many allometric relationships could influence associated constants. For example, uptake by the gill assumes diffusion across a constant surface area and binding to a constant amount of tissue (see Equation 3.32). However, gill surface area changes disproportionately with animal size (Hughes 1984; Muir 1969). Many respiratory processes that directly affect gill uptake display clear allometric trends. The amount of gut surface available for uptake per unit body mass also changes disproportionately with size (Ribble and Smith 1983). General body surface-to-volume ratio changes with size can also influence uptake associated with adsorption processes (Smock 1983). Allometric relationships also may affect elimination constants. For example, renal elimination, number of nephra in the kidneys, urine output, heartbeat, and gut beat are all linked disportionately to size in mammals (Adolph 1949). Although not always the case (Somero and Childress 1980), enzyme concentrations such as those associated with contaminant degradation (Walker 1987) display allometric relationships.

Anderson and Spear (1980a, 1980b) successfully used power models to describe the allometry associated with the simple one-compartment model constants (k_e and k_u in Equation 3.42). Copper pharmacokinetics in the pumpkinseed sunfish (*Lepomis gibbosus*) gill were clearly defined with these models. But rainbow trout (*Salmo gairdneri*) uptake showed only a slight scaling effect. For the sunfish, but not the rainbow trout, these relationships were extended to include size-dependent toxicity. Newman and Mitz (1988) also described size-dependent k_e and k_u for Zn accumulation in mosquitofish (*Gambusia holbrooki*, formerly *G. affinis*) using simple power functions. Incorporating these power relationships into Equation (3.42), they demonstrated that *b* values may not be constant for a species-metal combination. This conclusion was contrary to Boyden's early speculation (1974, 1977) and consistent with the results of many field studies, e.g., Cossa et al. (1980), Strong and Luoma (1981), and Lobel and Wright (1982). Using the same approach, clear power relationships were not found for Hg^{2+} kinetics in mosquitofish (Newman and Doubet 1989). Later, Newman and Heagler (1991) speculated that metabolic rate and gill surface-to-total fish volume ratio may both play key roles in determining these relationships for Zn accumulation. However, no strong evidence to support this speculation has been forwarded to date.

Rowland (1985) and Hayton (1989) also provide excellent and relevant treatments of allometry as related to PBPK modeling of mammals. Rowland's explanation begins with the observation that many biological functions can be described by the power model (Equation 3.95), with *B* replaced by some biological function, *F*(*B*)). The apparent volume of distribution and clearance are two such biological qualities pertinent to drug or toxicant pharmacokinetics that can be so described.

$$V_d = a_1 W^{b_1} \tag{3.102}$$

$$Cl = a_2 W^{b_2} \tag{3.103}$$

It follows by substitution that the $t_{1/2}$ can be expressed as a function of size-dependent V_d and *Cl*. (See Equation 3.51 and the text immediately below it for more information.)

$$t_{1/2} = \text{Ln}2\left[\frac{a_1}{a_2}\right]W^{b_1 - b_2} \tag{3.104}$$

As size decreases, $t_{1/2}$ will decrease if V_d decreases more rapidly than *Cl*. Smaller individuals will eliminate a substance faster than larger individuals.

Rowland extended his treatment to include drug elimination kinetics from a monoexponential compartment. The concentration at any time (C_t) is determined by the dose administered (D), apparent volume (V_d), and elimination rate constant (k_e) as described in Equation (3.105).

$$C_t = \frac{D}{V} e^{-k_e t} \tag{3.105}$$

This follows directly from Equation (3.2) and rearrangement of Equation (3.35).

The concentration at any time can be determined as a function of the scaling of the apparent volume of distribution and clearance by substitution for V_d (Equation 3.102), Cl (Equation 3.103, and k_e (Equation 3.3 combined with Equation 3.104) into Equation (3.105).

$$C_t = \frac{D}{a_1 W^{b_1}} e^{-[a_1/a_2][t/W^{b_1-b_2}]} \tag{3.106}$$

Equation (3.106) can be used with nonlinear regression methods to fit size-dependent data to this model. However, to facilitate linear plotting, Rowland rearranges Equation (3.107).

$$\frac{C_t}{D/W^{b_1}} = \left[\frac{1}{a_1}\right] e^{-[a_1/a_2][t/W^{b_1-b_2}]} \tag{3.107}$$

By taking the natural logarithm of both sides of Equation (3.107), one can derive a straight-line relationship.

$$Ln \frac{C_t}{D/W^{b_1}} \text{ vs. } \frac{t}{W^{b_1-b_2}} \tag{3.108}$$

An example showing good fit to such a plot is given by Rowland for methotrexate elimination from mice, rats, monkeys, dogs, and humans.

Except for ubiquitous data quality considerations mentioned previously, there is no reason that the above methods cannot be applied to ecotoxicology also. The techniques can be modified to accommodate an initial concentration scenario when a dose scenario is not pertinent. Substitution of body burden for dose would suffice. Incorporation of biexponential elimination would involve an expansion of Equation (3.106) similar to that with Equation (3.2) to Equation (3.8).

3.5 SUMMARY

Commonly used techniques for modeling bioaccumulation are presented in this chapter along with several potentially useful alternative approaches. Examples illustrate the more important approaches. The often blurry distinction between physical and kinetic compartments is made. Means of assessing the appropriateness of models are also demonstrated with examples. Clearance-based and physiologically based pharmacokinetic models are described, as is the fugacity approach to modeling bioaccumulation. Stochastic formulations of these models are mentioned only briefly. Also presented are model-free, statistical moments methods. Quantitative methods of modeling effects of two intrinsic qualities of organisms (lipid content and animal size) are presented.

This treatment highlights many subjects and approaches. This was done to provide the reader with a basic understanding of available methods. The reader is encouraged to explore the cited works listed below for a more comprehensive understanding of any particular methods. A rich selection of techniques is available to the worker who invests the time necessary to effectively implement them.

REFERENCES

Adolph, E.F. Quantitative relations in the physiological constitutions of mammals. *Science* 109:579–585 (1949).

Anderson, P.D., and P.A. Spear. Copper pharmacokinetics in fish gills. I. Kinetics in pumpkinseed sunfish, *Lepomis gibbosus*, of different body sizes. *Water Res.* 14:1101–1105 (1980a).

Anderson, P.D., and P.A. Spear. Copper pharmacokinetics in fish gills. II. Body size relationships for accumulation and tolerance. *Water Res.* 14:1107–1111 (1980b).

Arnot, J.A., and D. Mackay. Vignette 3.1. Fugacity and bioaccumulation. In *Fundamentals of ecotoxicology*, ed. M.C. Newman. 3rd ed. Boca Raton, FL: CRC Press, 2010.

Atkins, G.L. *Multicompartment models for biological systems.* London: Methuen & Co. LTD, 1969.

Bacci, E. *Ecotoxicology of organic contaminants.* Boca Raton, FL: CRC Press, 1994.

Bache, C.A., W.H. Gutenmann, and D.J. Lisk. Residues of total mercury and methylmercuric salts in lake trout as a function of age. *Science* 172:951–952 (1971).

Ballatori, N., and T.W, Clarkson. Developmental changes in the biliary excretion of methylmercury and glutathione. *Science* 216:61–63 (1982).

Barber, M.C., L.A. Suarez, and R.R. Lassiter. Modeling bioconcentration of nonpolar organic pollutants by fish. *Environ. Toxicol. Chem.* 7:545–558 (1988).

Barron, M.G., M.A. Mayes, P.G. Murphy, and R.J. Nolan. Pharmacokinetics and metabolism of triclopyrbutoxyethyl ester on coho salmon. *Aquat. Toxicol.* 16:19–32 (1990b).

Barron, M.G., G.R. Stehly, and W.L. Hayton. Pharmacokinetic modeling in aquatic animals. I. Models and concepts. *Aquat. Toxicol.* 17:187–212 (1990a).

Beauchamp, J.J., and J.S. Olson. Correction for bias in regression estimates after logarithmic transformation. *Ecology* 54(6):1403–1407 (1973).

Benson, A.A., and R.E. Summons. Arsenic accumulation in Great Barrier Reef invertebrates. *Science* 211:482–483 (1981).

Blaylock, B.G. Radionuclide data base available for bioaccumulation factors for freshwater biota. *Nucl. Saf.* 23(4):427–438 (1982).

Blaylock, B.G., and M.L. Frank. Distribution of tritium in a chronically contaminated lake. In *Behavior of tritium in the environment.* IAEA-SM-232/74. Vienna: International Atomic Agency, 1979, pp. 247–256.

Blaylock, B.G., M.L. Frank, and D.L. DeAngelis. Bioaccumulation of ^{95m}Tc in fish and snails. *Health Phys.* 42(3):257–266 (1982).

Boxenbaum, H.G., S. Reigelman, and R.M. Elashoff. Statistical estimations in pharmacokinetics. *J. Pharmacokinet. Biopharm.* 2(2):123–148 (1974).

Boyden, C.R. Trace element content and body size in mollusks. *Nature* 251:311–314 (1974).

Boyden, C.R. Effect of size upon metal content of shellfish. *J. Mar. Biol. Assoc. U.K.* 57:675–714 (1977).

Bozdogan, H. Akaike's information criterion and recent developments in information complexity. *J. Math. Psychol.* 44:62–91 (2000).

Campbell, F.L. Relative susceptibility to arsenic in successive instars of the silkworm. *Gen. Physiol. Notes* 9(6):727–733 (1926).

Chiou, C.T. Partition coefficients of organic compounds in lipid-water systems and correlations with fish bioconcentration factors. *Environ. Sci. Technol.* 19:57–62 (1985).

Christian, S.D., and E.E. Tucker. Least squares analysis with the microcomputer. Part 5. General least squares with variable weighting. *Am. Lab.* 18–32 (1984).

Connell, D.W. *Bioaccumulation of xenobiotic compounds.* Boca Raton, FL: CRC Press, 1990.

Connell, D.W., and D.W. Hawker. Use of polynomial expressions to describe the bioconcentration of hydrophobic chemicals by fish. *Ecotoxicol. Environ. Saf.* 16:242–257 (1988).

Cossa, D., E. Bourget, D. Pouliot, J. Piuze, and J.P. Chanut. Geographical and seasonal variations in the relationship between trace metal content and body weight in *Mytilus edulis*. *Mar. Biol. (Berl.)* 58:7–14 (1980).

Crist, R.H., K. Oberhoiser, D. Schwartz, J. Marzoff, D. Ryder, and D.R. Crist. Interactions of metals and protons with algae. *Environ. Sci. Technol.* 22(7):755–760 (1988).

Cutshall, N. Turnover of zinc-65 in oysters. *Health Phys.* 26:327–331 (1974).
De Bruijn, J., and J. Hermens. Uptake and elimination kinetics of organophosphorous pesticides in the guppy (*Poecilia reticulata*): Correlations with the octanol/water partition coefficient. *Environ. Toxicol. Chem.* 10:791–804 (1991).
Der, G., and B.S. Everitt. *Statistical analysis of medical data using SAS*. Boca Raton, FL: Chapman & Hall/CRC, 2006.
Edmonds, J.S., and K.A. Francesconi. Isolation and identification of arsenobetaine from the American lobster *Homarus americanus*. *Chemosphere* 10(9):1041–1044 (1981).
Ellgehausen, H., J.A. Guth, and H.O. Essner. Factors determining the bioaccumulation potential of pesticides in the individual compartments of aquatic food chains. *Ecotoxicol. Environ. Saf.* 4:134–157 (1980).
Erickson, R.J., and J.M. McKim. A simple flow-limited model for exchange of organic chemicals at fish gills. *Environ. Toxicol. Chem.* 9:159–165 (1990a).
Erickson, R.J., and J.M. McKim. A model of exchange of organic chemicals at fish gills: Flow and diffusion limitations. *Aquatic. Toxicol.* 18:175–198 (1990b).
Fagerström, T. Body weight, metabolic rate, and trace substance turnover in animals. *Oecologia (Berl.)* 29:99–104 (1977).
Foster, G.D., and D.G. Crosby. Xenobiotic metabolism of nitrophenol derivatives by the rice field crayfish (*Procambarus clarkii*). *Environ. Toxicol. Chem.* 5:1059–1070 (1986).
Fujita, M., and K. Hashizume. Status of uptake of mercury by the fresh water diatom, *Synedra ulna*. *Water Res.* 9:889–894 (1975).
Geyer, H., P. Sheehan, D. Kotzias, D. Freitag, and F. Korte. Prediction of ecotoxicological behavior of chemicals: Relationship between physico-chemical properties and bioaccumulation of organic chemicals in the mussel *Mytilus edulis*. *Chemosphere* 11(11):1121–1134 (1982).
Gibaldi, M. *Biopharmaceutics and clinical pharmacokinetics*. 4th ed. Philadelphia: Lea & Febiger, 1991.
Gibaldi, M., and D. Perrier. *Pharmacokinetics*. 2nd ed. New York: Marcel Dekker, 1982.
Giblin, F.J., and E.J. Massaro. Pharmacodynamics of methyl mercury in the rainbow trout (*Salmo gairdneri*): Tissue uptake, distribution and excretion. *Toxicol. Appl. Pharmacol.* 24:81–91 (1973).
Giesy, J.P., J.W. Bowling, and H.J. Kania. Cadmium and zinc accumulation and elimination by freshwater crayfish. *Arch. Environ. Contam. Toxicol.* 9:683–697 (1980).
Giesy, J.P., and D. Paine. Effects of naturally occurring aquatic organic fractions on ^{241}Am uptake by *Scenedesmus obliquus* (*Chlorophyceae*) and *Aeromonas hydrophila* (*Pseudomonadaceae*). *Appl. Environ. Microbiol.* 33(1):89–96 (1977).
Gobas, F.A.P.C., and D. Mackay. Dynamics of hydrophobic organic chemical bioconcentration in fish. *Environ. Toxicol. Chem.* 6:495–504 (1987).
Gobas, F.A.P.C., A. Opperhuizen, and O. Hutzinger. Bioconcentration of hydrophobic chemicals in fish: Relation with membrane permeation. *Environ. Toxicol. Chem.* 5:637–646 (1986).
Goldstein, R.A., and J.W. Elwood. A two-compartment, three-parameter model for the adsorption and retention of ingested elements by animals. *Ecology* 52(5):935–939 (1971).
Gordon, M.S. *Animal physiology: Principles and adaptations*. 2nd ed. New York: The Macmillan Company, 1972.
Greenblatt, D.J., and R.I. Shader. *Pharmacokinetics in clinical practice*. Philadelphia: W.B. Sanders Co., 1985.
Harrison, S.E., and J.F. Klaverkamp. Uptake, elimination and tissue distribution of dietary and aqueous cadmium by rainbow trout (*Salmo gairdneri* Richardson) and lake whitefish (*Coregonus clupeaformis* Mitchill). *Environ. Toxicol. Chem.* 8:87–97 (1989).
Hayton, W.L. Pharmacokinetic parameters for interspecies scaling using allometric techniques. *Health Phys.* 57:159–164 (1989).
Hayton, W.L. Personal communication, 1992.
Hayton, W.L., and M.G. Barron. Rate-limiting barriers to xenobiotic uptake by the gill. *Environ. Toxicol. Chem.* 9:151–157 (1990).
Hinton, T.G., F.W. Whicker, J.E. Pinder III, and S.A. Ibrahim. Comparative kinetics of ^{47}Ca, ^{85}Sr and ^{226}Ra in the freshwater turtle, *Trachemys scripta*. *J. Environ. Radioact.* 16:25–47 (1992).
Hirayama, K., and A. Yasutake. Sex and age differences in mercury distribution and excretion in methylmercury-administered mice. *J. Toxicol. Environ. Health* 18:49–60 (1986).
Hocking, R.R. The analysis and selection of variables in linear regression. *Biometrics* 32:1–49 (1976).
Holleman, D.F., J.R. Luick, and F.W. Whicker. Transfer of radiocesium from lichen to reindeer. *Health Phys.* 21:657–666 (1971).

Hughes, G.M. Measurement of gill area in fishes: Practices and problems. *J. Mar. Biol. Assoc. U.K.* 64:637–655 (1984).

Hunn, J.B. Role of calcium in gill function in freshwater fishes. *Comp. Biochem. Physiol.* 82A(3):543–547 (1985).

Jones, A.R., M. Taylor, and K. Simkiss. Regulation of calcium, cobalt and zinc by *Tetrahymena elliotti*. *Comp. Biochem. Physiol.* 78A(3):493–500 (1984).

Jorgensen, S.E. *Modelling in ecotoxicology.* New York: Elsevier, 1990.

Kac, M. Some mathematical models in science. *Science* 166:695–699 (1969).

Kinniburgh, D.G. General purpose adsorption isotherms. *Environ. Sci. Technol.* 20:895–904 (1986).

Koch, R.W., and G.M. Smillie. Bias in hydrologic prediction using log-transformed regression models. *Water Res.* 22(5):717–723 (1986).

Kolehmainen, S.E. The balances of ^{137}Cs, stable cesium and potassium of bluegill (*Lepomis macrochirus* Raf.) and other fish in White Oak Lake. *Health Phys.* 23:301–315 (1972).

Konemann, H., and K. Van Leeuwen. Toxicokinetics in fish: Accumulation and elimination of six chlorobenzenes by guppies. *Chemosphere* 9:3–19 (1980).

Kujawinski, E.B., J.W. Farrington, and J.W. Moffett. Importance of passive diffusion in the uptake of polychlorinated biphenyls by phagotrophic protozoa. *Appl. Environ. Microb.* 2000:1987–1993 (2000).

Landrum, P.F. Appendix 7. Derivation of units for simple bioaccumulation models. In *Fundamentals of ecotoxicology*, ed. M.C. Newman. 3rd ed. Boca Raton, FL: CRC Press, 2010.

Landrum, P.F., H. Lee II, and M.J. Lydy. Toxicokinetics in aquatic systems: Model comparisons and use in hazard assessment. *Environ. Toxicol. Chem.* 11:1709–1725 (1992).

Landrum, P.F., and M.J. Lydy. Toxicokinetics short course. Presented at 12th Meeting of the Society of Environmental Toxicology and Chemistry, Seattle, WA, November 3, 1991.

Landrum, P.F., and C.R. Stubblefield. Role of respiration in the accumulation of organic xenobiotics by the amphipod *Diporeia* sp. *Environ. Toxicol. Chem.* 10:1019–1028 (1991).

Les, A., and R.W. Walker. Toxicity and binding of copper, zinc, and cadmium by the blue-green alga, *Chroococcus paris*. *Water Air Soil Pollut.* 23:129–139 (1984).

Leversee, G.J., J.P. Giesy, P.F. Landrum, S. Gerould, J.W. Bowling, T.E. Fannin, J.D. Haddock, and S.M. Bartell. Kinetics and biotransformation of benzo(a)pyrene in *Chironomus riparius*. *Arch. Environ. Contam. Toxicol.* 11:25–31 (1982).

Lobel, P.B., H.P. Longerich, S.E. Jackson, and S.P. Belkhode. A major factor contributing to the high degree of unexplained variability of some elements concentrations in biological tissue: 27 elements in 5 organs of the mussel *Mytilus* as a model. *Arch. Environ. Contam. Toxicol.* 21:118–125 (1991).

Lobel, P.B., and D.A. Wright. Total body zinc and allometric growth ratios in *Mytilus edulis* collected from different shore levels. *Mar. Biol. (Berl.)* 66:231–236 (1982).

Lunde, G. The absorption and metabolism of arsenic in fish. *Fiskeridir. Skr. Ser. Teknol. Unders.* 5(12):1–16 (1972).

Luoma, S.N., C. Johns, N.S. Fisher, N.A. Steinberg, R.S. Oremland, and J.R. Reinfelder. Determination of selenium bioavailability to a benthic bivalve from particulate and solute pathways. *Environ. Sci. Technol.* 26:485–491 (1992).

Lyon, R., M. Taylor, and K. Simkiss. Ligand activity in the clearance of metals from the blood of the crayfish (*Austropotamobius pallipes*). *J. Exp. Biol.* 113:19–27 (1984).

Mackay, D. Correlation of bioconcentration factors. *Environ. Sci. Technol.* 16:274–278 (1982).

Mackay, D. *Multimedia environmental models: The fugacity approach.* Chelsea, MI: Lewis Publishers, 1991.

Mackay, D., and A.I. Hughes. Three parameter equation describing the uptake of organic compounds in fish. *Environ. Sci. Technol.* 18:439–444 (1984).

Mackay, D., and S. Paterson. Fugacity revisited. The fugacity approach to environmental transport. *Environ. Sci. Technol.* 116(12):654A–660A (1982).

Mallows, C.L. Some comments on C_p. *Technometrics* 15:661–675 (1973).

Mallows, C.L. More comments on C_p. *Technometrics* 37:362–372 (1995).

Matis, J.H., T.H. Miller, and D.M. Allen. Stochastic models of bioaccumulation. In *Metal ecotoxicology, concepts and applications*, ed. M.C. Newman and A.W. McIntosh. Chelsea, MI: Lewis Publishers, 1991, pp. 171–206.

Matis, J.H., and H.D. Tolley. On the stochastic modeling of tracer kinetics. *Fed. Proc.* 39(1):104–109 (1980).

Matis, J.H., and T.E. Wehrly. Stochastic models of compartmental systems. *Biometrics* 35:199–220 (1979).

Matis, J.H., T.E. Wehrly, and K.B. Gerald. The statistical analysis of pharmacokinetic data. In *Lecture notes in biomathematics. Tracer kinetics and physiologic modeling theory and practice*, ed. S. Levin. Berlin: Springer-Verlag, 1983, pp. 1–58.

Mayer, F.L. Residue dynamics of di-2-ethylhexylphthalate in fathead minnows (*Pimephales promelas*). *J. Fish. Res. Board Can.* 33:2610–2613 (1976).

McCallum, D.A., and P.M. Dixon. Reducing bias in estimates of the Richards growth function shape parameter. *Growth Dev. Aging* 54:135–141 (1990).

McCloskey, J.T., I.R. Schultz, and M.C. Newman. Estimating the oral bioavailability of methylmercury to channel catfish (*Ictalurus punctatus*). *Environ. Toxicol. Chem.* 17:1524–1529 (1998).

McKone, C.E., R.G. Young, C.A. Bache, and D.J. Lisk. Rapid uptake of mercuric ion by goldfish. *Environ. Sci. Technol.* 5:1138–1139 (1971).

Micromath Scientific Software. *RSTRIP polyexponential curve stripping/least squares parameter estimation*. Version 4. Salt Lake City, UT: Micromath Scientific Software, 1989.

Morgan, F. The uptake of radioactivity by fish and shellfish. I. ^{134}Caesium by whole animals. *J. Mar. Biol. Ass. U.K.* 44:259–271 (1964).

Morris, R.J., M.J. McCartney, A.G. Howard, M.H. Arbab-Zavar, and J.S. Davis. The ability of a field population of diatoms to discriminate between phosphate and arsenate. *Mar. Chem.* 14:259–265 (1984).

Muir, B.S. Gill dimensions as a function of fish size. *J. Fish. Res. Board. Can.* 26:165–170 (1969).

Myhill, J. Investigation of the effect of data error in the analysis of biological tracer data. *Biophys. J.* 7:903–911 (1967).

Neely, W.B., G.E. Blau, and G.L. Agin. The use of SIMUSOLV to analyze fish bioconcentration data. *Chemomet. Intellig. Lab. Sys.* 1:359–366 (1987).

Neely, W.B., D.R. Branson, and G.E. Blau. Partition coefficient to measure bioconcentration potential of organic chemicals in fish. *Environ. Sci. Technol.* 8(13):1113–1115 (1974).

Neter, J., W. Wasserman, and M.H. Kutner. *Applied linear statistical models in regression, analysis of variance, and experimental design*. Homewood, IL: Richard D. Irwin, 1990.

Newman, M.C. A statistical bias in the derivation of hardness-dependent metals criteria. *Environ. Toxicol. Chem.* 10:1295–1297 (1991).

Newman, M.C. Regression analysis of log-transformed data: Statistical bias and its correction. *Environ. Toxicol. Chem.* 12:1129–1133 (1993).

Newman, M.C., and W.H. Clements. *Ecotoxicology: A comprehensive treatment*. Boca Raton, FL: CRC Press, 2008.

Newman, M.C., and D.K. Doubet. Size-dependence of mercury(II) accumulation in the mosquitofish, *Gambusia affinis* (Baird and Girard). *Arch. Environ. Contam. Toxicol.* 18:819–825 (1989).

Newman, M.C., and M.G. Heagler. Allometry of metal bioaccumulation and toxicity. In *Metal ecotoxicology, concepts and applications*, ed. M.C. Newman and A.W. McIntosh. Chelsea, MI: Lewis Publishers, 1991, pp. 91–130.

Newman, M.C., and A.W. McIntosh. Lead elimination and size effects on accumulation by two freshwater gastropods. *Arch. Environ. Contam. Toxicol.* 12:25–29 (1983a).

Newman, M.C., and A.W. McIntosh. Slow accumulation of lead from contaminated food sources by the freshwater gastropods, *Physa integra* and *Campeloma decisum*. *Arch. Environ. Contam. Toxicol.* 12:685–692 (1983b).

Newman, M.C., and A.W. McIntosh. Appropriateness of aufwuchs as a monitor of bioaccumulation. *Environ. Pollut.* 60:83–100 (1989).

Newman, M.C., and S.V. Mitz. Size dependence of zinc elimination and uptake from water by mosquitofish *Gambusia affinis* (Baird and Girard). *Aquat. Toxicol.* 12:17–32 (1988).

Nichols, J.W., P.N. Fitzsimmons, F.W. Whiteman, T.D. Dawson, L. Babeu, and J. Jueneman. A physiologically based toxicokinetic model for dietary uptake of hydrophobic organic compounds in fish. *Toxicol. Sci.* 77:206–218 (2004).

Nichols, J.W., K.M. Jensen, J.E. Tietge, and R.D. Johnson. Physiologically based toxicokinetic model for maternal transfer of 2,3,7,8-tetrachlorodibenzo-p-dioxin in brook trout (*Salvelinus fontinalis*). *Environ. Toxicol. Chem.* 17:2422–2434 (1998).

Nichols, J.W., J.M. McKim, M.E. Andersen, M.L. Gargas, H.J. Clewell III, and R.J. Erickson. A physiologically based toxicokinetic model for the uptake and disposition of waterborne organic chemicals in fish. *Toxicol. Appl. Pharmacol.* 106:433–447 (1990).

Nichols, J.W., J.M. McKim, G.J. Lien, A.D. Hoffman, and S.L. Bertelsen. Physiologically based toxicokinetic modeling of three waterborne chloroethanes in rainbow trout (*Oncorhynchus mykiss*). *Toxicol. Appl. Pharmacol.* 110:374–389 (1991).

Nicoletto, P.F., and A.C. Hendricks. Sexual differences in accumulation of mercury in four species of centrachid fishes. *Can. J. Zool.* 66:944–949 (1988).

Paasivirta, J. *Chemical ecotoxicology.* Chelsea, MI: Lewis Publishers, 1991.

Pagenkopf, G.K. Gill surface interaction model for trace-metal toxicity to fishes: Role of complexation, pH, and water hardness. *Environ. Sci. Technol.* 17(6):342–347 (1983).

Pankow, J.F. *Aquatic chemistry concepts.* Chelsea, MI: Lewis Publishers, 1991.

Part, P., and O. Svanberg. Uptake of cadmium in perfused rainbow trout (*Salmo gairdneri*) gills. *Can. J. Fish. Aquat. Sci.* 38:917–924 (1981).

Penry, D.L. Applications of efficiency measurements in bioaccumulation studies: Definitions, clarifications, and a critique of methods. *Environ. Toxicol. Chem.* 17:1633–1639 (1998).

Pentreath, R.J. The accumulation of mercury from food by the plaice, *Pleuronectes platessa* L. *J. Exp. Mar. Biol. Ecol.* 25:51–65 (1976).

Peters, E.L., I.R. Schultz, and M.C. Newman. Rb and Cs kinetics and tissue distributions in channel catfish (*Ictalurus punctatus*). *Ecotoxicology* 8:287–300 (1999).

Phillips, D.L. Propagation of error and bias in half-life estimates based on two measurements. *Arch. Environ. Contam. Toxicol.* 18:508–514 (1989).

Piszkiewicz, D. *Kinetics of chemical and enzyme-catalyzed reactions.* New York: Oxford University Press, 1977.

Preston, E.M. The importance of ingestion in chromium-51 accumulation by *Crassostrea virginica* (Gmelin). *J. Exp. Mar. Biol. Ecol.* 6:47–54 (1971).

Raaijmakers, J.G.W. Statistical analysis of the Michaelis–Menten equation. *Biometrics* 43:793–803 (1987).

Reichle, D.E. Relation of body size to food uptake, oxygen consumption, and trace element metabolism in forest floor arthropods. *Ecology* 49(3):538–542 (1967).

Renwick, A.G. Toxicokinetics. In *Principles and methods of toxicology*, ed. A.W. Hayes. 5th ed. Boca Raton, FL: CRC Press, 2008.

Ribble, D.O., and M.H. Smith. Relative intestine length and feeding ecology of freshwater fishes. *Growth* 47:292–300 (1983).

Ricker, W.E. Linear regression in fishery research. *J. Fish. Res. Board. Can.* 30:409–434 (1973).

Rodgers, D.W. Tritium dynamics in juvenile rainbow trout, *Salmo gairdneri*. *Health Phys.* 50(1):89–98 (1986).

Rose, K.A., R.I. McLean, and J.K. Summers. Development and Monte Carlo analysis of an oyster bioaccumulation model applied to biomonitoring data. *Ecol. Modell.* 45:111–132 (1989).

Rowland, M. Physiologic pharmacokinetic models and interanimal species scaling. *Pharmacol. Ther.* 29:49–68 (1985).

SAS Institute. *SAS/STAT user's guide*. Release 6.03 ed. Cary, NC: SAS Institute, 1988.

Schultz, I.R., and M.C. Newman. Methyl mercury toxicokinetics in channel catfish (*Ictalurus punctatus*) and largemouth bass (*Micropterus salmoides*) after intravascular administration. *Environ. Toxicol. Chem.* 16:990–996 (1997).

Schultz, I.R., E.L. Peters, and M.C. Newman. Toxicokinetics and disposition of inorganic mercury and cadmium in channel catfish after intravascular administration. *Toxicol. Appl. Pharm.* 140:39–50 (1996).

Schulz-Baldes, M. Lead uptake from sea water and food, and lead loss in the common mussel *Mytilus edulis*. *Mar. Biol. (Berl.).* 25:177–193 (1974).

Scott, D.E., F.W. Whicker, and J.W. Gibbons. Effect of season on the retention of ^{137}Cs and ^{90}Sr by the yellow-bellied slider turtle (*Pseudemys scripta*). *Calif. J. Zool.* 64:2850–2853 (1986).

Simkiss, K. Lipid solubility of heavy metals in saline solutions. *J. Mar. Biol. Assoc. U.K.* 63:1–7 (1983).

Skrable, K.W., G.E. Chabot, C.S. French, M.E. Wrenn, J. Lipsztein, and T. Lo Sasso. Blood-organ transfer kinetics. *Health Phys.* 39:193–209 (1980).

Smock, L.A. Relationships between metal concentrations and organism size in aquatic insects. *Freshwater Biol.* 13:313–321 (1983).

Somero, G.N., and J.J. Childress. A violation of the metabolism-size scaling paradigm: Activities of glycolytic enzymes in muscle increase in larger-size fish. *Physiol. Zool.* 53(3):322–337 (1980).

Spacie, A., and J.L. Hamelink. Bioaccumulation. In *Fundamentals of aquatic toxicology*, ed. G.M. Rand and S.R. Petrocelli. New York: Hemisphere Publishing Corp., 1985, pp. 495–525.

Sposito, G. *The surface chemistry of soils.* New York: Oxford University Press, 1984.

Sprugel, D.G. Correction for bias in log-transformed allometric equations. *Ecology* 64(1):209–210 (1983).

Stary, J., and K. Kratzer. The cumulation of toxic metals on alga. *J. Environ. Anal. Chem.* 12:65–71 (1982).

Steele, G., P. Stehr-Green, and E. Welty. Estimates of the biologic half-life of polychlorinated biphenyls in human serum. *N. Engl. J. Med.* 314:926–927 (1986).

Strong, C.R., and S.N. Luoma. Variations in the correlation of body size with concentrations of Cu and Ag in the bivalve *Macoma balthica*. *Can. J. Fish. Aquat. Sci.* 38:1059–1064 (1981).

Stumm, W., and J.J. Morgan. *Aquatic chemistry. An introduction emphasizing chemical equilibria in natural waters.* New York: Wiley-Interscience, 1970.

Thomann, R.V. Equilibrium model of fate of microcontaminants in diverse aquatic food chains. *Can. J. Fish. Aquat. Sci.* 38:280–296 (1981).

Unlu, S.Y., and S.W. Fowler. Factors affecting the flux of arsenic through the mussel *Mytilus edulis*. *Mar. Biol.* 51:209–219 (1979).

Van Der Putte, I., and P. Part. Oxygen and chromium transfer in perfused gills of rainbow trout (*Salmo gairdneri*) exposed to hexavalent chromium at two different pH levels. *Aquat. Toxicol.* 2:31–45 (1982).

Van Liew, H.D. Semilogarithmic plots of data which reflect a continuum of exponential processes. *Science* 138:682–683 (1962).

Veith, G.D., D.L. DeFoe, and B.V. Bergstedt. Measuring and estimating the bioconcentration factor of chemicals in fish. *J. Fish. Res. Board Can.* 36:1040–1048 (1979).

Wagner, J.G. *Fundamentals of clinical pharmacokinetics*. Hamilton, IL: Drug Intelligence Publications, 1979.

Waiwood, B.A., V. Zitko, K. Haya, L.E. Burridge, and D.W. McLeese. Uptake and excretion of zinc by several tissues of the lobster (*Homarus americanus*). *Environ. Toxicol. Chem.* 6:27–32 (1987).

Walker, C.H. Species differences in microsomal monooxygenase activity and their relationship to biological half-lives. *Drug Metabol. Rev.* 7(2):295–323 (1978).

Walker, C.H. Kinetic models for predicting bioaccumulation of pollutants in ecosystems. *Environ. Pollut.* 44:227–240 (1987).

Wang, H.-K., and J.M. Wood. Bioaccumulation of nickel by algae. *Environ. Sci. Technol.* 18(2):106–109 (1984).

Wang, Y., and Q. Liu. Comparison of Akaike information criterion (AIC) and Bayesian information criterion (BIC) in selection of stock-recruitment relationships. *Fish. Res.* 77:220–225 (2006).

Watkins, B., and K. Simkiss. The effect of oscillating temperatures on the metal ion metabolism of *Mytilus edulis*. *J. Mar. Biol. Ass. U.K.* 68:93–100 (1988).

Whicker, F.W., and V. Schultz. *Radioecology: Nuclear energy and the environment.* Vol. II. Boca Raton, FL: CRC Press, 1982.

Williamson, P. Variables affecting body burdens of lead, zinc and cadmium in a roadside population of the snail *Cepaea hortensis* Muller. *Oecologia (Berl.)* 44:213–220 (1980).

Willis, J.N., and N.Y. Jones. The use of uniform labeling with zinc-65 to measure stable zinc turnover in the mosquitofish, *Gambusia affinis*. I. Retention. *Health Phys.* 32:381–387 (1977).

Yamaoka, K., T. Nakagawa, and T. Uno. Application of Akaike's information criterian (AIC) in the evaluation of linear pharmacokinetic equations. *J. Pharmacokinet. Biopharm.* 6:165–175 (1978).

Yang, R., V. Thurston, J. Neuman, and D.J. Randall. A physiological model to predict xenobiotic concentration in fish. *Aquat. Toxicol.* 48:109–117 (2000).

Zingaro, R.A. Biochemistry of arsenic: Recent developments. In *Arsenic: Industrial, biomedical, environmental perspectives*, ed. W.H. Lederer and R.J. Fensterheim. New York: Van Nostrand Reinhold Company, 1983, pp. 327–347.

CHAPTER 4

Lethal and Other Quantal Responses to Stress

Prior to the Renaissance period and extending well into that period, ... the art of poisoning [was brought to] its zenith.... The record of the city councils of Florence and particularly the infamous Council of Ten of Venice contain ample testimony of the political use of poisons. Victims were named, prices set, contracts recorded, and, when the deed was accomplished, payment made. The notation "factum", often appeared after the accomplishment of its transaction.

—Casarett (1975)

4.1 GENERAL

The U.S. Toxic Substances Control Act (TSCA), Resource Conservation and Recovery Act (RCRA), Federal Insecticide, Fungicide and Rodenticide Act (FIFRA), Water Pollution Control Act (WPCA), and National Environmental Policy Act (NEPA) all require that toxicity be addressed in standard ways for substances that may find their way into the environment. Unlike decisions based on mammalian toxicology, a discipline in which much basic research was done before enactment of such laws, decisions associated with ecotoxicology rely heavily on knowledge developed primarily within a regulatory framework (Dagani 1980). This has the advantage of focusing effort in the most efficient way possible to address very real regulatory needs. However, it also imposes a structure to the associated knowledge base that is decidedly not scientific, i.e., not organized on the basis of the most plausible explanatory principles as determined by the scientific methods (see Chapter 1). Also, there is an inherent delay in incorporating new information because the regulatory framework is built on the best information available at the time of enactment. Honesty requires that one also acknowledge the resistance from vested government and private institutions during most attempts at science-based change. Given the immediacy of our needs during the last several decades and the nature of institutions as they attempt to maintain their vested positions in a process, this regulatory context and behavior is understandable. However, for the sustained growth of sound knowledge in ecotoxicology, a framework based on scientifically sound explanatory principles must emerge in a timely manner as the dominant one. This chapter discusses the area of ecotoxicology with the strongest regulatory influence and, consequently, most in need of periodic and very careful reexamination.

Some toxicological consequences, such as the macabre one quoted above, are unambiguously "factum." Others are more difficult to demonstrate and, consequently, are prone to divergent interpretations. It is the intent of this chapter to outline and illustrate the use of candidate methods for quantifying lethal effects. Both well-established and alternate approaches will be discussed without regard to their acceptability for regulatory activities. Finally, the relative progress of regulatory and scientific ecotoxicology as applied to methods of quantifying lethal stress will be contrasted briefly.

The axiom from Bacon's *Novum Organum* (1620) that "truth is rightly named the daughter of time, not of authority," seems the most reliable guide to ascertaining which approach is, in fact, best.

4.2 DOSE-RESPONSE AT A SET ENDPOINT

4.2.1 General

Toxic response is a function of both the intensity (concentration or dose) and duration of exposure. Some techniques for measuring dose-response relationships control the duration of exposure and vary dose. Such techniques are defined herein as those for which a response is assessed at a final set time (endpoint). With many toxicity tests, the response is a quantal response, e.g., dead or alive, unresponsive to prodding or responsive, immobile or mobile. Operationally defined responses such as the last two are used if death is difficult to score. Often the proportion (P) of the total number of individuals that are dead after a specific exposure duration is used to summarize response in endpoint-based toxicity tests.

4.2.2 Metameters of Dose and Response

The frequency distribution of individual tolerances within an exposed population is often thought to be skewed. Consequently, mathematical transformations of the original measurements are commonly made to normalize the distribution of tolerances (Finney 1971) quantified in a variety of ways. For example, insect populations with subsets of individuals displaying extreme tolerances to a pesticide would be studied using logarithms of the exposure doses, instead of the arithmetic doses.

Measurements or measurement transformations used in the analysis of biological tests are referred to as metameters in the classic literature (Bacharach et al. 1942). The logarithm of the exposure dose or concentration is the most common dose or concentration metameter (Gaddum 1953). A variety of effect metameters (metameters for toxicant effect) are used in endpoint techniques. Probit, logit, and arcsine of the square root transformations of P are some of the most common. They are used to make linear the characteristically sigmoid cumulative response curves (Figure 4.1A) resulting from plotting the proportion of exposed individuals responding (P) versus log dose or log concentration.

The probit of P versus log dose (or log concentration) is the most common pair of transformations. The probit was introduced by Bliss (1935) using the concept of individual tolerance (measured as the smallest dose needed to kill a particular individual). When a particular dose is administered to many individuals in a population, those individuals with tolerances less than or equal to that dose will die. If groups of individuals from a population are exposed to a range of doses, a skewed frequency distribution of the proportion succumbing appears. Bliss (1935) suggested that logarithms of lethal doses be used to produce a normal distribution from such data. To obtain a straight line relationship, the normal equivalent deviations or N.E.D. (the proportion dying at each dose (P), expressed in terms of standard deviations from the mean of a normal distribution)* could be plotted against the logarithm of dose. The resulting curve is a linear form of the cumulative normal distribution for mortality. Bliss viewed the negative values of N.E.D. values below the mean as inconvenient. Consequently, he added 5 to the N.E.D. to make the occurrence of negative values rare. The effect metameter resulting from the addition of 5 to the N.E.D. is the probit or probability unit. It

* The N.E.D. is a proportion expressed in terms of the inverse of the unit normal cumulative distribution (normal curve with a mean = 0 and standard deviation = 1). The Excel functions, NORMINV(P,0,1) or NORMSINV(P), will generate it (N.E.D. or z) for proportion, P.

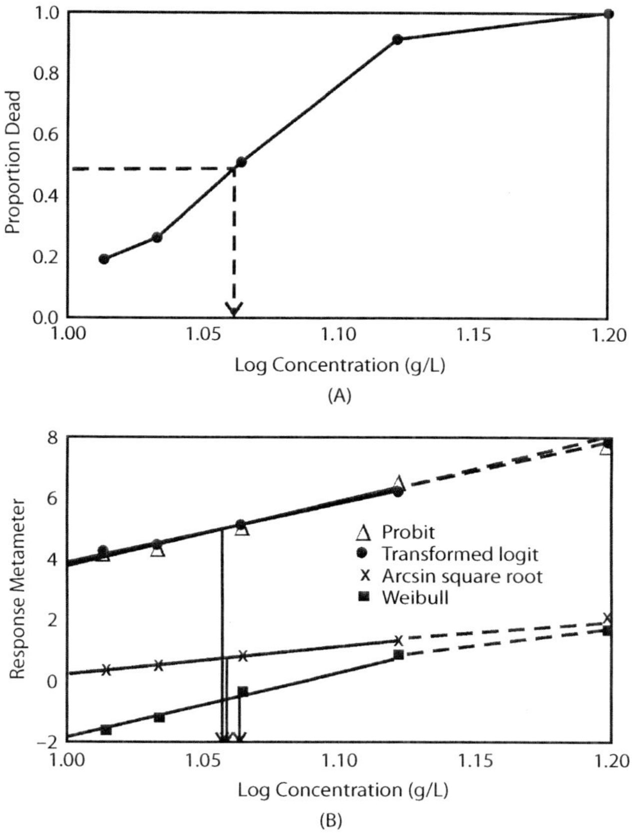

Figure 4.1 Sodium chloride toxicity to mosquitofish (*Gambusia holbrooki*). Panel A shows the sigmoidal plot of proportion dead versus log exposure concentration. Various linearizing metameters are used for these same data in panel B. Response metameters for the probit, transformed logit, arcsin square root, and Weibull functions were extracted from Appendix 5 of this book.

must be mentioned that the advantage of adding 5 to the N.E.D. is now dubious (see Finney 1978, p. 358). Indeed many software packages that execute probit analysis, such as SAS, use the N.E.D. without adding the superfluous 5. Only the pervasive use of probits instead of N.E.D. (i.e., z scores for estimated P values) transforms in toxicology argues for continued use for the sake of uniformity.

$$Probit = N.E.D. + 5 \tag{4.1}$$

Regardless, use of the probit or N.E.D. carries with it the assumption of a normal model (Equation 4.2) for the log dose-response relationship.

Normal cumulative distribution function:

$$F(x) = \frac{1}{\sqrt{2\pi\sigma}} \int_{-\infty}^{x} e^{-(x-\mu)^2/2\sigma^2} dx \tag{4.2}$$

where μ and σ = the mean and standard deviation, and $F(x)$ = theoretical proportion corresponding with a concentration metameter (P is the empirical proportion).

The probit or the N.E.D. can be obtained from tables such as Appendix 5 of this book, a table of probabilities for the normal curve, or by using special functions found in many statistical computer packages. The Excel® functions for producing these metameters are provided at the end of Appendix 5.

Berkson (1951) advocated the use of logits ("log odds units") instead of probits. Use of the logit carries with it the assumption that a logistic model, not a normal model, describes the log dose-response curve (Gaddum 1953). Berkson (1951) argued in favor of the logit metameter instead of the probit by pointing out that the probit transformation had no theoretical advantage over the logit and, unlike the logit, requires a statistical table. Today, the requirement of a table is irrelevant and the theoretical underpinnings for use of the logistic model remain no clearer than those for the normal model. Indeed, the lack of any theoretical superiority of either of these two transformations was common knowledge soon after their introduction (Gaddum 1953). Although the logistic model can be related to the Hill equation for enzyme kinetics (Christensen and Nyholm 1984) and the probit can be linked to adsorption models (Bliss 1935), they are used empirically to fit log dose-response data. Both metameters simply work, giving similar results (except at the tails of the curve) (Figure 4.1B).

$$Logit = Ln\left[\frac{P}{1-P}\right] \tag{4.3}$$

An alternate form of the logit is Equation (4.4) (see Finney 1978, p. 374, Equation 18.1.23). This transformation brings the logit values very close to those of the probit transformation except toward the tails of the log dose-response curve.

$$Transformed\ Logit = \frac{Logit}{2} + 5 \tag{4.4}$$

Both the logit and transformed logit can be found in tables such as Appendix 5 of this book. The third common response metameter is called by various names, such as the arcsine, angle sigmoid, or angular transformation (Armitage and Allen 1950).

$$Arcsine\ transform = arcsin\sqrt{P} \tag{4.5}$$

Note that the arcsine is expressed in radians, not degrees, and may also be denoted as $\sin^{-1}$. The inverse sine of x is the value of y that satisfies $x = \sin y$.

Gaddum (1953) and others suggested that the arcsine metameter not be used for proportions outside of the range of 0.20 to 0.80. Use of this metameter has the advantage of often making the variance for the metameter constant, and therefore any weight factor used in fitting will not vary with the value of P (Gaddum 1953; Finney 1978). Finney (1978) suggested that this metameter is less appealing than the probit or logit now because of the easy accessibility of computers. No theoretical model has been forwarded for its use in modeling the dose-response relationships.

Other less commonly used transformations are available. For example, Finney (1978) lists the Wilson-Worcester, Cauchy-Urban, and linear (or rectangular) sigmoid models. A Weibull model has been suggested for this purpose (Christensen 1984; Christensen and Nyholm 1984). The Weibull model (Equation 4.7) is one of several flexible generalizations of the exponential model (Equation 4.6). It may have positive or negative skewness depending on the shape parameter (λ) (Miller 1981).

Exponential cumulative distribution function:

$$F(x) = 1 - e^{-kx} \tag{4.6}$$

where k is greater than 0.

Weibull cumulative distribution function:

$$F(x) = 1 - e^{-(\propto x)^{\lambda}} \tag{4.7}$$

where α and λ are greater than 0.

The α and λ are scale and shape parameters, respectively. The Weibull model reduces to an exponential model when the shape parameter is 1.

Although used to fit log dose-response data empirically, the Weibull model can also be linked to models of carcinogen action (Christensen 1984). For example, when λ is an integer, the Weibull is analogous to a multiple-hit model. The Weibull transformation (U) defined by Equation (4.8) can be plotted against Ln of dose or concentration to yield a straight line.

$$U = Ln(-Ln(1-P)) \tag{4.8}$$

An example of such a plot is given in Figure 4.1B.

Example 4.1

What are the N.E.D., probit, logit, transformed logit, angular transform, and Weibull metameter values for 50, 16, and 84% mortalities?

All values can be extracted from Appendix 5. Alternatively, they may be generated with special functions and code such as the SAS code, NED = PROBIT(P), Prbt = PROBIT(P) + 5, Logit = LOG(P/(1-P)), Tlogit = Logit/2 + 5, Angular = ARCSIN(SQRT(P)), and Weibull = LOG(-LOG(1-P)), respectively. The Excel functions can also be used: NED=NORMINV(P,0,1), Prbt= NORMINV(P,0,1)+5, Logit=LOG(P/(1-P)), Tlogit=LOG(P/(1-P))+5, Angular=ASIN(SQRT(P)), and Weibull=LN(-LN(1-P)).

Proportion	N.E.D.	Probit	Logit	T. Logit	Angular	Weibull
0.50	0.000	5.000	0.000	5.000	0.785	−0.367
0.16	−0.994	4.005	−1.658	4.171	0.412	−1.747
0.84	0.994	5.994	1.658	5.829	1.159	0.606

Note that the angular transform is expressed in radians here. If one has the transform in degrees, it can easily be converted to radians. One radian is 180°/π or approximately 57.29578°. The angular transform of 0.50 is (0.7854 radians) (57.29578°/radian) or 45°.

4.2.3 The LC50/LD50

Trevan (1927) was the first to suggest the use of the LD50,* the lethal dose predicted to kill 50% of the exposed organisms, as a statistically reliable estimate of toxic effect instead of the more variable lethal threshold dose (Gaddum 1953). For animals exposed to concentrations in various media

* Note that the original notation was LD50, not LD_{50}, as is now commonly used. The original notation is used herein.

rather than doses, an LC50 (lethal concentration predicted to kill 50% of the exposed organisms) is estimated. For nonlethal or ambiguously lethal endpoints such as unresponsiveness to prodding or immobility of an invertebrate, the calculated concentration affecting 50% of the exposed organisms is often called the effective concentration or EC50. Graphical estimation of the LD50, LC50, or EC50 can be achieved using any of the response metameters (Appendix 5) described above to yield similar answers. For example, the values of probit (5.00000), transformed logit (5.00000), arcsine of the square root (0.78540), and Weibull (–0.36651) for $P = 0.50$ in Figure 4.1B correspond to concentration metameters of approximately 1.058, 1.057, 1.059, and 1.064, respectively. In turn, the antilogarithms of these concentration metameters correspond to 96 h LC50 values of approximately 11.43, 11.40, 11.46, and 11.59 g NaCl/L.

4.2.3.1 Litchfield–Wilcoxon Method

The above graphical method is incomplete because it fails to give a confidence interval for the estimated LC50 or EC50. Litchfield and Wilcoxon (1949) outline a straightforward, semigraphical procedure for estimating the LC50 and its 95% confidence interval. The following steps summarize their approach.

1. On log-probability paper (or plotting the log concentration and probit of the proportion responding), produce a preliminary plot of the data (P values versus the corresponding exposure concentrations). Because 0 and 100% mortality data cannot be plotted in this manner, omit such data from the plot. Draw a preliminary line through the points, "particularly those in the region of 40 to 60 per cent effect." Using this preliminary line, cull away points that fall below a predicted value of 0.01% or above a predicted value of 99.9%. Now draw a final line through the remaining data points.
2. Using the final line, extract the concentrations corresponding to the 16, 50, and 84% mortality (LC16, LC50, and LC84).
3. Calculate a slope function (S):

$$S = \frac{\frac{LC84}{LC50} + \frac{LC50}{LC16}}{2} \tag{4.9}$$

4. Determine the total number of individuals (N') tested between the 16 and 84% responses predicted by the line. Then calculate the 95% confidence interval (CI) as follows.

$$f_{LC50} = S^{2.77/\sqrt{N'}} \tag{4.10}$$

Upper limit of 95% CI:

$$Upper\ Limit = (LC50)(f_{LC50}) \tag{4.11}$$

Lower limit of 95% CI:

$$Lower\ Limit = LC50 / (f_{LC50}) \tag{4.12}$$

Example 4.2

Newman and Aplin (1992) exposed mosquitofish (*Gambusia holbrooki*) to a series of NaCl concentrations for 96 h. The resulting data are tabulated below. What are the estimates of LC50 and its 95% CI?

[NaCl] (g/L)	Log of [NaCl]	Numbers of Dead/Total	Proportion Dead	Probit
20.1	1.303	77/77	1.00	>7.576
15.8	**1.199**	**78/78**	**1.00**	**>7.576**
13.2	**1.121**	**69/76**	**0.91**	**6.341**
11.6	**1.064**	**40/77**	**0.52**	**5.050**
10.8	**1.033**	**22/79**	**0.28**	**4.417**
10.3	**1.013**	**16/76**	**0.21**	**4.194**
0.016	−1.796	0/79	0.00	<2.424

Optimum scaling of these or any data is difficult using log-probability paper. Such paper is becoming rare and difficult to find due to the ease of computer-based graphing. Regardless of the selection made for these data, the points will tend to be compressed into one small portion of the paper if the toxicant concentration range is narrow. Further, it is difficult to accurately place points between lines on such paper as the eye tends to place points with an arithmetic bias. As use of such paper is analogous to the probit approach discussed above and arithmetic paper can be easily used to properly scale the probit-log concentration data, these data will be graphed as the probit and log concentration metameters instead.

Figure 4.2 depicts these data. Only the boldface values in the table were plotted. The LC16, LC50, and LC84 can be extracted from this graph by taking the antilogarithm of the log concentration corresponding to the appropriate probit values for these percentages. Extract the probit values from Appendix 5. They are 4.00554, 5.0000, and 5.99446, respectively. (If you review the derivation of the probit, you might be confused by minor inaccuracies here. The probit values for ±1 standard deviation should be 4.00000 and 6.00000, not 4.00554 and 5.99446. These slight inaccuracies arise because 16 and 84% are the nearest whole percentage estimates of ±1 standard deviation from the mean.)

$$\begin{aligned} LC16 &= 10^{1.009} \text{ or } 10.21 \\ LC50 &= 10^{1.058} \text{ or } 11.43 \\ LC84 &= 10^{1.107} \text{ or } 12.79 \end{aligned}$$

$$\begin{aligned} S &= (LC84/LC50 + LC50/LC16)/2 \\ &= (12.79/11.43 + 11.43/10.21)/2 \\ &= 1.12 \end{aligned}$$

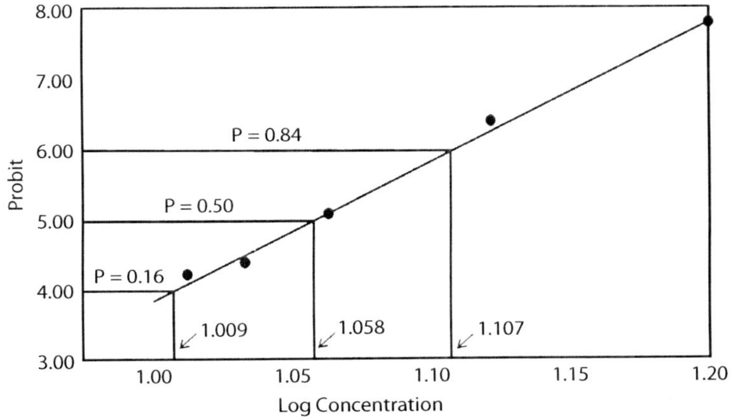

Figure 4.2 Sodium chloride toxicity data for mosquitofish using the probit-log concentration transformations. The probit values for the 0.16, 0.50, and 0.84 mortality proportions are used to estimate the LC50 and its 95% confidence interval (see Example 4.2, Litchfield–Wilcoxon method).

$$N' = 76 + 79 + 77 = 232$$

$$\begin{aligned} f_{LC50} &= S^{2.77/\sqrt{N'}} \\ &= 1.12^{2.77/\sqrt{232}} \\ &= 1.12^{0.182} \\ &= 1.021 \end{aligned}$$

95% CI for the LC50:

$$\begin{aligned} \text{LC50} \cdot f_{LC50} &= 11.43 \cdot 1.021 = 11.67 \\ \text{LC50}/f_{LC50} &= 11.43/1.021 = 11.19 \end{aligned}$$

4.2.3.2 Maximum Likelihood Method (Normal, Logistic, and Weibull Models)

The semigraphic Litchfield–Wilcoxon approach will yield slightly different answers depending on the person executing the procedures. A more formal approach to the estimation of LC50 is desirable.

Likely the first approach to come to the reader's mind is least-squares linear regression. This method fits data to a model by minimizing the sum of the squares of the deviations between the observed and predicted values of Y. Indeed, such methods, including weighted regression methods, can be used under certain conditions to fit this type of data. Then values can be predicted within the range of the dose metameter values used in the regression. Examples of such methods for unweighted and weighted regression are provided in Gad and Weil (1988, pp. 94–98) and Christensen and Nyholm (1984). However, such an approach cannot easily include 0 and 100% responses if they occur. To get around this problem, Berkson's substitutions of $1/2n$ and $1–1/2n$ could be used (Finney 1971) (n = the number of individuals in the exposure treatment associated with the P values of 0.00 or 1.00). Bliss (1935) provided a more involved means of coping with 0 and 100% mortality treatments. However, this "fudging" (Stephan 1977) can compromise results. Armitage and Allen (1950) describe an alternative, minimum χ^2 method for estimating LD50.

A maximum likelihood method can be used to estimate the LC50 instead. Neter et al. (1990) explain this method using a hypothetical probability function with one parameter, θ. If the population with the probability function $f(Y,\theta)$ is sampled, the joint probability function for the observations Y_1 to Y_n is (Equation 1.52a in Neter et al. 1990)

$$g(Y_1 \ldots Y_n) = \prod_{i=1}^{n} f(Y_i, \theta) \tag{4.13}$$

where θ = a single parameter of the example function. (The Π is similar to the Σ operation except that the product instead of the sum is taken for $f(Yi,\theta)$.)

The likelihood function for the observations is the following (Equation 1.52b in Neter et al. 1990).

$$L(\theta) = \prod_{i=1}^{n} f(y_i, \theta) \tag{4.14}$$

The maximum likelihood method maximizes the likelihood function to yield an estimate of θ. The maximum likelihood equations here have no explicit solutions, and consequently, solutions are reached by iteration. The equations are solved repeatedly using the estimates from the previous iteration as the initial values in the equations. The process is repeated until estimates from consecutive iterations are sufficiently similar to be acceptable for the intended use.

Estimates from maximum likelihood methods are biased, but according to Finney (1971), the bias in the estimation of LC50 is usually small, except if very few observations are taken. Also, the bias is often no worse than those of alternate techniques (Finney 1971), and the maximum likelihood procedure has very good precision relative to other estimation techniques. If there are less than two partial kills, the maximum likelihood methods cannot be used without adjustment of the $P = 0.00$ or $P = 1.00$ data such as described above for Berkson's substitutions. In summary, the maximum likelihood method produces a relatively precise but slightly biased estimate of the LC50. The iterative nature of the method means that calculations are normally done with a computer.

The normal (probit), logistic (logit), Weibull (Weibit), or other models can be fit using maximum likelihood methods. Weighted or unweighted maximum likelihood methods can be used (Gaddum 1953; Christensen and Nyholm 1984). Consequently, the question now becomes "Which model fits these data best using the maximum likelihood methods?" Goodness-of-fit estimates can be used to measure how closely a model fits the data set. The χ^2 statistics for the various models can be compared to select between the various models.

Example 4.3

Analysis of Newman and Aplin's (1992) data in Example 4.2 is extended to maximum likelihood fitting of normal, logistic, and Gompertz models. The fitting is done with the SAS PROC PROBIT (SAS 1988). Fitting the Gompertz model is similar in some ways to fitting a Weibull model because it is also a generalized exponential model. As outlined by Dixon (1992), the Gompertz model is the following:

$$P = 1 - e^{-e^{x'}b} \tag{4.15}$$

where P = effect metameter, x' = metameter for concentration, and b = an estimated parameter.

The Gompertz model as fit by the SAS PROC PROBIT is expressed as the following:

$$P = 1 - e^{-e^{a+bx}} \tag{4.16}$$

where a and b are maximum likelihood parameter estimates. Or

$$P = 1 - e^{-(e^{a}(e^{x})^{b})} \tag{4.17}$$

or, expressed as the Weibull model (see Equation 4.7),

$$P = 1 - e^{-\alpha Z^{\lambda}} \tag{4.18}$$

where $\alpha = e^{a}$, $\lambda = b$, and $Z = e^{x}$.

The SAS code for fitting these three models follows:

```
DATA SALT;
   INFILE CARDS EOF=EOF;
   INPUT DEAD TOTAL CONC @@;
   PROP = DEAD/TOTAL;
   /* PRODUCING CONCENTRATIONS TO BE PLOTTED        */
   /* LATER AGAINST PREDICTED VALUES.               */
   OUTPUT;
```

```
RETURN;
EOF: DO CONC = 10 TO 21 BY 0.50;
   OUTPUT;
   END;
   DATALINES;
16 76 10.30 22 79 10.80 40 77 11.60 69 76 13.20
78 78 15.80 77 77 20.10
;
/* PROBIT MODEL OF NACL-MOSQUITOFISH DATA         */
PROC PROBIT LOG10 INVERSECL LACKFIT;
   MODEL DEAD/TOTAL = CONC/D=NORMAL ITPRINT;
   OUTPUT OUT=P P=PROB STD=STD XBETA=XBETA;
RUN;
PROC PLOT;
   PLOT PROP*CONC="X" PROB*CONC="*"/OVERLAY;
TITLE 'PROBIT - OBSERVED AND PREDICTED VALUES';
RUN;
/* LOGIT MODEL OF NACL-MOSQUITOFISH DATA          */
PROC PROBIT LOG10 INVERSECL LACKFIT;
   MODEL DEAD/TOTAL = CONC/D=LOGISTIC ITPRINT;
   OUTPUT OUT=L P=LPROB STD=LSTD XBETA=LXBETA;
RUN;<*Example H1678Y3005P End Here*>
PROC PLOT;
   PLOT PROP*CONC="X" LPROB*CONC="*"/OVERLAY;
   TITLE 'LOGIT - OBSERVED AND PREDICTED VALUES';
RUN;
/* GOMPERTZ MODEL OF NACL-MOSQUITOFISH DATA       */
/* NOTE THAT THE NATURAL LOGARITHM IS USED IN     */
/* THIS MODEL TO MAKE IT EASY TO RELATE TO THE    */
/* WEIBULL MODEL.                                 */
PROC PROBIT LOG INVERSECL LACKFIT;
   MODEL DEAD/TOTAL = CONC/D=GOMPERTZ ITPRINT;
   OUTPUT OUT=G P=GPROB STD=GSTD XBETA=GXBETA;
RUN;
PROC PLOT;
   PLOT PROP*CONC="X" GPROB*CONC="*"/OVERLAY;
   TITLE 'GOMPERTZ - OBSERVED AND PREDICTED VALUES';
RUN;
```

THE NORMAL MODEL

The procedure converged within seven iterations. The Pearson goodness-of-fit χ^2 = 1.4112 (df = 4) with a probability = 0.8422 as generated with Excel CHIDIST(1.4112,4). (The Pearson χ^2 is used here. Explanations of this and other pertinent χ^2 statistics can be obtained from various sources, including Christensen and Nyholm 1984, SAS 1988, Berkson 1955, and Salsburg 1986, pp. 60–61.) The PROBIT procedure generated the following estimates of the LC50 and its 95% CI.

Estimated LC50 = 11.44 g NaCl/L
95% CI = 11.24 – 11.66 g NaCl/L

Buikema et al. (1982) and later Eaton and Gilbert (2008) urged that, in reporting acute toxicity data, the slope with its limits should be reported as well as the LC50, its 95% confidence limits, and the associated χ^2 statistic. The slope and the LC50 are required to fully describe the toxicity model. For example, two intersecting dose-response curves may have a common LC50 but very different slopes. Considering only the LC50 would not reveal the significant differences associated with these two curves. Although not provided in this summary of results, the slopes for these models are given in the analysis output.

The confidence limits presented here would be treated as fiducial limits. (Fiducial limits are "a statement that there is a 19 of 20 chance that the LC50 value falls within the specified limits ($P = 0.05$)") (Buikema et al. 1982). However, Fisher's fiducial limits concept was never generally adopted in the statistical community. Confidence intervals are not interpreted as stated above for fiducial limits. Finney (1971) and Buikema et al. (1982) provide further explanation of the conceptual difference between fiducial and confidence limits.

THE LOGISTIC MODEL

The procedure converged within seven iterations. The Pearson goodness-of-fit $\chi^2 = 2.0581$ (df = 4) with a probability = 0.7251.

Estimated LC50 = 11.43 g NaCl/L
95% CI = 11.23 – 11.64 g NaCl/L

THE GOMPERTZ MODEL

The methods converged within five iterations. The Pearson goodness-of-fit χ^2 (df = 4) = 0.2635 with a probability = 0.9920.

Estimated LC50 = 11.58 g NaCl/L
95% CI = 11.35 – 11.80 g NaCl/L

MODEL COMPARISON USING χ^2 RATIOS

A χ^2 value was estimated for each model. It is used to compare the expected proportions of dead test organisms as predicted by the model to the observed proportions dead. The larger the χ^2 value, the more extreme the failure of the method to model the tolerance distribution (Bliss 1935; Finney 1978). The χ^2-associated probability calculated for the three models showed no significant heterogeneity ($\alpha = 0.05$): there is no evidence for rejecting any of the three models.

However, the statistical significance of the χ^2 test is of little utility. The ratio of χ^2 values for the three models might be used to assess their relative goodness-of-fits (Christensen 1984). The χ^2 will decrease as the goodness-of-fit increases. Consequently, a ratio of χ^2 valued less than 1 indicates that the model associated with the χ^2 in the numerator fits the data set better than the model associated with the χ^2 in the denominator.

$$\chi^2_{Gompertz}/\chi^2_{probit} = 0.2635/1.4112 = 0.1867$$
$$\chi^2_{Gompertz}/\chi^2_{logit} = 0.2635/2.0581 = 0.1280$$
$$\chi^2_{probit}/\chi^2_{logit} = 1.4112/2.0581 = 0.6857$$

The ratios with the Gompertz χ^2 in the numerator were considerably less than 1, indicating that the Gompertz model is the best choice. Similarly, the probit model seemed better than the logit model. This conclusion can be confirmed by plotting the data against the predicted values for each from the models.

Pertinent to the Gompertz results, Christensen (1984) and Christensen and Nyholm (1984) explained the differences in results of using the Weibull model instead of the normal model to fit log dose-response data. Relative to the normal model, the Weibull model predictions at the concentration extremes bow upward from a straight line. This implied to Christensen (1984) that most organisms die above a certain concentration threshold. Generally, the Weibull model predicts higher mortality at lower concentrations than the probit model.

4.2.3.3 Trimmed Spearman–Karber Method

The maximum likelihood methods require the assumption of specific models. Use of the probit model assumes a (log) normal distribution and use of the logit assumes a (log) logistic model. Both the lognormal and log logistic curves are symmetrical about the mean population tolerance. The use of the Weibull metameter also implies a specific model, but the associated curve is not symmetrical about the mean. Unfortunately, the ability to identify the appropriateness of any model is often quite limited.

Another problem that may be encountered with maximum likelihood methods is associated with the process of convergence over iterations (Hamilton et al. 1977). For some data sets, the methods can fail to reach acceptable convergence. Also, it is possible that they can converge on inappropriate solutions depending on the initial values used to begin the iterative process. For this reason, it is always prudent to plot the predictions from a fitted model along with the original data to check for the reasonableness of the model estimates.

A modified Spearman–Karber method can be used instead of these methods if a symmetrical distribution can be assumed. A specific model need not be assumed in this approach. The procedure is described by Hamilton et al. (1977) in the following steps.

1. Let x_1 to x_k be the natural logarithms of k exposure concentrations with $x_1 < \ldots < x_k$. Also let n_1 to nk, and r_1 to r_k be the total number of individuals exposed at each concentration and the number that died after exposure to each concentration, respectively. If more than one concentration treatment had no mortalities, then the logarithm of the highest concentration giving no mortalities is x_1. There should be no mortalities in any concentrations below the x_1. Similarly, if more than one concentration resulted in 100% mortality, then the natural logarithm of the lowest concentration producing 100% mortality is x_k. All concentrations above x_k must have 100% mortality.
2. Let p_i be equal to r_i/n_i, the proportion of the exposed individuals dying by the endpoint. To perform the trimming, the p_i values need to be adjusted if the following conditions are not present, $p_1 \leq p_2 \leq \ldots \leq p_k$. This must be done as the assumption is made that the proportions increase monotonically, i.e., the true proportions of exposed individuals dying (P_i) increase with exposure concentration. New p_i values are calculated for adjacent p_i values that do not fulfill this requirement. For two inconsistent, adjacent p_i values, their new p values would be $(r_i + r_{i+1})/(n_i + n_{i+1})$.

   For example (from Table III of Hamilton et al. 1977), five tanks of increasing toxicant concentrations have r_i/n_i values of 0/20, 1/20, 0/20, 3/19, and 20/20. There is an inconsistency, as the second lowest concentration had one death but the next highest concentration had no mortality. The p_i values for these two tanks are replaced by a value intermediate between the two. The new p_i for these two tanks would be $(r_2 + r_3)/(n_2 + n_3) = (0 + 1)/(20 + 20) = 0.025$. If the r_i/n_i values were 0/20, 2/20, 1/20, 1/20, 3/19, and 20/20, then the process would be the following.

   a. Original p_i values: 0.00, 0.10, 0.05, 0.05, 0.15, 1.00
   b. $(r_2 + r_3 + r_4)/(n_2 + n_3 + n_4) = 4/60 = 0.067$
   c. Adjusted p_i values: 0.00, 0.067, 0.067, 0.067, 0.15, 1.00

   The process is continued for all inconsistent pairs until the p_i values are monotonically nondecreasing with concentration.
3. Next, the Ln concentrations (x_i) are plotted against the final p_i values.
4. Now the cumulative distribution curve (step 3) must be trimmed. A percentage (α) of the lower and upper portions of the curve is trimmed away to leave a graph of the central $100 - 2\alpha$ portion of the curve. The p scale is replaced by $_{\alpha}p = (p - \alpha/100)/(1 - 2\alpha/100)$, and these data ($_{\alpha}p_j$ versus x_i) are replotted. Any $_{\alpha}p$ values above 1 or below 0 are not used when replotting. Instead, the respective Ln concentrations predicted by the new 0.0 and 1.0 values on the $_{\alpha}p$ scale are used as the points at either end of the new curve.
5. The mean of the cumulative frequency curve is calculated as follows. Assume that $k = 6$, but with trimming, only 4 points remain ($_{\alpha}p_2$, $_{\alpha}p_3$, $_{\alpha}p_4$, $_{\alpha}p_5$) after step 4. The corresponding Ln concentration values are x_2, x_3, x_4, and x_5. There are now five intervals to be considered. Three intervals exist between the four x_i values. Two more exist on each end of the curve between the predicted x_i corresponding to $p = 0$ and x_2, and between x_5 and the predicted x_i corresponding to $p = 1.0$. Let an interval of Ln concentration be represented generally by (x_{j-1}, x_j) and an interval of $_{\alpha}p$ be represented by $(_{\alpha}p_{j-1}, {_{\alpha}p_j})$. The mean of the distribution is then calculated.

$$Mean = \sum_{i=1}^{j} ({}_{\alpha}p_i - {}_{\alpha}p_{i-1})\left(\frac{x_{i-1} + x_i}{2}\right) \quad (4.19)$$

where ${}_{\alpha}p_j - {}_{\alpha}p_{j-1}$ = the proportion of the total area under the curve in the interval, (x_{j-1}, x_j), and $(x_{j-1} + x_j)/2$ = the midpoint of the interval, (x_{j-1}, x_j).

The α-trimmed Spearman–Karber estimate of the LC50 is the antilog of the mean calculated with Equation (4.19).

6. Obviously, an α must be selected from the range $0 \le \alpha < 50$. Estimation of an appropriate α can be done using the recommendations of Hamilton et al. (1977). The method is always accurate when $\alpha \ge 100p_1$ (p_1 is the adjusted p_i associated with the lowest x_i) and $\alpha \ge 100(1 - p_k)$ (p_k is the adjusted p_i associated with the highest x_i). The α could then be set at the maximum of $100p_1$ and $100(1 - p_k)$. As α increases, the sensitivity of the method to anomalous values decreases. This would argue for selection of a large α. However, an upper limit is placed on the possible values of α, as the standard error increases as the α increases. Hamilton et al. (1977) suggested that an α of 10% is adequate when the highest concentration results in 95% mortality or more and the lowest concentration results in 5% mortality or less.
7. The variance (and 95% confidence interval) for the trimmed Spearman–Karber estimate can be estimated as outlined in the appendix of Hamilton et al. (1977). Simulations by these authors suggested that the variance and 95% confidence interval estimates are conservative.

Let $A = \alpha/100$, L = maximum$\{i: p_i \le A\}$ or MAX$\{i: p_i \le A\}$, U = minimum$\{i: p_i \ge 1 - A\}$ or MIN$\{p_i \ge 1 - A\}$, x_L = the largest Ln concentration value with an adjusted $p \le A$, x_U = the smallest Ln concentration value with an adjusted $p \ge 1 - A$, n_L = the number of individuals exposed to X_L, and n_U = the number of individuals exposed to X_U.

Necessary, intermediate calculations using the above-defined variables follow.

$$V_1 = \left[\frac{(x_{L+1} - x_L)(p_{L+1} - A)^2}{(p_{L+1} - p_L)^2}\right]^2 \left[\frac{p_L(1 - p_L)}{n_L}\right] \quad (4.20)$$

$$V_2 = \left[(x_L - x_{L+2}) + \frac{(x_{L+1} - x_L)(A - p_L)^2}{(p_{L+1} - p_L)^2}\right]^2 \left[\frac{p_{L+1}(1 - p_{L+1})}{n_{L+1}}\right] \quad (4.21)$$

$$V_3 = \sum_{i=L+2}^{U-2} (x_{i-1} - x_{i+1})^2 \frac{p_i(1 - p_i)}{n_i} \quad (4.22)$$

$$V_4 = \left[(x_{U-2} - x_U) + \frac{(x_U - x_{U-1})(p_U - 1 + A)^2}{(p_U - p_{U-1})^2}\right]^2 \frac{p_{U-1}(1 - p_{U-1})}{n_{U-1}} \quad (4.23)$$

$$V_5 = \left[\frac{(x_U - x_{U-1})(1 - A - p_{U-1})^2}{(p_U - p_{U-1})^2}\right]^2 \left[\frac{p_U(1 - p_U)}{n_U}\right] \quad (4.24)$$

$$V_6 = \left[\left[\frac{(x_U - x_{L+1})(1 - A - p_U)^2}{(p_U - p_{L+1})^2}\right] - \left[\frac{(x_{L+1} - x_L)(A - p_L)^2}{(p_{L+1} - p_L)^2}\right] + (x_L - x_U)\right]^2 \frac{p_{L+1}(1 - p_{L+1})}{n_{L+1}} \quad (4.25)$$

The variance about the mean can be estimated from the above equations and the following equations for various values of $U - L$.

If $U - L \geq 4$ then

$$Variance = \frac{(V_1 + V_2 + V_3 + V_4 + V_5)}{(2 - 4A)^2} \tag{4.26}$$

If $U - L = 3$ then

$$Variance = \frac{(V_1 + V_2 + V_4 + V_5)}{(2 - 4A)^2} \tag{4.27}$$

If $U - L = 2$ then

$$Variance = \frac{(V_1 + V_5 + V_6)}{(2 - 4A)^2} \tag{4.28}$$

If $U - L = 1$ then

$$Variance = (x_U - x_L)^2 \left[\left[\frac{(0.5 - p_U)^2}{(p_U - p_L)^4} \right] p_L \frac{(1 - p_L)}{n_L} + \left[\frac{(0.5 - p_L)^2}{(p_U - p_L)^4} \right] p_U \frac{(1 - p_U)}{n_U} \right] \tag{4.29}$$

The 95% confidence interval is twice the square root of the variance estimated with Equation (4.26), (4.27), (4.28), or (4.29).

Example 4.4

Calculations for the trimmed Spearman–Karber method will be demonstrated with the NaCl-mosquitofish data used in Examples 4.2 and 4.3. According to step 1, these data (concentration, k, x_i, r_i, n_i) are tabulated below in the first five rows. No adjustments as described in step 2 above are required. The data for the highest concentration are omitted as the next concentration (15.8 g NaCl/L) also resulted in a 100% mortality response. These data (p_i, x_i) are plotted in Figure 4.3A.

Concentration	20.1	15.8	13.2	11.6	10.8	10.3	0.016
k	—	6	5	4	3	2	1
x_i	—	2.760	2.580	2.451	2.380	2.332	–4.135
r_i	77	78	69	40	22	16	0
n_i	77	78	76	77	79	76	79
p_i	—	1.00	0.91	0.52	0.28	0.21	0.00
${}_{\alpha}p$	—	—	—	0.53	0.13	0.02	—

An α is estimated using step 6 above. Since $p_1 = 0.00$, α should be greater than or equal to $100p_1$, i.e., 0%. Since $p_k = p_6 = 1.00$, α should be greater than or equal to $100(1 - p_k)$, i.e., 0%. Both estimates result in the trivial recommendation that trimming of this data set should be greater than or equal to 0%. Hamilton et al. (1977) recommended an $\alpha = 10\%$ in such a case (step 6 above). For the sake of illustration, let's use the next p values ($k = 2$ and $k = 5$) to estimate α. The α should be the largest of the two values, 100(1.00 – 0.91) or 9% and 100(0.21) or 21%. For convenience, an α near the larger of these values (21%) is selected. Twenty percent will be trimmed from each tail of the curve (Figure 4.3A).

Now ${}_{\alpha}p$ values $((p - \alpha/100)/(1 - 2\alpha/100))$ are calculated and entered in the table above. As outlined in step 4, any ${}_{\alpha}p$ values greater than 1 or less than 0 are not plotted. Therefore, the values of 1.18 and –0.33 for $i = 5$ and 1 are discarded. Figure 4.3B is the line resulting from plotting ${}_{\alpha}p$ values against Ln concentration. The predicted data points corresponding to $p = 0.20$ and $p = 0.80$ have been scaled to ${}_{\alpha}p$ values of 0.00 and 1.00, respectively. The mean of the

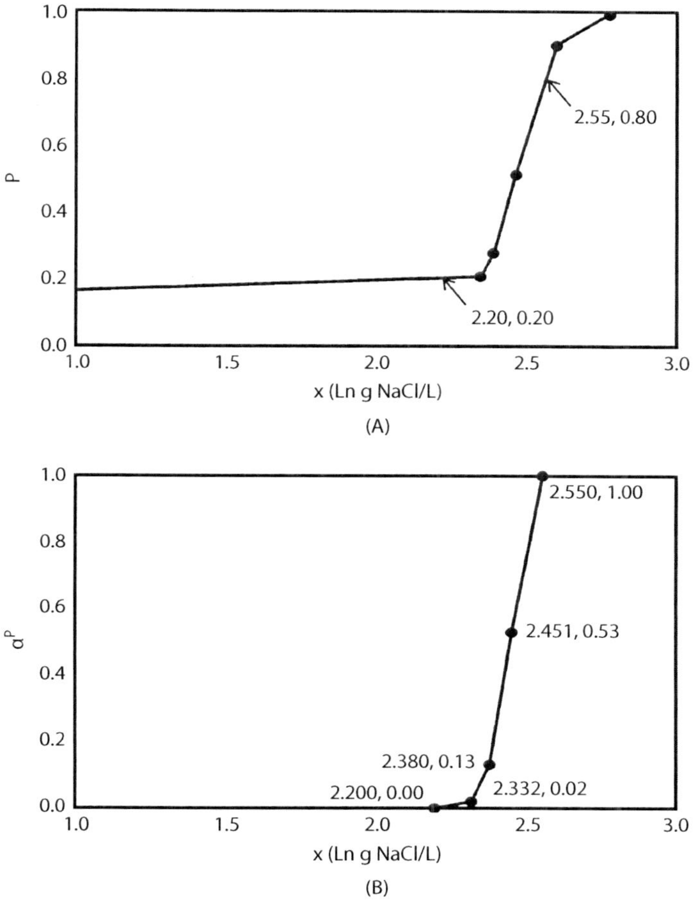

Figure 4.3 The trimmed Spearman–Karber method for estimating the LC50 and associated 95% confidence interval for the sodium chloride–mosquitofish mortality data. Panel A presents all pertinent data and the points to be trimmed. Panel B is the plot resulting from trimming and estimation of αP calculation. (See Example 4.4 for details.)

resulting distribution is estimated with Equation (4.19). The table below summarizes the associated calculations.

Interval $(j,j-1)$	$_{\alpha}p_j$	$_{\alpha}p_{j-1}$	x_j	x_{j-1}	Product
2,1	0.02	0.00	2.332	2.200	0.04
3,2	0.13	0.02	2.380	2.332	0.26
4,3	0.53	0.13	2.451	2.380	0.97
5,4	1.00	0.53	2.550	2.451	1.18
					Σ 2.45

The antilogarithm of 2.45 is the trimmed Spearman–Karber estimate of the LC50. That is, LC50 = $e^{2.45}$ = 11.58 g NaCl/L. This is close to estimates derived from the graphical (11.40 to 11.59 g NaCl/L) and maximum likelihood (11.43 to 11.58 g NaCl/L) methods in Examples 4.2 and 4.3.

The 95% confidence interval for the LC50 is estimated using Equations (4.20) to (4.24), and (4.26) as outlined in step 7.

$A = 0.20$ $U - L = 5 - 1 = 4$
$p_L = 0.00$ $p_U = 0.91$
$x_L = -4.135$ $X_U = 2.580$
$n_L = 79$ $n_U = 76$

With these data and those in the table above, V_1 to V_5 are calculated (Equations 4.20 to 4.24).

$V_1 = 0$ $V_2 = 0.000920118$
$V_3 = 0.000036137$ $V_4 = 0.000116697$
$V_5 = 0.000004764$

The variance is calculated using these estimates in Equation (4.26) as $U - L = 4$.

Variance = 0.00107716

The 95% confidence limits are then calculated.

95% confidence interval = 0.065640231
Lower limit = $e^{(2.45-0.065640231)}$ = 10.85 g NaCl/L
Upper limit = $e^{(2.45+0.065640231)}$ = 12.37 g NaCl/L

4.2.3.4 Binomial Method

Stephan (1977) argues that partial kills are not always necessary to estimate an LC50 and approximate 95% confidence intervals. He reasons that the LC50 must fall between the highest concentration giving 0% mortality (A) and the lowest concentration giving 100% mortality (B). Of course, this is strictly true only if sample error is assumed to be immaterial. The estimate of the LC50 is

$$LC50 = \sqrt{A \cdot B} \tag{4.30}$$

if A and B are expressed in concentration units. The interval between A and B can be used as the confidence interval for the LC50. This confidence interval is at least a 95% confidence interval if the number of individuals treated at a concentration interval (n) exceeds 5. For example, the coefficients for the confidence intervals are 93.8, 96.9, 98.4, and 99.2% for n = 5, 6, 7, and 8, respectively (Stephan 1977). If the same number of individuals were exposed at the A and B concentrations, the associated confidence coefficient is

$$Coefficient = 100[1 - 2(0.5)^n] \tag{4.31}$$

If the number of individuals exposed to concentration A (n_A) and the number exposed to concentration B (n_B) are different, Gelber et al. (1985) express the confidence coefficient for the interval between A and B in the form

$$Coefficient = 100[1 - 0.5^{n_A} - 0.5^{n_B}] \tag{4.32}$$

4.2.3.5 Moving Average Method

A simple moving average method can be used to estimate the LC50 (Gad and Weil 1988; Stephan 1977; Gelber et al. 1985) if the exposure concentrations are set in a geometric pattern, e.g., 2, 4, 8, 16 mg/L, and the same number of individuals are exposed to each concentration. Gelber et al. (1985) state that the restrictions on this method make it less appealing than the others mentioned previously. However, Stephan (1977) and Gad and Weil (1988) did not believe that the restrictions are so confining as to disfavor its use.

Gelber et al. (1985) presents the moving average method with the simple equation

$$\hat{p}_i = \sum_{j=i-\frac{s}{2}}^{i+\frac{s}{2}} w_j p_j \tag{4.33}$$

There are i exposure intervals with associated proportional kills of p_i. Because the concentrations are defined by a geometric series, all of the log concentrations have the same span (s) between them. Using the p_j and estimated weights (w_j), a moving rate of mortality is estimated with Equation (4.33).

The moving average method will mechanically produce results when there are fewer than two partial kills (Stephan 1977). Initially Stephan (1977) considered moving average estimates from data sets with one partial kill as statistically valid but later pointed out that such estimates should be considered approximate only (Stephan, personal communication, 1992). With no partial kills and the use of the logarithm of exposure concentrations, the moving average method produces the same approximation of the LC50 as Equation (4.30) (Stephan, personal communication, 1992).

Because of the restrictions imposed on this technique, no further discussion is given here. Instead the reader is referred to Gad and Weil (1988), who give a detailed explanation with an example and all necessary tables.

4.2.3.6 Up-and-Down (Sensitivity) Method

An alternative experimental design can be used to estimate LC50, or more often LD50, values from quantal data (Dixon 1965, 1991; Dixon and Massey 1969; Bruce 1985). Originally developed to test the mechanical force needed to detonate explosives (Dixon and Mood 1948), the sensitivity (also called up-and-down or staircase) design as applied to toxicity tests involves a sequence of exposures of one animal at a time. For example, a first animal is administered a dose of 55 mg/kg of body weight and survives a 48 h period of observation. A second animal is then given a higher dose and dies within 48 h. The next animal is given a dose between the first two doses, and so on.* An LD50 estimate and its 95% confidence limits are produced from such results (Figure 4.4).

The conditions are well established under which the up-and-down method for estimating a median effect dose (or concentration) is advantageous relative to the conventional approach. The reduced number of subjects used is especially advantageous if ethical, monetary, or animal husbandry issues are prominent, as is often the case with larger animals. Dixon and Mood (1948) commented about their original computational methods that a 30 to 40% reduction in the number of required test subjects was commonplace in sensitivity studies of explosives. The savings increased further with methodological improvements (e.g., Dixon 1965). Bruce (1985, 1987) reported that LD50 estimation with the up-and-down method required only 6 to 9 rats in contrast to the conventional test involving 40 to 50 rats. More recently, Lipnick et al. (1995) noted that the conventional test required roughly 30 rats but the up-and-down test required only 6 to 10 rats. Lipnick et al. (1995) documented clearly that both tests still produced similar results. The price of this advantage is the increased time needed to complete the series of exposures in the assay instead of a single simultaneous exposure of all individuals (Dixon 1991). Also, the planning and logistics

* In many protocols, a limit test is performed prior to the main test described in this section. For instance, the OECD (2006) limit test, intended to identify low-toxicity agents by dosing the fewest animals, administers a high dose to the first animal (i.e., 2,000 mg/kg). The main test is then carried out if that animal dies. If it survives, four additional animals are dosed in sequence. If three of these die, the main test is used to estimate the LD50. If three or more survive, the LD50 is reported as greater than 2,000 mg/kg. These OECD rules are somewhat different in the atypical situation in which 5,000 mg/kg is used as the initial dose of the limit test.

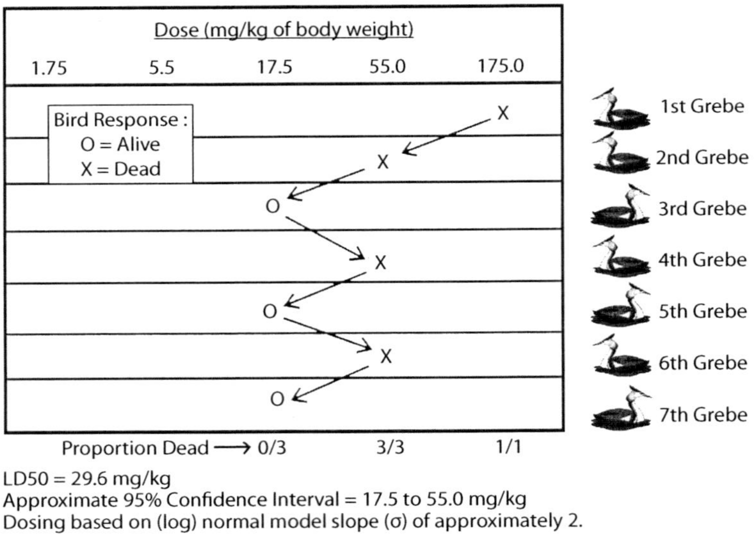

Figure 4.4 A depiction of the up-and-down method for the determination of the LD50. The first grebe was administered 175 mg/kg and died, resulting in a lower dosing of a second grebe. Because this second grebe also died, the dose given to the third grebe was even lower. With survival of the third grebe, the dose given to the fourth grebe was increased, resulting in death. The process was repeated until a stopping rule described in the text indicated that results for seven grebes were sufficient for estimation. The estimated proportions for the 17.5 and 55.0 results could be used in Equations (4.30) and (4.31) to produce a very rudimentary LD50 estimate of 31.0 with an associated confidence interval (17.5 to 55.0) that is at least a 75% confidence limit. The up-and-down method provides a better approach. Although illustrated here using a sequence of dosings of seven Western grebes, the data and illustration follow the OECD (2006) Table 4 example (assumed model slope of 2). The AOT425StatPgm shareware available from the EPA webpage http://www.epa.gov/oppfead1/harmonization was used to produce the LD50 and its associated 95% confidence interval.

become more complicated than for a conventional test, especially if the subjects are easily exposed together in large numbers. Such subjects would include small insects (Dixon and Mood 1948). Tests that estimate LC50 for small aquatic organisms exposed to carefully maintained concentrations in water are infrequently conducted under the up-and-down design for this reason. Another shortcoming with the up-and-down method is obvious from Figure 4.4: the method focuses on the median value so any need for accurate estimates away from the median becomes problematic (Rispin et al. 2002; Organization for Economic Cooperation and Development (OECD) 2006). This could be remedied with method modification if the assumption of a normal distribution can be justified. Consequences of deviations from distributional assumptions are more serious toward the tails (Dixon and Massey 1969). Another less obvious issue with implementing the up-and-down is the need to have some prior estimate the standard deviation for the (log) normally distributed data (Dixon and Massey 1969). As discussed below, the intervals chosen between administered doses are based on the standard deviation. Such information can be estimated in some instances from past studies or other information.

Methods for estimating LD50 from data sets with small numbers of observations (Dixon 1965) are described here, although other methods are available (Dixon and Mood 1948) for data sets with a large number of observations. The steps provided below are those given in Dixon's original description. A few important clarifications are added from the more recent guidelines of the OECD (2006), although specific discussion of the OECD (2006) methods and the associated software will also be given later in Example 4.6.

1. A series of doses is defined with the spacing between doses being equal to the standard deviation (σ)* from the probit model. Although the dose intervals should be σ units apart, Dixon (1965) indicates that the intervals can be off by as much as 50% with little consequence to the estimate. A smaller standard error for the estimated LD50 can be achieved if one has enough knowledge of the LD50 prior to testing such that the intervals chosen can be narrower than σ. One rule for establishing intervals is the ½σ oral dosing intervals recommended by the OECD (2006). "The dose progression factor should be chosen to be the antilog of 1/(the estimated slope of the dose-response curve) and should remain constant throughout the test (a progression of 3.2 corresponds to a slope of 2)" (OECD 2006).
2. A series of doses is defined, with the first being close to (slightly below) the suspected LD50. Established rules determine if the dose is dropped or increased after each outcome.
3. The toxicity assay continues for a total of N' tests, which depends on N. The N is "the total number of trials [N'] reduced by one less than the number of like responses at the beginning of the series" (Dixon 1965). Letting O = survival and X = death after dosing, a sequence of results might be OOOXOXOXO. The N' would be 9 and N would be 9 – (3–1) = 7.
4. The maximum likelihood solution for LD50 is obtained using Dixon's (1965) Table 1 (reorganized and split into two tables in Appendix 6, and used in Equation 4.34) (Dixon and Massey 1969; Dixon 1991),

$$LD50 = X_f + kd \tag{4.34}$$

where X_f = the last administered dose (expressed in log units), k = a tabulated maximum likelihood value from Appendix 6, and d = the interval between doses levels (expressed in log units).

If $N > 6$, Appendix 6 might not be used and a modified computation (Equation 2 of Dixon 1965) is used instead. The values in Appendix 6 are the maximum likelihood estimates of the particular results assuming a normal distribution, that is, $P(000X0X0X0|X_0,\mu,\sigma)$.

5. Assuming a normal distribution, the standard error of the estimated LD50 is approximately $\sigma\sqrt{2/N}$ (Dixon 1965). In some cases depending on the initial series, a further adjustment (C) is needed to this estimate (Dixon 1965; Dixon and Massey 1969). A general estimate of the standard error for each test sequence is tabulated below.

Applying these methods to the example given in Figure 4.4, the estimated LD50 would be the following:

$$\log LD50 = \log(17.5) + (0.5)(0.458) = 1.472$$

The 0.5 was the dosing interval in logarithm units and 0.458 is the value from Appendix 6 associated with a first part (XX) and second part (OXOXOX). (Note separate tabulations are provided for series with first parts starting with X or O.) The antilogarithm of 1.472 is 29.6 mg/kg. The σ = 1/2 (slope of 2) was assumed during the original dosing interval selection here and applied to estimate the standard error of the LD50:

$$\sigma\sqrt{2/N} = 1/2\sqrt{(2/6)} = 0.28$$

Adjusted estimates of the standard error for the (logarithm of) predicted LD50 (μ) are given in Table 1 in Dixon (1965) also. Dixon's tabulated estimates include a potential additional adjustment as explained in Dixon (1965) and Dixon and Massey (1969).

Total Number of Trials (N)	Estimated Standard Error
2	0.88σ
3	0.76σ
4	0.67σ
5	0.61σ
6	0.56σ

* The σ is the reciprocal of the slope of the conventional probit (response) versus log dose curve.

Because $N = 6$ here, the estimated standard error from the above estimates from Dixon's table would be 0.56σ or 0.28. The 95% confidence interval would be the antilogarithms of 1.472 ± (1.96) (0.28), that is, 8.4 to 104.9 mg/kg.

Example 4.5

The OECD (2006) protocol for up-and-down estimation of the LD50 prescribes specific steps for selecting dosing intervals and judging at what point dosing should be stopped. A shareware package, AOT425statpgm, is available from EPA (http://www.epa.gov/oppfead1/harmonization) to implement this approach. The dosing intervals are based initially assuming a σ/2 interval between doses and a slope of the dose-response curve of 2, although other slopes can be considered, e.g., Table 1 in OECD (2006). The σ/2 interval on a log dose scale corresponds to a 3.2 factor increase between doses on an arithmetic scale.

AOT425STATPGM (VERSION 1.0): TEST RESULTS AND RECOMMENDATIONS
Acute Oral Toxicity (OECD Test Guideline 425) Statistical Program

Test Seq.	Animal ID	Dose (mg/kg)	Short-Term Result	Long-Term Result
1	First	175	X	X
2	Second	55	X	X
3	Third	17.5	O	O
4	Fourth	55	X	X
5	Fifth	17.5	O	O
6	Sixth	55	X	X
7	Seventh	17.5	O	O

Stopping criteria met: 5 reversals in 6 tests.
Statistical estimate based on long-term outcomes:

Estimated LD50 = 29.57 (Based on an assumed sigma of 0.5).
Approximate 95% confidence interval is 17.5 to 55.

The first point to note above is the statement of the stopping rule for the dosing series. OECD (2006) sets the following as the appropriate rules. The (b) stopping rule was used in the above example series.

(a) 3 consecutive animals survive at the upper (dose) bound;
(b) 5 reversals occur in any 6 consecutive animals tested;
(c) at least 4 animals have followed the first reversal and the specified likelihood-ratios exceed the critical value.

The LD50 output above agrees with the calculations already illustrated for these example data. However, the approximate 95% confidence interval does not, being similar to that described from the binomial method instead. OECD estimates confidence intervals two ways. If three or more doses were tested and the dose in the middle has at least one animal that survived and one animal that died, the confidence interval might be estimated with specific formulae. But, like the binomial approach, an approximate confidence interval is estimated if all individuals survive below a dose and all individuals die at the next highest dose. The interval bounded by the doses with no and complete mortality is the approximate confidence interval. This is not a formal 95% confidence interval. This was the approach taken by the AOT425statpgm program with this example data set.

4.2.3.7 Coping with Control Mortalities

Sometimes a certain level of mortality in a control group is unavoidable and various rules of thumb exist for defining tolerable, albeit unwelcome, levels of control mortality. For example, if one is willing to make the optimistic assumption that the underlying cause of mortality associated with the controls does not influence the mortality associated with the toxicant (Stephan 1977), a range of control mortalities from 5% (Gaddum 1953) to 10% (Buikema et al. 1982; American Public Health Association (APHA) 1981) might be acceptable. A number of early authors (Gaddum 1953; Buikema et al. 1982; APHA 1981; Sprague 1969) recommended that Abbott's formula (Tattersfield and Morris 1924) be used to adjust the proportions dying in each exposure to account for this acceptable control mortality.

$$p_c = \frac{p - p_0}{1 - p_0} \tag{4.35}$$

where p_c, p, and p_0 = the corrected, original, and control mortality proportions, respectively.

Another approach would be to include natural or background mortality during maximum likelihood fitting of data to a model. As an illustration, the probit model specified above can be expressed in terms of the cumulative normal function, $\Phi()$,

$$P = \Phi(a + b(Log\ C)) \tag{4.36}$$

where P = proportion of the exposed individuals responding, and C = the exposure concentration (or dose).

Spontaneous (background or natural) mortality (P_S) can be incorporated into this model and estimated by maximum likelihood methods. In such a model, $P = P_S$ if C is 0.

$$P = P_S + (1 - P_S)(\Phi(a + b(Log\ C)) \tag{4.37}$$

Example 4.6

In a study of the lethal effects of PAH compounds released during oil spills, Unger et al. (2008) exposed grass shrimp (*Palaemonetes pugio*) to a range of naphthalene concentrations. The resulting mortality data are used here to illustrate how the spontaneous mortality (in the exposure system tanks) can be estimated. Equation (4.37) is the model fit to these data. The following SAS code fits and then graphs these data (Figure 4.5).

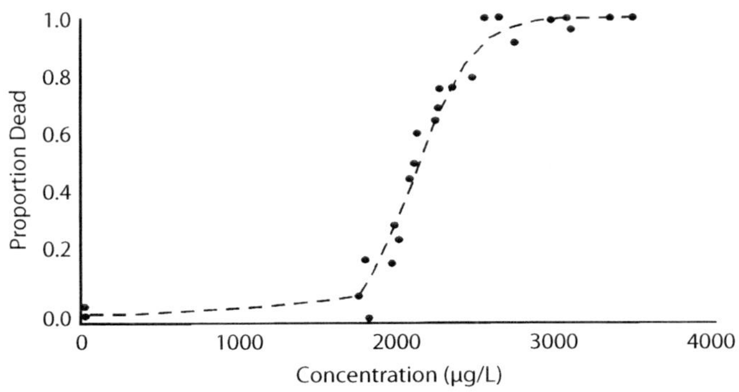

Figure 4.5 The observed data and resulting model (line) for naphthalene-exposed grass shrimp after 24 h (top panel) (Unger et al. 2008). These data were fit to Equation (4.37), which involved estimating the spontaneous, that is toxicant-unrelated, mortality in the exposure system.

```
GOPTIONS RESET=ALL BORDER;
DATA NAP; /* 24 HOUR DATA FOR GRASS SHRIMP PALAEMONETES PUGIO EXPOSED */
INPUT CONC DEAD TOTAL @@; /* TO DISSOLVED NAPHTHALENE IN SEAWATER     */
PROP=DEAD/TOTAL;
DATALINES;
0 1 26 0 0 26 0 0 26 0 0 26 0 1 26 0 0 26 1725 2 26 1792 5 26 1797 0 26
1972 8 25 1996 7 26 1940 5 26 2097 14 27 2057 12 26 2105 16 26 2204 17
26 2233 18 26 2250 20 26 2340 21 27 2448 21 26 2452 21 26 2530 26 26
2624 26 26 2726 24 26 2960 26 26 3090 25 26 3055 26 26 3332 26 26 3335
26 26 3483 26 26
;
/* THE OPTC OPTION ESTIMATES BASELINE MORTALITY THAT IS ASSUMED TO BE */
/* INDEPENDENT OF THAT RESULTING FROM NAPHTHALENE EXPOSURE            */
PROC PROBIT DATA=NAP LOG10 INVERSECL LACKFIT OPTC;
  MODEL DEAD/TOTAL=CONC/D=NORMAL ITPRINT;
OUTPUT OUT=P P=PROB STD=STD XBETA=XBETA;
  RUN;
SYMBOL1 COLOR=BLACK V=STAR;
SYMBOL2 COLOR=BLACK INTERPOL=JOIN V=NONE;
PROC GPLOT;
  PLOT PROP*CONC PROB*CONC/OVERLAY;
  RUN;
  QUIT;
```

The estimated spontaneous mortality (P_S) was 0.01 and has an associated 95% confidence limit ranging from –0.01 to 0.03. There was a very modest 1% mortality for the shrimp placed into the experimental system.

As an aside, Equation (4.36) can also be modified in a situation if a toxicant concentration threshold (C_T) exists below which no effect is observed,

$$P = \Phi[a + b(Log\ (C - C_T)] \tag{4.38}$$

4.2.3.8 Duplicate Treatments

Often it is important to determine if duplicate treatments are homogeneous. Stephan (1977) suggested that Fisher's exact test should be used for this purpose. Test results are suspect if the assumption of homogeneity is rejected for duplicates. This test, like a simple contingency χ^2test, can be set up with a 2 by 2 table such as the following:

	Number Dead	Number Surviving	Row Total
Tank A	a	b	$a + b$
Tank B	c	d	$c + d$
Column total	$a + c$	$b + d$	Grand total $a + b + c + d$

A χ^2 test could be applied to such a contingency table under certain conditions. The counts in each cell expected under the hypothesis of no difference are estimated as the product of the row total and column total divided by the grand total count. For example, assume the following results are obtained for two replicate tanks in a toxicity test:

	Number Dead	Number Surviving	Row Total
Tank A	1	19	20
Tank B	6	14	20
Column total	7	33	Grand total 40

The expected cell count for cell *b* (number surviving in tank A) would be ((20)(33))/40, or 16.5. Similarly, the expected counts for cells *a*, *c*, and *d* are 3.5, 3.5, and 16.5, respectively. The following formula could be used to calculate a χ^2 statistic for these data:

$$\chi^2 = \sum_{i=a}^{d} \frac{(O_i - E_i)^2}{E_i} \tag{4.39}$$

where O_i and E_i = the observed and expected counts in cell *i*.

The χ^2 statistic is 4.3290 in this example. The degrees of freedom for the contingency χ^2 test would be (number of columns – 1)(number of rows – 1), that is, (2–1)(2–1) = 1 df. A χ^2 table or a software function such as the Excel CHIINV(4.3290,1) would be used to get an associated *p* value of 0.0375. Some statisticians (e.g., Miller 1998; Piegorsh and Bailer 1997) incorporate Yates's continuity correction of $(a + b + c + d)/2$ (Yates 1934) into the above equation if small numbers are associated with cell counts,

$$\chi^2 = \frac{(a+b+c+d)\left(|ab-bc| - \left[\frac{(a+b+c+d)}{2}\right]\right)^2}{(a+b)(c+d)(a+c)(b+d)} \tag{4.40}$$

Regardless, the use of a χ^2 statistic will produce questionable χ^2 estimates if, as in this case, the expected count in any cell is less than 5. Miller (1998, p. 46) is less conservative than those who adhere to this commonly held rule, stating that the continuity corrected χ^2 statistic given in Equation (4.40) is adequate if the treatments (e.g., tanks A and B) have at least 10 observations each and the smallest cell count is 2 or higher. Here we will adhere to the conventional rule of using Fisher's exact test instead of a χ^2 test if any expected cell count is less than 5.

Fisher's exact test begins with the assumption (null hypothesis) that the four marginal totals have a hypergeometric conditional distribution (Miller 1998), and consequently, the probability of the four observed cell counts, given the four marginal totals, would be the following:

$$P(a,b,c,d|a+b,\ c+d,\ a+c,\ b+d) = \frac{\binom{a+b}{a}\binom{c+d}{c}}{\binom{a+b+c+d}{a+c}} \tag{4.41}$$

This equation can be expressed in terms of factorials for readers unfamiliar with the binomial coefficient notation in Equation (4.41). The binomial coefficient denotes the number of possible ways that subsets of size *j* may be drawn from a population of size *N*. (See Feller 1968, p. 34, or Pollard 1979, p. 4, for a detailed explanation of this notation.)

$$\binom{N}{j} = \frac{N!}{j!(N-j)!} = \frac{N(N-1)(N-2)\ldots(N-j+1)}{j!} \tag{4.42}$$

With Equation (4.42), Equation (4.41) can now be rewritten in terms of factorials (Miller 1998),

$$P(a,b,c,d|a+b,\, c+d,\, a+c,\, b+d) = \frac{(a+b)!(c+d)!(a+c)!(b+d)!}{(a+b+c+d)!a!b!c!d!} \tag{4.43}$$

Example 4.7

Assume that the following data were produced for water (A) and solvent control (B) tanks of a toxicity test. The question is whether the solvent tank mortality is higher than that of the water control tank, i.e., a one-sided test. The following code executed the Fisher's exact test and then produced two tables.

```
DATA TEST;
   INPUT TANK $ OUTCOME $ N;
   DATALINES;
A DEAD  1
A LIVE 19
B DEAD  6
B LIVE 14
;
RUN;
PROC FREQ DATA=TEST ORDER=DATA;
TABLES TANK*OUTCOME/EXACT;
WEIGHT N;
RUN;
```

TANK Frequency Percent Row Pct Col Pct	OUTCOME DEAD	LIVE	Total
A	1 2.50 5.00 14.29	19 47.50 95.00 57.58	20 50.00
B	6 15.00 30.00 85.71	14 35.00 70.00 42.42	20 50.00
Total	7 17.50	33 82.50	40 100.00

Fisher's Exact Test

Cell (1,1) Frequency (F)	1
Left-sided Pr <= F	**0.0457**
Right-sided Pr >= F	0.9958
Table Probability (P)	0.0416
Two-sided Pr <= P	0.0915

The last table gives the exact probabilities for left- and right-sided one-way tests, the table probability, and the exact two-sided probability. Small left-sided values draw support from the null hypothesis, that is, the count in the first cell is less than expected under the null hypothesis. As conventionally interpreted, the bolded p value suggests the probability of getting this or a more extreme table of data if the null hypothesis were true (as per Equation 4.43). The null hypothesis does not seem to be supported. Notice that the two-sided test p value is twice that of the one-sided p-value.

4.2.3.9 Summary of LC50 Methods

There are a variety of methods for estimating the LC50 (LD50 or EC50) and its associated 95% confidence interval. The simplest graphical techniques are convenient yet results are subject to inaccuracies from fitting a line by eye. The Litchfield–Wilcoxon method, a semigraphical method, can be used to estimate the LC50 and its 95% confidence interval, but it also is subject to a degree of error associated with fitting a line by eye. The maximum likelihood methods of fitting normal, logistic, or Weibull models assume specific models and require at least two partial kills unless adjustment is made to the proportions responding, e.g., Berkson's substitutions (Berkson 1955). The ratio of χ^2 values for candidate models and plots can be used to judge the relative appropriateness of each. The trimmed Spearman–Karber method provides a nonparametric alternative requiring only a monotonic increase in proportion responding and a symmetrical distribution. If there are no partial kills, then a binomial method might be applied. The simple moving average method can be applied if the exposure concentrations are set in a geometric pattern and the same numbers of individuals are exposed to each concentration. Finally, substantial reductions in number of subjects tested can be realized with the up-and-down method if it is appropriate.

The tedium of calculating these statistics can be minimized by using a variety of software packages. Pertinent procedures in the SAS package have been illustrated here. Straightforward shareware programs are available to estimate the LC50 and associated statistics using the probit (maximum likelihood method for a lognormal model) and Spearman–Karber methods. The code and manuals can be obtained from the U.S. Environmental Protection Agency as instructed in U.S. EPA (2002, Section 11.2.5.3) or http://www.epa.gov/EERD/stat2.htm. The AOT425statpgm shareware implementing the up-and-down method as formalized by the OECD (2006) can be obtained from the U.S. EPA (http://www.epa.gov/oppfead1/harmonization). (Mention of specific software throughout this book does not imply endorsement.)

4.2.3.10 Incipient LC50

The incipient (asymptotic, ultimate, or threshold) LC50 is the calculated concentration below which 50% of exposed individuals would live indefinitely relative to the lethal effects of the toxicant. Equally important to keep in mind is the procedural definition of the incipient LC50, which is the antilogarithm of the asymptotic log concentration of the (log) LC50 versus (log) time curve. (See Figure 4.6A for the general behavior of toxic response versus time. Although LT50 is used in this figure instead of LC50, the same general behavior is expected. The LC50 value decreases as exposure duration increases until an asymptotic LC50 is reached.) Often, mortality during the course of a toxicity test will be recorded at set intervals. The estimated LC50 is calculated for each time. The log of the LC50 for each time (X) is plotted against the log of time (Y) to produce a toxicity curve. The lethal threshold concentration (incipient LC50) (Sprague 1970) is estimated to be the concentration at which the curve begins to run parallel with the X axis.

Sprague (1969, 1970) provides a general discussion of quantitative methods for dealing with the incipient LC50. As pointed out by Chew and Hamilton (1985), there are some conceptual difficulties with this approach. First, LC50 values are often estimated at a series of times during a single

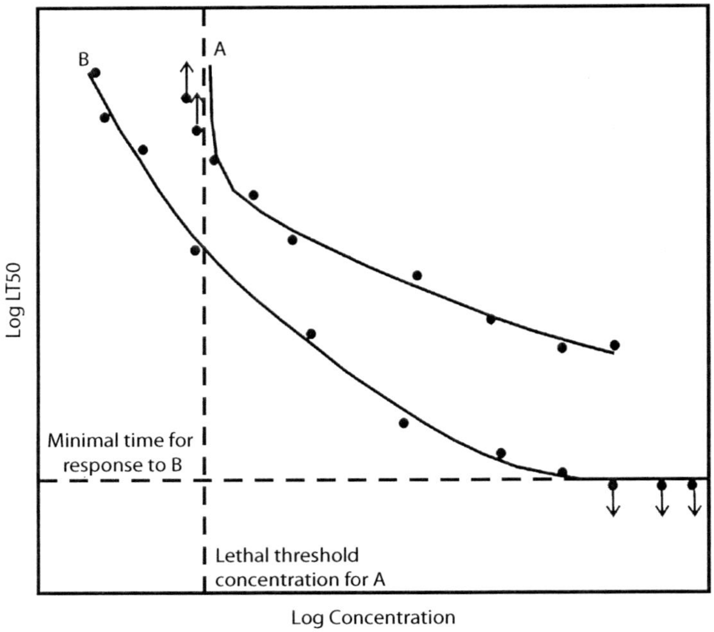

Figure 4.6 An illustration of the incipient lethal level or incipient threshold concentration (A). As the exposure concentration decreases, a concentration is reached below which 50% or more of the exposed individuals will live indefinitely relative to the effect of toxicant A. Some toxicants (B) may show no apparent lethal threshold concentration. A minimum time for the toxic response may also be apparent (B). Fifty percent of the individuals exposed survive for at least this length of time regardless of toxicant concentration.

toxicity test and treated as independent. Independence is a dubious assumption. Second, concentration is treated as an independent variable, yet in practice, it is set in the experimental design.

4.2.3.11 The Significance of the LC50

> Today, it is widely recognized that the LD50 is of marginal value as a measure of hazard, although it does provide a useful "ball park" indication of the relative hazard of a compound to cause serious, life-threatening poisoning from a single exposure.
>
> **—Eaton and Gilbert (2008)**

The LC50 does not indicate environmental safety of a specific concentration of toxicant (APHA 1981). Rather, it is a measure of toxicity that is best employed in a relative context, e.g., concentration X of chemical A has a higher toxic effect than concentration X of chemical B after 96 h of exposure, or the toxic effect is less under condition A than under condition B. It is quite dependent on conditions surrounding the toxic response. Furthermore, the time endpoint is often selected for convenience (Figure 4.7). As discussed earlier, the median value ($p = 0.50$) was selected because it was the most statistically reliable value. Despite the common acknowledgment of this fact, LC50 values are inappropriately used to imply environmental safety. An example might be the misinterpretation of the incipient LC50. Again, the incipient LC50 is the concentration below which 50% of the individuals will live indefinitely (relative to the toxicant effects) under the test conditions. The median value is nearly meaningless relative to the persistence of an endemic population in an ecological community: it is a statistically convenient value only. The inappropriate use of LC50 values

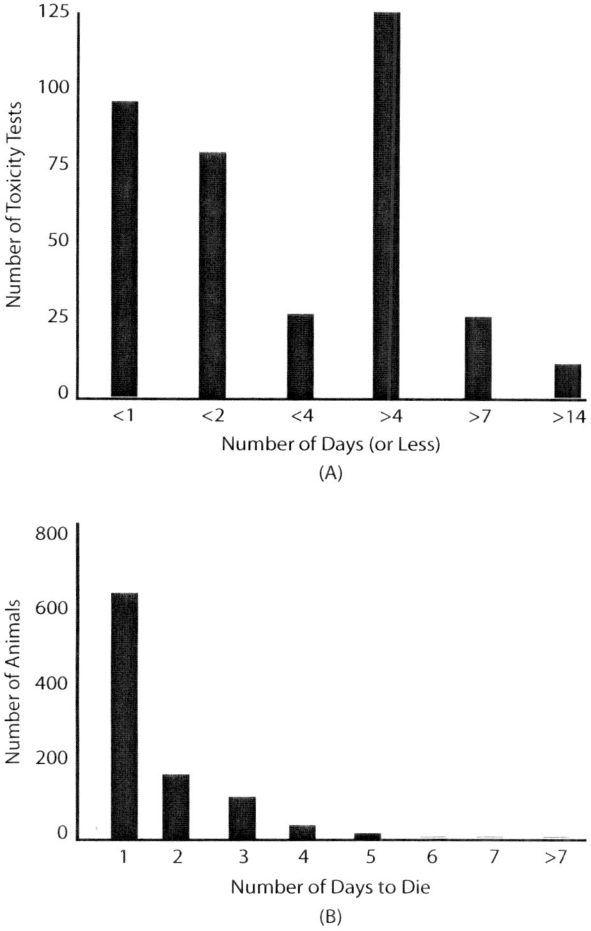

Figure 4.7 Two examples of published information used to justify selection of an exposure duration that encompasses acute lethality. The top panel displays the results of early LC50 tests compiled by Sprague (1969); however, nearly half of the tests required more than the 96 h being argued as adequate to include most acute mortality. Also, many tests were conducted without periodic refreshing of toxicant concentrations in the exposure water. (More detailed discussion of the shortcomings of these data is provided in Newman and Clements 2008.) The bottom panel was derived from Figure 1 in Bruce (1985), from which it was concluded that "the overwhelming majority of deaths occurred in the first few days." This evidence was used in a reasonable fashion to pragmatically define the duration between animal dosings needed in the up-and-down acute LD50 oral rat dosing test.

to suggest ecological risk also enters into the debate about the currently emerging species sensitivity distribution approach (see Posthuma et al. 2002, Chapters 1 and 2; Jagoe and Newman 1997).

Several additional complications emerge during the difficult task of using LC50 data to gauge species population fate under real-world exposure scenarios. Zhao and Newman (2004, 2006, 2007) explored relevant issues with the amphipod *Hyallela azteca* exposed to two contrasting toxicants, copper sulfate and sodium pentachlorophenol. The first was postexposure mortality. Although LD50 data were originally used by mammalian toxicologists to address questions in which postexposure mortality was considered irrelevant (e.g., Figure 4.7), postexposure mortality becomes potentially very relevant during the current application of these LC50 (or LD50) data to infer population consequences of field exposures. To explore some of these toxicodynamics features, Zhao and Newman

(2004, 2006) adjusted the conventional test scheme so that amphipods received a toxicant pulse as might be expected in many real-world acute exposures. The first point to make from their work is that some, but not all, toxicant exposures had substantial postexposure mortality of stressed individuals. The cumulative mortality for amphipods exposed to sodium pentachlorophenol did not change appreciably after exposure ended, but substantial postexposure mortality occurred for amphipods exposed to copper sulfate (Figure 4.8). If all mortality, including postexposure mortality, was considered at the LC50 concentration, the percentage dying of the copper-exposed amphipods would

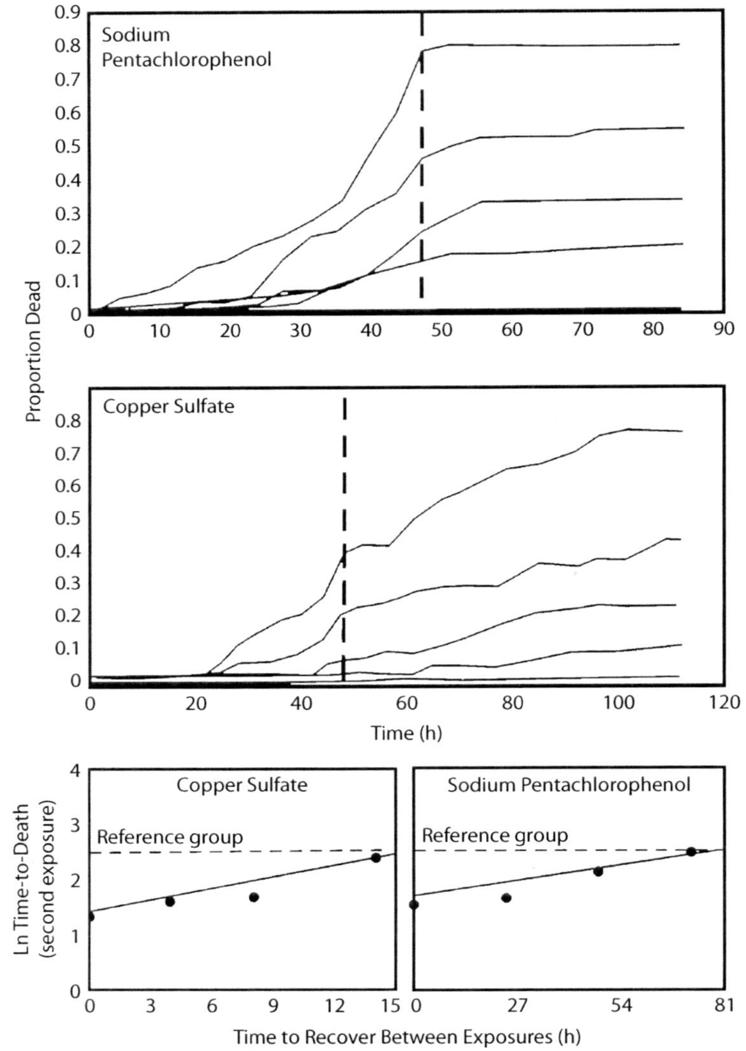

Figure 4.8 Results from pulsed exposures of the amphipod *Hyalella azteca* to copper sulfate and sodium pentachlorophenol. The top two panels show the cumulative mortality during and postexposure for both toxicants (dashed line indicates time that exposure ended). Substantial mortality occurred postexposure for amphipods exposed to copper sulfate but not amphipods exposed to sodium pentachlorophenol (Zhao and Newman 2004). As shown in the bottom panels, amphipods exposed to a pulse of these toxicants and allowed to recover for different durations before a second pulse was delivered showed that the recovery period between pulses influenced their mortality rates relative to previously unexposed amphipods (reference group) (Zhao and Newman 2006).

be 65 to 85%, not 50% (Figure 4.9). The conventional LC50 was a poor predictor of lethal consequences to copper sulfate-exposed organisms. Zhao and Newman (2004) argued that a "complete LC50," including mortality occurring during and after exposure, must be calculated for toxicants such as copper sulfate. Zhao and Newman (2006) also examined amphipod response as influenced by the amount of time between pulsed or intermittent toxicant exposures (bottom of Figure 4.8), finding that there was a critical time needed to recuperate from a previous exposure before the population might be expected to respond like a previously nonexposed one. These important features of postexposure mortality and recuperation time are not incorporated into predictions of environmental consequences from conventional LC50 values.

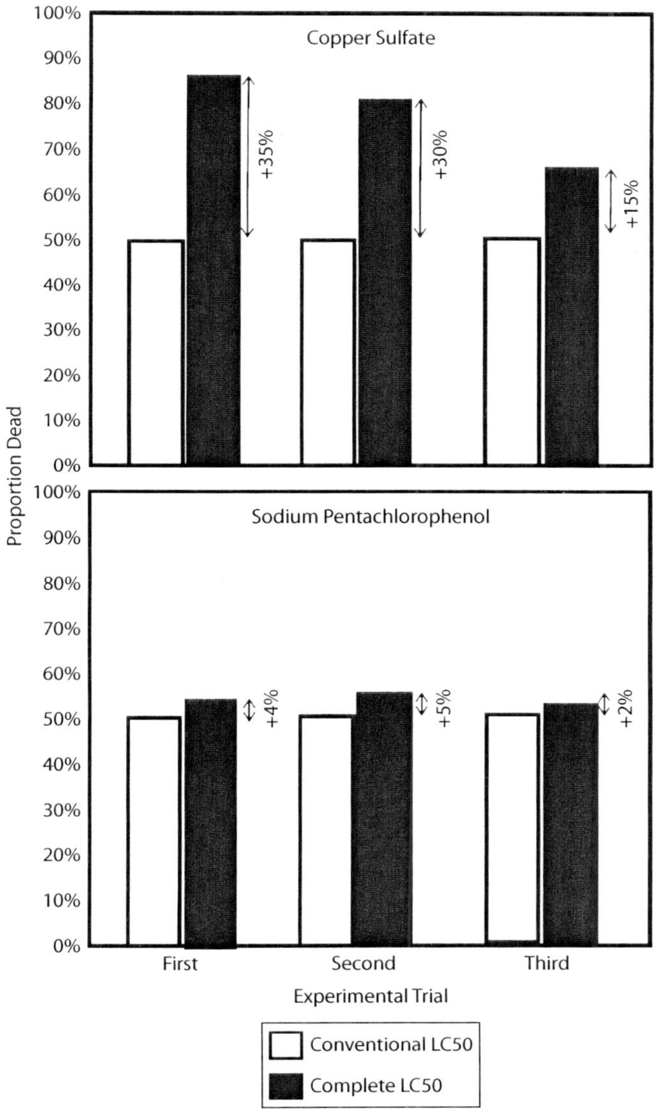

Figure 4.9 An illustration of the conventional and complete LC50 for the amphipod *Hyalella azteca* exposed to copper sulfate and sodium pentachlorophenol. (Modified from Zhao, Y., and M.C. Newman, *Environ. Toxicol. Chem.*, 23(9), 2147–2153, 2004, Figure 2.)

4.3 TIME TO DEATH

A definition of toxicity [is] "the accumulation of injury over short or long periods of time, which renders an organism incapable of functioning within the limits of adaptation." This definition implies that toxicity is a function of time in addition to the dose.

—Rozman and Doull (2000)

Protocols for estimating toxic endpoint (e.g. 96-h LC_{50}) dominate research in aquatic toxicology to the near exclusion of equally appropriate techniques such as survival modeling. By noting time-to-death during routine testing, statistical power, incorporation of covariates, prediction of toxic effect over time and linkage to life table characteristics can be improved. Consequently, prediction of ecological risk is enhanced.

—Newman and Aplin (1992)

The inclusion of exposure times in toxicology is not new (see for instance Bliss 1937, Sprague 1969) but its use in standard toxicity testing has been neglected until recently.

—Sanchez-Bayo (2009)

4.3.1 The Standard Approach

4.3.1.1 General

Another approach to analyzing toxicity data is to model time to death (resistance time), the duration of exposure prior to death of the individual. Sprague (1969) describes the straightforward analysis of time-to-death data sets. The cumulative proportions of exposed individuals dying are recorded at a series of time intervals. Originally, these data were then plotted on logarithm-probability paper (percentage mortality on the probability scale versus the log of time) to yield a line. If several concentrations were used in the test, one line may be generated for each toxicant concentration or level of the stressor. This approach is similar to that described above for dose-response plots assuming a normal distribution of P at a set endpoint when plotted against the logarithm of dose. From our previous discussion, it follows that the probit (or N.E.D.) of the proportion dying up to any time plotted against the logarithm of time should also produce a straight line if the assumption of a lognormal model is correct.

The cumulative mortality (that proportion of all individuals dying) at exposure duration t plotted against the logarithm of duration does not always produce a straight line. If there is a distinct change in slope for the time course of cumulative mortality, the resulting plot is called a split probit by convention. A split probit is often attributed to different mechanisms being dominant at different times or distinct subsets of individuals with different resistances to the stressor (Sprague 1969). However, inappropriate fitting of the data set to the lognormal model could be another reason for a split probit, as curve assessment usually ends at visual inspection and alternative models are rarely examined. Indeed, other historic conventions exist for including durations that are incongruent with the lognormal model. As examples, a single acute:chronic effect ratio over the pertinent scales (Kenaga 1982), or a simple reciprocal relationship between level of effect and duration of exposure in which a is the estimated incipient LC50 (Green 1965),

$$Ln\ C = a + b\frac{1}{T} \tag{4.44}$$

or perhaps the more involved relationship from Barata et al. (1999),

$$Ln\ C = \frac{Probit(P) - a - \dfrac{c}{Ln\ T}}{b} \tag{4.45}$$

where a, b, and c = estimated model parameters, T = duration of exposure, C = effect concentration (e.g., LC50) (Equation 4.44) or exposure concentration (Equation 4.45), and P = proportion of exposed individuals that die.

Just as the LC50 is useful in endpoint dose-response methods, the median response time is important in the simple time-to-death approach just described. The median lethal time (LT50, median period of survival, median resistance time, or median time to death) is the duration of exposure corresponding to a cumulative mortality of 50% for the exposed individuals. If a nonlethal or ambiguously lethal response is considered, median effective time (ET50) would be the appropriate term (Sprague 1969). The median lethal time is analogous to the LC50, and assuming that the normal model is appropriate for the logarithm of exposure duration-cumulative mortality data, it can be analyzed as described for the probit methods for the LC50. Shepard (1955) provides an excellent example of the median response time usage. In addition to discussing split probits and covariates such as acclimation, he also demonstrates an alternate method of estimating the asymptotic lethal concentration using the median response (resistance) time instead of the LC50 as described above. This approach will be discussed below in detail.

4.3.1.2 Litchfield Method for Estimating LT50

Litchfield (1949) described a rapid graphical method for estimation of the LT50 and its 95% confidence limits. It is very similar to the previously described Litchfield and Wilcoxon method published that same year. Indeed, with the exception of one exponent, the associated equations are identical to Equations (4.9) to (4.12).

Assume that all individuals die during the exposure period: all individuals have a quantified time to death. (If there were survivors, the data would be censored as defined in Chapter 2. The data set would be right censored because the observations censored are those associated with high values of time, i.e., values to the right of the distribution.) The cumulative percentage of individuals dying (number dead/total number exposed) is plotted against time (duration of exposure) on log-probability paper. The time corresponding to a cumulative percentage of 50% is the LT50.

Calculation of the 95% confidence limit for the LT50 is estimated with Equations (4.46) to (4.50). First, the slope function (equivalent to the standard deviation) is estimated by Equation (4.46).

$$S = \frac{\dfrac{LT84}{LT50} + \dfrac{LT50}{LT16}}{2} \tag{4.46}$$

where LT16, LT50, LT84 = the times corresponding to 16, 50, and 84% of the cumulative percent mortality.

The 95% confidence interval is estimated using f_{LT50}.

$$f_{LT50} = S^{1.96\sqrt{N}} \tag{4.47}$$

where N = the number of individuals exposed.

The upper and lower confidence limits are calculated by multiplying or dividing the estimated LT50 by f_{LT50}.

Upper limit of 95% CI:

$$Upper\ Limit = LT50 \cdot f_{LT50} \tag{4.48}$$

Lower limit of 95% CI:

$$Lower\ Limit = LT50 \,/\, f_{LT50} \tag{4.49}$$

If there are survivors, the equation used to estimate the f_{LT50} needs to be modified as outlined in Litchfield (1949). Values for E from Bliss's (1937) Table VIII are required for this adjustment. Pertinent material from this table is reproduced in Appendix 7. The E acts to adjust the N to account for the number of survivors. For censored data sets, the following equation becomes pertinent:

$$f_{LT50} = S^{1.96/\sqrt{N_2}} \tag{4.50}$$

where $N_2 = N/E$.

This value is used instead of that from Equation (4.47) in Equations (4.48) and (4.49) to estimate the 95% confidence limits for the LT50 of the censored data set.

Bliss (1937) estimated E using an x' value, the point of censoring expressed in terms of the standard deviation. This value is derived with the point of censoring, log LT50, and log standard deviation of the LT50. In this case, the log LT50 and log standard deviation are taken from the graph and Equation (4.46) as described above. Since these are right-censored data and Bliss's (1937) table is produced for left-censored data, the x' is obtained by plotting reaction times (1,000/t). The point of censoring is defined by inverting each survival time and subtracting 3 from the logarithm of each survival time. The process as outlined in Litchfield (1949) or Bliss (1937) to estimate x' can be simplified if one remembers that the N.E.D. expresses a proportion in terms of deviations from the mean. Consequently, the N.E.D. (Appendix 5) or z score for 1 minus the proportion dead at the end of the exposure can be used as x' in Appendix 7.

Example 4.8

Times to death of streptomycin-treated mice with tuberculosis were tabulated by Litchfield (1949) over a time course of 60 days. Fourteen mice died with the following times to death: 26, 29, 32, 32, 37, 39, 39, 42, 43, 43, 44, 47, 52, and 59 days. Six lived beyond 60 days. What were the LT50 and its 95% confidence interval for these mice?

Time (days)	Cumulative Number Dead	Number Dead/Total Number
26	1	0.05
29	2	0.10
32	4	0.20
37	5	0.25
39	7	0.35
42	8	0.40
43	10	0.50
44	11	0.55
47	12	0.60
52	13	0.65
59	14	0.70

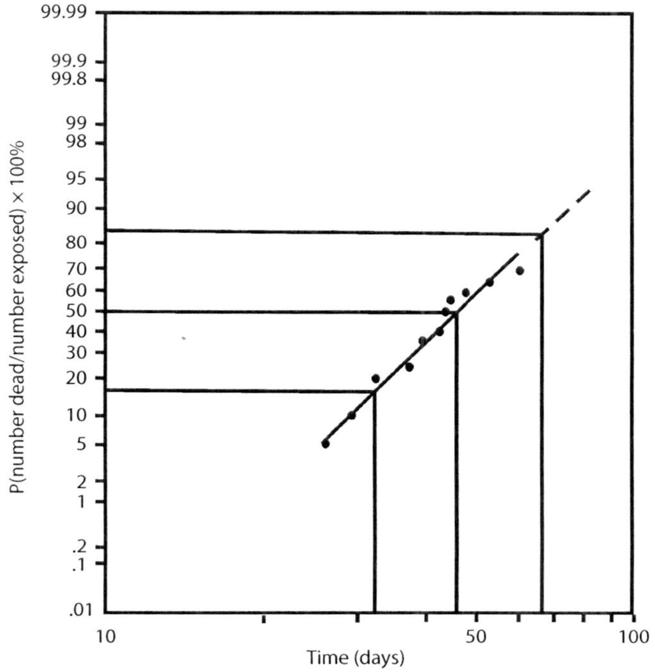

Figure 4.10 Mortality of streptomycin-treated mice with tuberculosis (Litchfield, J.T., Jr., *J. Pharmacol. Exp. Ther.*, 97, 399–408, 1949) is plotted using log-probability scales to estimate the LT50 and its 95% confidence interval (Example 4.8).

Using log-probability paper (Figure 4.10), an LT50 of 46 days is derived. Extrapolating and using the results in Equation (4.46),

$$S = [(66/46) + (46/32)]/2 = 1.44 \text{ days}$$

The 95% confidence intervals may be calculated using Equations (4.48) to (4.50), but x', E, N_2, and finally, f_{LT50} must be calculated first.

$$X' = \text{N.E.D. for } 1.00 - 0.70 = -0.52440$$
$$E = 1.1390 \text{ (Linearly interpolated from Table 7)}$$
$$N_2 = N/E = 20/1.1390 = 17.56$$
$$f_{LT\,50} = S^{1.96/\sqrt{N_2}} = 1.44^{1.96/4.190} = 1.44^{0.468} = 1.19$$
$$\text{Upper limit} = \text{LT50} \cdot (f_{\text{LT50}}) = 46(1.19) = 54.7 \text{ days}$$
$$\text{Lower limit} = \text{LT50}/f_{\text{LT50}} = 46/1.19 = 38.7 \text{ days}$$

The LT50 for these mice is 46 days with a 95% confidence interval of approximately 39 to 55 days.

4.3.1.3 Lethal Threshold Concentration

Sprague (1969) illustrated the use of plotting the logarithms of LT50 (Y) and concentration (X) to derive the lethal threshold concentration, the concentration at which 50% of the exposed individuals will survive indefinitely relative to the toxicant effects. When present (Figure 4.6A), this measure is analogous to the incipient LC50. However, there are data sets that appear to have no apparent threshold concentration (Figure 4.6B). Further, an asymptotic time may also be present (Figure 4.6B). The asymptotic time is the minimum exposure duration necessary for expression of a response (Sprague 1969). Notionally, no response is expected at any time less than the asymptotic time regardless of the toxicant concentration.

Gaddum (1953) and Lloyd (1960) quantify this curve using two formulae. Ostwald's equation (Lloyd 1960) is used when the logarithm of the concentration versus the logarithm of LT50 produces a straight line.

$$C^n t = K \tag{4.51}$$

where C = the toxicant concentration, t = LT50, and n, K = regression-derived constants.

If the plot is curvilinear, the equation of choice (Gaddum 1953; Lloyd 1960) is the following:

$$(C - C_0)^n (t - t_0) = K \tag{4.52}$$

where C_0, t_0 = threshold concentration and threshold time, respectively.

The initial estimates for the threshold concentration and time may be extracted by eye during the fitting process (Feller 1968). Gaddum (1953) suggests plotting C against $1/t$ for large t values to find the value of $1/t$ that equals 0. This is used as the estimate of C_0. Similarly, the plot of t versus $1/C$ for large C values can be used to estimate t_0.

Gaddum (1953) notes that the following formulae have been found by other workers to fit this type of data also.

$$(C - C_0)(1 - e^{-a(t-t_0)}) = K \tag{4.53}$$

$$(C - C_0)t = K\left(1 + \frac{t_0}{t}\right) \tag{4.54}$$

$$\left(\frac{C_0}{C} - 1\right)\left(\frac{t}{t_0} - 1\right) = K \tag{4.55}$$

Gaddum (1953) also points out that Equation (4.55) is simply another form of Equation (4.52).

None of the above methods for describing the time-response relationship can be definitively linked to any underlying mechanisms, although some attempt was made at linkage in the associated publications. In contrast, Chew and Hamilton (1985) derive a time-concentration-response curve based on pharmacokinetic principles. They assume that there exists a threshold amount of toxicant in some biological compartment such as the gill above which mortality occurs. They then use a one-compartment bioaccumulation model (see Equation 3.42 in Chapter 3) to estimate the time to reach that threshold given a series of specified exposure concentrations.

$$C_t = AC_W(1 - e^{-Bt}) \tag{4.56}$$

where C_t = the concentration in the compartment at time t, C_w = concentration in the source (water), and A, B = estimated constants.

Their model, as used to fit time-response data, takes the form

$$f(C_W) = \frac{\left[Ln\ C_W - Ln\left(C_W - \frac{\xi}{A}\right)\right]}{B} \tag{4.57}$$

where B = "the average rate of entry and elimination from the organism; B determines the curvature of $f(C_w)$," ξ/A = "the asymptotic LC50; that is, less than half the population of organisms would die due to the toxicant if the environmental concentration were less than ξ/A," and ξ = "the median concentration in the compartment at which half the organisms in the population will die."

Median survival time data for ammonia and cadmium toxicity to fish were fit to this model. (In health sciences, such a model that links pharmacokinetics to toxic or pharmacological action (pharmacodynamics) would be called a pharmacokinetic-pharmacodynamic model.) Nonlinear, least-squares fit of median time to death versus concentration was done with a weighting of the reciprocal of the median time-to-death values' estimated variance.

Data from concentration treatments with "too few deaths" were not used in the models. Chew and Hamilton (1985) suggested a rule of thumb for exclusion of treatments. Let k be the smallest integer such that

$$\alpha > 0.5^N \sum_{j=k}^{N} \binom{N}{j} \tag{4.58}$$

where α = the level of significance, and N = the number of organisms exposed to the concentration.

They suggest exclusion of any point from the analysis where the number of deaths (d) is less than or equal to k.

With the potential for linkage to bioaccumulation models, this approach of Chew and Hamilton (1985) has much appeal. It has the potential for elaboration to account for extrinsic and intrinsic factors affecting accumulation and toxicity kinetics. Further evaluations and enrichments of this approach, such as the work of Heming et al. (1989) with organochlorine pesticide toxicity, would be valuable contributions to the field.

4.3.2 The Survival Time Approach

4.3.2.1 General

> It is unfortunate that pollution biologists have tended to form a splinter group as far as toxicology is concerned. Standard techniques of analysis, developed by pharmacologists and statisticians for testing drugs, have too often been ignored.
>
> **—Sprague (1969)**

> The inclusion of exposure times in toxicology is not new..., but its use in standard toxicity testing has been neglected until recently.
>
> **—Sanchez-Bayo (2009)**

There was remarkably little attention paid to time-to-death (survival time) methods in the field of ecotoxicology (Dixon and Newman 1991; Newman and Aplin 1992; Sanchez-Bayo 2009) until recently, e.g., Crane et al. (2002), Barron et al. (2008), and Carriger et al. (2010). The reason is likely linked to the tendencies of ecotoxicologists as noted by Sprague in the above quote. This was unfortunate because a rich array of pertinent methods is available from the engineering and health sciences literature. For example, pertinent survival analysis methods continue to be described in detail in books such as Miller (1981), Cox and Oakes (1984), Altman (1991), Marubini and Valsecchi (1995), Woodward (2005), and Der and Everitt (2006). Also, most statistical software packages implement these methods.

Such techniques have been used in studies of covariate effects on toxicity (Diamond et al. 1989, 1991; Newman et al. 1989; Newman and Aplin 1992). However, as pointed out in Dixon and Newman (1991), Newman and Aplin (1992), and Crane et al. (2002), ecotoxicologists and ecological risk assessors have yet to take full advantage of them. It is the primary objective of this section to describe these methods with emphasis on their virtues. However, because of this stress on advantages, it must be emphasized here that the author is not implying that these techniques are superior in all ways to the endpoint methods described earlier. Survival time and endpoint methods both have their place in ecotoxicology. Sadly, there is still a gross imbalance in their application at present, although the need to address intermittent pulses of exposure has prompted increased interest, e.g., Ashauer et al. (2006) and Zhao and Newman (2006, 2007).

Several terms and concepts must be presented before discussing specific details of the various methods. Chief among these are the survival and hazard functions. Both are used in more in-depth analysis of time-to-death data.

Assume a time course of exposure with individuals dying over a period T. The mortality within the exposed group of individuals can be described as a probability density function ($f(t)$) or cumulative distribution function ($F(t)$). An estimate of the $F(t)$, that is, the probability of dying by t, would be the total number of individuals dead at time t, divided by the total number of exposed individuals.

$$F(t) = \frac{Number\ Dead_t}{Total\ Number\ Exposed} \tag{4.59}$$

The survival function ($S(t)$) is estimated by the number of individuals surviving to time t, divided by the total number of individuals exposed to the toxicant. This probability of surviving to t can be expressed in terms of $F(t)$.

$$S(t) = 1 - F(t) \tag{4.60}$$

The hazard rate or function ($h(t)$) is the probability of dying during a given very small interval. It has been interpreted as the force of mortality (Miller 1981), instantaneous mortality rate (Pinder et al. 1978), instantaneous failure rate, or proneness to fail (Nelson 1972) depending on the specific process being studied. It can be defined in terms of $f(t)$ and $F(t)$, or $S(t)$ (Nelson 1972; Miller 1981; Dixon and Newman, 1991).

$$h(t) = \frac{f(t)}{1 - F(t)} = \frac{f(t)}{S(t)} = \left[\frac{-1}{S(t)}\right]\left[\frac{dS(t)}{dt}\right] \tag{4.61}$$

The cumulative hazard function, $H(t)$, can also be defined in terms of cumulative mortality ($F(t)$) (Nelson 1972; Blackstone 1986).

$$H(t) = \int_{-\infty}^{t} h(t)dt = -Ln(1 - F(t)) \tag{4.62}$$

The cumulative survival ($S(t)$) is related to the cumulative hazard (Der and Everitt 2006),

$$S(t) = e^{-H(t)} \tag{4.63}$$

$F(t)$ and $f(t)$ can be expressed in terms of $H(t)$ and $h(t)$,

$$F(t) = 1 - e^{-H(t)} \tag{4.64}$$

$$f(t) = h(t)S(t) \tag{4.65}$$

With these preliminaries completed, we can now move to a more in-depth time-to-death analysis. It should be noted that although death (or some surrogate measure of death) is being used in these techniques, other events may also be analyzed with the methods described below. The only restriction here is that the discrete event scored for an individual can occur only once* during the time course, e.g., time to immobilization, time to first brood, time to recover after intoxication or stupefaction, time to remission, or time to develop cancer.

4.3.2.2 Nonparametric Methods

Two general nonparametric approaches can be used for time-to-death data, the life table and product-limit (Kaplan–Meier) methods (Miller 1981; Cox and Oakes 1984; SAS 1988). Neither requires a specific form for the underlying distribution describing survival. The familiar life or actuarial table places estimates of $S(t)$ in a fixed sequence of intervals, e.g., year age classes of humans. (See Chapter 6 for more details.) With the product-limit approach, the intervals may be of any length (Miller 1981). Harrell (1988) suggests that product-limit methods are preferable to life table methods because they have more resolution and carry fewer assumptions.

The product-limit estimate of $S(t)$ is defined in various sources (Miller 1981; SAS 1988; Cox and Oakes 1984; Blackstone 1986). The original description can be found in Kaplan and Meier (1958), and a description of the associated maximum likelihood method is given in Kalbfleisch and Prentice (1980). The notation here is that used by SAS (1988).

$$\hat{S}(t_i) = \prod_{j=1}^{i} \left[1 - \frac{d_j}{n_j} \right] \tag{4.66}$$

where i = the labels for the failure times, t_i, n_i = the number of individuals alive just before time t_i (the number of individuals alive and at risk of dying), and d_i = the number of individuals dying at t_i.

This product-limit estimate of $S(t)$ is undefined for times beyond the end of the exposure period (T) if survivors remain (Dixon and Newman 1991).

Greenwood's formula (Miller 1981; Harrell 1988; SAS 1988; Dixon and Newman 1991) can be used to estimate the variance for the product-limit estimate, $\hat{S}(t)$, if censoring only involved some individuals that survived beyond the duration of the exposure.

* Although irrelevant here, these methods can be applied if the event occurs more than once (Cox and Oakes 1984), as in the case of time to give birth. The event must still be discrete, i.e., did/did not give birth, or did/did not move out of a stream reach that received a runoff-related pulse of pesticide.

$$\hat{\sigma}^2 = \hat{S}^2(t_i)\sum_{j=1}^{i}\frac{d_j}{n_j(n_j - d_j)} \tag{4.67}$$

Dixon and Newman (1991) point out that Equation (4.67) reduces to Equation (4.68), the binomial variance, for all times before termination of the experiment if there has been no censoring before termination of the experiment.

$$\hat{\sigma}^2(t_i) = \frac{\hat{S}(t_i)[1 - \hat{S}(t_i)]}{N} \tag{4.68}$$

where N = the total number of individuals exposed.

The standard error for the $\hat{S}$ is estimated with Equation (4.69) (Marubini and Valsecchi 1995) by again assuming a Gaussian approximation for the binomial distribution and no censoring.

$$SE(t_i) = \sqrt{\frac{\hat{S}(t_i)\left[1 - \hat{S}(t_i)\right]}{N - 1}} \tag{4.69}$$

The SAS software package (SAS 1988) can estimate the approximate 95% confidence intervals using ±1.96 times the square root of the variance from Equation (4.68) (see Miller 1981, p. 52). According to Dixon and Newman (1991), such approximations might not be appropriate for small numbers of observations because these approximations assume that the survival estimate is normally distributed. More appropriate confidence intervals for the $\hat{S}$ can be generated using other methods described in sources such as Marubini and Valsecchi (1995, pp. 56–63) and Kalbfleisch and Prentice (1980, pp. 14–15).

The above methods can be used to estimate the $S(t)$ for an exposed group of individuals. Two or more survival curves can also be examined for equality using these methods. Such methods can be used to test if mortality data generated from replicate tanks can be assumed to be homogeneous. They can also be used to test this same assumption for survival data from different treatments. Both the log-rank and Wilcoxon methods test whether the null hypothesis is false that the observed times to death from two (or more) samples came from the same survival distribution (Dixon and Newman 1991). Both give similar results, but the Wilcoxon is more sensitive to deviations in deaths early in the exposure trial than the log-rank test. Both are described generally in Dixon and Newman (1991). The log-rank test is described in detail in Mantel (1966) and Cox (1972). The Wilcoxon test is described in Peto and Peto (1972). (An alternative weighted Kaplan–Meier statistic was described by Pepe and Fleming 1989. Although it will not be used here, the reader is referred to the above method if hazard plots cross each other for the two sets of data being compared.)

Example 4.9

The proportions of mosquitofish dying after 96 h of exposure to various concentrations of NaCl were used in Examples 4.2, 4.3, and 4.4. Actually, the proportion dying at each salt concentration was derived for pooled numbers from duplicate tanks for convenience of the examples. Further, time-to-death data were collected, not just 96 h mortality data. Was it appropriate to pool the duplicate tanks in the examples? Let's address this question (testing for homogeneity) using the null hypothesis that the observed times to death in duplicate tanks came from the same survival distribution. The SAS code to implement the nonparametric procedures for the Wilcoxon and log-rank tests is the following:

```
DATA TOXICITY;
   INFILE "B:TOXICITY.DAT";
   INPUT TTD 1-2 TANK $ 4-5 PPT 7-10 WETWT 12-16 STDLGTH 18-20;
   IF TTD>96 THEN FLAG=1;
      ELSE FLAG=2;
RUN;
PROC SORT;
   BY PPT TANK TTD;
RUN;
PROC LIFETEST;
   TIME TTD*FLAG(1);
   STRATA TANK;
   BY PPT;
RUN;
```

The results are tabulated below for the 10.3, 10.8, 11.6, 13.2, and 15.8 g NaCl/L duplicate tanks. Estimation of statistics for the 20.1 g NaCl/L treatment was not possible because the extremely rapid deaths resulted in insufficient points for the associated survival curve.

Treatment	Test	χ^2	df	Associated P
10.3	Log-rank	2.7926	1	0.0947
	Wilcoxon	2.8142	1	0.0934
10.8	Log-rank	0.6140	1	0.4333
	Wilcoxon	0.7154	1	0.3976
11.6	Log-rank	0.2320	1	0.6300
	Wilcoxon	0.1166	1	0.7327
13.2	Log-rank	2.8338	1	0.0923
	Wilcoxon	4.4549	1	0.0348
15.8	Log-rank	2.8358	1	0.0922
	Wilcoxon	2.8641	1	0.0906

With one exception in the 10 tests, the null hypothesis could not be rejected ($\alpha = 0.05$). The χ^2 for the Wilcoxon test using the duplicate 13.2 g NaCl/L treatments had an associated probability less than 0.05, although that for the log-rank had an associated probability greater than 0.05. This is understandable because the Wilcoxon test is sensitive to deviations in deaths early in the exposure trial and there was a slight inequality in salt concentrations as the tanks initially filled. However, beyond the first few points on the survival curves, these curves were close to each other. Given these facts and the results of the log-rank test, the duplicate tanks were deemed sufficiently homogeneous to pool.

4.3.2.3 Parametric and Semiparametric Methods

4.3.2.3.1 General

Parametric and semiparametric methods are also available for the analysis of time-to-death data. Indeed, survival time methods incorporating specific models are common in health sciences, engineering, and economics (Cox and Oakes 1984). Although their common origins with life table methods can be traced back to the European plague years of the 1660s (Blackstone 1986), the extensive use of survival (or failure) time methods did not begin until World War II (Miller 1981). Today, they are used to address diverse topics ranging from mechanical failure (Nelson 1969, 1972), to coronary disease risk factors (Pryor et al. 1983), to cancer mortality (Lew et al. 1983). They have more recently been introduced into ecological disciplines such as population genetics, e.g., Manly (1985, Chapter 5). Unfortunately, they are only now beginning to be used by ecotoxicologists.

Survival time models can take several forms. Cox and Oakes (1984, Table 2.1) and Miller (1981, Chapter 2) provide explicit survival and hazard functions for the more commonly modeled distributions. Proportional hazard models use the hazard of a reference group or type as a base hazard and then scale (make proportional) the hazard of other groups to that baseline hazard. For example, the hazard of contracting lung cancer for smokers might be compared to the baseline hazard for nonsmokers. In contrast, accelerated failure models use functions that describe the change in Ln time to death resulting from some change in covariates. Continuing the example, the effect of smoking on Ln time to death may be estimated with an accelerated failure model. Both forms of survival models are described below.

4.3.2.3.2 Proportional Hazard Models

4.3.2.3.2.1 Assuming a Specific Model — The general expression of a proportional hazard model is the following:

$$h(t,x_i) = e^{f(x_i)}h_0(t) \tag{4.70}$$

where $h(t,x_i)$ = the hazard at time t for a group or class x_i, $h_0(t)$ = the baseline hazard, and $e^{f(xi)}$ = a function that relates the $h(t,x_i)$ to the baseline hazard.

The $f(x_i)$ in Equation (4.70) is some function used to fit the data set. It can be used to fit a continuous variable, such as fish weight, or class variables, such as fish sex or type of treatment. A vector of coefficients and a matrix of covariates can be used if more than one covariate is incorporated into the model.

4.3.2.3.2.2 Cox Proportional Hazard Model — The proportional hazard models described above assume that a specific distribution describes the baseline hazard and that hazards between classes are proportional. If the baseline distribution is not apparent or if it is not desirable to select a specific distribution, a semiparametric method is available. The Cox proportional hazard model retains the assumption of proportional hazards but uses a family of Lehmann alternatives (see Kotz and Johnson 1983, pp. 598–601) to fit the baseline hazard (Newman and Aplin 1992). Cox proportional hazard models are common in clinical studies because, in many cases, the underlining distribution is not as important as the relative hazards for the classes of interest. They are uncommon in ecotoxicology but were applied to describe poisoning of mosquitofish with mercury (Heagler et al. 1993) and honeybees with pesticides (Dechaume Moncharmont et al. 2003).

4.3.2.3.3 Accelerated Failure Models

Another form of survival model is the accelerated failure time model,

$$Ln\ t_i = f(x_i) + \varepsilon_i \tag{4.71}$$

where t_i = the time to death, $f(x_i)$ = a function that relates Ln t_i to the covariate(s), and ε_i = the error term.

In this case, the Ln time to death, not the hazard, is modified by $f(x_i)$.

4.3.2.3.4 General Form for Survival Time Models

As described by Dixon and Newman (1991), the accelerated failure time model can be converted to the form of a hazard model.

$$h(t, x_i) = e^{f(x_i)} h_0(t, e^{f(x_i)}) \tag{4.72}$$

Dixon and Newman (1991) give examples of the common types of regression functions that can be used for $f(x_i)$. For example, they include such functions as a linear equation ($a + bX$) for a continuous variable. Linear models for log-transformed variables or polynomial models may be used also. For class variables such as sex or treatment type, the function can simply estimate the mean response for each class. Candidate functions used to model the error distribution include the exponential, Weibull, Gompertz, normal, lognormal, log logistic, and gamma distributions. Other, less common functions are tabulated on page 17 of Cox and Oakes (1984).

If the exponential or Weibull function is selected for the distribution, the model (Equation 4.72) can take the form of a proportional hazard model (Equation 4.70) (Dixon and Newman 1991). The only difference among these functions is the way in which time to death is incorporated.

Exponential:

$$Ln\ h(t) = a + bX \tag{4.73}$$

Weibull:

$$Ln\ h(t) = a + bX + c\ Ln(t) \tag{4.74}$$

where a, b, c = constants, and X = the independent variable.

The hazard is constant for the exponential model over the duration of exposure (time). For the Weibull distribution, the Ln of the hazard either increases ($+c$) or decreases ($-c$) linearly over the duration of exposure.

4.3.2.3.5 Selecting the Appropriate Model

Given the variety of candidate functions that can be selected for formulation of survival models, the question becomes: Which function is the best?

There are several means of addressing this question. Various linearizations are used for the exponential, Weibull, normal, lognormal, and log logistic models.

Model	**Y**	**X**
Exponential	Ln $S(t)$	t
Weibull	Ln(–Ln $S(t)$)	Ln t
Normal	Probit ($F(t)$)	t
Lognormal	Probit ($F(t)$)	Ln t
Log logistic	Ln($S(t)/F(t)$)	Ln t

The transformed data are plotted for the candidate models. If a model is appropriate, the pertinent plot will produce a straight line. If, for example, there are several classes within the data set, e.g., fish exposed to seven different salt concentrations, curves for each of the seven subsets of fish can be plotted. If a range of individuals differing by a continuous variable is exposed, e.g., fish of different sizes, lines for arbitrary groupings (very small, small, medium, large) can be plotted. This approach can be limited if the number of individuals in the particular group being plotted is small. In that case, the variability in the plot can make it difficult to assess the linearity of the plots.

The log-likelihood statistic generated by statistical packages that fit survival models can also be used in model selection. The log-likelihood statistic can be used directly if the number of parameters estimated for each candidate model is the same. For example, the Weibull, lognormal, and log logistic have the same number of parameters to be estimated, and consequently, the associated log-likelihood estimates for these three models could be compared directly, much as the sums of squares are compared for models fit with least-squares methods. The model with the largest log-likelihood value fits the data best. This same approach could be used also in testing various transformations of covariates to select the best transformation.

Harrell (1988) and Dixon and Newman (1991) describe the likelihood ratio test for more formally assessing improvement of fit for nested models. Twice the difference in the log-likelihood values for two models approximates a χ^2 distribution for samples with large numbers of observations. (Harrell 1988 states that this χ^2 statistic is roughly equivalent to the Pearson χ^2 statistic.) The χ^2 value for the models being assessed is compared to a critical χ^2 under the null hypothesis of no significant difference. (The associated degrees of freedom are calculated as the difference in the number of estimated parameters between the two models being compared.) Dixon and Newman (1991) point out that this likelihood ratio test is only valid if one of the models is nested in the other. They gave an example of its use for deciding if a linear (oxygen concentration used as a continuous variable) or quadratic (oxygen concentration and the square of the oxygen concentration both used as continuous variables) model for $f(x)$ was the most appropriate for describing survival of oxygen-stressed fish (Shepard 1955).

Akaike's information criterion (AIC) can be used to compare the appropriateness of a series of candidate models varying in complexity, i.e., number of parameters to be fit (Atkinson 1980; Harrell 1988). It adjusts the log-likelihood values to account for the differences in the number of parameters available to fit the data set. This criterion, as applied here, has a different form than that given in Chapter 3.

$$AIC = -2\left(Log\ likelihood\right) + 2P \tag{4.75}$$

where P = the number of parameters being fit with the model.

The AIC values are compared for the various models; the model with the lowest AIC has the best fit.

It must be noted that the methods described above indicate the relative fits of different models to the data set. It is possible that none of the candidate models fit the data. The adequacy of the selected model should be examined by plotting the predicted and observed data together.

Example 4.10

The sodium chloride toxicity data generated by Newman and Aplin (1992) can now be fully analyzed using survival time modeling methods. Both covariates (fish wet weight and salt exposure concentration) will be incorporated. To illustrate various aspects of the process, the data will be fit with untransformed covariates and then with transformed covariates.

UNTRANSFORMED COVARIATES

What underlying distribution should be assumed? The series of linearizations listed earlier in this chapter can be used during the first attempts at addressing this question. $F(t)$ and $S(t)$ are estimated by the cumulative mortality (number dead/total number) and cumulative survival (number surviving/total number) for all times up to the termination of the exposure. The appropriate transformations of these metameters and time (duration of exposure) are then plotted. Figure 4.11 shows the results for the Weibull model linearization. This linearization appeared

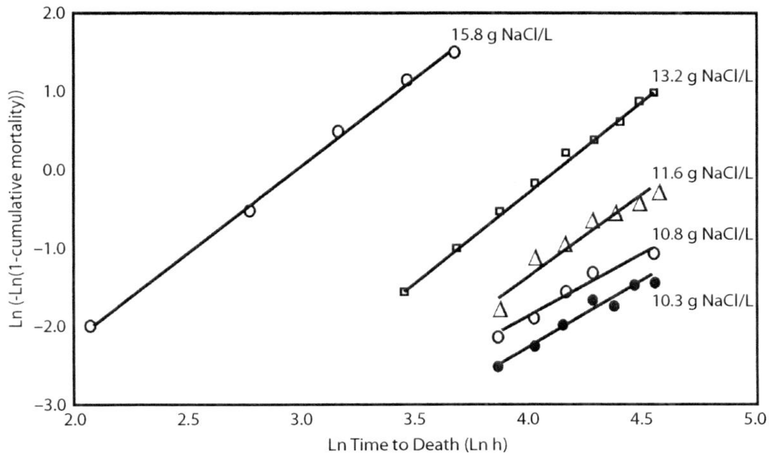

Figure 4.11 Linear transformation of times to death for mosquitofish exposed to five concentrations of sodium chloride (Example 4.9). The other exposure concentrations (<1 and 20.1 g NaCl/L) produced insufficient numbers of observations to be used in this plot. The generally straight and parallel lines suggest that the Weibull distribution was adequate for describing these data. (Modified from Newman, M.C., and M. Aplin, *Aquatic Toxicol.*, 23, 85–96, 1992. With permission from Elsevier.)

best as judged by visual comparison with those for the other distributions. (Note that no customary linearization is available for the gamma distribution.)

The log-likelihood statistic can also be used to compare candidate models if they are estimating the same number of parameters. As the models vary in the number of parameters being estimated, the AIC (Equation 4.75) is used. The following SAS code fits these data to survival models assuming exponential, Weibull, lognormal, log logistic, and gamma distributions.

```
DATA TOXICITY;
     INFILE "B:TOXICITY.DAT";
     INPUT TTD 1-2 TANK $ 4-5 PPT 6-10 WETWGT 12-16;
     IF PPT>0;
     IF TTD>96 THEN FLAG=1;
ELSE FLAG=2;
RUN;
PROC SORT;
   BY PPT WETWGT TTD;
RUN;
PROC LIFEREG;
   MODEL TTD*FLAG(1)  = PPT WETWGT/DISTRIBUTION=EXPONENTIAL;
   MODEL TTD*FLAG(1)  = PPT WETWGT/DISTRIBUTION=WEIBULL;
   MODEL TTD*FLAG(1)  = PPT WETWGT/DISTRIBUTION=LNORMAL;
   MODEL TTD*FLAG(1)  = PPT WETWGT/DISTRIBUTION=LLOGISTIC;
   MODEL TTD*FLAG (1)= PPT WETWGT/DISTRIBUTION=GAMMA;
RUN;
```

The models generated the log-likelihood values tabulated below. The AIC was calculated using the log-likelihood statistic and the number of parameters being estimated. (How one determines the number of estimated parameters will be discussed.)

Distribution	Log-Likelihood	Number of Parameters	AIC
Exponential	–385	3	776
Weibull	–194	4	396

Distribution	Log-Likelihood	Number of Parameters	AIC
Lognormal	–202	4	412
Log logistic	–198	4	404
Gamma	–193	5	396

The AIC is smallest for the Weibull and gamma distributions, suggesting that they both provide better fits to the data set than the other distributions. Both the Weibull and gamma distributions are generalized exponential functions. However, as the Weibull distribution results in a proportional hazard model, it will be used here to illustrate several points. The resulting proportional hazard (Weibull) model follows.

Variable	df	Estimate	Standard Error	χ^2	$P > \chi^2$
Intercept, μ	1	7.8579	0.0853	8,487	<0.0001
[NaCl] β_s	1	–0.2953	0.0052	3,258	<0.0001
Wet wgt β_w	1	0.0602	0.2566	17	0.0001
Scale, σ	1	0	0.3046	0.0137	

For this model, four parameters (μ, β_s, β_w, σ) are estimated. The χ^2 estimates for β_s and β_w suggest that the salt concentration and fish wet weight have significant effects on time to death ($\alpha = 0.05$). The negative sign associated with β_s indicates that the time to death decreases as the salt concentration increases. The positive sign for β_w indicates that time to death is shortened as fish wet weight decreases.

The median time to death can be estimated with this proportional hazard model (Equation 4.70).

$$MTTD = e^{\mu} e^{\beta_W Wgt + \beta_S [NaCl]} e^{\sigma W} \tag{4.76}$$

where Wgt = wet weight of fish (g), $[NaCl]$ = salt concentration (g NaCl/L), and W = response metameter for Weibull with $P = 0.50$ (Appendix 5).

The median time to death can be estimated with this model for any salt concentration or fish wet weight within the ranges used to generate the model (Figure 4.12). The median time to death

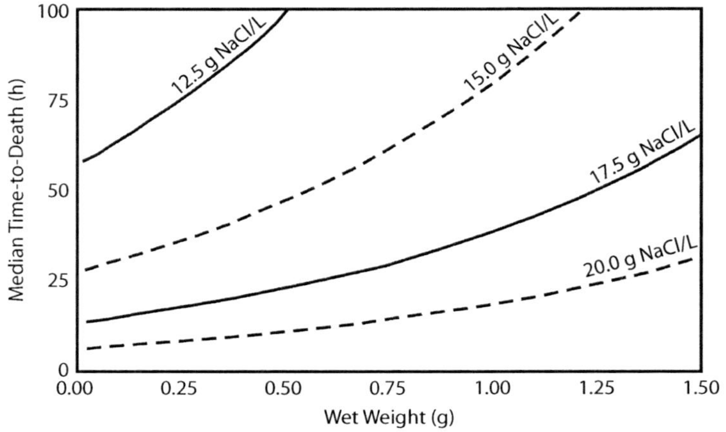

Figure 4.12 Predicted median times to death for mosquitofish exposed to various salt concentrations. Predictions are made over a wide range of fish sizes (wet weight). See Example 4.9 for further details. (Modified from Newman, M.C., and M. Aplin, *Aquatic Toxicol.,* 23, 85–96, 1992. With permission from Elsevier.)

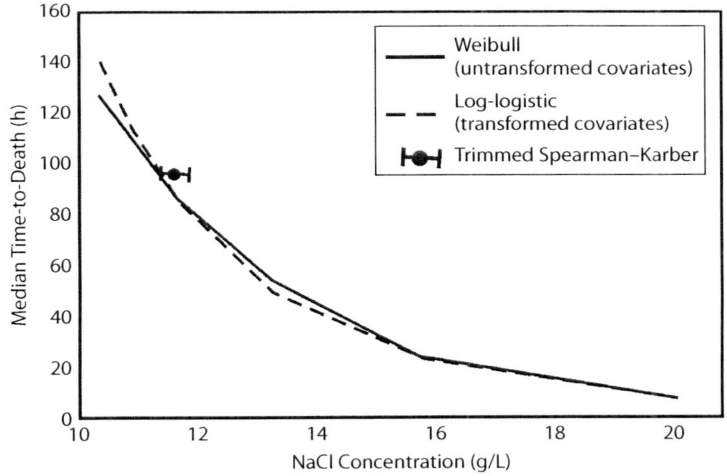

Figure 4.13 Comparison of the trimmed Spearman–Karber estimate of 96 h LC50 to concentrations corresponding to predicted median times to death of 96 h using the Weibull model with untransformed variable (salt concentration and wet weight) and log logistic model with transformed variables (Ln salt concentration, Ln wet weight). See Example 4.9 for further details. (Modified from Newman, M.C., and M. Aplin, *Aquatic Toxicol.*, 23, 85–96, 1992. With permission from Elsevier.)

(and the associated standard error) for the average size fish (0.136 g) can be calculated within the range of salt concentrations tested (Figure 4.13). The resulting curve of median time to death versus salt concentration is consistent with the trimmed Spearman–Karber 96 h LC50 estimate calculated in Example 4.4. Values for *W* for other proportions can be used to estimate times to death other than that of the median.

The 96 h LC50 may also be estimated directly from the model.

$$96hLC50 = \frac{Ln96 - \mu - \beta_W Wgt - \sigma W}{\beta_S} \tag{4.77}$$

For a fish of average weight (0.136 g), the 96 h LC50 is estimated to be 11.26 g NaCl/L.

With this proportional hazard model, relative risk can be estimated for fish of various sizes, or fish exposed to various salt concentrations (Dixon and Newman 1991). Newman and Aplin (1992) gave the following example (incorrectly). The relative risk of a 0.10 g fish to that of a 1.0 g fish is $e^{-\beta_W(\Delta Wgt)/\sigma}$, where Δ^{Wgt} is the difference in weight of the fish. With the Weibull model, the relative risk is $e^{-1.0602(0.9\ g)/0.3046}$, or 22.9. The smaller fish has a risk 22.9 times higher than the larger fish. Similarly, for an increase in salt concentration of 10 g/L, the relative risk for mosquitofish is $e^{-\beta_S(\Delta[NaCL])/\sigma} = 16{,}231.053$. Risk increases approximately 16,231 times with a 10 g/L increase in salt concentration. Further, the risk of a 0.1 g fish at 20 g/L is 22.7·16,231, or 371,690 times greater than a 1.0 g fish at 10 g/L.

TRANSFORMED COVARIATES

The data analysis in Newman and Aplin (1992) ended with the above calculations. The covariates, salt concentration, and fish wet weight remained untransformed for purposes of illustration only. However, from the material covered to this point, one could object that a better concentration metameter to use might have been Ln of concentration. Also, as will be discussed soon, one could easily have argued for the use of Ln of wet weight, not wet weight. Let's use these arguments to illustrate how one might assess appropriate transformations of covariates using the AIC.

The SAS LIFEREG code given above was run three additional times using various transformation combinations of wet weight and salt concentration: Ln wet weight and g NaCl/L,

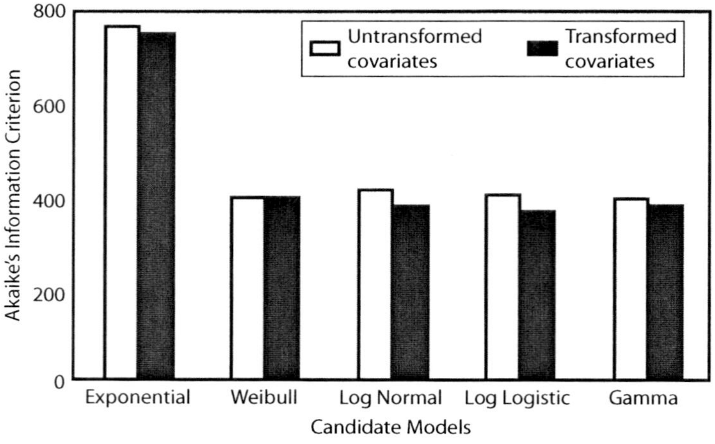

Figure 4.14 Model selection for sodium chloride–mosquitofish toxicity data using the Akaike's information criterion. See Example 4.10 for more details.

wet weight and Ln g NaCl/L, and Ln wet weight and Ln g NaCl/L. The first two combinations yielded no improvement in the AIC values relative to the models using wet weight and g NaCl/L. However, the AIC values were improved through the use of logarithms for both covariates (exponential: 758, Weibull: 396, lognormal: 380, log logistic: 372, gamma: 380) (Figure 4.14). The log logistic and lognormal models both had AIC values that dropped approximately 30 units when the transformed variables were used. The log logistic model incorporating the transformed variables had the lowest AIC of all models examined. Although the original Weibull model was adequate for the intended illustration, the AIC indicates that other models, e.g., the log logistic model using transformed covariates, are more appropriate for these data. Indeed, the predicted median times to death for a 0.136 g mosquitofish over a range of NaCl concentrations (Figure 4.13) compares more favorably for the log logistic model with transformed covariates than those for the Weibull model to the widely accepted, trimmed Spearman–Karber 96 h LC50 value. (Equations 4.76 and 4.77 can be used provided the transformed covariates are used and the appropriate W parameter for the logistic model is used.)

Although only mentioned briefly above, more biologically derived functions that link accumulation kinetics and toxicodynamics can be developed and hold great promise for predicting toxicological consequences under different exposure scenarios. Kooijman and coworkers have worked for some time to insert such an approach into the current regulatory framework, producing some interesting techniques, e.g., DEB model (Jager et al. 2006). The general toxicokinetic-toxicodynamic approach is based on the premise set out in the pleasingly simple equation (Rozman and Doull 2000)

$$\frac{dT}{dE} = \frac{dT}{dD}\frac{dD}{dK}\frac{dK}{dE} \tag{4.78}$$

where T = toxicity, D = toxicodynamics, K = toxicokinetics, and E = exposure.

Toxic effect is a consequence of exposure, not simply external concentration or ingested dose. It involves delivery of a substance through time to the target tissue(s) or site(s), and the consequent toxicodynamics, including damage remaining postexposure. Many, but not all, applications of such an approach in ecotoxicology focus on a critical body burden. The critical body burden is defined simply as the concentration that must be reached in the body or a specific target location in the body in order for a lethal effect to occur. In contrast, Legierse et al. (1999) model aquatic biota uptake of organophosphorus pesticide and link effect to consequent irreversible inhibition of acetylcholinesterase by their phase I oxon metabolites. They compare this critical target site/receptor occupation model to the more conventional critical body residue model as a means of incorporating intensity and duration of exposure into predictions of organophorus pesticide toxic consequences. Ashauer et al. (2006) link exposure duration and intensity in models involving toxicant that manifest lethal effect under other contexts, such as toxic effects linked to threshold hazard or cumulative damage. Exact formulation of these toxicokinetic-toxicodynamic survival models depends on toxic mode, which can range from simple general damage, as in the case of general oxidative damage by a metal, to narcosis of nonpolar organic compounds, to biocides with reversible or irreversible receptor binding, to hydrogen bond donor acidity-linked lethality (Ashauer et al. 2006). Mode of action will determine whether a critical level in the body, cumulative AUC, or cumulative or instantaneous hazard is the best exposure metric to relate to lethal consequence. As summarized in Figure 4 in Ashauer et al. (2006), the number of parameters requiring estimation varies widely among candidate survival time models. The potential user must judge the level of detail required to answer the question being addressed during the survival modeling effort.

4.4 QUANTIFYING THE EFFECTS OF EXTRINSIC FACTORS*

4.4.1 Overview

Quantitative treatment of some of the more common external modifiers of toxic response will be presented in this section. Such factors can be incorporated into survival time analyses as demonstrated earlier with salt concentration. Although done infrequently, they can also be incorporated into endpoint methods such as probit and logit models fit with maximum likelihood methods. For example, the probit procedure in the SAS package (SAS 1988) readily incorporates covariates. However, the ability to effectively incorporate covariates is limited with endpoint models relative to survival time models because less information is extracted from the mortality trial. For example, only six data pairs (proportion dead at 96 h versus salt concentration) were used in the probit procedure to estimate the LC50 for the mosquitofish toxicity data, but hundreds of data pairs (time to death versus salt concentration) were used in analyzing the results of this toxicity test with survival time models.

Many of the factors discussed influence toxic impact by shifting the distribution of the total amount of toxicant among different chemical species and physical phases. Here, they are presented in the context of toxic impact.

4.4.2 Inorganic Toxicants

4.4.2.1 Ammonia

The influence of pH and temperature on ammonia toxicity provides a clear illustration of toxic impact modification by extrinsic factors. Shifts in pH strongly influence ammonia speciation by modifying the equilibrium concentrations of nonionized (NH_3) and ionized ammonia (NH_4^+).

$$NH_3 + nH_2O \leftrightarrow NH_3 \cdot nH_2O \leftrightarrow NH_4^+ + OH^- + (n-1)H_2O$$

* The qualifiers, intrinsic and extrinsic, are being used here only to separate factors that are characteristics of the organism or not. Size of an organism is an intrinsic factor but the lipophilicity of a toxicant is an extrinsic factor.

Temperature and ionic strength also influence the distribution defined by this equation, but to a lesser extent (Hillaby and Randall 1979; Thurston et al. 1981; Gersich and Hopkins 1986).

The nonionized ammonia appears to be the most toxic form, notionally because of its ease of diffusion across gills. Ionized ammonia is subject to much slower transport.

An increase in pH shifts the equilibrium to the left in the above equation and, consequently, increases toxic impact. In the range of pH values found in freshwater, an increase of one pH unit is estimated to increase the nonionized ammonia concentration 10-fold. Emerson et al. (1975) provide the following convenient method for deriving an estimate of nonionized ammonia for various pH and temperature conditions at zero salinity.

The ionization constant (K_a) for the abbreviated equation, $NH_4^+ \leftrightarrow H^+ + NH_3$, is the following:

$$K_a = \frac{[H^+][NH_3]}{[NH_4^+]} \tag{4.79}$$

Empirical fit of data yielded the following regression line for pK_a (Emerson et al. 1975). (Remember that $pKa = -\log Ka$.)

$$pK_a = 0.09018 + \frac{2729.92}{T} \tag{4.80}$$

where T = temperature in degrees Kelvin (degrees Celsius + 273.15).

The percentage of the total ammonia ($NH_3 + NH_4^+$) present as un-ionized ammonia is estimated at a specific pH by using the pK_a from Equation (4.80) in Equation (4.81).

$$Percent = \frac{100}{10^{pK_a - pH} + 1} \tag{4.81}$$

Equations (4.80) and (4.81) are appropriate for estimating percent of total ammonia present as nonionized ammonia within the temperature range of 0 and 30°C, and pH range of 6.0 to 10.0. As Equation (4.80) is an empirical relationship, this method cannot be used outside of conditions under which the regression line was generated.

Example 4.11

What is the change in the percentage of total ammonia present as nonionized ammonia if pH is increased from 7.0 to 7.9 and temperature is increased from 15°C to 28°C?

a. Percentage at pH = 7.0 and temperature = 15°C. The pKa estimation using Equation (4.80):

$$pK_a = 0.09018 + 2729.92/(15 + 273.15) = 9.56413$$

Estimation of percentage nonionized ammonia from Equation (4.81):

$$\text{Percent} = 100/(10^{9.56413-7.0} + 1) = 0.272\%$$

b. Percentage at pH = 7.9 and temperature = 28°C:

$$pK_a = 0.09018 + 2729.92/(28 + 273.15) = 9.15516$$
$$\text{Percent} = 100/(10^{9.15516-7.9} + 1) = 5.264\%$$

c. Difference in percentages: The percentage of nonionized ammonia shifted upward from 0.272% to 5.264%, or approximately 5%, with the change in pH and temperature.

Under the assumption that nonionized ammonia is the primary toxic species, total ammonia concentrations can be transformed to nonionized ammonia concentrations prior to use as a metameter in the methods described above. In doing so, the conditions under which the relationships were fit must be satisfied. Also, Thurston et al. (1981) indicate that ionized ammonia can have a small, but significant, impact on ammonia toxicity.

4.4.2.2 Metals

4.4.2.2.1 Water

Many factors can modify dissolved metal and metalloid toxicity. Examples include dissolved oxygen (Clubb et al. 1975), major cations (Muller 1980), phosphate (Freedman et al. 1980), and other ligands. Understanding metal speciation is central for understanding and potentially quantifying the influence of many factors. However, an understanding of speciation is not always sufficient, as will be discussed in the next section. According to the free ion activity model (FIAM), the aquated or free metal ion is assumed to be the most toxic form of a metal (Campbell and Tessier 1996; Di Toro et al. 2001), although other metal-ligand complexes may also be toxic (Andrew et al. 1977; Borgmann 1983). Consequently, speciation is estimated in recent studies of metal toxicity with emphasis on the aquo metal ion. Current practice also dictates that the chemical composition of waters used for toxicity tests be clearly defined so that speciation can be estimated later. Several programs have been developed to facilitate estimation of speciation. Loehle et al. (1986) reviewed the general approach and accuracy of several of the more common models. Most are valuable tools if their limitations relative to bioavailability and toxicity are understood. Currently, a Visual Basic version of the most widely applied MINTEQ program can be obtained from Jon Petter Gustafsson (KTH, Department of Land and Water Resources Engineering, Stockholm, Sweden, gustafjp@kth.se). Alternatively, analytical methods may be employed to quantify species, e.g., specific ion electrodes such as that for the cupric ion. These data also can provide valuable information regarding the influence of water quality on toxic impact of other inorganic toxicants.

Many aspects of metal and metalloid speciation remain extremely difficult to predict accurately. The poorly quantified role of natural organic ligands, incompletely defined chemical processes occurring at biological surface microlayers, and the ambiguous importance of chemical kinetics in some cases remain problematic. Realizing that a complete knowledge of speciation and associated ramifications is extremely difficult to obtain, empirical methods have been developed to quantify the influence of extrinsic factors on metal toxicity. Today, these methods are effectively used in a variety of applications, including water quality criteria and standard development. Their eventual linkage with models based on first principles remains a major goal in ecotoxicology, e.g., Meyer (1999).

Perhaps the best example of an empirical relationship of this type is that between water hardness and metal toxicity. Hardness has a strong effect on the toxic impact of dissolved metals such as beryllium (Slonim and Slonim 1973), cadmium (Carroll et al. 1979; Wright and Frain 1981; Pascoe et al. 1986), copper (Howarth and Sprague 1978), and zinc (Zitko and Carson 1976). Hardness is thought to influence metal toxicity by one or several mechanisms, including hardness metal (Ca^{2+} and Mg^{2+}) competition with the toxic metal for biological binding sites,* modification of biological processes such as ion regulation, or modification of speciation or solubility. As hardness is often correlated with alkalinity and ionic strength, indirect effects such as shifts in buffering capacity or ligand complexation have also been suggested.

Empirical relationships between metal toxic effect and water hardness commonly involve logarithmic transformations of hardness and toxic impact, e.g., see Slonim and Slonim (1973), Howarth

* This is the central premise of the biological ligand model (BLM); that is, metal bioactivity manifests if the amount of metal-ligand complexes exceeds a critical concentration (Di Toro et al. 2001).

and Sprague (1978), and Nelson et al. (1986). This approach is incorporated into water quality criteria for several metals; e.g., see the U.S. EPA criteria documents for copper, cadmium, or zinc (U.S. EPA 1985a, 1985b, 1987). Least-squares regression is used to fit the line,

$$Log\ LC50 = bLog\ H + Log\ a \tag{4.82}$$

where H = hardness, b = the regression slope, and $Log\ a$ = the regression intercept.

Next, the model is backtransformed to the form

$$LC50 = aH^b \tag{4.83}$$

However, as discussed in Chapter 3, such a backtransformed model generates biased predictions of mean toxic effect (mean LC50) (Newman 1991). This bias arises because the associated error term in the regression model was neglected during the backtransformation. The complete model is

$$Log\ LC50 = bLog\ H + Log\ a + \epsilon \tag{4.84}$$

where ϵ = the random error term in the model.

The error term is incorporated into the complete backtransformed model.

$$LC50 = aH^b 10^{\epsilon} \tag{4.85}$$

The term 10^{ϵ} can be estimated using the model mean square error if the residuals are believed to be normally distributed.

$$10^{\epsilon} = 10^{2.302(MSE/2)} \tag{4.86}$$

where MSE = the model mean square error, and 2.302 = Ln 10.

If the residuals are not normally distributed, a "smearing estimate" of bias (Koch and Smillie 1986) can be used.

$$LC50 = aH^b \left[\frac{1}{N} \sum_{i=1}^{N} 10^{r_i} \right] \tag{4.87}$$

where N = the number of observation pairs, and r_i = the ith regression residual.

Newman (1991) provides examples of this bias using data from U.S. EPA criteria documents. The bias ranges from 2 to 57% for the 12 sets of data examined. (An example of backtransformation bias correction is given in Chapter 3 relative to anthracene elimination kinetics.)

Unless the error term ($\in$) is 0, the LC50 predicted with Equation (4.83) will not be the mean predicted value of the median lethal concentration (LC50). As discussed in Chapter 3, the median of the median lethal concentration is predicted with Equation (4.83) (Miller 1984). The bias discussed above is not pertinent if prediction of the median is acceptable or desirable. However, it should be clear that such a decision has been made because the tendency is to assume that the mean value is predicted. The bias should be estimated if prediction of the mean response is desired. Minimally, the error mean square should be included in any publication reporting such a regression model. This allows future correction if desired.

The mechanism underlying the influence of hardness on metal toxicity was modeled successfully by Meyer (1999), who included H^+, Ca^2, a gill surface-associated ligand group ($\equiv L_G$), and a divalent metal ion, M^{2+}, in his illustrative model. The divalent metal ion competes with Ca^{2+} and H^+ for the gill-associated ligand, and also speciates as influenced by alkalinity components (i.e., OH^-, HCO_3^-, and CO_3^{2-}). The influence of the speciation and competition on the metal LC50 (expressed as total dissolved metal) are defined by Equation (4.88).

$$LC50 \cong k' \frac{[Ca^{2+}]}{[Ca \equiv L_G]}\left(1+\frac{k''}{10^{-pH}}+k''' \cdot A\right) \tag{4.88}$$

where k', k'', and k''' = constants, A = alkalinity ($CaCO_3$ equivalents per L), $[Ca^{2+}]$ = concentration of Ca^{2+} ion, and $[Ca \equiv L_G]$ = concentration of metal binding sites that are bound with Ca.

Models have also been produced to quantify intermetal trends in bioactivity, including lethality, based on differences in metal coordination chemistry (Newman and McCloskey 1998; McCloskey et al. 1996; Newman and Clements 2008; Newman et al. 1998; Tatara et al. 1998). Simply put, a metal's coordination chemistry influences bioavailability, internal transformations, distribution among internal pools, elimination, and adverse effect. Most models use metrics of binding tendencies as described by the hard and soft acids and bases (HSAB) theory. In this context, the metal and ligand donor atom are Lewis acids and bases, respectively. Class A metals are hard Lewis acids and B metals are soft Lewis acids. The descriptors *hard* and *soft* refer to how readily the metal's outer valence shell deforms during metal-ligand interaction. Metrics of qualities such as metal ion softness have been refined, e.g., Jones and Vaughn (1978) and Kinraide (2009), to such an extent that they can be used for predicting trends in toxicology. One such model for 20 mono-, di-, and trivalent metal ions' effect on bacterial bioluminescence is depicted in Figure 4.15 (top panel). The EC50 data from McCloskey et al. (1996) were regressed against a recently refined softness index (Kinraide 2009) to show that effect increases as metal softness increases. Publications describing such QICARs appear at increasing frequency in the literature.

Example 4.12

The QICAR model depicted in Figure 4.15 was generated with the SAS code below.

```
/* This code models metal ion softness index against EC50 values for  */
/* all metals from Microtox toxicity data of (McCloskey et al. 1996). */
/* TOTLEC is the log (base 10) of the EC50 (uM/L) after 15 minutes    */
/* exposure and is expressed as total dissolved metal, not the free   */
/* ion. This also includes Hg for which the neutral chloride species  */
/* were not considered. Note that SOFTCON is the computed softness    */
/* index Sigma Con Comp obtained from Kinraide 2009 Env. Tox. Chem.   */
/* 28:525-533, Table 2 value.                                         */

DATA QICAR;
   INPUT METAL $ TOTLEC SOFTCON @@;
   DATALINES;
HG2+ -0.037 1.16 CA2+ 4.976 -0.99 CD2+ 1.424 0.17 CU2+ 0.208 0.65
MG2+ 4.941 -1.02 MN2+ 3.196 -0.20 NI2+ 2.753 0.29 PB2+ 0.061 0.46
ZN2+ 1.547 -0.09 CO2+ 2.942 0.27 CR3+ 2.265 0.02 FE3+ 2.009 0.34
CS1+ 5.606 -0.63 K1+ 5.796 -0.73 SR2+ 5.372 -0.88 LI1+ 5.469 -0.97
NA1+ 5.603 -0.80 BA2+ 4.980 -0.76 LA3+ 3.229 -0.53 AG1+ -0.034 0.84
;
RUN;
PROC GLM;
        MODEL TOTLEC=SOFTCON;
RUN;
```

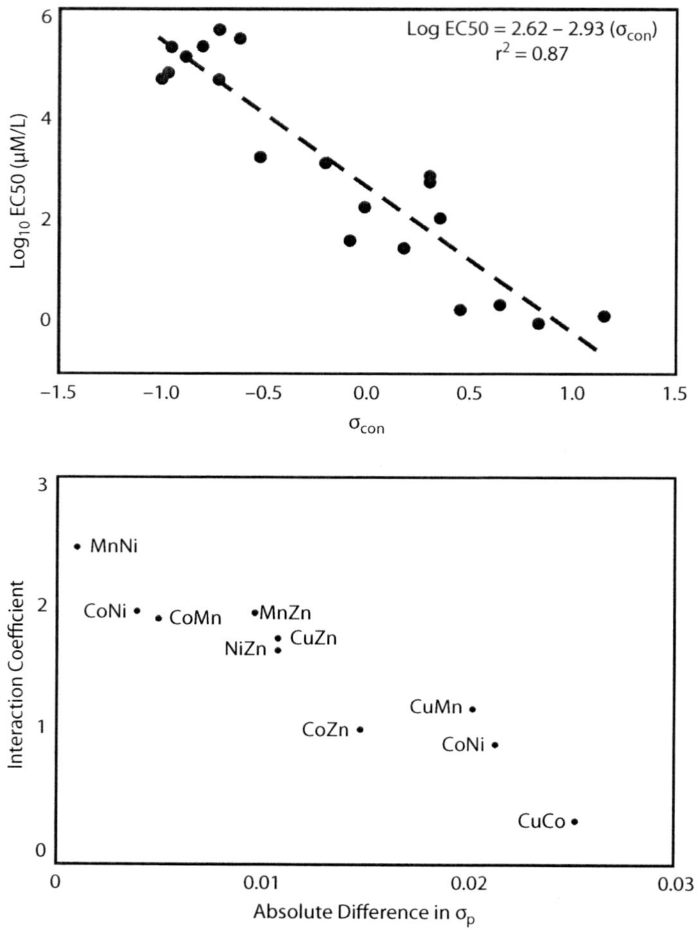

Figure 4.15 Quantitative ion characteristic–activity relationship (QICAR) for predicting intermetal ion toxicity trends (top panel) and interactions between paired metal ions (bottom panel). Bacterial bioluminescence inhibition 15 min EC50 data for 20 mono-, di-, and trivalent metal ions (McCloskey et al. 1996) are modeled with Kinraide's softness index, σ_{con} (Kinraide 2009). The bottom panel is one of two making up Figure 9.8 in Newman and Clements (2008). These data were originally from Ownby and Newman (2003) and fit to the equation shown in Example 4.15, which is the same as Equation (4.105) except an interaction coefficient (ρ) is added. Deviations from perfect independent action for binary mixtures of metals (as indicated from $\rho \neq 1$) are predictable from differences in paired metal softnesses. Specific paired metals in mixtures are indicated next to each data point, e.g., MnNi = a binary mixture of Mn^{2+} and Ni^{2+}.

The following output was produced (r^2, slope, and intercept bolded):

```
                         The GLM Procedure

Dependent Variable: TOTLEC
                                              Sum of Mean
     Source            DF       Squares          Square       F Value     Pr > F
Model                   1    74.07573967    74.07573967     116.03      <.0001
Error                  18    11.49167253     0.63842625
Corrected Total        19    85.56741220

R-Square    Coeff Var     Root MSE     TOTLEC Mean
0.865700     25.64812     0.799016      3.115300

Source      DF     Type I SS      Mean Square     F Value     Pr > F
SOFTCON      1    74.07573967     74.07573967      116.03     <.0001

Source      DF    Type III SS     Mean Square     F Value     Pr > F
SOFTCON      1    74.07573967     74.07573967      116.03     <.0001

Parameter               Estimate          Error         t Value    Standard Pr > |t|
Intercept             2.616991813     0.18455730        14.18          <.0001
SOFTCON(Slope)       -2.931224627     0.27212376       -10.77          <.0001
```

Before moving our discussion to sediment-associated toxicity, it must be reiterated that many other extrinsic factors influence the toxic impact of dissolved metals and metalloids. Temperature (Cairns et al. 1975), pH (Newman and Jagoe 1993), and salinity (Fales 1978; MacInnes and Calabrese 1979) are three important examples. Much work remains to be done to quantify such effects in terms other than simple chemical speciation.

4.4.2.2.2 Sediments

Sediment toxicity received much attention during the 1980–1990s; e.g., see Cairns et al. (1984), Nebeker et al. (1984), DeWitt et al. (1989), and Burton (1991). Speciation in particles and interstitial water has received much deserved attention as sediment toxicity protocols and methods were developed. Despite these efforts, major obstacles impede our understanding of sediment chemistry. Luoma (1989) acknowledged this lack of essential information regarding sediment speciation and the consequences of this ignorance to prediction of sediment toxicity. Improved methods for analyzing metal distribution among sediment components, improved computational methods for predicting sediment-water exchange, and a better understanding of processes controlling bioaccumulation from solution and food are identified as areas where the necessary knowledge is clearly lacking.

Although our present knowledge is limited, several quantitative approaches suggest themselves to a researcher interested in quantifying toxic effects of metal-contaminated sediments. Several procedurally defined sediment fractions modify bioavailability of sediment-bound metals in surficial or oxic sediments. Easily extracted iron (1 N HCl or an equivalent extractant) or hydrous iron oxides tend to decrease bioavailability of silver, arsenic, cobalt, lead, and zinc (Luoma and Jenne 1977; Luoma and Bryan 1978; Cook et al. 1979; Langston 1980; Tessier et al. 1984; Campbell et al.

1988; Young and Harvey 1991). Manganese oxides can also lessen metal bioavailability (Young and Harvey 1991). Although the organic content of sediment can decrease bioavailability (Crecelius et al. 1982), the extent to which bioavailability is influenced varies widely between the benthic species-metal combinations. However, in some cases, metal concentrations can be normalized to sediment organic carbon content (Ogendi et al. 2007). Finally, when sequential extractions of sediments are performed, the most readily extracted sediment fractions are generally found to be the most available fractions (Tessier et al. 1984; Young and Harvey 1991).

As a consequence of these studies, empirical relationships between sediment-bound metals and bioaccumulation in the associated fauna have been developed between accumulated metal and some sediment fraction, e.g., metal in an EDTA extract (Ray et al. 1981) or exchangeable fraction (Rule and Alden 1990). Alternatively, since certain sediment qualities influence bioavailability, an empirical relationship can be developed for accumulated metal and normalized sediment metal concentrations, e.g., amount of metal divided by the amount of 1 N HCl extractable iron (Jenne and Luoma 1977; Luoma and Bryan 1978; Langston 1980). Although these techniques often improve quantification of metal bioavailability from oxic sediments, it is important to understand that they are empirical relationships; i.e., extrapolation beyond the data set used to generate the relationship is not strictly valid. Also, there is considerable variability in the literature regarding these methods. Although these methods often generate concentration metameters superior to total concentrations of metals, they remain imperfect tools at this time.

Under anoxic conditions, most of the metals in sediments are present generally as highly insoluble sulfides (Patrick et al. 1977; Di Toro et al. 1990). Consequently, toxicities of metals in anoxic or near-anoxic sediments are strongly influenced by S^{2-} (Patrick et al. 1977; Bryan 1985; Bjornberg et al. 1988). For this reason, normalization of metal concentrations to sulfide concentration has been used for estimation of potential toxic impact in anoxic sediments (Di Toro et al. 1990; Ankley et al. 1991; Carlson et al. 1991). Solid phase sulfides in sediments are estimated with acid volatile sulfides (AVSs). Procedurally the AVSs are sulfides extracted with 1 N cold HCl. They are believed to be predominately iron sulfides, especially metastable amorphous FeS, mackinawite (FeS), and greigite (Fe_3S_4) (Morse and Rickard 2004; Campbell et al. 2006). They could also include manganese sulfides. The molar concentration of metals extracted in cold HCl (simultaneously extracted metals (SEMs)) can be divided by the molar concentration of AVSs to yield a normalized estimate of metal available in anoxic sediments to have a toxic impact. For stoichiometric reasons, values of SEM/AVS < 1 were used initially to imply that all of the metal is precipitated with the sulfides and unavailable for toxic action. (The metal in the interstitial water is assumed to be the metal available to have a toxic impact. Those associated with solid phases are not considered. Ingestion and consequent bioaccumulation of solid phase metals have not been considered fully yet in this approach.) More recent applications use SEM/AVS > 0 to imply potential toxicity instead of SEM/AVS > 1. These methods can be used to enhance quantification of toxic effect of metals and metalloids in sediments. However, they are empirical relationships employing procedurally defined quantities. As such, they must be used with care (Long et al. 1998; Lee et al. 2000; Morse and Rickard 2004; Campbell et al. 2006). Considerable work remains to be done in this area of ecotoxicology. Much of the underlying chemistry remains only generally defined. For example, some benthic organisms create an oxic microenvironment via water pumping, and prediction based on anoxic geochemistry might be inadequate for such species (Campbell et al. 2006). Also, many species ingest sediment, and consequently, particulate-associated metals gain entry to organisms via ingestion in addition to uptake from interstitial waters.

4.4.2.3 Organic Toxicants

Chemical factors such as water hardness and pH, and also physical factors, can modify organic toxicant effects (Nimmo 1985). For example, light can modify the toxic impact of photolabile, organic contaminants that degrade to toxic products, e.g., anthracene (Bowling et al. 1983). Such phototoxicity of a chemical can occur in the presence of light due to the production of toxic photolysis products. Perhaps the most extensively quantified, extrinsic factors examined to date are the qualities of organic toxicants themselves. The associated models, quantitative structure-activity relationships (QSARs), focus on differences in structure (or some property consequent to structure) of a class of organic compounds and quantitatively relate them to bioactivity. Such QSARs range from simple, but still common, models relating a single molecular quality (e.g., lipophility) to bioactivity, to more complex computer-driven models relating the three-dimensional structures of molecules to their bioactivity (e.g., ligand structure–receptor site complementarity) (McKinney et al. 2000). The applicable molecular qualities can be broken down into measures of lipophilicity, steric conformation, molecular volume, ionization, polarity, and reactivity. In ecotoxicology, the most commonly used involve lipophility: the correlation between toxic effect and K_{ow} is one very familiar example in ecotoxicology. Simple empirical models have been generated to predict K_{ow} using known qualities of a parent and substituent structures in an organic chemical (Hansch et al. 1968, 1972). These relationships are very useful in predicting the toxic impact of many classes of organic compounds, although other factors, e.g., electrophilic qualities of the toxicant (McKinney et al. 2000) or attachment to electron-withdrawing functions on the molecule (Hansch et al. 1972; McKinney et al. 2000), often must be included as well.

More generally, any relationship linking the molecular structure and physical properties of a compound to its biological activity is called a structure-activity relationship (SAR). Quantitative structure-activity relationships are SARs that express relationships in quantitative terms. The most generally applied "are statistical models that are nearly always obtained by regressing values of a common test endpoint for a series of chemicals against one or more quantifiable properties of the chemicals" (Suter 1993). Commonly used properties are hydrophobicity (e.g., K_{ow}), topological (e.g., molecular connectivity index), electrical (e.g., Hammett constants and ionization potentials), and steric (e.g., total molecular surface area). The QSAR approach, used for many years in drug development, is now producing invaluable empirical relationships in ecotoxicology. Attempts are at hand to link such relationships to even more basic molecular qualities (Borman 1990).

One illustrative approach to quantifying the relationship between toxicity and structure of organic compounds is the classic additive constitutive approach. This approach estimates the influence of structural components on the lipophilic (hydrophobic) qualities of the toxicant. The lipophilicity is assumed to dictate narcotic activity through its influence on membrane penetration, bioaccumulation in target organs, and binding to proteins and other cell constituents (Borman 1990). The contribution of a structural constituent of the toxicant is estimated using the logarithms of the K_{ow} values for the parent molecule and the parent molecule with the structural component added. A parameter (π_X) is estimated (Hansch and Leo 1979).

$$\pi_X = Log\ K_{OWX} - Log\ K_{OWH} \tag{4.89}$$

where π_X = the logarithm of the partition coefficient for the structure added to the parent molecule (Hansch et al. 1972), K_{owX} = the K_{ow} for the derivative compound, and K_{owH} = the K_{ow} of the parent compound.

Tabulations of these π_X values can be used to estimate the K_{ow} of a specific compound within a class of compounds sharing a common parent structure. Hansch and Leo (1979) provide tabulations of π_X values. A K_{owH} is found from these or other tables for the parent chemical. The π_X values are found for each substituent added to that parent. According to additivity principles, the logarithm of K_{ow} for toxicant can then be estimated as the logarithm of the K_{ow} for the parent molecule plus the sum of the pertinent substituent π_X values.

$$Log\ K_{OW} = Log\ K_{OWH} + \sum_{i=1}^{n} \pi_i \tag{4.90}$$

The substituent addition to the parent compound favors the water phase if the Hansch π_X parameter has a negative sign. A positive parameter shifts the distribution in favor of the octanol phase.

This estimate of K_{ow} can be correlated to toxic activity (Laughlin et al. 1985) or bioavailability (Neely et al. 1974). Toxicity for untested compounds within the class used to establish the QSAR can be predicted with such relationships.

Similarly, molecular topology, electronic, or steric parameters may be used to establish empirical structure-activity relationships (Suter 1993; Borman 1990). Several characteristics may be used together in more complex QSARs. For example, a combined approach incorporating measures of hydrophobicity and degree of disassociation of an active group may be used for transport and toxic action of a weak electrolyte, a substance in equilibrium between nonionized and ionized forms (Lipnick 1985). Hansch and Leo (1979) provided general explanations of the basic types of QSARs. Suter (1993) provided a table of QSARs used by the U.S. EPA Office of Toxic Substances. It is strikingly apparent from Suter's table that hydrophobicity is used as the sole or primary quality of interest in most of these QSARs.

Laughlin et al. (1985) provide an illustration of QSAR application for a class of similar compounds with a notionally identical mode of action. They produced sound regression models of constituent π_X sums versus toxicity to mud crab zoeae for di- and triorganotin compounds. Ninety-four to 95% of the variance in toxicity among organotins is explained by the sum of the constituent π_X values. Computer estimations of total molecular surface area for these compounds were used to produce QSARs also. Ninety-three to 95% of the variance in toxic impact was explained by differences in the total surface area of the various organotins. These models using total surface areas and Hansch π parameters suggested to Laughlin et al. (1985) that partitioning was the best predictor of toxic impact for the organotin antifouling agents. This confirmed earlier work by these authors suggesting that electronic factors had little influence on toxicity. (Hansch and Leo 1979 provide a general introduction to electronic parameters such as the σ constants used by Laughlin et al. 1985.)

Example 4.13

A study of K_{ow} effects on chlorinated phenol toxicity to freshwater fish produces the relationship log (1/LC50) = 0.58log K_{ow} – 3.20 (Table 7.1 in Suter 1993). Predict the 96 h LC50 for 2,4-dichlorophenol using this relationship. First, calculate the predicted K_{ow} for 2,4-dichlorophenol assuming constituent additivity and a phenol parent structure to which the two Cl atoms are bound. Then use the above QSAR to estimate the 96 h LC50.

The median K_{ow} for phenol calculated from values in Appendix II of Hansch and Leo (1979) is 1.485. A π_{Cl} (0.71) is taken from Table VI-1 in Hansch and Leo (1979). These values are used in Equation (4.80) to estimate the log K_{ow} for this compound. The sum is 1.485 + 0.71 + 0.71, or 2.905. The K_{ow} for this compound is predicted to be the antilogarithm of 2.905. (Aside: The median of the three values given in Appendix II of Hansch and Leo (1979) for log K_{ow} of 2,4-dichlorophenol is 3.08.)

The LC50 can now be estimated using the calculated K_{ow}.

$$\text{Log}\ (1/\text{LC50}) = 0.58 \cdot \log K_{ow} - 3.20 = 0.58 \cdot 2.905 - 3.20$$
$$\text{Log}\ (1/\text{LC50}) = -1.5151$$
$$1/\text{LC50} = 10^{-1.5151} \text{ or } 0.03054$$
$$\text{LC50} = 32.74\text{, or approximately } 33\ \mu\text{moles/L}$$

Assessment of the toxic impact of organic compounds in sediments may also take advantage of QSARs. Di Toro et al. (1991) used K_{ow} as a measure of hydrophobicity of nonionic organic chemicals in their treatment of sediment toxicity. They assumed pragmatically that interstitial water concentrations determine the availability and toxicity of associated chemicals. The partitioning of an organic compound among the interstitial water, organic carbon in the sediment solid phases, and the organism determined its toxic impact. The toxicity of the organic compound in the interstitial water (as determined by partitioning) was shown to be similar to that of the chemical dissolved in the overlying water. The practical consequence is that sediment toxicity could be predicted from routine toxicity data for the dissolved compound if partitioning between the sediment organic carbon phase and the interstitial waters is quantified (Equation 4.91).

$$K_p = \frac{C_s}{C_d} = f_{oc}K_{oc} \tag{4.91}$$

where K_p = partition coefficient, C_s = concentration in the sediment, C_d = concentration in the interstitial water, K_{oc} = partition coefficient for the organic carbon phase of the sediments (assumed equal to K_{ow}), and f_{oc} = the mass fraction of organic carbon in the sediments.

The toxic impact of the nonionic organic compound can be estimated with knowledge of the sediment organic carbon content, toxicity of the organic compound when dissolved in water, and its K_{ow}. Implied by Equation (4.91) is the potential for toxic impact normalization based on the organic carbon content of sediments. Relative to toxic impact, sediment concentrations of nonionic organic compounds in sediments can be more accurately expressed in terms of amount/unit organic carbon than in terms of amount/unit sediment mass. Such a concentration metameter or a transformation of this metameter could be used in any of the above-described endpoint and survival time methods.

4.5 QUANTIFYING EFFECTS OF INTRINSIC FACTORS

4.5.1 Overview

Only two examples of common intrinsic factors will be discussed for the sake of brevity. Other important factors, such as stress hormone response to photoperiod or circadian rhythms (McLeay and Munro 1979), will not be discussed in balance with their importance, as they are normally controlled rather than quantified. Although the effects of an individual's sex have been quantified (Anderson and Weber 1975; Dixon and Newman 1991; Newman et al. 1989) and can be striking (Pallotta et al. 1962), they will not be discussed per se as they were discussed briefly in Example 4.10.

4.5.2 Acclimation

The term *acclimation*, as used by aquatic toxicologists, may have several different meanings. Physiological acclimation refers to an adaptive change in the individual in response to a change in environmental conditions. More specifically, physiological acclimation often refers to shifts taking place under controlled laboratory conditions, and acclimatization refers to those taking place under natural conditions. Acclimation may also be used to refer to the time period that an organism spends in an exposure system prior to toxicant addition, i.e., acclimation to test conditions. Herein, acclimation is used to "refer to a nonlethal exposure and the physiological responses thereto regardless of the effect of these factors on tolerance to subsequent exposure" (Chapman 1985).

Several approaches have been taken to quantify acclimation. For example, Shepard (1955) quantified physiological acclimation of trout to low oxygen conditions. Acclimation involved changes in the oxygen binding capacity of the blood. The incipient lethal level (*ILL*, mg O_2/L) was linearly

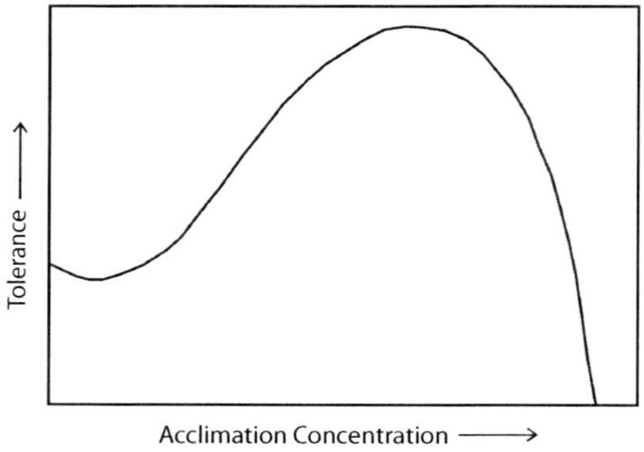

Figure 4.16 The influence of toxicant acclimation concentration on toxicant tolerance. Toxicant tolerance is expressed here as the difference between the copper incipient lethal level of nonacclimated and acclimated individuals. (From Bliss, C.I., *J. Exp. Biol.*, 13, 95–110, 1936, Figures 2 and 3. With permission.)

related to the acclimation level of oxygen (A, mg O_2/L): $ILL = 0.086A + 0.87$. He estimated the ultimate incipient lethal level (the lowest concentration to which fish may be acclimated and still experience 50% or less mortality) by solving this relationship for $ILL = A$.

Chapman (1985) described the general toxic response model incorporating acclimation to metals (Figure 4.16). This figure was based primarily on LC50 data. There is a general zone similar to that described above for oxygen-acclimated trout within which increasing acclimation concentrations increases the LC50. There is a maximum acclimation concentration above which tolerance decreases. There is also a lower limit below which tolerance decreases. This curve depicts the net results of decreasing tolerance due to damage and response to the metal that enhances tolerance. Induction of changes that lessen toxic impact can have a threshold concentration below which no response occurs, although damage still occurs. This produces a decrease in tolerance at low acclimation concentrations. The upper limit of tolerance is associated with the increasingly significant damage incurred during exposure to acclimation concentrations (Chapman 1985). Dixon and Sprague (1981a) observed such a curve when examining copper acclimation by rainbow trout (*Oncorynchus mykiss*). They used the change in the incipient lethal level for acclimated and nonacclimated trout as the response variable.

Dixon and Sprague (1981a) also included acclimation time in their treatment of copper toxicity to rainbow trout. By varying acclimation time (0 to 21 days) as well as acclimation concentration, they derived a response surface for change in tolerance. They described a response surface resembling a ski jump (Figure 4.16, with the added dimension of time with tolerance increasing to a plateau at higher acclimation times). They described the surface quantitatively with a multiple regression model. It incorporated log acclimation concentration (A), a transformation of acclimation time (T), A^2, A^3, T^2, T^3, AT, A^2T, and AT^2.

As suggested by Figure 4.16, not all preexposures produce enhanced tolerance. Although acclimation to arsenic enhanced tolerance to that metalloid, Dixon and Sprague (1981b) found a decreasing cyanide tolerance after 7 to 14 days of cyanide preexposure. They attributed this to significant kidney damage at all sublethal, preexposure concentrations.

4.5.3 Size

Early efforts in drug administration saw drug dosage (the amount of a substance given per unit weight or some other normalizing unit) calculated as directly proportional to an individual's mass, e.g., 3 mg drug/kg of body weight. Moore, as cited in Bliss (1936), argued that dosage should be adjusted to the amount of surface available for drug adsorption. He suggested use of the 2/3 power of body weight to facilitate scaling to absorptive surface. Rhomberg and Wolff (1998) reviewed the mammalian toxicology literature, stating that the ¾ power was more pervasive today, but different powers are estimated for many toxicological scaling relationships. Earlier, Bliss (1936) had argued against use of a constant factor and developed means of determining empirical constants for each situation. His work remains a foundation paper for quantification of size effects on toxicant or drug action.

Bliss used Campbell's time-to-death data for arsenic given orally to silkworm larvae (Campbell 1926) to formulate his approach. First, to combine data from various dosages, he minimized the variance among dosages by converting times to death to rates of toxic action (1,000/min of survival). Next, he used analysis of covariance to demonstrate that toxic action was influenced by both animal size (weight) and dosage. The smaller animals were more sensitive to the toxic action of arsenic. He then proceeded to examine various functions of dose (the amount of a substance given to an individual) and body weight that provided convenient linear relationships.

$$f(t) = a + b_1 f(m) - b_2 f(w) \tag{4.92}$$

where $f(t)$ = some function of survival time, t, $f(m)$ = some function of the mass of arsenic in the dose, $f(w)$ = some function of the individual weight, and a,b_1,b_2 = regression constants.

To linearize the effect of dose, the logarithm of dose was plotted against the rate of toxic action (Figure 4.17A). He judged that the relationship was linear despite significant variablity. The residuals were judged to be normally distributed after visual inspection. The residuals (deviations in rate) from the regression model of log dose versus rate of toxic action were then plotted against log body weight. A linear function was visually determined as adequate to describe the effect of body weight on toxic action (Figure 4.17B). The following model for size effects was generated.

$$y = a + b_1 Log\ m - b_2 Log\ w \tag{4.93}$$

where y = 1,000/time to death.

This model can be rearranged to estimate the size-specific dosage.

$$y = a + b_1\left[Log\ m - \frac{b_2}{b_1} Log\ w\right] \tag{4.94}$$

$$y = a + b_1 Log \frac{m}{w^h} \tag{4.95}$$

where $h = b_2/b_1$.

The size factor (h) was relatively constant over several silkworm instars.

Bliss's work remains unquestionably a major contribution; however, several points should be made regarding its general application. Several toxicologists (e.g., Rall and North 1953; Lemanna et al. 1955; Pallotta et al. 1962; Lemanna and Hart 1968) have used examples to point out that other types of relationships occur. They advised that caution be used before assuming any general law during data analysis. Misapplication can increase variability and confound data interpretation (Rall and North 1953; Pallotta et al. 1962). Although Bliss's relationship can be used effectively (Anderson and Weber 1975; Hedtke et al. 1982), other relationships should be considered.

Often, smaller individuals are more sensitive than larger individuals (Heit and Fingerman 1977; Newman et al. 1989), but there are toxicant-species pairs for which larger animals are most sensitive (Hogan et al. 1987) or for which there is no size dependence (Angelakos 1960). Further, size can have an effect on one measure of toxic effect but not on another, e.g., size-dependent rate of mortality but not size-independent incipient LC50 (Adelman et al. 1976).

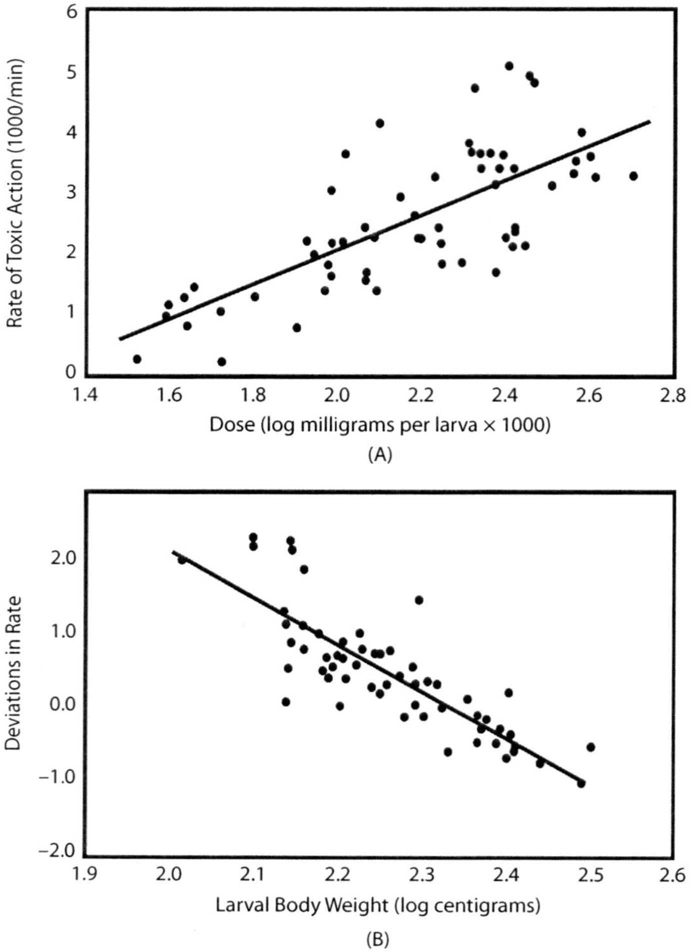

Figure 4.17 Bliss developed an approach to incorporating animal size into expressions of toxic effects using arsenic toxicity to silkworm larvae. Panel A is a plot and regression line derived from dose rate of toxic action data. Panel B is a plot and associated regression line for the residuals from the first regression line (dose versus rate of toxic action) versus body weight. See the text for more details. (Reproduced and adapted from Bliss, C.I., *J. Exp. Biol.,* 13, 95–110, 1936, Figures 1 and 2. With permission of the Company of Biologists, Ltd.)

Today, the imperative to transform data to produce linear relationships is unnecessary. In attempting to linearize the relationships, some of Bliss's judgments regarding residual distributions may have been optimistic, especially when the variability about the predicted values is considered. Several of these points can be addressed with computational software such as that used in Example 4.10. Even if found true for arsenic toxicity to silkworm larvae, the assumption of a normal distribution for the distribution of values for toxic action for all size-toxic impact relationships remains untested. To be of general use, the methods advocated by Bliss must also be modified to accommodate censoring (survivors or individuals removed during the experiment). Considering the magnitude of size effects and the variation in size among and within populations, it is surprising that so much refinement of methods remains to be done.

Anderson and Weber (1975) extended Bliss's approach to prediction of time endpoints for fish exposed to toxicants dissolved in water. Toxic responses of guppies (*Poecilia reticulata*) to dieldrin, potassium pentachlorophenate, potassium cyanide, copper chloride, zinc chloride, and nickel chloride were used. Beginning with Equation (4.95), they used endpoint data (probits of proportion dead) as the dependent variable. The concentration of toxicant dissolved in the water was used as m. Wet weight (w) was used as a measure of animal size. Multiple regression was used as described above to estimate h.

They also connected this approach with the common allometric model of LC50 for animal size. The allometric equation begins with the transformed regression model.

$$Log\ C = a + b\ Log\ w \tag{4.96}$$

where C = LC50 or LD50, w = animal weight, and a,b = regression constants.

As mentioned previously, relationships such as that described by Equation (4.96) are often back-transformed to their original units.

$$C = aw^b \tag{4.97}$$

Equation (4.97) can be rearranged.

$$Log(C / w^b) = Log\ a \tag{4.98}$$

In terms of Equation (4.95), the intercept ($log\ a$) in Equation (4.96) is that value of log (m/w^h) for which y is the metameter for 50% mortality. The b in Equation (4.96) is equal to the h in Equation (4.95). In this manner, Anderson and Weber (1975) linked Bliss's methods to power relationships commonly used to describe animal size effects on LC50 or LD50. They also recommended that their multivariate approach employing log-transformed values of toxicant concentration and animal size be generally used to scale toxic effects.

Example 4.14

As suggested earlier in this chapter and by Example 4.10, survival analysis methods can model allometric effects on toxicity. This suggestion will be examined more thoroughly here. Mosquitofish were exposed again to sodium chloride for 96 h. Three tests were conducted with fish of various sizes. In test 1, wet weights ranged from 0.005 to 0.047 g (mean: 0.016; standard deviation: 0.006 g). All 280 fish in this test were juveniles. In test 2, wet weights ranged from 0.033 to 0.223 g (mean: 0.091 g; standard deviation: 0.039 g). The 278 fish in this test were a mixture of juveniles and sexually mature females.

In the final test 3, wet weights ranged from 0.108 to 1.985 g (mean: 0.516 g; standard deviation: 0.330). All 208 fish in the third test were adult females. The overall mean and standard deviation for the 766 fish employed in these tests were 0.179 and 0.271 g, respectively. Through the process described in Example 4.10, log wet weight and log salt concentration were selected in the various models examined. This is consistent with Bliss's model (Equation 4.93). Using the SAS procedure LIFEREG (see Example 4.10), exponential, Weibull, lognormal, normal, log logistic, and gamma distributions were selected for description of the error function. With the transformations described above, the normal model is similar to Bliss's model (Equation 4.93) except time to death is used instead of 1,000/time to death. The other options were not assessed by Bliss and remain unexplored in most subsequent studies.

The following SAS code performs these analyses of pooled data from all three tests and added test ("run") as a class variable. With minor modification, the code can be used to analyze each test separately.

```
DATA TOXICITY;
INFILE "B:MORT.DAT";
```

```
   INPUT RUN 1 UTTD 3-4 LTTD 6-7 WT 9-13 LENGTH 15-16 SALT 18-22;
   LWGT = LOG10(WT);
   LSALT = LOG10(SALT);
RUN;
   /* UNLIKE PREVIOUS SAS PROGRAMS IN EXAMPLES USING LIFEREG,          */
   /* INTERVAL CENSORING IS USED IN THIS EXAMPLE. THE INTER-           */
   /* VALS USED IN THIS EXPERIMENT WERE SUFFICIENTLY LARGE             */
   /* RELATIVE TO THE TOTAL LENGTH OF EXPOSURE THAT UPPER (UTTD)       */
   /* AND LOWER (LTTD) LIMITS OF EACH SAMPLING INTERVAL WERE           */
   /* USED. IN PREVIOUS EXAMPLES, IT WAS ASSUMED THAT THE SAMP-        */
   /* ING INTERVALS WERE SUFFICIENTLY SMALL THAT THE TIME OF           */
   /* SAMPING WAS AN ADEQUATE MEASURE OF TIME-TO-DEATH. HERE,          */
   /* THE SAMPLING TIME IS THE UPPER PORTION OF THE INTERVAL IN        */
   /* WHICH THE FISH DIED AND THE PREVIOUS TIME OF EXAMINING FOR       */
   /* DEATH WAS USED AS THE LOWER LIMIT. FOR THE FIRST INTERVAL        */
   /* WITH DEATHS, THE LTTD IS MISSING. SIMILARLY, FOR THE FINAL       */
   /* SAMPLING, THE UTTD IS MISSING.                                   */
PROC SORT;
   BY RUN LSALT LWT;
RUN;
PROC LIFEREG;
   CLASS RUN;
   MODEL (LTTD,UTTD) = RUN LSALT LWT/D=EXPONENTIAL;
   MODEL (LTTD,UTTD) = RUN LSALT LWT/D=WEIBULL;
   MODEL (LTTD,UTTD) = RUN LSALT LWT/D=LNORMAL;
   MODEL (LTTD,UTTD) = RUN LSALT LWT/D=NORMAL;
   MODEL (LTTD,UTTD) = RUN LSALT LWT/D=LLOGISTIC;
   MODEL (LTTD,UTTD) = RUN LSALT LWT/D=GAMMA;
RUN;
```

The AIC values for each model were the following:

Model	Run 1	Run 2	Run 3	Overall
Exponential	1,546	1,044	544	3,124
Weibull	1,270	798	452	2,580
Lognormal	1,236	824	456	2,576
Normal	1,324	816	464	2,658
Log logistic	1,240	798	452	2,558
Gamma	1,236	798	452	2,562

In this example, the normal model never provided the best fit (lowest AIC). This observation is consistent with our previous discussions of survival time metameters. The above analysis and the studies cited above indicate that several candidate distributions should be examined prior to modeling size effects on time to death. This conclusion can also be extended to toxic endpoint methods because, as outlined above, the endpoint methods of Anderson and Weber (1975) are linked to Bliss's approach. Other metameters should be explored in addition to probits suggested by Anderson and Weber (1975).

4.6 TOXICANT MIXTURES

4.6.1 Methods Based on Additivity

Prediction of the joint effects of toxicants has been an important goal of ecotoxicologists because toxicants are commonly present together in contaminated systems (Monsoon 2005). Toxicants may interact and influence adsorption, binding to plasma components, tissue distributions, elimination, action at receptor sites, and toxicant metabolism (Calamari and Alabaster 1980). Sprague (1970) used the following example to define the possible joint effects of toxicants. He uses the toxic unit scale. In his example, the lethal threshold concentrations and dissolved metal concentrations are used but other metameters may also be used.

$$Toxic\ Unit = \frac{[Toxicant]}{[Lethal\ Threshold]} \tag{4.99}$$

where $[Toxicant]$ = the concentration of the toxicant in the exposure solution, and $[Lethal\ Threshold]$ = lethal threshold concentration of the toxicant.

Assume that one-half of a toxic unit of toxicant A and one-half of a toxic unit of B are administered together. Notionally, the combination is additive if the mixture of the two half toxic units produces exactly one unit of toxic response. An additive response is one in which the response is the summation of the responses for each toxicant administered separately. A combined effect may also be less than additive or more than additive. Brown (1968), in noting that additivity was often observed, suggested that additivity can be justified initially with the concept of Selyean stress (see Chapter 1). Each toxicant acts as a Selyean stressor and contributes equally to the degree of experienced shock. (It is ambiguous to this author whether Selyean stress or damage is being discussed in Brown's explanation.) In contrast, Sprague (1970) stated that no theoretical basis can be claimed for additivity, and it is applied for empirical reasons only. Enserink et al. (1991) suggested that additivity may be a consequence of similar modes of action for the toxicants in the mixture.

If the toxicants act additively, any combination of toxic units of A and B summing to 1 will have a measured joint toxic effect of 1. Joint effects of arsenic, cadmium, chromium, copper, mercury, lead, nickel, and zinc to *Daphnia magna* were found to be approximately additive by Enserink et al. (1991). Survival during acute exposure, and changes in reproduction and body weight during chronic exposure were used as effects during this study. Toxicant antagonism is present if more than 1 toxic unit is needed with A in the presence of B to have the effect. For example, copper was found to be antagonistic to the toxic action of methylmercury to blue gourami (Roales and Perlmutter 1974a). Zinc is also antagonistic to cygon (O,O-dimethyl-S-(N-methylcarbamoylmethyl) phosphorodithioate) toxicity to zebrafish (Roales and Perlmutter 1974b). If the presence of B enhances the toxic impact of A, the interaction is called synergism. Babich and Stotzky (1983) reported potentiation between nickel and copper for microbial growth.

Calamari and Alabaster (1980) defined a similar set of terms in their treatment of toxicant mixtures. However, they added the distinction between toxicant interactions at similar or dissimilar sites of action. For example, two toxicants can have a similar, synergistic effect. Interpretation and prediction are greatly enhanced if information is available for making this distinction.

In an early attempt to predict joint effects of mixtures of toxicants in ecotoxicology, Marking and Dawson (1975) developed a scale for assessing toxicant interactions. They used the LC50 as the normalizing concentration in their computations.

$$S = \frac{LC50_{AC}}{LC50_{AI}} + \frac{LC50_{BC}}{LC50_{BI}} \tag{4.100}$$

where $LC50_{AC}$ = LC50 for A in the presence of B, $LC50_{AI}$ = LC50 for A in the absence of B, $LC50_{BC}$ = LC50 for B in the presence of A, and $LC50_{BI}$ = LC50 for B in the absence of A.

Values of $S < 1$ indicated a greater than additive response and values of $S > 1$ suggested a less than additive response. However, the degree of deviation from additivity is not similar for values of the same magnitude in the negative and positive directions from $S = 1$. Marking and Dawson (1975) used a modification of S to linearize the scale and used 0 as the reference point (additivity) with this modified index. The modified index was called the additive index or Marking's additive index. For $S \leq 1$, the additive index is $1/S - 1$. It is $-S + 1$ for $S > 1$. This additive index scale is linear with strict additivity indicated by a value of 0. Significant deviation from an additive index of 0 was tested by use of the LC50 95% CIs to estimate the acceptable range of additive indices (Marking 1985). The lower 95% confidence limits for $LC50_{AI}$ and $LC50_{BI}$ and upper limits for $LC50_{AC}$ and $LC50_{BC}$ were substituted into Equation (4.100) to estimate the lowest value for the additive index. Upper limits of $LC50_{AI}$ and $LC50_{BI}$ and lower limits of $LC50_{AC}$ and $LC50_{AC}$ and $LC50_{BC}$ are used in Equation (4.100) to estimate the highest value of the additive index. Inclusion of 0 in the resulting interval suggests a lack of any statistically significant deviation from additivity. Unfortunately, inclusion of 0 has also been used to conclude that toxicant mixtures are additive. For example, Thompson et al. (1980) stated: "The value of the [additive index] calculated from the results of this study was +0.218 with a range from −0.64 to +1.30. Since 0 is contained within these limits, the toxicity of Zn-Cu mixtures to bluegills was additive under the conditions of these tests." Although the reader can quickly ascertain the true conclusion that the authors were making, the conclusion as stated is inaccurate. This problem stems from some of Marking's own statements, such as "Mixtures that resulted in ranges for the additive index that overlapped zero were judged to be only additive in toxicity; ranges that did not overlap zero were either greater or less than additive in toxicity" (Marking 1985). His meaning is clear but the conclusion of additivity if ranges include 0 is statistically invalid and must be avoided.

Marking (Marking and Dawson 1975; Marking 1985) also expressed additive indices in terms of magnification factors. For example, additive indices of 9, 1, 0, and −9 have magnification factors of 10, 2, 1, and 0.10 times, respectively. An index of 9 suggests that the combined toxic effect of A and B is 10 times greater than that of A or B alone.

This approach is central in much work in regulatory ecotoxicology, although it lacks the theoretical soundness of the mode-of-action-based methods described in the next section. Di Toro et al. (1990) incorporated additivity into methods for assessing sediment quality for metals. Under the assumption of addition, Di Toro et al. (1990) suggested that the sum of the metal concentrations in the SEM fraction divided by the AVS should not exceed 1 when several toxic metals are present at high concentrations in the sediment. Similarly, Spehar and Fiandt (1986) examined metal interactions in the context of water quality criteria.

The above approach can be enriched considerably using factorial experimental designs (Calamari and Alabaster 1980; Piegorsh et al. 1988) and response surface methods, e.g., Parker's (1979) analysis of combined metal toxicity to protozoa or Voyer et al.'s (1982) examination of metal interactions on flounder larvae. Voyer and Heltshe (1984) reanalyzed models developed by Parker (1979) and Voyer et al. (1982) and used the results to distinguish between various types of additivity. For Parker's (1979) and Voyer et al.'s (1982) data, the following models were defined and tested for goodness-of-fit.

$$\hat{y} = a_x + b_y \tag{4.101}$$

where $\hat{y}$ = the response variable, a_x = the measured response to the administered level of toxicant x when the concentration of toxicant $y = 0$, and b_y = the measured response to the administered level of toxicant y when the concentration of toxicant $x = 0$.

$$\hat{y} = a + b_1 x_1 + b_2 x_2 \tag{4.102}$$

where x_1 = concentration of toxicant 1, x_2 = concentration of toxicant 2, b_1 = estimated coefficient for toxicant 1, b_2 = estimated coefficient for toxicant 2, and a = estimated parameter.

$$\hat{y} = a + b_1x_1 + b_2x_2 + b_{11}x_1^2 + b_{22}x_2^2 + b_{12}x_1x_2 \tag{4.103}$$

where b_{11} = estimated coefficient for quadratic effect associated with toxicant 1, b_{22} = estimated coefficient for quadratic effect associated with toxicant 2, and b_{12} = estimated coefficient for interactive effect.

Equations (4.101), (4.102), and (4.103) describe simple additive, linear additive, and quadratic models for toxicant interactions, respectively. Voyer and Heltshe (1984) suggested from their analyses that assessment of alternate models such as Equations (4.102) and (4.103) can enhance prediction of effect for toxicant mixtures. Protocol is outlined for conducting factorial experimentation for developing appropriate models. (See Box and Draper 1987, especially Chapter 7, and Chapter 9 in Neter et al. 1990 for general discussions of this approach.) It is worthwhile to note that, if a factorial design is chosen as in Example 4.15, it is not always necessary to have responses for all possible combinations of concentrations$_A$ by concentrations$_B$. A specific design called a fixed-ratio mixture ray design can be used that requires only a subset of the possible concentration combinations (Casey et al. 2004; Coffey et al. 2005).

4.6.2 Methods Based on Mode of Action

Toxicants can have similar joint action, in which case they act by the same mechanism (i.e., have identical modes of action) (Finney 1942), and "one component can be substituted at a constant proportion for the other … toxicity of a mixture is predictable directly from that of the constituents if their relative proportions are known" (Finney 1947). In contrast, independent joint action of toxicants exists if each toxicant produces an effect independent of the other and by a different mode of action (Finney 1942, 1947). Relative to quantifying joint action of such chemicals, Finney (1942) makes an important distinction: "In mixtures whose constituents act similarly any quantity of one constituent can be replaced by proportionate amount of any other without disturbing the potency, but for mixtures whose constituents act independently the mortalities, not the doses, are additive." This distinction is central in the following discussions.

Models for joint action of toxicants in mixture build upon the straightforward probit models, e.g.,

$$Probit(P) = a + b\ Log\ C + \in \tag{4.104}$$

where C = the exposure concentration, a = an estimated regression intercept, b = an estimated regression parameter accounting for the influence of exposure concentration, and $\in$ = error term.

If one applied the N.E.D. instead of the probit transformation, a rearranged form of this model is specified in Equation (4.36), that is, $P = \Phi(a + b(Log\ C))$.

If two independently acting chemicals, A and B, were combined at specific concentrations (or doses) in an exposure solution, the proportion dying of individuals exposed to the mixture (P_{A+B}) can be predicted from the proportion that would have died of individuals exposed to A alone at the specified concentration (P_A) and the proportion that would have died of individuals exposed to B alone at the specified concentration (P_B) (Finney 1942).

$$P_{A+B} = P_A + P_B(1 - P_A) = P_A + P_B - P_AP_B \tag{4.105}$$

The reason P_{A+B} is not simply the sum of P_A and P_B is clear if one thinks of a proportion as the probability of an individual dying at that specific concentration of that toxicant. If an outcome can result from two independent processes with associated probabilities of P_A and P_B, the probability of the event occurring (Equation 4.105) is a direct outcome of a fundamental rule of probability (see p. 59 in

Hacking 2001 for details). The term $-P_A P_B$ is needed to adjust for the reality that if A kills an organism, that organism is not alive to be killed by B. In theory, this model can be expanded to include many toxicants, although the amount of work needed to validate the model might be extremely difficult to derive.

$$P_{A+B+C+...} = 1-(1-P_A)(1-P_B)(1-P_C)... \tag{4.106}$$

Toxicants that display similar action require a different model. To model joint action of such toxicants, Finney (1947) noted that toxicity curves are parallel for toxicants with similar action.* The influence of each toxicant alone is modeled with Equation (4.36), and then the two models combined, as shown below, to predict the joint effect. Let Equations (4.107) and (4.108) be the models for the two toxicants, A and B, alone.

$$Probit(P_A) = Intercept_A + Slope(Log\ C_A) \tag{4.107}$$

$$Probit(P_B) = Intercept_B + Slope(Log\ C_B) \tag{4.108}$$

The log of the relative potency of A and B can be estimated by combining these two models and rearranging the resulting model.

$$Log\ \rho_B = \frac{Intercept_B - Intercept_A}{Slope} \tag{4.109}$$

This relative potency measure can now be used to include both toxicants in a mixture model.

$$Probit(P_{A+B}) = Intercept_A + Slope \cdot Log(C_A + \rho_B C_B) \tag{4.110}$$

More similarly acting toxicants can be included in the model using the appropriate relative potencies. Perhaps the best example of modeling many similarly acting toxicants is the toxic equivalent (TEQ) approach taken for mixtures of dibenzo-p-dioxins, dioxin-like PCBs, and dibenzofurans. These toxicants, often present in mixture, have aryl hydrocarbon (AH) receptor interaction as the first stage of their toxic action. The total toxicity of their mixtures is determined using toxic equivalency factors (TEFs) (Van der Berg et al. 1998). The TEF for each compound expresses that compound's toxicity relative to the reference toxicant, 2,3,7,8-tetrachlorodibenzo-p-dioxin (2,3,7,8-TCDD). The reference toxicant, 2,3,7,8-TCDD, is assigned a TEF value of 1, and other similarly acting compounds are given experimentally derived TEF scaled to this TEF. For a mixture of these compounds in an environmental sample, the products of concentration times TEF for each AH receptor compound are summed to estimate the TEQ of the mixture.

Example 4.15

The bacterial bioluminescence QICAR studies described in Example 4.12 and Figure 4.15 were expanded in Newman and Clements (2008) to include binary metal ion mixtures. Although modeling was done from the vantages of similar and independent joint action, only the approach for modeling mixture effects based on similar action is illustrated here. Ten binary metal ion mixtures were used (Figure 4.15 bottom panel). Five concentrations (including 0) of each metal ion were

* It is important to emphasize that Finney suggested this rule but added crucial conditions beyond similar mode of action. For the approach described here to be valid (i.e., parallel slopes), the toxicants should also have sufficiently similar toxicokinetics. This condition is often overlooked in mixture studies.

combined in mixture with five concentrations (including 0) of the other metal ion in a full factorial design, that is, 2 metals × 5 concentrations. The P_A and P_B are predicted with models produced using data for each metal in combination with a 0 concentration of the other. The corresponding P_{A+B} was the response proportion observed directly for each binary mixture. The predicted P_A and P_B, and observed P_{A+B} were combined for all mixture combinations as shown in the code below.

```
/* LA AND CE ARE THE CONCENTRATIONS OF LANTHANUM AND CERUM.          */
/* PAPB IS THE MEASURED BIOLUMINESCENCE AFTER 15 MINUTES OF EXPOSURE. */
DATA LACE;
INPUT LA CE PAPB @@;
PAPB=100*((PAPB-.372)/(1-.372)); NORMZ=100;
DATALINES;
    0 0 .372      0 3.125 .359      0 6.25 .385      0 12.50 .481      0 25.00 .662
 3.125 0 .333  3.125 3.125 .370  3.125 6.25 .447  3.125 12.50 .533  3.125 25.00 .684
 6.250 0 .368  6.250 3.125 .419  6.250 6.25 .449  6.250 12.50 .568  6.250 25.00 .747
12.50  0 .500 12.50  3.125 .548 12.50  6.25 .569 12.50  12.50 .629 12.50  25.00 .761
25.00  0 .667 25.00  3.125 .725 25.00  6.25 .708 25.00  12.50 .757 25.00  25.00 .821
;
DATA LAN; SET LACE; IF CE=0; RUN;
PROC PROBIT LOG10 INVERSECL LACKFIT DATA=LAN; /* PROC PROBIT A */
  MODEL PAPB/NORMZ=LA/D=NORMAL ITPRINT;
  OUTPUT OUT=PLAN P=PPROB;
RUN;
DATA LAN2; SET PLAN; PAPB=PAPB/100; RUN;
DATA CEN; SET LACE; IF LA=0; RUN;
PROC PROBIT LOG10 INVERSECL LACKFIT DATA=CEN; /* PROC PROBIT B */
  MODEL PAPB/NORMZ=CE/D=NORMAL ITPRINT;
  OUTPUT OUT=PCEN P=PPROB;
RUN;
DATA NEW;
  SET LACE;
  IF LA NE 0; IF CE NE 0;
  PAPB=PAPB/100;
  LCE=LOG10(CE); LLA=LOG10(LA);
  INTERLA=-3.5687+2.4985*LLA; /* Resulting Model from PROC PROBIT A */
  PLA=PROBNORM(INTERLA);
  INTERCE=-4.3893+3.0872*LCE; /* Resulting Model from PROC PROBIT B */
  PCE=PROBNORM(INTERCE);
  EXPECT=PLA+PCE;
  RUN;
PROC GLM DATA=NEW;
  MODEL PAPB=PLA PCE PLA*PCE/CLPARM;
RUN;
```

These data (P_A, P_B, P_{A+B}) for each mixture could then be fit to a modification of Equation (4.104) that incorporates an interaction coefficient (ρ),

$$P_{A+B} = P_A + P_B - \rho P_A P_B$$

The interaction coefficient would be 1 with independent joint action but would increase as the action of the paired metal ions deviated from independent joint action.

4.6.3 Graphical Depictions Using Isobolograms

A graphical approach has also been applied to depict joint effects of binary mixtures (Figure 4.18). Deviations from a straight line predicted for strict additivity are used in this approach to infer antagonism and synergism. Typical might be the fitting of data to the logistic model,

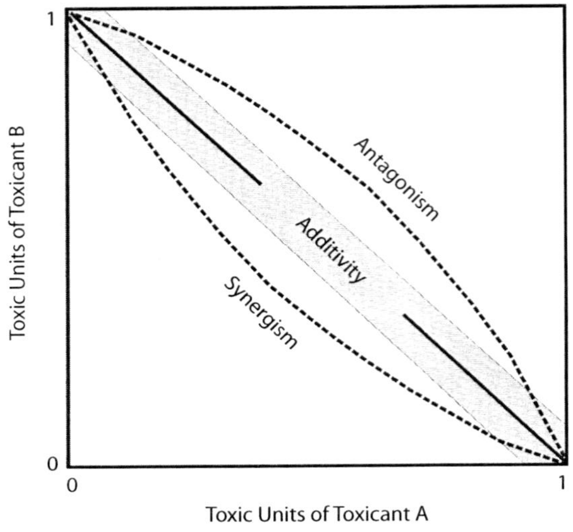

Figure 4.18 A toxic unit-based isobologram applied to binary mixtures of toxicants (A and B here) showing the diagonal line for simple additivity and its 95% confidence interval. Deviations away from this line suggest antagonism or synergism (dashed curves).

$$Ln\left[\frac{p}{1-p}\right] = a + b_A C_A + b_B C_B + b_{AB} C_A C_B \tag{4.111}$$

where $Ln[p/(1 - p)]$ = logarithm of the odds of responding, a, b_A, b_B, b_{AB} = estimated model parameters, and C_A and C_B = concentrations (or doses) of toxicants A and B, respectively.

Advancements in this approach have been made, including that of Chen and Pounds (1998), who expanded it for use with responses that are not bound by 0 and 1.

$$y = y_{minimum} + \frac{y_{maximum} - y_{minimum}}{1 + e^{-(a + b_A C_A + b_B C_B + b_{AB} C_A C_B)}} \tag{4.112}$$

or

$$Ln\left[\frac{y - y_{minimum}}{y_{maximum} - y}\right] = a + b_A C_A + b_B C_B + b_{AB} C_A C_B \tag{4.113}$$

Instead of 0 and 1, the $y_{minimum}$ and $y_{maximum}$ are the minimum and maximum values that the response (y) can take. As in Equation (4.111), the magnitude of b_{AB} suggests the magnitude of interaction of the two toxicants.

4.7 SUMMARY

Two goals were set during the development of this chapter. The first was to present the significant progress made in regulatory ecotoxicology to quantify lethal stress. Valuable methods are described for use in quantifying lethal and other quantal responses to stress. Most grew out of practical or regulatory applications. En masse, they are based on a sound statistical foundation.

However, they were developed to their present state more to meet regulatory goals than to elucidate explanatory principles.

The second goal of the chapter was to make apparent our contrastingly slow progress in the context of scientific ecotoxicology. Explanatory principles underlying methods selection, e.g., use of probit versus Weibull metameters, or toxicological definitions, e.g., acute versus chronic toxicity, are often ignored or discussed superficially. Relatively little effort has been spent in linkage to underlying principles. Very notable exceptions include development and application of metal speciation models and QSARs in ecotoxicology. Movement toward their linkage to physiologically based pharmacokinetic (PBPK) and pharmacodynamic models remains an exciting exception also.

In the context of scientific ecotoxicology, too much effort is spent on the "description of the diversity of phenomena" (Rapport et al. 1985) instead of determination of underlying themes or theory reduction. This is surprising as regulatory ecotoxicologists have provided such a richness of methodologies and information to be bent to these goals. Further, alternative metameters, toxicity endpoints, and methodologies are ignored in favor of methods with no apparent conceptual superiority. Instead, methods are selected for their regulatory significance, e.g., use of a probit estimate of a 96 h LC50 instead of a survival time model.

Traditional methodologies are not built upon or modified as efficiently as possible because of an overzealous adherence to standard methods and a general lack of emphasis on strong scientific inference. A good example is the delay in modifying Bliss's methods for incorporating size into models of toxic impact. Remember that these important techniques were developed initially with an emphasis on linear transformation. Such an approach is understandable if one keeps in mind the general inconvenience of executing more involved methods in the 1930s. Regardless, why should these historical constraints still have such a strong influence on present-day exploration of appropriate models?

REFERENCES

Adelman, I.R., L.L. Smith Jr., and G.D. Siesennop. Effect of size or age of goldfish and fathead minnows on use of pentachlorophenol as a reference toxicant. *Water Res.* 10:685–687 (1976).

Altman, D.G. *Practical statistics for medical research.* Boca Raton, FL: Chapman & Hall/CRC, 1991.

American Public Health Association. *Standard methods for the examination of water and wastewater.* 15th ed. Washington, DC: American Public Health Association, 1981.

Anderson, P.D., and L.J. Weber. Toxic response as a quantitative function of body size. *Toxicol. Appl. Pharmacol.* 33:471–483 (1975).

Andrew, R.W., K.E. Biesinger, and G.E. Glass. Effects of inorganic complexing on the toxicity of copper to *Daphnia magna. Water Res.* 11:309–315 (1977).

Angelakos, E.T. Lack of relationship between body weight and pharmacological effect exemplified by histamine toxicity in mice. *Proc. Soc. Exp. Biol. Med.* 103:296–298 (1960).

Ankley, G.T., G.L. Phipps, E.N. Leonard, D.A. Benoit, V.R. Mattson, P.A. Kosian, A.M. Cotter, J.R. Dierkes, D.J. Hansen, and J.D. Mahony. Acid-volatile sulfide as a factor mediating cadmium and nickel bioavailability in contaminated sediments. *Environ. Toxicol. Chem.* 10:1299–1307 (1991).

Armitage, P., and I. Allen. Methods of estimating the LD50 in quantal response data. *J. Hyg.* 48:298–322 (1950).

Ashauer, R., A. Boxall, and C. Brown. Predicting effects on aquatic organisms from fluctuating or pulsed exposure to pesticides. *Environ. Toxicol. Chem.* 25:1899–1912 (2006).

Atkinson, A.C. A note on the generalized information criterion for choice of a model. *Biometrika* 67:413–418 (1980).

Babich, H., and G. Stotzky. Synergism between nickel and copper in their toxicity to microbes: Mediation by pH. *Ecotoxicol. Environ. Saf.* 7:576–587 (1983).

Bacharach, A.L., M.E. Coates, and T.R. Middleton. A biological test for vitamin P activity, *Biochem. J.* 36:407–412 (1942).

Bacon, F. *Advancement of learning and novum organum.* New York: Wiley Book Co., 1620; reprinted, 1944.

Barata, C., D.J. Baird, and S.J. Markich. Comparing metal toxicity among *Daphnia magna* clones: An approach using concentration-time-response surfaces. *Arch. Environ. Contam. Toxicol.* 37:326–331 (1999).

Barron, M.G., S. Raimondo, C. Russom, D.N. Vivian, and S.H. Yee. Accuracy of chronic aquatic toxicity estimates determined from acute toxicity data and two time-response models. *Environ. Toxicol. Chem.* 27:2196–2205 (2008).

Berkson, J. Why I prefer logits to probits. *Biometrics* 7(4):327–339 (1951).

Berkson, J. Maximum likelihood and minimum χ^2 estimates of the logistic function. *J. Am. Stat. Assoc.* 50:130–162 (1955).

Bjornberg, A., L. Hakanson, and K. Lundbergh. A theory of the mechanisms regulating bioavailability of mercury in natural waters. *Environ. Pollut.* 49:53–61 (1988).

Blackstone, E.H. Analysis of death (survival analysis) and other time-related events. In *Current status of clinical cardiology*, ed. F.J. Macartney. Boston: MTP Press Limited, 1986.

Bliss, C.I. The calculation of the dosage-mortality curve. *Ann. Appl. Biol.* 22:134–307 (1935).

Bliss, C.I. The size factor in the action of arsenic upon silkworm larvae. *J. Exp. Biol.* 13:95–110 (1936).

Bliss, C.I. The calculation of the time-mortality curve. *Ann. Appl. Biol.* 24:815–852 (1937).

Borgmann, U. Metal speciation and toxicity of free metal ions to aquatic biota. In *Aquatic toxicology*, ed. J.O. Nriagu. New York: John Wiley & Sons, 1983.

Borman, S. New QSAR techniques eyed for environmental assessments. *Chem. Eng. News* 19:20–23 (1990).

Bowling, J.W., G.J. Leversee, P.F. Landrum, and J.P. Giesy. Acute mortality of anthracene-contaminated fish exposed to sunlight. *Aquat. Toxicol.* 3:79–90 (1983).

Box, G.E.P., and N.R. Draper. *Empirical model-building and response surfaces.* New York: John Wiley & Sons, 1987.

Brown, V.M. The calculation of the acute toxicity of mixtures of poisons to rainbow trout. *Water Res.* 2:723–733 (1968).

Bruce, R.D. An up-and-down procedure for acute toxicity testing. *Fund. Appl. Toxicol.* 5:151–157 (1985).

Bruce, R.D. A confirmatory study of the up-and-down method for acute oral toxicity testing. *Fund. Appl. Toxicol.* 8:97–100 (1987).

Bryan, G.W. Bioavailability and effects of heavy metals in marine species. In *Wastes in the sea: Near shore waste disposal*, ed. B. Ketchum, J. Capuzzo, W. Burt, I. Duedall, P. Park, and D. Kester. Vol. 6. New York: John Wiley & Sons, 1985.

Buikema, A.L., Jr., B.R. Niederlehner, and J. Cairns Jr. Biological monitoring: Part IV. Toxicity testing. *Water Res.* 16:239–262 (1982).

Burton, G.A., Jr. Assessing the toxicity of freshwater sediments. *Environ. Toxicol. Chem.* 10:1585–1627 (1991).

Cairns, J., Jr., A.G. Heath, and B.C. Parker. Temperature influence on chemical toxicity to aquatic organisms. *J. Water Pollut. Control Fed.* 47(2):267–280 (1975).

Cairns, M.A., A.V. Nebeker, J.H. Gakstatter, and W.L. Griffis. Toxicity of copper-spiked sediments to freshwater invertebrates. *Environ. Toxicol. Chem.* 3:435–445 (1984).

Calamari, D., and J.S. Alabaster. An approach to theoretical models in evaluating the effects of mixtures of toxicants in the aquatic environment. *Chemosphere* 9:533–538 (1980).

Campbell, F.L. Relative susceptability of arsenic in successive instars of the silkworm. *J. Gen. Physiol.* 9(6):727–733 (1926).

Campbell, P.G.C., P.M. Chapman, and B.A. Hale. Risk assessment of metals in the environment. *Issues Environ. Sci. Technol.* 21:102–131 (2006).

Campbell, P.G.C., A.G. Lewis, P.M. Chapman, A.A. Crowder, W.K. Fletcher, B. Imber, S.N. Luoma, P.M. Stokes, and M. Winfrey. *Biologically available metals in sediments*. No. 27694. Ottawa, Ontario: National Research Council Canada, 1988.

Campbell, P.G.C., and A. Tessier. Ecotoxicology of metals in the aquatic environment: Geochemical aspects. In *Ecotoxicology. A hierarchical treatment*, ed. M.C. Newman and C.H. Jagoe. Boca Raton, FL: CRC Press/Lewis Publishers, 1996.

Carlson, A.R., G.L. Phipps, V.R. Mattson, P.A. Kosian, and A.M. Cotter. The role of acid-volatile sulfide in determining cadmium bioavailability and toxicity in freshwater sediments. *Environ. Toxicol. Chem.* 10:1309–1319 (1991).

Carriger, J.F., T.C. Hoang, and G.M. Rand. Survival time analysis of least killifish (*Heterandria formosa*) and mosquitofish (*Gambusia affinis*) in acute exposures to endosulfan sulfate. *Arch. Environ. Contam. Toxicol.* 58:1015–1022 (2010).

Carroll, J.J., S.J. Ellis, and W.S. Oliver. Influences of hardness on the acute toxicity of cadmium to brook trout (*Salvelinus fontinalis*). *Bull. Environ. Contam. Toxicol.* 22:575–581 (1979).

Casarett, L.J. Origin and scope of toxicology. In *Toxicology. The basic science of poisons*, ed. L.J. Cassarett and J. Doull. New York: MacMillan Publishing Co., 1975.

Casey, M., C. Gennings, W.H. Carter Jr., V.C. Moser, and J.E. Simmons. Detecting interaction(s) and assessing the impact of component subsets in a chemical mixture using fixed-ratio mixture ray designs. *Agric. Biol. Environ. Stat.* 9(3):339–361 (2004).

Chapman, G.A. Acclimation as a factor influencing metal criteria. In *Aquatic toxicology and hazard assessment: Eighth symposium*, ed. R.C. Bahner and D.J. Hansen. ASTM STP 891. Philadelphia: American Society for Testing and Materials, 1985, pp. 119–136.

Chen, D.G., and J.G. Pounds. A nonlinear isobologram model with Box-Cox transformation to both sides for chemical mixtures. *Environ. Health Perspect.* 106(Suppl. 6):1367–1371 (1998).

Chew, R.D., and M.A. Hamilton. Toxicity curve estimation: Fitting a compartment model to median survival times. *Trans. Am. Fish. Soc.* 114:403–412 (1985).

Christensen, E.R. Dose-response functions in aquatic toxicity testing and the Weibull model. *Water Res.* 18(2):213–221 (1984).

Christensen, E.R., and N. Nyholm. Ecotoxicological assays with algae: Weibull dose-response curves. *Environ. Sci. Technol.* 18(9):713–718 (1984).

Clubb, R.W., A.R. Gaufin, and J.L. Lords. Synergism between dissolved oxygen and cadmium toxicity in five species of aquatic insects. *Environ. Res.* 9:285–289 (1975).

Coffey, T., C. Gennings, J.E. Simmons, and D.W. Herr. D-optimal experimental designs to test for departure from additivity in a fixed-ratio mixture ray. *Toxicol. Sci.* 88(2):467–476 (2005).

Cook, M., G. Nickless, R.E. Lawn, and D.J. Roberts. Biological availability of sediment-bound cadmium to the edible cockle, *Cerastoderma edule*. *Bull. Environ. Contam. Toxicol.* 23:381–386 (1979).

Cox, D.R. Regression models and lifetables (with discussion). *J. R. Stat. Soc. B* 34:187–200 (1972).

Cox, D.R., and D. Oakes. *Analysis of survival data.* New York: Chapman & Hall, 1984.

Crane, M., M.C. Newman, P.F. Chapman, and J. Fenlon (eds.). *Risk assessment with time to event models.* Boca Raton, FL: CRC/Lewis Publishers, 2002.

Crecelius, E.A., J.T. Hardy, C.I. Bobson, R.L. Schmidt, C.W. Apts, J.M. Gurtisen, and S.P. Joyce. Copper bioavailability to marine bivalves and shrimp: Relationship to cupric ion activity. *Mar. Environ. Res.* 6:13–26 (1982).

Dagani, R. Aquatic toxicology matures, gains importance. *Chem. Eng. News* 30:18–23 (1980).

Dechaume Moncharmont, F.-X., A. Decourtye, C. Hennequet-Hantier, O. Pons, and M.-H. Pham-Delegue. Statistical analysis of honeybee survival after chronic exposure to insecticides. *Environ. Toxicol. Chem.* 22:3088–3094 (2003).

Der, G., and B.S. Everitt. *Statistical analysis of medical data using SAS.* Boca Raton, FL: Chapman & Hall/CRC Press, 2006.

DeWitt, T.H., R.C. Swartz, and J.O. Lamberson. Measuring the acute toxicity of estuarine sediments. *Environ. Toxicol. Chem.* 8:1035–1048 (1989).

Diamond, S.A., M.C. Newman, M. Mulvey, P.M. Dixon, and D. Martinson. Allozyme genotype and time to death of mosquitofish, *Gambusia affinis* (Baird and Girard), during acute exposure to inorganic mercury. *Environ. Toxicol. Chem.* 8:613–622 (1989).

Di Toro, D.M., H.E. Allen, H.L. Bergman, J.S. Meyer, P.R. Paquin, and R.C. Santore. Biotic ligand model of aquatic toxicity of metals. *Environ. Toxicol. Chem.* 20:2383–2396 (2001).

Di Toro, D.M., J.D. Mahony, D.J. Hansen, K.J. Scott, M.B. Hicks, S.M. Mayr, and M.S. Redmond. Toxicity of cadmium in sediments: The role of acid volatile sulfide. *Environ. Toxicol. Chem.* 9:1487–1502 (1990).

Di Toro, D.M., C.S. Zarba, D.J. Hansen, W.J. Berry, R.C. Swartz, C.E. Cowan, S.P. Pavlou, H.E. Allen, N.A. Thomas, and P.R. Paquin. Technical basis for establishing sediment quality criteria for nonionic chemicals using equilibrium partitioning. *Environ. Toxicol. Chem.* 10:1541–1583 (1991).

Dixon, D.G., and J.B. Sprague. Acclimation to copper by rainbow trout (*Salmo gairdneri*)—A modifying factor in toxicity. *Can. J. Fish. Aquat. Sci.* 38:880–888 (1981a).

Dixon, D.G., and J.B. Sprague. Acclimation-induced changes in toxicity of arsenic and cyanide in rainbow trout, *Salmo gairdneri* Richardson. *J. Fish Biol.* 18:579–589 (1981b).

Dixon, P.M. Personal communication, 1992.

Dixon, P.M., and M.C. Newman. Analyzing toxicity data using statistical models for time-to-death: An introduction. In *Metal ecotoxicology, concepts and applications*, ed. M.C. Newman and A.W. McIntosh. Chelsea, MI: Lewis Publishers, 1991.

Dixon, W.J. The up-and-down method from small samples. *J. Am. Stat. Assoc.* 60(312):967–978 (1965).

Dixon, W.J. Staircase bioassay: The up-and-down method. *Neurosci. Biobehav. R.* 15:47–50 (1991).

Dixon, W.J., and F.J. Massey Jr. *Introduction to statistical analysis*. New York: McGraw-Hill Book Co., 1969.

Dixon, W.J., and A.M. Mood. A method for obtaining and analyzing sensitivity data. *J. Am. Stat. Assoc.* 43(241):109–126 (1948).

Eaton, D.L., and S.G. Gilbert. Principles of toxicology. In *Toxicology. The basic science of poisons*, ed. C.D. Klaassen. 7th ed. New York: McGraw-Hill Co., 2008.

Emerson, K., R.C. Russo, R.E. Lund, and R.V. Thurston. Aqueous ammonia equilibrium calculations: Effect of pH and temperature. *J. Fish. Res. Board Can.* 32:2379–2383 (1975).

Enserink, E.L., J.L. Maas-Diepeveen, and C.J. Van Leeuwen. Combined effects of metals: An ecotoxicological evaluation. *Water Res.* 25(6):679–687 (1991).

Fales, R.R. The influence of temperature and salinity on the toxicity of hexavalent chromium to the grass shrimp *Palaemonetes pugio* (Holthuis). *Bull. Environ. Contam. Toxicol.* 20:447–450 (1978).

Feller, W. *An introduction to probability theory and its application.* New York: John Wiley & Sons, 1968.

Finney, D.J. The analysis of toxicity tests on mixtures of poisons. *Ann. Appl. Biol.* 29:82–94 (1942).

Finney, D.J. *Probit analysis. A statistical treatment of the sigmoid response curve*. Cambridge: Cambridge University Press, 1947.

Finney, D.J. *Probit analysis*. 3rd ed. London: Cambridge at the University Press, 1971.

Finney, D.J. *Statistical method in biological assay*. 3rd ed. London: Charles Griffin and Company, 1978.

Freedman, M.L., P.M. Cunningham, J.E. Schindler, and M.J. Zimmerman. Effect of lead speciation on toxicity. *Bull. Environ. Contam. Toxicol.* 25:389–393 (1980).

Gad, S.C., and C.S. Weil. *Statistics and experimental design for toxicologists.* Caldwell, NJ: Telford Press, 1988.

Gaddum, J.H. Bioassays and mathematics. *Pharmacol. Rev.* 5:87–134 (1953).

Gelber, R.D., P.T. Lavin, C.R. Mehta, and D.A. Schoenfeld. Statistical analysis. In *Fundamentals of aquatic toxicology*, ed. G.M. Rand and S.R. Petrocelli. New York: Hemisphere Publishing Corp., 1985.

Gersich, F.M., and D.L. Hopkins. Site-specific acute and chronic toxicity of ammonia to *Daphnia magna* Straus. *Environ. Toxicol. Chem.* 5:443–447 (1986).

Green, R.H. Estimation of tolerance over an indefinite time period. *Ecology* 46:887 (1965).

Hacking, I. *An introduction to probability and inductive logic*. Cambridge: Cambridge University Press, 2001.

Hamilton, M.A., R.C. Russo, and R.V. Thurston. Trimmed Spearman–Karber method for estimating median lethal concentrations in toxicity bioassays. *Environ. Sci. Technol.* 11(7):714–719 (1977).

Hansch, C., and A. Leo. *Substituent constants for correlation analysis in chemistry and biology.* New York: John Wiley & Sons, 1979.

Hansch, C., A. Leo, and D. Nikaitani. On the additive-constitutive character of partition coefficients. *J. Org. Chem.* 37(20):3090–3092 (1972).

Hansch, C., J.E. Quinlan, and G.L. Lawrence. The linear free-energy relationship between partition coefficients and the aqueous solubility of organic liquids. *J. Org. Chem.* 33(1):347–350 (1968).

Harrell, F.E., Jr. *Survival and risk analysis.* Durham, NC: Duke University Medical Center, 1988.

Heagler, M.G., M.C. Newman, M. Mulvey, and P.M. Dixon. Allozyme genotype in mosquitofish, *Gambusia holbrooki*, during mercury exposure: Temporal stability, concentrations effects and field verification. *Environ. Toxicol. Chem.* 12:385–395 (1993).

Hedtke, J.L., E. Robinson-Wilson, and L.J. Weber. Influence of body size and developmental stage of coho salmon (*Oncorhynchus kisutch*) on lethality of several toxicants. *Fund. Appl. Toxicol.* 2:67–72 (1982).

Heit, M., and M. Fingerman. The influences of size, sex and temperature on the toxicity of mercury to two species of crayfishes. *Bull. Environ. Contam. Toxicol.* 18(5):572–580 (1977).

Heming, T.A., A. Sharma, and Y. Kumar. Time-toxicity relationships in fish exposed to the organochlorine pesticide methoxychlor. *Environ. Toxicol. Chem.* 8:923–932 (1989).

Hillaby, B.A., and D.J. Randall. Acute ammonia toxicity and ammonia excretion in rainbow trout (*Salmo gairdneri*). *J. Fish. Res. Board Can.* 36:621–629 (1979).

Hogan, G.R., B.S. Cole, and J.M. Lovelace. Sex and age mortality responses in zinc acetate-treated mice. *Bull. Environ. Contam. Toxicol.* 39:156–161 (1987).

Howarth, R.S., and J.B. Sprague. Copper lethality to rainbow trout in waters of various hardness and pH. *Water Res.* 12:455–462 (1978).

Jager, T., E.H.W. Heugens, and S.A.L.M. Kooijman. Making sense of ecotoxicological test results: Toward application of process-based models. *Ecotoxicology* 15:305–314 (2006).

Jagoe, R., and M.C. Newman. Bootstrap estimation of community NOEC values. *Ecotoxicology* 6:293–306, 1997.

Jenne, E.A., and S.N. Luoma. Forms of trace elements in soils, sediments, and associated waters: An overview of their determination and biological availability. In *Biological implications of metals in the environment*, ed. R.E. Wildung and H. Drucker. CONF-750929. Springfield, VA: National Technical Information Service, 1977, pp. 110–143.

Jones, M.M., and W.K. Vaughn. HSAB theory and acute metal ion toxicity and detoxification processes. *J. Inorg. Nucl. Chem.* 40:2081–2088 (1978).

Kalbfleisch, J.D., and R.L. Prentice. *The statistical analysis of failure time data.* New York: John Wiley & Sons, 1980.

Kaplan, E.L., and P. Meier. Nonparametric estimation from incomplete observations. *J. Am. Stat. Assoc.* 53:457–481 (1958).

Kenaga, E.E. Predictability of chronic toxicity from acute toxicity of chemicals in fish and aquatic invertebrates. *Environ. Toxicol. Chem.* 1:347–358 (1982).

Kinraide, T.B. Improved scales for metal ion softness and toxicity. *Environ. Toxicol. Chem.* 28:525–533 (2009).

Koch, R.W., and G.M. Smillie. Bias in hydrologic prediction using log-transformed regression models. *Water Res.* 22:717–723 (1986).

Kotz, S., and N.L. Johnson. *Encyclopedia of statistical sciences: Icing the tails to limit theorems.* Vol. 4. New York: John Wiley & Sons, 1983.

Langston, W.J. Arsenic in U.K. estuarine sediments and its availability to benthic organisms. *J. Mar. Biol. Ass. U.K.* 60:869–881 (1980).

Laughlin, R.B., Jr., R.B. Johannesen, W. French, H. Guard, and F.E. Brinckman. Structure-activity relationships for organotin compounds. *Environ. Toxicol. Chem.* 4:343–351 (1985).

Lee, B., S.B. Griscom, J. Lee, H.J. Choi, C. Koh, S.N. Luoma, and N.S. Fisher. Influences of dietary uptake and reactive sulfides on metal bioavailability from aquatic sediments. *Science* 287:282–284 (2000).

Legierse, K.C.H.M., H.J.M. Verhaar, W.H. Vaes, J.H.M. De Bruijn, and J.L.M. Hermens. Analysis of the time-dependent acute toxicity of organophosphorus pesticides: The critical target occupation model. *Environ. Sci. Technol.* 33:917–925 (1999).

Lemanna, C., and E.R. Hart. Relationship of lethal toxic dose to body weight of the mouse. *Toxicol. Appl. Pharmacol.* 13:307–315 (1968).

Lemanna, C., W.I. Jensen, and I.D.J. Bross. Body weight as a factor in the response of mice to botulinal toxins. *Am. J. Hyg.* 62:21–28 (1955).

Lew, A.A., C.L. Day, T.J. Harrist, W.C. Wood, and M.C. Mihm Jr. Multivariate analysis. Some guidelines for physicians. *J. Am. Med. Assoc.* 249(5):641–643 (1983).

Lipnick, R.L. A perspective on quantitative structure-activity relationships in ecotoxicology. *Environ. Toxicol. Chem.* 255–257 (1985).

Lipnick, R.L., J.A. Cotruvo, R.N. Hill, R.D. Bruce, K.A. Stitzel, A.P. Walker, I. Chu, M. Goddard, L. Segal, J.A. Springer, and R.C. Myers. Comparison of the up-and-down conventional LD_{50}, and fixed-dose acute toxicity procedures. *Food Chem. Toxic.* 33(3):223–231 (1995).

Litchfield, J.T., Jr., and F. Wilcoxon. A simplified method of evaluating dose-effect experiments. *J. Pharm. Exp. Ther.* 96:99–113 (1949).

Litchfield, J.T., Jr. A method for rapid graphic solution of time-per cent effects curves. *J. Pharmacol. Exp. Ther.* 97:399–408. (1949).

Lloyd, R. The toxicity of zinc sulphate to rainbow trout. *Ann. Appl. Biol.* 48(1):84–94 (1960).

Loehle, C., P. Bertsch, and G. Mills. An evaluation of chemical speciation in the MEXAMS metal transport model. *Environ. Software* 1(2):106–112 (1986).

Long, E.R., D.D. MacDonald, J.C. Cubbage, and C.G. Ingersoll. Predicting the toxicity of sediment-associated trace metals with simultaneously extracted trace metal:acid-volatile sulfide concentrations and dry weight-normalized concentrations: A critical comparison. *Environ. Toxicol. Chem.* 17:972–974 (1998).

Luoma, S.N. Can we determine the biological availability of sediment-bound trace elements? *Hydrobiologia* 176/177:379–396 (1989).

Luoma, S.N., and G.W. Bryan. Factors controlling the availability of sediment-bound lead to the estuarine cockle, *Scrobicularia plana. J. Mar. Biol. Ass. U.K.* 58:793–802 (1978).

Luoma, S.N., and E.A. Jenne. The availability of sediment-bound cobalt, silver, and zinc to a deposit-feeding clam. In *Biological implications of metals in the environment*, ed. R.E. Wilding and H. Dricker. CONF-750929. Springfield, VA: National Technical Information Service, 1977, pp. 213–230.

MacInnes, J.R., and A. Calabrese. Combined effects of salinity, temperature, and copper on embryos and early larvae of the American oyster, *Crassostrea virginica*. *Arch. Environ. Contam. Toxicol.* 8:553–562 (1979).

Manly, B.F.J. *The statistics of natural selection on animal populations*. New York: Chapman & Hall, 1985.

Mantel, N. Evaluation of survival data and two new rank statistics arising in its consideration. *Cancer Chemother. Rep.* 50:163–170 (1966).

Marking, L.L. Toxicity of chemical mixtures. In *Fundamentals of aquatic toxicology*, ed. G.M. Rand and S.R. Petrocelli. New York: Hemisphere Publishing Corp., 1985, chap. 7.

Marking, L.L., and V.K. Dawson. Method for assessment of toxicity or efficacy of mixtures of chemicals. *U.S. Fish Wildlife Serv. Invest. Fish Control* 67:1–8 (1975).

Marubini, E., and M.G. Valsecchi. *Analyzing survival data from clinical trials and observational studies*. Chichester, UK: John Wiley & Sons, 1995.

McCloskey, J.T., M.C. Newman, and S.B. Clark. Predicting the relative toxicity of metal ions using ion characteristics: Microtox® bioluminescence assay. *Environ. Toxicol. Chem.* 15:1730–1737 (1996).

McKinney, J.D., A. Richard, C. Waller, M.C. Newman, and F. Gerberick. The practice of structure activity relationships (SAR) in toxicology. *Toxicol. Sci.* 56:8–17 (2000).

McLeay, D.J., and Munro, J.R. Photoperiodic acclimation and circadian variations in tolerance of juvenile rainbow trout (*Salmo gairdneri*) to zinc. *Bull. Environ. Contam. Toxicol.* 23:552–557 (1979).

Meyer, J.S. A mechanistic explanation for the ln(LC50) vs ln(hardness) adjustment equation for metals. *Environ. Sci. Technol.* 33:908–912 (1999).

Miller, D.M. Reducing transformation bias in curve fitting. *Am. Stat.* 38(2):124–126 (1984).

Miller, R.G., Jr. *Survival analysis*. New York: John Wiley & Sons, 1981.

Miller, R.G., Jr. *Beyond ANOVA. Basics of applied statistics*. Boca Raton, FL: Chapman & Hall/CRC Press, 1998.

Monosoon, E. Chemical mixtures: Considering the evolution of toxicological and chemical assessment. *Environ. Health Perspect.* 113:383–390 (2005).

Morse, J.W., and D. Rickard. Chemical dynamics of sedimentary acid volatile sulfide. *Environ. Sci. Technol.* 38:131A–136A (2004).

Muller, H.-G. Acute toxicity of potassium dichromate to *Daphnia magna* as a function of the water quality. *Bull. Environ. Contam. Toxicol.* 25:113–117 (1980).

Nebeker, A.V., M.A. Cairns, J.H. Gakstatter, K.W. Malueg, G.S. Schuytema, and D.F. Krawczyk. Biological methods for determining toxicity of contaminated freshwater sediments to invertebrates. *Environ. Toxicol. Chem.* 3:617–630 (1984).

Neely, W.B., D.R. Branson, and G.E. Blau. Partition coefficients to measure bioconcentration potential of organic chemicals in fish. *Environ. Sci. Technol.* 8(13):1113–1115 (1974).

Nelson, H., D. Benoit, R. Erickson, V. Mattson, and J. Lindberg. *The effects of variable hardness, pH, alkalinity, suspended clay, and humic on the chemical speciation and aquatic toxicity of copper*. EPA/600/3-86/023. Springfield, VA: NTIS, 1986.

Nelson, W. Hazard plotting for incomplete failure data. *J. Qual. Technol.* 1(1):27–52 (1969).

Nelson, W. Theory and applications of hazard plotting for censored failure data. *Technometrics* 14(4):945–966 (1972).

Neter, J., W. Wasserman, and M.H. Kutner. *Applied linear statistical models. Regression, analysis of variance and experimental designs*. Boston: Irwin, 1990.

Newman, M.C. A statistical bias in the derivation of hardness-dependent metals criteria. *Environ. Toxicol. Chem.* 10:1295–1297 (1991).

Newman, M.C., and M. Aplin. Enhancing toxicity data interpretation and prediction of ecological risk with survival time modeling: An illustration using sodium chloride toxicity to mosquitofish (*Gambusia holbrooki*). *Aquatic Toxicol.* 23:85–96 (1992).

Newman, M.C., and W.H. Clements. *Ecotoxicology: A comprehensive treatment*. Boca Raton, FL: CRC Press, 2008.

Newman, M.C., S.A. Diamond, M. Mulvey, and P. Dixon. Allozyme genotype and time to death of mosquitofish, *Gambusia affinis* (Baird and Girard) during acute toxicant exposure: A comparison of arsenate and inorganic mercury. *Aquatic Toxicol.* 15:141–156 (1989).

Newman, M.C., and C.H. Jagoe. Ligands and the bioavailability of metals in aquatic environments. In *A mechanistic understanding of bioavailability: Physical-chemical interactions*, ed. J. Hamelink and W. Benson. Chelsea, MI: Lewis Publishers, 1993.

Newman, M.C., and J.T. McCloskey. Predicting relative toxicity and interactions of divalent metal ions: Microtox® bioluminescence assay. *Environ. Toxicol. Chem.* 15:275–281 (1998).

Newman, M.C., J.T. McCloskey, and C.P. Tatara. Using metal-ligand binding characteristics to predict metal toxicity: Quantitative ion character-activity relationships (QICARS). *Environ. Health Perspect.* 106:1419–1425 (1998).

Nimmo, D.R. Pesticides. In *Fundamentals of aquatic toxicology: Methods and applications*, ed. G.M. Rand and S.R. Petrocelli. New York: Hemisphere Publishing Corp., 1985, pp. 335–373.

Ogendi, G.M., W.G. Brumbaugh, R.E. Hannigan, and J.L. Farris. Effects of acid-volatile sulfide on metal bioavailability and toxicity to midge (*Chironomus tentans*) larvae in black shale sediments. *Environ. Toxicol. Chem.* 26:325–334 (2007).

Organization for Economic Cooperation and Development. *Guidelines for the testing of chemicals/Section 4: Health effects test no. 425: Acute oral toxicity: Up-and-down procedure. Adopted March 23, 2006*. Paris, 2006.

Ownby, D.R., and M.C. Newman. Advances in quantitative ion character-activity relationships (QICARs): Using metal-ligand binding characteristics to predict metal toxicity. *QSAR Comb. Sci.* 22:241–246 (2003).

Pallotta, A.J., M.G. Kelly, D.P. Rall, and J.W. Ward. Toxicology of acetoxycycloheximide as a function of sex and body weight. *J. Pharmacol. Exp. Ther.* 136:400–405 (1962).

Parker, J.G. Toxic effects of heavy metals upon cultures of *Uronema marinum* (Ciliophora: Uronematidae). *Mar. Biol.* 54:17–24 (1979).

Pascoe, D., S.A. Evans, and J. Woodworth. Heavy metal toxicity to fish and the influence of water hardness. *Arch. Environ. Contam. Toxicol.* 15:481–487 (1986).

Patrick, W.H., Jr., R.P. Gambrell, and R.A. Khalid. Physiochemical factors regulating solubility and bioavailability of toxic heavy metals in contaminated dredged sediment. *J. Environ. Sci. Health* A12(9):475–492 (1977).

Pepe, M.S., and T.R. Fleming. Weighted Kaplan–Meier statistics: A class of distance tests for censored survival data. *Biometrics* 45:497–507 (1989).

Peto, R., and J. Peto. Asymptotically efficient rank-invariant test procedures. *J. R. Stat. Soc. A* 135:185–207 (1972).

Piegorsh, W.W., and A.J. Bailer. *Statistics for environmental biology and toxicology*. London: Chapman & Hall, 1997.

Piegorsh, W.W., C.R. Weinberg, and B.H. Margolin. Exploring simple independent action in multifactorial tables of proportions. *Biometrics* 44:595–603 (1988).

Pinder, J.E., III, J.G. Wiener, and M.H. Smith. The Weibill distribution: A new method of summarizing survivorship data. *Ecology* 59(1):175–179 (1978).

Pollard, J.H. *A handbook of numerical and statistical techniques with examples mainly from the life sciences*. New York: Cambridge University Press, 1979.

Posthuma, L., G.W. Suter II, and T.P. Traas (eds.). *Species sensitivity distributions in ecotoxicology*. Boca Raton, FL: CRC/Lewis Publishers, 2002.

Pryor, D.B., F.E. Harrell Jr., K.L. Lee, R.M. Califf, and R.A. Rosati. Estimating the likelihood of significant coronary artery disease. *Am. J. Med.* 75:771–780 (1983).

Rall, D.P., and W.C. North. Consideration of dose-weight relationships. *Proc. Soc. Exp. Biol. Med.* 83:825–827 (1953).

Rapport, D.J., H.A. Regier, and T.C. Hutchinson. Ecosystem behavior under stress. *Am. Nat.* 125(5):617–640 (1985).

Ray, S., D.W. McLeese, and M.R. Peterson. Accumulation of copper, zinc, cadmium and lead from two contaminated sediments by three marine invertebrates—A laboratory study. *Bull. Environ. Contam. Toxicol.* 26:315–322 (1981).

Rhomberg, L.R., and S.K. Wolff. Empirical scaling of single oral lethal doses across mammalian species based on a large database. *Risk Anal.* 18(6):741–753 (1998).

Rispin, A., D. Farrar, E. Margosches, K. Gupta, K. Stitzel, G. Carr, M. Greene, W. Meyer, and D. McCall. Alternative methods for the median lethal dose (LD_{50}) test: The up-and-down procedure for acute oral toxicity. *ILAR J.* 43(4):233–243 (2002).

Roales, R.R., and A. Perlmutter. Toxicity of methylmercury and copper, applied singly and jointly, to the blue gourami, *Trichogaster trichopterus*. *Bull Environ. Contam. Toxicol.* 12(5):633–639 (1974a).

Roales, R.R., and A. Perlmutter. Toxicity of zinc and Cygon, applied singly and jointly, to zebrafish embryos. *Bull. Environm. Contam. Toxicol.* 12(4):475–480 (1974b).

Rozman, K.K., and J. Doull. Dose and time variables of toxicity. *Toxicology* 144:169–178 (2000).

Rule, J.H., and R.W. Alden III. Cadmium bioavailability to three estuarine animals in relationship to geochemical fractions in sediments. *Arch. Environ. Contam. Toxicol.* 19:878–885 (1990).

Salsburg, D.S. *Statistics for toxicologists.* New York: Marcel Dekker, 1986.

Sanchez-Bayo, F. From simple toxicological models to prediction of toxic effects in time. *Ecotoxicology* 18:343–354 (2009).

SAS Institute. *Additional SAS/STAT procedures.* SAS Technical Report P-179, Release 6.03. Cary, NC: SAS Institute, 1988.

Shepard, M.P. Resistance and tolerance of young speckled trout (*Salvelinus fontinalis*) to oxygen lack, with special reference to low oxygen acclimation. *J. Fish. Res. Board. Can.* 12(3):387–446 (1955).

Slonim, C.B., and A.R. Slonim. Effect of water hardness on the tolerance of the guppy to beryllium sulfate. *Bull. Environ. Contam. Toxicol.* 10(5):295–301 (1973).

Spehar, R.L., and J.T. Fiandt. Acute and chronic effects of water quality criteria-based metal mixtures on three aquatic species. *Environ. Toxicol. Chem.* 5:917–931 (1986).

Sprague, J.B. Measurement of pollutant toxicity to fish. I. Bioassay methods for acute toxicity. *Water Res.* 3:793–821 (1969).

Sprague, J.B. Measurement of pollutant toxicity to fish. II. Utilizing and applying bioassay results. *Water Res.* 4:3–32 (1970).

Stephan, C.E. Methods for calculating an LC_{50}. In *Aquatic toxicology and hazard evaluation, ASTM STP 634*, ed. F.L. Mayer and J.L. Hamelink. Philadelphia: American Society for Testing and Materials, 1977.

Stephan, C.E. Personal communication, 1992.

Suter, G.W., III. *Ecological risk assessment.* Chelsea, MI: Lewis Publishers, 1993.

Tatara, C.P., M.C. Newman, J.T. McCloskey, and P.L. Williams. Use of ion characteristics to predict relative toxicity of mono-, di- and trivalent metal ions: *Caenorhabditis elegans* LC50. *Aquat. Toxicol.* 42:255–269 (1998).

Tattersfield, F., and H.M. Morris. An apparatus for testing the toxic values of contact insecticides under controlled conditions. *Bull. Entomol. Res.* 14:223–233 (1924).

Tessier, A., P.G.C. Campbell, J.C. Auclair, and M. Bisson. Relationships between the partitioning of trace metals in sediments and their accumulation in the tissues of the freshwater mollusc *Elliptio complanata* in a mining area. *Can. J. Fish. Aquat. Sci.* 41:1463–1472 (1984).

Thompson, K.W., A.C. Hendricks, and J. Cairns Jr. Acute toxicity of zinc and copper singly and in combination to the bluegill (*Lepomis macrochirus*). *Bull. Environ. Contam. Toxicol.* 25:122–129 (1980).

Thurston, R.V., R.C. Russo, and G.A. Vinogradov. Ammonia toxicity to fishes. Effect of pH on the toxicity of the un-ionized ammonia species. *Environ. Sci. Technol.* 15(7):837–840 (1981).

Trevan, J.W. The error of determination of toxicity. *Proc. R. Soc. London B* 101:483–514 (1927).

Unger, M.A., M.C. Newman, and G.G. Vadas. Predicting survival of grass shrimp, *Palamonetes pugio*, exposed to naphthalene, fluorine, and dibenzothiophene. *Environ. Toxicol. Chem.* 27(8):1802–1808 (2008).

U.S. Environmental Protection Agency. *Ambient water quality criteria for copper—1984.* EPA 440/5-84-031. Springfield, VA: National Technical Information Service, 1985a.

U.S. Environmental Protection Agency. *Ambient water quality criteria for cadmium—1984.* EPA 440/5-84-032. Springfield, VA: 1985b.

U.S. Environmental Protection Agency. *Ambient aquatic life water quality criteria for zinc.* EPA 440/5-87-003. Springfield, VA: 1987.

U.S. Environmental Protection Agency. *Methods for measuring the acute toxicity of effluents and receiving waters to freshwater and marine organisms.* EPA-821-R-02-012. Washington, DC, 2002.

Van den Berg, M., L. Birnbaum, A.T.C. Bosveld, B. Brunstrom, P. Cook, M. Feeley, G.P. Giesy, A. Hanberg, R. Hasagawa, S.W. Kennedy, T. Kubiak, J.C. Larsen, F.X.R. van Leeuwen, A.K.D. Liem, C. Nolt, R.E. Peterson, L. Poellinger, S. Safe, D. Schrenk, D. Tillitt, M. Tysklind, M. Younes, F. Waerm, and T. Zacharewski. Toxic equivalency factors (TEFs) for PCBs, PCDDs, PCDFs for human and wildlife. *Environ. Health Perspect.* 106:775–792 (1998).

Voyer, R.A., J.A. Cardin, J.F. Heltshe, and G.L. Hoffman. Viability of embryos of the winter flounder *Psuedopleuronectes americanus* exposed to mixtures of cadmium and silver in combination with selected fixed salinities. *Aquat. Toxicol.* 2:223–233 (1982).

Voyer, R.A., and J.F. Heltshe. Factor interactions and aquatic toxicity testing. *Water Res.* 18(4):441–447 (1984).

Woodward, M. *Epidemiology. Study design and data analysis.* 2nd ed. Boca Raton, FL: Chapman & Hall/CRC, 2005.

Wright, D.A., and J.W. Frain. The effect of calcium on cadmium toxicity in the freshwater amphipod, *Gammarus pulex* (L.).*Arch. Environ. Contam. Toxicol.* 10:321–328 (1981).

Yates, F. Contingency tables involving small numbers and the χ^2 test. *J. Brit. R. Stat. Soc. Suppl.* 1:217–235 (1934).

Young, L.B., and H.H. Harvey. Metal concentrations in chironomids in relation to the geochemical characteristics of surficial sediments. *Arch. Environ. Contam. Toxicol.* 21:202–211 (1991).

Zitko, V., and W.G. Carson. A mechanism of the effect of water hardness on the lethality of heavy metals to fish. *Chemosphere* 5:299–303 (1976).

Zhao, Y., and M.C. Newman. Shortcomings of the laboratory-derived median lethal concentration for predicting mortality in field populations: Exposure duration and latent mortality. *Environ. Toxicol. Chem.* 23(9):2147–2153 (2004).

Zhao, Y., and M.C. Newman. Effects of exposure duration and recovery time during pulsed exposures. *Environ. Toxicol. Chem.* 25(5):1298–1304 (2006).

Zhao, Y., and M.C. Newman. The theory underlying dose-response models influences predictions for intermittent exposures. *Environ. Toxicol. Chem.* 26(3):543–547 (2007).

CHAPTER 5

Statistical Tests for Detection of Chronic Lethal and Sublethal Stress

Scientific research is a process of guided learning. The objective of statistical methods is to make that process as efficient as possible.

—**Box et al. (1978)**

5.1 GENERAL

Predictive models such as those described in Chapter 4 can be applied to acute toxicity data and, in many cases, to chronic lethal or sublethal effects data (e.g., Newman 2008). However, as effects become more subtle and difficult to model, predictive models are routinely replaced by statistical tests that notionally test for the presence of a statistically significant effect. Also, statistical testing methods are very often applied to chronic lethal or sublethal effects data for no reason other than presumed regulatory fiat. This chapter will examine statistical testing as applied to chronic toxicity and sublethal indicators of stress.

Page limits rule out detailed discussion of quantitative applications for particular chronic lethal or sublethal responses, although such discussion is obviously desirable. Only general methods will be treated. Fortunately, excellent reviews of such techniques already exist (Adams 1990; McCarthy and Shugart 1990; Huggett et al. 1992; OECD 2006a, 2006b).

5.2 METHOD SELECTION

5.2.1 The U.S. EPA Scheme

The original U.S. EPA methods establish (Weber et al. 1988, 1989) and their revisions (U.S. EPA 2002a, 2002b) describe statistical methods applicable to regulatory assessment of sublethal and chronic lethal effects. The U.S. EPA approach as diagrammed for fathead minnow larval survival (Figure 5.1) is used initially to frame the first part of this chapter, with supplemental material included as warranted. Most of the supplemental material comes from a review of postanalysis of variance methods by Day and Quinn (1989) and an OECD guidance document (OECD 2006a). Most of the techniques associated with the leftmost branch in Figure 5.1 ("Point Estimation") are described in detail in Chapter 4. The statistical testing techniques represented in the large, central portion of Figure 5.1 will be discussed and expanded in this chapter.

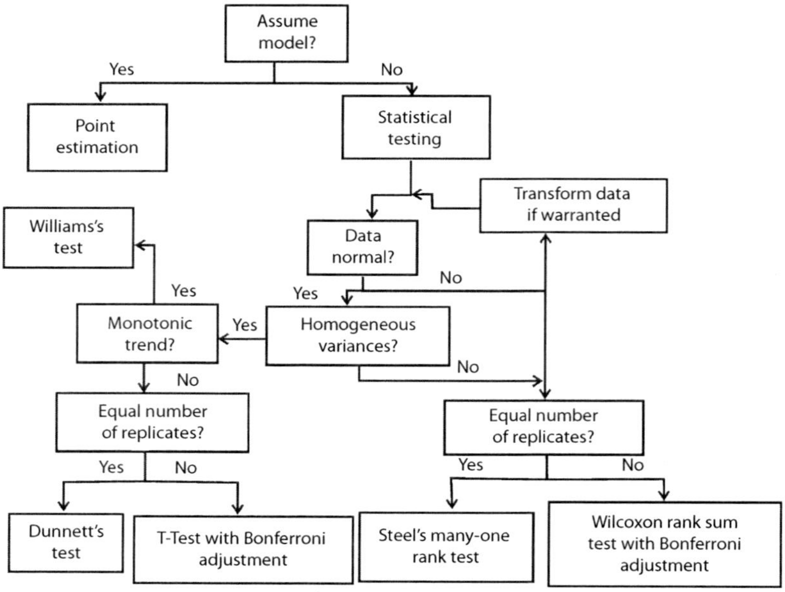

Figure 5.1 U.S. EPA scheme for analysis of chronic survival and sublethal effects as recommended by Weber et al. (1989) for regulatory testing. This scheme has been used for more than two decades (see U.S. EPA 2002b, Figure 2). (From Newman, M.C., and W.H. Clements, *Ecotoxicology: A Comprehensive Treatment*, CRC Press, Boca Raton, FL, 2008, Figure 10.4.)

5.2.2 The OECD Scheme

The published OECD scheme (OECD 2006a) differs from that of the U.S. EPA (2002b) in several notable ways. The OECD makes considerable effort to distinguish among methods best applied to quantal (e.g., dead or alive), continuous (e.g., growth rate), or discrete* (e.g., number of young or number of leaves per individual) response data.† More to the point, the OECD statistical testing scheme identifies best methods for quantal or continuous data, sketching out these methods as shown in Figures 5.2 and 5.3 (OECD 2006a). Also, a focused attempt is made in the OECD document to identify the best statistical approach as distinct from that most commonly applied in regulatory ecotoxicity testing. More attention is paid to the test power and making use of any underlying biology associated with the response data. Also, the document discusses in detail the calculation of EC10 or other point estimates using methods like those described in Chapter 4.

* The OECD document differentiates among nominal, ordinal, and interval discrete data. Nominal data are data for which a value is assigned for qualities that are not associated in a particular order, e.g., adult male, adult female, or juvenile. Ordinal or ranked data involve relative magnitudes, such as low, moderate, high, and severe. Interval data are discontinuous data for discrete counts such as the number of young born.

† More generally, data are continuous or discontinuous. Continuous, but not discontinuous, data can be measured at infinitely smaller scales. Discontinuous data are countable (Levine et al. 2001). Discontinuous data might be defined as discrete data, as discussed by OECD (2006a), or as meristic data, as often discussed in fields such as taxonomy (Sokal and Rohlf 1981). Other data types needing careful consideration during statistical testing include ratios, proportions, percentages, or quotients (Zar 1999).

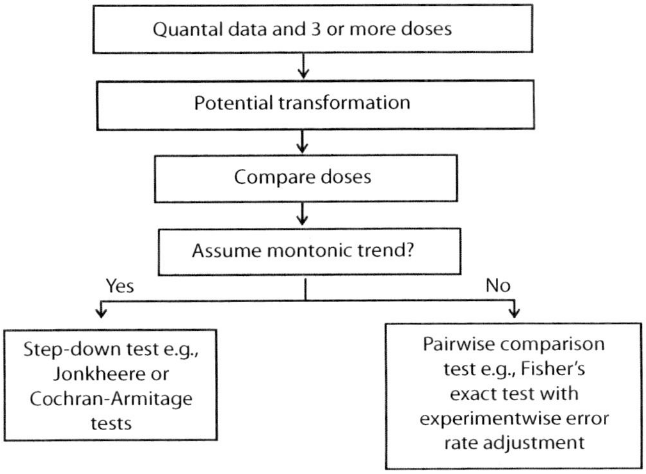

Figure 5.2 Modified OECD scheme for hypothesis testing involving quantal data (see OECD 2006a, Figure 5.1). If there are only two doses (control and a single toxicant concentration), then a test such as Fisher's exact test would be relevant. A two-dose test might be used for limit tests such as those discussed in Chapter 4 relative to the up-and-down LC50 technique. Limit tests are intended to infer simply whether the toxicant at the high experimental concentration is nontoxic. The conclusion drawn from a nonsignificant limit test might be that the toxicant has very low potential for adverse effect.

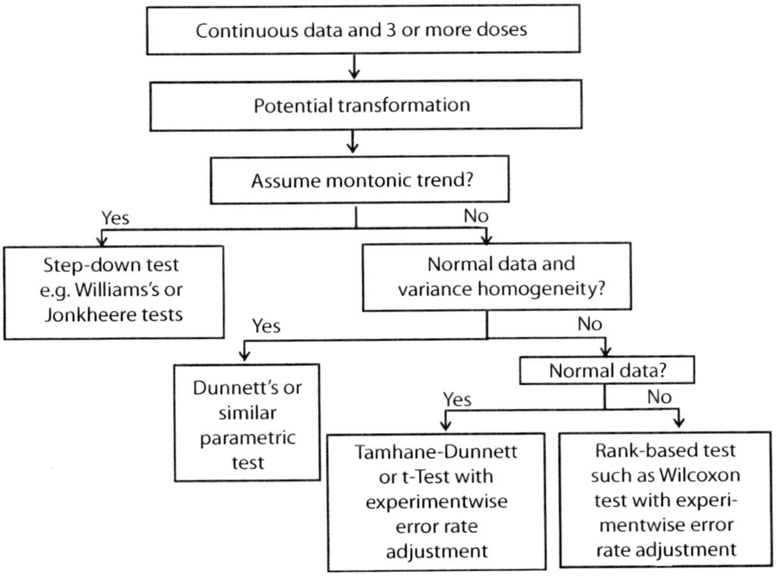

Figure 5.3 Modified OECD scheme for hypothesis testing involving continuous data (see OECD 2006, Figures 5.2 and 5.3). If there are only two doses (control and a single toxicant concentration), then a test such as Fisher's exact test might be relevant.

5.3 ONE-WAY ANALYSIS OF VARIANCE

5.3.1 General Design

A general experimental design leading to a one-way analysis of variance (ANOVA) is assumed in much of the following discussion, although in some instances, such as Dunnett's test, an initial ANOVA might not be required before statistical testing. In an experiment leading to one-way ANOVA, a series of treatments is established, including a control or reference treatment, with replicates at each treatment level. At the end of the experiment, observations from replicates for each treatment are made for a response variable. An ANOVA might be conducted to determine how probable it would be to obtain the observed data if the null hypothesis was true of no difference in means among treatments. With a significant outcome, a post-ANOVA test might be done to suggest which treatments might differ.

5.3.2 Assumption of Independence

The assumption of independence is central to ANOVA. In the case of ecotoxicity testing, this involves careful attention to assignment of individual subjects (sampling units) to exposure groups. For example, a formal random assignment of individual fish to tanks within the experiment is essential to producing valid inferences. This should involve random, not haphazard, assignment. The Excel® functions RAND() and RANDBETWEEN() can easily be applied to randomly assign individuals to treatments. However, more than this might be required. Spatial placement of tanks, cages, or beakers might be important depending on the physical configuration of the experiment and the experimental design. Despite efforts to avoid confounding factors, there might be differences in light intensity, temperature, humidity, or concentration in the experimental area that produce violations of independence among replicates. As a hypothetical example, assume that five sets of five replicate tanks containing 10 fish were to receive five different concentrations (0, 64, 128, 256, and 512 μg/L) for sodium pentachlorophenol. Physical restrictions might require that the 25 tanks be distributed among five constant temperature water tables that might vary slightly during testing in temperature or some other confounding factor. Each table has a flow of cooling water entering at one end and draining from the other. The flow across a table could produce slight temperature differences on the table despite one's best efforts. A simple randomized assignment of tanks to tables and positions on tables could create dependencies by chance alone. A Latin square design could be applied instead, such that each concentration treatment had one tank on each table and each position. The 5 × 5 Latin square interspersion of pentachlorophenol treatments shown below assures this will be the case. Any differences associated with water table and positions on a table are distributed uniformly among rows and columns.

	Position on Table (μg/L)				
Table	**At Inlet**	**Near Inlet**	**Center**	**Near Outlet**	**At Outlet**
A	0	256	512	128	64
B	256	128	64	512	0
C	512	0	128	64	256
D	64	512	256	0	128
E	128	64	0	256	512

The importance of thoughtfully assigning individuals to treatments or replicates, and ensuring independence among measured units such as tanks of exposed fish, cannot be overemphasized. Statistically significant outcomes can as easily result from violations of independence as from true differences among treatments.

The experimental unit (replicate) is the smallest unit of experimental material to which a treatment can be allocated independently of all other units. By definition, experimental units (e.g., aquariums, beakers, or plant pots) must be able to receive different treatments. Each experimental unit may contain multiple sampling units (e.g., fish, daphnia or plants) on which measurements are taken. Within each experimental unit, sampling units may not be independent. However, in some special case situations, individual organisms (housed in common units) can be treated as the experimental units: these special cases require some proof or strong argument of independence of organisms.

—OECD (2006a)

Pseudoreplication is central in our discussion of independence. In his widely read paper, Hurlbert (1984) defines pseudoreplication as "the use of inferential statistics to test for treatment effects with data from experiments where either treatments are not replicated (though samples may be) or replicates are not statistically independent." Pseudoreplication was present in 48% of ecological studies reviewed by Hurlbert (1984). As an example involving ecotoxicity testing, the standard proportional dilutor used in many flow-through aquatic toxicity tests has replicate tanks that receive spiked waters from the same dilution tank. Are these tanks truly replicates? Further, replicate tanks for a concentration treatment often sit physically close to each other in a constant temperature water trough: a condition could easily emerge that Hurlbert discusses as inadequate treatment interspersion. Gradients in temperature and other confounding factors should be explored and minimized to avoid dependence. Intentional designs such as randomized block or Latin square designs can reduce the influence of such issues.* The survival time experiments of Newman and coworkers described in Chapter 4 are a second ecotoxicology example. Tanks, but not individual fish, might be considered the true replicates if times to death were noted for individual fish within a single tank and duplicate tanks of fish were used per toxicant concentration. Individual fish were the sampling units for which a response was measured, but one could argue that, without careful documentation of no confounding factors influencing the duplicate tanks, the tanks were the true replicated experimental units. Justification including formal demonstration needs to be provided in order to interpret the responses (survival differences) as those of fish, not tanks (see Example 4.9). How much deviation from perfect replication (in this case, tank effects) and independence is acceptable to meet experimental goals must be determined thoughtfully because absolute replication and complete independence are unrealistic expectations for most experiments (Hurlbert 1984). Realizing the need to minimize the risk of confounding by uncontrolled experimental conditions, Newman and coworkers modified their more recent survival test designs by placing individuals in separate containers. The cost of doing so was increased effort to maintain dosing consistency and animal husbandry.

5.3.3 Other Assumptions

Formal requirements of one-way ANOVA techniques to test for equal means among treatments include a common variance for observations of all treatments and normal distributions for the observations. As an example, counts often display a Poisson distribution so the mean and variance would be expected to be equal. Consequently, variance would increase with an increase in mean response. Transformations of quantal data (e.g., dead/alive), proportions (e.g., percent hatched), or other expressions of response are often necessary to satisfy requirements of homogeneity of variance and normality. The reciprocal of a variable might be used to this end (Weber et al. 1989). Often logarithm or square root transformations (Miller 1986; Gelber et al. 1985; Weber et al. 1989) are used, although other power functions are also available (Miller 1986).

* For detailed discussion of randomized block and Latin square designs, see Cochran and Cox (1957), Hicks (1973), Box et al. (1978), Neter et al. (1990), or Davis (2000). Many software packages have modules to help generate such designs. For example, the SAS program has an ADX module that includes a library of design templates but also allows creation of customized designs.

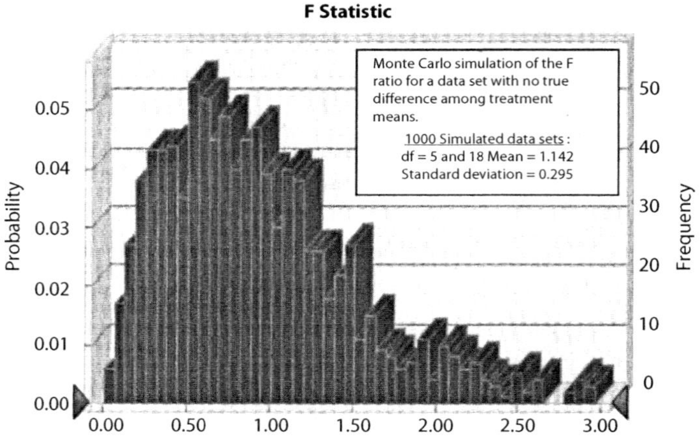

Figure 5.4 The mean sum of $\text{squares}_{\text{among}}$ and mean sum of $\text{squares}_{\text{within}}$ both reflect the common variance if there is no difference among means of treatments. The ratio of these two estimates (*F* ratio) will become larger than that expected by chance alone if treatment means do differ. The *F* distribution is defined by the degrees of freedom of the numerator and denominator, and consequently, the median *F* ratio expected under the null hypothesis of no treatment difference will vary with the degrees of freedom, approaching 1 if both *df* become very large. It follows that the critical *F* ratio corresponding with a particular α depends on the df_{among} and df_{within}. In this simulated illustration, the design used later to generate the sodium pentachlorophenol–fathead minnow data set is applied, but with random values inserted into the 24 cells (6 concentrations × 4 replicates = 24; mean = 1.142; standard deviation = 0.295). One thousand data sets were simulated and *F* ratios estimated for each. The *F* ratio corresponding to an α of 0.05 on the resulting plot would be that with only 5% of the area under the curve to its right; i.e., only 50 of the simulated 1,000 *F* ratios had this or a more extreme value. That being the case, an *F* ratio greater than 2.77 (df_{among} = 5, df_{within} = 18) would be grounds for rejection of the null hypothesis if α = 0.05.

The most common transformation for quantal or proportional data is the arcsine square root transformation.* This transformation is done to stabilize variances under the assumption that the data come from a binomial distribution. Weber et al. (1989) and others modify the procedure slightly by recommending that arcsine $\sqrt{1/(4n)}$ and (1.5708 – radians for the proportion $P = 0$) be used if the proportions are 0.0 and 1.0, respectively. (The 1.5708 – radians for $P = 0$ is the arcsine square root transformation of $(n - 1/4)/n$.) Here n is the number of animals exposed in the specific group for which the transformation is being done. Snedecor and Cochran (1973) recommend this modification if n is less than 50; however, they also discuss a more accurate alternative for data sets with small numbers of observations. The purpose of these transformations is to modify quantal or proportion data to minimize the likelihood of violating the assumption of variance homogeneity required by several of the more powerful, subsequent methods. Also, the distributions for the transformed proportions are more likely to be normal than those for the untransformed proportions (Gelber et al. 1985).

In an ANOVA, the estimated total sum of squares is broken down into the among- and within-treatment sums of squares. The mean sum of $\text{squares}_{\text{within}}$ is an estimate of the sampling or error variance, and the mean sum of $\text{squares}_{\text{among}}$ is an estimate of the variance existing among treatments. Both the within and among mean sums of squares are estimates of a common variance (σ_o^2) if the null hypothesis of no significant difference among treatment means is true. But, if the null hypothesis is not true, the mean sum of $\text{squares}_{\text{among}}$ will be some value larger than the common variance. Consequently, the mean sum of $\text{squares}_{\text{among}}$/mean sum of $\text{squares}_{\text{within}}$, or *F* ratio, can be used to test the null hypothesis of equal means between treatments (Figure 5.4). If the *F* ratio

* The transform is expressed in radians, not degrees. The arcsine square root transformation prescribed for proportion survival models was already discussed in Chapter 4. Appendix 5 provides arcsine square root transformations for proportions.

(named after R.A. Fisher) is larger than a specified critical value, this indicates a significant deviation from the null hypothesis of equal means. However, the particular treatments that differ from the others must be determined with post-ANOVA testing.

Example 5.1

Weber et al. (1989) provide the following fathead minnow (*Pimephales promelas*) larval survival data for sodium pentachlorophenol (NaPCP) exposure. Ten fish were exposed in each tank and proportions of the total number alive at the end of the exposure reported. Arcsine square root transformed data were analyzed here using a one-way ANOVA.

Toxicant Concentration (μg/L)	Replicate—Proportion Surviving			
	A	B	C	D
0	1.0	1.0	0.9	0.9
32	0.8	0.8	1.0	0.8
64	0.9	1.0	1.0	1.0
128	0.9	0.9	0.8	1.0
256	0.7	0.9	1.0	0.5
512	0.4	0.3	0.4	0.2

These quantal data are transformed using values from Appendix 5. For 1.0 (all 10 fish survived the exposure), the transformation used is that recommended by Weber et al. (1989), or 1.5708 – arcsine $\sqrt{1/(4n)}$ = 1.5708 – 0.1588 = 1.412. Substituting these values into the table,

Toxicant Concentration (μg/L)	Replicate			
	A	B	C	D
0 (treatment 1)	1.412	1.412	1.249	1.249
32 (treatment 2)	1.107	1.107	1.412	1.107
64 (treatment 3)	1.249	1.412	1.412	1.412
128 (treatment 4)	1.249	1.249	1.107	1.412
256 (treatment 5)	0.991	1.249	1.412	0.785
512 (treatment 6)	0.685	0.580	0.685	0.464

Following the general notation of Dixon and Massey (1969), there are six treatments (q = 6), including a control or reference treatment (0 μg/L). The six treatments or rows have means $\mu_1, \mu_2, \ldots, \mu_6$. The variance is assumed to be the same (σ_o^2) for all six means. (Normally, this assumption would be tested prior to an ANOVA.) The treatments have random samples drawn from them of size $n_1, n_2, \ldots, n_6$. In the table above, there are four replicate tanks for each treatment. To develop a working table for an ANOVA, let X_{ij} be the jth observation (replicate) out of z = 4 from treatment i.

	j					
i	1	2	3	4	Total (T_i)	Mean ($\bar{X}_i$)
1	1.412	1.412	1.249	1.249	5.322	1.330
2	1.107	1.107	1.412	1.107	4.733	1.183
3	1.249	1.412	1.412	1.412	5.485	1.371
4	1.249	1.249	1.107	1.412	5.017	1.254
5	0.991	1.249	1.412	0.785	4.437	1.109

i	j = 1	2	3	4	Total (T_i)	
6	0.685	0.580	0.685	0.464	2.414	0.604
					Grand total (T_G)	**27.408**
					Grand mean ($\bar{X}_G$)	**1.142**

The total or overall variance can be estimated by s_T^2, the mean total sum of squares.

$$s_T^2 = \frac{\sum_{i=1}^{q}\sum_{j=1}^{z}(X_{ij} - \bar{X}_G)^2}{N-1} \tag{5.1}$$

where N = the total number of observations over all treatments ($N = n_1 + n_2 + \ldots + n_6$).

The numerator is the overall or total sum of squares and the denominator is the total degrees of freedom.

The variance among treatment means is estimated by the mean sum of squares, s_A^2.

$$s_A^2 = \frac{\sum_{i-1}^{q}\frac{T_i^2}{n_i} - \frac{T_G^2}{N}}{q-1} \tag{5.2}$$

The numerator and denominator are the sum of squares among the treatment means and the treatment degrees of freedom, respectively.

The within-treatment (or error or among-replicate) variance is estimated by the mean error sum of squares, s_W^2.

$$s_w^2 = \frac{\sum_{i=1}^{q}\sum_{j=1}^{z}(X_{ij} - \bar{X}_i)^2}{N-q} \tag{5.3}$$

The numerator and denominator are the within-treatment (or error) sum of squares and the error degrees of freedom, respectively. An ANOVA table of variance is then constructed.

Variance	**Sum of Squares**	***df***	**Mean Sum of Squares**
Among	1.574	5	0.315 (s_A^2)
Within	0.426	18	0.024 (s_W^2)
Total	2.000	23	0.087 (s_T^2)

The null hypothesis that the treatment means are the same can be tested with an F ratio. The calculated F ratio for later comparison to tabulated values is the ratio of the mean sums of squares (MSS). In this case, MSS_{among}/MSS_{within} = 0.315/0.024 = 13.125. An F ratio is extracted from a table given α, $q - 1$, and $N - q$ (e.g., Table 16 in Rohlf and Sokal 1981) or with a function such as the Excel FINV(.05,5,18). In this case, $F_{0.05}(5,18)$ is 2.77. There is a probability of 0.05 that the calculated F would be larger than the tabulated value of 2.77 by chance alone if all μ values were the same. In the present example, the null hypothesis is rejected because 13.125 is greater than the critical F ratio of 2.77. However, as the analysis stands, it remains ambiguous which particular treatment means are not equal. Tukey's or Scheffe's tests could be used to resolve this ambiguity. They would test for significant differences among all pairs of means. The significance probability (α) would be based on all possible comparisons of means, that is, $[q(q-1)]/2$ or 15 comparisons.

The following SAS code performs the calculations described above, including Tukey's and Scheffe's tests:

```
DATA FATHEAD;
   INPUT ARC CONC @@;
   DATALINES;
   1.4120 0 1.4120 0 1.2490 0 1.2490 0
   1.1071 32 1.1071 32 1.4120 32 1.1071 32
   1.2490 64 1.4120 64 1.4120 64 1.4120 64
   1.2490 128 1.2490 128 1.1071 128 1.4120 128
   0.9912 256 1.2490 256 1.4120 256 0.7854 256
   0.6847 512 0.5796 512 0.6847 512 0.4636 512
   ;
PROC GLM;
   CLASS CONC;
   MODEL ARC=CONC;
   MEANS/TUKEY SCHEFFE;
RUN;
```

The wide range of post-ANOVA methods to compare treatments was reviewed by Day and Quinn (1989) and Tukey (1991). Some are concerned only with comparison of a specific pair of means, but others involve comparisons among all means. Still others compare one mean such as the control mean to the other means. Depending on the specific method, one of several kinds of errors rates may be pertinent. Consequently, a brief review of statistical errors pertinent to these post-ANOVA hypothesis tests follows.

A type I error occurs if the null hypothesis is rejected when it is true. The type I error rate (number of incorrect statements/total number of statements if the experiment was repeated many times) can be expressed as either a significance probability (α) or a confidence probability or coefficient ($1-\alpha$) (Miller 1966). In contrast, a type II error occurs if the null hypothesis is accepted when it is untrue. Its associated probability is designated β. The power of a test is the probability of rejecting the null hypothesis when it is untrue, i.e., $1-\beta$. The type I error rate for post-ANOVA comparisons can be estimated for a specific comparison or for an entire set of comparisons comprising an experiment. The definition for error rate given above is sufficient without elaboration for the case of a single comparison. An experimentwise or familywise type I error rate is used in the case involving many comparisons. The experimentwise error rate is "the proportion of experiments in which at least one false rejection of the null hypothesis is made" (Steel 1960). A clear understanding of the distinction between the experimentwise type I error rate and the type I error rate for a specific comparison of paired means (pairwise type I error rate) is important to understanding the procedures in the rest of this chapter.

Sheskin (2004) provided a series of equations that explain the most common way of coping with experimentwise error rates. The error rate used for each comparison of two means is adjusted downward so that the experimentwise error rate is maintained at a specified probability. The experimentwise error rate (α) can be related to the error rate used for testing each pair in the c number of comparisons in the entire experiment (α'),

$$\alpha = 1-(1-\alpha')^c \tag{5.4}$$

Equation 5.4 defines the probability of an event (type I error) occurring with c trials based on basic probability theory, that is, probability(A and B) = probability(A) × probability(B) if A and B are independent. Here, the c events have the same probability and are assumed to be independent of

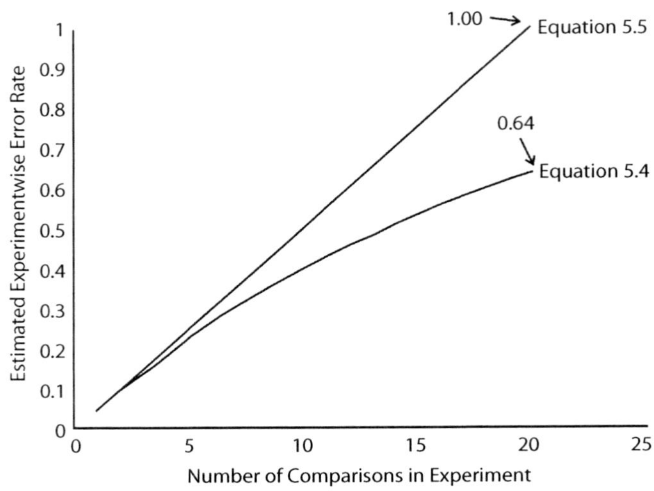

Figure 5.5 An illustration of the differences in experimentwise error rates estimated with Equation (5.4) and the approximate Equation (5.5). The experimentwise error rates were calculated using a set pairwise α of 0.05 and different numbers of comparisons. Estimates diverge as the number of comparisons increases, with the true experimentwise error rate inflation being reflected in the curve for Equation (5.4). The theoretical error rate with 20 comparisons is 0.64, yet Equation (5.4) gives an estimate of 1. The biased experimentwise type I error rates estimated with Equation (5.4) are the reason why the related Bonferroni adjustment will tend to be conservative, particularly with a large number of comparisons.

one another.* As an example, the probability of not making a type I error with two trials is $(1 - 0.05) \times (1 - 0.05) = (0.95)^2$. Equation (5.4) is a generalization for the probability of making a type I error in c trials (tests). A simpler, but conservative approximation is also in common use for this purpose,

$$\alpha = (c)(\alpha') \tag{5.5}$$

That Equation (5.5) is conservative is illustrated in Figure 5.5. The predicted experimentwise error rates for Equation (5.5) become increasingly biased upward as the number of comparisons increases. Most adjustments of the c pairwise error rate (α') in an experiment are based on Equation (5.4) or (5.5),

$$\alpha' = 1 - \sqrt[c]{1-\alpha} \quad \text{or} \quad 1-(1-\alpha)^{1/c} \tag{5.6}$$

$$\alpha' = \frac{\alpha}{c} \tag{5.7}$$

Equation (5.7) is the Bonferroni adjustment referred to at the bottom of Figures 5.1 to 5.3.† Again, the Bonferroni adjustment will be slightly conservative because it is based on Equation

* Events are independent if the outcome of one does not influence the outcome of the other(s).

† The Bonferroni-Holm adjustment used by OECD (2006a) is distinct in that the adjustments and statistical comparisons are made in a sequence beginning with the pair of means with the lowest unadjusted p value. Pairwise comparisons continue for the pair with the next lowest p value and stop with the first nonsignificant outcome. More specifically, the c unadjusted p values for the pairwise comparisons are ranked from smallest to largest and the first is compared to α/c. If it is significant, the next p value (p_i) is compared to $\alpha/(c - i + 1)$. The process is continued until a nonsignificant outcome occurs. That and all pair differences with higher p values are considered statistically nonsignificant.

(5.5): the actual experimentwise error rate will be slightly less than that derived using basic probability theory (Equation 5.6). The Dunn–Šidák adjustments discussed later are based on Equation (5.6). If pairs of tests in the experiment are independent, the adjusted error rates are exact. Otherwise, they too are slightly conservative but less so than the Bonferroni adjusted error rates (Sokel and Rohlf 1995). Sheskin (2004) emphasizes an important concern with any adjustments to pairwise error rates to ensure a certain value (or, in the case of Equation 5.7, upper limit to the value) of the experimentwise error rate, "although a reduction in the value of [the pairwise error rate] reduces the likelihood of committing a[n experimentwise] Type I error, it increases the likelihood of committing a Type II error." All else being equal, the smaller the type I error rate, the larger the type II error rate will be. Put more pragmatically, smaller type I error rates will reduce statistical power (Figure 5.6). He goes on to state that experimentwise error rate adjustments are

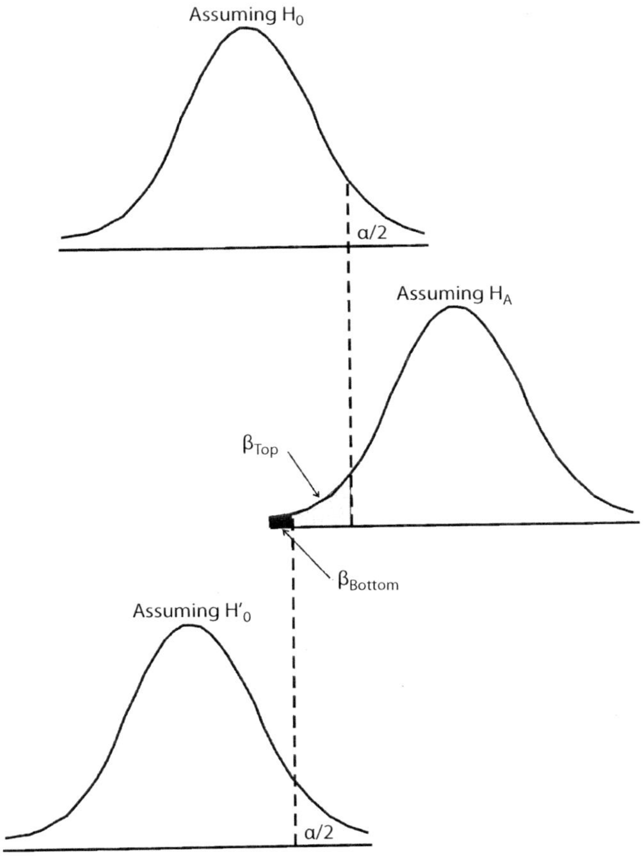

Figure 5.6 A two-sided test illustration of the relationships between type I (α) and type II (β) error rates, and also effect size. The top and middle curves are hypothetical normal distributions under the null hypothesis (H_0) and alternate hypothesis (H_A), respectively. A dotted line is drawn from the point on the H_0 curve area where the area to its right is $\alpha/2$. The area on the H_A curve to the left of that line (filled with grey and black) is the β_{Top}. The area to its right is the power, $1 - \beta_{Top}$. The area to the line's right under the H_A curve will decrease if the line is moved to decrease the $\alpha/2$ area under the H_0 curve: power decreases if α is decreased. The middle and bottom curves can be used to further show how the magnitude of the difference in means influences power. The bottom curve (H'_0) has a mean that is less than the original H_0 mean. Applying the same $\alpha/2$, it is clear that the power (nonblack area under H_A) increased relative to that for H_0. The ability to reject a false null hypothesis decreases as α decreases and increases with the magnitude of the true difference increases.

generally recommended if the experimenter sets an *a priori* number of pairwise comparisons in the experiment, as is the case with laboratory toxicity testing relevant here. These issues will be discussed in more detail later in this chapter.

5.4 TEST OF NORMALITY: SHAPIRO–WILK'S TEST

The assumption of normality is addressed prior to use of a conventional ANOVA and post-ANOVA tests, including Dunnett's test, a *t*-test with Bonferroni's adjustment, Dunn–Šidák's *t*-test, or Williams's test. This may be done using one of a variety of tests, although the power associated with these methods is often poor (Zar 1999). Graphical methods, including probability or quantile-quantile plotting, can be used to visually assess the normality of data. Sokal and Rohlf (1981, 1995), Miller (1986), and Levine et al. (2001) provide excellent overviews with examples for several of these techniques. Miller (1986) recommends visual inspection of probit, normal equivalent deviations (N.E.D.), or z score plots (untransformed or transformed data), as outlined in Chapter 4, for detecting nonnormality. The n values are ordered from smallest ($i = 1$) to largest ($i = n$). Next, the probit or N.E.D. for $i/(n + 1)$ is plotted against the observation value (or its transform if appropriate).* If the points produce a straight line, the data are judged to be sufficiently normal. "If a deviation from normality cannot be spotted by eye on probit paper, it is not worth worrying about" (Miller 1986). Miller (1986) suggests that particular attention be paid to deviations at the two ends of the line, as they are more important than values toward the center of the line for tests comparing means, e.g., an ANOVA and related tests. The reader wishing fuller discussion, including the specific effects of kurtosis and skewness on ANOVA, is referred to pages 5–16 in Miller (1986).

The assumption of normality can be assessed more formally with Shapiro–Wilk's (Shapiro and Wilk 1965), χ^2, or Kolmogorov–Smirnov (Sokal and Rohlf 1981) tests. Miller (1986) argues against the use of the latter two tests because they are less sensitive to deviations at extremes than to the less crucial deviations at the distribution center. The Shapiro–Wilk's test will be used here to assess the null hypothesis that the distribution from which observations were taken is normal. If the null hypothesis is rejected, the distribution is judged to be nonnormal at a specified α. It is important to note that failure to reject the null hypothesis does not prove that a distribution is normal. Indeed, with a sufficiently large sample, the null hypothesis of normality would be rejected for any data set. Fortunately, several of these tests, such as ANOVA itself (Dixon and Massey 1969; Miller 1986; Snedecor and Cochran 1973), produce acceptable results despite moderate violations of normality or homogeneity of variance. Monte Carlo studies indicate that probabilities derived from an ANOVA are close to the real probabilities if the underlying distribution is at least symmetrical and the variances for the treatments are within threefold of each other (Salsburg 1986).

Shapiro–Wilk's test orders and then compares the ordered observations to expected order statistics for a normal distribution. To begin, the unordered observations ($x_1, x_2, \ldots, x_n$) are ordered and relabeled such that $y_1 \leq y_2 \leq \ldots \leq y_n$. Then, using the n ordered observations and coefficients from Appendix 8, the following calculations are performed.

$$s^2 = \sum_{i=1}^{n}(y_i - \overline{y})^2 = \sum_{i=1}^{n}(x_i - \overline{x})^2 \tag{5.8}$$

If n is an even number, define $k = n/2$ and estimate b with Equation (5.9).

* The van der Waerden scaling transformation (Equation 2.6) is suggested here but Tukey (Equation 2.7) or Blom (Equation 2.8) scaling transformations may also be used.

$$b = \sum_{i=1}^{k} (a_{n-i+1})(y_{n-i+1} - y_i) \tag{5.9}$$

where a_{n-i+1} = values taken from Appendix 8 ($n \leq 50$).

If n is an odd number, define $k = (n - 1)/2$ and estimate b with Equation (5.10).

$$b = a_n(y_n - y_1) + \ldots + a_{k+2}(y_{k+2} - y_k) \tag{5.10}$$

The s^2 from Equation (5.8) and b as estimated with either Equation (5.9) or (5.10) are then used to calculate W.

$$W = \frac{b^2}{s^2} \tag{5.11}$$

Appendix 9 gives critical values of W for $n \leq 50$. An estimated W value smaller than the critical W value from Appendix 9 indicates that the H_0 of a normal distribution is rejected.

Example 5.2

Was the assumption of normality for the data used in Example 5.1 appropriate? The observations in the 128 μg/L treatment will be used to illustrate each detail of this method. Then the assumption of normality for all 24 observations will be estimated. (Note: These calculations are quite tedious. Fortunately, the SAS procedure PROC UNIVARIATE can perform them if the number of observations is less than 2,000 (SAS Institute 1988a). PROC UNIVARIATE will also perform the Kolmogorov–Smirnov, Anderson–Darling, and Cramer–van Mises tests of normality. The statement PROC UNIVARIATE NORMAL PLOT also produces a quantile-quantile probability plot.)

The 128 μg/L treatment ($n = 4$):

$$x_1,\ldots x_4 = 1.249, 1.249, 1.107, 1.412 \text{ so } y_1,\ldots y_4 = 1.107, 1.249, 1.249, 1.412$$

$$\bar{x} = 1.254$$

$$s^2 = (1.107 - 1.254)^2 + (1.249 - 1.254)^2 + (1.249 - 1.254)^2 + (1.412 - 1.254)^2 = 0.0466$$

Since n is even, b is estimated with Equation (5.8) and $k = 2$.

$$b = a_4(y_4 - y_1) + a_3(y_3 - y_2) = 0.6872(1.412 - 1.107) + 0.1677(1.249 - 1.249) = 0.2096$$

$$b^2 = 0.2096^2 = 0.0439$$

$$W = 0.0439/0.0466 = 0.9421$$

The critical value from Appendix 9 for W at $\alpha = 0.05$ and $n = 4$ is 0.748. Consequently, there is insufficient evidence for rejecting the null hypothesis of a normal distribution.

ALL OBSERVATIONS

The pertinent treatment mean must be subtracted from each observation value within a treatment before W can be estimated. The resulting values, adjusted to remove any treatment influence by subtracting the pertinent treatment mean from each value, are then arranged from smallest to largest and used according to the procedures described above.

i	Adjusted Value	$(y_i - \bar{y})^2$	a_{n-i+1}	$a_{n-i+1}(y_{n-i+1} - y_i)$
1	−0.324	0.102	0.4493	0.282
2	−0.147	0.020	0.3098	0.116
3	−0.140	0.018	0.2554	0.076
4	−0.122	0.014	0.2145	0.044
5	−0.118	0.013	0.1807	0.036
6	−0.081	0.006	0.1512	0.024
7	−0.081	0.006	0.1245	0.020
8	−0.076	0.005	0.0997	0.012
9	−0.076	0.005	0.0764	0.009
10	−0.076	0.005	0.0539	0.006
11	−0.024	0.000	0.0321	0.001
12	−0.005	0.000	0.0107	0.000
13	−0.005	0.000		**b = 0.626**
14	0.014	0.000		**b^2 = 0.392**
15	0.041	0.001		
16	0.041	0.001		
17	0.041	0.001		
18	0.081	0.006		
19	0.081	0.006		
20	0.082	0.006		
21	0.082	0.006		
22	0.158	0.023		
23	0.229	0.050		
24	0.303	0.088		
		s^2 = 0.382		

The $\bar{y}$ for the adjusted values is −0.005. The s^2 is the summation of the 24 $(y_i - \bar{y})^2$ values, or 0.382. The products, $a_{n-i+1}(y_{n-i+1} - y_i)$, are summed, and that sum is squared to derive b^2. The W is estimated as $b^2/s^2 = 0.392/0.382 = 1.026$. The critical value from Appendix 9 for W (α = 0.05, n = 24) is 0.916. There is no indication of significant deviation from normality for these adjusted values.

5.5 TEST FOR HOMOGENEITY OF VARIANCES: BARTLETT'S TEST

Homogeneity of variances is routinely examined because it is a formal requirement for ANOVA. Bartlett's test can be used under the assumption that the observations are normally distributed. The null hypothesis is equal variances for the treatments. The resulting statistic is compared to a critical χ^2 with associated degrees of freedom, $(q - 1)$ and α. The calculations are the following.

$$M = \left(\sum_{i=1}^{q} df_i\right) \ln \bar{s}^2 - \sum_{i=1}^{q} df_i \ln s_i^2 \tag{5.12}$$

where $\bar{s}^2$ = the mean of all of the individual treatment variances, and df_i = the degrees of freedom for treatment i (e.g., $n_i - 1$).

$$C = 1 + \left[\frac{1}{3(q-1)}\right]\left[\sum_{i=1}^{q}\frac{1}{df_i} - \frac{1}{\sum_{i=1}^{q} df_i}\right] \quad (5.13)$$

The M/C is compared to a critical χ^2.

Example 5.3

Let's apply Bartlett's method to test for variance homogeneity in the data for Examples 5.1 and 5.2.

i	Treatment (µg/L)	s_i^2	$\ln s_i^2$	$df_i(n_i - 1)$
1	0	0.008856	−4.727	3
2	32	0.023256	−3.761	3
3	64	0.006642	−5.014	3
4	128	0.015541	−4.164	3
5	256	0.076770	−2.567	3
6	512	0.011099	−4.501	3
		Σ	−24.734	18

$\bar{s}^2 = 0.023694$ and $q = 6$
$M = (18)(\ln 0.023694) - (3(-24.734)) = 6.837$
$C = 1 + [1/15][6/3 - 1/18] = 1.130$
$M/C = 6.05$

A critical value of 11.07 is obtained from a table of χ^2 values ($\alpha = 0.05$, $df = 5$), e.g., Table 14 in Rohlf and Sokal (1981) or from a software function such as the Excel CHIINV(0.05,5). The null hypothesis of equal variances is not rejected because M/C is less than this critical χ^2. Strictly speaking, this indicates no statistically significant deviation from homogeneity of variances. It does not prove that the variances are equal.

Barlett's and Shapiro–Wilk's tests can be implemented with SAS for the example data set using the following code.

```
DATA FATHEAD;
     INPUT ARC CONC @@;
     DATALINES;
     1.4120   0 1.4120   0 1.2490   0 1.2490   0
     1.1071  32 1.1071  32 1.4120  32 1.1071  32
     1.2490  64 1.4120  64 1.4120  64 1.4120  64
     1.2490 128 1.2490 128 1.1071 128 1.4120 128
     0.9912 256 1.2490 256 1.4120 256 0.7854 256
     0.6847 512 0.5796 512 0.6847 512 0.4636 512
     ;
/* SHAPIRO-WILK'S TEST FOR NORMALITY OF TREATMENT MEAN        */
/* ADJUSTED RESIDUALS.                                        */
DATA RESID;
   SET FATHEAD;
   IF CONC=0 THEN RESID=ARC-1.33050;
   IF CONC=32 THEN RESID=ARC-1.18332;
```

```
  IF CONC=64 THEN RESID=ARC-1.37125;
  IF CONC=128 THEN RESID=ARC-1.254275;
  IF CONC=256 THEN RESID=ARC-1.109400;
  IF CONC=512 THEN RESID=ARC-0.603150;
PROC UNIVARIATE NORMAL PLOT;
  VAR RESID;
RUN;
/* NEXT BARTLETT'S TEST FOR HOMOGENEITY OF VARIANCE IS DONE */
PROC GLM DATA=FATHEAD;
    CLASS CONC;
    MODEL ARC=CONC;
    MEANS CONC/HOVTEST=BARTLETT;
RUN;
```

The results for tests of normality pasted below do not result in rejecting the assumption of normality for all of the data after treatment means are subtracted from each observation.

```
Test                  --Statistic---      -----p Value------
Shapiro-Wilk          W       0.973332    Pr < W      0.7491
Kolmogorov-Smirnov    D       0.128914    Pr > D      >0.1500
Cramer-von Mises      W-Sq    0.049473    Pr > W-Sq   >0.2500
Anderson-Darling      A-Sq    0.324368    Pr > A-Sq   >0.2500
```

Although slightly different from the results of the above hand calculations, SAS results for Bartlett's test provide no evidence to reject the assumption of homogeneity of variances.

```
Bartlett's Test for Homogeneity of ARC Variance

Source  DF  Chi-Square  Pr > ChiSq
CONC    5   6.0439      0.3020
```

Both tests for normality and homogeneity of variance are done before an ANOVA to assess underlying assumptions. Further, they are useful in evaluating transformations. The tests are run before and after transformations such as those described above to determine the effectiveness of the transformations.

5.6 TREATMENT MEANS COMPARED TO THE CONTROL MEAN

5.6.1 Dunnett's Test

5.6.1.1 Equal Number of Observations

Often the null hypothesis that treatment means are the same as a control or reference group mean is tested instead of the null hypothesis of equivalent means for all treatments. In contrast to Example 5.1, in which Tukey's or Scheffe's tests were mentioned as tests for significant differences among all means, the means for treatments exposing fish to various concentrations of NaPCP might be compared only to that for unexposed fish. Methods such as Dunnett's test (Dunnett 1955, 1964) accommodate multiple comparisons of this nature and provide a more powerful method of testing this hypothesis than application of Tukey's or Scheffe's test.* Dunnett's test is based on the assump-

* The increase in power is a consequence of only $q - 1$ comparisons being made instead of $[q(q - 1)]/2$ comparisons. Consequently, the pairwise error rates do not require as much downward adjustment to maintain an experimentwise error rate.

tions of normality and equal variances. In the example used above, the differences between the mean of the 0 μg/L and the five other treatments would be tested jointly with an experimentwise α as follows.

Let p be the number of treatments to be compared to the control ($p = q - 1$). The estimated treatment means are designated $\bar{x}_1, \ldots, \bar{x}_p$, and the estimated control mean is designated $\bar{x}_0$. The estimated standard deviation for the q sets of observations is s. The number of observations in each treatment is designated as n_i.

$$s^2 = \frac{\sum_{i=0}^{p}\sum_{j=1}^{n_i}(x_{ij} - \bar{x}_i)^2}{\left(\sum_{i=0}^{p} n_i\right) - (p+1)} \tag{5.14}$$

Notice that Equation (5.14) is the same as Equation (5.3) for the estimated within-treatment variance except minor changes in notation were made to conform to Dunnett's original description of the technique (Dunnett 1955). Often, an ANOVA is performed prior to implementation of Dunnett's test to facilitate estimation of this variance. An ANOVA can also be used with more complicated experimental designs to ensure that interaction terms are insignificant. Dunnett (1964) discusses estimation if interactions are important.

An allowance is estimated with Equation (5.15) if the number of observations is the same for all treatments including the control.

$$A = ts\sqrt{\frac{2}{n}} \tag{5.15}$$

This allowance is used to establish a confidence limit for the differences between the two means beyond which the difference is significant. Appendices 10 and 11 may be used for one- and two-sided tests, respectively.

Example 5.4

The fathead minnow toxicity data used in the examples above will be analyzed with Dunnett's procedures. A one-sided test with a lower limit is appropriate because the toxicant is expected to lower, rather than raise, survival. A two-sided test will be discussed briefly only to illustrate the associated calculations. Note that the i used here is equivalent to $i - 1$ in previous examples.

i	$\bar{x}_i$	$\Sigma(x_{ij} - \bar{x}_i)^2$
0	1.330	0.02657
1	1.183	0.06977
2	1.371	0.01993
3	1.254	0.04662
4	1.109	0.23031
5	0.604	0.03330
	Σ	**0.4265**

$\Sigma n_i = 24$
$s^2 = 0.4265/(24 - 6) = 0.02369$ (within treatment mean sum of squares in Example 5.1)
$s = 0.1539$

ONE-SIDED TEST (LOWER LIMIT)

The allowance (A) is the allowable difference in treatment means. In other words, A estimates the lower confidence limit for the differences, $\bar{x}_i - \bar{x}_0$.

From Appendix 10, t for a one-sided test ($\alpha = 0.05$, $df = 18$, $p = 5$) is 2.41.

$$A = (2.41)(0.1539)(\sqrt{2/4}) = 0.2623$$

i	$\bar{x}_i - \bar{x}_0$
1	1.183 − 1.330 = −0.147
2	1.371 − 1.330 = 0.041
3	1.254 − 1.330 = −0.076
4	1.109 − 1.330 = −0.221
5	0.604 − 1.330 = −0.726

Only $\bar{x}_5$ is significantly different ($\alpha = 0.05$) from $\bar{x}_0$ because $|-0.726|$ is a larger difference for $\bar{x}_5 - \bar{x}_0$ than 0.2623. No other differences are significant at the chosen α.

TWO-SIDED TEST

Although not pertinent to this specific example, a two-sided test for these data might involve estimation of both upper and lower confidence limits. The A would be estimated with a t taken from Appendix 11 instead of Appendix 10.

$$t = 2.76$$
$$A = (2.76)(0.1539)(\sqrt{2/4}) = 0.3004$$

The A would be compared to both negative and positive differences between means.

The substitution of the following SAS code for the last five lines of code in Example 5.1 allows calculation of two-sided and one-sided (lower-limit) Dunnett's tests.

```
PROC GLM;
   CLASS CONC;
   MODEL ARC=CONC;
   MEANS CONC/DUNNETT("0") DUNNETTL("0");
RUN;
```

5.6.1.2 Unequal Number of Observations

By design or mishap, the numbers of observations in the various treatments may not be equal. In fact, there are sound reasons for having more observations in the control than in the exposure treatments (Dunnett 1955; Williams 1972; Box et al. 1978). For an α of 0.05 or less, the optimum allocation of observations between the control and treatments can be estimated (Dunnett 1955; Williams 1972).

$$\frac{n_0}{n_i} = \sqrt{q - 1} \quad (5.16)$$

In our example, the optimum number of observations (replicate tanks) in the control would be 2.24 times that in an exposure treatment, e.g., nine control tanks and four replicate tanks for each exposure treatment. Dunnett's test may be used with unequal n values for the control and treatments by substituting the A estimated in Equation (5.17) for that estimated by Equation (5.15).

$$A = ts\sqrt{\frac{1}{n_0} + \frac{1}{n_i}} \tag{5.17}$$

However, t values in the associated tables were developed under the condition of equal observation numbers or, more specifically, a correlation of ½ (Dunnett 1955, 1964). An estimate of this correlation if the number of observations differs is given in Equation (5.18).

$$\rho_{ij} = \frac{1}{\frac{n_0}{n_i} + 1} \tag{5.18}$$

Increasing the number of control observations (n_0) relative to the number in the exposure treatments (n_i) will lower the value of ρ_{ij} below 1/2. This will decrease the probability associated with the confidence limits below the assumed 0.95 or 0.99. Consequently, use of the approach described above for a one-sided test will produce an approximate (conservative) test if observation numbers are unequal. However, the values of P (probability) do not differ very much from the tabulated values for a wide range of ρ_{ij} values on either side of 1/2. Day and Quinn (1989) recommend Dunnett's test using Kramer's modification (Kramer 1956) (Equation 5.17) as the best method for control-treatment comparisons with unequal observation numbers.

For two-sided tests, Dunnett (1964) provided a correction factor for adjusting the tabulated t values for unequal observation numbers between the control and the other treatments. First, $1 - (n_i/n_0)$ is calculated. The result is multiplied by the superscript in the appropriate table of t values (Appendices 10 and 11). This factor is the percentage by which the tabulated t value should be increased to allow for the greater number of control observations. If there were nine control replicates in the above example, the tabulated t value (2.76) would be adjusted by (3.6(1 – 4/9)), or 2%. The adjusted t value would be (2.76)(1.02), or 2.815. More precise estimates for one- and two-sided tests when the correlation is not 1/2 can be generated by statistical programs such as SAS®. For example, the SAS algorithm used in Example 5.4 will perform these calculations when observation numbers are unequal between the control and exposure treatments, and among exposure treatments. When the numbers of observations are different among experimental treatments, the correlation is estimated by the SAS program using the harmonic mean of the observation numbers.

5.6.2 *t*-Test with Bonferroni's Adjustment

Weber et al. (1989) recommend that Bonferroni's adjusted t-test be used instead of Dunnett's test if the numbers of observations are unequal because the confidence limit for Dunnett's test is approximate with unequal numbers of observations. However, for the two-sided test, adjustment is easily accomplished for Dunnett's test as discussed above. Further, this is not a general restriction as some computer applications adjust ρ_{ij} accordingly. The SAS procedure GLM (SAS Institute 1988a) adjusts the ρ_{ij} and can be used as a more powerful tool than the recommended t-test with Bonferroni's adjustment. (The difference in power results from the fact that the t-test with Bonferroni's adjustment sets an upper limit (α) on the experimentwise error rate, as already discussed, but Dunnett's test fixes α for the experimentwise error rate.)

Generally, Bonferroni's adjustment of the *t*-test modifies the α associated with the statistical test for each pair of means to accommodate the multiple comparisons. An upper limit for the experimentwise α (the probability of making at least one type I error in a series of comparisons) is defined, e.g., 0.05, and an adjusted α' is estimated for use in testing each individual pair in the experiment so as to maintain this upper limit for the experimentwise error rate. With Bonferroni's adjustment, α' is estimated by α/p (see Equation 5.7, where *c* was used to denote the number of comparisons in the test), and the experimentwise α divided by the number of pairs being tested in the entire experiment. For example, the α' used for testing the individual pairs in Example 5.1 would be 0.05/5, or 0.01. The α' is used to extract pertinent *t* values used for the *p* comparisons. Obviously, such a procedure would require *t* values for adjusted α values. Such tables of *t* values are provided in Appendices 12 to 15.* The experimentwise α, *df*, and *p* are used to extract a critical *t* from these tables. The *t* values in the columns for $p = 1$ (mean of the control versus the mean of one treatment only) are simply Student's *t* statistics because $\alpha/p = \alpha/1 = \alpha$.

Under the assumptions of normality and homogeneity of variance, t_i is estimated with Equation (5.19) for each pair of means. The *s* is estimated with the square root of the within-treatment mean sum of squares (Equation 5.3).

$$t_i = \frac{\bar{X}_0 - \bar{X}_i}{s\sqrt{\dfrac{1}{n_0} + \dfrac{1}{n_i}}} \tag{5.19}$$

Each value is compared to tabulated Bonferroni's *t* values. Appendices 11 ($\alpha = 0.01$) and 13 ($\alpha = 0.05$) are used for one-sided tests, and Appendices 14 ($\alpha = 0.01$) and 15 ($\alpha = 0.05$) for two-sided tests. Values in Appendices 12 and 13 were generated with the SAS function TINV using as arguments $1\text{-}\alpha/p$ as the quantile, *df*, and 0 as the distribution center (SAS Institute 1988b).

5.6.3 Dunn–Šidák *t*-Test

An alternative to the *t*-test with Bonferroni's adjustment is the Dunn–Šidák *t*-test (Ury 1976), which has slightly more power than the *t*-test with Bonferroni's adjustment (Day and Quinn 1989). (Less conservative estimates of the probability, *P*, based on the Bonferroni adjustment are also available, e.g., Wright 1992, and provide more power than the commonly used Bonferroni adjustment described above.)

Although the Dunn–Šidák *t*-test also adjusts the α to get an exact or, under some conditions, an upper limit for the experimentwise error rate, the adjustment is different from Bonferroni's adjustment. The use of an adjusted α (α′) as defined below results in an experimentwise error rate ≤ α for the *p* comparisons (Sokal and Rohlf 1981).

$$\alpha' = 1 - (1 - \alpha)^{1/p} \tag{5.20}$$

The *t* values associated with the resulting unconventional values of α' cannot be taken directly from most Student's *t* tables but can be easily generated from software functions such as that described in the footnote. If software is unavailable, tables such as Appendix 16 (one-sided test) or 17 (two-sided test) are available. (Values in Appendix 16 were generated with the SAS function

* The Bonferroni or Dunn-Šidák t could easily be produced with Excel TINV(α′,*df*). The df here is 18, as described in the example. TINV(0.01,18) yields 2.878 for a two-sided test after Bonferroni adjustment of α′ to 0.05/5. This is the same as that given in Appendix 15 for $df = 18$ and $p = 5$. Although TINV provides estimates for two-tailed t distributions, TINV can also generate *t* values for a one-sided test by using 2α′, that is, TINV(0.02,18) or 2.552.

TINV using as arguments $(1 - \alpha)^{1/p}$ as the quantile, *df*, and 0 as the distribution center (SAS Institute 1988b).) Use of two-tailed tables of Dunn–Šidák's *t* values for one-sided tests by halving the experimentwise α has been recommended by Rohlf and Sokal (1981). However, a minor inaccuracy occurs during such use (Dixon 1993), and tables such as Appendix 16 should be used instead. Note that there is no column for $p = 1$ comparison in these tables. As described with Bonferroni's *t* tables, the Dunn–Šidák *t* values are simply Student's *t* values when $p = 1$. The adjustment on the *t* values as defined by Equation (5.20) becomes simply $\alpha' = \alpha$ when $p = 1$. Consequently, the *t* values provided for $p = 1$ in Bonferroni's *t* tables (Student's *t* statistic) can be used as the Dunn–Šidák *t* for $p = 1$.

Example 5.5

Both a *t*-test with Bonferroni's adjustment and a Dunn–Šidák test will be used to analyze the fathead minnow survival data.

i	$\bar{X}_0 - \bar{X}_i$	t_i
1	0.147	1.352
2	−0.041	−0.377
3	0.076	0.699
4	0.221	2.032
5	0.726	6.676

***t*-TEST WITH BONFERRONI'S ADJUSTMENT**

The *t* value for $\alpha = 0.05$, $q - 1 = p = 5$, $df_{within} = 18$ is 2.552 (Appendix 13). The t_i for treatment 5 (512 µg/L) was larger than this critical *t* value, suggesting according to statistical testing convention that the mean for this treatment was significantly different from that of the control (0 µg/L). No other treatment means were significantly different from that of the control.

***t*-TEST BY DUNN-ŠIDÁK'S METHOD**

The *t* value for an experimentwise $\alpha = 0.05$, $p = 5$, and $df_{within} = 18$ is 2.543 (Appendix 16). The t_i calculated above for treatment 5 is greater than 2.543, leading to rejection of the null hypothesis of equal means. Again, none of the other means differed significantly from the control mean.

5.7 MONOTONIC TREND: WILLIAMS'S TEST

Williams (1971, 1972) noted that the methods described above might not make full use of all of the information available to the researcher. If one assumes that response changes monotonically with increasing dose, the more powerful isotonic regression approach may be developed. Williams's approach has a null hypothesis that mean responses are equal among treatment doses and an alternate hypothesis that mean responses monotonically increase or decrease with treatment dose. The test is conducted in two stages. In the first stage, the presence or absence of a significant deviation from the null hypothesis is tested using the most extreme comparison (control versus highest concentration or dose). If the most extreme difference is significant, the lowest dose producing a significant mean response is identified in the second stage.

Williams's procedures with equal numbers of observations (n) among treatments can be summarized as follows. In the design, there is a control ($i = 0$) and p treatments (dose levels, $i = 1$ to p) with dose increasing with i. The mean responses ($\bar{X}_i$ values) are assumed to be independent and normal. The observations have a common variance (σ^2) estimated by s^2, the mean sum of $\text{squares}_{\text{within}}$.

Next the $\bar{X}_i$ values are used to produce a series of maximum likelihood estimates of expected mean responses assuming the alternate hypothesis of monotonic ordering with dose. The estimates are $M_0 \le M_1 \le M_2 \le M_3 \le \ldots \le M_p$ if the mean responses increase with dose. If mean responses are expected to decrease, the signs are changed on the means ($\bar{X}_i$) before use. One of the inequalities between M_i values must be strict, i.e., all cannot be equal. Adjustment must also be made if the $\bar{X}_i$ values (or $-\bar{X}_i$ values) do not satisfy the series $\bar{X}_0 \le \bar{X}_i \le \bar{X}_{i+1} \le \bar{X}_{i+2} \le \ldots \le \bar{X}_p$. The adjustment process involves the following steps.

Assume that $\bar{X}_0 \le \bar{X}_i > \bar{X}_{i+1} \le \bar{X}_{i+2} \le \ldots \le \bar{X}_p$. The $\bar{X}_i > \bar{X}_{i+1}$ must be adjusted to satisfy the inequality series for the M_i values. The two means are replaced by a single estimated mean.

$$\bar{X}_{i,i+1} = \frac{w_i\bar{X}_i + w_{i+1}\bar{X}_{i+1}}{w_i + w_{i+1}} \tag{5.21}$$

The weights (w_i, w_{i+1}) are proportional to the number of observations used to produce each mean. Equation (5.21) reduces to Equation (5.22) because n was the same for all treatments.

$$\bar{X}_{i,i+1} = \frac{\bar{X}_i + \bar{X}_{i+1}}{2} \tag{5.22}$$

This estimated mean ($\bar{X}_{i,i+1}$) replaces the two means ($\bar{X}_i$, $\bar{X}_{i+1}$) in the series. This process is repeated until the series of inequalities is satisfied. For every such replacement, the number of unique means in the series is reduced by one.

The largest M in the final series (M_p) is compared to the $\bar{X}_0$ to test the null hypothesis. The test statistic ($\bar{t}_p$) is estimated with Equation (5.23).

$$\bar{t}_p = \frac{M_p - \bar{X}_0}{s\sqrt{\frac{2}{n}}} \tag{5.23}$$

The null hypothesis is rejected if the estimated $\bar{t}_p$ is greater than a critical $\bar{t}_{p,\alpha}$ (Appendix 18 for $\alpha = 0.01$ or Appendix 19 for $\alpha = 0.05$). But no further conclusion save that a significant response is present at the highest M may be made at this point. More information is extracted during a second stage in which the lowest dose with a significant response is determined. A $\bar{t}_{p-1}$ is calculated with Equation (5.23) for M_{p-1} instead of M_p and compared to a critical value of $\bar{t}_{p-1,\alpha}$. The null hypothesis that the M_{p-1} is the same as the M_0 is rejected if the estimated $\bar{t}_{p-1}$ is greater than the critical $\bar{t}_{p-1,\alpha}$. The process is repeated for M_{p-2}, M_{p-3}, …, M_1 to determine the lowest dose at which a significant response was detected.

Example 5.6

The fathead minnow survival data will be used to illustrate Williams's test as applied to treatments with equal numbers of observations ($n = 4$). From previous examples, the following information is available: $s^2 = 0.024$, $df = 18$, $p = 5$, $q = 6$.

$\bar{X}_i$	$-\bar{X}_i$	M_i	df'',p''	$\bar{t}_p$	$\bar{t}_{p,\alpha}$
1.330	−1.330	−1.330			
1.183	−1.183	−1.277	18,1	0.484	1.734

$\bar{X}_i$	$-\bar{X}_i$	M_i	df'',p''	$\bar{t}_p$	$\bar{t}_{p,\alpha}$
1.327	−1.371	−1.277	18,2	0.484	1.818
1.254	−1.254	−1.254	18,3	0.694	1.845
1.109	−1.109	−1.109	18,4	2.017	1.859
0.604	−0.604	−0.604	18,5	6.627	1.867

Is there a significant response? Because the assumed response is a decrease in survival with increasing dose, the sign of the mean responses must be changed before preceding (column 2 above). Notice that $-\bar{X}_1$ is greater than $-\bar{X}_2$. This pair must be modified as per Equation (5.22) to generate the associated M_i.

$$\bar{X}_{1,1} = \frac{(-1.183 + (-1.371))}{2} = -1.277$$

The M_i values for all treatment means are provided in column 3 above. Using Equation (5.23), $\bar{t}_p$ is estimated from the largest M_i.

$$\bar{t}_p = \frac{-0.604 - (-1.330)}{\sqrt{\frac{2(0.024)}{4}}} = 6.627$$

From Appendix 19, a critical value of $\bar{t}_{5,\alpha=0.05}$ (one-sided test) is found to be 1.867. Since 6.627 is greater than 1.867, the null hypothesis is rejected. (The value for $p = 1$ in the above table is taken from a Student's t distribution table for $\alpha = 0.05$, $df = 18$ for a one-sided test. Alternatively, the same value can be obtained from a table for a two-sided test by using 2α or 0.10 as the α for that table. Bonferroni adjusted t values for $p = 1$ within Appendices 12 to 15 are Student's t values that may be used for this purpose.)

What is the lowest concentration with a significant response?

The process is repeated for all M_i values with critical $\bar{t}_{p,\alpha}$ taken from Appendix 19 using the df and p listed in column 4 of the table above. The results are provided in columns 5 and 6 above. The lowest concentration with a significant response was 256 μg/L. Although SAS does not implement Williams's test directly, the SAS code below does produce such a test.

```
DATA FATHEAD;
     INPUT ARC CONC @@;
     DATALINES;
     1.4120    0 1.4120    0 1.2490    0 1.2490    0
     1.1071   32 1.1071   32 1.4120   32 1.1071   32
     1.2490   64 1.4120   64 1.4120   64 1.4120   64
     1.2490  128 1.2490  128 1.1071  128 1.4120  128
     0.9912  256 1.2490  256 1.4120  256 0.7854  256
     0.6847  512 0.5796  512 0.6847  512 0.4636  512
     ;
RUN;
/* WILLIAMS'S TEST IS DONE AFTER ADJUSTING THE TREATMENT MEANS SO       */
/* THEY ARE ISOTONIC. THE PROBMC OPTION AND SOME ADDITIONAL CODE        */
/* ARE USED. THIS APPROACH WORKS FOR WILLIAMS'S TEST WITH EQUAL         */
/* NUMBERS OF OBSERVATIONS. THE MSE (0.02367466) HAS BEEN CALCULATED    */
/* PRIOR TO RUNNING THIS CODE. THE ASSOCIATED DF=24-6=18,THE            */
/* PROBABILITY OF THE T FOR EACH TESTED PAIR IS ESTIMATED FOR           */
/* "P"=5,4,3,2,1. Tp=(MEANp-MEANcontrol)/(SQRT(2*MSE/N)WHERE            */
/* N=THE NUMBER OF OBSERVATIONS IN EACH TREATMENT.                      */
DATA ISOTON;
INPUT PAIR LABEL $ @@;
```

```
/* CALCULATED TBARp FOR EACH PAIR TESTED */
IF PAIR=5 THEN TCALC=(-.604+1.330)/SQRT(2*.02367466/4);
IF PAIR=4 THEN TCALC=(-1.109+1.330)/SQRT(2*.02367466/4);
IF PAIR=3 THEN TCALC=(-1.254+1.330)/SQRT(2*.02367466/4);
IF PAIR=2 THEN TCALC=(-1.277+1.330)/SQRT(2*.02367466/4);
IF PAIR=1 THEN TCALC=(-1.277+1.330)/SQRT(2*.02367466/4);
ASSOCIATED WITH THE CALCULATED T ABOVE */
PROB=1 - PROBMC("WILLIAMS",TCALC,.,18,5);
/* RESULTS SIGNIFICANT AT 0.05? */
IF PROB<0.05 THEN SIGN = "YES"; ELSE SIGN = " NO";
DATALINES;
5 512vs0 4 256vs0 3 128vs0 2 64vs0 1 32vs0
;
RUN;
PROC PRINT NOOBS;
TITLE "ONE-SIDED WILLIAMS'S TEST WITH ALPHA=0.05";
  VAR LABEL TCALC Talpha PROB SIGN;
RUN;
```

This code produces the following results:

```
ONE-SIDED WILLIAMS'S TEST WITH ALPHA=0.05

LABEL     TCALC     Talpha     PROB     SIGN

512vs0  6.67283  1.86684  0.00000  YES
256vs0  2.03126  1.85894  0.03610  YES
128vs0  0.69853  1.84547  0.33396  NO
64vs0   0.48713  1.81781  0.42536  NO
32vs0   0.48713  1.73406  0.42536  NO
```

Although the tabulated TCALC values are slightly different from the $\bar{t}_p$ estimated by hand calculator above, the conclusions from the tests are the same. Notice that the SAS PROBMC("WILLIAMS", 0.95,df, PAIR) could be used instead of the appendix tables of critical $\bar{t}_{i,\infty}$ for Williams's test. (The SAS PROBMS() function could also have been used to produce for one- and two-sided Dunnett's tests for Example 5.4, although Dunnett's test is more easily executed with the SAS code shown in that example.)

Williams (1972) provides a modification of this approach if the number of observations is not equal between the control and experimental treatments. If the number of control observations (n_0) is equal to or greater than the number of observations in each experimental treatment (n_i), $\bar{t}_i$ can be estimated. Again, there are sound reasons for n_0 to be greater than n_i. Like Dunnett's calculations (Dunnett 1955), Williams's power calculations suggested that n_0/n_i (= w) should be equal to or slightly higher than $\sqrt{q-1}$.

$$\bar{t}_i = \frac{M_i - \bar{X}_0}{s\sqrt{\dfrac{1}{n_i} + \dfrac{1}{n_0}}} \tag{5.24}$$

The critical $\bar{t}_{i,\alpha}$ must now be adjusted for n_0/n_i. Recognizing that the $\bar{t}_{i,\alpha}$ decreases linearly with $1/w$, Williams estimated a series of β_t values for extrapolating within Appendices 19 and 20 to account for this effect. They are placed as exponents in the tables. The value of $\bar{t}_{i,\alpha}$ with a w of 1 (i.e., $\bar{t}_{i,\alpha(1)}$) is modified with w and β_t using Equation (5.25).

$$\bar{t}_{i,\alpha(w)} = \bar{t}_{i,\alpha(1)} - 10^{-2}\beta_t\left(1 - \frac{1}{w}\right) \tag{5.25}$$

The modified $\bar{t}_{i,\alpha(w)}$ is the critical value compared to the $\bar{t}_p$ values estimated with Equation (5.23). These extrapolated values are correct to 0.01 if $w \leq 4$, or 0.02 if $w = 5$ and 6 (Williams 1972). Williams (1972) provides the following better formulae if the control is known, that is, w approaches ∞.

$$\bar{t}_{i,w(\infty)} = \bar{t}_{i,\alpha(1)} - 10^{-2}(\beta + 2) \tag{5.26}$$

Example 5.7

Let's assume that there were eight control observations and four experimental (noncontrol) observations in a data set similar to the fathead minnow data used previously (w = 8/4 = 2.00). Assume that the s^2 is 0.024. The df would be 28 – 6, or 22. The $\bar{X}_0$ and M_p are –1.330 and –0.604, respectively. Equation (5.23) is used to estimate a $\bar{t}_i$ of 7.653.

From Appendix 19 (α = 0.05, df = 22, p = 5), $\bar{t}_{i,\alpha(1)}$ and β_t are found to be 1.846 and 5, respectively.

$$\bar{t}_{i,\alpha(2)} = \bar{t}_{i,\alpha(1)} - 10^{-2}(\beta_t)\left(1 - \left(\frac{1}{w}\right)\right)$$
$$= 1.846 - 10^{-2}(5)(1 - 0.5) = 1.821$$

The null hypothesis is rejected since $\bar{t}_5$ is greater than $\bar{t}_{i,\alpha(2)}$.

Williams's test can still be used if unequal numbers of observations occur among noncontrol treatments. Williams (1972) used the maximum likelihood estimators previously described (Equation 5.21) with the number of observations for each treatment used as weights.

Equation (5.27) is a formal expression of the procedure.

$$M_i = MAX_{1 \leq u \leq i} MIN_{i \leq v \leq p} \frac{\sum_{j=u}^{v} n_i \bar{X}_i}{\sum_{j=u}^{v} n_i} \tag{5.27}$$

Equation (5.27) is used to generate the maximum likelihood estimates of the means. Appendices 18 and 20 can still be used provided there are only moderate differences in n_i values. Williams (1972) reports that these tables may be applied with confidence if $0.80 \leq n_i/n_p \leq 1.25$ for all $1 \leq i \leq p - 1$.

Example 5.8

Williams's test is performed here on a modified, fathead minnow survival data set. One additional observation will be added to the 0 (transform = 1.412), 32 (transform = 1.107), 64 (transform = 1.249), and 256 (transform 0.991) μg/L treatments to produce this modified data set. Thus, these four treatments now have five observations each and the remaining two treatments

(128 and 512 μg/L) have four observations each. (The new s is estimated to be 0.145. The data passed tests for variance homogeneity and normality at $\alpha = 0.05$. As all n_i/n_p are less than or equal to 1.25, Appendix 19 can be used.) Only values from Appendix 19 for the two treatments with observation numbers different from that of the control (128 and 512 μg/L) need to be adjusted with Equation (5.24) before tabulation below.

$\bar{X}_i$	$-\bar{X}_i$	M_i	df'',p''	$\bar{t}_p$	$\bar{t}_{p,\alpha}$
1.347	−1.347	−1.347			
1.168	−1.168	−1.257	22,1	0.982	1.717
1.347	−1.347	−1.257	22,2	0.982	1.798
1.254	−1.254	−1.254	22,3	0.957	1.817
1.086	−1.086	−1.086	22,4	2.848	1.838
0.604	−0.604	−0.604	22,5	7.644	1.836

The null hypothesis is rejected for the last two treatments. The lowest concentration displaying a mean survival significantly less than that of the control is the 256 μg/L treatment.

If it is ambiguous whether the monotonic trend will involve an increase or decrease, a two-sided Williams' test is appropriate. Tables such as Appendices 20 ($\alpha = 0.01$) and 21 ($\alpha = 0.05$) can be used for this purpose. The testing process is analogous to that described above for the one-sided tests.

5.8 STEEL'S MULTIPLE TREATMENT-CONTROL RANK SUM TEST

Formally, the parametric methods described above should not be used if the assumption of normality is rejected for the data set (Figure 5.1). Less powerful nonparametric tests such as Steel's rank sum tests (Steel 1959, 1960, 1961) would be used instead. Although equal variance is a formal requirement for these tests, they are believed to be relatively robust to variance heterogeneity (Steel 1959). They carry the assumption of a continuous distribution for the measured variable. Steel's test as described initially requires equal numbers of observations for all treatments and the control. Later, a modification for unequal numbers of observations in the control versus the experimental treatments is discussed.

The formal null hypothesis is that all observations come from the same population regardless of treatment. This population is described by the cumulative distribution function, F. Therefore, the null hypothesis can be written as $F_0 = F_1 = \ldots = F_p$, where F_0 is the distribution of the control observations and $F_1, \ldots, F_p$ (or F_i values) are those for the experimental treatments 1 to p. It is tested using a specified experimentwise error rate. The alternative hypothesis is that one or more F_i is "stochastically larger" (Steel 1959) than F_0, i.e., $F_0 < F_i$, $F_0 > F_i$, or $F_0 \neq F_i$. If $F_0 < F_i$, the median and other percentiles of the experimental treatment are larger than those of the control. The opposite is true if $F_0 > F_i$. Steel (1959) describes the locations of the control and treatment distributions as "different" in the case $F_0 \neq F_i$.

To test the null hypothesis, the observations for each control–experimental treatment pair are pooled and then ranked from smallest to largest. The average rank is used in the case of ties for each of the tied values. Next, the ranks are summed for the experimental treatment observations. This rank sum for the experimental treatment is designated T_i. Next, the rank sum is calculated for the observations from the control (T_i') using the convenient relationship

$$T_i' = (2n+1)n - T_i \tag{5.28}$$

where n = number of replicates in each treatment.

The ranking and summing process is repeated for each of the p control–experimental treatment pairs. For the results from each pair, the minimum of T_i and T_i' (MIN(T_i,T_i')) is compared to a critical value from Appendix 23 (two-sided test). With a one-sided test, whether T_i or T_i' is used as the minimum value for a particular pair for comparison to the critical value (Appendix 22) will determine if the treatment is significantly smaller or greater than the control.

Example 5.9

The transformed fathead minnow survival data tabulated in Example 5.1 will be used to illustrate Steel's multiple treatment-control rank sum test. The ranking process will be illustrated with the 0 μg/L-32 μg/L pair.

Transformed Survival	Treatment (μg/L)	Rank
1.107	32	2
1.107	32	2
1.107	32	2
1.249	0	4.5
1.249	0	4.5
1.412	0	7
1.412	0	7
1.412	32	7

$T_1 = 2 + 2 + 2 + 7 = 13 \qquad T_1' = ((2)4 + 1)4 - 13 = 23$

The results of T_i and T_i' calculations for the five treatment pairs as illustrated above for the 0 μg/L-32 μg/L pair are summarized.

i	Pair	T_i	T_i'	MIN(T_i,T_i')	Critical T Value
1	0–32	13	23	13	10
2	0–64	20	16	16	10
3	0–128	15	21	15	10
4	0–256	14	22	14	10
5	0–512	10	26	10	10

For the one-sided test, the treatment decreases survival so T_i is used. The T_i values are compared to the tabulated critical Steel's rank sums T values. Since T_5 is equal to the critical T value from Appendix 22 (one-sided test, $\alpha = 0.05$, $n = 4$, $p = 5$), the null hypothesis is rejected. The null hypothesis is not rejected for any of the other pairs, as their associated T_i values are greater than this critical T value. Only the 512 μg/L concentration had significantly elevated mortality. This is the same conclusion as that reached with Dunnett's test.

For a two-sided test, the MIN(T_i,T_i') would be compared to a critical value from Appendix 23. However, for this low number of observations and number of comparisons, no critical value can be estimated for this test statistic.

Appendices 22 and 23 are limited to 2 to 9 treatment comparisons and 4 to 20 observations per comparison. To extend these tables, Steel (1959) suggests that Dunnett's t (Appendices 10 to 11) may be used. The critical T is estimated using three approximations.

$$\mu_T = \frac{n(2n+1)}{2} \tag{5.29}$$

$$\sigma_T^2 = \frac{n^2(2n+1)}{12} \tag{5.30}$$

$$T = Integer\ portion\ of\ (\mu_T - t\sigma_T) \tag{5.31}$$

The T in Equation (5.31) is obtained from the appropriate Dunnett's t table for $df = 4$. In using Dunnett's tables for one-sided tests, the difference between the assumed correlation ($\rho_{ij} = 1/2$) as discussed earlier for Dunnett's test and the true correlation (approximately $n/(2n + 1)$) is ignored.

Miller (1966) describes a procedure for using Steel's test if the number of observations in the control (n_0) is different from the number in the experimental treatments but $n_1 = n_2 = n_3 = \ldots = n_p$. All observation numbers for treatments (n_i) are equal to n but different from n_0. Let R_{ij} be the rank of treatment i values' jth observation and $\rho = n/(n + n_0 + 1)$.

$$R_i = \sum_{j=1}^{n} R_{ij} \tag{5.32}$$

For a one-sided test, the null hypothesis is rejected if MAX(R_1 … R_p) ≥ r^α. The critical r^α is estimated using Equation (5.33).

$$r^\infty \approx \frac{n(n+n_0+1)}{2} + \frac{1}{2} + m^\infty_{p(\rho)}\sqrt{\frac{nn_0(n+n_0+1)}{12}} \tag{5.33}$$

where $m\alpha_{p(\rho)}$ = value obtained from tables of Gupta (1963).

For a two-sided test, the null hypothesis is rejected if $R^* = \text{MAX}(R_1^* \ldots R_p^*) \geq r_*^\alpha$. The R_i^* are the MAX(R_i, $n(n + n_0 + 1) - R_i$). The r_*^α is estimated with Equation (5.34).

$$r_*^\alpha \approx \frac{n(n+n_0+1)}{2} + \frac{1}{2} + |m|^\alpha_{p(\rho)}\sqrt{\frac{nn_o(n+n_0+1)}{12}} \tag{5.34}$$

where $|m|^\alpha_{p(\rho)}$ = value from Dunnett's tables (see Appendix) for $df = \infty$.

Day and Quinn (1989) recommend that unless the correlation is large, Fligner's (1984) modification of Steel's test can be used when observation numbers are unequal between the control and experimental treatments, or between experimental treatments. The sum of ranks associated with the ith treatment (R_i) is estimated.

$$R_i = \sum_{b=1}^{n_0}\sum_{c=1}^{n_i}\Psi(X_{ic} - X_{0b}) + \frac{n_i(n_i+1)}{2} \tag{5.35}$$

where $\Psi(a) = 1$ if $a > 0$, 0.5 if $a = 0$, and 0 if $a < 0$. An r_i is estimated as the larger of R_i and $n_i(n_i + n_0 + 1) - R_i$.

$$r_i = MAX(R_i,\ n_i(n_i + n_0 + 1) - R_0) \tag{5.36}$$

The r_i values are compared to critical values estimated using Bonferroni's adjustment to generate an experimentwise error rate. Remember that because the Bonferroni adjustment results in an upper bound for the experimentwise error rate, this modified test will have less power than the unmodified test.

5.9 WILCOXON RANK SUM TEST WITH BONFERRONI'S ADJUSTMENT

Weber et al. (1989) recommend use of the Wilcoxon rank sum test with Bonferroni's adjustment for the experimentwise error rate if the number of observations varies among treatments. For each experimental treatment-control pair, observations are combined and ranked. In the one-sided procedure, the values are ranked from smallest to largest if the treatment effect is thought to decrease the value of the variable relative to that of the control. If the treatment effect is thought to increase the values of the variable, the signs of the values are changed prior to ranking, that is, multiplied by −1. For the two-sided test, the minimum of the treatment rank sum and control rank sum is compared to the test statistic. In the case of ties, each of the tied values is replaced by the average rank. Next, the ranks are summed for the experimental treatment observations (R_T) and for the control observations (R_C). This process is repeated for all treatment-control pairs. The rank sums (R_T and/or R_C) resulting from the p pairs are compared to critical values from Appendix 24 (one-sided test) or 25 (two-sided test).

The critical values in Appendices 24 and 25 were generated from probability tables for the Wilcoxon statistic (Beyer 1968; Kokoska and Nevison 1989) using experimentwise α values and the Bonferroni adjustment to estimate α'. Values for $m = 9$ or 10 in Appendix 24 were taken directly from Table 18 of Kokoska and Nevison (1989). The Mann–Whitney U values were estimated from Table 18 of Kokoska and Nevison (1989) for $m = 9$ and 10 and used to calculate values of R for $p = 9$ and 10 (Equation 5.36). For other combinations of experimental and control replicate numbers (p and m, respectively), Mann–Whitney U values were extracted from Beyer's (1968) Table X.3 using the adjusted α values (α' in Appendices 24 and 25) and the number of observations in the treatment (p) and control (m). These U values were then used to produce critical rank sums for the experimental treatment with the relationship (Noether 1971)

$$R = U + \frac{p(p+1)}{2} \tag{5.37}$$

If the calculated rank sum is less than or equal to the critical rank sum value in the one-sided test table, the null hypothesis of no difference between the treatment and control would be rejected. For a two-sided test, the MIN(R_T, R_C) is compared to the critical rank sum value in the two-sided test table. (The R_C is the calculated rank sum for the control.)

Example 5.10

The transformed fathead toxicity data as modified in Example 5.8 will be used to illustrate the Wilcoxon rank sum test with Bonferroni's adjustment for a one-sided test.

Toxicant Concentration (μg/L)	Replicate A	Replicate B	Replicate C	Replicate D	Replicate E
0	1.412	1.412	1.249	1.249	1.412
32	1.107	1.107	1.412	1.107	1.107
64	1.249	1.412	1.412	1.412	1.249
128	1.249	1.249	1.107	1.412	
256	0.991	1.249	1.412	0.785	0.991
512	0.685	0.580	0.685	0.464	

The rank sums are calculated and compared to critical values in Table 24 (one-sided test with $\alpha = 0.05$).

Pair	R_T	R_C	m	p	Critical Values One-Sided Test
0–32	18.5	36.5	5	5	16
0–64	27.5	27.5	5	5	16
0–128	15.5	29.5	5	4	10
0–256	19.5	35.5	5	5	16
0–512	10.0	35.0	5	4	10

Because the R_T for the control-512 μg/L treatment comparison is equal to the critical value, the null hypothesis is rejected for this pair. None of the other control-treatment pairs were significantly different because the associated R_T values were not less than or equal to the critical rank sum value from Appendix 24. This contrasts with results of the more powerful Williams's test that indicated significant effects at both the 256 and 512 μg/L treatments in this data set (Example 5.8). The SAS program can also facilitate these tests. As an example, the following code generates estimates for the control versus 32 μg/L treatment.

```
DATA FIRST;
     INPUT ARC CONC @@;
     CARDS;
     1.4120  0 1.4120  0 1.2490  0 1.2490  0 1.412  0
     1.1071 32 1.1071 32 1.4120 32 1.1071 32 1.107 32
RUN;
PROC NPAR1WAY DATA=FIRST CORRECT=YES;
     CLASS CONC;
     VAR ARC;
     EXACT WILCOXON;
RUN;
```

The following summary of scores and estimated p values is the generated output.

```
Wilcoxon Scores (Rank Sums) for Variable ARC
Classified by Variable CONC

       Sum of Expected   Std Dev   Mean
CONC  N Scores Under H0   Under H0  Score

0     5  36.50   27.50    4.564355  7.30
32    5  18.50   27.50    4.564355  3.70
```

Three approaches (normal approximation, t-test, and exact test) produce similar p values for the one-way test, i.e., –0.0313, 0.0477, and 0.0357. Because the experiment involves five comparisons, the pairwise error rate for this test is 0.05/5 = 0.01. There is no evidence supporting rejection of the null hypothesis. This same procedure would be performed for all five pairs in the experiment to complete the analyses.

```
Wilcoxon Two-Sample Test Statistic (S) 36.5000

Normal Approximation
Z 1.8623
One-Sided Pr >  Z           0.0313
Two-Sided Pr > |Z|          0.0626

t Approximation
One-Sided Pr >  Z           0.0477
Two-Sided Pr > |Z|          0.0955
```

```
Exact Test
One-Sided Pr >=  S              0.0357
Two-Sided Pr >=  |S - Mean|     0.0714
```

Assuming a monotonic trend, the OECD (Figures 5.2 and 5.3) recommends the Jonckherre(-Terpstra) test as the nonparametric equivalent of Williams's test. The SAS code below implements this test as just done with the other two nonparametric tests. Only the two highest treatments were tested in the code because the second highest (256 μg/L) was not significant. Appendix 26 lists code that was written for one-way increasing or decreasing responses, and also two-way test of responses.

```
DATA FATHEAD;
     INPUT ORDR $ CONC ARC @@;
     DATALINES;
     ONE    0 1.4120 ONE    0 1.4120 ONE    0 1.2490 ONE    0 1.2490
     TWO   32 1.1071 TWO   32 1.1071 TWO   32 1.4120 TWO   32 1.1071
     THR   64 1.2490 THR   64 1.4120 THR   64 1.4120 THR   64 1.4120
     FOR  128 1.2490 FOR  128 1.2490 FOR  128 1.1071 FOR  128 1.4120
     FIV  256 0.9912 FIV  256 1.2490 FIV  256 1.4120 FIV  256 0.7854
     SIX  512 0.6847 SIX  512 0.5796 SIX  512 0.6847 SIX  512 0.4636
     ;
RUN;
DATA FATHEAD5; SET FATHEAD; RUN; /* ALL TREATMENTS INCLUDED */
PROC FREQ ORDER=DATA;
TABLES ORDR*ARC/JT;
OUTPUT OUT=FINAL JT;
RUN;
DATA FATHEAD4; SET FATHEAD; IF CONC<512; RUN; /*512 UG/L OMITTED */
PROC FREQ ORDER=DATA;
TABLES ORDR*ARC/JT;
OUTPUT OUT=FINAL JT;
RUN;
```

5.10 A SECOND LOOK AT STATISTICAL TESTING

Rituals seem to be indispensable for self-definition of social groups ... and there is nothing wrong with them. However, they should be the subject rather than the procedure of social sciences. Elements of social rituals include (i) the repetition of the same action, (ii) a focus on special numbers or colors, (iii) fears about serious sanctions for rule violations, and (iv) wishful thinking and delusions that virtually eliminate critical thinking ... the null [hypothesis significance testing] ritual has each of these four characteristics: incremental repetition of the same procedure; the magical 5 percent number; fear of sanctions by editors or advisors; and wishful thinking about the outcome, the p-value, which blocks researcher's intelligence.

—Gigerenzer (2008)

5.10.1 General

Bayesian methods were believed to be the best means of making statistical inferences prior to the 1920s (Howson and Urbach 1989). Discomfort with the subjective aspect of Bayesian probabilities prompted attempts to formulate more objective methods during that same decade. The statistical testing approach just described* emerged during the 1930s as common practice, eventu-

* The specific approach described to this point is null hypothesis significance testing (NHST). Reflecting the needs expressed by regulators, the approach discussed so far would be labeled even more specifically as nil hypothesis significance testing because the tested hypothesis is no difference or correlation (see Nickerson 2000): the tested effect size (ES) is 0.

ally becoming the gold standard of inference in many sciences. Initiating the process, Sir Ronald A. Fisher (1922, 1935) established significance testing as *the* objective* tool for inference (Morrison and Henkel 1970; Howson and Urbach 1989). His aim was logical falsification of a hypothesis under the premise that a low probability of obtaining data under an assumed hypothesis suggests that that hypothesis itself ought to be rejected as false.† Improbable implies implausible in the practice of significance testing. *P* values, the probability of getting the outcome or a more extreme outcome if the null hypothesis were true, are generated using a normal, binomial, *F* ratio, *t*, χ^2, or some other statistic. This practical falsification approach requires a hypothesis to be nullified. Fisher describes that null hypothesis as "the hypothesis that the phenomenon to be demonstrated is in fact absent." Eventually, an error rate below which the *p* value prompts rejection was judged necessary for logical falsification, although the error rate (α) might vary depending on the situation. As Fisher states, "[No researcher] has a fixed level of significance at which ... he rejects hypotheses: he rather gives his mind to each particular case in light of his evidence and his ideas" (Fisher 1956). Regardless, a 0.05 error rate convention eventually emerged (Gigerenzer 2004).‡

Failure to reject the null hypothesis does not imply the null hypothesis is true: only falsification of a hypothesis moves inferences ahead. Such reasoning follows directly from Popper's partial solution to the problem of deduction; that is, you can prove something is false but not that something is true (Grattan-Guinness 2004). Importantly, there is no alternative hypothesis during a Fisherian significance test (Biau et al. 2010; Goodman 1993; Gigerenzer 2004; Newman 2008). If the null hypothesis is falsified that a chemical dose has no lethal effect, there is no alternate hypothesis that the dose has a lethal effect. Nor does logic somehow dictate that the dose must be toxic. Oddly, most applications of these methods are motivated by the desire to reject a null hypothesis because the researcher suspects an effect (Biau et al. 2010).

Jerzy Neyman and Egon Pearson proposed another approach called hypothesis testing (Neyman and Pearson 1928) soon after Ronald Fisher introduced significance testing. They argued that Fisher's logical refutation context was inappropriate in most instances,§ and instead, a preferable, more objective¶ context was that allowing one to decide the best of two or more hypotheses based on data-derived probabilities. The logical falsification of Fisher's significance testing of a single hypothesis was abandoned in favor of hypothesis testing based on decision making about two or more hypotheses. Inferences modifying belief were replaced by decisions resulting in the highest chance of being correct (Howson and Urbach 1989). As Neyman and Pearson (1933) explain, "Without hoping to know whether each separate hypothesis is true or false, we may search for rules to govern our behavior with regard to them, in following which we insure that, in the long run of experience, we shall not often be wrong." Hypothesis testing requires that a second error rate (β) and an effect size (ES) be defined *a priori*. The type II error rate reflects the probability of falsely rejecting the alternate hypothesis. The ES defines the magnitude of the difference or correlation that the practitioner deems important to be able to detect if present. Unlike Fisher's significance testing,

* The claim of objectivity has been scrutinized by many authors (e.g., Newman 2008) who point out that the choices of a null hypothesis, a specific test, and α are subjective. Strictly thinking, practical falsification is also questionable because rare events do occur.

† Not only is this approach unsound logically, but it has been made clear by numerous authors that the probability of getting observations under the assumption of the null hypothesis is an unreliable estimate of the probability of the null hypothesis being true given the observations. The skeptical reader is referred to Berger and Sellke (1987) and Trafimow and Rice (2009) for quantitative illustration.

‡ According to Hurlbert and Lombardi (2009), the 0.05 convention began with Karl Pearson's unpublished opinion that observed data beyond three probable errors of the mean, that is, 0.0456, might be viewed as significant. Gosset (Student 1908) eventually published this opinion as a good rule of thumb. Gosset states in his acknowledgments that Karl Pearson provided "constant advice and criticism" of the paper, suggesting again Pearson's influence in the eventual establishment of the 0.05 convention.

§ Such opinions remain common. As an example, Hogben (1970) states, "In my own experience so far, cases where Professor Fisher's theory would have been suitable have not been very frequent."

¶ A moment of reflection makes it clear that choices of α, β, ES, hypotheses, and the testing technique are also done subjectively.

Neyman–Pearson hypothesis testing attempts to minimize the probability of deciding in favor of a false hypothesis and of deciding against a true hypothesis (Hacking 2001). It is crucial to emphasize that it is naïve to apply a Neyman–Pearson interpretation if α, β,* and ES are not established *a priori*. Using the dosing example again, it would be an invalid to decide that evidence favors the alternate hypothesis that the dose was lethal if β and ES had not been integrated into the testing *a priori* (Newman 2008).

During the decades following the 1930s, the originators of significance and hypothesis testing engaged in strong, and sometimes quite personal, criticism of each other that distracted scientists from the important features and relative values of these tests (Ziliak and McCloskey 2008). Proponents and opponents of the two methods gradually codified and added detail to produce the current statistical testing approach. The methods described in the first half of this chapter are typical of the resulting hybrid approach. Presently, omission of such methods from any submitted ecotoxicology manuscript or regulatory report would prompt reservations from reviewers, editors, or review panel members about conclusions. Yet, this approach has been described in the literature with the following phrasing:

"A kind of essential mindlessness in the conduct of research" (Bakan 1966)
"More fantasy than fact" and "not only useless but it is also harmful" (Carver 1978)
"Machinery for producing phoney corroborations" (Lakatos 1978)
"The ritual of null hypothesis testing" (Cohen 1994)
"Insignificance of statistical significance" (McCloskey 1995)
"The most bone-headly misguided procedure ever institutionalized" (Rozeboom 1997)
"Retards the growth of scientific knowledge" (Schmidt and Hunter 1997)
"The ritual ... with dichotomous decisions around a sacred 0.05 criterion" (Germano 1999)
"The *p* value fallacy" (Goodman 1999)
"Explicitly denounced by most eminent and most experienced scientists" (Lecoutre et al. 2001)
"Mindless statistics" (Gigerenzer 2004)
"Ill-founded strategy" (Ioannidis 2005)
"[Comparing a] 'trivial' null hypothesis and a single alternative" (Stephens et al. 2005)
"Statistical ritual ... cargo cult science" (Guthery 2008)
"Can lead to unjustified interpretations" (Cumming 2012)

The irritation evident in these comments will not be expressed again in this chapter, although the author does empathize with these critics. Dawkins' "good humoured" exasperation strategy taken in *The Extended Phenotype* (Dawkins 1982) will be adopted instead. Sources of the criticisms will be examined neutrally and two possible ways of avoiding them in ecotoxicology discussed.

> It is, I think, a delicate issue for individual statisticians to expose genuine difficulties of interpretation without becoming a source of negative thinking and discouragement.
>
> **—Cox (2009)**

Unquestionably, statistical testing has grown to play a prominent role in most sciences. As one illustration, only 17% of published psychology papers applied these methods in the decade between 1917 and 1929, but 94% of such publications were applying them by the 1990s (Nickerson 2000). Anderson et al. (2000) found nil† hypothesis testing to be pervasive in ecology and wildlife management journals. Surveys of conservation biology journal articles found that 92 and 78% applied nil hypothesis tests in 2001 and 2005, respectively (Fidler 2006). The majority of environmental

* Or power, $1 - \beta$, which has meaning in Neyman–Pearson hypothesis testing but not Fisherian significance testing (Hacking 2001).

† A null hypothesis of no difference or correlation is a nil hypothesis.

toxicology and chemistry journal articles surveyed by Newman (2008) relied on null (predominantly nil) hypothesis testing. Especially relevant to ecotoxicological practice are the comments of Salsburg (2001):

> A simplified version of … hypothesis testing can now be found in all elementary statistics textbooks…. Since it has been codified, this version of the formulation is exact and didactic…. This rigid approach to hypothesis testing has been accepted by regulatory agencies like the U.S. Food and Drug Administration and the Environmental Protection Agency.

Unfortunately, statistical testing has developed some inferential slang that, like obtuse verbal slang, often confuses understanding. Contemporary practice of statistical testing can be summarized to be one favoring a conventional α (often 0.05) in combination with an undefined β and an ES of 0. Outcomes are deemed either significant or not. Sometimes outcomes are categorized as not significant ($p > 0.05$), significant ($0.01 < p \leq 0.05$), or highly significant ($p \leq 0.01$). Concerns summarized by Newman (2008) about this approach are repeated below along with possible ways of reducing common errors.

Concern 1. Confusing $p(E|H_0)$ with $p(H_0|E)$. Data are used to generate a test statistic that is then associated by some function with a p value. If test assumptions are met, the p value is the probability of obtaining the observations or more extreme observations if the null hypothesis is correct, $p(E|H_0)$. A pervasive misconception (Goodman 1999; Nickerson 2000) is that the $p(E|H_0)$ reflects the probability of the null hypothesis being true given the evidence, $p(H_0|E)$. This misconception is so common that it is given names such as the inverse problem (Gigerenzer 2000) or the less charitable "'odds-against-chance' fantasy" (Carver 1978). Illustrating the pervasiveness of this misconception, Gigerenzer (2004) described a survey given to German psychology students, lecturers, and professors showing that 80% of professors teaching statistics and 100% of students misunderstood what a p value meant. This book's author (Newman) obtained similar results after administering a derived version of that survey to academic or professional audiences in China (Xiamen University and Hua Zhong Normal University), India (Cochin University), Spain (Society of Toxicology and Chemistry Conference, Seville), the United States (Virginia Institute of Marine Science), and Vietnam (Society of Toxicology and Chemistry Meeting, Ho Chi Minh City). Clearly, misinterpretations of p values are common and widespread.

That $p(E|H_0) \neq (p(H_0|E)$ is evident from Bayes's theorem, a theorem requiring minimal understanding of probability theory (Hacking 2001; review also Section 1.6 of Chapter 1),

$$p(H_0 \mid E) = \frac{p(H_0)p(E \mid H_0)}{p(E)} \tag{5.38}$$

where $p(H_0)$ = prior probability of H_0, that is, probability of H_0 prior to the evidence being gathered, and $p(E)$ = probability of the evidence with no condition of H_0 being true. Obviously, $p(H_0)$ and $p(E)$ are needed before $p(H_0|E)$ can be estimated from a p value.

The widespread counterargument is also false that $p(E|H_0)$ and $p(H_0|E)$ are close enough for the sound inference in most cases. Berger and Sellke (1987) provide a detailed explanation, concluding that "p-values can be highly misleading measures of the evidence provided by the data against the null hypothesis." Trafimow and Rice (2009) conducted simulations to test the argument that a strong correlation exists between $p(E|H_0)$ and $p(H_0|E)$, concluding "the correlation … is unimpressive and fails to provide a compelling justification for computing p values." Figure 5.7 illustrates this point for p values ($p(E|H_0)$) of 0.01 and 0.05. The p value underestimates the $p(H_0|E)$, especially at low values for $p(E)$.

If an alternative hypothesis is specified, a related and common misconception is a p value implies that the probability of the alternative hypothesis being true, $p(H_A)$, is approximately 1 minus the p value (Nickerson 2000). Again, the error in such an inference can be illustrated with Bayes's theorem. With the inclusion of an alternate hypothesis, Bayes's theorem becomes Equation (5.39) or (5.40), depending on whether the probability for H_0 or H_A is the issue.

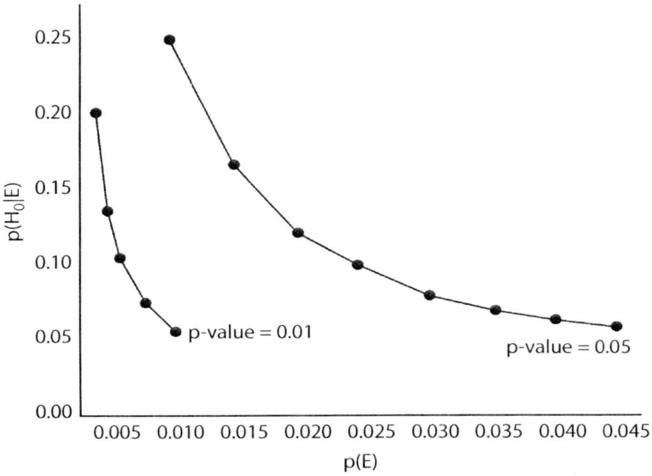

Figure 5.7 Illustration that the p value ($p(E|H_0)$ in Equation 5.38) does not reliably reflect the probability of the null hypothesis being true given the evidence ($p(H_0|E)$ in Equation 5.38). The probability of the hypothesis being true prior to collecting the evidence ($p(H_0)$) was arbitrarily set to 0.05, and the range of 0.005 to 0.045 was selected for $p(E)$. Especially at low probabilities of getting the evidence ($p(E)$), the p value consistently understates the actual probability of the null hypothesis being true given the evidence ($p(H_0|E)$. No simple correlation exists between $p(E|H_0)$ and $p(H_0|E)$, and misinterpretation of the p value will result in overstating the evidence against the null hypothesis.

$$p(H_0 \mid E) = \frac{p(H_0)p(E \mid H_0)}{p(E \mid H_0)p(H_0) + p(E \mid H_A)p(H_A)} \tag{5.39}$$

$$p(H_A \mid E) = \frac{p(H_A)p(E \mid H_A)}{p(E \mid H_0)p(H_0) + p(E \mid H_A)p(H_A)} \tag{5.40}$$

Clearly, $1 - p(E|H_0)$ is not equal to $p(H_A|E)$. A p value cannot be used alone to infer the probability of the alternate hypothesis being true. As a useful aside to be applied later, the relative probabilities for H_0 and H_A (i.e., likelihood ratio for the two hypotheses) can be expressed with Equation (5.41) (Howson and Urbach 1989).

$$\frac{p(H_0 \mid E)}{p(H_A \mid E)} = \frac{p(E \mid H_0)p(H_0)}{p(E \mid H_A)p(H_A)} \tag{5.41}$$

Notice that the last three equations, unlike Equation (5.38), do not require one to know $p(E)$.

Concern 2. It is often unclear what is being inferred from a significant outcome. Inference from a positive significance test context is that the null hypothesis should be rejected because the evidence had a sufficiently low probability of occurring by chance if the null hypothesis was true (yet see concern 1 above). Early in the codification of statistical testing, the logical extension was made that the alternative hypothesis must then be true. As evidence, Berkson (1938) states, "My view is *that there is never any valid reason for rejection of the null hypothesis except on the willingness to embrace an alternative one.*"* But a significant outcome does not mean that an alternate hypothesis is true because there is no such alternate hypothesis with Fisherian significance testing. For the Neyman–Pearson hypothesis testing vantage, the decision can be made to favor the alternate

* Berkson's original italics retained.

hypothesis based on the evidence; however, the β and ES must have been specified *a priori* in order to do so. Otherwise, the Neyman–Pearson vantage cannot be applied.

Linked to this confusion is the common use of a "roving α" (Goodman 1993), that is, labeling outcomes as nonsignificant ($p > 0.05$), significant ($0.01 < p \leq 0.05$), or highly significant ($p \leq 0.01$). This approach is nonsensical from the current Fisherian context* of falsifying a null hypothesis: something cannot be highly true as opposed to merely true. From a Neyman–Pearson vantage, α and β are set prior to gathering evidence and a decision is made within the context of those error rates. You decide to favor either one hypothesis or the other so the roving α approach is nonsensical here too.

Concern 3. Balance type I and II error rates. Current practice typified in the methods described in the first half of this chapter has a set α in combination with a vaguely defined β. This practice emerges from the central falsification feature of Fisherian significance testing. If an alternate hypothesis becomes part of statistical inference (Neyman–Pearson's hypothesis testing), common practice now suggests that the researcher need only be concerned with avoiding a type I error. Type II error is unimportant and may be allowed to range within the dictates of some standard test design. This suggestion is based on recent convention, not careful thought. In the context of this chapter, setting α but allowing β to rove inside a range constrained loosely by experimental design suggests that the only error with serious consequences is that of falsely deciding there is an effect (Suter 1993; Newman 2008). Falsely deciding there is no effect is less important. There is no logical defense of this behavior in ecotoxicity testing other than an argument based on recent precedence that is now being questioned openly in many other sciences.

The relative importance of type I and II errors can be incorporated into statistical testing with error rate quotients (Cohen 1988). The quotient, α/β, reflects the practitioner's perception of the relative seriousness of making these errors. For example, the standard design of ecotoxicity tests described in this chapter produces a β in the range of 0.2 (Van der Hoeven 1998) when a conventional α of 0.05 is employed. The result is a design in which the consequences of deciding incorrectly there is a toxic effect are four times worse than those of deciding incorrectly there is no toxicity effect, i.e., $\alpha/\beta = 0.05/0.20 = 0.25$. This imbalance is hard to defend in ecotoxicology, although it was much more reasonable for early agricultural applications of statistical testing. The error discussed in Section 5.3.3 about assuming that a decreasing α results in more rigorous test results is germane to this issue. Decreasing α to enhance testing rigor carries the unintended decision that consequences of type II error are much less serious than those of type I error.

Concern 4. Inattention to effect size (ES). Fisher's significance testing established the term *null hypothesis* to denote the hypothesis to be nullified logically. A pervasive misinterpretation exists that null means nil, i.e., no difference or correlation. Problems with neglecting ES by defaulting to zero are described throughout the scientific literature (Anderson et al. 2000; Berkson 1938; Ziliak and McCloskey 2004; Stephens et al. 2005; Nakagawa and Cuthill 2007). This inattention is integrated into many, but a thankfully decreasing number of, statistical packages. That fundamental problems emerge from this misunderstanding can be illustrated by pointing out that any nil hypothesis will be rejected given enough observations or replicates because absolute conformity to the hypothesized distribution/model is an unreasonable expectation from real world data.

Why is this problematic in day-to-day applications of statistical testing? A partial answer comes from the regulatory ecotoxicology literature, in which a key perceived advantage that statistical testing has over regression-based prediction is that statistical testing removes the difficulty of determining what is a biologically meaningful effect size (Stephan and Rogers 1985). This strategy eschews the responsibility of applying subject knowledge to determining what level of effect is important and stresses statistical significance instead. Although a common mistake during statistical testing (McCloskey 1995), consensus among thoughtful statisticians is that this behavior inhibits sound decisions from evidence. Tests are designed and executed without sufficient understanding of the ES that could be detected. For example, Jennions and Møller (2003) found that most tests employed in behavioral ecology publications had quite low power ($1 - \beta = 0.40$ to 0.47) to detect a medium ES.†

* The original approach of Fisher would accommodate the roving α more than the current methods that Cohen (1994, 1995) calls the "usual reject-H_0-confirm-the-theory approach."

† Jennions and Møller (2003) examined effect relative with the correlation coefficients, r. The percentage is that of total variance explained by the hypothesized relationship ($100 \times r^2$). A medium effect was approximately 9%.

Crane and Newman (2000) examined chronic toxicity fish growth tests, estimating that ES of 10 to 34% would be required for the examined tests to produce a significant outcome, and that estimated ES ranged widely among published studies. Conventional *Ceriodaphnia* reproduction tests with an $n = 10$ animals per treatment concentration were able to detect only effects in the range of 31 to 100% inhibition (Oris and Bailer 1993). Van der Hoeven (1998) concluded about three ecotoxicity protocols used in an OECD ring test that "a reduction inhibition of 20% will be observed in 80% of the *D. magna*, and in only 18% of the tests with either *E. fetida* or *F. candida* [assuming a one-sided Dunnett's test]." It remains ambiguous in all such studies whether an effect below these specified magnitudes would be biologically important. Common ecotoxicity protocols do not require that experimental design ensure that a biologically important effect size could be detected.

Concern 5. Publication bias favoring significant findings. A documented bias exists that favors publication of statistically significant over nonsignificant outcomes (Bakan 1966; Nakagawa 2004). Ioannidis (2005) points out that this bias causes difficulty with estimating the prestudy (prior) probability of a relationship for use in equations like Equations (5.38) to (5.41). Bayesian interpretation of results and meta-analysis are compromised as discussed later.

Published criticisms of statistical test application in ecotoxicology science and practice (de Bruijn and Hof, 1997; Chapman et al. 1996; Crane and Newman 2000; Hoekstra and van Ewijk 1993a, 1993b; Kooijman 1996; Newman 2008, 2010; Newman and Clements 2008; Stephen and Rogers 1985; Van Dam et al. 2012; Van der Hoeven 1997; Warne and van Dam 2008; Landis and Chapman 2011) have urged change, but the codified methods remain deep-rooted. A recent OECD committee (OECD 2006a) recommendation to gradually phase out effect metrics based on statistical testing (NOEC) is especially encouraging. The question is whether the rate of change is acceptable or too slow for responsible stewardship of the environment.

Two contrasting ways forward are relevant to ecotoxicologists wishing to avoid the problems described above. The first way (Hurlbert and Lombardi 2009) rejects Neyman–Pearson and paleo-Fisherian methods. A neoFisherian approach is presented as the best for scientific inquiry. The second (Newman 2008) advocates more attention to the Neyman–Pearson vantage as a way of best gathering evidence that can be used later for Bayesian inferences and meta-analyses. Both approaches contribute evidence-based insight if outcomes are interpreted thoughtfully in combination with subject knowledge.

Hurlbert and Lombardi (2009) propose neoFisherian significance testing (NFST). The essence of the neoFisherian approach is the following. First, the critical α is not specified and the terms *significant* and *not significant* are abandoned. Second, high p values only infer that suspension of judgment should occur. Third, a "three-valued logic" should be adopted. This three-valued logic was envoked from early work by Lehmann (1950), Bahadur (1952), Kaiser (1960), and more recently, Harris (1997). With a two-sided test, three-valued logic renders to either (1) reserving judgment or judging an effect in the positive (2) or negative (3) direction. As a pertinent example, the data collected after chemical exposure of a group of individuals either (1) provided no useful insight, (2) provided evidence of increased mortality, or (3) provided evidence of decreased mortality. Fourth, the neoFisherian approach recognizes effect size as an essential feature to include in all inferences. Fifth, it recognizes that the frequent use of confidence intervals in lieu of statistical testing is useful but often unnecessary. At the onset of Hurlbert and Lombardi's paper, they reject the Neyman–Pearson hypothesis testing, opining that "Neyman and Pearson seem not to have intended their framework to be used in scientific work in a rigid manner adopted in many quarters" and "[the Neyman–Pearson approach] is suited for situations, such as industrial quality control or 'commercial specifications' … where different actions will be taken according to whether $P \leq \alpha$ or $P > \alpha$."*

* Arguably, the context of this chapter is one in which decisions are being made about a toxicant dose or concentration being harmful or not with the intent of guiding regulatory action. The Neyman–Pearson framework does seem very relevant. Further, the information generated within such an approach would more readily facilitate subsequent Bayesian and meta-analysis.

The second approach (Newman 2008; review also Section 1.6 of Chapter 1) urges more attention to the Neyman–Pearson vantage but does so with the intent of supporting Bayesian inference and future meta-analyses. First, it states that confidence intervals avoid many of the issues described above and should be applied instead of statistical testing if possible. Confidence intervals provide insight simultaneously about the ES, statistical significance, and variability in the data. If statistical tests are to be used, several changes should be made to the conventional (Hurlbert and Lombardi's paleoFisherian) approach. First, type I and II error rates are defined and justified *a priori*. Second, the ES is also defined and justified *a priori*. Justification should be based on subject knowledge, not simply statistical or experimental convention. Third, the distinction between $p(E|H_0)$ and $p(H_0|E)$ should be kept in mind during interpretation of test outcome. Fourth, it is critical to design tests permitting positive predictive value estimation. The advantage of doing this will be discussed shortly. Fifth, practitioners are encouraged to publish negative results. Sixth, null hypotheses that are not nil should be favored. Seventh, definitive inferences from isolated tests should be avoided. Only a series of logically linked experiments or observation sets permit sound decisions. Finally, test power must be estimated *a priori*, not *post priori*. As already described, power, like α, is useful for evidence-based decisions only if established *a priori*. This important point is detailed very clearly in Hoenig and Heisey (2001). Common metrics from statistical software packages such as the minimum significant difference or observed power can be misinterpreted if this point is misunderstood. These *post priori* metrics indicate the magnitude of the difference that could have been detected *given the experimental design and outcome*. As an important point, observed power depends on the calculated p value (Hoenig and Heisey 2001). As useful as these metrics are for designing future experiments, they should not be confused with $1 - \beta$ established *a priori*. They are deceptive if used to suggest the ES that would have been detected after a nonsignificant outcome was generated for an experiment.

5.10.2 Deference to Power, Effect Size, and Balanced Error Rates

Methods exist for incorporating β and ES into experimental or sampling design. Cohen (1988) produced a textbook for this purpose and Dattalo (2008) provides a practical guide to current methods, including bootstrap and Monte Carlo techniques. The SAS package has many procedures applicable to a range of designs, including most of those illustrated in this chapter. Example 5.11 provides the SAS code as one example. Less expensive packages such as Minitab®* also have power calculators. The excellent shareware G*Power 3, developed by the University of Dusseldorf's Institute of Experimental Psychology, can be downloaded from http://www.psycho.uni-duesseldorf.de/aap/projects/gpower/ for this purpose. Another shareware package, PS V9, produced by DuPont and Plummer (Vanderbilt University Department of Biostatistics), can be downloaded from http://biostat.mc.vanderbilt.edu/wiki/Main/PowerSampleSize. The U.S. EPA DEFT is useful for this purpose and can be obtained from http://www.epa.gov/esd/databases/deft/install.htm. Two shareware packages estimate sample size or power within the context of generating a spatial sampling plan such as a field survey. The Pacific National Laboratory's Visual Sampling Plan (VSP) and its manual can be downloaded from http:/vsp.pnl.gov and http://vsp.pnl.gov/docs/PNNL%2019915.pdf, respectively. The SADA software developed at the University of Tennessee–Knoxville can be downloaded from http://www.tiem.utk.edu/~sada/secondary_sampling.shtml. Appendix 27 describes and illustrates those methods most relevant to the tests discussed in this chapter.

* Mention of software does not suggest endorsement by the author.

Example 5.11

Power estimation can be illustrated with the PCP toxicity data. The number of replicates required for a one-way ANOVA *F*-test assuming an α of 0.05 can be explored with the treatment means, overall standard deviation, and practical limits to the number of replicates. Although not done here, particular treatment comparisons could be specified. The power $(1 - b)$ is estimated for each potential number of replicates in a final, definitive test.

```
PROC POWER;
   ONEWAYANOVA
   TEST=OVERALL
   GROUPMEANS = 1.33| 1.183 | 1.371 | 1.254 | 1.109 | 0.604
   STDDEV = 0.17
   ALPHA = 0.05
   NPERGROUP = 2 TO 8
   POWER=.;
RUN;
```

A table is generated for estimated power for the requested number of replicates per treatment,

Index	Computed Power N per Group	Power
1	2	0.753
2	3	0.987
3	4	>.999
4	5	>.999
5	6	>.999
6	7	>.999
7	8	>.999

Three or more replicates are sufficient to get satisfactory power for the *F*-test. A judgment of equally serious consequences for type I and II errors ($a = b = 0.05$) would be supported. Use of more replicates might result in wasted resources and unnecessary killing of animals.

But the question of power could also be addressed for the *t*-tests used to compare treatments. Power is estimated here for comparing two treatments with a pairwise α of 0.01. It is examined for differences in means (ES) of 0.2, 0.3, and 0.4. The NPERGROUP refers to the number of replicates in each of the two treatments.

```
PROC POWER;
     TWOSAMPLEMEANS TEST=DIFF
     MEANDIFF= .2 .3 .4
     STDDEV = 0.17
     ALPHA=0.01 /* BONFERRONI CORRECTED */
     POWER = .
     SIDES=1
     NPERGROUP = 2 TO 25;
     PLOT;
RUN;
```

SAS generated tabular and graphic output, but only the graph is shown here (Figure 5.8). Power is approximately 0.97, 0.80, and 0.42 with a total of eight replicates for differences between means of 0.4, 0.3, and 0.2, respectively. Power would be 0.60, 0.35, and 0.16 for differences of 0.4, 0.3, and 0.2 if four replicates were used. What ES is important to detect if present should be determined by knowledge of the biological consequences of a certain magnitude of effect. A definitive experiment would be designed based on what was practical and what was important to be able to detect with the justified type I and II error rates.

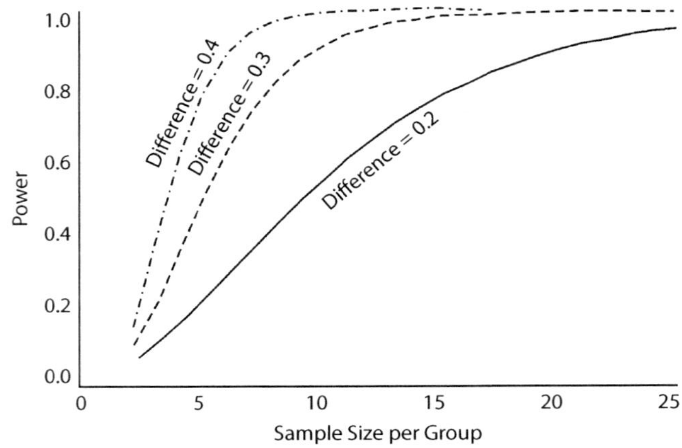

Figure 5.8 Illustration of power analysis for the PCP toxicity data. The power analyses are for *t*-test after the pairwise α was adjusted downward to obtain an experimentwise α of 0.05. Results are shown for three hypothetical differences in treatment means (0.2, 0.3, and 0.4) and different total numbers of replicates per treatment/group.

The null hypothesis difference was 0 by default, but adding NULLDIFF=0.2 to the TWOSAMPLEMEANS statement in the above code would allow another null difference (0.2 in this example) to be used.

5.10.3 Positive Predictive Value (PPV) and Negative Predictive Value (NPV)

> … what the vulgar call chance is nothing but a secret and conceal'd cause. That species of probability, therefore, is what we must chiefly examine.
>
> **—Hume (1739)**

So how does one make the desired statistical inferences if the *p* value from a conventional statistical test is such a poor tool? The answer has already been sketched out (Equations 5.38 to 5.40). Here we will begin answering this question using the context of a diagnostic test that produces either a positive or negative result for tested individuals of a particular population. There can be true positive (TP), false positive (FP), false negative (FN), and true negative (TN) test outcomes. The proportion of positive tests that are correct (that is, truly reflect the presence of the condition or disease) is called the test sensitivity, and the proportion of negative tests that are correct is called the test specificity.

$$Sensitivity = \frac{TP}{TP + FN} \tag{5.42}$$

$$Specificity = \frac{TN}{FP + TN} \tag{5.43}$$

$$Prevalence\ of\ Condition = \frac{TP + FN}{Number\ of\ Individuals\ Tested} \tag{5.44}$$

The positive predictive value (PPV) of a test is the proportion of all the tested individuals that are correctly diagnosed as having the condition. Conversely, the negative predictive value (NPV) is the proportion of individuals that are correctly diagnosed as not having the condition. The PPV and NPV for testing of a specific population can be expressed in terms of TP, FP, FN, and TN (Equations 5.45 and 5.47), or in terms of conditional probabilities (Equations 5.46 and 5.48) (Altman 1991; Gaddis and Gaddis 1990).

$$PPV = \frac{TP}{TP + FP} \tag{5.45}$$

$$P(Condition \mid Positive\ Test) = \frac{P(Positive\ Test \mid Condition)P(Condition)}{P(Positive\ Test)} \tag{5.46}$$

$$NPV = \frac{TN}{FN + TN} \tag{5.47}$$

$$P(No\ Condition \mid Negative\ Test) = \frac{p(Negative\ Test \mid No\ Condition)p(No\ Condition)}{p(Negative\ Test)} \tag{5.48}$$

Equations (5.45) and (5.47) are relevant to testing a specific population and must be modified to accommodate populations with other prevalences for the condition (Altman 1991; Altman and Bland 1994). Prevalence is the estimated prior probability discussed earlier. The PPV and NPV can also be expressed in terms of prevalence, sensitivity, and specificity:

$$PPV = \frac{Sensitivity \times Prevalence}{Sensitivity \times Prevalence + (1 - Specificity) \times (1 - Prevalence)} \tag{5.49}$$

$$NPV = \frac{Specificity\ x\ (1 - Prevalence)}{(1 - Sensitivity) \times Prevalence + Specificity \times (1 - Prevalence)} \tag{5.50}$$

A careful comparison of Equations (5.49) and (5.50) to Equations (5.38) to (5.40) will make it clear that this approach is directly relevant to interpreting outcomes of statistical tests.

Example 5.12

Sterne and Davey Smith (2001) provide an instructive example of applying these metrics to evaluate epidemiological results. Their example involved a collection of 1985 published studies of 300 risk factors for coronary heart disease. The proportion of such null hypotheses that were found in the past to be false was approximately 10% of the studies ($p(H_0) = 0.1$), so 90% of studies found no effect ($p({}^{-}H_0) = 0.9$). Drawing upon past analyses of such studies, power was estimated to be only 0.50 with a typical α of 0.05.

Equation (5.50) or (5.39) can be adjusted to accommodate this information to determine false positive value (*FPV*) of these tests of heart disease risk factors.

$$FPV = \frac{\alpha(1 - p(H_0))}{(1 - \beta)(p(H_0)) + \alpha(1 - p(H_0))} = \frac{0.05 * 0.90}{0.50 * 0.10 + 0.05 * 0.90} = \frac{0.045}{0.095} = 0.47 \tag{5.51}$$

The FPV indicates that roughly half of the statistically significant studies would be false positives. The combination of the prior probability of getting a true relationship and the typically poor power of these kinds of studies resulted in very low ability to infer that a relationship truly exists between heart disease and the risk factor based on a significant test outcome.

The FPV for significance tests was illustrated in Example 5.12. Referring to the FPV as a false positive result probability (FPRP) of a statistically significant hypothesis test outcome, Wacholder et al. (2004) and Sterne and Davey Smith (2001) use a more general form of Equation (5.51) for its calculation

$$FPRP = \frac{\alpha(1-\pi)}{\alpha(1-\pi)+(1-\beta)\pi} \tag{5.52}$$

where π = the prior probability, that is, $P(H_A)$. In the case of the example given above, the H_A is that a relationship truly exists between coronary heart disease and the risk factor. Calculation of FPRP for Example 5.12 is the following:

$$FPRP = \frac{\alpha(1-\pi)}{\alpha(1-\pi)+(1-\beta)\pi} = \frac{(0.05)(1-0.1)}{(0.05)(1-0.1)+(1-0.5)(0.1)} = \frac{0.045}{0.095} = 0.47$$

The probability of a false significant outcome is nearly the same as that for a true significant outcome. Similarly, the probability of a false negative outcome (false negative result probability (FNRP)) can be calculated. Figure 5.9 illustrates that a simple diagram using natural frequencies can also be used to estimate PPV, NPV, FPRP, and FNRP. Notice the relationships in the figure: $FPRP = 1 - PPV$ and $FNRP = 1 - NPV$.

The overall accuracy of such a test is expressed in the field of information retrieval with a single metric, the F measure,

$$F = 2\,\frac{PPV(1-\beta)}{PPV+(1-\beta)} \tag{5.53}$$

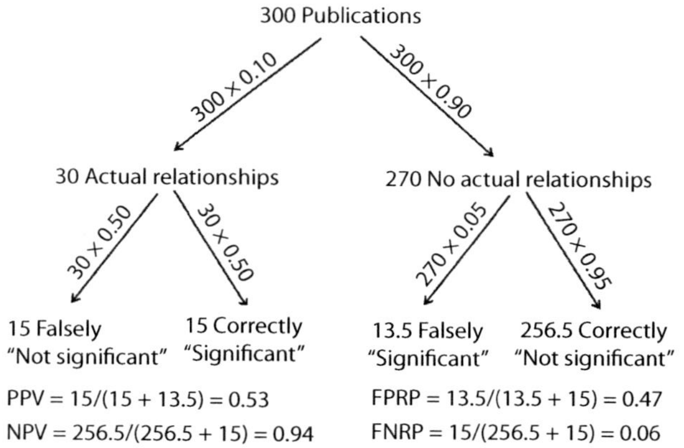

Figure 5.9 Graphical illustration of PPV and FPV estimation for 300 epidemiological studies attempting to identify risk factors for coronary heart disease. (From Sterne and Davey Smith, 2001.)

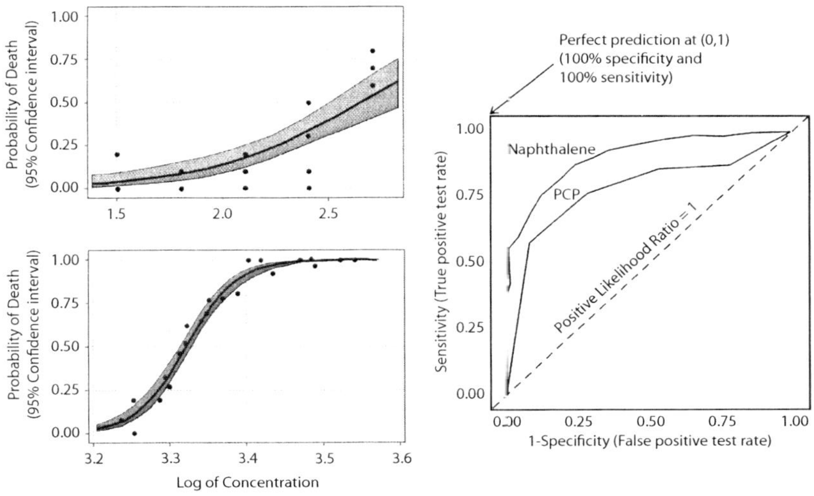

Figure 5.10 ROC curves for naphthalene and pentachlororpherol toxicity data modeled with a binary logistic model, Ln $(p/(1-p)) = a + b(\log_{10}$ toxicant concentration). The two panels on the left show the observed proportions dying at each $\log_{10}$ concentration, a line depicting model predictions, and a gray area indicting the approximate 95% confidence interval for the estimated probabilities. In the ROC curve (right panel), the LR is plotted for different values of $\log_{10}$ concentration. The area under the ROC curve reflects how well each logistic model based on $\log_{10}$ concentration discriminates between exposed individuals that will or will not die. The ideal would be a line that rose steeply toward the upper left corner (1,0) of the graph.

All of these probabilities become available if hypothesis testing is adopted in the manner recommended in Newman (2008). The results from this approach can be used in further Bayesian inferences. Hurlbert and Lombardi's neoFisherian significance testing/assessment (NFST) seems appropriate for some investigations, but not ecotoxicity testing, which has the goal of deciding what is or is not toxic, and then acting on that decision. An approach including prior probabilities* and thoughtfully chosen decision error rates seems most appropriate.

Example 5.13

Returning for a moment in this example to regression modeling (Chapter 4), specificity and sensitivity can be informative for gauging the accuracy of model predictions of death or survival for exposed individuals. The naphthalene (Example 4.6) and pentachlorophenol (Example 5.1) toxicity data can be fit to a binary logistic model, Ln $(p/(1-p)) = a + b(\log_{10}$ toxicant concentration) to illustrate this point. Notice that now predictions are not being made for a single diagnostic test but from a continuous variable, $\log_{10}$ concentration. The following SAS code fits the models and produces receiver operator characteristic (ROC) curves for each (Figure 5.10).

```
DATA FATHEAD;
   INPUT DEAD TOTAL CONC @@;
   LCONC=LOG10(CONC);
   DATALINES;
   2 10   32 2 10   32 0 10   32 2 10   32
   1 10   64 0 10   64 0 10   64 0 10   64
   1 10 128 1 10 128 2 10 128 0 10 128
```

* If no information or sound justification exists for selecting a particular prior probability, a noninformative prior probability might be chosen until more evidence is produced to improve the estimate. A noninformative prior probability might be a 0.50 chance of the solution being toxic or not.

```
   3 10 256 1 10 256 0 10 256 5 10 256
   6 10 512 7 10 512 6 10 512 8 10 512
   ;
ODS GRAPHICS ON; /* PLOT FILES ARE SAVED TO THE PROGRAM DIRECTORY */
PROC LOGISTIC DATA=FATHEAD PLOTS=EFFECT PLOTS=ROC;
     MODEL DEAD/TOTAL=LCONC/OUTROC=ROCOUT;
     OUTPUT OUT=ESTIMATED PREDICTED=ESTPROB L=LOWER95 U=UPPER95;
RUN;
DATA NAP;
     INPUT CONC DEAD TOTAL @@;
     LCONC=LOG10(CONC);
     DATALINES;
     0 1 26 0 0 26 0 0 26 0 0 26 0 1 26 0 0 26 1725
     2 26 1792 5 26 1797 0 26 1972 8 25 1996 7 26
     1940 5 26 2097 14 27 2057 12 26 2105 16 26
     2204 17 26 2233 18 26 2250 20 26 2340 21 27 2448
     21 26 2452 21 26 2530 26 26 2624 26 26 2726 24 26
     2960 26 26 3090 25 26 3055 26 26 3332 26 26 3335
     26 26 3483 26 26
;
PROC LOGISTIC DATA=NAP PLOTS=EFFECT PLOTS=ROC;
       MODEL DEAD/TOTAL=LCONC/OUTROC=ROCOUT;
       OUTPUT OUT=ESTIMATED PREDICTED=ESTPROB L=LOWER95 U=UPPER95;
RUN;
ODS GRAPHICS OFF;
```

The ROC chart is a plot of the sensitivity and 1 – specificity for different values of $\log_{10}$ concentration. Sensitivity is the probability of correctly predicting mortality based on $\log_{10}$ toxicant concentration. Specificity is the probability of correctly predicting survival based on $\log_{10}$ concentration of the toxicant. It follows that 1 – specificity is the probability of incorrectly predicting lethality for exposed individuals. Sensitivity/(1 – specificity) is the positive Likelihood Ratio (LR). The diagonal (line of no discrimination) in the figure is the line for which the positive LR is 1. The line of no discrimination reflects the situation in which the numbers of true positives and false negatives are the same; therefore, a positive LR of 1 indicates a predictive reliability akin to a flip of a fair coin. The area under the line of no discrimination is 0.5, so an AUC of 0.5 would indicate a model with useless predictive value. Predictive value for models with AUC of 0.70 to 0.80, 0.80 to 0.90, and 0.90 to 1.00 might be described as fair, good, and excellent, respectively. The areas under the curve (AUCs) for the naphthalene and pentachlorophenol curves were 0.90 and 0.78, respectively, so the accuracy of prediction for the naphthalene model was excellent, but that for pentachlorophenol was only fair.

Different models could also have been applied to a data set and the AUC for each model used to estimate the relative value of each model for generating accurate predictions. An example in which three models might be compared is the situation involving two explanatory variables. The models would include variable 1 alone, variable 2 alone, and both variables 1 and 2 together.

5.10.4 The Virtues of Confidence Intervals

> Significance testing unquestionably dominates epidemiology today. In attempting to refrain from the practice over the past 17 years, I have often been expected, assumed, encouraged, and sometimes even forced to engage in it by editors, reviewers, colleagues, professors, students, funding sources, regulators, attorneys, and journalists. It is not easy to be a non-tester in a testing world.
>
> **—Poole (2001)**

Persuasive arguments have been made to avoid the difficulties associated with current statistical tests by using confidence intervals where appropriate (Altman et al. 2000; Di Stefano 2004; Fidler et al. 2004; Newman 2008; Poole 2001). The argument normally begins by recounting the numerous errors in the currently applied statistical testing. It is then pointed out that confidence intervals simultaneously convey the magnitude of the effect, the precision of the estimated effect, and the statistical significance of the observed effect (Di Stefano 2004; Fidler et al. 2004; Nakagawa and Cuthill 2007). Conventional statistical tests give short shrift to the first two features that, with a moment of objective reflection, are clearly as, or more, important to an ecotoxicologist as statistical significance (Figure 5.11). Cumming and Finch (2005) suggest three rules for confidence interval use: (1) select error bars associated directly with the relevant effect, (2) presentation should be sensitive to the experimental design, and (3) the confidence intervals should be thoroughly interpreted. Much more guidance can be found in Altman et al. (2000) and Cumming (2012), who also provide shareware for calculating confidence intervals. Dattalo (2008) and others provide methods for confidence power interval estimation.

Several issues must be kept in mind when applying confidence intervals. The nature of confidence interval is often confused by practitioners and misrepresented in the published literature. By strict definition, if one were to generate many such intervals, 95% of those intervals would contain the measured effect difference or correlation coefficient. It is not strictly correct to claim the probability is 0.95 that a particular interval includes the difference or correlation coefficient. Also, the 95% value for confidence intervals, like the type I error rate of 0.05, is a convention, and other values might be more appropriate depending on circumstances or goals of a study. Incautious application of confidence intervals can lead to some of the misinterpretations already discussed relative to conventional statistical testing (Poole 2001).

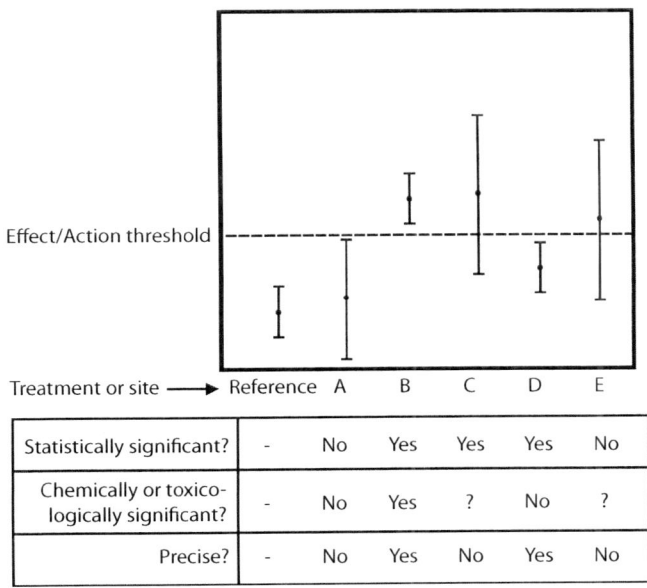

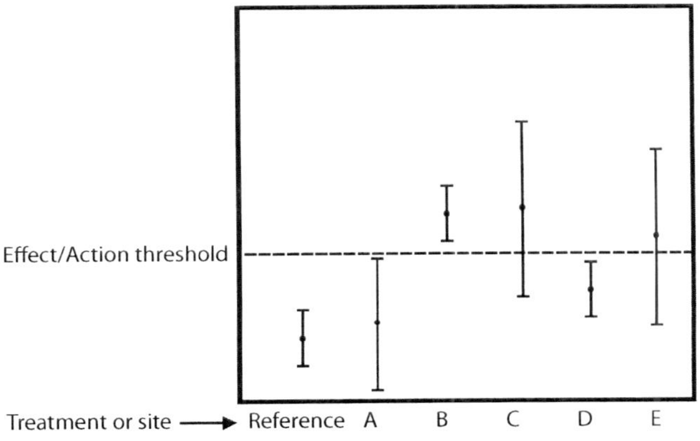

Statistically significant?	-	No	Yes	Yes	Yes	No
Chemically or toxicologically significant?	-	No	Yes	?	No	?
Precise?	-	No	Yes	No	Yes	No

Figure 5.11 An illustration of 95% confidence intervals (CIs; the interval $\bar{x} - t_{n-1,95}SE \le \mu \le \bar{x} + t_{n-1,95}SE$, where SE = standard error) for treatment means from an experiment that includes hypothetical adverse effects for different treatments (or sites). A toxicological effect large enough to be of concern is shown by the effect/action threshold line. Notice that sole reliance on statistical significance could result in B, C, and D being treated identically. Yet differences in ES and precision are obviously important in deciding how best to act toward each of these results. Judgment to act about B would be easier than about the imprecise C. A reasonable decision about C might be to gather more or different information before acting. Also, both A and E could be treated identically as not statistically significant, yet the argument could easily be made that E warrants further scrutiny.

Example 5.14

Estimation of the number of samples needed to get a confidence interval of a desired precision can be illustrated using SAS PROC POWER. The ES in this context is the distribution half width, the distance from the point estimate to the mean. Analogous to power is the PROBWIDTH, which is the probability that the desired confidence interval precision (or better) will be obtained at the specified half width. In this example, a confidence interval is estimated with a t statistic. An α of 0.05 for a one-sided test is specified, although some other value could have been chosen. The desired ES (HALFWIDTH) is 0.1, and the standard deviation (STDEV) is given as 0.15. The probability of getting the specified precision (PROBWIDTH) is set at 0.95. In addition to estimating the required number of observations to meet these specifications, a graph of number of observations for ES values from 0.1 to 0.5 is also requested (Figure 5.12).

```
PROC POWER;
   ONESAMPLEMEANS CI=T
      ALPHA = 0.05
      HALFWIDTH = 0.1
      STDDEV = 0.15
      PROBWIDTH = 0.95
      SIDES = 1
      NTOTAL = .;
   PLOT X=EFFECT MIN=0.1 MAX = 0.5;
RUN;
```

The output below indicates that 13 observations would be needed to meet the specified requirements. The probability of getting the specified precision is actually slightly better (0.959) than the requested 0.95.

```
    Confidence Interval for Mean
      Fixed Scenario Elements

Distribution                   Normal
Method                          Exact
Number of Sides                     1
Alpha                            0.05
CI Half-Width                     0.1
Standard Deviation               0.15
Nominal Prob(Width)              0.95
Prob Type                 Conditional
```

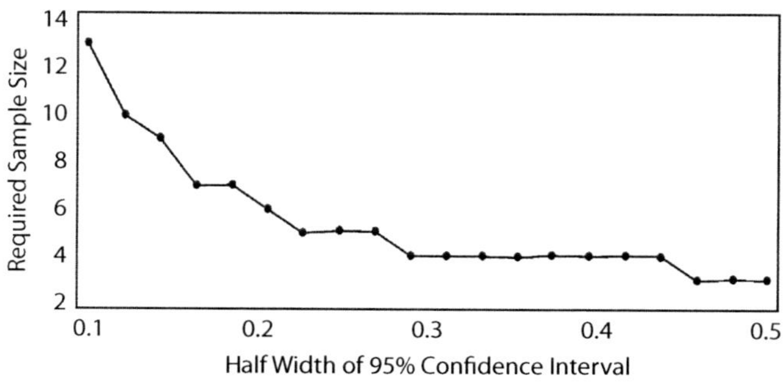

Figure 5.12 Illustration of estimating the number of observations needed to meet a specified precision for a one-sided 95% confidence interval.

```
  Computed N Total
Actual  Prob        N
(Width)         Total
0.959              13
```

5.11 INFERRING BIOLOGICAL SIGNIFICANCE FROM STATISTICAL SIGNIFICANCE

There was an ISO resolution (ISO TC147/SC5/WG10 Antalya 3) as well as an OECD workshop recommendation (OECD, 1998) that the NOEC should be phased out from international standard.... However, the NOEC is still required in many regulatory standards.... Therefore guidance will be provided on the statistical methods for determination of the NOEC.

—OECD (1996a)

Difficulties beyond the statistical ones just discussed make it difficult to infer material harm using significance testing. The conventional methods described above test statistical significance, not biological significance. As pointed out by Salsburg (1986), the conventional sense of the term *statistically significant* is that a test simply "signified that something has happened different from the proposed [null] hypothesis." It is thought to imply the plausibility of the null hypothesis, not importance of any effect (Box et al. 1978). Nonetheless, judgments are made at present regarding biological importance and standard regulatory methods have been created to imply biological significance from results of statistical tests. OECD (2006a) applies NOEC from a proof of hazard context, that is, the chemical being tested is assumed to be nontoxic at a specific concentration/dose unless there is sufficient evidence to the contrary. The context of the OECD suggests that a statistically significant effect at a tested concentration is sufficient evidence of toxicity. The OECD context does not encompass an alternative hypothesis, although such consideration is noted to have promise for future refinement of the methods (OECD 2006a, p. 34). As described already, conclusions arising from such methods have many serious shortcomings that should be recognized when making inferences. The following conventional definitions as used in a regulatory process can be used to highlight some of the more important challenges.

No observed effect concentration or level (NOEC or NOEL)

"The highest dose for which the difference with the control group is not statistically significant" (Hoekstra and Van Ewijk 1993a).

"The highest concentration of a material in a toxicity test that has no statistically significant adverse effect on the exposed population of test organisms as compared with the controls. When derived from a life cycle or partial life cycle test, it is numerically the same as the lower limit of the MATC" (Hoekstra and Van Ewijk 1993a). This definition is also used by Weber et al. (1989).

Lowest observed effect concentration or level (LOEC or LOEL)

"The lowest test dose at which the response is significantly different from the control group" (Hoekstra and Van Ewijk 1993a).

"The lowest concentration of a material used in a toxicity test that has a statistically significant adverse effect on the exposed population of test organisms as compared to the controls. When derived from a life cycle or partial life cycle test, it is numerically the same as the upper limit of the MATC [maximum acceptable toxicant concentration]" (Rand and Petrocelli 1985). This definition is also used by Weber et al. (1989).

Maximum acceptable toxicant concentration (MATC)

"An undetermined concentration within the interval bounded by the NOEC and LOEC that is presumed safe by virtue of the fact that no statistically significant adverse effect was observed" (Weber et al. 1989).

Safe concentration
"The highest concentration of toxicant that will permit normal propagation of fish and other aquatic life in receiving waters. The concept of a 'safe concentration' is a biological concept, whereas the 'no observed effect concentration' is a statistically defined concentration" (Weber et al. 1989).

Consideration of the definitions for LOEL and NOEL suggests an immediate difficulty. Sound decisions regarding the consequences of toxicant release to an aquatic system require more than statistical testing. A lack of any ecological or temporal context for these rudimentary effects (survival, weight gain, maturation, and reproduction) detected during highly structured and temporally deficient experiments often precludes logically defensible decision making.

The definitions of acceptable or unacceptable contaminant levels are based on statistical testing as described in the first half of this chapter. However, all of the problems discussed in the second half of the chapter argue against this being a reliable approach. Other approaches exist, such as confidence intervals, bioequivalence testing, modeling, or testing according to the scheme described above. If all assumptions are met, confidence or credible intervals for predictive models as described in the previous chapter could be used to define acceptable and unacceptable levels more effectively (Suter et al. 1987). Dixon and Garrett (1993) and Denton et al. (2011) argue that bioequivalence tests for insignificant difference in responses of exposure treatments and the control treatment may be appropriate in some cases. A design involving α, β, ES, and prior probabilities, as discussed above, might also provide a way ahead in many situations.

The tendency to interpret α values too strictly or improperly is a substantial concern with the current approach. A biologically critical effect with an associated p value = 0.06 should not be ignored if derived from a properly designed study. On the other hand, a highly statistically significant effect might be biologically trivial if the variable measured has no material effect on an organism's fitness. The traditional α of 0.05 or 0.01 used in conventional statistical testing is arbitrary and insufficient without β and ES; this must be understood if they are to be used effectively in inferring biological importance. Readers wishing for more detail about this can review the referenced materials above and also Box et al. (1978, p. 109), Snedecor and Cochran (1973, p. 27), Noether (1971, p. 64), Salsburg (1986, p. 84), or Green (1978, p. 9).

5.12 SUMMARY

A wide range of statistical methods exist for testing or quantifying chronic lethal and sublethal effects on biota (Figures 5.1 to 5.3). Unfortunately, the strength of inference derived from the present methods is questionable. Indeed, Platt's (1964) comment, "The mathematical box is a beautiful way of wrapping up a problem, but it will not hold the phenomena unless they have been caught in a logical box to begin with" seems very pertinent here.

There are a variety of means to strengthen inference. Newman (2008) suggests the set of recommendations described above. Methods allowing estimation of PPV or those based on confidence intervals, if appropriate, would enhance inferential strength. Refocusing on statistical methods and experimental designs with the necessary power is essential at this point.* For example, statistical power may be enhanced by increasing sample size or choosing among tests thoughtfully. The methods toward the left-hand side of Figure 5.1 tend to have the highest power. In contrast to the

* The critical reader will have noticed that hypothesis tests criticized in this chapter are to be found throughout the other chapters of this book. Power calculations allowing *a priori* establishment of desired α, β, and effect size need to be made for those tests to be conducted in a manner consistent with the recommendations outlined here. For confidence intervals, a similar set of calculations is needed to determine the sample size required to obtain the desired precision. Appendix 27 details with examples the means of doing these calculations for the most important methods applied in this book. Software packages to implement the methods conveniently are also discussed.

tendency suggested in Figure 5.1, the number of control observations could be increased with a consequent increase in statistical power. Certainly, the strength of inference associated with "no significant effect" is clarified if an estimate of power and ES are presented also. (For further discussion of the importance of power in ecological studies, the reader is referred to Toft and Shea 1983, Rotenberry and Weins 1985, Gerrodette 1987, Peterman and Bradford 1987, and Peterman 1990). Second, clearly classifying the nature of the measured effect, (e.g., Selyean stress, effect with hormesis, preadaptive stress, damage, ambiguous effect, or neutral effect) enhances selection of the most appropriate ES to consider during test design. The type of effect will strongly affect method selection (e.g., more powerful one-sided versus two-sided test, or avoidance of a test assuming monotonic trend if hormesis was expected). Failure to do so weakens inference and muddles conclusions regarding biological significance. Third, clearly state the limitations of inferences derived from the test results. For example, conclusions from tests using time endpoints, e.g., 96-hour LC50, should not be used to imply biological consequences in a field situation involving longer exposure durations. Fourth, these methods should be used in tandem with those addressing higher- and lower-order effects on other ecological components to strengthen inferences about biological consequences.

REFERENCES

Adams, S.M., ed. Biological indicators of stress in fish. In *American Fisheries Society Symposium 8*. Bethesda, MD: American Fisheries Society, 1990.

Altman, D.G. *Practical statistics for medical research*. Boca Raton, FL: Chapman & Hall/CRC Press, 1991.

Altman, D.G., and J.M. Bland. Diagnostic tests 2: Positive values. *BMJ* 309:102 (1994).

Altman D., D. Machin, T.N. Bryant, and M.J. Gardner. *Statistics with confidence: Confidence intervals and statistical guidelines*. 2nd ed. London: British Medical Journal Books, 2000.

Anderson, D.R., K.P. Burnham, and W.L. Thompson. Null hypothesis testing: Problems, prevalence, and an alternative. *J. Wildlife Manag.* 64(4):912–923 (2000).

Bailey, B.J.R. Tables of the Bonferroni t statistic. *J. Am. Stat. Assoc.* 72:469–478 (1977).

Bahadur, R. R. A property of the t-test. *Sankhya* 12: 79–88 (1952).

Bakan, D. The test of significance in psychological research. *Psychol. Bull.* 66:423–437 (1966).

Berger, J.O., and T. Sellke. Testing a point null hypothesis: The irreconcilability of p values and evidence. *J. Am. Stat. Assoc.* 82:112–122 (1987).

Berkson, J. Some difficulties of interpretation encountered in the application of the chi-squared test. *J. Am. Stat. Assoc.* 33:526–536 (1938).

Beyer, W.H. *Handbook of tables for probability and statistics*. 2nd ed. Cleveland, OH: CRC Press, 1968.

Biau, D.J., B. Jolles, and R. Porcher. P value and the theory of hypothesis testing. *Clin. Orthop. Relat. Res.* 468:885–892 (2010).

Box, G.E.P., W.G. Hunter, and J.S. Hunter. *Statistics for experimenters. An introduction to design, data analysis, and model building*. New York: John Wiley & Sons, 1978.

Carver, R.P. The case against statistical significance testing. *Harvard Educ. Rev.* 48:378–399 (1978).

Chapman, P.M., R.S. Caldwell, and P.F. Chapman. A warning: NOECs are inappropriate for regulatory use. *Environ. Toxicol. Chem.* 15(2):77–79 (1996).

Chen, J.J., and R.L. Kodell. Quantitative risk assessment for teratological effects. *J. Am. Stat. Assoc.* 84:966–971 (1989).

Cochran, W.G., and G.M. Cox. *Experimental designs*. New York: John Wiley & Sons, 1957.

Cohen, J. *Statistical power analysis for the behavioral sciences*. 2nd ed. Mahwah, NJ: Lawrence Erlbaum Associates, 1988.

Cohen, J. The earth is round (p<0.05). *Am. Psychol.* 49:997–1003 (1994).

Cohen, J. The earth is round (p<0.05): Rejoinder. *Am. Psychol.* 50:1104 (1995).

Cox, C. Threshold dose-response models in toxicology. *Biometrics* 43:511–523 (1987).

Cox, D.R. Commentary: Smoking and lung cancer: Reflections on a pioneering paper. *Int. J. Epidemiol.* 38:1192–1193 (2009).

Crane, M., and M.C. Newman. What level of effect is a no observed effect? *Environ. Toxicol. Chem.* 19:516–519 (2000).

Crump, K.S. A new method for determining allowable daily intakes. *Fundam. Appl. Toxicol.* 4:854–871 (1984).

Cumming, G. *Understanding the new statistics*. New York: Routledge, 2012.

Cumming, G., and S. Finch. Inference by eye. *Am. Psychol.* 60(2):170–180 (2005).

Dattalo, P. *Determining sample size*. New York: Oxford University Press, 2008.

Davis, B. *Introduction to agricultural statistics*. Albany, NY: Delmar Thomson Learning, 2000.

Dawkins, R. *The extended phenotype*. New York: Oxford University Press, 1982.

Day, R.W., and G.P. Quinn. Comparisons of treatments after an analysis of variance in ecology. *Ecol. Monogr.* 59(4):433–463 (1989).

de Bruijn, J.H., and Hof, M. How to measure no effect. Part IV. How acceptable if the ECx from an environmental policy point of view? *Environmetrics* 8:263–267 (1997).

Denton, D. L., J. Diamond, and L. Zheng. Test of significant toxicity: A statistical application for assessing whether an effluent or site water is truly toxic. *Environ. Toxicol. Chem.* 30: 1117–1126 (2011).

Di Stefano, J. A confidence interval approach to data analysis. *Forest Ecol. Manag.* 187:173–183 (2004).

Dixon, P.M. Personal communication, 1993.

Dixon, P.M., and K.A. Garrett. Statistical issues for field experimenters. In *Wildlife toxicology and population modeling integrated studies of agroecosystems*, ed. R.J. Kendall and T. Lacher. Chelsea, MI: Lewis Publishers, 1993, pp. 67–73.

Dixon, W.J., and F.J. Massey Jr. *Introduction to statistical analysis*. New York: McGraw-Hill Book Co., 1969.

Dunnett, C.W. A multiple comparison procedure for comparing several treatments with a control. *J. Am. Stat. Assoc.* 50:1096–1121 (1955).

Dunnett, C.W. New tables for multiple comparisons with a control. *Biometrics* 20:482–491 (1964).

Fidler, F., M.A. Burgman, G. Cumming, R. Buttrose, and N. Thomason. Impact of criticism of null-hypothesis significance testing on statistical reporting practices in conservation biology. *Conserv. Biol.* 20(5):1539–1544 (2006).

Fidler, F., N. Thomason, G. Cumming, S. Finch, and J. Leeman. Editors can lead researchers to confidence intervals, but can't make them think. *Psychol. Sci.* 15(2):119–126 (2004).

Fisher, R.A. On the mathematical foundations of theoretical statistics. *Philos. Trans. R. Soc.* A222:309–368 (1922).

Fisher, R.A. Statistical tests. *Nature* 136:474 (1935).

Fisher, R.A. *Statistical methods and scientific induction*. Edinburgh: Oliver & Boyd, 1956.

Fligner, M.A. A note on two-sided distribution-free treatment versus control multiple comparisons. *J. Am. Stat. Assoc.* 79:208–211 (1984).

Gaddis, G.M., and M.L. Gaddis. Introduction to biostatistics: Part 3, sensitivity, specificity, predictive value, and hypothesis testing. *Ann. Emerg. Med.* 19:591–597 (1990).

Gelber, R.D., P.T. Lavin, C.R. Mehta, and D.A. Schoenfeld. Statistical analysis. In *Fundamental of aquatic toxicology. Methods and applications*, ed. G.M. Rand and S.R. Petrocelli. Washington, DC: Hemisphere Publishing Corporation, 1985, pp. 267–312.

Germano, J.D. Ecology, statistics, and the art of misdiagnosis: The need for a paradigm shift. *Environ. Rev.* 7:167–190 (1999).

Gerrodette, T. A power analysis for detecting trends. *Ecology* 68(5):1364–1372 (1987).

Gigerenzer, G. *Adaptive thinking. Rationality in the real world.* Oxford: Oxford University Press, 2001.

Gigerenzer, G. Mindless statistics. *J. Socio-Econ.* 33:587–606 (2004).

Gigerenzer, G. *Rationality for mortals. How people cope with uncertainty.* Oxford: Oxford University Press, 2008.

Goodman, S.N. P values, hypothesis tests, and likelihood: Implications for epidemiology of a neglected historical debate. *Am. J. Epidemiol.* 137(5):485–495, 1993.

Goodman, S.N. Toward evidence-based medical statistics. 1: The p value fallacy. *Ann. Intern. Med.* 130:995–1004 (1999).

Grattan-Guinness, I. Karl Popper and 'the problem of induction': A fresh look at the logic of testing scientific theories. *Erkenntnis* 60:107–120 (2004).

Green, R.H. *Sampling design and statistical methods for environmental biologists*. New York: John Wiley & Sons, 1978.

Gupta, S.S. Probability integrals of multivariate normal and multivariate *t*. *Ann. Math. Stat.* 34:792–828 (1963).

Guthery, F.S. Statistical ritual versus knowledge accrual in wildlife science. *J. Wildlife Manag.* 72(8):1872–1875 (2008).

Hacking, I. *An introduction to probability and inductive logic*. Cambridge: Cambridge University Press, 2001.

Harris, R. J. Significance tests have their place. *Psychol. Sci.* 8: 8–11 (1997).

Hicks, C.R. *Fundamental concepts in the design of experiments*. New York: Holt, Rinehart and Winston, 1973.

Hoekstra, J.A., and P.H. Van Ewijk. Alternatives for the no-observed-effect level. *Environ. Toxicol. Chem.* 12:187–194 (1993a).

Hoekstra, J.A., and P.H. Van Ewijk. The bounded effect concentration as an alternative to the NOEC. *Sci. Total Environ. Suppl.* 705–711 (1993b).

Hoenig, J.M., and D.M. Heisey. The abuse of power: The pervasive fallacy of power calculations for data analysis. *Am. Stat.* 55(1):1–6 (2001).

Hogben, L. The contemporary crisis or the uncertainties of uncertain inference. In *The significance test controversy*, ed. D.L. Morrison and R.E. Henkel. New Bunswick, NJ: Transaction Publishers, 1970, pp. 8–40.

Howson, C., and P. Urbach. *Scientific reasoning: The Bayesian approach*. La Salle, IL: Open Court, 1989.

Huggett, R.J., R.A. Kimerle, P.M. Mehrle Jr., and H.L. Bergman. *Biomarkers. Biochemical, physiological, and histological markers of anthropogenic stress*. Chelsea, MI: Lewis Publishers, 1992.

Hume, D. *A treatise of human nature*. New York: Penguin Books, 1739; reprinted, 1985.

Hurlbert, S.H. Pseudoreplication and design of ecological field experiments. *Ecol. Monogr.* 54(2):187–211 (1984).

Hurlbert, S.H., and C.M. Lombardi. Final collapse of the Neyman–Pearson decision theoretic framework and rise of the neoFisherian. *Ann. Zool. Fennici* 46:311–349 (2009).

Ioannidis, J.P.A. Why most published research findings are false. *PLoS Med.* 2(8):e124 (2005). DOI: 10.1371/journal.pmed.0020124.

Jennions, M.D., and A.P. Møller. A survey of the statistical power of research in behavioral ecology and animal behavior. *Behav. Ecol.* 14(3):438–445 (2003).

Kaiser, H. F. Directional statistics decisions. *Psychol. Rev.* 67: 167 (1960).

Kokoska, S., and C. Nevison. *Statistical tables and formulae*. New York: Springer-Verlag, 1989.

Kooijman, S.A.L.M. An alternative for NOEC exists, but the standard model has to be abandoned first. *Oikos* 75:310–316 (1996).

Kramer, C.Y. Extension of multiple range tests to group means with unequal numbers of replications. *Biometrics* 12(3):307–310 (1956).

Lakatos, I. Falsification and the methodology of scientific research programmes. In *The methodology of scientific research programmes: Imre Lakatos' Pholisophical Papers*. Vol. 1. Cambridge: Cambridge University Press, 1978.

Landis, W.P., and P.M. Chapman. Well past time to stop using NOELs and LOELs. *Integr. Environ. Assess. Manag.* 7:vi–viii (2011).

Lecoutre, B., M.-P. Lecoutre, and J. Poitevineau. Uses, abuses and misuses of significance tests in the scientific community: Won't the Bayesian choice be unavoidable? *Int. Stat. Rev.* 69:399–417 (2001).

Lehmann, E. L. Some principles of the theory of testing hypotheses. *Annals of Mathematical Statistics* 21: 1–26 (1950).

Levine, D.M., P.P. Ramsey, and R.K. Smidt. *Applied statistics for engineers and scientists*. Upper Saddle River, NJ: Prentice Hall, 2001.

McCarthy, J.F., and L.R. Shugart, eds. *Biomarkers of environmental contamination*. Chelsea, MI: Lewis Publishers, 1990.

McCloskey, D.N. The insignificance of statistical insignificance. *Am. Sci.* 272:32–33 (1995).

Miller, R.G., Jr. *Simultaneous statistical inference*. New York: McGraw-Hill Book Company, 1966.

Miller, R.G., Jr. *Beyond ANOVA, basics of applied statistics*. New York: John Wiley & Sons, 1986.

Morrison, D.E., and R.E. Henkel. *The significance test controversy*. New Brunswick, NJ: Transaction Publishers, 1970.

Nakagawa, S. A farewell to Bonferroni: The problems of low statistical power and publication bias. *Behav. Ecol.* 15(6):1044–1045 (2004).

Nakagawa, S., and I.C. Cuthill. Effect size, confidence interval and statistical significance: A practical guide for biologists. *Biol. Rev.* 82:591–605 (2007).

Neter, J., W. Wasserman, and M.H. Kutner. *Applied linear statistical models*. 3rd ed. Homewood, IL: Richard D. Irwin, 1990.

Newman, M.C. What exactly are you inferring? A closer look at hypothesis testing. *Environ. Toxicol. Chem.* 27:1013–1019 (2008).

Newman, M.C. *Fundamentals of ecotoxicology*. 3rd ed. Boca Raton, FL: Taylor & Francis/CRC Press, 2010.

Newman, M.C., and W.H. Clements. *Ecotoxicology: A comprehensive treatment.* Boca Raton, FL: CRC Press, 2008.

Neyman, J., and E.S. Pearson. On the use and the interpretation of certain test criteria for purposes of statistical inference. *Biometrika* 20:175–240, 263–294 (1928).

Neyman, J., and E.S. Pearson. On the problem of the most efficient tests of statistical hypotheses. *Philos. Trans. R. Soc.* A231:289–337 (1933).

Nickerson, R.S. Null hypothesis significance testing: A review of an old and continuing controversy. *Psychol. Methods* 5(2):241–301 (2000).

Noether, G.E. *Introduction to statistics. A fresh approach.* New York: Houghton Mifflin Company, 1971.

Organization for Economic Cooperation and Development (OECD). *Current approaches in the statistical analysis of ecotoxicity data: A guidance to application.* OECD Series on Testing and Assessment No. 54, ENV/JM/MONO(2006)18. Paris: OECD, 2006a.

Organization for Economic Cooperation and Development (OECD). Annexes. In *OECD Current approaches in the statistical analysis of ecotoxicity data: A guidance to application.* OECD Series on Testing and Assessment No. 54, ENV/JM/MONO(2006)18/ANN. Paris: OECD, 2006b.

Oris, J.T., and A.J. Bailey. Statistical analysis of the *Ceriodaphnia* toxicity test: Sample size determination for reproductive effects. *Environ. Toxicol. Chem.* 12:85–90 (1993).

Peterman, R.M. The importance of reporting statistical power: The forest decline and acidic deposition example. *Ecology* 71(5):2024–2027 (1990).

Peterman, R.M., and M.J. Bradford. Statistical power of trends in fish abundance. *Can. J. Fish. Aquat. Sci.* 44:1879–1889 (1987).

Platt, J.R. Strong inference. *Science* 146:347–353 (1964).

Poole, C. Low p-values or narrow confidence intervals: Which are more durable? *Epidemiology* 12(3):291–294 (2001).

Rand, G.M., and S.R. Petrocelli. *Fundamentals of aquatic toxicology. Methods and applications.* Washington, DC: Hemisphere Publishing Corp., 1985.

Rohlf, F.J., and R.R. Sokal. *Statistical tables.* 2nd ed. New York: W.H. Freeman and Company, 1981.

Rotenberry, J.T., and J.A. Wiens. Statistical power analysis and community-wide patterns. *Am. Nat.* 125:164–168 (1985).

Rozeboom, W.W. Good science is abductive, not hypothetico-deductive. In *What if there were no significance tests?* ed. L.L. Harlow, S.A. Mulaik, and J.H. Steiger. Hillsdale, NJ: Erlbaum, 1997, pp. 335–392.

Salsburg, D.S. *Statistics for toxicologists.* New York: Marcel Dekker, 1986.

Salsburg, D.S. *The lady tasting tea.* New York: Henry Holt and Company, 2001.

SAS Institute. *SAS procedures guide, release 6.03.* Cary, NC: SAS Institute, 1988a.

SAS Institute. *SAS language guide for personal computers, release 6.03.* Cary, NC: SAS Institute, 1988b.

Schmidt, F.L., and J.E. Hunter. Eight common but false objections to the discontinuation of significance testing in the analysis of research data. In *What if there were no significance tests?* ed. L.L. Harlow, S.A. Mulaik, and J.H. Steiger. Hillsdale, NJ: Erlbaum, 1997, pp. 37–67.

Shapiro, S.S., and M.B. Wilk. An analysis of variance test for normality (complete samples). *Biometrika* 52:591–611 (1965).

Sheskin, D.J. *Handbook of parametric and nonparametric statistical procedures.* 3rd ed. Boca Raton, FL: Chapman & Hall/CRC Press, 2004.

Snedecor, G.W., and W.G. Cochran. *Statistical methods.* 6th ed. Ames: Iowa State University Press, 1973.

Sokal, R.R., and F.J. Rohlf. *Biometry. The principles and practice of statistics in biological research.* 2nd ed. New York: W.H. Freeman and Company, 1981.

Sokal, R.R., and F.J. Rohlf. *Biometry. The principles and practice of statistics in biological research.* 3rd ed. New York: W.H. Freeman and Company, 1995.

Steel, R.G.D. A multiple comparison rank sum test: Treatments versus control. *Biometrics* 15:560–572 (1959).

Steel, R.G.D. A rank sum test for comparing all pairs of treatments. *Technometrics* 2(2):197–207 (1960).

Steel, R.G.D. Some rank sum multiple comparison tests. *Biometrics* 17:539–552 (1961).

Stephan, C.E., and J.W. Rogers. Advantages of using regression analysis to calculate results of chronic toxicity tests. In *Aquatic toxicology and hazard assessment: Eighth symposium*, ed. R.C. Bahner and D.J. Hansen. ASTM STP 891. Philadelphia: American Society for Testing and Materials, 1985.

Stephens, P.A., S.W. Buskirk, G.D. Hayward, and C. Martínez del Rio. Information theory and hypothesis testing: A call for pluralism. *J. Appl. Ecol.* 42:4–12 (2005).

Sterne, J.A., and G. Davey Smith. Sifting the evidence—What's wrong with significance tests? *BMJ* 322:226–231 (2001).

Student (W.S. Gosset). The probable error of a mean. *Biometrika* 5:1–25 (1908).

Suter, G.W., II. *Ecological risk assessment.* Chelsea, MI: Lewis Publishers, 1987.

Suter, G.W., II, A.E. Rosen, E. Linder, and D.F. Parkhurst. Endpoints for responses of fish to chronic toxic exposures. *Environ. Toxicol. Chem.* 6:793–809 (1987).

Toft, C.A., and P.J. Shea. Detecting community-wide patterns: Estimating power strengthens statistical inference. *Am. Nat.* 122:618–625 (1983).

Trafimow, D., and S. Rice. A test of the null hypothesis significance testing procedure correlation argument. *J. Gen. Psychol.* 136(3):261–269 (2009).

Tukey, J.W. The philosophy of multiple comparisons. *Stat. Sci.* 6(1):100–116 (1991).

Ury, H.K. A comparison of four procedures for multiple comparisons among means (pairwise contrasts) for arbitrary sample sizes. *Technometrics* 18(1):89–97 (1976).

U.S. Environmental Protection Agency (EPA). *Methods for measuring the acute toxicity of effluents and receiving waters to freshwater and marine organisms.* 5th ed., EPA-821-R-02-012. Washington, DC, U.S. EPA Office of Water, 2002a.

U.S. Environmental Protection Agency (EPA). *Methods for measuring the chronic toxicity of effluents and receiving waters to freshwater organisms.* 4th ed., EPA-821-R-02-013. Washington, DC, U.S. EPA Office of Water, 2002b.

Van Dam, R. A., A. J. Harford, and M. St. J. Warne. Time to get off the fence: The need for definitive international guidance on statistical analysis of ecotoxicity data. *IEAM* 8: 242–245 (2012).

Van der Hoeven, N. How to measure no effect. Part III: Statistical aspects of NOEC, ECx and NEC estimates. *Environmetrics* 8:255–261 (1997).

Van der Hoeven, N. Power analysis for the NOEC: What is the probability of detecting small toxic effects on three different species using the appropriate standardized test protocols? *Ecotoxicology* 7:355–361 (1998).

Warne, Michael St. J., and R. van Dam. NOEC and LOEC data should no longer be generated or used. *Australian J. Ecotoxicol.* 14:1–5 (2008).

Wacholder, S., S. Chanock, M. Garcia-Closas, L. El ghormli, and N. Rothman. Assessing the probability that a positive report is false: An approach for molecular epidemiology studies. *J. Natl. Cancer J.* 96(6): 434–442 (2004).

Weber, C.I., W.I. Horning, D.J. Claim, T.W. Neiheisel, P.A. Lewis, E.L. Robinson, J. Menkeclick, and F. Kessler. *Short-term methods for estimating the chronic toxicity of effluents and receiving waters to marine and estuarine organisms.* EPA/600/4–87/028. Cincinnati: Environmental Monitoring Systems Laboratory, Environmental Protection Agency, 1988.

Weber, C.I., W.H. Peltier, T.J. Norberg-King, W.B. Horning II, F.A. Kessler, J.R. Menkedick, T.W. Neiheisel, P.A. Lewis, D.J. Klemm, Q.H. Pickering, E.L. Robinson, J.M. Lazorchak, L.J. Wymer, and R.W. Freyberg. *Short-term methods for estimating the chronic toxicity of effluents and receiving waters to freshwater organisms.* EPA/600/4–89/001. Cincinnati, OH: Environmental Monitoring Systems Laboratory, Environmental Protection Agency, 1989.

Williams, D.A. A test for differences between treatment means when several dose levels are compared with a zero dose control. *Biometrics* 27:103–117 (1971).

Williams, D.A. The comparison of several dose levels with a zero dose control. *Biometrics* 28:519–531 (1972).

Wright, S.P. Adjusted P-values for simultaneous inference. *Biometrics* 48:1005–1013 (1992).

Zar, J.H. *Biostatistical analysis.* 4th ed. Upper Saddle River, NJ: Prentice Hall, 1999.

Ziliak, S.T., and D.N. McCloskey. Significance redux. *J. Socio-Econ.* 33:665–675 (2004).

Ziliak, S.T., and D.N. McCloskey. *The cult of statistical significance. How the standard error costs us jobs, justice, and lives.* Ann Arbor, MI: University of Michigan Press, 2008.

CHAPTER 6

Population and Metapopulation Effects

Pollutants matter because of their effects on populations, and so, indirectly, on communities too, but pollutants act by their effects on individual organisms.

—Moriarty (1983)

The populational nature of ecological toxicology is of fundamental importance.

—Bezel and Bolshakov (1990)

The major task of applied ecotoxicological studies is to provide information on potential deleterious effects of chemicals on populations and ecosystems.

—Laskowski (2001)

The plethora of ecological problems generated by both industrial and domestic contaminants calls for an effective means to reliably assess population- and community-effects.

—Emlen and Springman (2007)

The primary goal of most ecotoxicologists is ensuring persistence and vitality of populations within ecological communities.

—Newman (2010)

6.1 GENERAL

6.1.1 Population Level Research

The importance of populations as pointed out in the above quotes dictates that more research addressing questions at the population level is essential to sound environmental stewardship. The wealth of information that currently exists about effects on individuals also suggests the advantage of more insightful linkage of effects measured at the individual level to those at the population level (Calow and Sibley 1990; Forbes et al. 2001; Raimondo and McKenny 2006). The current unreliable direct extrapolation of individual-based effect metrics to population consequences needs to be deemphasized as argued by an increasing number of ecotoxicologists (e.g., Calow et al. 1997; Chandler et al. 2004; Forbes and Calow 1999; Forbes et al. 2001, 2008; Jensen et al. 2001; Raimondo and McKenney 2005a, 2005b; Stark and Banks 2003; Walthall and Stark 1997). As early evidence of this need, Daniels and Allan (1981) found that 48 h LC50 values for two zooplankton species exposed to dieldrin were much higher than the pesticide concentrations that actually led to a population growth rate below zero.*

Population biology and genetics are extremely rich in conceptual and mathematical models, yet despite the recent surge in studies applying molecular technologies, they still remain relatively wanting for appropriate data sets with which to apply them. Intelligent blending of these models and theories with the data-rich field of ecotoxicology holds certain promise of exciting contributions in the near future. (Equally certain during this infusion of ideas and methods will be the occasional misapplication of concepts and methods.) It is the goal of this chapter to foster appropriate application of these useful concepts and methods. Because it is impossible to provide sufficient detail on all pertinent materials in one chapter, the reader is strongly encouraged to explore the cited sources, and Chapters 12 to 19 in Newman and Clements (2008).

6.1.2 Definition of Population

A population might be defined as a group of individuals of a species occupying a defined space at a particular time (Reid 1961; Krebs 1972) or "a collective group of organisms of the same species (or some other groups within which individuals may exchange genetic information) occupying a particular space" (Odum 1971). It has emergent qualities not characteristic of the individuals that make up the population (Haddon 2001). Although the concept of population is easily grasped, indistinct temporal and spatial boundaries often make operational definition of a study population frustratingly arbitrary. And critical features of populations often include exchange of individuals within a network of subpopulations (metapopulation). Collections of subpopulations can range from a simple source-sink configuration (e.g., mainland → island), to subpopulations exchanging at varying rates among patches differing in their capacities to support the species, to a highly fragmented situation with very little exchange among subpopulations (Harrison 1992). The spatial distribution of individuals within the subpopulations can also shift through time. From the vantage of population genetics, the concept of population is given slightly different emphasis than in demography. For example, Cavalli-Sforza and Bodmer (1971) defined a Mendelian population as a population of interbreeding individuals sharing a common gene pool. Such populations may be divided into demes ("semi-isolated subpopulations with various patterns of genetic exchange or migration between them" (Hartl and Clark 1989)).† The definition of consequent genetic clines (a steady change in allele frequency across a defined region of study) and related phenomena then become germane to sound genetic studies of populations. Obviously, sound conclusions regarding effects at the population level require a clear definition of the study population.

6.2 EPIDEMIOLOGY

6.2.1 Basic Principles and Metrics

Epidemiology is the science of the cause, incidence, prevalence, temporal changes, and spatial distribution of infectious or noninfectious disease in populations. To distinguish it from the study of environmental toxicant effects on humans (environmental epidemiology), ecological epidemiology

* Although generally disregarded by practitioners of the species sensitivity distribution approach, it follows from this conclusion that any method based on collections of individual-based effects metrics has dubious value for inferring consequences to ecological communities or species assemblages (Forbes and Calow 2010; Newman and Clements 2008).

† Nee (2007) defines two kinds of metapopulations. A Levins metapopulation exists in a landscape made up of habitat patches differing in suitability of maintaining the species. Empty patches are colonized by migration, and patch populations go locally extinct as a function of patch quality and interpatch exchange. The second metapopulation is composed of subpopulations exchanging among patches without patch subpopulations going extinct. This second metapopulation is most similar to the subpopulations or demes of the population geneticist.

is the moniker given specifically to the study of disease risk to nonhuman species inhabiting contaminated locations (Suter 1993).

6.2.2 Causation

Assigning causation is a difficult task when dealing with situations having high dimensionality and minimal replication, such as a population inhabiting a contaminated habitat. The habitat could be contaminated with one or more chemicals, and the population will also be exposed to a range of factors affecting individual fitness. Rules of thumb have been established and currently are the tools of choice in many studies of disease causation in populations. Perhaps the most popular are Hill's nine aspects of disease association (Hill 1965), which are routinely covered in epidemiology textbooks (e.g., Gordis 2004). More recent and narrowly focused rules are those published by Evans (1976) and Fox (1991). All are qualitative rules of thumb founded on rudimentary concepts of causality and common sense (for more detail see Newman and Clements 2008). Their qualitative nature makes them appealing tools* but also makes them more susceptible than quantitative methods to bias and cognitive errors. Fortunately, quantitative methods for assigning probabilities to candidate causes of a disease or pathology in a population exist, as discussed in Section 6.6, explicitly formulated in Equations (1.1) to (1.5), and illustrated with Example 1.1 of Chapter 1. Equation (5.41) and associated text are also relevant to the example below. Discussions in earlier chapters will be expanded here to include a specific example and exploration of Bayesian factors.

Expanding on our previous applications of Bayesian equations, the previous relationships for competing hypotheses can be expressed specifically as the odds for competing causes or explanations for some outcome, such as death or presence of a pathological condition in individuals in a population. In the case of two possible causes (see Equations 1.4, 1.5, and 5.41),

$$\frac{P(H_1 \mid D)}{P(H_2 \mid D)} = \frac{p(H_1)}{p(H_2)} \cdot \frac{p(D \mid H_1)}{p(D \mid H_2)} \tag{6.1}$$

where H_1 and H_2 = the hypothesized causes 1 and 2, respectively, and D = the data or evidence. $p(H_1 \mid D)/p(H_2 \mid D)$ is the posterior odds of the two competing hypotheses conditional on the observed data; $p(H_1)/p(H_2)$ is the prior odds of the two hypotheses, and $p(D \mid H_1)/p(D \mid H_2)$ is the likelihood ratio for D given H_1 or H_2. $(p(D \mid H_1)/p(D \mid H_2))$ is also called a Bayes factor in this context (Gelman et al. 1997; Goodman 1999; O'Hagan 2006) and is used to generate the posterior odds for discrete hypotheses (e.g., H_1, H_2) when multiplied by the prior odds. A Bayes factor reflects the support provided to the hypotheses by the observed data. Bayes factors are being used with increasing frequency as alternatives to statistical test p values.

Evidence or data of different kinds can be combined in this approach. The posterior odds derived by multiplying the prior odds by the data-derived Bayes factor can then be used as a new prior odds that can be combined with another Bayes factor generated with additional data or evidence. This process can be repeated to produce posterior odds until a data-based judgment can be made about the relative plausibilities of the competing causal hypotheses. The process can be generalized to several competing causal hypotheses if warranted (see Equation 1.5).

* Perhaps these qualitative methods are also favored due to motivations like those underlying one of the symptoms of pathological science, that is, aversion to high-risk testing of a favored scientific hypothesis (Rousseau (1992) in Chapter 1). Here the aversion is to applying quantitative methods to assign causality that might put one's favored causal theory at more risk than would a less explicit qualitative set of rules. Regardless, Popper's conclusions (Popper (1968) in Chapter 1) that quantitative measurement is generally superior to qualitative methods would seem to hold true for causal inference as well. Bayesian quantitative methods also allow direct application of Chamberlin's method of multiple hypotheses (see Chapter 1).

Example 6.1

A colony of endangered grebe nests is monitored for 48 days each year on a small lake that has high background levels of aryl hydrocarbon receptor (Ahr)-mediated toxicants (dioxins and furans) from diffuse sources. The chance of a chick dying during the 48 days after hatching due to these Ahr-mediated toxicants is 1 in 10. A tanker truck accident spills industrial solvent into the lake resulting for 1 day in concentrations high enough to kill one in eight chicks from the lake population. Fortunately the solvent drops to sublethal concentrations after that day. (The one in eight estimate was derived from published studies of a past solvent spill and a similar grebe species that inhabits a lake in this same industrialized region.) Ornithologists studying the endangered species document a grebe chick death during the day of the spill. Before fining the trucking company, the regulatory authorities ask, "How likely is it that the spilled solvent, and not the high background levels of dioxins and furans, caused the death of the endangered grebe chick?" They understand that a "preponderance of evidence," and not a "beyond a reasonable doubt," context is relevant when seeking monetary compensation.

Prior odds of death under the two scenarios (spilled solvent or high levels of Ahr-mediated pollutants) need to be estimated with the information provided above. There is a 1 in 10 chance of a chick dying due to the high background level of high Ahr-mediated pollutants. There is a one in eight chance of chick death in the presence of the background Ahr-mediated pollutants plus the spilled solvent. So the chance of death due to the spilled solvent alone is the following:

$$\frac{1}{8}-\frac{1}{10}=\frac{5}{40}-\frac{4}{40}=\frac{1}{40}$$

Given this and the fact that 4 out of 40 chicks would die due to the high Ahr-mediated pollutants, 1 of 5 chicks dying would be a consequence of the spilled chemical and 4 of 5 chicks dying would be a consequence of the high levels of Ahr-mediated pollutants.

The timing of the chick death is useful information too. The chance of the chick dying the day of the spill if the Ahr-mediated pollutants caused the death was 1 day/48 days, but the chance of the chick dying that day if the solvent caused the death was 1 day/1 day. The posterior odds for the two causes (H_1 = solvent, H_2 = Ahr-mediated pollutants) can be calculated with information on the baseline odds of dying from the two causes and the timing data.

$$\frac{P(H_1 \mid D_{Base\ Rate\ 1}, D_{Timing\ 1})}{P(H_2 \mid D_{Base\ Rate\ 2}, D_{TTiming\ 2})}=\frac{p(H_1 \mid D_{Base\ Rate\ 1})}{p(H_2 \mid D_{Base\ Rate\ 2})}\cdot\frac{p(D_{Timing\ 1} \mid H_1, D_{Base\ Rate\ 1})}{p(D_{Timing\ 2} \mid H_2, D_{Base\ rate\ 2})}$$

$$\frac{P(H_1 \mid D_{Base\ Rate\ 1}, D_{Timing\ 1})}{P(H_2 \mid D_{Base\ Rate\ 2}, D_{Timing\ 2})}=\frac{1/5}{4/5}\cdot\frac{1/1}{1/48}=\frac{12}{1}$$

So the data suggest that the probabilities are 12/13 and 1/13 that death occurred from the solvent exposure or Ahr-mediated pollutants, respectively. The calculated probability from these data that the solvent caused the chick death was 12/13, or 0.92. The strength of evidence is substantial relative to a preponderance of evidence requirement for legal action ($p > 0.5$), although the addition of more data could modify this state of belief. (This example was derived directly from one in Lane et al. 1987 for ascertaining the probability of an effect being caused by an adverse drug reaction. The reader is referred to that publication and Hutchinson and Lane 1989 for more background and detail.)

The computational approach just described would be extremely useful in instances such as the recent confused and acrimonious assessment of the cause of fishkills along the mid-Atlantic coast (Stow and Borsuk 2003; Borsuk et al. 2004; Belousek 2004). The interested reader is directed to books by Almond (1995), Pearl (2000), and Neapolitan (2004), and publications by Garbolino and

Taroni (2002) and Uusitalo (2007) for more details on this approach. Software packages such as Netica® (Norsys 2006) are available to facilitate these types of analyses.

6.2.3 Prevalence and Incidence

In the simplest context, prevalence is the proportion or percentage of individuals in a population with the pathological condition at a particular time, and incidence is the number of new cases of the condition during a time period. In situations with low prevalences, communication can be made clearer by expressing prevalence as the number of individuals per some standard size, such as 5 cases in 100,000 individuals, instead of a very small proportion or percentage. Ahlbom (1993) defines the incidence rate as the number of new cases in a population normalized to some standard measure of exposure time period, such as 2 years or 1,000 person-years.* One thousand individual-years might be 1,000 individuals in 1 year or 200 individuals for 5 years. Estimates might even involve many individuals followed for different time durations. Incidence describes the number of previously normal individuals in the at-risk population that acquire the pathological condition during a specified period. An incidence rate might be expressed as the number of new cases per 1,000 individual-years. Incidence is one measure of risk in a specific at-risk population or subpopulation.

In this simplest of forms, Woodward (2005) provides the following estimates of prevalence and incidence rate that are normalized to the population size at mid-year of a study:

$$Prevalance = \frac{Number\ of\ Individuals\ with\ the\ Condition\ at\ Mid\text{-}year}{Mid\text{-}year\ Population\ Size} \tag{6.2}$$

$$Incidence = \frac{Number\ of\ New\ Cases\ that\ Year}{Mid\text{-}year\ Population\ Size} \tag{6.3}$$

The prevalence can be estimated from the incidence and the duration of interest if the disease dynamics are relatively constant through time, prevalence = incidence × duration. The estimated incidence rate $(\hat{I})$ is the following:

$$\hat{I} = \frac{N}{T} \tag{6.4}$$

where N = the number of individuals with the condition, and T = the total time the population was exposed expressed in individual-years, or simply individuals during a specified period of time. Obviously, the computations are correct only if time is included consistently. The error for the number of individuals with the pathological condition (N) is assumed to conform to a Poisson distribution (Equation 6.5) characterized by the single parameter, λ. The assumption of a Poisson distribution means that the distribution mean and variance for N are identical, as will be discussed in more detail later in this chapter. The probability of a specific number of individuals with the pathology (i) is estimated as follows:

$$P_i = \frac{\lambda^i e^{-\lambda}}{i!} \quad \text{or} \quad e^{-\lambda}\left(\frac{\lambda^i}{i!}\right) \tag{6.5}$$

* Given the subject of this book, the common phrasing *person-years* will be replaced by *individual-years* here.

where P_i = the probability of i individuals with the pathology and λ = the Poisson parameter reflecting the expected mean rate. The estimated N would be $\hat{I} \cdot T$ according to Equation (6.4). The variance is the following:

$$\sum_{0}^{\infty} \frac{\lambda^i e^{-\lambda}}{i!} i^2 - \lambda^2 \tag{6.6}$$

One method for generating a 95% confidence interval for the estimated number of cases is suggested by Equation (6.5) (Ahlbom 1993). This equation is used iteratively as follows to get the upper (Equation 6.7) and lower (Equation 6.8) 95% confidence limits. The λ values are changed in these equations until the two sides of each equation are close enough to be considered equivalent.

$$\sum_{i=0}^{N} \frac{e^{-\lambda_{upper}} \lambda_{upper}^{i}}{i!} = 0.025 \tag{6.7}$$

$$\sum_{i=N}^{\infty} \frac{e^{-\lambda_{lower}} \lambda_{lower}^{i}}{i!} = 0.025 \tag{6.8}$$

Some of the tedium of this approach might be alleviated by using the Excel® POISSON function to do the associated calculations. The 0.025 on the right of the equations is changed if another confidence interval is desired, e.g., 0.05 for a 90% two-sided confidence interval. The asymmetry of the Poisson distribution decreases as N increases, so Ahlbom (1993) suggests that a normal approximation method be used instead if N is in the range of 15 to 20 or higher. One consequence of the Poisson-based approach is that a simple table of 95% confidence limits can be generated for N up to 20 to avoid the tedium of iteratively solving the equations. Table 6.1 does this, but if needed, Ahlbom (1993) and Beyer (1991) provide more extensive tabulations.

Another consequence is that the simpler computations provide an approximate 95% confidence interval based on a normal distribution if N is greater than 20.

$$Number\ of\ Individuals\ with\ the\ Pathology = \hat{N} \pm 1.96\sqrt{\hat{N}} \tag{6.9}$$

Confidence limits for $\hat{I}$ can be generated by dividing both confidence limits for N by T. A very similar approach can be used to produce confidence limits for prevalence (Ahlbom 1993). In that case, a binomial distribution is assumed instead of a Poisson distribution when applying the iterative approach described above. The normal approximation for 95% confidence limits is the following (Ahlbom 1993; Woodward 2005):

$$Prevalance = \frac{\hat{N}}{N_{total}} \pm 1.96\sqrt{\frac{\frac{\hat{N}}{N_{total}}\left(1 - \frac{\hat{N}}{N_{total}}\right)}{N_{total}}} \tag{6.10}$$

Often estimated spatial or temporal differences in incidences ($\widehat{IRD}$) or prevalences ($\widehat{PD}$) become the focus of epidemiological analyses, requiring calculation of their confidence intervals (Ahlbom 1993; Woodward 2005).

Table 6.1 95% Confidence Intervals for a Poisson Distribution with N = 0 to 20

N	Lower Limit	Upper Limit
0	0.0	3.7
1	0.1	5.6
2	0.2	7.2
3	0.6	8.8
4	1.1	10.2
5	1.6	11.7
6	2.2	13.1
7	2.8	14.4
8	3.4	15.8
9	4.1	17.1
10	4.8	18.4
11	5.5	19.7
12	6.2	21.0
13	6.9	22.3
14	7.7	23.5
15	8.4	24.7
16	9.1	26.0
17	9.9	27.2
18	10.7	28.4
19	11.4	29.7
20	12.2	30.9

$$\widehat{IRD} = \frac{\hat{N}_1}{T_1} - \frac{\hat{N}_2}{T_2} \pm Z_{\alpha/2}\sqrt{\frac{\hat{N}_1}{T_1^2} + \frac{\hat{N}_2}{T_2^2}} \tag{6.11}$$

$$\widehat{PD} = P_1 - P_2 \pm Z_{\alpha/2}\sqrt{\frac{\frac{\hat{N}_1}{N_{total\ 1}}\left(1 - \frac{\hat{N}_1}{N_{total\ 1}}\right)}{N_{total\ 1}} - \frac{\frac{\hat{N}_2}{N_{total\ 2}}\left(1 - \frac{\hat{N}_2}{N_{total\ 2}}\right)}{N_{total\ 2}}} \tag{6.12}$$

Example 6.2

Mulvey and coinvestigators published a series of papers describing the population genetic effects of PAH contamination on fish (*Fundulus heteroclitus*) in the Elizabeth River (Virginia) (Mulvey et al. 2002, 2003; Ownby et al. 2002). The prevalence of liver cancer and other lesions was high in some subpopulations (Vogelbein 2010). Unpublished liver lesion data from the survey described by Mulvey et al. (2002, 2003) and Ownby (2002) included data on preneoplastic altered hepatocelluar lesions, including the following information (site identifications identical as those given in the above publications):

AW site (high PAH): 52 of 57 sampled fish had altered loci
CH site (low PAH): 2 of 60 sampled fish had altered loci

$$Prevalance_{AW} = \frac{52}{57} \pm 1.96\sqrt{\frac{\frac{52}{57}\left(1 - \frac{52}{57}\right)}{57}} = 0.91 \pm 0.07$$

$$Prevalance_{CH} = \frac{2}{60} \pm 1.96\sqrt{\frac{\frac{2}{60}\left(1-\frac{2}{60}\right)}{60}} = 0.03 \pm 0.04$$

$$\widehat{PD} = 0.91 - 0.03 \pm 1.96\sqrt{\frac{\frac{52}{57}\left(1-\frac{52}{57}\right)}{57} - \frac{\frac{2}{60}\left(1-\frac{2}{60}\right)}{60}} = 0.88 \pm 0.06$$

The 95% confidence interval for $\widehat{PD}$ does not overlap with 0, suggesting a large and material difference in lesion prevalence between these two locations. The PAH concentrations (ng/g dry sediment) were in the range of 2,000 to 7,000 at the CH site and 200,000 to 350,000 at the AW site (Mulvey et al. 2002, 2003).

Populations can be compared using differences in these metrics; however, ratios of incidence rates can also be useful in this regard. Often the rate ratio (*RR*) is produced with one incidence being that for a reference population ($\hat{I}_0$),

$$\widehat{RR} = \frac{\hat{I}_1}{\hat{I}_o} \tag{6.13}$$

The Ln of *RR* is often estimated along with its associated confidence limits (6.15),

$$Variance\ of\ Ln(RR) \approx \frac{1}{N_1} + \frac{1}{N_0} \tag{6.14}$$

$$Ln\ \widehat{RR} \pm Z_{\alpha/2}\sqrt{\frac{1}{N_1} + \frac{1}{N_0}} \tag{6.15}$$

The antilogarithms of the Ln $\widehat{RR}$ and its confidence limits can be used to express results in arithmetic units.

The presence of a pathological condition in a sampled group can also be expressed as the odds of having the pathology (odds = number with pathology/number without pathology), giving rise to the odds ratio as another metric for comparing disease in two groups. The odds ratio (OR) for two groups experiencing a pathological condition suspected as being related to exposure to some factor is based on the numbers of sampled individuals with the condition who were (N_a) and were not (N_b) exposed to the factor, and those without the condition who were (N_c) and were not (N_d) exposed to the agent,

$$\widehat{OR} = \frac{\frac{N_a}{N_b}}{\frac{N_c}{N_d}} \quad \text{or} \quad \frac{N_a N_d}{N_b N_c} \tag{6.16}$$

A large OR suggests association between the agent and pathological condition. The Ln of OR is used to express results of such studies because it often conforms to a normal distribution. It (Equation 6.17) and associated confidence intervals (Equation 6.19) can be estimated as follows (Ahlbom 1993; Sahai and Khurshid 1996; Woodward 2005):

$$Ln\ \widehat{OR} = Ln\frac{N_a}{N_1 - N_a} - Ln\frac{N_c}{N_0 - N_c} \tag{6.17}$$

$$Variance\ of\ Ln\ \widehat{OR} \approx \frac{1}{N_a} + \frac{1}{N_b} + \frac{1}{N_c} + \frac{1}{N_d} \tag{6.18}$$

$$Ln\ \widehat{OR} \pm Z_{\alpha/2}\sqrt{\frac{1}{N_a} + \frac{1}{N_b} + \frac{1}{N_c} + \frac{1}{N_d}} \tag{6.19}$$

where N_1 (= $N_a + N_b$) and N_0 (= $N_c + N_d$) are the number of sampled individuals in the exposed and unexposed groups, respectively. The antilogarithms of the Ln $\widehat{OR}$ and its confidence limits could be used to express results in arithmetic units. Formal hypothesis test for the odds ratio might be done by the methods described in Mantel and Haenszel (1959) or Woodward (2005), including those for experiments with confounding factors.

6.2.4 Logistic Regression Models

Models involving epidemiological information are diverse, but logistic regression is a very common one due to the nature of the data just described. More specifically, binary logistic regression would often be applied in the case of the disease presence/absence data described to this point. The natural logarithm of the odds of the pathology being present is related to one or more risk factors.

$$Ln\left[\frac{p_{pathology}}{1 - p_{pathology}}\right] = a + bx \tag{6.20}$$

where x is some factor potentially influencing $p_{pathology}$, and a and b are estimated parameters. The situation might involve two groups differing in whether or not they were exposed to a chemical. It might also involve individuals whose exposure was characterized by some continuous function. Several risk factors might incorporated into an expanded form of Equation (6.20). Interpretation and useful statistics emerging from logistic regression differ depending on such qualities.

Example 6.3

The liver lesion data used in Example 6.2 can be extended to include logistic regression to determine the potential change in risk with changes in PAH concentrations in sediments. Mulvey et al. (2002, 2003) did not report in the literature liver lesion information generated by coinvestigator W. Vogelbein's team, but that information for 16 Elizabeth River sites is used to illustrate logistic regression here. The prevalence of altered loci in the *F. heteroclitus* liver is examined with the sum of PAH in sediments being a possible explanatory variable. The PAH concentrations measured in duplicate composite sediment samples from each site were averaged and then summed as per Horness et al. (1998). This concentration was normalized to sediment organic carbon content to yield units of mg/kg (dry sediment)/(total organic carbon content/100). The model is that shown by Equation (6.20),

$$Ln\left[\frac{p_{Lesion}}{1 - p_{Lesion}}\right] = a + b(PAH) + \varepsilon$$

as fit using the following SAS code:

```
DATA LESION; /* AF2 = PRESENCE IN ANY OF ALL ALTERED FOCI TYPES         */
   INPUT SITE $ A N PAH @@; /* A = NUMBER WITH LESION & N = SAMPLE SIZE*/
   DATALINES;
   AW 52 57 4153 CC 0 60 19 CF 0 60 283 CH 58 60 178 CI 56 60 14
   CS 37 60 2503 JC 0 60 456 KY 0 60 12 NC 9 60 420 NM 0 60 100
   PC 0 60 101 PP 0 60 68 QY 1 60 19 RN 5 60 231 RS 51 60 1197
   SC 5 60 254
   ;
   RUN;
ODS GRAPHICS ON; /* ROC PLOT FILES SAVED TO THE PROGRAM DIRECTORY       */
        /* BINARY LOGISTIC MODEL FOR LESION AF2 VERSUS SEDIMENT PAH     */
PROC LOGISTIC DATA=LESION PLOTS=ROC(ID=PROB);
     MODEL A/N = PAH/CLPARM=PL CLODDS=PL ALPHA=0.05;
     OUTPUT OUT=PAH PREDICTED=PHAT LOWER=PLOW UPPER=PUP;
RUN;
PROC MEANS DATA=PAH NOPRINT;
   VAR PAH PLOW PHAT PUP;
   BY SITE;
OUTPUT OUT=PAH2 MEAN=PAH PLOW PHAT PUP;
RUN;
PROC PRINT DATA=PAH2;
   VAR SITE PAH PLOW PHAT PUP;
   RUN;
PROC GPLOT DATA=PAH2;
     PLOT PLOW*PAH PHAT*PAH PUP*PAH/OVERLAY;
RUN;
ODS GRAPHICS OFF;
```

The output was the following:

```
                    Model Fit Statistics
                                          Intercept
                             Intercept          and
             Criterion            Only   Covariates
             AIC              1148.134      953.118
             SC               1152.997      962.845
             -2 Log L         1146.134      949.118

            Analysis of Maximum Likelihood Estimates
                                 Standard          Wald
Parameter    DF    Estimate         Error    Chi-Square    Pr > ChiSq
Intercept     1     -1.5866        0.0956      275.3937        <.0001
PAH           1     0.00102      0.000092      121.7106        <.0001

                    Odds Ratio Estimates
                          Point          95% Wald
            Effect     Estimate     Confidence Limits
            PAH           1.001      1.001       1.001

       Parameter    Estimate    95% Confidence Limits
       Intercept     -1.5866      -1.7774        -1.4024
       PAH           0.00102     0.000844        0.00121
```

The results including the ROC curve are also shown in Figure 6.1. The AIC suggests that the addition of PAH to the model resulted in a better model than one including only the intercept.

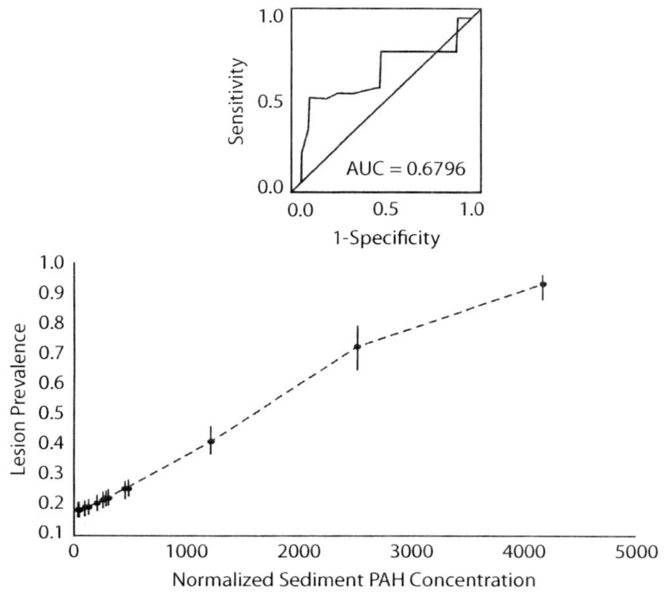

Figure 6.1 Logistic regression model for liver lesions in Elizabeth River *F. heteroclitus* including sediment PAH concentration as an explanatory variable.

Although the coefficient for PAH (0.00102; 95% confidence interval, 0.00084 to 0.00121) was statistically significant, the ROC AUC was only 0.68, which, using the general interpretative scheme from Chapter 5 (Example 5.13), indicates only fair predictive value for this PAH concentration-based model.

The metric and methods described in this short section have been employed successfully in a wide range of ecotoxicological studies. Of course, their customary use is in human disease study, as in the case of thyroid cancer incidence in subpopulations resulting notionally from the Chernobyl nuclear accident (Ron 2007). But prevalence surveys for histopathological markers such as those in fishes (Horness et al. 1998) are also common. Another area of study in which these techniques were applied to advantage involves the impact of the antifouling agent tributyltin (TBT). This antifouling agent has become widely distributed and, unfortunately, is an endocrine disruptor at very low concentrations. The worldwide appearance of imposex, especially in prosobranch mollusks, has been carefully documented using these metrics, e.g., Bryan and Gibbs (1991), Barreiro et al. (2001), de Wolf et al. (2004), Oehlmann et al. (2007), Chan et al. (2008), Sousa et al. (2009), Swennen et al. (2009), and Cheung et al. (2010).

6.3 POPULATION SIZE

6.3.1 General

Determining the effects of environmental factors on population size is an integral part of ecology. Liebig's law of the minimum* (Liebig 1840) is one excellent example that linked nutritional

* Population size is limited by some essential factor in its environment that is scarce relative to the amounts of other essential factors.

requirements of individuals to limits on population size. Shelford's law of tolerance* (Shelford 1911, 1913) spawned innumerable comparisons of laboratory-derived tolerance limits for species to population size or distribution in the field. Refinement of such concepts included the development of tolerance polygons that incorporated acclimation into the pertinent environmental factors.

Concepts linking individual tolerances to population distributions and sizes remain central to ecotoxicological predictions for exposed populations. Laboratory studies provide tolerance limits for various toxicants within a pertinent milieu, and these limits are used with field estimates of bioavailable toxicant to explain differences in population size among contaminated and uncontaminated sites.

6.3.2 Measurement of Population Size

6.3.2.1 General

Size is perhaps the most commonly measured field population attribute. It is often expressed as either relative abundance or density, although size estimates can also be complete counts of all individuals in a population. Relative abundances are expressed in terms such as catch per unit effort, annual number of individuals on a typical wintering ground, or sightings per Christmas bird count. They have arbitrary units of relative population size and, as such, are often called population indices instead of population size estimates (Nichols 1992). Densities are expressed as number of individuals or biomass per unit of space (area or volume). Densities may be absolute (estimates of the population size in a given area) or relative (estimates of relative densities among populations). The number of individuals of a particular species per cm^2 of Hester-Dendy sampler surface would be an example of a relative density, for example. Further, densities may be crude or ecological. Crude densities are numbers per unit of space and ecological densities are numbers per unit of suitable habitat space.

Techniques used to determine the size of field populations are categorized as quadrat, mark-recapture, or removal-based methods. Quadrat methods predict population size with density estimates for sample plots (quadrats) within the population area and the overall area that the entire population occupies. Mark-recapture (capture-recapture, tag-recapture, or Peterson) methods employ a precensus marking and release of individuals followed by recapture. The proportion of the total number captured during the census that were marked individuals is used to estimate the population size. Removal methods involve multiple resamplings without replacement and use the decrease in capture rate to estimate population size. Removal-based estimates may involve few or many resamplings. Multiple resampling methods fit a line to catch per unit effort versus total catch-to-date data and extrapolate the total catch to date downward to a catch per unit effort of 0.

6.3.2.2 Quadrat Estimates

The total area (or volume) of interest is divided into subunits or quadrats. The subunits are chosen of an appropriate, albeit arbitrary, size, e.g., 1 m by 1 m square plots along a transect. Quadrat methods are most conveniently applied to small, sedentary organisms that are easily counted and unlikely to avoid "capture" in the quadrat during counting (Pielou 1974). Of course, methods such as aerial photography of a nesting colony of birds can make the approach useful for some mobile species as well. To generate reliable counts within quadrats, three conditions should be met (Krebs 1972). First, the population within each quadrat should be known accurately. Second, the quadrat area and total area occupied by the population must be known. Third,

* The tolerance of individuals of a species over an environmental gradient or series of environmental gradients can determine the species' geographical distribution or sizes of local populations.

the quadrats must be representative of the whole area of interest. The third condition will be discussed in more detail in Section 6.4. Wiegert's (1962) and Hendricks' (1956) methods for estimating optimal quadrat size are described with examples by Krebs (1999). Although providing software to apply both methods, Krebs (1999) states his preference for Wiegert's method because it makes fewer simplifying assumptions.

Using the formulations outlined by Pielou (1974), the total number of individuals in the population is N and the number of quadrats (sampling units) within which the entire population is contained is Y. The mean number of individuals per quadrat ($\bar{N}$) is obviously N/Y. This $\bar{N}$ is estimated as the sum of the number of individuals counted in each quadrat (Σn) divided by the number of quadrats counted (y) or $\bar{n} = \Sigma n/y$. The predicted, total population size ($\hat{N}$) is then $Y\bar{n}$. The precision of this estimate depends on the number of quadrats counted and the evenness of the distribution of individuals among the quadrats. The variance associated with this estimate is calculated as the following (Pielou 1974):

$$Variance\ of\ \hat{N} = \frac{Y^2}{y(y-1)}\left[1-\frac{y}{Y}\right]\left[\sum_{i=1}^{y} n_i^2 - \frac{\left(\sum_{i=1}^{y} n_i\right)^2}{y}\right] \tag{6.21}$$

Equation (6.21) is appropriate if, during random selection of quadrats, a quadrat could only be selected once (i.e., sampling without replacement). Pielou (1974) replaced $1 - y/Y$ with 1 in Equation (6.21) if a quadrat can be picked for sampling more than once (sampling with replacement).

The standard error (square root of the variance, or $s_{\hat{N}}$) of the total population size, associated degrees of freedom ($df \le y - 1$ with $df = y - 1$ if the y is specified prior to the census), and a t statistic from a function such as the Excel TINV() function or a table such as Appendix 15 ($\alpha = 0.05$, $p = 1$ column) can be used to generate a 95% confidence limit for the population size estimate. Note that the probability argument in Excel TINV() is that for a two-tailed t-distribution.

$$95\%\ Confidence\ Interval:\ \hat{N} \pm t s_{\hat{N}} \tag{6.22}$$

Pielou (1974) also presents estimates of population density. As mentioned, the estimate of the average population density ($\hat{N}_a$) is $\bar{N}$. The average population density has a variance defined by Equation (6.23).

$$Variance\ of\ \hat{N}_a = \frac{1}{Y^2}[Variance\ of\ \hat{N}] \tag{6.23}$$

The 95% confidence limits of the mean population density ($\hat{N}_a$) are defined by Equation (6.24).

$$95\%\ Confidence\ Limit:\ \hat{N}_a \pm t\hat{N}_a \tag{6.24}$$

Example 6.4

A population of clams on a beach was decimated after an oil spill. To monitor the reestablishment of the population, a baseline estimate of the decimated population size is obtained for a 500 m^2 area of beach. The number of clams in each of 50, randomly selected quadrats (1 m × 1 m squares) is counted. Each quadrat is sampled without replacement.

Clams/Quadrat	Frequency of Quadrat Count
0	12
1	24
2	6
3	3
4	2
5	1
6	1
7	1
≥8	0

a. Estimation of $\hat{N}$:

$$\bar{n} = \frac{\sum n}{y} = \frac{24 + 12 + 9 + 8 + 5 + 6 + 7}{50} = 1.42 \frac{clams}{m^2 \ quadrat}$$

$$\hat{N} = \bar{n} \cdot Y = 1.42 \frac{clams}{m^2} \cdot 500 \ m^2 = 710 \ clams$$

b. Estimation of associated variance and 95% confidence interval:

$$\begin{aligned}
\text{Variance} &= [500^2/(50 \cdot 49)] \cdot [1 - (50/500)] \cdot [217 - (5{,}041/50)] \\
&= [102.0408] \cdot [0.9] \cdot [116.1800] \\
&= 10{,}669.59 \\
\text{Standard error} &= 103.29 \\
\text{95\% confidence limits} &= 710 \pm t \cdot 103.29 \\
&= 710 \pm 2.0086 \cdot 103.29 \\
&= 710 \pm 207 \\
\text{95\% CI} &= 494 \text{ to } 908 \text{ clams}
\end{aligned}$$

c. Estimation of $\hat{N}_a$:

$$\hat{N}_a = \bar{N} \approx \bar{n} = 1.42 \frac{clams}{m^2 \ quadrat}$$

d. Estimation of associated variance and 95% confidence interval:

$$\begin{aligned}
\text{Variance} &= \left[\frac{1}{Y^2}\right] \text{Variance of } \hat{N} \\
&= [1/500^2] \cdot [10669.59] \\
&= 0.04268 \\
\text{95\% Confidence Limits} &= 1.42 \pm 0.086 \cdot 0.04268 \\
&= 1.42 \pm 0.09 \\
\text{95\% C.I.} &= 1.33 \text{ to } 1.51 \text{ clams/m}^2 \text{ quadrat}
\end{aligned}$$

Krebs (1999) provides software that implements similar computational methods. The Excel Add-in PopTools also implements these calculations under its Sampling → Jolly-Seber Method options.

6.3.2.3 Mark-Recapture Estimates

Often the characteristics of a species or its habitat make quadrat methods inappropriate or impractical. For example, placing open quadrats over an area of pond and counting the number of sunfish within each quadrat generates meaningless results because of the avoidance behavior of this mobile species. Often, mark-recapture methods can be used in such situations. In a precensus trapping, individuals are marked and then released so that they can enter back into the population. A census is then done by retrapping. The numbers of marked and unmarked individuals are used to estimate the total population size.

Such estimates assume several conditions (Pielou 1974; Emmel 1976; Lebreton et al. 1993; Efford et al. 2004). First, marking or tagging of an individual should not alter its behavior or risk of mortality. Second, marked individuals must be scored accurately. This condition is violated if tags are lost or marks become indistinguishable. Third, marked animals should randomly mix within the population so that their likelihoods of recapture are the same as those of unmarked individuals. Estimates can be compromised if there is insufficient time between marking and recapture for the marked individuals to mix thoroughly within the population. Learned behavior associated with being trapped initially (trap addiction or avoidance) may also invalidate this condition. Finally, it is assumed that there are insignificant numbers of births, deaths, and migrants between the initial marking and recapture dates. Such a population that remains constant during sampling is referred to as a closed population.*

Again, using the formulations and notation of Pielou (1974), the intuitive Lincoln's estimator of population size is calculated from mark-recapture data according to Equation (6.25). (The Lincoln's estimator is also called the Petersen-Lincoln estimator.)

$$N = \frac{nM}{m} \tag{6.25}$$

where N = total population size (number of individuals), M = the total number of marked individuals, n = the number of individuals caught at censusing, and m = the number of marked individuals caught at censusing.

Equation (6.25) overestimates N because m is not independent of n. Various modifications of Lincoln's estimator have been generated to minimize this bias. For example, Emmel (1976) recommends Bailey's modification (Equation 6.26) as a less biased estimator of N. (Bailey 1951 provides more detail on this method.) Equation (6.28) may be used to calculate the variance for this estimate (Seber 1970).

$$\hat{N} = \frac{M(n+1)}{m+1} \tag{6.26}$$

Seber (1970), Pielou (1974), and White et al. (1982) recommend Chapman's (1951) modification, which produces an estimate with negligible bias if $m \geq 7$.

$$\hat{N} = \frac{(M+1)(n+1)}{m+1} - 1 \tag{6.27}$$

* Although not discussed, methods for open populations do exist (see Huggins et al. 2003; McDonald and Amstrup 2001; Nichols 1992; White et al. 1982; White and Burnham 1999; Williams et al. 2002). Open populations will only be discussed here in the general context of migration and metapopulation dynamics.

$$Variance\ of\ \hat{N} = \frac{(M+1)(n+1)(M-m)(n-m)}{(m+1)^2(m+2)} \tag{6.28}$$

The 95% confidence limits are the following:

$$95\%\ Confidence\ Limit = \hat{N} \pm 2\sqrt{Variance\ of\ \hat{N}} \tag{6.29}$$

Example 6.5

A question arose during a mesocosm study of trophic transfer regarding how much of a pollutant was present in a prey species (sunfish) for possible transfer to largemouth bass. An essential piece of information for the associated calculation was the estimated sunfish population size. In a precensus survey, 42 sunfish were tagged and released back into the pond. During a follow-up censusing, 17 tagged and 21 untagged sunfish were caught ($M = 42$, $n = 17 + 21 = 38$, $m = 17$).

a. N is estimated using Chapman's modification (Equation 6.27):

$$\begin{aligned}\hat{N} &= [(42+1)(38+1)]/(17+1)-1 \\ &= 92.17 \text{ or } 92 \text{ sunfish}\end{aligned}$$

b. Variance in $\hat{N}$ (Equation 6.28) and 95% confidence interval (Equation 6.29):

$$\begin{aligned}\text{Variance} &= [(42+1)(38+1)(42-17)(38-17)]/[(17+1)^2(17+2)] \\ &= [43 \cdot 39 \cdot 25 \cdot 21]/[324 \cdot 19] = 143.02 \\ \text{Standard error} &= 11.96 \approx 12 \\ 95\%\ \text{CL} &= 92 \pm 2 \cdot 12 \\ &= 92 \pm 24 \\ 95\%\ \text{CL}&: 68 \text{ to } 116 \text{ sunfish}\end{aligned}$$

Krebs (1999) software does these calculations with slightly different formulations. The Excel Add-in PopTools also does these estimations. Many studies involve more than two samplings, and associated methods are described in Nichols (1992).

6.3.2.4 Removal-Based Estimates

Numbers of individuals collected during consecutive catches may also be used to estimate population size. Individuals from previous catches are removed from the population and the decrease in catch is used to estimate size. Removal can be by physical removal or by placing some distinguishing mark on captured individuals prior to release back into the environment. For example, seines may be placed as obstructing curtains upstream and downstream of a stream segment in which one wishes to estimate the population size of a fish species. After capture, the individuals may be removed by returning them to the stream below the location of the downstream seine. They are physically removed from the population being estimated. Alternatively, turtles from a pond may be captured, counted, marked unique by shell notching, and released back into the pond. They are not counted if caught again.

The specific equations used to provide an estimate of population size and the associated variance depend on the number of times the population is sampled. Those outlined by Seber and Le Cren (1967) and Pielou (1974) for two- and three-catch samplings will be presented in detail. Then, methods based on many samplings will be discussed. Regardless of the number of catches, these methods are based on the assumption that all individuals in the closed population have the same probability of capture, and this probability does not change between periods of capture. The number of individuals captured each time is assumed to substantially decrease the size of the remaining population. The error structure is assumed to be binomial (Seber and Le Cren 1967).

For methods based on two periods of capture, the maximum likelihood estimator of population size and its variance are defined by Equations (6.30) and (6.31).

$$E(\hat{N}) = N + \frac{q(1+q)}{p^3} = N + b \tag{6.30}$$

$$Variance\ of\ \hat{N} = \frac{Nq^2(1+q)}{p^3} + \frac{2q(1-p^2-q^3)}{p^5} - b^2 \tag{6.31}$$

where $E(\hat{N})$= the expected value of the population size estimate, N = population size, p = probability of capture, q = probability of escaping capture, and b = bias.

Equations (6.32) and (6.33) are approximations based on Equations (6.30) and (6.31). (Note that the expected values of $C_1 = Np$ and $C_2 = Np(1 - p)$ or Npq.)

$$\hat{N} = \frac{C_1^2}{(C_1 - C_2)} \tag{6.32}$$

$$Variance\ of\ \hat{N} = \frac{(C_1C_2)^2(C_1 + C_2)}{(C_1 - C_2)^4} \tag{6.33}$$

where $\hat{N}$ = predicted population size (moment estimate), C_1 = the number caught during the first trapping, and C_2 = the number caught during the second trapping.

The probability of capture (p) is estimated using Equation (6.34). The probability of escaping capture (q) is $1 - p$.

$$\hat{p} = \frac{C_1 - C_2}{C_1} \tag{6.34}$$

where $\hat{p}$ = predicted probability of capture.

Equations (6.32) and (6.33) are adequate if $Np^3 > 16q^2(1 + q)$. Otherwise, the bias (b) in these approximations should be assessed (Seber and Le Cren 1967). The bias is estimated by Equation (6.35).

$$b = \frac{q(1+q)}{p^3} \tag{6.35}$$

Estimates of population size and its associated variance can be adjusted for this bias.

$$\hat{N}_{ub} = \hat{N} - b \tag{6.36}$$

where $\hat{N}_{ub}$ = unbiased, predicted population size.

$$Var_{ub} = Variance\ of\ \hat{N} - \frac{2q(1-p^2-q^3)}{p^5} + b^2 \tag{6.37}$$

where var_{ub} = unbiased variance of $\hat{N}$.

If the bias divided by the standard error (calculated by taking the square root of the variance from Equation 6.33) is greater than 0.1, Seber and Le Cren (1967) suggested that the bias not be corrected in the variance estimate.

Pielou (1974) also provides general equations for estimating N from three samplings. Obviously, the addition of more catches should improve the precision of the estimate. Her estimates of N from three catches (C_1, C_2, C_3) and the associated variance are generated with the following equations. (Note that the expected values are $C_1 = Np$, $C_2 = Npq$, and $C_3 = Npq^2$. The q and p are as defined previously.) The population size is estimated by Equation (6.38).

$$\hat{N} = \frac{C_1 + C_2 + C_3}{1 - q^3} \tag{6.38}$$

First, an estimate ($\hat{q}$) of q must be derived. Equations (6.39) and (6.40) are used to calculate Q for this purpose.

$$T = C_1 + C_2 + C_3 \tag{6.39}$$

$$Q = \frac{C_2 + 2C_3}{T} \tag{6.40}$$

The Q from Equation (6.40) is placed into quadratic Equation (6.41) to get $\hat{q}$.

$$(2-Q)q^2 + (1-Q)q - Q = 0 \tag{6.41}$$

This is done by using the following solution to the simple quadratic equation of the form given in Equation (6.42).

$$ax^2 + bx + c = 0 \tag{6.42}$$

Equation (6.42) can be solved for x.

$$x = \frac{-b \pm \sqrt{b^2 - 4ac}}{2a} \tag{6.43}$$

Letting $2 - Q = a$, $1 - Q = b$, and $-Q = c$, q can be calculated:

$$q = \frac{-(1-Q) \pm \sqrt{(1-Q)^2 - 4(2-Q)(-Q)}}{2(2-Q)} \tag{6.44}$$

Equation (6.44) can be simplified to Equation (6.45).

$$q = \frac{(Q-1) \pm \sqrt{1+6Q-3Q^2}}{2(2-Q)} \tag{6.45}$$

Although Equation (6.45) has two roots, only the probability within the range of 0 to 1 is pertinent. The $\hat{q}$ is placed into Equation (6.38) in lieu of q to provide an estimate ($\hat{N}$) of N. The variance and 95% confidence interval for $\hat{N}$ are calculated with Equations (6.46) and (6.47).

$$Variance\ of\ \hat{N} = \frac{\hat{N}T(\hat{N}-T)}{T^2 - \dfrac{9\hat{N}(\hat{N}-T)(1-\hat{q})^2}{\hat{q}}} \tag{6.46}$$

$$95\%\ Confidence\ Limit = \hat{N} \pm 2\sqrt{Variance\ of\ \hat{N}} \tag{6.47}$$

Example 6.6

Population size of trout in a small, glacial lake experiencing acidification is estimated. Adult trout are caught during consecutive samplings, marked, and returned to the lake. The lake is then resampled. The brief time between samplings and the relative isolation of the lake are judged sufficient to satisfy the assumption of a closed population. (Data used here are those simulated for $N = 625$ and $p = 0.40$ by White et al. 1982, Table 4.2, pp 104.)

Two samplings (C_1 and C_2) yielded 260 and 141 unmarked trout:

a. $\hat{N}$ is estimated with Equation (6.32):

$$\hat{N} = 2602/(260 - 141)$$
$$= 568.07, \text{ or } 568 \text{ adult trout}$$

b. Variance of $\hat{N}$ is estimated with Equation (6.33):

$$\text{Variance} = [(260 \cdot 141)^2 \cdot (260 + 141)]/(260 - 141)^4$$
$$= [1{,}343{,}955{,}600 \cdot 401]/200{,}533{,}921 = 2{,}687.46$$
$$\text{Standard error} = 51.84 \text{ or } 52 \text{ trout}$$

c. p and q are estimated with Equation (6.34):

$$\hat{p} = [260 - 141] / 260 = 0.46$$
$$\hat{q} = 1 - 0.46 = 0.54$$

d. Bias is estimated with Equation (6.35):

$$b = [0.54 \times (1 + 0.54)]/0.46^3 = 0.8316/0.0973 = 8.54$$

e. Unbiased estimate of N (Equation 6.36): If $Np^3 > 16q^2(1 + q)$, then bias correction would be made on this estimate of N. The calculated $\hat{N}$, $\hat{p}$, and $\hat{q}$ are used to test this inequality.

$$Np^3 \approx 568 \cdot 0.46^3$$
$$\approx 55.287$$
$$16q^2 \approx 4.666$$

Since the estimate of Np^3 is greater than the estimate of $16q^2$, the bias correction should be made.

$$\begin{aligned}\hat{N}_{ub} &= 568 - 9\\ &= 559 \text{ adult trout}\end{aligned}$$

f. The unbiased estimate of the variance: If b/standard error is greater than 0.1, Seber and Le Cren (1967) suggested that no bias correction be made for the variance estimate. Using the above estimates of b and standard error,

$$\begin{aligned}b/SE &\approx 8.54/51.84\\ &\approx 0.16\end{aligned}$$

No bias correction is recommended. However, for the sake of illustration in this example, Equation (6.37) and estimates of p, q, and b will be used to estimate the unbiased variance.

$$\begin{aligned}Var_{ub} &= \text{Variance of } \hat{N} - [2 \cdot 0.54 \cdot (1 - 0.46^2 - 0.54^3)/0.46^5] + 8.54^2\\ &= 2{,}687.46 - 33.08 + 72.93\\ &= 2{,}727.31\\ &\text{(Unbiased standard error} = 52.22)\end{aligned}$$

Three samplings (C_1, C_2, and C_3): Assume that three samplings were made with counts of 260, 141, and 97 unmarked trout. The methods described above can be used to provide another estimate.

Equation (6.39) is used to calculate T:

$$T = 260 + 141 + 97 = 498 \text{ adult trout}$$

Equation (6.40) is used to estimate Q:

$$\begin{aligned}Q &= [141 + 2 \cdot 97]/498\\ &= 0.6727\end{aligned}$$

Equation (6.45) is used to estimate q:

$$q = \frac{(0.6727 - 1) \pm \sqrt{1 + 60.6727 - 30.6727^2}}{2(2 - 0.6727)}$$

$$\begin{aligned}q &= 0.5992\\ &\approx 0.60\end{aligned}$$

g. $\hat{N}$ estimation: These estimates of q and T are used in Equation (6.38) to calculate the population size.

$$\begin{aligned}\hat{N} &= (260 + 141 + 97)/(1 - 0.60^3)\\ &= 635 \text{ adult trout}\end{aligned}$$

h. Variance of $\hat{N}$ is estimated with Equation (6.46):

$$\begin{aligned}\text{Variance} &= \frac{635 \cdot 498(635 - 498)}{498^2 - \dfrac{9 \cdot 635(635 - 498)(1 - 0.60)^2}{0.60}}\\ &= 43{,}323{,}510/39{,}216\\ &= 1{,}104.74\\ &\text{Standard error} = 33.24 \text{ or } 33 \text{ trout}\end{aligned}$$

The Krebs (1999) software will do similar calculations for three or more samplings, but not two samplings. The calculations are done under the mark-recapture → catch-effort models for Exploited Populations options.

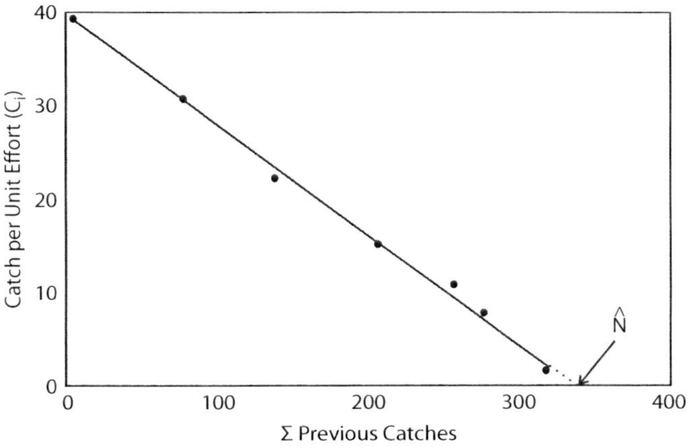

Figure 6.2 Leslie–De Lury regression method for estimation of $\hat{N}$.

With multiple catch methods, the catch per unit effort (e.g., fish caught per day) may be used in linear regression against the total number of individuals caught prior to that sampling. Extrapolation of total number caught downward to a catch per unit effort of 0 allows estimation of the total population size (x-intercept) (Figure 6.2). This method (Leslie–De Lury method) is detailed by Seber and Le Cren (1967) as follows.

Let the consecutive samplings (C_1, C_2, …, C_K or C_i) be characterized by the same probability of capture (p). Assume constant effort between removals, e.g., catch per unit effort. Plot or fit to a regression line C_i versus the sum of all catches up to, but not including, C_i.

$$C_i = Slope\left(\sum_{j=1}^{i-1} C_i\right) + Intercept \tag{6.48}$$

To this point, the assumption of equal p values between sampling periods could not be tested. This is one of the disadvantages of using only one or two samplings to estimate p. However, with the regression method, the line would be straight if p is constant between samplings ($p_1 = p_2 \ldots = p_K$). More formal methods are available for determining if p probabilities are equal (Zippin 1958; Seber and Le Cren 1967; White et al. 1982).

Example 6.7

Assume that four consecutive catches of sizes 260, 141, 97, and 50 were made from the trout population described in Example 6.6. The population size and associated probability of capture can be estimated.

Linear regression can be performed manually or with the following SAS code to fit the catch per unit effort (C_i, dependent variable) versus sum of all previous catches (independent variable). In the SAS code, CATCH is the number of trout caught during a particular sampling and TOTAL is the total number of trout caught up to that sampling. Four pairs of data are used (CATCH/TOTAL: 260/0, 141/260, 97/401 and 50/498).

```
DATA CATCH;
   INPUT CATCH TOTAL @@;
   DATALINES;
   260 0 141 260 97 401 50 498
```

```
   ;
   RUN;
PROC GLM;
   MODEL CATCH=TOTAL;
   OUTPUT OUT=PREDICT P=CATHAT R=RESID;
RUN;
```

The r^2 associated with the regression results was 0.995. The slope was –0.4149. The probability of capture can be estimated to be the slope multiplied by –1, e.g., approximately 0.41 (standard error of estimate = 0.02). This is very close to the p (0.40) set by White et al. (1982) during generation of these simulated data. The regression model can be solved for CATCH = 0 to estimate the population size.

```
CATCH = -0.4149×TOTAL + 257.2119
    0 = -0.4149×TOTAL + 257.2119
TOTAL = -257.2119/-0.4149 = 620 trout
```

The estimate of the total number of trout ($\hat{N}$ = 620) is slightly less than the actual total number of 625 trout set during generation of these simulated data.

White et al. (1982) recommend a maximum likelihood method instead of the regression method described here. More detail regarding this method and other mark-recapture or removal methods for closed populations can be obtained from Otis et al. (1978), White et al. (1982), Krebs (1989), and Skalski and Robson (1992). Several programs, including CAPTURE and TRANSECT (Laake et al. 1979) and MARK (White et al. 2006), do these calculations. Also, shareware for these purposes can be obtained from the Utah Cooperative Research Unit, Utah State University, Logan, Utah 84322. Shareware applying methods described in Krebs (1999) is available from http://nhsbig.inhs.uiuc.edu/wes/krebs.html.

6.3.3 Simple Population Growth

6.3.3.1 The Exponential Growth Model

Descriptions of population dynamics traditionally begin with density-independent exponential (also called geometric or Malthusian) growth, e.g., May (1976), Vandermeer and Goldberg (2003), Rockwood (2006), or May and McLean (2007). The unrestrained trajectory of population size (dN/dt) is expressed as a function of the population size (N) and r. The constant r is referred to as the per capita growth rate, per capita average growth rate, Malthusian constant, intrinsic rate of increase, or instantaneous rate of increase. Notice that it is growth rate expressed per individual in the population; e.g., $r = 1$ indicates one individual was added to the population per one individual in the population in the specified unit of time. It is the difference between the birth (b) and death (d) rates expressed as per unit of time. Often, r_m is used to designate the maximum rate at which a population can grow if completely unrestrained; that is, r_m is a physiological feature of individuals making up the population. The symbol r is then reserved for description of population growth that might not be completely unrestrained, e.g., measured growth in a field population experiencing logarithmic growth.

$$\frac{dN}{dt} = bN - dN = (b - d)N = rN \tag{6.49}$$

This equation is transformed to Equation (6.50) to predict population size at any time.

$$N_t = N_0 e^{rt} \tag{6.50}$$

where N_t = population size at time t, and N_0 = population size at time 0.

Note that these equations are similar to those presented in Chapter 3 for elimination and uptake. Indeed, Equation (6.50) is the negative exponential model for loss if r is negative, i.e., the population is decreasing in size. Consequently, several pertinent aspects of curve fitting will not be discussed here because a review of Chapter 3 and associated examples provide such detail.

It is important to realize that additional features such as migration can be included to these models. For convenience, let's express the relationship in terms of individuals. If B, D, I, and E are the number of individuals born, dying, immigrating, and emigrating during period N_t to N_{t+1} (Rockwood 2006),

$$N_{t+1} = N_t + (B - D) + (I - E) \tag{6.51}$$

or

$$N_{t+1} = N_t + (B + I) - (D + E) \tag{6.52}$$

In terms of the previous discussion in which b and d were used, the per individual immigration and emigration rates can be expressed as i and e, and Equation (6.53) is then relevant.

$$N_{t+1} = N_t[(b + i) - (d + e)] \tag{6.53}$$

The values of b, i, d, and e often are not constant. They can be dependent on the density of the population being modeled and also that of any population(s) linked to the modeled population by migration. (Example 6.9 describes one such case.)

Ignoring migration for the moment, Equation (6.50) may be made linear during data fitting.

$$Ln\ N_t = rt + Ln\ N_0 \tag{6.54}$$

A semilogarithmic plot (Ln N_t, t) will produce a straight line with a slope of r.

Estimates of r from Equation (6.54) may be used to calculate doubling times (the time, expressed in the same unit as the growth rate, in which the number of individuals doubles in the population, i.e., t for which $N_t/N_0 = 2$).

$$t_d = \frac{Ln\left(\frac{1}{0.5}\right)}{r} = \frac{Ln\ 2}{r} \tag{6.55}$$

If necessary, two points on the growth curve can be used to estimate growth rate. The growth during a specified period (λ) is estimated under the assumption of exponential growth. This λ is expressed per individual in the population at the time. Like r, it is the difference between the birth (b) and death (d) rates, but for the time period, not per unit of time. The λ is referred to as the finite rate of increase, although May (1976) refers to λ as the multiplicative growth factor per generation and considers the term *finite rate of increase* a misnomer. Stated concisely, $N_{t+1} = \lambda N_t$. The λ calculated from the population sizes at two times might be used to estimate r. (Recommendations given for estimation of half-life from two points in Chapter 3 are suggested here for estimation of λ with only two points.)

$$\lambda = \frac{N_t}{N_0} = e^r \tag{6.56}$$

Slightly different constants may be used with the same basic mathematics for microbial growth dynamics (Stanier et al. 1970). Microbial growth by division is also expressed as an exponential growth rate constant, k. Units are generations or divisions per unit of time.

$$N_t = N_0 2^{kt} \tag{6.57}$$

$$Ln \frac{N_t}{N_0} = kt \tag{6.58}$$

The k can be generated by determining numbers of cells at two times during exponential growth. The instantaneous or specific growth rate constant (μ in units such as day^{-1} or h^{-1}) can be calculated from k using the relationship $\mu = \mathrm{Ln}\, 2 \cdot k$. The specific growth rate expressed as μ is calculated with Equation (6.59).

$$\mu = \frac{Ln \frac{N_{t_2}}{N_{t_1}}}{t_2 - t_1} \tag{6.59}$$

Similar to Equation (6.50),

$$N_t = N_0 e^{\mu t} \tag{6.60}$$

A series of specific growth rates may be determined for some microbial population over time. The change in the maximum specific growth rates may then be used as a measure of pollutant effect, e.g., algal growth (Christensen and Nyholm 1984).

Example 6.8

Cladoceran reproduction was used by Hamilton (1986) to assess the effects of a toxicant. The cumulative number of young produced by a control female over 7 days was monitored during a static renewal toxicity test. (These data were extracted from Figure 1 of Hamilton 1986.)

Day	Cumulative Number of Young	Ln N_t (including female)
0	0	0
1	0	0
2	0	0
3	0	0
4	4	1.609
5	14	2.639
6	14	2.639
7	28	3.332

Either linear regression (Equation 6.54) or nonlinear regression (Equation 6.50) can be used to estimate r from these data. Here, an r is estimated by linear regression and used as an initial r in an iterative, nonlinear regression method.

```
DATA GROWTH;
   INPUT N T @@;
   LNN = LOG(N);
   DATALINES;
   1 0 1 1 1 2 1 3 5 4 15 5 15 6 29 7
   ;
RUN;
PROC GLM;
   MODEL LNN=T;
   OUTPUT OUT=PREDICT P=LNNHAT R=RESID;
RUN;
PROC PLOT;
   PLOT LNN*T="*" LNNHAT*T="P"/OVERLAY;
RUN;
/* NONLINEAR FIT USING THE ESTIMATE GENERATED WITH */
/* THE ABOVE LINEAR METHODS, E.G., 0.56 AS THE */
/* INITIAL ESTIMATE OF r. */
PROC NLIN DATA=GROWTH;
   PARMS R=0.56;
   BOUNDS 0<R;
   DER.R=T*EXP(R*T);
   MODEL N=1*EXP(R*T);
   OUTPUT OUT=CURVE P=NHAT R=CRESID;
RUN;
PROC PLOT;
   PLOT N*T="*" NHAT*T="P"/OVERLAY;
   PLOT CRESID*T="*"/VREF=0;
RUN;
```

The initial estimate of r from the linear regression was 0.56 ± 0.09. This estimate was used in the nonlinear regression to produce a final estimate of 0.48 ± 0.01 (asymptotic 95% confidence interval: 0.45 to 0.50). The doubling time for this clone would be approximately (Ln 2)/0.48, or 1.44 days.

6.3.3.2 The Logistic Growth Model

Populations do not continue unrestrained growth indefinitely. The equations given above are modified to steadily decrease the growth rate by some factor as the population density increases. This factor K is linked to a constant carrying capacity, the maximum population size (density) that available resources in a particular environment are capable of supporting indefinitely.

$$r_{dd} = r\left[\frac{K-N}{N}\right] \tag{6.61}$$

where r_{dd} = the population density-dependent r.

The density-dependent growth rate decreases toward 0 as N approaches K. The change in N is expressed for logistic growth by the differential Equation (6.62).

$$\frac{dN}{dt} = rN\left[\frac{K-N}{K}\right] = rN\left[1-\frac{N}{K}\right] \tag{6.62}$$

Another term (I) can be incorporated into Equation (6.62) if some factor such as a toxicant has an adverse effect on population growth that is independent of population density. The I can be expressed as a loss from the population, $E_{Toxicant} \cdot N$.*

$$\frac{dN}{dt} = rN\left[1 - \frac{N}{K}\right] - I = rN\left[1 - \frac{N}{K}\right] - E_{Toxicant}N \quad (6.63)$$

Example 6.9

Adding migration of individuals among subpopulations to the above phenomenological models suggests distinct features of contaminant influence on populations. An individual moves in a habitat mosaic to optimize acquisition of resources and overall Darwinian fitness (Grunbaum 2011; de Jager et al. 2011). The qualities of the available patches within the mosaic influence this movement in complex ways. Relevant to differences among patches created by contamination is the ideal free distribution (IFD) theory (Fretwell and Lucas 1970; Abrahams 1986). This theory holds that the probability of an individual moving from one patch to another is influenced by perceived differences in patch quality. Patches having more seemingly unexploited resources will be more attractive than those with fewer. Relative to the impact of population density on available resources, migration rates will be influenced by the size of subpopulations relative to the resources available to support the subpopulation on each patch (i.e., patch carrying capacity, K). Polluted patches could draw in many individuals from unpolluted patches if the presence of a pollutant in a habitat patch adversely impacts death and birth rates in the subpopulation so much that the available resources in the polluted patch are perceived as high by potential immigrants.

Subpopulations living in two patches that exchange individuals will be used to illustrate how migration between patches changes population dynamics (Figure 6.3). One of the patches will be polluted with consequently lower i and higher d values than the adjacent clean patch. The subpopulations are allowed to grow through time according to the simple model

$$N_{t+1} = N_t\left\{1 + [(b+i) - (d+e)]\left[1 - \frac{N_t}{K}\right]\right\} \quad (6.65)$$

For the simulations incorporating no IFD features, the migration rates (i and e) are constant. Population size projections are made for both patches (Figure 6.3, bottom panel) and the combined patches (top panel) through time. As just mentioned, these migration rates can be influenced by the perceived amount of unexploited resources in the two patches. For simplicity, these IFD-based dynamics are incorporated into the model as the following:

$$e_{polluted\ patch} = i_{clean\ patch} = e_{maximum\ for\ polluted\ patch}\left[1 - \frac{N_{t\ clean\ patch}}{K_{clean\ patch}}\right] \quad (6.66)$$

$$i_{polluted\ patch} = e_{clean\ patch} = i_{maximum\ for\ polluted\ patch}\left[1 - \frac{N_{t\ polluted\ patch}}{K_{polluted\ patch}}\right] \quad (6.67)$$

Equations (6.66) and (6.67) allow simple IFD features to be added to Equation (6.65) for both patch subpopulations. The carrying capacity was set to 1,000 and simulations began with

* This treatment of density-independent effects on population growth is directly analogous to fisheries modeling of industry take from or yield from fisheries due to human harvesting. See Haddon (2001) for more details.

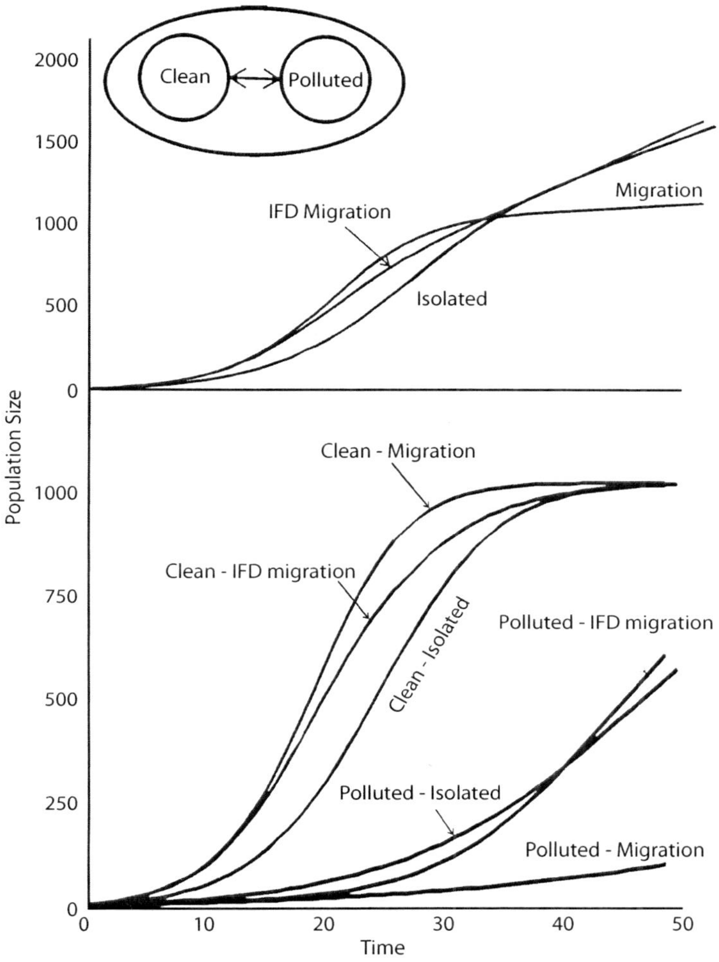

Figure 6.3 Dynamics of two subpopulations occupying either a polluted or clean patch with ("Migration" or "IFD Migration") or without ("Isolated") exchange through migration. Immigration and emigration are either constant ("Migration") or dependent on the perceived available resources in the other patch ("IFD Migration"). The top panel displays the dynamics for the sum of the number of individuals in both patches and the bottom panel displays the same of each patch ("Clean" or "Polluted") alone.

10 individuals for each patch. The i and e were set as maxima for patches if IFD features were included. The d and e rates or their maxima are higher for the polluted patch than the clean patch. The following vital rates were used:

IFD Features Not Included	IFD Features Included
Polluted	
$b = 0.3$	$b = 0.3$
$i = 0.05$	$i_{max} = 0.05$
$d = 0.2$	$d = 0.2$
$e = 0.1$	$e_{max} = 0.1$

IFD Features Not Included	IFD Features Included
Clean	
$b = 0.3$	$b = 0.3$
$i = 0.1$	$i_{max} = 0.1$
$d = 0.1$	$d = 0.1$
$e = 0.05$	$e_{max} = 0.05$

The time course of population size for the combined patches (Figure 6.3, top panel) is influenced clearly by density-dependent (IFD-driven) migration. The total number of individuals in both patches increases fastest in the model that includes IFD-based migration dynamics, but is slowest for the model with density-independent migration dynamics.

In isolation, the polluted patch population increases very slowly and the clean patch population increases to reach its carrying capacity before time = 50. The propensity of individuals in the polluted patch to leave the patch more readily than individuals coming into it from the clean patch results in a material deceleration of the polluted patch population growth. In both patches, IFD-based migration fosters faster population growth than does simple migration. With IFD migration favoring movement of individuals into patches with lower population densities, the presence of the clean patch improves population growth in the contaminated patch. The opposite was true with simple migration.

Note: This simulation is intended only to illustrate the importance of considering migration dynamics in projections of population dynamics. Different dynamics could be expected for other combinations for vital rates or formulations for IFD processes.

The N at any time is determined for populations with overlapping generations with Equation (6.64), which describes the classic logistic curve.

$$N_t = \frac{K}{1 + \left[\frac{K - N_0}{N_0}\right] e^{-rt}} \tag{6.64}$$

The logistic equations may be modified to make them more flexible. For example, Gilpin and Ayala (1973) add a parameter (θ) that allows different growth curve symmetries.

$$\frac{dN}{dt} = rN\left[1 - \left(\frac{N}{K}\right)^{\theta}\right] \tag{6.68}$$

The differential Equation (6.62) and integrated Equation (6.64) are often used for populations undergoing continuous growth with overlapping generations (May 1976; May and Oster 1976). For fitting data, an alternate form of Equation (6.64) is convenient.

$$N_t = \frac{K}{1 + e^{a - rt}} \tag{6.69}$$

The a is an integration constant defining the position of the curve relative to the origin (Pielou 1969; Krebs 1972). Odum (1971) shows that a is equal to $\mathrm{Ln}((K - N)/N)$ at time = 0. (At time = 0, $N = N_0$ so a becomes $\mathrm{Ln}((K - N_0)/N_0)$. Substitution of this for a in Equation (6.69) yields Equation (6.64).) Equation (6.69) may be converted to the linear form,

$$Ln\left[\frac{K-N}{N}\right]=a-rt \tag{6.70}$$

The K is estimated by eye from a graph of N versus time and then used to estimate $\text{Ln}((K-N)/N)$ for each t. A line is then fit to $\text{Ln}((K-N)/N)$ versus t. The slope is $-r$ and the intercept is a.

Example 6.10

The population growth of a marine amphipod on a clean sediment and sediment spiked with crude oil is monitored for 160 days. The r and K for the amphipods growing in the clean and spiked sediments are estimated. (The data below are fabricated based on those of Gause 1931.)

	Population Size (Number/Aquarium)	
Time (days)	**Clean Sediment**	**Spiked Sediment**
0	2	2
10	50	50
30	200	240
40	300	230
60	1,100	500
85	1,250	480
100	1,800	600
120	1,600	580
140	1,850	600
160	1,500	580

These data are graphed (N versus t) and estimates of K for each sediment are extracted from the graph. These estimates are then used to calculate $\text{Ln}((K-N)/N)$. Linear regression models of $\text{Ln}((K-N)/N)$ versus t were used to generate estimates of parameters a (clean: 3.00, spiked: 1.60) and r (clean: 0.04, spiked: 0.03). The estimates of K and r were then used in a nonlinear regression procedure to fit these data to Equation (6.64). The estimates and their associated asymptotic standard errors and asymptotic 95% confidence intervals are given below.

Parameter	**Clean Sediment**	**Spiked Sediment**
	r	
Estimate (SE)	0.12 ± 0.01	0.14 ± 0.01
95% CI	0.11–0.14	0.12–0.16
	K	
Estimate (SE)	1611 ± 78	561 ± 27
95% CI	1,430–1,791	500–623

The crude oil spike had a very clear effect of lowering the carrying capacity of the sediments with K dropping from approximately 1,611 to 561 after spiking. The effect on population r was not as clear.

The SAS code used to perform these calculations follows. The program listed below was used in two phases. First, the linear regression procedures (PROC GLM) were performed to estimate r. Next, the nonlinear regression procedures (PROC NLIN) were added. Estimates of K and r from the PROC GLM procedures were used as initial values in the iterative fitting procedure to produce the final parameter values.

```
DATA LOGISTIC;
   INPUT SEDIMENT $ N T @@;
      IF SEDIMENT="CLEAN" THEN K=1900;
      IF SEDIMENT="CLEAN" THEN NINIT=2;
      IF SEDIMENT="SPIKED" THEN K=650;
      IF SEDIMENT="SPIKED" THEN NINIT=2;
      Y=(K-N)/N;
      LNY=LOG(Y);
      DATALINES;
         CLEAN 2 0 CLEAN 50 10 CLEAN 200 30 CLEAN 300 40 CLEAN 1100 60
         CLEAN 1250 85 CLEAN 1800 100 CLEAN 1600 120 CLEAN 1850 140
         CLEAN 1500 160 SPIKE 2 0 SPIKE 50 10 SPIKE 240 30 SPIKE 230 40
         SPIKE 500 60 SPIKE 480 85 SPIKE 600 100 SPIKE 580 120 SPIKE 600
         140 SPIKE 580 160
      ;
   RUN;
PROC SORT;
      BY SEDIMENT;
   RUN;
PROC GLM;
      MODEL LNY=T;
      BY SEDIMENT;
   RUN;
   DATA CLEAN;
      SET LOGISTIC;
      IF SEDIMENT="CLEAN";
   RUN;
   DATA SPIKE;
      SET LOGISTIC;
      IF SEDIMENT="SPIKE";
   RUN;
   PROC NLIN DATA=CLEAN;
      PARMS R=0.04 KAY=1900;
      BOUNDS 0<R;
      BOUNDS 0<KAY<2500;
      MODEL N=KAY/(1+((KAY-NINIT)/NINIT)*EXP(-R*T));
   RUN;
   PROC NLIN DATA=SPIKE;
      PARMS R=0.04 KAY=1900;
      BOUNDS 0<R;
      BOUNDS 0<KAY<2500;
      MODEL N=KAY/(1+((KAY-NINIT)/NINIT)*EXP(-R*T));
   RUN;
```

Difference equations are used for populations with nonoverlapping generations or discrete growth such as those of many insects and annual plants. The following difference equations are commonly used for discrete populations (May 1976; May and Oster 1976; May and McLean 2007).

Growth unrestrained by population density:

$$N_{t+1} = \lambda N_t \tag{6.71}$$

Density-dependent (restrained) growth:

$$N_{t+1} = N_t e^{r(1-N_t/K)} \tag{6.72}$$

or

$$N_{t+1} = \lambda N_t[1 + aN_t]^{-\beta} \tag{6.73}$$

or

$$N_{t+1} = \frac{\lambda N_t}{1 + [aN_t]^b} \tag{6.74}$$

or

$$N_{t+1} = N_t e^{r(1-N_t/K)^\theta} \tag{6.75}$$

Equation (6.72) is the classic Ricker model (Ricker 1954) with an equilibrium N of K. Equation (6.75) is a modified Ricker or θ-Ricker model (Thomas et al. 1980) that allows more flexible growth symmetry in the model, i.e., more flexible curve shape. It is Gilpin and Ayala's (1973) modification of the logistic model that also has an equilibrium N of K. Equation (6.73) from Hassell et al. (1976) includes a and β, constants that define the density dependence of the population. The equilibrium N for Equation (6.73) is $[\lambda^{1/\beta} - 1]/a$. The similar Equation (6.74) from Watkinson (1992) has an equilibrium population size of $[\lambda - 1]^{1/b}/a$.

Data may be fit to the model described by Equation (6.72) by linear regression with the following equation (Krebs 1972). The estimated values of A and B are used to calculate r and K.

$$\frac{N_{t+1} - N_t}{N_t} = A - BN_{t+1} \tag{6.76}$$

where N_t, N_{t+1} = population densities at times t and $t + 1$, A = a constant equal to $e^{rt} - 1$, and $B = A/K$. Hassell (1974) describes similar procedures for fitting data to the model described by Equation (6.73).

These equations represent only a few of the most common models used to describe simple, density-dependent growth. The interested reader is referred to May and Oster (1976) for a tabulation of additional models. None of these equations has any clear superiority to another except for the purely historical dominance of the Ricker model. Indeed, even the general value of the logistic model for description of population growth remains in debate, e.g., Watkinson (1992), Szathmary (1991), and Ginzburg (1992, 1993).

In the equations pertinent to overlapping generations (continuous growth), the population responds instantaneously to density effects. This is not realistic in many instances. Hutchinson (1948) introduced a delay differential equation that incorporates a lag in response (decrease in population growth rate) to density effects. The change in growth rate is in response to population density at some time prior to N_t.

$$\frac{dN_t}{dt} = rN_t\left[1 - \frac{N_{t-T}}{K}\right] \tag{6.77}$$

where T is the time lag in response.

More complicated models incorporating lags are common. Krebs (1972) and Odum (1971) both present models with two lags. The first (T or reaction time lag) is that incorporated into Equation (6.77). It is the time lag before a negative effect of crowding is realized. The second (g or reproductive lag) is the time to respond (begin increasing) to a favorable change in the environment.

$$\frac{dN_t}{dt} = rN_{t-g}\left[1 - \frac{N_{t-T}}{K}\right] \tag{6.78}$$

For the equations pertinent to discrete populations (nonoverlapping generations), there is a time delay or lag implicit in their structure (May 1976). For example, the increase in N from t to t + 1 may be the change in population density per generation. Consequently, incorporation of lag terms into these simplistic models analogous to the lag terms in the continuous growth models is not required. Regardless, time delays in both the continuous and discrete logistic models strongly influence the equilibrium stability as discussed in the next section.

6.3.3.3 Pollutant Take and Sustainable Yield

Fisheries and wildlife biologists have for many years estimated maximum sustainable yields, that is, the largest number of individuals that can be harvested from a population without causing that population to eventually become locally extinct (Murray 1993). Models applied to this issue are formulated as fixed-quota or fixed-effect harvest formulations (Donovan and Welden 2002). Equation (6.63) is a straightforward modification of a fixed-effect harvest model that only shifted the context slightly from density-independent take by fishing to take by toxicant action, i.e.,

$$\frac{dN}{dt} = rN\left[1 - \frac{N}{K}\right] - I = rN\left[1 - \frac{N}{K}\right] - E_{Toxicant}N$$

The equivalent difference model would be the following:

$$N_{t+1} = N_t + rN_t\left[\frac{K - N_t}{K}\right] - E_{Toxicant}N_t \tag{6.79}$$

Donovan and Welden (2002) provide explanation of this relationship in terms of fisheries harvesting (Chapter 29, Equation 3). This particular model (Equations 6.63 and 6.79) assumes that toxicants do not materially impact natality. It predicts that the maximum sustainable yield will occur if the population size is one-half of the carrying capacity and will drop off on either side of that maximum (Beddington and Kirkwood 2007) (Figure 6.4). This is true regardless of the agent whose action is reflected in I. The qualities of some populations might produce different maxima, in which case a modification of Equation (6.63) incorporating a shape factor like that in Equation (6.68) can be used to accommodate this refinement.

$$\frac{dN}{dt} = rN\left[1 - \left(\frac{N}{K}\right)^{\theta}\right] - E_{Toxicant}N \tag{6.80}$$

Equation (11.11) in Beddington and Kirkwood (2007) can be reparameterized to produce the same for a toxicant-impacted population with discrete growth dynamics,

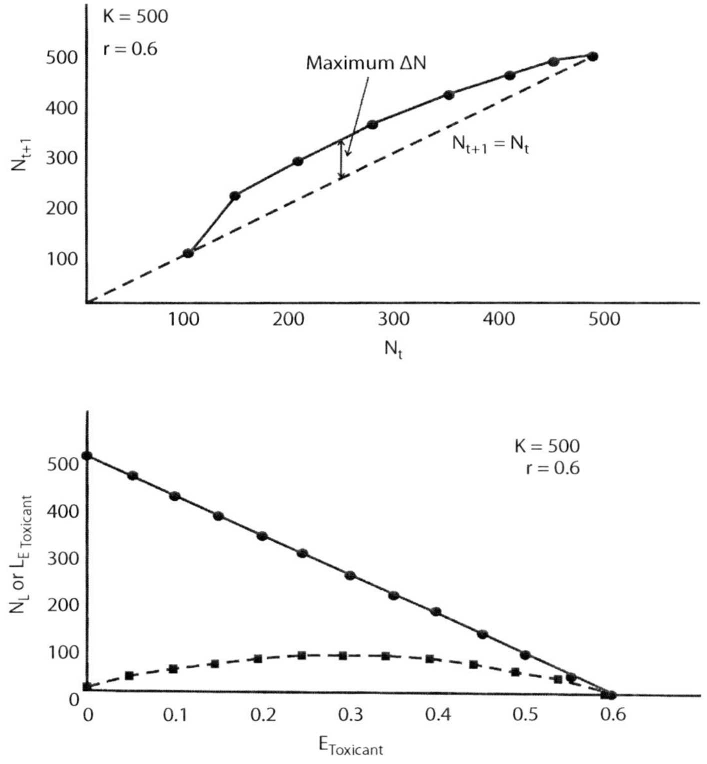

Figure 6.4 The dashed line in the top panel indicates the condition of equilibrium, that is, each individual in the population is exactly replaced by one new individual for each time step. Any N_{t+1} observation above this line denotes that more individuals are being added to the population than required for simple replacement. The maximum production of individuals in a population according to Equation (6.79) (with $E_{Toxicant} = 0$) will occur at a population size of one half the carrying capacity. The bottom panel (solid line) depicts predictions of Equation (6.82) that the conditional carrying capacity (N_L) will decrease toward 0 as $E_{Toxicant}$ approaches K. The dashed line in the lower panel depicts the number of individuals lost to the toxicant at various values of $E_{Toxicant}$ up to the $r(L_{E_{Toxicant}})$.

$$N_{t+1} = rN_t\left[1-\left(\frac{N_t}{K}\right)^{\theta}\right] - E_{Toxicant}N_t \tag{6.81}$$

Modifying Murray's (1993) formulation for fisheries take to address toxicant take (Equation 6.63), the new carrying capacity of a population experiencing I (N_L) will be the following (Newman and Clements 2008):

$$N_L = K\left[1-\frac{E_{Toxicant}}{r}\right] \tag{6.82}$$

under the condition that r is larger than $E_{Toxicant}$. This new carrying capacity, N_L, will drop to zero as $E_{Toxicant}$ approaches r. The number of individuals removed from the population due to the toxicant ($L_{E_{Toxicant}}$) is estimated with Equation (6.83). This equation multiplies the new carry capacity as defined in Equation (6.82) by the $E_{Toxicant}$ to estimate the number lost.

$$L_{E_{Toxicant}} = E_{Toxicant} K \left[1 - \frac{E_{Toxicant}}{r} \right] \tag{6.83}$$

So, these models project that the carrying capacity of an impacted population will decrease but growth rate could increase or decrease as I increases.

Although straightforward to understand when used to judge impact on experimental populations, growth rate is a useful metric for field population effect only when interpretation involves other qualities of the population. It would be questionable to simply imply population effect by only observing the change in number of individuals in a contaminated field population and comparing that to the change in numbers for a reference field population (Moriarty 1983).

6.3.3.4 Population Stability

In Chapter 1, Popper (1968) was quoted as follows: "The empirical basis of objective science has thus nothing 'absolute' about it. Science does not rest upon solid bedrock. The bold structure of its theory rises, as it were, above a swamp, but not down to any natural or 'given' base; and if we stop driving the piles deeper, it is not because we have reached firm ground. We simply stop when we are satisfied that the piles are firm enough to carry structure, at least for the time being." This quote is extremely pertinent to the topic of population stability. In the last several decades, it was necessary to drive the piles deeper into the swamp as the scientific structure above the foundation became more and more substantial.

The concept that population density always approaches a single equilibrium point is an unrealistic depiction based on expediency and history. It emerged from the simplistic balance in nature concept that has a long history in Western society (Ehrlich and Birch 1967; Pool 1989), and it remains at the foundation of many present-day arguments or decisions in ecotoxicology. For example, the concept (expectation) of an equilibrium population density permeates Sarokin and Schulkin's (1992) excellent discussion of pollution-related population disturbances. In their review, a "surprisingly meager" amount of information forced the authors to assume that any large deviation in population size implies dysfunction. This assumption is probably indefensible for many populations.

For populations displaying continuous growth, Equation (6.77) may be used to incorporate a lag in response (change in the rate of growth). Although treated as such to this point, this model does not necessarily describe a gradual growth to K. The dynamics of this model are dependent on the relative sizes of T and the characteristic return time (T_R).

The r may be used to estimate a characteristic return time, a parameter reflecting the rate at which the population will return toward K (or any analogous equilibrium density). The T_R is the inverse of r: $T_R = 1/r$. As T increases relative to T_R, population density becomes increasingly unstable (May 1976). The population density tends to overshoot and then to undershoot K as T increases relative to T_R. Figure 6.5 illustrates three general regions of population stability described by May (1976).

$0 < rT < e^{-1}$	Damped stable point
$e^{-1} < rT < 0.5\pi$	Oscillatory, damped, stable point
$0.5\pi < rT$	Stable limit cycles

Damped stable point dynamics for population density involve convergence at a single point such as K. Oscillatory, damped, stable point dynamics include swings back and forth with eventual convergence on K. A population displaying stable limit cycles will oscillate with a fixed period and amplitude about K. The cycles might involve oscillation between two or more points. (See Schaffer and Kot 1986.)

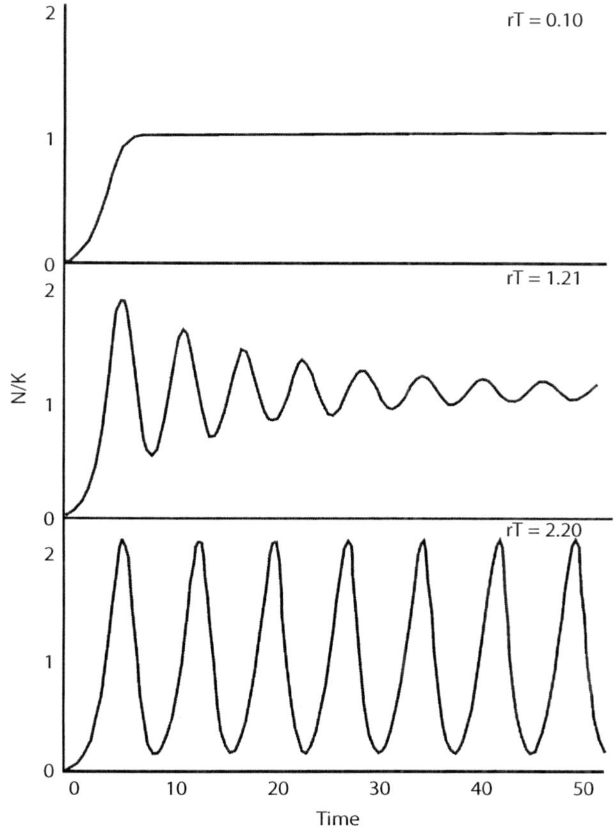

Figure 6.5 Population dynamics for a continuous growth model (Equation 6.77) with increasing values of *r*.

Discrete growth models also are capable of surprisingly complex dynamics. The dynamics of Equation (6.72) become increasingly complex as *r* increases (May 1974, 1976). May (1974) provides stability qualities for various ranges of *r* in Equation (6.72).

$0 < r < 2.000$	Stable point
$2.000 < r < 2.526$	Stable, oscillating cycle between 2 points
$2.526 < r < 2.656$	Stable, oscillating cycle among 4 points
$2.656 < r < 2.685$	Stable, oscillating cycle among 8 points
$2.685 < r < 2.692$	Stable, oscillating cycle among 16 to 64+ points
$2.692 < r$	Chaotic dynamics

Figure 6.6 illustrates the associated dynamics using N/K to scale population densities relative to the carrying capacity. Stable point (convergence to one population density) and stable cycle (stable oscillations about a point) dynamics have been previously described. Chaotic behavior is characterized as apparently random dynamics arising from a deterministic basis such as Equation (6.72) with specific parameter values. With chaotic dynamics, the population density cannot be predicted at any time in the future.

The remarkable work of May (1974) suggests that, under some conditions, it might be impossible to predict N even for simple models such as Equation (6.72). This stimulated several decades of reexamination of population dynamics. Using the model described by Equation (6.73) to fit insect

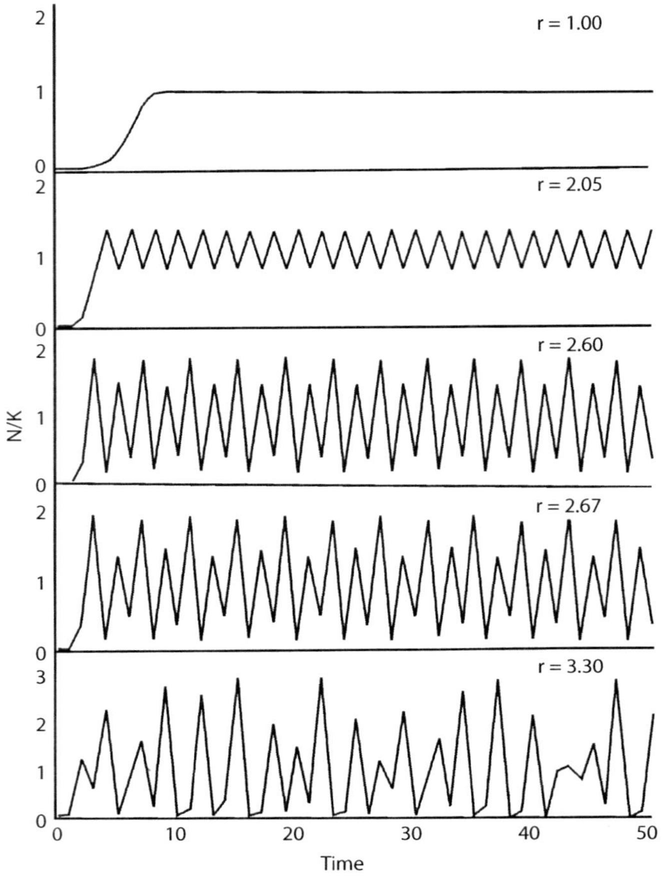

Figure 6.6 Population dynamics for a discrete growth model (Equation 6.72) with increasing values of *rT*.

population data, it was quickly discovered that *r* values begetting chaotic dynamics were unrealistically high (Hassell et al. 1976). Only one laboratory population had a sufficiently high *r* to display chaotic dynamics. This fostered scientific opinions that field populations likely do not display chaotic dynamics. Thomas et al. (1980) and Philippi et al. (1987) found no indication of chaotic behavior in laboratory populations of *Drosophila* fit to Equation (6.75) and also concluded that populations are unlikely to exhibit chaotic behavior. May (1980, 1987) suggested that the addition of more details to these models fosters chaotic behavior. Other criticisms (Renshaw 1991) focused on the loss of many qualities of chaos in population models when stochastic components are incorporated. The suggestion was that natural populations experiencing random perturbations would be unlikely to display any detectable chaotic dynamics. The substance of this criticism was strengthened by May's own conclusions (May 1989) regarding the extreme difficulty of discerning chaotic behavior in the presence of stochastic components.

This aspect of population behavior remains an exciting area of research; however, the consideration of complex dynamics has become a central theme in population ecology. Unfortunately, it has yet to be applied as conscientiously in population ecotoxicology. This is surprising because Allen (1990) suggested more than 20 years ago that periodic forcing of a population such as that associated with weather can increase the likelihood of chaos and instability. Many contaminant effects on populations share qualities similar to those described by Allen for periodic forcing by weather.

Further, the logic leading to Pool's suggestion (Pool 1989) that insect populations sprayed with pesticides may be a fertile area for detecting chaos can be applied equally well to nontarget species of interest to ecotoxicologists.* In addition to the references provided above, the interested reader is directed toward Haddon (2001), Holden (1986), Glass and Mackey (1989), Kot et al. (1988), May and Anderson (1983), May and McLean (2007), May et al. (1974), Nychka et al. (1990), Olsen and Schaffer (1990), Rockwood (2006), Schaffer and Kot (1986), and Vandermeer and Goldberg (2003) for further discussions of pertinent concepts and methods.

The stochastic and deterministic complexities of laboratory and field populations ensure that the debate regarding whether or not chaos occurs in populations will continue for many years to come. Regardless of the outcome, Robert May made a major contribution to population biology by shifting attention away from stable points, such as a constant K, and toward the variation that should be expected for populations.

6.4 DEMOGRAPHY

6.4.1 General

In our discussions to this point, a population is described simply as a group of homogeneous individuals. This is an obvious oversimplification. A population is characterized by individuals differing in age and sex. As a consequence, there will be important differences between individuals relative to their contribution of offspring to the population or likelihood of dying at any particular time. These differences will influence the dynamics and qualities of the population through time. Indeed, r and K are really summary statistics arising from the dynamics of age- and sex-structured populations. To understand populations more completely, the consequences of differences in age- and sex-specific recruitment and loss should be examined. A brief sketch of demography (the quantitative study of birth, death, migration, age, and sex within populations) will be provided here for that purpose. A simple life table approach will be presented first and then expanded to include a matrix approach to structured populations.

6.4.2 Life Tables

6.4.2.1 General

Demographers developed life tables to describe age-specific birth and death in populations. Some life tables include only mortality data (l_x schedules). Sex-specific mortality is frequently calculated. More comprehensive tables might include age-specific natality ($l_x m_x$ schedules). Often, males are not tabulated as their contributions to natality are considerably more difficult to quantify than those of females for many species.

Life tables may be developed with different types of data. Cohort (horizontal or dynamic) life tables follow a group of individuals born at the same time. The age-specific mortality and natality of all or a specific number of individuals, such as 1,000, are followed in that cohort. A dynamic-composite life table may be constructed if cohort results from a population followed during different years are composited into a single table. A time-specific (vertical, static, stationary, or current) life table examines the individuals in a population at a specific moment. The population might include many cohorts. Importantly, cohort and time-specific tables are equivalent only if birth and mortality

* Note that, as mentioned earlier, many of these equations are similar to those used for bioaccumulation kinetics. There is no reason to assume that such complex dynamics are not present under certain situations involving bioaccumulation. This might be an interesting area to be addressed in the near future.

rates are constant and independent of which cohort is contributing to population parameter estimates (Krebs 1972; Stearns 1992), e.g., the population is at equilibrium and the environment does not change between samplings.

Time intervals used to construct life tables differ among studies. Several years may be used for long-lived organisms, whereas months or weeks may be more appropriate for short-lived species. Regardless, repeated samples such as those used in cohort table construction must be taken at consistent intervals and times, e.g., annually during spawning season.

Life table analyses enjoy relatively consistent terminology; however, formulations do vary. An age interval is often designated as X. An x then becomes the subscript for a series of age-specific variables, e.g., the number of survivors that are present at the beginning of interval X is n_x and the proportion of the original group of individuals surviving to the beginning of interval $X + 1$ is l_x. The number dying during X is d_x. The mortality rate in interval X is q_x. The estimated average number of offspring produced by a female in the interval X is m_x. Specific formulations can involve quotients or numbers of individuals.

With these quantities defined for each interval, several statistics can be calculated to aid in understanding the dynamics of a population. The mean expected life span for an individual alive at age x (e_x) can be estimated with survivorship data. Rate of increase (r) and mean generation time (T_c) may be estimated with survival and natality data.

6.4.2.2 Death

Deevey (1947) describes three general survivorship curves (Figure 6.7, top panel). To a certain age, the probability of survival is high with species conforming to the type I curve. Survival rates drop drastically after that age. Human populations in developed countries conform to the type I curve. For type II curves, there is a steady decrease in survival probability through time with the rate of

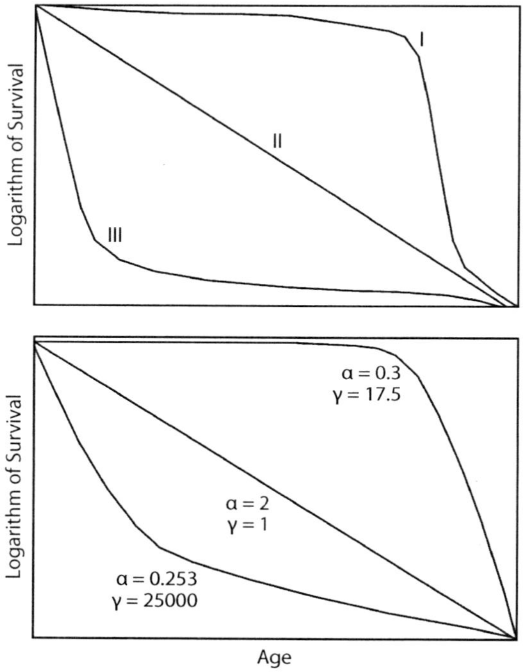

Figure 6.7 Deevey (1947) described Type I, II, and III survival curves as shown the top panel. Pinder et al. (1978) demonstrated that a Weibull distribution can be used to describe these and intergrading survival curves (bottom panel).

survival remaining constant. Many adult bird populations display such survivorship curves. With type III curves, there is high initial mortality followed by a slow decrease in survival probability. Oyster populations can display type III curves with very high mortality at early life stages. These survivorship curves and many intergrading curves can be better described by the Weibull model (Pinder et al. 1978) or other generalized exponential models, such as the gamma model (see Chapter 4 for more details). Figure 6.7 (bottom panel) illustrates the use of the Weibull model (Equation 6.84) to produce curves similar to the descriptive curves of Deevey (1947).

$$Survival\ Probability = e^{-(\alpha t)^{\gamma}} \tag{6.84}$$

The Weibull model might be fit to survivorship data and used to estimate l_x, d_x, or q_x. Fox (1989), Lebreton et al. (1993), Pinder et al. (1978), and Rago and Dorazio (1984) demonstrated the specific application of the Weibull and related models to life tables. Associated statistical methods were discussed in Chapter 4.

Count data are used directly in many life table analyses. Age-specific mortalities and life expectancies are calculated from these counts of individuals.

Example 6.10

The mortality rates and life expectancies in a fish population from a water body receiving heated effluent are required for an environmental assessment. One thousand fish were marked and their numbers monitored until there were no survivors (6 years). The survival data (n_x) were used to estimate the mortality in numbers of individuals, $d_x = n_{x+1} - n_x$. Standardized survivorship (l_x) is n_x/n_0 and age-specific survivorship (g_x) is l_{x+1}/l_x. The proportion dying of the individuals present at the beginning of the interval (q_x) is d_x/n_x. From these data, L_x was estimated as $(l_x + l_{x+1})/2$ and T_x was estimated by summing the L_x values from the table bottom up to the pertinent age class. The mean expected life span (e_x in years) was then calculated to be T_x/l_x for each age interval.

Age (years)	n_x	d_x	l_x	q_x	L_x	T_x	e_x
0–1	1,000	800	1.000	0.800	0.600	0.810	0.81
1–2	200	122	0.200	0.610	0.139	0.210	1.05
2–3	78	50	0.078	0.641	0.053	0.071	0.91
3–4	28	24	0.028	0.857	0.016	0.018	0.64
4–5	4	4	0.004	1.000	0.002	0.002	0.50
5–6	0						

Fish in the first year had a mean life expectancy (e_x) of 0.81 years. The life expectancy increased slightly for the next 2 years but then decreased until the last year of life when it was only 0.50 year.

These data could also have been analyzed using time-to-death techniques described in Chapter 4. Significant differences between survival curves could also have been judged.

6.4.2.3 Birth and Death

Natality can be added to survival tables to estimate additional population characteristics. Normally, only females are included in these tables because of the ease of determining female contributions to productivity relative to those of males. The mean number of females born per female in age class X (m_x, age-specific birth rate) is used as a measure of age-specific productivity or fecundity. Once the age-specific birth rate is estimated, the net reproductive rate (R_o) can be calculated assuming exponential growth. The net reproductive rate is the expected number of females produced during the lifetime of a newborn female. The products $l_x m_x$ are summed over all the X classes up to the final one (ω) to estimate the R_o.

$$R_0 = \sum_{X=0}^{\omega} l_x m_x \tag{6.85}$$

The mean generation time (T_c) is then estimated by dividing the sum of the $xl_x m_x$ products by R_o. (Note that the midpoint of each interval is used for x in this calculation, e.g., 0.5 for the interval 0 to 1 year old.) The instantaneous rate of increase (r) for the population is approximately Ln R_o/T_c.

$$T_c = \frac{\left[\sum_{x=0}^{\omega} x_{midpoint} l_x m_x\right]}{R_0} \tag{6.86}$$

$$r \approx \frac{Ln\ R_0}{T_c} \tag{6.87}$$

This estimate of r is only an approximation that Donovan and Welden (2002) opine is usually within 10% of that produced with the Euler correction detailed below.

Example 6.11

A cohort of female crayfish was followed for 6 years in a contaminated mesocosm. The table below summarizes the calculations necessary to estimate the R_o and T_c for these decapods.

Age (x)	l_x	m_x	$l_x m_x$	$xl_x m_x$
0–1 or 0.5	1.000	0.000	0.000	0.000
1–2 or 1.5	0.312	2.238	0.698	1.047
2–3 or 2.5	0.095	1.390	0.132	0.330
3–4 or 3.5	0.037	0.410	0.015	0.053
4–5 or 4.5	0.003	0.400	0.001	0.005
5–6 or 5.5	0.000		$\Sigma = 0.846$	$\Sigma = 1.435$

The R_o is 0.846 females produced during the lifetime of a female. The R_o is below a rate of simple replacement ($R_o = 1$). Assuming that the results were generated from a population with a stable structure, this low R_o suggests that the population size will slowly decline with time. The generation time is 1.435/0.846, or 1.70 years. The estimated r is approximately (Ln 0.846)/1.70, or –0.098.

These and more involved calculations are performed most effectively with any of a variety of shareware packages. Perhaps the most valuable and widely used at the moment is the PopTools Add-in to Excel. It can be downloaded from a variety of sites at this time, including http://www.poptools.org/. Another package written in R is popbio (Stubben and Milligan 2007). Attention to the matrix approach described below is recommended at this point to the reader contemplating a demographic study of populations exposed to pollutants. The matrix approach began with the classic publications by Leslie (1945, 1948) and later Lefkovitch (1965), and has grown steadily to become a central tool in population ecology (Caswell 2001).

Equation (6.87) provides only an approximation of r; however, the Euler–Lotka equation (Equation 6.88) may be used to determine r if the population has a stable structure. Stearns (1992) argues that moderate violations of this assumption of a stable population will have little effect on the results of using Euler's correction of the estimated r. The population is assumed to be changing exponentially, and to have constant birth and death rates.

$$\sum_{x=0}^{\omega} l_x m_x e^{-rx} = 1 \tag{6.88}$$

The Euler–Lotka equation is applied by repeated substitution of estimates of r until the left side of Equation (6.88) is sufficiently close to 1. The r estimated using the life table in Example 6.11 could be used as the first approximation of r in Equation (6.88). The value of the left side of the equation will be larger or smaller than 1. The r is adjusted slightly so that the value of the left side is closer to 1. For example, if insertion of r results in the left side of Equation (6.88) being 1.20, the r is increased slightly and used again in Equation (6.88). The process is repeated until the final r results in the solution to the left side of the equation that is close enough to 1 for the purposes of the demographic study. Programs such as PopTools provide a convenient means of doing these calculations. For example, an r of -0.137 is estimated if PopTools is used to do this with the results of Example 6.11. (The corresponding λ estimate would be $e^r = e^{-0.137} = 0.872$.)

In the above paragraph, comment without explanation was made regarding the stability of population structure. Given a specific r or λ, a population will eventually establish a constant or stable distribution of individuals among the age classes. This structure is referred to as the stable age structure of the population. This distribution of individuals between age classes can be predicted if age-specific births and death rates are relatively constant in a population with exponential dynamics. (See Donovan and Welden 2002, Krebs 1972, Rockwood 2006, Stearns 1992, or Smith 1980) who provide much more detail.)

$$C_x = \frac{\lambda^{-x} l_x}{\sum_{i=0}^{\omega} \lambda^{-1} l_i} \tag{6.89}$$

or, because $\lambda = e^r$,

$$C_x = \frac{e^{-rx} l_x}{\sum_{x=0}^{\omega} e^{-rx} l_x} \tag{6.90}$$

where C_x = the proportion of all individuals in age class x.

If the population size is constant, the stable age distribution is called the stationary or life table age distribution (Krebs 1972; Stearns 1992).

Example 6.12

Estimate the stable structure for a population of fish with an $r = 0.113$ ($\lambda = 1.12$) and the l_x values listed below.

x (years)	l_x	λ^{-x}	$\lambda^{-x}l_x$	C_x
0	1.000	1.000	1.000	0.574
1	0.513	0.893	0.458	0.263
2	0.291	0.797	0.232	0.133
3	0.074	0.712	0.053	0.030
4	0.000	0.636	0.000	0.000
			$\Sigma = 1.743$	$\Sigma = 1.000$

Respectively, 57.4, 26.3, 13.3, and 3.0% of the fish will be in the 0-, 1-, 2-, and 3-year classes.

The approach described above to estimate r for an age-structured population may be used to assess the effect of contaminants on populations. For example, Daniels and Allan (1981) calculated r for zooplankton populations exposed to pesticides. A distinct drop in the population parameter, r, was noted for both a copepod and cladoceran species when dieldrin concentrations exceeded specific levels. They argued that, with the same amount of effort, a more ecologically realistic measure of population health (r) could be generated than by standard toxicity testing techniques. Demographic analyses would be valuable additions to the conventional methods recommended by Mount and Norberg (1984) and Hamilton (1986) for analysis of zooplankton toxicity test data. Munzinger and Guarducci (1988) also found that standard methods were poor indicators of effects on populations exposed to metals and recommended demographic methods as being superior. Similarly, Pesch et al. (1991) examined the effects of sediment contamination on λ for laboratory populations of the polychaete, *Neanthes arenaceodentata*. In using this approach to compare populations under various contamination regimes, the precautions of Rago and Dorazio (1984) regarding the distributions to be expected for λ should be kept in mind. They point out that the distributions of λ are often skewed and normal statistics could be inappropriate for their analysis.

The contribution that a female makes to the population depends on the combination of her potential production of young and her survival probabilities at different age classes. Natality estimates alone do not provide optimum insight about a female's contribution. To better understand this quality, the expected number of offspring produced by a female of age class X during her lifetime (R_x) can be calculated. The R_x adjusts m_x based on the probability of survival to age y, l_y/l_x. Stearns (1992) calls this the reproductive value of a stationary population.

$$R_x = \sum_{y=x}^{\omega} \frac{l_y}{l_x} m_y \tag{6.91}$$

The R_x for age class 0 is the net reproductive rate. The R_x formulation could be changed to get the expected number during the rest of her lifetime (Donovan and Welden 2002),

$$R_x = \sum_{y=x+1}^{\omega} \frac{l_y}{l_x} m_y \tag{6.92}$$

Fisher's reproductive value better reflects the expected reproduction from the female's current age onward, which can be estimated with the more involved equation (Stearns 1992; Lanciani 1998; Rockwood 2006)

$$V_x = \frac{\sum_{y=x}^{\omega} e^{-ry} l_y m_y}{e^{-rx} l_x} \quad \text{or} \quad \frac{\sum_{i=0}^{\omega} \lambda^{-1} B_{x+1}}{S_x} \tag{6.93}$$

where S_x and B_{x+1} = the number of females surviving to age X and total number of daughters born to all females at age $X + 1$. Fisher (1930) introduced it as a measure of "impact on long-term population dynamics of individuals living at the same time."

Example 6.13

The R_x and V_x will be calculated for a fish population with the qualities tabulated below.

x (years)	l_x	m_x	R_x (Eq. (6.91))	V_x
0	1.000	0.000	2.130	0.695
1	0.513	3.005	4.160	1.899
2	0.291	2.018	2.035	1.008
3	0.074	0.068	0.068	0.034
4	0.000	0.000		

R_x: $R_0 = ((1/1)0) + ((0.513/1)3.005) + ((0.291/1)2.018) + ((0.074/1)0.068) = 2.130$
$R_1 = ((0.513/0.513)3.005) + ((0.291/0.513)2.018) + ((0.074/0.513)0.068) = 4.160$
$R_2 = ((0.291/0.291)2.018) + ((0.074/0.291)0.068) = 2.035$
$R_3 = ((0.074/0.074)0.068) = 0.068$

The R_0 indicates that each new female will more than replace herself during her life. The PopTools Add-in to Excel was used to do the more involved computations for V_x. In PopTools, you begin by entering the following matrix into a spreadsheet with m_x in the top row and l_{x+1}/l_x in the subdiagonal.

0	3.005	2.018	0.068
0.513	0	0	0
0	0.567	0	0
0	0	0.254	0

The V_x estimates (left eigenvector) were calculated by PopTools with matrix methods described next. From the PopTools menu, select Matrix Tools → Basic Analysis (check option "Scale Eigenvectors such that <wv> = 1?"). Select the matrix and then an Excel cell to receive the output. Results: $V_0 = 0.695$, $V_1 = 1.899$, $V_2 = 1.008$, and $V_3 = 0.0338$.

Additional features and further explanation of R_x and V_x are available in the substantial demography literature. For example, V_x has meaning relative to natural selection and life history strategies (Stearns 1992). The interested reader is directed to Cushing and Yicang (1994), Grafen (2006), Lanciani (1998), Pianka and Parker (1975), and Chapter 4 of Rockwood (2006) for other formulations.

Despite its obvious value, reproductive value remains underutilized for predicting potential age-dependent effects on populations from contaminated sites. (Barnthouse et al. 1987 do use a similar approach.) For example, a toxic event that left alive most of the 3-year-old fish described in Example 6.13, but none of the younger fish, would have a devastating impact on the population because of the low V_A for 3-year-old fish. Based on this type of analysis, Petersen and Petersen (1988) question the common assumption that because young organisms are more sensitive than older organisms to toxicants, there will be a large effect on a population via loss of young. Clearly, V_x should be explored when making such judgments. In the next section, elasticity analysis will be discussed as a potentially better means of doing such analyses.

6.4.2.4 Other Considerations

In this discussion of demography, several important factors were ignored out of necessity. Obviously, the sex ratio is important. With age-specific sex ratios, males can be incorporated more fully into life table analyses. Emmel (1976) and Smith (1980) describe the general effect of age-specific sex ratios on demography. Migration can also contribute to demographic qualities of populations. More complex models, including those involving genetic components, incorporate migration.

6.4.3 Matrix Methods

6.4.3.1 General Concepts and Basic Methods

Populations can be envisioned as either age- or stage-structured (Figure 6.8). Only age-structured population analyses have been described to this point. Age-structured models specify the probability of living to the next age (P, estimated by l_{X+1}/l_X) and natality (F). As will be discussed in detail, this age-specific information can be placed into a matrix with F_x in the top row and P_x in the matrix subdiagonal. Stage-structured population analyses assign individuals to different life cycle stages instead of ages. Examples might include seed-seedling-juvenile-adult plant, caterpillar-pupae-butterfly, and nestling-fledgling-subadult-adult bird. Stage-specific models specify survival probability (now denoted G_x) and natality (F_x), and also the probability of remaining at a stage during a time step (P_x). This information can also be placed into a matrix as shown at the top of Figure 6.8.

For the rest of this section, matrix methods will be applied to age- and stage-structured population data. The PopTools shareware will be used for this purpose. The reader interested in learning more about PopTools after reading this section should download the help tool for the current PopTools release for a modest fee from the link given on www.poptools.org/download, i.e., http://scinergy.hoodwards.com/. Relevant features and operations involving matrices and vectors are reviewed in Appendix 27 for readers either unfamiliar with or forgetful of matrix algebra.

6.4.3.2 Leslie Matrix Age-Based Approach

More than half century ago, Leslie (1945, 1948) took natality and survival probabilities from life tables and placed them into a matrix. In the resulting Leslie matrix (**L**), the probability (P_x) of a female alive in age class X being alive to enter age class $X + 1$ is placed in the matrix subdiagonal, $P_x = l_{x+1}/l_x$. The numbers of daughters (F_x) born in the time interval x to $x + 1$ per female in that age class were placed in the top row of the Leslie matrix. The only conditions required for these elements were that $0 < P_x < 1$ and $F_x \geq 0$. The other matrix elements of **L** were zeros.

$$L = \begin{bmatrix} 0 & F_1 & F_2 & F_3 \\ P_0 & 0 & 0 & 0 \\ 0 & P_1 & 0 & 0 \\ 0 & 0 & P_2 & 0 \end{bmatrix} \tag{6.94}$$

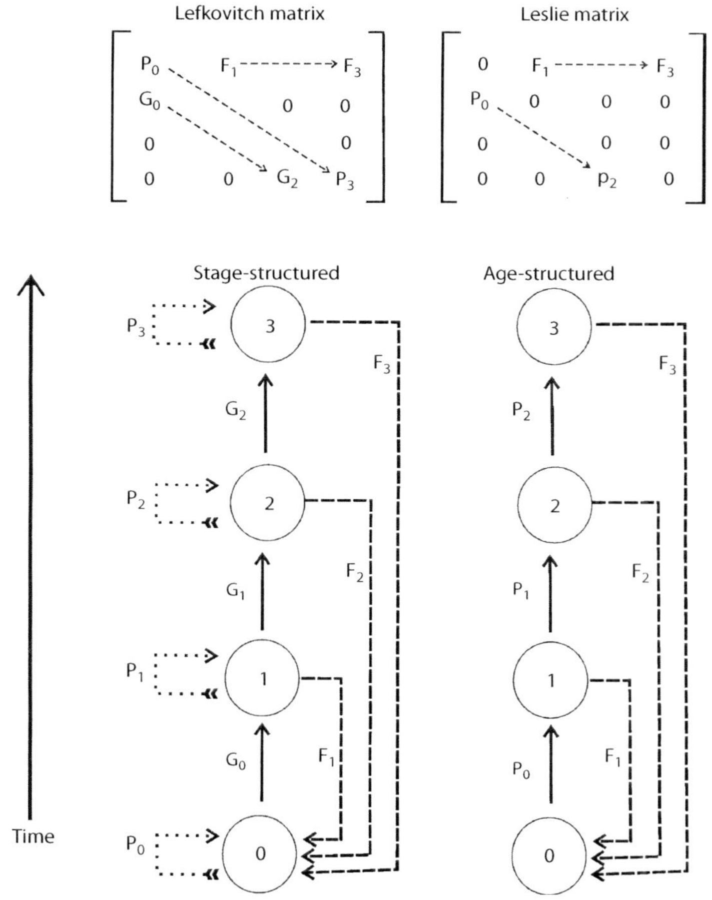

Figure 6.8 Life history diagrams of age- and stage-structured populations. The matrices used to analyze each type are shown above in each diagram.

An initial (time t) population state vector can be multiplied by the Leslie matrix to produce another vector containing the estimated number of individuals in each age class at time $t + 1$. This process can be repeated to generate a series of vectors of numbers of individuals in each age class through time.

$$\begin{bmatrix} 0 & F_1 & F_2 \\ P_0 & 0 & 0 \\ 0 & P_1 & 0 \end{bmatrix} \begin{bmatrix} n_{0,t} \\ n_{1,t} \\ n_{2,t} \end{bmatrix} = \begin{bmatrix} n_{0,t+1} \\ n_{1,t+1} \\ n_{2,t+1} \end{bmatrix}$$

$$\begin{bmatrix} 0 & F_1 & F_2 \\ P_0 & 0 & 0 \\ 0 & P_1 & 0 \end{bmatrix} \begin{bmatrix} n_{0,t+1} \\ n_{1,t+1} \\ n_{2,t+1} \end{bmatrix} = \begin{bmatrix} n_{0,t+2} \\ n_{1,t+2} \\ n_{2,t+2} \end{bmatrix}$$

Etc.

An example using the rules explained in Appendix 27 is the following:

$$t=0:\begin{bmatrix} 0 & 1 & 4 \\ 0.5 & 0 & 0 \\ 0 & .25 & 0 \end{bmatrix}\begin{bmatrix} 200 \\ 0 \\ 0 \end{bmatrix}=\begin{bmatrix} 0\cdot 200+1\cdot 0+4\cdot 0 \\ 0.5\cdot 200+0\cdot 0+0\cdot 0 \\ 0\cdot 200+0\cdot 0+0\cdot 0 \end{bmatrix}=\begin{bmatrix} 0 \\ 100 \\ 0 \end{bmatrix}$$

$$t=1:\begin{bmatrix} 0 & 1 & 4 \\ 0.5 & 0 & 0 \\ 0 & .25 & 0 \end{bmatrix}\begin{bmatrix} 0 \\ 100 \\ 0 \end{bmatrix}=\begin{bmatrix} 0\cdot 0+1\cdot 100+4\cdot 0 \\ 0.5\cdot 0+0\cdot 100+0\cdot 0 \\ 0\cdot 0+0.25\cdot 100+0\cdot 0 \end{bmatrix}=\begin{bmatrix} 100 \\ 0 \\ 25 \end{bmatrix}$$

Example 6.14

Population projection methods based on a Leslie matrix can be demonstrated with the information in Example 6.11. In that example, a crayfish population was analyzed with a life table. The table in that example is reproduced here with changes in column headings and computation of P_x.

Age (x)	l_x	F_x	$P_x = l_{x+1}/l_x$
0–1 or 0.5	1.000	0.000	0.312/1.000 = 0.31200
1–2 or 1.5	0.312	2.238	0.095/0.312 = 0.30449
2–3 or 2.5	0.095	1.390	0.037/0.095 = 0.38947
3–4 or 3.5	0.037	0.410	0.003/0.037 = 0.08108
4–5 or 4.5	0.003	0.400	
5–6 or 5.5	0.000		

This information can be entered as Leslie matrix elements. A starting population can be defined with a vector $\mathbf{v}_0$.

$$L=\begin{bmatrix} 0 & 2.238 & 1.390 & 0.410 & 0.400 \\ 0.312 & 0 & 0 & 0 & 0 \\ 0 & 0.30449 & 0 & 0 & 0 \\ 0 & 0 & 0.38947 & 0 & 0 \\ 0 & 0 & 0 & 0.08108 & 0 \end{bmatrix}$$

$$v_0=\begin{bmatrix} 0 \\ 10 \\ 10 \\ 10 \\ 10 \end{bmatrix}$$

Matrix **L** might be placed into an array of cells A1:E5 of an Excel spreadsheet and the initial population state vector $\mathbf{v}_0$ in A7 to A11. PopTools is invoked from the Add-in selection of the Excel top menu bar. From the PopTools menu bar, "Matrix Tools" and then "Matrix Projection" are selected. The matrix and state vector cells are entered as requested, and the cell to the top left corner of the array of output cells is selected, e.g., A13. The number of time steps is specified, e.g., 10, and the "Go" button is clicked to generate the projected number of individuals in each age class for the specified number of time steps. The following lines will be generated (with the leftmost totals being added as the rounded sum of the values in the generated cells to its left).

Time	Stage 1	Stage 2	Stage 3	Stage 4	Stage 5	Total
0	0	10	10	10	10	40
1	44.38	0	3.0449	3.8947	0.8108	52
2	6.153558	13.84656	0	1.185897	0.315782	22
3	31.60113	1.91991	4.216139	0	0.096153	38
4	10.19565	9.859553	0.584593	1.64206	0	22
5	23.55151	3.181044	3.002135	0.227682	0.133138	30
6	11.43875	7.348071	0.968596	1.169242	0.01846	21
7	18.2781	3.56889	2.237414	0.377239	0.094802	25
8	11.28977	5.702769	1.086691	0.871406	0.030587	19
9	14.64281	3.522408	1.736436	0.423234	0.070654	20
10	10.49858	4.568556	1.072538	0.67629	0.034316	17

If conditions specified in the Leslie matrix persist, the population will slowly decline through time. This was also the conclusion in Example 6.11.

The population information derived earlier from life tables can also be obtained from a Leslie matrix. The left and right eigenvectors are the reproductive values (R_x values) and stable age structure of the population. The eigenvalue for the right eigenvector is an estimate of the population growth rate, λ or e^r. The eigenvectors and eigenvalues can be calculated step by step with Excel as explained very clearly by Donovan and Welden (2002). However, the details will not be shown here based on the assumption that most readers will elect to use software instead of doing the laborious calculations. PopTools and the crayfish population described in Example 6.11 will be used in Example 6.15 to illustrate this approach.

Example 6.15

We can begin this example with the Leslie matrix and initial population state vector placed into an Excel spreadsheet in Example 6.14. Start PopTools, select "Matrix Tools," and then "Reproductive Value." Enter the Leslie matrix cells and also the cell that will be the topmost one for the vector of output, e.g., I13. Click "Go" to produce the following reproductive value estimates.

Reproductive Value
0.151887
0.451406
0.258298
0.072888
0.065521

These estimates are normalized to add up to 1. The same process can be repeated to generate a vector of the stable age structure by simply selecting "Age Distribution" instead of "Reproductive Value." Excel cell I20 might be chosen as the top cell for the output.

The age structure estimates are also normalized to add up to 1.

Age Structure
0.667811

```
Age Structure

0.224701
0.073786
0.030992
0.00271
```

The following menu selection from PopTools will produce the eigenvalue λ that is an estimate of the finite population growth rate from the Leslie matrix, "Matrix Tools" and "Finite Rate of Increase." The Leslie matrix is selected and cell I27 chosen to produce an estimate of 0.927.

There is also a simple PopTools "Basic Analysis" option under "Matrix Tools" that generates these and several other population metrics. (Age structure and reproductive value are expressed so that they both sum to 100% with this option.) It produces the following output for the Leslie matrix.

OUTPUT PANEL A

Eigenvalues		Eigenvectors (R&L)	
Real	Imaginary	Age/stage struct	Reprod val
0.927262	0	66.8%	15.2%
-0.03109	-0.11314	22.5%	45.1%
-0.03109	0.113143	7.4%	25.8%
-0.12742	0	3.1%	7.3%
-0.73765	0	0.3%	6.6%

OUTPUT PANEL B

r	-0.075519409	(rate of increase)
Ro	0.846677204	(expected number of replacements)
T	2.203880618	(generation time - time for increase of Ro)
mu1	2.196050121	(mean age of parents of offspring of a cohort)

OUTPUT PANEL C

```
R (expected lifetime production)
          0.846677204  2.713708987  1.562313991  0.442432  0.4
```

These estimates can be compared to those produced in Example 6.11 based on simple life table computations. The simple λ produced with the "Finite Rate of Increase" and also "Basic Analysis" options were identical (0.927). For the "Basic Analysis" output, it can be found at the top of the OUTPUT TABLE A eigenvalues which is the position of the dominant eigenvalue. The approximate r in Example 6.11 (before application of the Euler-Lotka correction) was lower (−0.098) than that from PopTools (−0.076 in OUTPUT PANEL B), which was estimated as the natural logarithm of the dominant eigenvalue, Ln(0.927) = −0.0758. The expected lifetime production, R_0, estimates were identical (0.847 in first column of OUTPUT PANEL C). The generation times were slightly different for the simple life table (1.70) and matrix technique (2.20 in OUTPUT PANEL B).

Previously the issue was broached about deciding which stage- or age-specific vital rate might be the most critical relative to the impact of environmental toxicants. The common assumption is that early/young individuals will be the most sensitive of all life history stages; however, this does not mean that the most sensitive stage of an individual is the most critical to maintaining population viability. To make that kind of judgment validly, it must be determined how many changes to each feature of an individual's life history impacts overall population growth rate. Sensitivity and elasticity analysis facilitate such an assessment (Caswell 2001). To begin, the right (**w**) and left (**v**) eigenvectors are estimated. Sensitivity of λ to changes in life stage vital rates can be assessed with **w** and **v**. Sensitivity is defined by Equation (6.95) (de Kroon et al. 1986; Caswell 2001; Caswell et al. 2004), where v_i and w_j are elements of **w** and **v**,

$$\frac{w_j v_i}{\mathbf{vw}} \tag{6.95}$$

Sensitivities calculated for survival and reproduction will have different units, making comparison of the relative importance of the various elements in a sensitivity matrix difficult. To resolve this problem, Caswell (2001) introduced elasticities that have the same units for survival and reproduction. Elasticity is the "rate of change in the log of λ with respect to the log of an element of [**L**]" (Vandermeer and Goldberg 2003):

$$e_{ij} = \frac{\partial(Ln\ \lambda)}{\partial(Ln\ p_{ij})} = \frac{p_{ij}}{\lambda}\frac{\partial\lambda}{\partial p_{ij}} = \frac{p_{ij}}{\lambda} v_i w_j \tag{6.96}$$

where p_{ij} = the element of interest such as survival of a specific age class. Because the sum of all of the elasticities for the entire matrix is 1, "e_{ij} is the proportional sensitivity of λ to changes in p_{ij}" (Vandermeer and Goldberg 2003), as will be illustrated in Example 6.16 for the crayfish population used in the last several examples.

Example 6.16

Elasticity analysis will be done for the crayfish population Leslie matrix using the PopTools Add-in to Excel. The same matrix used in the last examples will be used again. The "Matrix Elasticity" option in "Matrix Tools" is selected. The Leslie matrix (A1:E5) is selected in addition to a cell (e.g., G30) to be the top left corner of the output array. The following elasticity matrix is generated:

0	0.367148	0.07488	0.009277	0.000791
0.4521	0	0	0	0
0	0.084948	0	0	0
0	0	0.010068	0	0
0	0	0	0.000791	0

The highest elasticities (0.4521 and 0.3671) were associated with survival of 0- to 1-year-old females and natality of 1- to 2-year-old females. These demographic features were the most crucial to determining the change in growth rate of the population. Roughly 82% (that is, 100(0.45 + 0.37)) of the value of λ was determined by these two demographic features.

Only a few ecotoxicological applications of sensitivity and elasticity have been published to date. Examples include Jensen et al. (2001), Tanaka and Nakanishi (2001), Salice and Miller (2003), Forbes and Calow (2010), and Forbes et al. (2010).

6.4.3.3 Lefkovitch Stage-Based Approach

As shown in Figure 6.8, demographic formulations can involve structures based on life stage instead of age (Caswell 2001; Donovan and Welden 2002; Vandermeer and Goldberg 2003). The matrix approach, like that of the Leslie matrix, is applied to stage-structured populations. The matrix is called a Lefkovitch matrix (Lefkovitch 1965). The Lefkovitch matrix contains information about fertility for each stage (F), survival probability from one stage to the next (G), and the probability of an individual remaining at a particular stage (P):

$$\begin{bmatrix} P_0 & F_1 & F_2 & F_3 \\ G_0 & P_1 & 0 & 0 \\ 0 & G_1 & P_2 & 0 \\ 0 & 0 & G_2 & P_3 \end{bmatrix} \tag{6.97}$$

Like the Leslie matrix approach to age-structured populations, population projections can be made by multiplying the Lefkovitch matrix by a population state vector. The λ and other population metrics can also be estimated with a Lefkovitch matrix. Like the Leslie matrix approach, the calculations tend to be tedious if done manually, so programs such as PopTools are routinely applied to analyze and make projections.

6.5 SPATIAL DISTRIBUTION OF INDIVIDUALS

6.5.1 General

To this point, it has been assumed that individuals were randomly distributed within the area or volume containing the population. However, deviations from such randomness lead to very different behaviors at the population level. This is particularly true for sessile or sedentary species existing within relatively large areas (Pielou 1969). Consequently, the distribution of individuals within a population or study area can be a critical quality to assess. This section will present general methods applicable to this task. The associated descriptions and terminology are extracted from Pielou (1969, 1974) and Ludwig and Reynolds (1988). The reader is encouraged to review these and similar references for more detail. Various software will be applied here, including the shareware that complements Krebs's textbook (Krebs 1999) and associated software written by John Brzustowski. (See http://nhsbig.inhs.uiuc.edu/wes/krebs.html for explanation.)

The three general distribution patterns are uniform, random, and clumped (aggregated, clustered, or patchy) (Figure 6.9). A uniform distribution of individuals within an area is the least common pattern. More likely, individuals will be randomly distributed or clumped. Further, clumps themselves may be distributed in any of these three patterns. Patterns can also differ in intensity and grain (Figure 6.10). A coarse-grained pattern has large gaps between large clusters; e.g., the pattern in Figure 6.10A has a more coarse grain than that of Figure 6.10B. A high-intensity pattern has large differences in densities, e.g., the pattern in Figure 6.10A has a higher intensity than that of Figure 6.10C. In Figure 6.10, the distribution shown in panel A has both coarser grain and higher intensity than that in panel D.

Pielou (1969) distinguishes between patterns associated with arbitrary sampling units and discrete habitat sites or sampling units (SUs). Examples of patterns for discrete sampling units are the pattern of crayfish numbers under each of 1,000 rocks, that for number of individuals for an insect species per stream snag along a length of stream, or that for the distribution of individuals for a particular parasitic species among 200 hosts. The sampling units (rocks, stream snags, and hosts) in which the individuals are distributed are discrete. In contrast, units of space may be arbitrary if the individuals are contained within a continuous habitat, e.g., the number of clams per m^2 of sediment surface or number of a species of zooplankton per m^3 of water. Approaches used for testing and describing spatial patterns are not the same for discrete and arbitrary sampling units.

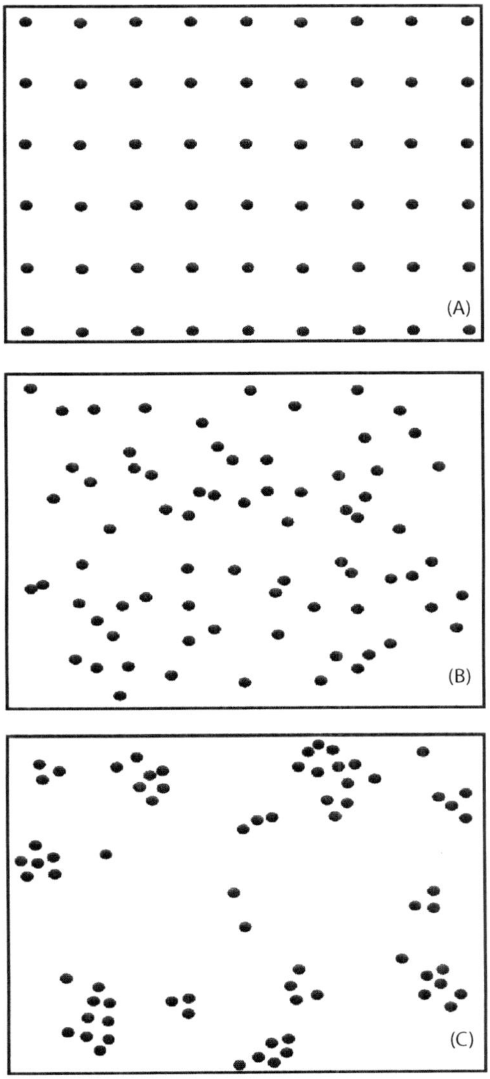

Figure 6.9 Three types of spatial patterns for individuals: (A) = uniform, (B) = random, and (C) = clumped.

6.5.2 Indices for Discrete Sampling Units

If individuals are randomly distributed among sampling units, their frequency distribution can be described by a Poisson distribution. Because the variance and mean of any Poisson distribution are equal, the $s^2/\bar{x}$ estimate of σ^2/μ may be used as a measure of general conformity to the Poisson distribution. Indeed, the variance divided by the mean is called the index of dispersion. A Poisson distribution is indicated if σ^2 is equal to μ, implying a random pattern. Similarly, a σ^2 less than μ indicates a negative binomial distribution and implies a clumped pattern. A σ^2 greater than μ suggests a positive binomial distribution and implies a uniform distribution. Pielou (1969) provides a χ^2 test of fit to the Poisson distribution based on this quotient of variance to mean. Ludwig and Reynolds (1988) give similar tests for small ($N < 30$) and large ($N \geq 30$) numbers of SUs. Krebs

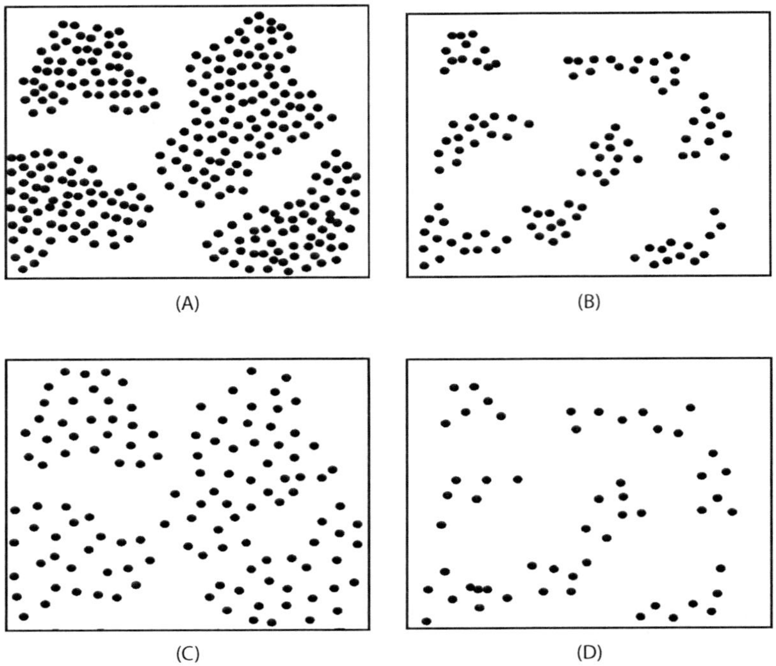

Figure 6.10 An example of grain and intensity. (Modified and reorganized from Pielou, E.C., *Population and Community Ecology: Principles and Methods*, Gordon and Breach Science Publishers, New York, 1974.)

(1999) provides software as well as an excellent explanation of these methods.* As will be shown, this quotient is used as a test statistic and as a descriptive statistic for spatial distribution.

Pielou (1969), Krebs (1989, 1999), and Ludwig and Reynolds (1988) provide straightforward χ^2 tests for deviations from the Poisson or negative binomial distributions. Tests for uniform distributions are not provided by these authors because this distribution is uncommon. Rejection with these tests is thought to suggest inadequacy of the distributions for description of the observed spatial patterns; e.g., the pattern is not random (rejection of Poisson model) or the pattern is not clumped (rejection of the negative binomial model). Again, in a strict sense, failure to reject the null hypothesis based on a p value does not prove that the distribution is appropriate. Failure to reject the null hypothesis can arise from testing inadequacies, e.g., lack of statistical power, as well as the qualities of the distribution. Ambiguity also exists with rejection of the null hypothesis for reasons discussed in the last chapter.

The null hypothesis (H_0) that the Poisson distribution describes the data (implying a random pattern) is tested with a minimum of 30 SUs (Ludwig and Reynolds 1988). The probabilities of counts of x of 1, 2, 3, …, r individuals per SU is described by Equation (6.98) for a Poisson distribution.

$$P_i = \frac{\lambda^i e^{-\lambda}}{i!} \approx \frac{\bar{x}^i e^{-\bar{x}}}{i!} \tag{6.98}$$

where P_i = probability of i individuals per SU (i = 0, 1, 2, …, r), λ = mean of the number of individuals per SU, and $\bar{x}$ = estimated mean number of individuals per SU.

* The reader should note that Hurlbert (1990) argues that this quotient is not a good measure of departure from randomness.

The observed frequencies (F_i) of counts of individuals per SU are tabulated first. The expected frequencies (E_i) corresponding to each of the $r + 1$ observed frequencies are estimated with the total number of individuals observed on all SUs (N) and the estimated P_i from Equation (6.98).

$$E_i = \frac{\bar{x}^i e^{-\bar{x}}}{i!} N \tag{6.99}$$

A χ^2 statistic can then be calculated using the observed and expected frequencies for the data.*

$$\chi^2 = \sum_{i=0}^{r} \frac{[F_i - E_i]^2}{E_i} \tag{6.100}$$

The H_0 is rejected if this χ^2 value is greater than a tabulated value. The critical χ^2 value can be obtained from Table 14 in Rohlf and Sokal (1981) with an associated $\alpha = 0.05$ and $df = (r + 1) - 1 = r$. The $df = r$ is used because this is the number of frequency classes minus the number of estimated parameters. Alternatively, the Excel CHIINV(χ^2,df) function will produce the associated p value.

Ludwig and Reynolds (1988) describe a very similar method for testing the H_0 that the spatial pattern can be described by a negative binomial distribution. Acceptance of the H_0 implies, but certainly does not prove, that these data have a clumped pattern. Let μ be the mean number of individuals per SU and k be a measure of the degree of clumping. Again, a table of observed frequencies of numbers of individuals per SU (F_i) is constructed. The expected probabilities are then calculated under the assumption of a negative binomial distribution.

$$P_i = \left[\frac{\mu}{\mu + k}\right]^i \left[\frac{(k+i-1)!}{[i!(k-1)!]}\right]\left[1 + \frac{\mu}{k}\right]^{-k} \tag{6.101}$$

Ludwig and Reynolds (1988) provide the following forms of Equation (6.101) to facilitate calculations.

$$P_0 = \left[1 + \frac{\bar{x}}{\hat{k}}\right]^{-k} \tag{6.102}$$

$$P_1 = \left[\frac{\bar{x}}{(\bar{x} + \hat{k})}\right]\left[\frac{\hat{k}}{1}\right] P_0 \tag{6.103}$$

$$\vdots$$

$$P_r = \left[\frac{\bar{x}}{\bar{x} + \hat{k}}\right]\left[\frac{\hat{k} + r - 1}{r}\right] P_{r-1} \tag{6.104}$$

The μ in Equation (6.01) is estimated by $\bar{x}$ and the k is estimated by the following procedure. First, Equation (6.105) is used for an initial estimate of k.

* Krebs (1999) notes that the χ^2 statistic can be estimated directly from the index of dispersion (I) and number of counted sampling units (N), $\chi^2 = I(N - 1)$. He also provides a graphical method based on this estimated χ^2 versus degrees of freedom to suggest tendency toward uniform, random, and clumped distributions.

$$\hat{k} = \frac{\bar{x}^2}{s^2 - \bar{x}} \quad (6.105)$$

where s^2 = the variance estimated from all of the SUs.

If the original $\bar{x} > 4$ and the above estimated $\hat{k} > 4$, this estimate of k is used in Equations (6.102) to (6.104). Otherwise, the following iterative process is used to estimate k in Equations (6.102) to (6.104). The $\hat{k}$ from Equation (6.105) is inserted into Equation (6.106) and both sides of Equation (6.106) calculated.

$$Log \frac{N}{N_0} = \hat{k}\ Log\left[1 + \frac{\bar{x}}{\hat{k}}\right] \quad (6.106)$$

where N = the total number of SUs, and N_0 = the number of SUs with 0 individuals.

If the value on the right side of Equation (6.106) is smaller than the value on the left side of this equation, the $\hat{k}$ is increased slightly and the right side of the equation solved again. The $\hat{k}$ is decreased if the value of the right side is larger than that of the left side. This process is repeated until both sides of Equation (6.106) are sufficiently similar for the intent of the study. The final $\hat{k}$ from this process is used in Equations (6.102) to (6.104) to estimate the P_i values.

As described for the Poisson distribution, a χ^2 statistic is generated with a resulting table of expected and observed frequencies (Equation 6.100). The estimated χ^2 is compared to a table of χ^2 with an associated α or a χ^2 produced with Excel CHIINV(probability, df). The p value could also be calculated with the Excel CHIDIST(χ^2,df) function. The df would be $r + 1 - 2$, or $r - 1$ in this instance, because the number of parameters being estimated is now 2.

Example 6.17

The gregarious behavior of a particular benthic species results in a clumped distribution of individuals under intertidal rocks. This behavior is maintained under normal laboratory conditions as well. The species behavior is examined in mesocosms spiked with petroleum industry waste. The distribution of individuals among 109 rocks was measured. The H_0 is posed that the distribution can be described by a negative binomial function—the associated implication being that a negative binomial is a good model for the usual clumped distribution of individuals. The observed frequencies (F_i) for I values of 0 to 11 individuals per rock are tabulated in the leftmost columns (1 and 2) below.

i (number/rock)	F_i (Observed P)	$E_i(P_iN)$ (Expected P)	$(F_i - E_i)^2/E_i$
0	5	1.53	12.04
1	6	5.89	<0.01
2	5	11.99	48.86
3	15	16.79	3.20
4	21	18.31	7.24
5	30	16.46	183.33
6	10	12.75	7.56
7	5	8.72	13.84
8	4	5.34	1.80
9	4	3.05	0.90
10	2	1.64	0.13
11	2	0.76	1.54
			$\Sigma = 28.59$

The mean number of individuals per rock ($\bar{x}$) is 4.59 with a variance of 5.37 (N = 109). The initial estimate of $\hat{k}$ is found with Equation (6.105) to be $4.59^2/(5.37 - 4.59) = 27.01$. Because $\bar{x} > 4$ and $\hat{k} > 4$, this initial estimate is adequate for use in further calculations. Using $\bar{x}$ and $\hat{k}$ in Equations (6.102) to (6.104), the expected probabilities for the $r + 1$ counts are calculated.

$$P_0 = \left[1 + \frac{4.79}{27.01}\right]^{-27.01} = 0.014$$

$$P_1 = \left[\frac{4.59}{4.59 + 27.01}\right][27.01][0.014] = 0.054$$

Similarly, P_2 to P_{11} are calculated to be 0.110, 0.154, 0.168, 0.151, 0.117, 0.080, 0.049, 0.028, 0.015, and 0.007, respectively. These P_i values are multiplied by the total number of sampling units (109 rocks) to get the predicted frequencies (column 3 in above table). The differences between the 11 pairs of observed and expected frequencies are squared (column 4) and summed to get the χ^2 value of 28.59. This statistic is compared to a χ^2 with $\alpha = 0.05$ and $df = r - 1 = 10$. A value of 18.307 is found in Table 14 of Rohlf and Sokal (1981) or with Excel CHIINV(0.05,10). The Excel CHIDIST(28.59,1) generates a p value of 0.001451. The negative binomial model is rejected for description of the species pattern under these contaminated conditions.

The index of dispersion is also used as a descriptive statistic with values ranging from 0 (extreme uniformity) to 1 (random) to N (extreme clumping). Often this index is modified by subtraction of 1 to produce the index of clumping (David and Moore 1954). The index allows comparison between populations with different means but identical total numbers of individuals sampled (Pielou 1969).

$$Index\ of\ Clumping = \frac{s^2}{\bar{x}} - 1 \tag{6.107}$$

The w statistic is used for this purpose (Pielou 1969). The index of clumping is $s_j^2/\bar{x}_j - 1$, with j = 1 and 2 for the two sets of data to be compared (Pielou 1969). Notionally, the indices from the two data sets differ at $\alpha = 0.05$ if w is beyond $\pm\ 2.5/\sqrt{n-1}$.

$$w = -0.5\ Ln\left[\frac{\frac{s_1^2}{\bar{x}_1}}{\frac{s_2^2}{\bar{x}_2}}\right] \tag{6.108}$$

Pielou (1969) makes a pertinent observation relative to the use of this index. She suggests that if deaths occurred for some reason, and the original and surviving populations were sampled, this comparison could be used to test whether deaths were random. Obviously, the mean will drop with mortality. Perhaps not so obvious is the fact that the index of clumping will also decrease if deaths were random relative to the differing densities within the sample area. The index (I) will decrease by $\Theta * I$, where Θ is the proportion of the original population surviving. Deviations from this predicted decrease would suggest nonrandom (density-dependent) mortality.

The values for the indices of dispersion and clumping are dependent on the total number of individuals sampled (n). Green's index is a modification of these indices that corrects for this dependence on n. Consequently, Green's index allows comparisons between samples varying in total number of individuals sampled (n), mean number of individuals per SU ($\bar{x}$), and number of SUs (N). Ludwig and Reynolds (1988) and Krebs (1999) give the extreme range of Green's index to be $-1/(n$

– 1) (completely uniform) to 0 (random) to 1 (maximum clumping). Continuing the above argument of Pielou, any significant change in Green's index with loss of a portion of the population suggests nonuniform loss of individuals relative to spatial density.

$$Green's\ Index = \frac{\frac{s^2}{\bar{x}} - 1}{n - 1} \tag{6.109}$$

Many other indices for discrete sampling units have been developed. The reader is directed to Pielou (1969, 1974), Krebs (1999), or Ludwig and Reynolds (1988) for more discussion of these indices.

6.5.3 Indices for Arbitrary Sampling Units

6.5.3.1 Indices Based on Quadrats

Often arbitrary sampling units are quadrats such as those used earlier to estimate population densities. The approach mentioned in the footnote on page 323 could be applied or the variance to mean quotient for quadrats differing in size can be used to assess clumping of individuals. This can be seen readily using Figure 6.9. If randomly positioned quadrats of increasing size are placed over a uniform pattern, the variance to mean ratio will remain relatively constant. With a random pattern, the ratio will fluctuate in a random fashion. Clear peaks (or trends) in the ratio will emerge with clumped distributions.

Alternatively, the variance to mean quotient for contiguous pairs of quadrats of increasing size may be used to assess spatial patterning (Pielou 1974). (Often a row of quadrats is sampled and adjacent quadrats are sequentially pooled to get blocks of quadrats of increasing size.) The likelihood of smaller pairs of blocks being within the same clump or gap between clumps is higher than that for larger pairs. Consequently, the variance to mean quotient from a clumped pattern will increase as block size increases. This approach may be formulated as an ANOVA if desired. A peak in the variance to mean quotient (or mean square for number of quadrats per block) will occur at roughly the mean clump size (Pielou 1974). Blocks of 2^n (e.g., 1, 2, 4, 8, 16, 32) contiguous quadrats are often plotted against the logarithm of mean squares in such exercises.

Example 6.18

Eight 1 × 1 m contiguous quadrats are sampled in a row across a beach at the mean low tide line. The numbers of species A found in the eight quadrats are 5, 4, 8, 12, 14, 4, 2, and 4 across the beach. The mean for the entire sample of 53 individuals in the eight quadrats is 6.62 individuals per quadrat. Estimate the variance-to-mean ratio for quadrat blocks of sizes 1 through 4.

Block = 1 quadrat: Four block (= quadrat) pairs are available for estimating the quotient. The block pair means are the total number of individuals in all pertinent quadrats (number in quadrat 1 + number in quadrat 2 of the pair in this case) divided by the total number of quadrats in the pair (2 in this case). The pair variances are

$$\frac{\left[\sum (x - \bar{x})^2\right]}{2}$$

Pair 1(5,4):	$\bar{x} = (5 + 4)/2 = 4.5$	$s^2 = [(5 - 4.5)^2 + (4 - 4.5)^2]/2 = 0.25$
Pair 2(8,12):	$\bar{x} = 10.00$	$s^2 = 4.00$
Pair 3(14,4):	$\bar{x} = 9.00$	$s^2 = 25.00$
Pair 4(2,4):	$\bar{x} = 3.00$	$s^2 = 1.00$

The average s^2 for blocks of 1 quadrat is (0.25 + 4.00 + 25.00 + 1.00)/4 or 7.56. The block average ($\bar{x}_b$) is 53/8 blocks of two 1 × 1 m quadrats, or 6.62. The $s^2/\bar{x}_b$ is 7.56/6.62, or 1.14.

Block = 2 quadrats: Two pairs of contiguous blocks can be made by combining adjacent quadrats, (5 + 4, 8 + 12) and (14 + 4, 2 + 4).

Pair 1(9,20): $\bar{x} = (9 + 20)/4 = 7.25$ $s^2 = [(9 - 7.25)^2 + (20 - 7.25)^2]/2 = 82.81$

Pair 2(18,8): $\bar{x} = (18 + 6)/4 = 6.00$ $s^2 = [(18 - 6)^2 + (8 - 6)^2]/2 = 74.00$

The average s^2 for blocks of two quadrats is (82.81 + 70.56)/2, or 76.69. The block average ($\bar{x}_b$) is 53 individuals/4 blocks, or 13.25. The $s^2/\bar{x}_b$ is 74.00/13.25, or 5.58.

Block = 4 quadrats: One pair of blocks (5 + 4 + 8 + 12, 14 + 4 + 2 + 4) can be used.

Pair 1(29,24): $\bar{x} = (29 + 24)/8 = 6.62$ $s^2 = [(29 - 6.62)^2 + (24 - 6.62)^2]/2 = 401.46$

The average s^2 for blocks of four quadrats is 401.46. The block average ($\bar{x}_b$) is 53 individuals/2 blocks, or 26.50. The $s^2/\bar{x}_b$ is 401.46/26.50 or 15.15.

These $s^2/\bar{x}_b$ are plotted against the blocks sizes to get an estimate of clumping.

As mentioned, an ANOVA using one, two, and four quadrat blocks can be performed for this purpose also. The mean squares for the block classes will be in the same proportion as the relative $s^2/\bar{x}_b$ estimated above, i.e., approximately 1.14 to 5.79 to 15.15 (1:5.08:13.29). Clearly, the mean squares could be plotted against block size instead of the $s^2/\bar{x}_b$ to reveal the clumping in the data.

Ludwig and Reynolds (1988) presented additional quadrat-based methods, including a refinement of the method described above. This paired-quadrat variance method estimates the variance between pairs of fixed-size quadrats at specific distances between the (paired) quadrats. The variance for a spacing of 1 is calculated with Equation (6.110).

$$s^2 = \left[\frac{1}{N-1}\right]\left[\left(\frac{x_1 - x_2}{2}\right)^2 + \left(\frac{x_2 - x_3}{2}\right)^2 + \ldots + \left(\frac{x_{N-1} - x_N}{2}\right)^2\right] \tag{6.110}$$

where N = the total number of quadrats from which the counts were obtained.

The s^2 values for subsequent spacings are obtained as illustrated here for spacings of 2.

$$s^2 = \left[\frac{1}{N-2}\right]\left[\left(\frac{x_1 - x_3}{2}\right)^2 + \left(\frac{x_2 - x_4}{2}\right)^2 + \ldots + \left(\frac{x_{N-2} - x_N}{2}\right)^2\right] \tag{6.111}$$

For larger spacings, the denominator in the first term in Equation (6.111) and the intervals between quadrats used in each variance estimate are adjusted accordingly. The variance values are then plotted against spacings between quadrats. Ludwig and Reynolds (1988) recommend that a conservative estimate of the maximum number of spacings to be calculated is 10% of N. However, they also state that 20% of N might be acceptable.

Example 6.19

The quadrat data from the previous example are reanalyzed using the paired-quadrat variance method just described. The 1 × 1 m contiguous quadrats (x_1, x_2, x_3, x_4, x_5, x_6, x_7, and x_8) contain 5, 4, 8, 12, 14, 4, 2, and 4 individuals, respectively. For illustrative purposes, we will ignore Ludwig and Reynolds's (1988) recommendation for calculating variances for spacings up to 10% of N, e.g., 0.1*8 = 0.8 or 1 m spacing. Instead, an arbitrary number of 4 spacings will be used here for illustrative purposes only.

s^2 for spacing = 1:

$$\left[\frac{1}{8-1}\right]\left[\left(\frac{5-4}{2}\right)^2+\left(\frac{4-8}{2}\right)^2+\left(\frac{8-12}{2}\right)^2+\left(\frac{12-14}{2}\right)^2+\left(\frac{14-4}{2}\right)^2+\left(\frac{4-2}{2}\right)^2+\left(\frac{2-4}{2}\right)^2\right]$$

$$= 5.18$$

s^2 for spacing = 2:

$$\left[\frac{1}{8-2}\right]\left[\left(\frac{5-8}{2}\right)^2+\left(\frac{4-12}{2}\right)^2+\left(\frac{8-14}{2}\right)^2+\left(\frac{12-4}{2}\right)^2+\left(\frac{14-2}{2}\right)^2+\left(\frac{4-4}{2}\right)^2\right]=13.21$$

s^2 for spacing = 3:

$$\left[\frac{1}{8-3}\right]\left[\left(\frac{5-12}{2}\right)^2+\left(\frac{4-14}{2}\right)^2+\left(\frac{8-4}{2}\right)^2+\left(\frac{12-2}{2}\right)^2+\left(\frac{14-4}{2}\right)^2\right]=18.25$$

s^2 for spacing = 4:

$$\left[\frac{1}{8-4}\right]\left[\left(\frac{5-14}{2}\right)^2+\left(\frac{4-4}{2}\right)^2+\left(\frac{8-2}{2}\right)^2+\left(\frac{12-4}{2}\right)^2\right]=11.31$$

Plotting of the variances with distances would show a distinct clumping at 3 m spacings.

Many of these methods are sensitive to quadrat size. A quadrat too small will result in many 0 counts, but a quadrat too big will obscure spatial detail. In Upton and Fingleton's (1985) review of the literature, they suggest quadrat sizes resulting in an average count of 1.0 to 4.0 individuals per quadrat. Consequently, quadrat techniques become inefficient if individuals are widely dispersed. The distance-based methods described below are more appropriate in such situations.

6.5.3.2 Indices Based on Distance

Numerous distance techniques have been developed. Distances can be between individuals and their nearest neighbors, e.g., the classic work of Clark and Evans (1954). Distances to the second, third, fourth, or further nearest neighbors can also be determined and used to calculate spatial patterning. Pielou (1969, 1974) and Cressie (1991) provide detailed discussion and illustration of these techniques. Krebs (1999) provides description and programs for doing nearest-neighbor analysis.

Alternatively, the distance from a randomly selected point to the nearest individual can be compared to the distance between that individual and its nearest neighbor. The *T*-square index is one such technique recommended by Ludwig and Reynolds (1988) and Cressie (1991). Ludwig and Reynolds's (1988) treatment of the technique is provided here as an example. Krebs's (1999) code mentioned previously will compute the associated statistics, including tests of significance.

First, a point O is randomly selected from within the area containing the individuals. The distance x is measured between point O and the nearest individual (P). A perpendicular line intersecting line OP at P is then drawn. This perpendicular line defines two half planes, one of which contains the original randomly selected point O. The nearest neighbor (Q) to point P is identified on the other side of the perpendicular line, i.e., on the half plane opposite of that containing point O. The distance (y) between P and Q is measured. This process is repeated for N sample points.

Cressie (1991) indicates that the number of random sampling points should be less than or equal to the number of individuals in the area divided by 10. Normal approximations are appropriate for testing if N is approximately 10 or more.

The C statistic is calculated with Equation (6.112).

$$C = \frac{\sum_{i=1}^{N} \frac{x_i^2}{x_i^2 + 0.5y_i^2}}{N} \tag{6.112}$$

C is 0.5 for a random pattern. It decreases as the pattern tends toward uniformity and increases as the pattern tends toward clumping. Significant deviations from 0.5 can be tested with z.

$$z = \frac{C - 0.5}{\sqrt{\frac{1}{12N}}} \tag{6.113}$$

This calculated z is compared to a tabulated z for a standard normal distribution or Excel NORMSINV(), e.g., 1.96 for $\alpha = 0.05$.

Example 6.20

Normally, individuals of clam species X will distribute themselves randomly in a homogeneous substrate. Random patches of fly ash-contaminated sediments are formed within the otherwise clean sediment of a mesocosm. One hundred clams are then placed randomly into the mesocosm and allowed 1 month to move about in the sediments. Did the spatial pattern of clams remain random? (Control mesocosms show that clams maintain a random pattern in the absence of contaminated patches. One month is adequate for extensive movement about the mesocosm by the clams.) Ten random points (N = 10) are used to generate x_i, y_i data pairs. The C is then estimated with Equation (6.112).

x_i (cm)	x_i^2	y_i (cm)	y_i^2	$x_i^2/(x_i^2 + 0.5y_i^2)$
63	3,969	15	225	0.9724
37	1,369	5	25	0.9910
56	3,136	30	900	0.8745
71	5,041	6	36	0.9964
58	3,364	34	1,156	0.8534
63	3,969	23	529	0.9375
72	5,184	31	961	0.9152
75	5,625	21	441	0.9623
59	3,481	23	529	0.9294
64	4,096	29	841	0.9069
				Σ = 9.3390

From Equation (6.112), C = 9.3390/10 = 0.9339.

From Equation (6.113),

$$z = \frac{0.9339 - 0.50}{\sqrt{\frac{1}{(12 * 10)}}} = 4.7530$$

Since z (4.7530) is greater than 1.96, the H_0 that the clams are randomly distributed is rejected. The large value of C suggests significant clumping in the heterogeneous habitat. The Excel NORMSDIST(z) can also be applied to get the associated p value for this two-sided test, e.g., 2(1-NORMDIST(1.96)) = 2(1 – 0.975002) ≈ 0.05.

6.5.4 Consequences of Spatial Heterogeneity

Spatial heterogeneity was mentioned only generally in the discussions of population density, demographics, and dynamics. However, uneven distributions of individuals within a network of habitat patches of varying quality can influence overall population qualities (see Example 6.9). The local populations in excellent habitat patches might have birth rates exceeding death rates, and those in poor patches may have higher death rates than birth rates. Immigration and emigration rates will also differ for the various patches. The consequent large differences in fitnesses of individuals within different patches might produce sources and sinks; i.e., some areas produce surplus offspring that move into the poorer habitat patches (Lewin 1986, 1989). Alternatively, migration of individuals among similar patches can also involve less extreme situations. Regardless of which situation exists, it was an unfortunate reality that ecologists' early preoccupation with homogeneous populations "[had been] increasingly conditioning our perception of reality" (Weins 1976). Pulliam and Danielson (1991) demonstrated the importance of understanding the unique processes occurring in heterogeneous populations that were downplayed in most ecological models until the last few decades. This short section is intended to broaden ecotoxicological discussion of populations to consider features of populations occupying landscapes unevenly.

General context and some definitions from Hanski and Gilpin (1991) are needed at the onset of our discussions of metapopulations. There are three relevant scales here: local, metapopulation, and geographical scales. A local scale is one in which "individuals [of the same species] move and interact with each other in the course of their routine feeding and breeding activities." The models discussed to this point are relevant to this scale. At the other extreme, the geographical scale is one encompassing the entire geographical range of the species. The scale relevant to the present discussion is the metapopulation scale, that is, one "at which individuals [of a species] infrequently move from one place (population) to another, typically across habitat types which are not suitable for their feeding and breeding actitivities."

6.5.4.1 Metapopulations

Levins was the first to use the term *metapopulation* to define a population comprised of smaller populations occupying a network of habitat patches among which exchange of individuals was possible. The description of metapopulation scale given above should make it clear that the probability of movement of individuals among patches is lower than that for movement within a patch. Because the earliest publications on this topic (i.e., Levins 1969) focused on metapopulation consequences in pesticide application programs, it is difficult to explain why metapopulation concepts have flourished since 1969 in ecology and conservation biology, but not ecotoxicology.

6.5.4.2 Quantifying Metapopulation Dynamics

The most prominent elucidation of metapopulation dynamics has come from Ilkka Hanski at the University of Helsinki. For this reason, the notations here conform to those in his publications, i.e., Hanski (1991, 1996, 1998), Hanski and Gilpin (1991), and Hanski and Ovaskainen (2003).

Levins's original model assumed that a population in each patch within a mosaic of patches was either extinct or at its local carrying capacity (K),* and that individuals were equally likely to migrate from any patch to another. Initially, these assumptions eliminated the need to consider features such as the ideal free distribution discussed in Example 6.9. The rate or probability of patch colonization was determined by the fraction of all patches that were occupied (p). The extinction (e)† and colonization/immigration (m) parameters were integrated into the Levins model to predict the change in proportion of available patches occupied through time.

$$\frac{dp}{dt} = mp(1-p) - ep \tag{6.114}$$

The estimated p at equilibrium ($\hat{p}$) for this simple model is $1 - e/m$. The metapopulation will collapse unless $1 - e/m$ is a positive, nonzero number. Hanski and Gilpin (1991) and Hanski (1996) point out that this model can be expressed in a form similar to a logistic model (i.e., Equation 6.62),

$$\frac{dp}{dt} = (m-e)p\left[1 - \frac{p}{1-\left(\frac{e}{m}\right)}\right] \tag{6.115}$$

The bracketed $m - e$ part of this model is the intrinsic rate of increase of p for a metapopulation when p is small. So, the $m - e$ and $1 - e/m$ terms are analogous to the r and the population carrying capacity, K, of the logistic model.

A decrease in average patch size or decrease in exchange among patches will diminish the likelihood of metapopulation persistence. Metapopulation $\hat{p}$ will decrease with a decrease in the average size of a patch, as might happen with pollution-related loss of suitable habitat area. But, even with an adequate average patch size, the degree of patch isolation can strongly influence metapopulation persistence. Hanski (1991) provides the following formulae that make m and e functions of average patch area (A) and average patch isolation (D), and include these features in predictions of metapopulation $\hat{p}$.

$$m = m_0 \mathrm{e}^{-aD} \tag{6.116}$$

$$e = e_0 \mathrm{e}^{-bA} \tag{6.117}$$

$$\hat{p} = 1 - \left(\frac{e_0}{m_0}\right)\mathrm{e}^{-bA+aD} \tag{6.118}$$

The m and e in the model presented to this point (Equation 6.114) are not influenced by the probability of a nearby patch being occupied. Yet the chance of movement between two adjacent

* Such models that focus on predicting whether or not patches are occupied are called occupancy models. Metapopulation models that focus on the distribution of population sizes among patches are called structured metapopulation models (Hanski and Gilpin 1991).

† The italicized e in these equations is the model extinction parameter, but the nonitalicized e is the mathematical constant e (inverse of the natural logarithm).

patches is much higher than that between patches situated a considerable distance from each other.* Hanski's modification to Levins's model changes this by making e dependent on the p. A rescue effect is included in which the increased probability of emigrants from nearby occupied patches reduces the probability of the patch going locally extinct.

$$\frac{dp}{dt} = mp(1-p) - ep(1-p) \tag{6.119}$$

The p will approach 1 at equilibrium if $m > e$, i.e., $\hat{p} = 1$. Gotelli (1991) explains that the immigration rate is determined by the quadratic function $mp(1-p)$ in both Equations (6.114) and (6.119). Extinction rate in Equation (6.114) increases linearly as determined by the term ep, but the extinction rate in Equation (6.119) is defined by yet another quadratic function, $ep(1-p)$. In Equation (6.119), the extinction rate increases as more patches are occupied when p is small, but as p becomes larger (>0.5), the extinction rate begins to decrease as more patches are occupied. So Equation (6.119) incorporates the rescue effect into a basic occupancy model. But this modified equation does not incorporate what is called the propagule rain effect. The propagule rain effect occurs in situations in which extinction rate is independent of nearby patch occupancy. Such a situation might occur if there was a large seedbank in the soil or a large reservoir of resting stage individuals or eggs in sediments of a water body. The reseeding of the patch would not depend on p in such cases. Equation (6.120) (Gotelli 1991) defines the situation in which the extinction rate is independent of p.

$$\frac{dp}{dt} = m(1-p) - ep \tag{6.120}$$

The equilibrium p ($\hat{p}$) for this model is $m/(m + e)$. Both rescue effect and propagule rain can be included in these models using the following formulation:

$$\frac{dp}{dt} = m(1-p) - ep(1-p) \tag{6.121}$$

which has an equilibrium p ($\hat{p}$) of 1 if $m > e$. The general behavior of these four models (Equations 6.114, 6.119, 6.120, and 6.121) can be illustrated assuming $m = 0.3$ and $e = 0.1$ (Figure 6.11). The p is predicted to increase steadily to $1 - e/m$ with the Levins model (Equation 6.114), but the addition of a rescue effect afforded by the Hanski model (Equation 6.119) changes the curve shape and $\hat{p}$ to 1. The addition of propagule rain alone (Equation 6.120) allows p to increase quickly to $m/(m + e)$; however, the inclusion of both the rescue effect and propagule rain features in Equation (6.121) results in a very rapid increase to a $\hat{p}$ of 1.

The minimum viable metapopulation size concept should be mentioned before ending our discussion of basic metapopulation models. This concept is analogous to the minimum viable population (MVP) size, which is the smallest size that a species population can be and still have a high probability of remaining extant; e.g., x number of individuals are required for the population to have a 95% probability of persistence for 100 years. Gurney and Nisbet (1978, Equations 10 and 11) and Nisbet and Gurney (1982, Equations 6.8.10 and 6.8.11) used a stochastic version of Levins's model for metapopulations with various total numbers of suitable patches (H) and fractions of occupied patches at equilibrium ($\hat{p}$). The expected time to metapopulation extinction (T_M) was then related to

* Movement of individuals into a patch from a distant patch that requires migration into and out of a series of patches is called stepping-stone dispersal.

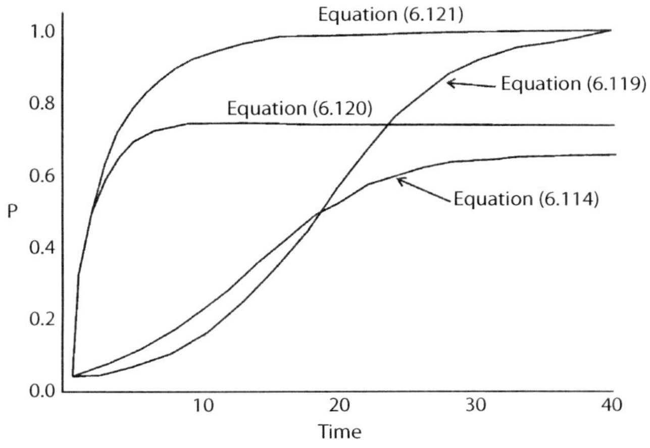

Figure 6.11 Trajectories of a simple metapopulation (m = 0.3 and e = 0.1) under the four models: Levins (6.114), Hanski with rescue effect (6.119), model including propagule rain (6.120) and model including rescue effect and propagule rain (6.121).

the expected time to extinction for a patch in the metapopulation (T_L).* The following relationship between T_M and T_L provided an approximate linkage:

$$T_M = T_L e^{(H\hat{p}^2)/(2(1-\hat{p}))} \tag{6.122}$$

Roughly speaking, a metapopulation will be persistent indefinitely if the following relationship is true:

$$\hat{P}\sqrt{H} \geq 3 \tag{6.123}$$

The positive influence of average patch population size (ℵ) on p is the final feature that we will discuss for occupancy models (Hanski 1991). This influence was added to the Levins model (Equation 6.114) by modifying the way extinction rate is handled. An e_0 and a were added for this purpose.

$$\frac{dp}{dt} = mp(1-p) - e_0 e^{-ap} p \tag{6.124}$$

The dynamics of this model are more complex than any described to this point and would require considerable space here to describe adequately. The interested reader is directed to Hanski (1991). Regardless of the details, the reader should understand from Equation (6.122) that average patch population size influences metapopulation qualities.

6.5.4.3 Consequences of Contamination

What are some ecotoxicological implications of metapopulation ecology? Several issues are suggested from the Levins model. Although e will not change, a metapopulation that becomes fragmented or loses patches due to pollution will have a decreased m because patch connectivity is

* See also Hanski and Gilpin (1997) and Rockwood (2006) for further details.

reduced. Fragmentation due to contaminant presence can result in the demise of a metapopulation. Beyond the Levins model context, extinction rate (e) could increase if average patch population size decreased due to pollution (Equation 6.122). So both m and e can be adversely impacted and result in higher risk of metapopulation extinction.

Some pollutant effects on a metapopulation might be ameliorated by the propagule rain or rescue effect. A metapopulation with some patches rendered useless by pollution might persist (Equations 6.122 and 6.123). However, the persistence predicted by these simple metapopulation models might come at a price. Harding and McNamara (2002) explain the relevant antirescue effect in which extinction rate of a patch is actually increased by an increase in immigration. In their publication, the disadvantage was associated with harmful parasites or diseases carried by immigrants. Also relevant might be migration between patches by ecomorphs that have substantial differences in fitness in the new patch. From the ecotoxicological vantage, immigrants with high pollutant body burdens or damage from exposure in another patch could contribute to an antirescue effect. Migrating individuals might also have substantially reduced natality rates due to previous toxicant exposure in a contaminated patch. Analogous to the ecomorph example might be a situation in which migrating individuals differ in resistance to chemical stressors due to past exposure histories. (Spromberg et al. (1998) might have been unaware of the antirescue effect when they postulated this possible impact in populations exposed to pollutants. They proposed the moniker of "action at a distance hypothesis.")

Nee (2007) discussed the concept of pathogen eradication threshold that has features relevant to metapopulation persistence of nontarget species during pollutant exposure. Essentially this concept states that a disease need not be eliminated in all individuals (individuals are disease patches in the sense of metapopulation patches) in order to be eradicated. Only a certain proportion of the susceptible population need be vaccinated to eliminate the disease from the population. We can change the situation slightly to consider how many patches must be made vacant by a chemical pollutant in order for the metapopulation to collapse. The epidemiology literature has ample models for this purpose, including some similar to Equation (6.123). The point to be understood is that metapopulations can slowly disappear even if some patches are uncontaminated or remediated.

Ecotoxicology research taking a metapopulation vantage appears infrequently in the recent literature. Publishing in the applied ecology literature, Sherratt and Jepson (1993) explored metapopulations of nontarget invertebrates subjected to chronic pesticide exposure. Lahr (1997) took the metapopulation vantage to study toxicant impact in temporary ponds. Spromberg et al. (1998) did simple three-patch simulations to make a lucid argument for a metapopulation approach to ecotoxicology. Even more recently, Ares (2003) and Chaumot et al. (2003) reviewed metapopulation issues pertinent to ecotoxicology. Metapopulation concepts were included in Relyea and Hoverman's 2006 review of ecology themes in ecotoxicology. Hopefully, the low frequency at which metapopulation-based ecotoxicology studies have appeared in the literature during the last two decades will increase during the next decade.

Metapopulation qualities also influence population genetics (e.g., Hanski 1998). As a specific example, Bell and Gonzalez (2011) recently concluded that local dispersal contributed materially to adaptation during yeast metapopulation exposure to salt stress. This linkage should be kept in mind while reading the next chapter section.

6.6 POPULATION GENETICS

6.6.1 Basic Concepts

6.6.1.1 Evolution by Natural Selection

The theory of evolution by natural selection is based on several assumptions. First, it is assumed that populations are capable of producing surplus offspring. Second, individuals within a population occupying a specific environment will differ in their abilities to survive and reproduce, i.e., their Darwinian fitness. Third, these differences in fitness have a hereditary basis. As a consequence of the surplus offspring and heritable differences in fitness, the genes of more fit individuals will be overrepresented in each successive generation. This "process by which genotypes with greater fitness leave, on the average, more offspring than do less fit genotypes" is natural selection (Hartl and Clark 1989). It is the mechanism suggested by Darwin for evolutionary change.

6.6.1.2 Hardy–Weinberg Equilibrium

The genotype frequencies within populations remain stable through time if certain conditions exist: (1) the population is a large population of randomly mating, diploid species with overlapping generations, (2) no natural selection is occurring, and (3) mutation and migration rates are negligible. The Hardy–Weinberg (or Castle–Hardy–Weinberg) principle defines the expected frequencies of genotypes for such a population. Let two alleles (100, 165) be present for a gene at frequencies of p and q, respectively. Predicted genotype frequencies for a two-allele locus (e.g., 100 and 165) at Hardy–Weinberg equilibrium are 100/100: p^2, 100/165: $2pq$, and 165/165: q^2.

Conformity of genotype frequencies to the Hardy–Weinberg model can be assessed with a χ^2 test. The observed number of individuals of each genotype (e.g., $O_{100/100}$, $O_{100/165}$, $O_{165/165}$) are compared to the expected number of individuals of that genotype under the Hardy–Weinberg model (e.g., $E_{100/100}$, $E_{100/165}$, $E_{165/165}$). For a polymorphic locus with two alleles, the χ^2 value is estimated as

$$\chi^2 = \sum_{i=1}^{3} \frac{(O_i - E_i)^2}{E_i} \tag{6.125}$$

where O_i = the number of individuals observed for the three possible genotypes, and E_i = the expected number of individuals for the three possible genotypes.

This χ^2 is compared to a tabulated critical χ^2 value with a specified α and $df = 1$ or one generated with the Excel CHIINV(p,df) function. (The df is the number of genotypes minus the two parameters in the data (p and q).) A calculated χ^2 greater than the critical χ^2 would lead to rejection of the H_0 and indicate that significant deviation from Hardy–Weinberg expectations has occurred.

Alternatively, the p value for the calculated χ^2 could be estimated with the Excel CHIDIST(χ^2,df) function.

Example 6.21

A population in a contaminated area experienced a decimating exposure to a toxicant. The population was examined the next year before breeding began. Among the parameters measured is the frequency of genotypes for an electrophoretic trait in 1,250 individuals (N). The genotype frequencies met Hardy–Weinberg expectations during several preexposure samplings. Is the population still in Hardy–Weinberg equilibrium relative to this trait?

Genotype	Observed Numbers	Expected Numbers
100/100	201	167.445
100/165	513	580.110
165/165	536	502.445

The expected frequencies for the 100/100, 100/165, and 165/165 genotypes are predicted from the observed allele frequencies. Each diploid individual has two alleles; i.e., there is a total of 2(1,250), or 2,500, alleles in the 1,250 individuals. Of these alleles, 2(201) + 1(513), or 915, of the alleles present are the 100 allele. The frequency for the 100 allele (p) is 915/2,500, or 0.366. The frequency of the 165 allele (q) is 1 – 0.366, or 0.634. The expected frequency of the 100/100 genotype is $q^2(N) = 0.366^2(1{,}250) = 167.445$ Similarly, the expected frequencies of the 100/165 ($2 \times 0.366 \times 0.634 \times 1{,}250$) and 165/165 ($0.634^2 \times 1{,}250$) genotypes are 580.110 and 502.445, respectively.

$$\chi^2 = \frac{(201-167.445)^2}{167.445} + \frac{(513-580.110)^2}{580.110} + \frac{(536-502.445)^2}{502.445} \approx 16.73$$

The critical χ^2 ($\alpha = 0.05$, $df = 1$) is 3.84; therefore, the H_0 is rejected. The genotype frequencies are significantly different from Hardy–Weinberg expectations. This deviation may be a consequence of one or several violations of the assumptions underlying the Hardy–Weinberg model. There may have been selection against a sensitive genotype (the heterozygote, 100/165). The removal of individuals from the habitat could have resulted in a rapid movement of migrants into the area. The general deficit of heterozygotes favors, but certainly does not prove, the possibility that significant migration produced the disequilibrium. (See Wahlund effect below.)

The assessment of Hardy–Weinberg equilibrium for two alleles can be extended to three alleles at a locus. Predicted frequencies of the six associated genotypes can be made if frequencies for the three alleles are known. For example, frequencies of the 66 (g), 100 (p), and 165 (q) alleles would be used to predict the following six genotype frequencies.

Genotype	Frequency
66/66	g^2
66/100	$2gp$
66/165	$2gq$
100/100	p^2
100/165	$2pq$
165/165	q^2

The six pairs of observed and expected genotype frequencies would be used in Equation (6.125) with $i = 6$. The degrees of freedom associated with the χ^2 would be the number of genotypes minus the number of alleles, i.e., 6 – 3, or 3 df.

6.6.1.3 Genetic Drift

6.6.1.3.1 General

Hardy–Weinberg expectations of no change in genotype frequencies through time are based on specific assumptions. The violation of the assumption of a large (effectively infinite) population has many ramifications pertinent to ecotoxicology. With a reduction in population size, genetic drift (random change in allele frequencies) can be accelerated. Thus, drift may be large enough to result in fixation or loss of an allele at a locus. (The probability of fixation for an allele is equal to its frequency.)

6.6.1.3.2 Effective Population Size

The rate of genetic drift is related to the number of individuals contributing genes to the next generation, the effective population size (N_e). Obviously, because not all individuals contribute genes to the next generation, the N_e will most often be some number less than the total number of individuals in the population. (If females store sperm, the N_e could be larger than the total number of individuals because absent males may be contributing via stored sperm.) Further, N_e changes through generations as the demographic composition and total number of individuals fluctuates in the population.

The effective population size, N_e, for a population with nonoverlapping generations may be estimated over many generations of changing population size as follows (Hartl and Clark 1989):

$$\frac{1}{N_e} = \left[\frac{1}{t}\right]\left[\frac{1}{N_1} + \frac{1}{N_2} + \ldots + \frac{1}{N_t}\right] \tag{6.126}$$

where t = time in generations, and N_i = population size at time i.

This harmonic mean estimate of N_e has the advantage of weighting the small N observations more heavily than the large N observations. This weighting accommodates the greatly increased probability of change in allele frequency with decreasing N. (Indeed, a very low N such as that associated with a drop to a low population size after a toxic event can produce a genetic bottleneck. The bottleneck reflects the decreased number of individuals available to carry an allele into the next generation.)

If the number of males and females contributing genes to the next generation is unequal, the effective population size (population with nonoverlapping generations) may be estimated with Equation (6.127) (Hartl and Clark 1989).

$$N_e = \frac{4N_m N_f}{N_m + N_f} \tag{6.127}$$

In populations with overlapping generations, estimation of N_e becomes more difficult. Hartl and Clark (1989) provided a general ($N_e \approx N/2$, where N = number of individuals in the population) and a more detailed estimate (Equation 6.128) for this purpose.

$$N_e = \frac{4N_a L}{\sigma_n^2 + 2} \tag{6.128}$$

where N_a = the natality within a unit of time, L = the mean generation time, and σ^2_n = the variance in brood size.

Crow and Kimura (1970) estimated the influence of N_e and the initial frequency (p) of a neutral allele (an allele not experiencing selection) on the rate at which it will be lost ($p \rightarrow 0.0$) or driven to fixation ($p \rightarrow 1.0$). Excluding cases for which the allele is lost, the average time expressed in generations to fixation is estimated by Equation (6.129).

$$\bar{t}_1 = -\frac{1}{p}[4N_e(1-p)Ln(1-p)] \tag{6.129}$$

In cases for which alleles do not go to fixation, the average time to loss for a neutral allele as a function of N_e is defined by Equation (6.130).

$$\bar{t}_0 = -4N_e\left[\frac{p}{1-p}\right]Ln\ p \tag{6.130}$$

Figure 6.12 summarizes the importance of p and N_e on the time to fixation or loss for a neutral allele.

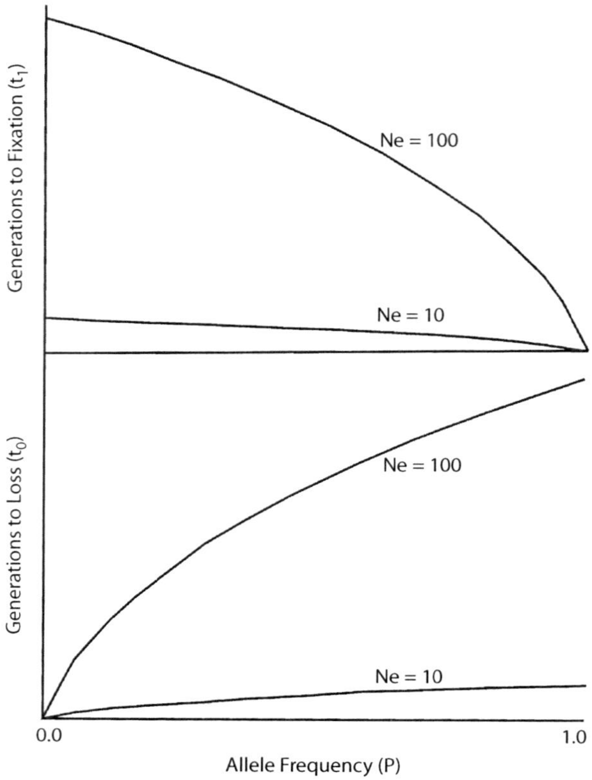

Figure 6.12 The time expressed in generations to fixation (t_1) or loss (t_0) of a neutral allele as functions of N_e and initial allele frequency (p). Note the accelerating influence of a drop of N_e from 100 to 10 individuals.

Several important points can be made from Equations (6.126) to (6.130). First, although it is a common assumption in ecotoxicology (Nevo et al. 1978, 1981; Lavie and Nevo 1986), change in allele frequency associated with pollution does not necessarily indicate selection. It might arise from accelerated genetic drift as a consequence of a generally reduced N_e or an abrupt shift associated with a genetic bottleneck. Differences in survival between sexes may exacerbate this condition (Equation 6.127). Demographic shifts also modify N_e (Equation 6.128) and, under certain situations, influence genetic drift. As the N_e is lowered by a toxic event, the rates at which a rare allele ($p < 0.5$) is lost or a common allele ($p > 0.5$) becomes fixed are greatly accelerated. Additionally, the probability of a loss of total genetic diversity would be increased if contamination reduces N_e.

6.6.1.4 *Wahlund Effect*

The Hardy–Weinberg model also assumes that there is no significant migration into a population. If this is not true, the observed genotype frequencies measured in the population are composites of frequencies of the initial population and the population of migrants. The Wahlund effect can occur if these separate populations are themselves in Hardy–Weinberg equilibrium. The Wahlund effect is simply that immediately after combination but before breeding, the composite population such as that just described will have a deficit of heterozygotes (Hartl and Clark 1989). (This may have been the reason for the decrease in heterozygotes noted in Example 6.21.) Endler (1986) noted that the Wahlund effect can befuddle detection of selection in natural populations if the population dynamics are inadequately understood. He further suggested that if the sampling area changes between sampling periods, the Wahlund effect can occur due to undetected differences in genetic structure over the various areas sampled.

One generation after mixing and mating, the resulting population will come to Hardy–Weinberg equilibrium and the frequencies of homozygotes in the resulting population will be lower than those in the separate populations. This reduction in homozygotes after mixing and breeding is called the Wahlund principle. (For richer detail, see Hartl and Clark 1989.)

6.6.1.5 Natural Selection

Deviation from Hardy–Weinberg expectations could also indicate selection. As discussed, selection acts on differences in fitness between individuals. Specifically, fitness differences exist in mating or fertilization ability, fertility, fecundity, or survival (Endler 1986). Selection may be directional. For a quantitative trait (a "continuous" trait affected by several loci), this may involve a directional shift in the distribution of individual qualities, e.g., toward individuals able to produce the most stress protein. Directional selection can occur for a "Mendelian" trait involving one locus if one homozygote has higher fitness than the other genotypes at the pertinent locus. Alternatively, selection may be normalizing. Fitness may be highest for individuals toward the middle of a distribution for a quantitative character or for the heterozygote for a simple, Mendelian character. In contrast, with disruptive selection, fitness is lowest for the individuals toward the center of a distribution or for the heterozygote.

Selection associated with mortality (viability) can be measured by examining the genotype frequencies of adults and comparing these frequencies to Hardy–Weinberg expectations. For example, let a two-allele locus have expected genotype frequencies of 100/100 = p^2, 100/165 = $2pq$, and 100/165 = q^2. With selection, the measured frequencies are w_1p^2, w_22pq, and w_3q^2, respectively. The values of w_1, w_2, and w_3 are the relative fitnesses of the three genotypes.

Example 6.22

Mosquitofish populations from areas experiencing chronic exposure to agricultural pesticides are found to possess high frequencies of a tolerance allele, T. Controlled matings of laboratory-reared fish produce 200 TT, 200 TS, and 200 SS fish. Both the T (p) and S (q) allele frequencies are 0.50. After exposure to pesticide, the surviving numbers of the three genotypes were TT = 195, TS = 123, and SS = 14. The relative fitnesses for the TT (w_1), TS (w_2), and SS (w_3) genotypes were estimated.

Genotype	Initial Number	Surviving Number	Proportion Alive	Normalized to TT (w)
TT	200	195	0.975	1.000
TS	200	123	0.615	0.631
SS	200	14	0.070	0.072

The relative fitnesses of the SS homozygote (w = 0.072) and TS heterozygote (w = 0.631) are considerably lower than that of the TT homozygote.

Natural selection is extremely difficult to demonstrate convincingly in natural populations. Estimation of power is critical in such studies because the absence of statistically significant differences in fitness between genotypes does not prove that there are no biologically significant differences in fitness. Lewontin and Cockerham (1957) present statistical methods of testing for significant natural selection and for estimation of associated power. It is illustrative to note that, with differences in fitnesses as large as w = 1.0 for genotype bb and w = 0.5 for genotype BB, sample sizes of 378 individuals are required to ensure detection of selection in 90% of all studied cases

(Table 1 in Lewontin and Cockerham 1957). With $w = 0.8$ for bb and $w = 1.0$ for BB, the required sample size increases to 3,405! These are impressive sample sizes if one keeps in mind that a selection coefficient in the range of a 1% difference can have a very significant effect on the change in allele frequencies (Hartl and Clark 1989). (The relationship between fitness and the selection coefficient is $s = 1 - w$. It is the relative decrease with selection (Wilson and Bossert 1971).) Hartl and Clark (1989) suggested that studies lacking the power to detect differences of 10% or less are of questionable value.

Inherent in the task of detecting natural selection are many logistical problems as tabulated by Endler (1986, pp. 99, 108, 116). Further, approaches taken in many cases are inferentially weak. Endler (1986) cited three flaws in most such studies. First, many studies only focus on one component of fitness and fail to estimate the lifetime fitness of individuals. For example, survival under stress is examined but reproductive fitness is ignored. Second, only one or a few traits are studied with little attention to other traits or trait interactions. Third, the function of the trait being studied is often vague or undefined. The reason or mechanism for differential fitness might remain undetermined.

This last point is particularly damning in ecotoxicology, where understanding of the mechanism is often as important as detecting the presence of selection (Newman et al. 1989). Fitness is a term linked to a specific set of conditions. As conditions shift, the relative fitnesses may also change. Without knowledge of the underlying mechanisms, inference about selection under differing conditions is questionable. "The time has passed for 'quick and dirty' studies of natural selection" (Endler 1986).

There is insufficient space here to describe detailed methods used to test for directional, disruptive, or stabilizing selection. The reader is directed to Chapter 6 of Endler (1986), Chapter 5 and other chapters in Manly (1985), and Arnold and Wade (1984) for more information.

6.6.1.6 Quantitative Genetics

Some traits are influenced by many genes and their expression may be strongly influenced by the environment. The time to death for an organism is an excellent example of such a trait. Such quantitative traits are not often analyzed as Mendelian traits. Methods developed by agricultural scientists and breeders are applied instead.

Often regression analysis is used to examine such traits. A large set of paired measurements for a trait expressed in an offspring and its parent may be used. A significant slope suggests a relationship; however, an environment shared by the parent and offspring may also result in a significant slope. (The phenotype, not the genotype, is being measured.) Mitchell-Olds and Shaw (1987) provide a useful review of regression techniques in this area.

Wilson and Bossert (1971), Hartl and Clark (1989), and Stearns (1992) define phenotypic variance with Equation (6.131). This equation is based on the assumption of no significant genetic-environmental interactions.

$$P = \mu + G + E \tag{6.131}$$

where P = the phenotype of an individual, μ = the population mean of the trait, G = the deviation from μ resulting from the individual's genotype, and E = the deviation from μ resulting from the individual's microenvironment or consequences of development.

The total phenotype variance in a population (σ_p^2) is the sum of the variance due to the genotype (σ_g^2) and variance due to the environment (σ_e^2). The σ_g^2 is composed of the additive effects of all pertinent genes (σ_a^2), epistatic variance due to gene interactions (σ_i^2), and dominance variance (σ_d^2). The heritability of a trait may be described in these terms.

Narrow sense heritability (h^2) is simply the sum of the variance due to additive genetic factors divided by the total phenotype variance, σ_a^2/σ_p^2 (Hartl and Clark 1989; Stearns 1992). Narrow sense heritability can be calculated from the slope (b) of the regression model described above.

$$b = \frac{h^2}{2} \tag{6.132}$$

Broad sense heritability includes narrow sense heritability plus variance due to environmental and other genetic effects. It is defined as σ_g^2/σ_p^2 (Hartl and Clark 1989; Stearns 1992). An analysis of covariance (ANCOVA) may be used to determine if significant broad sense heritability is present.

Example 6.23

Lee et al. (1992) exposed mother and offspring mosquitofish to inorganic mercury and measured times to death. After birth, individuals from each brood were placed into identical 100 L pools and allowed to mature before toxic exposure to mercury. Their times to death were noted during acute exposure. Individuals within each brood shared a common mother and environment during maturation. All broods shared similar rearing conditions. Mothers were also exposed to inorganic mercury and their times to death recorded. Narrow and broad sense heritabilities for mercury tolerance were examined.

Narrow sense heritability: Ten mother-offspring pairs were used to build a regression model. Mother times to death, offspring median times to death, and offspring wet weight were included in this model. The slope was not statistically significant at $\alpha = 0.05$ (H_0: slope = 0, $t = 0.64$, p value = 0.54), indicating no detectable narrow sense heritability. The addition of covariates such as fish wet weight, log of fish wet weight, or fish standard length failed to improve the model. (Although insignificant narrow sense heritability was noted here, it has been detected by others for mercury (Blanc 1973) or lead (Burger 1974) exposure in other fish species. Klerks and Levinton (1989) and Posthuma et al. (1993) also calculated high heritability for metal tolerance in oligochaetes and springtails.)

Broad sense heritability: Seventeen broods of offspring were used in an ANCOVA with fish wet weight included to adjust for brood size differences. (See Stearns 1992 for more detail.) The SAS code pertinent to the analysis was the following:

```
PROC GLM;
  CLASS BROOD;
  MODEL KIDTTD = BROOD KIDWGT;
RUN;
```

Source	df	Sums of Squares	Mean Square	F	Prob. > F
Model	17	8877.07	522.18	2.98	0.0003
Error	115	20,134.32	175.08		
Corrected total	132	29011.39			
Brood	16	8,290.88	518.18	2.96	0.0040
Wet weight	1	586.19	586.19	3.35	0.0699

The model clearly indicated a significant effect of brood on time to death for the 133 fish (p value = 0.0040).

The lack of detectable narrow sense, but detectable broad sense, heritability relative to time to death suggests that components of variance involving nonadditive genetic, genetic-environmental, or environmental factors during maturation were important in determining time to death.

In ecotoxicology, it is also important to understand clearly the relationship (reaction norm) in which the phenotype for a particular genotype varies as a continuous function of an environmental factor (Stearns 1989). It defines the ability of a particular genotype to express different phenotypes over a gradient of some environmental factor such as temperature or contaminant level. Figure 6.13 illustrates the reaction norm concept. In the top and middle panels, for a particular genotype, a

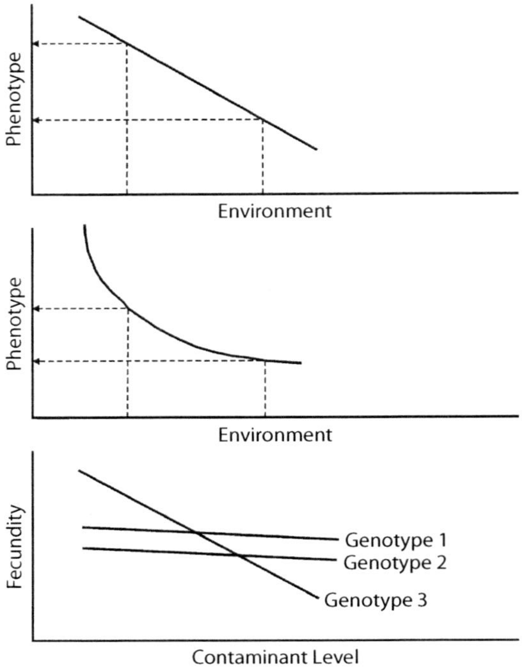

Figure 6.13 An illustration of reaction norms. The top panel shows the phenotypes expressed by a particular genotype over the range of an environmental factor. The middle panel shows a nonlinear reaction norm for a genotype. The bottom panel illustrates parallel (no significant environment by genotype interactions for genotypes 1 and 2) and nonparallel (significant environment by genotype interactions, genotype 3 versus the other two genotypes) reaction norms.

range of phenotypes is expressed over a range of intensities for some environmental factor. These phenotypes can be some surrogate measure of fitness, e.g., growth rate, fecundity, or stress protein production. The relationship can be a straight line (top panel) or a curved line (middle panel).

Reaction norms for a series of genotypes may be compared to reveal important detail, as shown in the bottom panel of Figure 6.13 for fecundity, as influenced by contaminant level. Genotypes 1 and 2 have parallel reaction norms suggesting no obvious genetic-environment interactions in the context of fecundity. Their response is similar over the range of environmental conditions, although the fecundity of genotype 2 remained consistently lower than that of genotype 1. Nonparallel reaction norms suggest genetic-environment interactions. In the context of genotype 3 versus the other two genotypes, there was a significant genetic-environment interaction. The fecundity of genotype 3 dropped much faster with increasing contaminant level than those of genotypes 1 and 2. It is critical to note that at the point at which genotypes 1 and 3 cross, there would be no detectable difference in fecundity. A study examining genotype effects at this contaminant level would lead one to the incorrect conclusion that genotype had no effect on fecundity under contaminant exposure. In fact, at lower levels of contamination, genotype 3 could have the highest fecundity of the three genotypes, yet it would have the lowest fecundity at higher levels.

Reaction norms are important to keep in mind as genetic-environmental interactions are common (Stearns 1992). The reaction norm will influence the rate of response to a selective agent, and the reaction norms themselves can be subject to natural selection (Hartl and Clark 1989). Stearns (1989, 1992) provides a thorough discussion of associated methods. Westcott (1986) discusses other methods of analyzing genetic-environmental interactions.

6.6.2 Lethal Stress (Viability)

Genetic differences in survival can be assessed using techniques described in Chapters 4 and 5. Manly (1985) suggests that the time-to-death approach described in Chapter 4 can be used very effectively for this purpose. Time-to-death methods are also beginning to be used in ecotoxicology (Dixon and Newman 1991; Diamond et al. 1989; Newman et al. 1989; Benton and Guttman 1992a, 1992b; Newman and Aplin 1992) and will be illustrated here. Other methods, such as those described in Chapters 4 and 5 for endpoint tests, are used for these purposes, e.g., Nevo et al. (1981), Lavie et al. (1984), Lavie and Nevo (1986), Hughes et al. (1992), and Kopp et al. (1992). However, statistical power is often undefined and lower with such methods, and as previously discussed, power is critically important in selection studies (Lewontin and Cockerham 1957).

An illustration of the time-to-death approach is provided in Example 6.24 for a single genetic locus. Multiple locus effects on survival (Lavie and Nevo 1988; Benton and Guttman 1992a, 1992b) may be similarly assessed by defining classes based on multiple locus combinations for incorporation into the model.

Example 6.24

Diamond et al. (1989) exposed mosquitofish of various genotypes to inorganic mercury. Their data are reanalyzed here with omission of all genetic data except those for the PGI-2 locus. The times to death were used to model and test for the potential effects of PGI-2 genotype on time to death. Fish were sexed and weighed prior to electrophoretic determination of their genotypes. The SAS PROC LIFEREG was used to produce the following output based on a Weibull distribution model:

Variable	*df*	Estimate (SE)	χ^2	Prob. of χ^2	Label
Intercept (μ)	1	4.134(0.130)	1004.1	0.0001	
Sex (β)	1		61.5	0.0001	
	1	0.358(0.046)	61.5	0.0001	Female
	0	0	—	—	Male
Wet weight (β)	1	3.157(0.361)	76.3	0.0001	
PGI-2 genotype (β)	5		17.33	0.0039	
	1	0.370(0.117)	9.90	0.0017	100/100
	1	0.468(0.117)	16.06	0.0001	100/66
	1	0.362(0.123)	8.66	0.0033	100/38
	1	0.389(0.125)	9.64	0.0019	66/66
	1	0.339(0.141)	5.79	0.0161	66/38
	1	0	—	—	38/38
Scale (σ)	1	0.514(0.091)			

The wet weight and sex of a mosquitofish have a significant (α = 0.05) effect on time to death, as indicated by the probability of getting a χ^2 as large as the calculated χ^2 by chance alone. As discussed in Chapter 4, the estimates of β for the two sexes indicate that males were more sensitive than females during mercury exposure. The positive β for wet weight also suggests that smaller fish were more sensitive than larger fish.

The overall χ^2 for PGI-2 genotype indicates a significant genotype effect on time to death. The reference genotype (38/38) had a time to death significantly shorter than those of all of the other genotypes, as indicated by the associated χ^2 values. For example, χ^2 = 9.90 was calculated for the H_0 that the β for the 100/100 genotype was not significantly different from that of the 38/38 genotype. (Because the 38/38 genotype was the reference genotype, its β was set at 0.)

The relative risks of the various genotypes can be used to estimate relative fitnesses. As described in Chapter 4, relative risk for a class variable such as genotype or sex is $e^{-\beta/\sigma}$.

Genotype	β	Relative Risk	w
100/100	0.370	$e^{-0.370/0.514} = 0.487$	0.82
100/66	0.468	$e^{-0.468/0.514} = 0.402$	1.00
100/38	0.362	$e^{-0.362/0.514} = 0.494$	0.81
66/66	0.389	$e^{-0.389/0.514} = 0.469$	0.86
66/38	0.339	$e^{-0.339/0.514} = 0.517$	0.78
38/38	0	$e^{-0/0.514} = 1.000$	0.40

These relative risks (column 3) are interpreted as described previously. For example, the risk of death for the 38/38 homozygote is roughly twice (1.000/0.487) that of the 100/100 homozygote. These relative risks can be converted to relative fitnesses (*w* values). This is done by making the *w* for the most fit genotype (100/66) equal to 1 (0.402/0.402) and all other genotypes' relative fitnesses some amount less than 1 (right column above) by dividing 0.402 by the relative risk of the genotype in question. These *w* values may now be used in conventional population genetics models.

The number of scored loci for which an individual is heterozyous has been correlated with a variety of surrogate measurements of fitness (Samallow and Soule 1983; Garton et al. 1984; Koehn and Gaffney 1984; Danzmann et al. 1986; Ferguson 1986; Mitton et al. 1986; Leary et al. 1987; Pemberton et al. 1988). For example, Diamond et al. (1989) and Newman et al. (1989) examined the number of heterozygous loci over eight scored loci as a measure of heterozygosity and found correlation with mosquitofish time to death during acute mercury or arsenate exposure. These correlations between measures of fitness and heterozygosity may be a product of heterosis such as that described by Watt (1983). (Heterosis is the superior performance of heterozygotes.) Trehan and Gill (1987) support this mechanism by demonstrating superior acid phosphatase activity for heterozygous *Drosophila*. Inbreeding depression and overdominance may also provide mechanisms for this multiple locus heterozygosity advantage (Turelli and Ginzburg 1983; Smouse 1986). However, in some cases, an apparent heterozygosity effect is simply an artifact arising from the additive effects of individual loci (Newman et al. 1989). In still other cases, there is no detectable multiple locus heterozygosity advantage.

A multiple locus heterozygosity effect on survival has been tentatively identified in some studies of pollutant effects on aquatic populations (Diamond et al. 1989; Newman et al. 1989; Kopp et al. 1992). However, as just mentioned, Newman et al. (1989) demonstrated that multiple locus heterozygosity effects can be artifacts. Furthermore, Kopp et al. (1992) found lower levels of heterozygosity at contaminated (low pH/high aluminum) sites than at clean sites, results inconsistent with their laboratory studies that predicted high heterozygosity in survivors of contaminated conditions. Field surveys (Battaglia et al. 1980; Kopp et al. 1992) also note drops in multiple locus heterozygosity with increased pollution. Selection against heterozygotes at several loci is the most often invoked mechanism for explanation of this drop in multiple locus heterozygosity (Battaglia et al. 1980; Kopp et al. 1992). However, several of the mechanisms described above (e.g., genetic bottlenecks, accelerated drift associated with low N_e, Wahlund effect) seem probable but untested explanations. Furthermore, interpretation of such findings is confounded by selection involving other unexamined components of fitness. Indeed, Nadeau and Baccus (1981) and Clegg et al. (1978) suggest that selection for reproductive components is more common than survival-related selection and can be in the opposite direction as survival selection. Thus, there is a high risk of inaccurate conclusions based only on survival differentials. Unfortunately, these studies are too preliminary to provide any strong inferences at this time. However, they do suggest an additional tool with which population responses may be assessed.

In addition to relationships between individual heterozygosity and fitness, averaged over all individuals, heterozygosity describes population genetic diversity. A drop in mean heterozygosity due to any contaminant-related mechanism can indicate a decrease in the overall genetic diversity of the population. As the long-term viability of the population (evolutionary potential) depends on genetic diversity, these changes are important regardless of any implied selection based on multiple locus heterozygosity.

In addition to heterozygosity, several indices also describe genetic diversity. These include counts of polymorphic loci (e.g., 9 of 12 loci were polymorphic) often expressed as proportions (e.g., $P = 0.66$), or mean number of alleles per scored locus (e.g., 1.75 alleles/scored locus). Leberg (1992) suggested that the number of polymorphic loci or average number of alleles per locus provide the best measures of genetic diversity if genetic bottlenecks are suspected.

The average heterozygosity and associated within-population variance for a sample of n individuals could be estimated (Weir 1990).

$$\bar{H}_l = \frac{1}{n}\sum_{j=1}^{n} x_{jl} \tag{6.133}$$

$$Variance\ \bar{H}_l = \frac{1}{n}\bar{H}_l(1-\bar{H}_l) \tag{6.134}$$

where $x_{jl} = 1$ if the individual is a heterozygote, or 0 if the individual is a homozygote.

Equation (1.133) estimates average heterozygosity for a locus (l) by simply dividing the sum of x_{jl} values by n. H_l values may be averaged for several (m) loci.

This average heterozygosity fails to consider genetic variance information associated with different homozygotes. Weir (1990) suggests that for inbred populations that have few heterozygotes but several different homozygotes, other indices of genetic diversity such as Equation (6.135) are better than Equation (6.133) as a measure of genetic variability. The genetic diversity (Equation 6.136) averaged over m loci can also be used to estimate genetic variation.

$$D_l = 1 - \sum_{u=1}^{f} \bar{P}_{lu}^2 \tag{6.135}$$

$$\bar{D} = 1 - \frac{1}{m}\sum_{l=1}^{m}\sum_{u=1}^{f} \bar{P}_{lu}^2 \tag{6.136}$$

where $\bar{P}_{lu}$ = the measured frequency of the uth allele at the lth locus, and f = the total number of alleles at the lth locus.

6.6.3 Selection Components

Differential survival (viability or zygotic selection) is only one of several important components of selection. (Figure 6.14 shows the major selection components in the context of the life cycle of a sexual organism.) Viability selection begins at zygote formation and continues through the lifetime of the individual. It may be examined at various ages or stages, e.g., Christiansen et al. (1974). Viability selection can also affect reproductive success because an individual living longer than

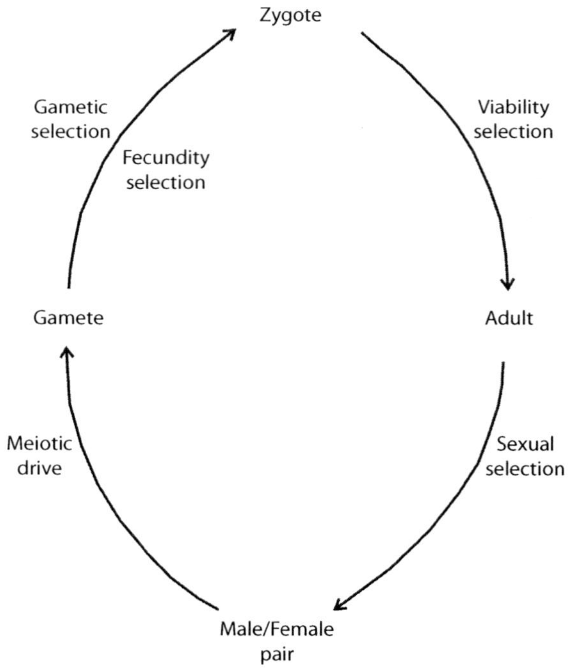

Figure 6.14 Components of selection. (Based on Hartl, D.L., and A.G. Clark, *Principles of Population Genetics*, 2nd ed., Sinauer Associates, Sunderland, MA, 1989, Chapter 4, Figure 1.)

another can have more opportunity to mate and produce more offspring. Consequently, it is important to understand the demographic qualities of a species population when interpreting selection components (Nadeau and Baccus 1981; Nadeau et al. 1981). Sexual selection is the differential mating success of individuals with specific genotypes. It may involve mechanisms such as the classic male-male competition for access to females or female mate selection. Meiotic drive, the differential production of gametes by heterozygote parents, may also occur (Christiansen and Frydenberg 1973; Hartl and Clark 1989). Gametic selection involves the differential success of gametes produced by heterozygotes (Hartl and Clark 1989). In the classic method papers of Christiansen and coworkers (Bungaard and Christiansen 1972; Christiansen and Frydenberg 1973; Siegismund and Christiansen 1985), these last two selection components are tested together as gametic selection. In their methods, gametic selection is defined as a distorted segregation in heterozygotes. Finally, more offspring (zygotes) may be produced from matings involving different pairs of genotypes (fecundity selection). Fecundity selection is expressed through mating pairs, not individuals. Selection at these various stages in combination with the demographic qualities of a population can produce distinct differences in the rate of allele frequency change (Hartl and Clark 1989). Also, several components can act to balance each other with an epiphenomenal maintenance of polymorphism (Hartl and Clark 1989) that would be inexplicable if only one component of fitness were quantified.

Prout (1965) sampled individuals in successive generations to examine selection components. Christiansen and Frydenberg (1973) developed a selection component analysis scheme that can be applied to populations of sexual, live-bearing species. Genotypes are determined for adults, gravid females, and one offspring per gravid female. In their analysis, expected and observed frequencies are tested with χ^2 statistics for the following series of sequential hypotheses. It is important to note that the hypotheses below are nested and, upon rejection of a hypothesis, no further hypotheses can be tested. The sequence of hypotheses from Table IV of Christiansen and Frydenberg (1973) is outlined below.

H_1: Half of the offspring from heterozygous females are heterozygous; i.e., there is no selection among female gametes. Rejection of this hypothesis suggests gametic selection in heterozygous females.

H_2: The frequency of transmitted male gametes is independent of female genotype. The male gametes are transmitted identically among all female genotypes. Provided H_1 was not rejected, rejection of H_2 suggests nonrandom mating with female sexual selection.

H_3: The frequency of transmitted male gametes is equal to the frequency in adult males; i.e., all males mate with equal success. Rejection implies differential male mating success and gametic selection in males if H_1 and H_2 were not rejected prior to this test.

H_4: The frequencies of genotypes are equal among gravid and nongravid (mature) females. Rejection implies differential female mating success if previous hypotheses were not rejected.

H_5: Equal frequencies of genotypes between adult males and females. Rejection suggests that zygotic selection is not the same for males and females if H_1 to H_4 were not previously rejected.

H_6: The adult population frequency is the same as that estimated for the zygotic population. Rejection suggests zygotic (survival or viability) selection.

Acceptance of all six hypotheses implies no detected selection at any component, i.e., no evidence that the gene is not neutral.

The space available here is insufficient for providing additional details of this approach. More information can be derived from Christiansen and Frydenberg (1973), Hartl and Clark (1989), and Weir (1990). Nadeau and Baccus (1981) and Nadeau et al (1981) describe a modification that uses genotypes of all offspring for each female to remove the interdependence of the sequence of hypotheses outlined above. By performing this additional work, most hypotheses can then be tested independently. Williams et al. (1990) provide a comprehensive selection component technique using linear modeling methods.

Example 6.25

Genetic changes were noted in mosquitofish (*Gambusia holbrooki*) populations exposed to inorganic mercury in 7,250 L mesocosms (fictitious example derived from Mulvey et al. 1995). Fish were harvested from the mesocosms (duplicate mesocosms A and B for each treatment) and their genotypes at eight loci determined by starch gel electrophoresis. One late-stage embryo was removed from each gravid female for electrophoresis. The results (probabilities associated with the χ^2 tests) from selection component analysis of data for the PGI-2 locus are shown below. Three allozymes were scored for this locus. For ease of analysis, the rare PGI-2 alleles (38, 66) were pooled in these analyses yielding three different genotypes, 100/100, 100/—, or —/—. The — indicates either a 38 or 66 allele.

	Treatment			
	0 μg/L		**20 μg/L**	
Hypothesis Implying	**A**	**B**	**A**	**B**
Female gametic selection	0.54	0.64	0.07	0.71
Random mating	0.76	0.96	0.51	0.91
Male reproductive selection	0.70	0.88	0.07	0.73
Female sexual selection	0.55	0.52	0.01	0.09
All components listed above	0.76	0.85	0.01	0.54
Zygotic selection equal in sexes	0.54	0.18	<0.01	0.26
Zygotic selection	0.68	0.19	0.42	0.58
All zygotic selection components	0.62	0.16	0.02	0.35
Total fit to random mating and neutrality	0.78	0.61	<0.01	0.51

There was no evidence of selection at the PGI-2 locus in the 0 μg/L mesocosms. In the 20 μg/L mesocosms, there was significant ($\alpha = 0.05$) deviation from the expected equality of genotypes between gravid and nongravid females in duplicate A. Female sexual selection was implied. Note that testing stops at this hypothesis and no further hypotheses (male sexual selection or zygotic selection) can be tested.

Which genotypes did best relative to the female sexual selection? The percentage of mature females that were gravid are the following:

PGI-2 Genotype	0 μg/L	20 μg/L
100/100	69.6%	43.3%
100/—	68.8%	70.7%
—/—	67.8%	68.0%

Clearly, the PGI-2 100/100 homozygotes had lower reproductive success (female sexual selection component) under mercury exposure than the other genotypes. An ANCOVA including female size and genotype versus fecundity (the number of late-stage embryos) indicated that the 100/100 genotype also produced significantly fewer young than the other genotypes under mercury stress.

6.6.4 Tolerance

The term *tolerance* has been used to imply physiological acclimation and the term *resistance* used to imply genetic adaptation. Conforming to Weis and Weis (1989), the term *tolerance* will be used here to cover both acclimation and genetic adaptation. Distinction will be made by using the terms *acclimation* and *genetic adaptation*, not *tolerance* and *resistance*. The focus in this portion of the chapter will be genetic adaptation (to toxicants) as expressed through the phenotype, "a change in a phenotype that occurs in response to a specific environmental signal and has a clear functional relationship to that signal that results in an improvement in growth, survival, or reproduction" (Stearns 1992).

Adaptation to pollutants can involve a wide range of mechanisms, including detoxification (Fabacher and Chambers 1972; Chambers and Yarbrough 1979; Angus 1983; Klerks and Bartholomew 1991), transport (Yarbrough 1974; Wood and Wang 1983), essential element regulation (Beeby 1991), and behavior (Kynard 1974). Adaptation can occur rapidly. For example, Klerks and Levinton (1989) estimate heritabilities of 0.59 to 1.08 for metal tolerance in an oligochaete species that result in clear adaptation within one to four generations. Once selection pressure is removed, loss of tolerance may be equally rapid due to the costs of maintaining the tolerance mechanism (Mulvey and Diamond 1991).

The rate of tolerance acquisition is modified by many factors as tabulated by Mulvey and Diamond (1991). Generally, selection will be faster for a trait under monogenic control than for a trait under polygenic control. Tolerance associated with a dominant gene will be selected for much more rapidly in early generations than tolerance associated with a recessive gene. The rate of adaptation is generally faster with shorter generation times and higher population growth rates. The presence of uncontaminated patches acting as refugia or a significant influx of migrants will lower the rate of adaptation.

An extremely important but generally neglected aspect of adaptation is the concept of adaptive constraint. Stearns (1992) provides the following explanations of constraint: "any pattern or state that can be attributed to phylogeny, as opposed to recent microevolution within the currently existing population," or "genes act through proteins that change the properties of cells, that cells are the key players in development, and that cells interact in processes constrained by physics in ways that cannot be changed by simple gene substitutions—steered a bit, modified slightly, but not fundamentally changed. Only a small portion of phenotype space is then available for exploration by gene substitutions." Although considerable effort is expended in detection and quantification of population adaptation to pollutants, the limits of such adaptation often remain vaguely defined. Klerks and Weis (1987) and Mulvey and Diamond (1991) correctly point out that observed enhanced tolerance in extant populations from contaminated environments likely represents the exceptional outcome of species fate. The general loss of species reflects the other, more common fate of species as a consequence of adaptive constraints.

Such adaptive constraints can beget selection constraints: selection can enhance toxicant tolerance only so far. Further, strong selection can lead to reduced genetic diversity and loss of alleles. Selection constraints can become more severe with reductions in population size and genetic variation within the population (Stearns 1992). Consequently, reduction in population size or genetic variation by natural or pollution-related causes can restrict the population's ability to adapt to future stressors.

6.7 SUMMARY

The objective of this chapter is to provide a population framework for understanding and measuring pollution effects. The majority of the chapter presents basic ecological concepts and methods directly applicable to ecotoxicology. Examples are used to demonstrate the pertinence of each method to ecotoxicology. Basic methods are outlined to assess population size, dynamics, and demographics. Important concepts regarding the stability of population densities are discussed and related to ecotoxicology. Finally, population genetics concepts and methods are illustrated and their relevance to ecotoxicology argued. Fortunately, much progress has been made in this area since the first edition of this book (e.g., see Forbes 1999; Newman 2001; Newman and Clements 2008, Chapters 16 to 18). The importance of examining all selection components is discussed.

REFERENCES

Abraham, M.V. Patch choice under perceptual constraints: A cause for departures from an ideal free distribution. *Behav. Ecol. Sociobiol.* 19:409–415 (1986).

Ahlbom, A. *Biostatistics for epidemiologists.* Boca Raton, FL: Lewis Publishers, 1993.

Allen, J.C. Factors contributing to chaos in population feedback systems. *Ecol. Modell.* 51:281–298 (1990).

Almond, R.G. *Graphical belief modeling.* London: Chapman & Hall, 1995.

Angus, R.A. Phenol tolerance in populations of mosquitofish from polluted and nonpolluted waters. *Trans. Am. Fish. Soc.* 112:794–799 (1983).

Ares, J. Time and space issues in ecotoxicology: Population models, landscape pattern analysis, and long-range environmental chemistry. *Environ. Toxicol. Chem.* 22:945–957 (2003).

Arnold, S.J., and M.J. Wade. On measurement of natural and sexual selection: Theory. *Evolution* 38:709–719 (1984).

Bailey, N.T.J. On estimating the size of mobile populations from capture-recapture data. *Biometrika* 38:293–306 (1951).

Barnthouse, L.W., G.W. Suter II, A.E. Rosen, and J.J. Beauchamp. Estimating responses of fish populations to toxic contaminants. *Environ. Toxicol. Chem.* 6:811–824 (1987).

Barreiro, R., R. Gonzalez, M. Quintela, and J.M. Ruiz. Imposex, organotin bioaccumulation and sterility of female *Nassarius reticulates* in polluted areas of NW Spain. *Mar. Ecol.-Prog. Ser.* 218:203–212 (2001).

Battaglia, B., P.M. Bisol, V.U. Fossato, and E. Rodino. Studies on the genetic effects of pollution in the sea. *Rapp. P. V. Réun. Cons. Int. Explor. Mer.* 179:267–274 (1980).

Beddington, J.R., and G.P. Kirkwood. Fisheries. In *Theoretical ecology. Principles and applications*, ed. R.M. May and A.R. McLean. Oxford: Oxford University Press, 2007, pp. 148–157.

Beeby, A. Toxic metal uptake and essential metal regulation in terrestrial invertebrates: A review. In *Metal ecotoxicology. Concepts and applications*, ed. M.C. Newman and A.W. McIntosh. Chelsea, MI: Lewis Publishers, 1991, pp. 65–89.

Bell, G., and A. Gonzalez. Adaptation and evolutionary rescue in metapopulations experiencing environmental deterioration. *Science* 332:1327–1330 (2011).

Belousek, D.W. Scientific consensus and public policy: The case of *Pfiesteria*. *J. Philos. Sci. Law* 4:1–33 (2004).

Benton, M.J., and S.I. Guttman. Allozyme genotype and differential resistance to mercury pollution in the caddisfly, *Nectopsyche albida*. I. Single-locus genotypes. *Can. J. Fish. Aquat. Sci.* 49:142–146 (1992a).

Benton, M.J., and S.I. Guttman. Allozyme genotype and differential resistance to mercury pollution in the caddisfly, *Nectopsyche albida*. II. Multilocus genotypes. *Can. J. Fish. Aquat. Sci.* 49:147–149 (1992b).

Beyer, W.H. *CRC standard probability and statistics. Tables and formulae*. Boca Raton, FL: CRC Press, 1991.

Bezel, V.S., and V.N. Bolshakov. Population ecotoxicology of mammals. In *Bioindications of chemical and radioactive pollution*, ed. D.A. Krivolutsky. Boca Raton, FL: CRC Press, 1990, pp. 141–186.

Blanc, J.M. Genetic aspects of resistance to mercury poisoning in steelhead trout *(Salmo gairdneri)*. MS thesis, Oregon State University, 1973.

Borsuk, M.E., C.A. Stow, and K.H. Reckhow. A Bayesian network of eutrophication models of synthesis, prediction, and uncertainty analysis. *Ecol. Model.* 173:219–239 (2004).

Bryan, G.W., and P.E. Gibbs. Impact of low concentrations of tributyltin (TBT) on marine organisms: A review. In *Metal ecotoxicology. Concepts and applications*, ed. M.C. Newman and A.W. McIntosh. Chelsea, MI: Lewis Publishers, 1991, pp. 323–361.

Bungaard, J., and F.B. Christiansen. Dynamics of polymorphism. I. Selection components in an experimental population of *Drosophila melanogaster*. *Genetics* 71:439–460 (1972).

Burger, C.V. Genetic aspects of lead toxicity in laboratory populations of guppies *(Poecilia reticulata)*. MS thesis, Oregon State University, 1974.

Calow, P., and R.M. Sibly. A physiological basis of population processes: Ecotoxicological implications. *Funct. Ecol.* 4:283–288 (1990).

Calow, P., R.M. Sibly, and V. Forbes. Risk assessment on the basis of simplified life-history scenarios. *Environ. Toxicol. Chem.* 16:1983–1989 (1997).

Caswell, H. *Matrix population models. Construction, analysis and interpretation*. 2nd ed. Sunderland, MA: Sinauer Assoc., 2001.

Caswell, H., T. Takada, and C.M. Hunter. Sensitivity analysis of equilibrium in density-dependent matrix population models. *Ecol. Lett.* 7:380–387 (2004).

Cavalli-Sforza, L.L., and W.F. Bodmer. *The genetics of human populations*. San Francisco: W.H. Freeman and Co., 1971.

Chambers, J.E., and J.D. Yarbrough. A seasonal study of microsomal mixed-function oxidase components in insecticide-resistant and susceptible mosquitofish, *Gambusia affinis*. *Toxicol. Appl. Pharmacol.* 48:497–507 (1979).

Chan, K.M., K.M.Y. Leung, K.C. Cheung, M.H. Wong, and J.-W. Qiu. Seasonal changes in imposex and tissue burden of butyltin compounds in *Thais clavigera* populations along the coastal area of Mirs Bay, China. *Mar. Pollut. Bull.* 57:645–61 (2008).

Chandler, G.T., T.L. Cary, A.C. Bejarano, J. Pender, and J.L. Ferry. Population consequences of fipronil and degradates to copepods at field concentrations: An integration of life cycle testing with Leslie matrix population modeling. *Environ. Sci. Technol.* 38:6407–6414 (2004).

Chapman, D.G. Some properties of the hypergeometric distribution with applications to zoological samples censuses. *Univ. Calif. Publ. Stat.* 1(7):131–160 (1951).

Chaumot, A., S. Charles, P. Flammarion, and P. Auger. Ecotoxicology and spatial modeling in population dynamics: An illustration with brown trout. *Environ. Toxicol. Chem.* 22:958–969 (2003).

Cheung, M.A., H.Y.M. Leung, and K.M.Y. Leung. Vignette 8.2: The use of neogastropods as an indicator of tributyltin contamination along the south China coast. In *Fundamentals of ecotoxicology*, ed. M.C. Newman. 3rd ed. Boca Raton, FL: Taylor & Francis/CRC Press, 2010, pp. 222–226.

Christensen, E.R., and N. Nyholm. Ecotoxicological assays with algae: Weibull dose-response curves. *Environ. Sci. Technol.* 18(9):713–718 (1984).

Christiansen, F.B., and O. Frydenberg. Selection component analysis of natural polymorphisms using population samples including mother-offspring combinations. *Theor. Popul. Biol.* 4:425–445 (1973).

Christiansen, F.B., O. Frydenberg, A.O. Gyldenholm, and V. Simonsen. Genetics of *Zoarces* populations VI. Further evidence, based on age group samples, of a heterozygote deficit in the EstIII polymorphism. *Hereditas* 77:225–236 (1974).

Clark, P.J., and F.C. Evans. Distance to nearest neighbor as a measure of spatial relationships in populations. *Ecology* 35:445–453 (1954).

Clegg, M.T., A.L. Kahler, and R.W. Allard. Estimation of life cycle components of selection in an experimental plant population. *Genetics* 89:765–792 (1978).

Cressie, N.A.C. *Statistics for spatial data.* New York: John Wiley & Sons, 1991.

Crow, J.F., and M. Kimura. *An introduction to population genetics theory.* Minneapolis: Burgess Publishing Company, 1970.

Cushing, J.M., and Z. Yicang. The net reproductive value and stability in matrix population models. *Nat. Resour. Model.* 8:297–333 (1994).

Daniels, R.E., and J.D. Allan. Life table evaluation of chronic exposure to a pesticide. *Canadian J. Fish. Aquat. Sci.* 38:485–494 (1981).

Danzmann, R.G., M.M. Ferguson, F.W. Allendorf, and K.L. Knudsen. Heterozygosity and developmental rate in a strain of rainbow trout (*Salmo gairdneri*). *Evolution* 40:86–93 (1986).

David, F.N., and P.G. Moore. Notes on contagious distributions in plant populations. *Ann. Bot. (Lond.)* 18:47–53 (1954).

Deevey, E.S. Life tables for natural populations. *Q. Rev. Biol.* 22:283–314 (1947).

de Jager, M., F.J. Weissing, P.M. Herman, B.A. Nolet, and J. von de Koppel. Levy walks evolve through interaction between movement and environmental complexity. *Science* 332:1551–1553 (2011).

de Kroon, H., A. Plaisier, J. van Groenenadael, and H. Caswell. Elasticity: The relative contribution of demographic parameters to population growth rate. *Ecology* 67:1427–1431 (1986).

de Wolf, H., C. Handa, T. Backeljau, and R. Blust. A baseline survey of intersex in *Littorina littorea* along the Scheldt estuary, the Netherlands. *Mar. Pollut. Bull.* 48:587–603 (2004).

Diamond, S.A., M.C. Newman, M. Mulvey, and D. Martinson. Allozyme genotype and time to death of mosquitofish, *Gambusia affinis* (Baird and Girard), during acute exposure to inorganic mercury. *Environ. Toxicol. Chem.* 8:613–622 (1989).

Dixon, P.M., and M.C. Newman. Analyzing toxicity data using statistical models for time-to-death: An introduction. In *Metal ecotoxicology. Concepts and applications*, ed. M.C. Newman and A.W. McIntosh. Chelsea, MI.: Lewis Publishers, 1991, pp. 207–242.

Donovan, T.M., and C.W. Welden. *Spreadsheet exercises in conservation biology and landscape ecology.* Sunderland, MA: Sinauer Assoc., 2002.

Efford, M.G., D.K. Dawson, and C.S. Robbins. DENSITY: Software for analyzing capture-recapture data from passive detector arrays. *Anim. Biodiv. Conserv.* 27:217–228 (2004).

Ehrlich, P.R., and L.C. Birch. The "balance of nature" and population control. *Am. Nat.* 101(918):97–107 (1967).

Emlen, J.M., and K.R. Springman. Developing methods to assess and predict the population level of effects of environmental contaminants. *Integr. Environ. Assess. Manag.* 3:157–165 (2007).

Emmel, T.C. *Population biology.* New York: Harper and Row Publishers, 1976.

Endler, J.A. *Natural selection in the wild.* Princeton, NJ: Princeton University Press, 1986.

Evans, A.S. Causation and disease: The Henle-Koch postulates revisited. *Yale J. Biol. Med.* 49:175–195 (1976).

Fabacher, D.L., and H. Chambers. Rotenone tolerance in mosquitofish. *Environ. Pollut.* 3:139–141 (1972).

Ferguson, M.M. Developmental stability in rainbow trout hybrids: Genomic coadaptation or heterozygosity? *Evolution* 40:323–330 (1986).

Fisher, R.A. *The genetic theory of natural selection*. New York: Dover, 1930.

Forbes, V.E. *Genetics and ecotoxicology*. Philadelphia: Taylor & Francis, 1999.

Forbes, V.E., and P. Calow. Is the per capita rate of increase a good measure of population-level effects in ecotoxicology? *Environ. Toxicol. Chem.* 18:1544–1556 (1999).

Forbes, V.E., and P. Calow. Vignette 10.2: Contaminant effects on population demographics. In *Fundamentals of ecotoxicology*, ed. M.C. Newman. 3rd ed. Boca Raton, FL: Taylor & Francis/CRC Press, 2010, pp. 293–297.

Forbes, V.E., P. Calow, and R.M. Sibly. Are current species extrapolation models a good basis for ecological risk assessment? *Environ. Toxicol. Chem.* 20:442–447 (2001).

Forbes, V.E., P. Calow, and R.M. Sibly. The extrapolation problem and how population modeling can help. *Environ. Toxicol. Chem.* 27:1987–1994 (2008).

Forbes, V.E., M. Olsen, A. Palmqvist, and P. Calow. Environmentally sensitive life-cycle traits have low elasticity: Implications for theory and practice. *Ecol. Appl.* 20:1449–1455 (2010).

Fox, G.A. Life tables and statistical inferences. *Bull. Ecol. Soc. Am.* 70:229–230 (1989).

Fox, G.A. Practical causal inference for ecoepidemiologists. *J. Toxicol. Environ. Health* 33:359–373 (1991).

Fretwell, S.D., and H.L. Lucas Jr. On territorial behavior and other factors influencing habitat distribution in birds. *Acta Biotheor.* 19:16–36 (1970).
Garbolino, P., and F. Taroni. Evaluation of scientific evidence using Bayesian networks. *Forensic Sci. Int.* 125:149–155 (2002).
Garton, D.W., R.K. Koehn, and T.M. Scott. Multiple locus heterozygosity and physiological energetics of growth in the coot clam, *Mulinin lateralis*, from a natural population. *Genetics* 108:445–455 (1984).
Gause, G.F. The influence of ecological factors on the size of population. *Am. Nat.* 65:70–76 (1931).
Gelman, A.B., J.S. Carlin, H.S. Stern, and D.B. Rubin. *Bayesian data analysis*. Boca Raton, FL: Chapman & Hall/CRC, 1997.
Gilpin, M.E., and F.J. Ayala. Global models of growth and competition. *Proc. Natl. Acad. Sci. USA* 70:3590–3593 (1973).
Ginzburg, L.R. Evolutionary consequences of basic growth equations. *Trends Ecol. Evol.* 7:133–134 (1992).
Ginzburg, L.R. Reply from L. Ginzburg. *Trends Ecol. Evol.* 8:69–70 (1993).
Glass, L., and M.C. Mackey. *From clocks to chaos. The rhythms of life.* Princeton, NJ: Princeton University Press, 1989.
Goodman, S.N. Toward evidence-based medical statistics. 2. The Bayes factor. *Ann. Intern. Med.* 130:1005–1013 (1999).
Gordis, L. *Epidemiology*. 3rd ed. Philadelphia: Elsevier Saunders, 2004.
Gotelli, N.J. Metapopulation models: The rescue effect, the propagule rain, and the core-satellite hypothesis. *Am. Nat.* 138:768–776 (1991).
Grafen, A. A theory of Fisher's reproductive value. *J. Math. Biol.* 53:15–60 (2006).
Grunbaum, D. Why did you Levy? *Science* 332:1514–1515 (2011).
Gurney, W.S.C., and R.M. Nisbet. Single-species population fluctuations in patchy environments. *Am. Nat.* 112:1075–1090 (1978).
Haddon, M. *Modeling and quantitative methods in fisheries*. Boca Raton, FL: Chapman & Hall/CRC Press, 2001.
Hamilton, M.A. Statistical analysis of the cladoceran reproductivity test. *Environ. Toxicol. Chem.* 5:205–212 (1986).
Hanski, I. Single-species metapopulation dynamics: Concepts, models and observations. *Biol. J. Linn. Soc.* 42:17–38 (1991).
Hanski, I. Metapopulation ecology. In *Population dynamics in ecological space and time*, ed. O.E. Rhodes Jr., R.K. Chesser, and M.H. Smith. Chicago: University of Chicago Press, 1996.
Hanski, I. Metapopulation dynamics. *Nature* 396:41–49 (1998).
Hanski, I., and M. Gilpin. Metapopulation dynamics: Brief history and conceptual domain. *Biol. J. Linn. Soc.* 42:3–16 (1991).
Hanski, I., and M. Gilpin. *Metapopulation biology. Ecology, genetics, and evolution*. San Diego: Academic Press, 1997.
Hanski, I., and O. Ovaskainen. Metapopulation theory from fragmented landscapes. *Theor. Popul. Biol.* 64:119–127 (2003).
Harding, K.C., and J.M. McNamara. A unifying framework for metapopulation dynamics. *Am. Nat.* 160:173–185 (2002).
Harrison, S. Local extinction in a metapopulation context. In *Metapopulation dynamics: empirical and theoretical investigations*, ed. I. Hanski and M. Gilpin. San Diego: Academic Press, 1992, pp. 89–103.
Hartl, D.L., and A.G. Clark. *Principles of population genetics*, 2nd ed. Sunderland, MA: Sinauer Associates, 1989.
Hassell, M.P. Density-dependence in single-species populations. *J. Anim. Ecol.* 44:283–296 (1974).
Hassell, M.P., J.H. Lawton, and R.M. May. Patterns of dynamical behavior in single-species populations. *J. Anim. Ecol.* 45:471–486 (1976).
Hendricks, W.A. *The mathematical theory of sampling*. New Brunswick: Scarecrow Publishers, 1956.
Hill, A.B. The environment and disease: Association or causation? *Proc. R. Soc. Med.* 58:295–300 (1965).
Holden, A.V. *Chaos*. Princeton, NJ: Princeton University Press, 1986.
Horness, B.H., D.P. Lomax, L.L. Johnson, M.S. Myers, S.M. Pierce, and T.K. Collier. Sediment quality thresholds: Estimates from hockey stick regression of liver lesion prevalence in English sole (*Pleuronectes vetulus*). *Environ. Toxicol. Chem.* 17:872–882 (1998).
Huggins, R., H.-C. Yang, A. Chao, and P.S.F. Yip. Population size estimation using local sample coverage for open populations. *J. Stat. Plan. Infer.* 113:699–714 (2003).
Hughes, J.M., M.W. Griffths, and D.A. Harrison. The effects of an organophosphate insecticide on two enzyme loci in the shrimp *Caradina* sp. *Biochem. Syst. Ecol.* 20:89–97 (1992).

Hurlbert, S.H. Spatial distribution of the montane unicorn. *Oikos* 58:257–271 (1990).

Hutchinson, G.E. Circular causal systems in ecology. *Ann. NY Acad. Sci.* 50:221–246 (1948).

Hutchinson, T.A., and D.A. Lane. Assessing methods for causality assessment of suspected adverse drug reactions. *J. Clin. Epidemiol.* 42:5–16 (1989).

Jensen, A., V.E. Forbes, and E. Davis Parker Jr. Variation in cadmium uptake, feeding rate, and life-history effects in the gastropod *Potamopyrgus antipodarum*: Linking toxicant effects on individuals to the population level. *Environ. Toxicol. Chem.* 20:2503–2513 (2001).

Klerks, P.L., and P.R. Bartholomew. Cadmium accumulation and detoxification in a Cd-resistant population of the oligochaete *Limnodrilus hoffmeisteri*. *Aquat. Toxicol.* 19:97–112 (1991).

Klerks, P.L., and J.S. Levinton. Rapid evolution of metal resistance in a benthic oligochaete inhabiting a metal-polluted site. *Biol. Bull.* 176:135–141 (1989).

Klerks, P.L., and J.S. Weis. Genetic adaptation to heavy metals in aquatic organisms: A review. *Environ. Pollut.* 45:173–205 (1987).

Koehn, R.K., and P.M. Gaffney. Genetic heterozygosity and growth rate in *Mytilus edulis*. *Mar. Biol.* 82:1–7 (1984).

Kopp, R.L., S.I. Guttman, and T.E. Wissing. Genetic indicators of environmental stress in central mudminnow (*Umbra limi*) populations exposed to acid deposition in the Adirondack Mountains. *Environ. Toxicol. Chem.* 11:665–676 (1992).

Kot, M., W.M. Schaffer, G.L. Truty, D.J. Graser, and L.F. Olsen. Changing criteria for imposing order. *Ecol. Modell.* 43:75–110 (1988).

Krebs, C.J. *Ecology. The experimental analysis of distribution and abundance.* New York: Harper and Row Publishers, 1972.

Krebs, C.J. *Ecological methodology.* New York: Harper Collins Publishers, 1989.

Krebs, C.J. *Ecological methodology.* Menlo Park, CA: Addison-Wesley Educational Publishers, 1999.

Kynard, B. Avoidance behavior of insecticide susceptible and resistant populations of mosquitofish to four insecticides. *Trans. Am. Fish. Soc.* 3:557–561 (1974).

Laake, J.L., K.P. Burnham, and D.R. Anderson. *User's guide for program TRANSECT.* Logan UT: Utah State University Press, 1979.

Lahr, J. Ecotoxicology of organisms adapted to life in temporary freshwater ponds in arid and semi-arid regions. *Arch. Environ. Contam. Toxicol.* 32:50–57 (1997).

Lanciani, C.A. A simple equation for presenting reproductive value to introductory biology and ecology classes. *Bull. ESA* 79:192–193 (1998).

Lane, D.A., M.S. Kramer, T.A. Hutchinson, J.K. Jones, and C. Naranjo. The causality assessment of adverse drug reactions using a Bayesian approach. *Pharmaceut. Med.* 2:265–283 (1987).

Laskowski, R. Why short-term bioassays are not meaningful—Effects of a pesticide (Imidacloprid) and a metal (cadmium) on pea aphids (*Acyrthosiphon pisum* Harris). *Ecotoxicology* 10:177–183 (2001).

Lavie, B., and E. Nevo. Genetic selection of homozygote allozyme genotypes in marine gastropods exposed to cadmium pollution. *Sci. Total Environ.* 57:91–98 (1986).

Lavie, B., and E. Nevo. Multilocus genetic resistance and susceptibility to mercury and cadmium pollution in the marine gastropod, *Cerithium scabridum*. *Aquat. Toxicol.* 13:291–296 (1988).

Lavie, B., E. Nevo, and U. Zoller. Differential viability of phosphoglucose isomerase allozyme genotypes of marine snails in nonionic detergent and crude oil-surfactant mixtures. *Environ. Res.* 35:270–276 (1984).

Leary, R.F., F.W. Allendorf, and K.L. Knudsen. Differences in inbreeding coefficients do not explain the association between heterozygosity at allozyme loci and developmental stability in rainbow trout. *Evolution* 41:1413–1415 (1987).

Leberg, P.L. Effects of population bottlenecks on genetic diversity as measured by allozyme electrophoresis. *Evolution* 46:477–494 (1992).

Lebreton, J.-D., R. Pradel, and J. Clobert. The statistical analysis of survival in animal populations. *Trends Ecol. Evol.* 8(3):91–95 (1993).

Lee, C.J., M.C. Newman, and M. Mulvey. Time to death of mosquitofish (*Gambusia holbrooki*) during acute inorganic mercury exposure: Population structure effects. *Arch. Environ. Contam. Toxicol.* 22:284–287 (1992).

Lefkovitch, L.P. The study of population growth in organisms grouped by stages. *Biometrics* 21:1–18 (1965).

Leslie, P.H. On the use of matrices in certain population mathematics. *Biometrika* 33:183–212 (1945).

Leslie, P.H. Some further notes on the use of matrices in population mathematics. *Biometrika* 35:213–245 (1948).

Levins, R. Some demographic and genetic consequences of environmental heterogeneity for biological control. *Bull. Entomol. Soc. Am.* 15:237–240 (1969).

Lewin, R. Supply-side ecology. *Science* 234:25–27 (1986).

Lewin, R. Sources and sinks complicate ecology. *Science* 243:477–478 (1989).

Lewontin, R.C., and Cockerham, C.C. The goodness-of-fit for detecting natural selection in random mating populations. *Evolution* 13:561–564 (1957).

Liebig, J. *Chemistry in its application to agriculture and physiology*. London: Taylor and Walton, 1840.

Ludwig, J.A., and J.F. Reynolds. *Statistical ecology. A primer on methods and computing*. New York: John Wiley & Sons, 1988.

Manly, B.F.J. *The statistics of natural selection on animal populations*. New York: Chapman & Hall, 1985.

Marubini, E., and M. Grazia Valsecchi. *Analyzing survival data from clinical trials and observational studies*. New York: John Wiley & Sons, 1995.

May, R.M. Biological populations with nonoverlapping generations: Stable points, stable cycles, and chaos. *Science* 186:645–647 (1974).

May, R.M. *Theoretical ecology. Principles and applications*. Philadelphia: W.B. Saunders Co., 1976.

May, R.M. Nonlinear phenomena in ecology and epidemiology. *Ann. N.Y. Acad. Sci.* 357:267–281 (1980).

May, R.M. Chaos and the dynamics of biological populations. *Proc. R. Soc. Lond.* A413:27–24 (1987).

May, R.M. The chaotic rhythms of life. *New Sci.* 18:37–41 (1989).

May, R.M., and R.M. Anderson. Epidemiology and genetics in the coevolution of parasites and hosts. *Proc. R. Soc. Lond.* B219:281–313 (1983).

May, R.M., G.R. Conway, M.P. Hassell, and T.R.E. Southwood. Time delays, density-dependence and single-species oscillations. *J. Anim. Ecol.* 43:747–770 (1974).

May, R.M., and A. McLean. *Theoretical ecology. Principles and applications*. 3rd ed. Oxford: Oxford University Press, 2007.

May, R.M., and G.F. Oster. Bifurcations and dynamic complexity in simple ecological models. *Am. Nat.* 110(974):573–599 (1976).

McDonald, T.L., and S.C. Amstrup. Estimation of population size using open capture-recapture models. *J. Agric. Biol. Environ. Stat.* 6:206–220 (2001).

Mitchell-Olds, T., and R.G. Shaw. Regression analysis of natural selection: Statistical inference and biological interpretation. *Evolution* 41:1149–1161 (1987).

Mitton, J.B., C. Carey, and T.D. Kocher. The relation of enzyme heterozygosity to standard and active oxygen consumption and body size of tiger salamanders, *Ambystoma triginum*. *Physiol. Zool.* 59:574–582 (1986).

Moriarty, F. *Ecotoxicology. The study of pollutants in ecosystems*. London: Academic Press, 1983.

Mount, D.I., and T.J. Norberg. A seven-day life-cycle cladoceran toxicity test. *Environ. Toxicol. Chem.* 3:425–434 (1984).

Mulvey, M., and S.A. Diamond. Genetic factors and tolerance acquisition in populations exposed to metals and metalloids. In *Metal ecotoxicology. Concepts and applications*, ed. M.C. Newman and A.W. McIntosh. Chelsea, MI: Lewis Publishers, 1991, pp. 301–322.

Mulvey, M., M.C. Newman, A. Chazal, M.M. Keklak, M.G. Heagler, and S. Hales. Genetic and demographic changes in mosquitofish (*Gambusia holbrooki*) populations exposed to mercury, *Environ. Toxicol. Chem.* 14:1411–1418 (1995).

Mulvey, M., M.C. Newman, W. Vogelbein, and M.A. Unger. Genetic structure of *Fundulus heteroclitus* from PAH-contaminated and neighboring sites in the Elizabeth and York Rivers, *Aquat. Toxicol.* 61:195–209 (2002).

Mulvey, M., M.C. Newman, W.K. Vogelbein, M.A. Unger, and D.R. Ownby. Genetic structure and mDNA diversity of *Fundulus heteroclitus* populations from polycyclic aromatic hydrocarbon-contaminated sites, *Environ. Toxicol. Chem.* 22:671–677 (2003).

Munzinger, A., and M.-L. Guarducci. The effect of low zinc concentrations on some demographic parameters of *Biomphalaria glabrata* (Say), mollusca: Gastropoda. *Aquat. Toxicol.* 12:51–61 (1988).

Murray, J.D. *Mathematical biology*. Berlin: Springer-Verlag, 1993.

Nadeau, J.H., and R. Baccus. Selection components of four allozymes in natural populations of *Peromyscus maniculatus*. *Evolution* 35:11–20 (1981).

Nadeau, J.H., K. Dietz, and R.H. Tamarin. Gametic selection and the selection component analysis. *Genet. Res. Camb.* 37:275–284 (1981).

Neapolitan, R.E. *Learning Bayesian networks*. Upper Saddle River, NJ: Pearson/Prentice Hall, 2004.

Nee, S. Metapopulations and their spatial dynamics. In *Theoretical ecology. Principles and applications*, ed. R. May and A. McLean. 3rd ed. Oxford: Oxford University Press, 2007, pp. 35–45.

Nevo, E., T. Perl, A. Beiles, and D. Wool. Mercury selection of allozyme genotypes in shrimps. *Experientia* 37:1152–1154 (1981).

Nevo, E., T. Shimony, and M. Libni. Pollution selection of allozyme polymorphisms in barnacles. *Experientia* 34:1562–1564 (1978).

Newman, M.C. *Population ecotoxicology*. Chichester, UK: John Wiley & Sons, 2001.

Newman, M.C. *Fundamentals of ecotoxicology*. 3rd ed. Boca Raton, FL: CRC Press, 2010.

Newman, M.C., and M.S. Aplin. Enhancing toxicity data interpretation and prediction of ecological risk with survival time modeling: An illustration using sodium chloride toxicity to mosquitofish (*Gambusia holbrooki*). *Aquat. Toxicol.* 23:85–96 (1992).

Newman, M.C., and W.H. Clements. *Ecotoxicology. A comprehensive treatment*. Boca Raton, FL: CRC Press, 2008.

Newman, M.C., S.A. Diamond, M. Mulvey, and P. Dixon. Allozyme genotype and time to death of mosquitofish, *Gambusia affinis* (Baird and Girard) during acute toxicant exposure: A comparison of arsenate and inorganic mercury. *Aquat. Toxicol.* 15:141–156 (1989).

Nichols, J.D. Mark-recapture models. *Bioscience* 42:94–102 (1992).

Nisbet, R.M., and W.S.C. Gurney. *Modelling fluctuating populations*. Chichester, UK: John Wiley & Sons, 1982.

Norsys. *Netica*. Version 3. Vancouver, BC:, Norsys Software Copororation, 2006.

Nychka, D., S. Ellner, D. McCaffrey, and A.R. Gallant. Statistics for chaos. *Statistical Computing and Statistical Graphics Newsletter*, 4–11 (1990).

Odum, E.P. *Fundamentals of ecology*. 3rd edition. Philadelphia: W.B. Saunders Company, 1971.

Oehlmann, J., P. Di Benedetto, M. Tillmann, M. Duft, M. Oetken, and U. Schulte-Oehlmann. Endocrine disruption in prosobranch molluscs: Evidence and ecological prevalence. *Ecotoxicology* 16:29–43 (2007).

O'Hagan, A. Bayes factors. *Significance* 2006:184–186 (2006).

Olsen, L.F., and W.M. Schaffer. Chaos versus noisy periodicity: Alternative hypotheses for childhood epidemics. *Science* 249:499–504 (1990).

Otis, D.L., K.P. Burnham, G.C. White, and D.R. Anderson. Statistical inference from capture data on closed animal populations. *Wildl. Monogr.* 62:1–135 (1978).

Ownby, D.R., M.C. Newman, M. Mulvey, W.K. Vogelbein, M.A. Unger, and L.F. Arazayus. Fish (*Fundulus heteroclitus*) populations with different exposure histories differ in tolerance of creosote-contaminated sediments. *Environ. Toxicol. Chem.* 21:1897–1902 (2002).

Pearl, J. *Causality. Models, reasoning, and inference*. Cambridge: Cambridge University Press, 2000.

Pemberton, J.M., S.D. Albon, F.E. Guinness, T.H. Clutton-Brock, and R.J. Berry. Genetic variation and juvenile survival in red deer. *Evolution* 42:921–93 (1988).

Pesch, C.E., W.R. Munns, and R. Gutjahr-Gobell. Effects of a contaminated sediment on life history traits and population growth rate of *Neanthes arenaceodentata* (Polychaeta: Nereidae) in the laboratory. *Environ. Toxicol. Chem.* 10:805–815 (1991).

Petersen, R.C., Jr., and L.B.-M. Petersen. Compensatory mortality in aquatic populations: Its importance for interpretation of toxicant effects. *Ambio* 17(6):381–386 (1988).

Philippi, T.E., M.P. Carpenter, T.J. Case, and M.E. Gilpin. *Drosophila* population dynamics: Chaos and extinction. *Ecology* 68(1):154–159 (1987).

Pianka, E.R., and W.S. Parker. Age-specific reproductive tactics. *Am. Nat.* 109:453–464 (1975).

Pielou, E.C. *An introduction to mathematical ecology*. New York: John Wiley & Sons, 1969.

Pielou, E.C. *Population and community ecology. Principles and methods*. New York: Gordon and Breach Science Publishers, 1974.

Pinder, J.E., III, J.G. Weiner, and M.H. Smith. The Weibull distribution: A new method of summarizing survivorship data. *Ecology* 59(1):175–179 (1978).

Pool, R. Ecologists flirt with chaos. *Science* 243:310–313 (1989).

Popper, K.R. *The logic of scientific discovery*. London: Hutchinson and Company, 1968.

Posthuma, L., R.F. Hogervorst, E.N.G. Joose, and N.M. Van Straalen. Genetic variation and covariation for characteristics associated with cadmium tolerance in natural populations of springtail *Orchesella cincta* (L.). *Evolution* 47:619–631 (1993).

Prout, T. The estimation of fitness from genotypic frequencies. *Evolution* 19:546–551 (1965).

Pulliam, H.R., and B.J. Danielson. Sources, sinks, and habitat selection: A landscape perspective on population dynamics. *Am. Nat.* 137: S50–S66 (1991).

Rago, P.J., and R.M. Dorazio. Statistical inference in life-table experiments: The finite rate of increase. *Can. J. Fish. Aquat. Sci.* 41:1361–1374 (1984).

Raimondo, S., and C.L. McKenney Jr. Projected population-level effects of thiobencarb exposure on the mysid, *Americamysis bahia*, and extinction probability in a concentration-decay exposure system. *Environ. Toxicol. Chem.* 24:564–572 (2005a).

Raimondo, S., and C.L. McKenney Jr. Projecting population-level responses of mysids exposed to an endocrine disrupting chemical. *Integr. Comp. Biol.* 45:151–157 (2005b).

Raimondo, S., and C.L. McKenney Jr. From organisms to populations: Modeling aquatic toxicity data across two levels of biological organization. *Environ. Toxicol. Chem.* 25:589–596 (2006).

Reid, G.K. *Ecology of inland waters and estuaries.* New York: Van Nostrand Reinhold Company, 1961.

Relyea, R., and J. Hoverman. Assessing the ecology in ecotoxicology: A review and synthesis in freshwater systems. *Ecol. Lett.* 9:1157–1171 (2006).

Renshaw, E. *Modelling biological populations in space and time.* Cambridge: Cambridge University Press, 1991.

Ricker, W.E. Stock and recruitment. *J. Fish. Res. Board Can.* 11:559–623 (1954).

Rockwood, L.L. *Introduction to population ecology.* Malden, MA: Blackwell Publishing, 2006.

Rohlf, F.J., and R.R. Sokal. *Statistical tables.* 2nd ed. New York: W.H. Freeman and Company, 1981.

Ron, E. Thyroid cancer incidence among people living in areas contaminated by radiation from the Chernobyl accident. *Health Phys.* 93:502–511 (2007).

Sahai, H., and A. Khurshid. *Statistics in epidemiology. Methods, techniques and application.* Boca Raton, FL: CRC Press, 1996.

Salice, C.J., and T.J. Miller. Population-level responses to long-term cadmium exposure in two strains of the freshwater gastropod *Biomphalaria glabrata*: Results from life-table response experiment. *Environ. Toxicol. Chem.* 22:678–688 (2003).

Samallow, P.B., and M.E. Soule. A case of stress related heterozygote superiority in nature. *Evolution* 37:646–649 (1983).

Sarokin, D., and J. Schulkin. The role of pollution in large-scale population disturbances. Part 1: Aquatic populations. *Environ. Sci. Technol.* 26(8):1476–1484 (1992).

Schaffer, W.M., and M. Kot. Chaos in ecological systems: The coals that Newcastle forgot. *Trends Ecol. Evol.* 1(3):58–63 (1986).

Seber, G.A.F. The effects of trap response on tag recapture estimates. *Biometrics* 26:13–22 (1970).

Seber, G.A.F., and E.D. Le Cren. Estimating population parameters from catches large relative to the population. *J. Anim. Ecol.* 36:631–643 (1967).

Shelford, V.E. Physiological animal geography. *J. Morphol.* 22:551–618 (1911).

Shelford, V.E. *Animal communities in temperate America.* Chicago: University of Chicago Press, 1913.

Sherratt, T.N., and P.C. Jepson. A metapopulation approach to modeling the long-term impact of pesticides on invertebrates. *J. Appl. Ecol.* 30:696–705 (1993).

Siegismund, H.R., and F.B. Christiansen. Selection component analysis of natural polymorphisms using population samples including mother-offspring combinations, III. *Theor. Popul. Biol.* 27:268–297 (1985).

Skalski, J.R., and D.S. Robson. *Techniques for wildlife investigations. Design and analysis of capture data.* New York: Academic Press, 1992.

Smith, R.L. *Ecology and field biology.* 3rd ed. New York: Harper and Row Publishers, 1980.

Smouse, P.E. The fitness consequences of multiple-locus heterozygosity under the multiplicative overdominance and inbreeding depression models. *Evolution* 40:946–957 (1986).

Sousa, A., F. Laranjeiro, S. Takahashi, S. Tanabe, and C.M. Barroso. Imposex and organotin prevalence in a European post-legislative scenario: Temporal trends from 2003 to 2008. *Chemosphere* 77:566–573 (2009).

Spromberg, J.A., B.M. John, and W.G. Landis. Metapopulation dynamics: Indirect effects and multiple distinct outcomes in ecological risk assessment. *Environ. Toxicol. Chem.* 17:1640–1649 (1998).

Stanier, R.Y., M. Doudoroff, and E.A. Adelberg. *The microbial world.* 3rd ed. Englewood Cliffs, NJ: Prentice-Hall, 1970.

Stark, J.D., and J.E. Banks. Population-level effects of pesticides and other toxicants on arthropods. *Annu. Rev. Entomol.* 48:505–519 (2003).

Stearns, S.C. The evolutionary significance of phenotypic plasticity. *Bioscience* 39:436–445 (1989).

Stearns, S.C. *The evolution of life histories.* Oxford: Oxford University Press, 1992.

Stow, C.A., and M.E. Borsuk. Enhacing causal assessment of estuarine fishkills using graphical models. *Ecosystems* 6:11–19 (2003).

Stubben, C., and B. Milligan. Estimating and analyzing demographic models using the popbio package in R. *J. Stat. Softw.* 22:1–23 (2007).

Suter, G.W., Jr. *Ecological risk assessment*. Boca Raton, FL: CRC/Lewis Press, 1993.
Swennen, C., U. Sampantarak, and N. Ruttanadakul. TBT-pollution in the Gulf of Thailand: A re-inspection of imposex incidence after 10 years. *Mar. Pollut. Bull.* 58:526–532 (2009).
Szathmary, E. Simple growth laws and selection consequences. *Trends Ecol. Evol.* 6:366–370 (1991).
Tanaka, Y., and J. Nakaishi. Life history elasticity and the population-level effect of p-nonylphenol on *Daphnia galeata*. *Ecol. Res.* 16:41–48 (2001).
Thomas, W.R., M.J. Pomerantz, and M.E. Gilpin. Chaos, asymmetric growth and group selection for dynamical stability. *Ecology* 61(6):1312–1320 (1980).
Trehan, K.S., and K.S. Gill. Subunit interaction: A molecular basis of heterosis. *Biochem. Genet.* 25:855–862 (1987).
Turelli, M., and L.R. Ginzburg. Should individual fitness increase with heterozygosity? *Genetics* 104:191–209 (1983).
Upton, G.J., and B. Fingleton. *Spatial data analysis by example. Point pattern and quantitative data.* Vol. 1. New York: John Wiley & Sons, 1985.
Uusitalo, L. Advantages and challenges of Bayesian networks in environmental modeling. *Ecol. Model.* 203:312–318 (2007).
Vandermeer, J.H., and D.E. Goldberg. *Population ecology. First principles*. Princeton, NJ: Princeton University Press, 2003.
Vogelbein, W.K. Vignette 7.2: Polycyclic aromatic hydrocarbons and liver cancer in fish. In *Fundamentals of ecotoxicology*, ed. M.C. Newman. 3rd ed. Boca Raton, FL: Taylor & Francis/CRC Press, 2010, pp. 206–210.
Walthall, W.K., and J.D. Stark. A comparison of acute mortality and population growth rate as endpoints of toxicological effect. *Ecotox. Environ. Safe.* 37:45–52 (1997).
Watkinson, A. Intuition and the logistic equation. *Trends Ecol. Evol.* 7(9):314–315 (1992).
Watt, W.B. Adaptation at specific loci. II. Demographic and biochemical elements in the maintenance of the *Colias* PGI polymorphism. *Genetics* 103:691–724 (1983).
Weins, J.A. Population responses to patchy environments. *Annu. Rev. Ecol. Syst.* 7:81–120 (1976).
Weir, B.S. *Genetic data analysis. Methods for discrete population genetic data.* Sunderland, MA: Sinauer Associates, 1990.
Weis, J.S., and P. Weis. Tolerance and stress in a polluted environment. The case of the mummichog. *Bioscience* 39:89–95 (1989).
Westcott, B. Some methods of analyzing genotype-environment interaction. *Heredity* 56:243–253 (1986).
White, G.C., D.R. Anderson, K.P. Burnham, and D.L. Otis. *Capture-recapture and removal methods for sampling closed populations.* LA-8787-NERP. Los Alamos, NM: Los Alamos National Laboratory, 1982.
White, G.C., and K.P. Burnham. Program MARK: Survival estimation from populations of marked animals. *Bird Study* 46(Suppl.):S120–S139 (1999).
White, G.C., W.L. Kendall, and R.T. Barker. Multistate survival models and their extensions in program MARK. *J. Wildlife Manag.* 70:1521–1529 (2006).
Wiegert, R.G. The selection of an optimum quadrat size for sampling the standing crop of grasses and forbs. *Ecology* 43:125–129 (1962).
Williams, B.K., J.D. Nichols, and M.J. Conroy. *Analysis and management of animal populations.*San Diego: Academic Press, 2002.
Williams, C.J., W.W. Anderson, and J. Arnold. Generalized linear modeling methods for selection component experiments. *Theor. Popul. Biol.* 37:389–423 (1990).
Wilson, E.O., and W.H. Bossert. *A primer of population biology.* Sunderland, MA: Sinauer Associates, 1971.
Wood, J.M., and H.-K. Wang. Microbial resistance to heavy metals. *Environ. Sci. Technol.* 17:582A–590A (1983).
Woodward, M. *Epidemiology. Study design and data analysis.* 2nd ed. Boca Raton, FL: Chapman & Hall/CRC, 2005.
Yarbrough, J.D. Insecticide resistance in vertebrates. In *Survival in toxic environments*, ed. M.A.Q. Khan and J.P. Bederka Jr. New York: Academic Press, 1974, pp. 373–397.
Zippin, C. The removal method of population estimation. *J. Wildl. Manag.* 22:82–90 (1958).

CHAPTER 7

Community Effects

Ecotoxicology is a three-legged stool that, to date, has been balanced on two legs: chemical fate and single-species toxicology. These are necessary but insufficient components of the field. The third leg, that of organism interactions, has been slower to develop as a tool for assessing how an ecosystem is affected by and recovers from chemical stresses.

—**Taub (1989)**

7.1 GENERAL

In studies of community ecotoxicology, as in many endeavors believed to involve high complexity, there is a tendency to retreat to a descriptive approach to knowledge accrual. Although descriptive studies are unquestionably essential, predominance of such an approach to the detriment of experimental approaches eventually begets a weak inferential structure to the associated knowledge base. (See Chapter 1 for discussion of strength of inference.) Fortunately, a pronounced shift has occurred in the last decade to include more experimentation and quantification in community ecotoxicology. Microcosm, mesocosm, and whole lake studies have been especially valuable in this regard, although they fell unjustifiably out of favor with regulatory agencies for a period.

A brief quantitative overview of toxicant influence on species interactions will be presented here. Discussion will begin with two-species interactions and end with metacommunities occupying broad geographical regions. The influence of toxicants on species niche integrity will be emphasized initially. The intent is to demonstrate that the integrity of a species' role, interactions, and habitat in an ecosystem or region (herein referred to as niche integrity) is as critical as its physiological or demographic integrity in determining the final outcome of toxicant exposure in an ecological arena. Various measures of community integrity will then be discussed relative to the effects of toxicants. Finally, trophic transfer of toxicants will be discussed, including theoretical models, and then a practical regression method based on stable nitrogen isotopes.

Throughout this chapter, the term *niche* will be used in the Hutchinsonian context (Hutchinson 1957). Wetzel (1983) clearly presents the Hutchinsonian fundamental and realized niches as

> a certain biological activity space in which an organism exists in a particular habitat. This space is influenced by the physiological and behavioral limits of a species and by the effects of environmental parameters (physical and biotic, such as temperature and predation) acting on it. Each of these parameters can be ordinated on an axis, and can be thought of as a dimension in space. The fundamental niche, then, can be viewed as an n-dimensional space or hypervolume, with each of its n axes or dimensions corresponding to the range of an environmental parameter over which the organism can exist. Since many physical and biotic factors interact, each species occupies only a portion of its fundamental niche; this portion can be referred to as the species realized niche.

In terms of individual fitness, the niche is the volume within *n*-dimensional space where fitness is positive (Green 1971), although as we will discuss relative to metacommunities, it might in some dimensions extend slightly further outward from that space with a positive fitness.

7.2 SIMPLE SPECIES INTERACTIONS

7.2.1 Predator–Prey Interactions

One of the best-known examples of natural selection in the wild is industrial melanism, the gradual increase to predominance of melanic forms in industrialized areas. The increase in dark morphs is correlated with a general darkening of habitat with soot and loss of lichens. With this darkening, the advantage due to camouflage shifts to favor dark over light forms with an epiphenomenal increase in the frequency of darker individuals. The mechanism was demonstrated in Kettlewell's (1955) classic study of differential bird predation on color morphs of the peppered moth (*Biston betularia*), in which predator–prey interactions were carefully quantified in enclosure (aviary) and mark-release field studies.

Although Kettlewell's (1955) enclosure and field studies remain an integral lesson during the training of every ecologist, including the ecotoxicologist (see Moriarty 1983, pp. 79–84), studies of pollution effects on predator–prey interactions remain biased toward highly structured laboratory assays (Atchison et al. 1987; Henry and Atchison 1991). Fewer studies than warranted (Clements et al. 1989; Newman and Clements 2008) use enclosure or field designs for predator–prey studies. This neglect compromises our understanding of toxicant effects because predator–prey interactions can be a crucial component of individual fitness and, in the special case of cannibalism (Schneider et al. 1980), population demography. Furthermore, if prey with highest body burdens of toxicants are most prone to predation due to their weakened state (Kania and O'Hara 1974), such information can also be important in fully understanding trophic transfer (Goodyear 1972). Indeed, Goodyear (1972) and Atchison et al. (1987) suggest that results of traditional acute or chronic mortality tests are less relevant to assessing the effects of pollutants than those focused on ecological mortality (toxicant-related diminution of fitness within an ecosystem context of sufficient magnitude as to be equivalent to physiological death, e.g., death due to a compromised ability to avoid predation).

The most common approach to quantifying toxicant influence on predator–prey interactions involves laboratory experiments in which the prey alone or both predator and prey are exposed to toxicant. Predation rate at intervals throughout the trial or the intensity of predation measured at the end of a trial period is compared between control and exposed animals. For example, Kania and O'Hara (1974) placed mercury-exposed and unexposed mosquitofish (*Gambusia holbrooki*, formerly *G. affinis*) into a chamber with a narrow shelf acting as a refuge from a predator (largemouth bass, *Micropterus salmoides*). After 60 h, the numbers of exposed and nonexposed mosquitofish falling prey to the bass were compared with a simple χ^2 test. For all exposures above 0.005 μg/L of inorganic mercury, there was a statistically significant increase in predation as a consequence of the compromised ability of the prey to avoid the predator. (With more complicated designs, many of the methods described in Chapter 4 or 5 could be used to analyze this type of data.) Goodyear (1972) obtained similar results for the effects of γ radiation on mosquitofish predation by bass; however, the percentage of fish surviving predation is noted at intervals over the entire time course. This type of data is amenable to nonparametric, semiparametric, and parametric methods described in Chapter 4 for time-to-death studies. Techniques in Chapter 4 have the advantage of allowing model development as well as hypothesis testing. Other approaches share this advantage. For example, in studying cadmium effects on fish predation, Sullivan et al. (1978) quantify the magnitude of effect in addition to testing for a significant effect.

Example 7.1

Effects of an insecticide (mirex or dodecachlorooctahydro-1,3,4-metheno-2H-cyclobuta[cd] pentalene) on predation of grass shrimp (*Palaemonetes vulgaris*) by pinfish (*Lagodon rhomboides*) were quantified by Tagatz (1976). After 13 days of shrimp exposure to sublethal concentrations of mirex, pinfish were placed into each exposure tank. Predation was measured in control and mirex-exposed tanks during the next 3 days. The results for day 3 were the following: control, 42 of 177 shrimp survived predation; mirex exposed, 5 of 115 shrimp survived predation. (These test statistics may differ slightly from those in the original paper because they are estimated from total counts and the survival percentages in the original paper.) A χ^2 test can be used to test for a significant deviation from expected intensity of predation if mirex had no effect.

To begin, a contingency table is generated. The observed and expected scores are placed into the table along with marginal column and row totals. The expected scores are placed in brackets within the table. The expected scores of dead and living animals in the table below are generated with the marginal column totals (expected = (column total × row total)/grand total). These expected scores are used under the assumption that the overall predation results reflected in the marginal column totals are independent of treatment category (mirex exposure or control). Consequently, one would expect to see them reflected also in results for the mirex-exposed and control treatments. These expected scores are then used together with the observed scores to calculate a χ^2 statistic.

	Alive	**Dead**	**Row Total**
Control	42(28.5)	135(148.5)	177
Mirex	5(18.5)	110(96.5)	115
Column total	47	245	292

Expected number of control shrimp alive: (47 × 177)/292 = 28.5
Expected number of control shrimp dead: (245 × 177)/292 = 148.5
Expected number of mirex-exposed shrimp alive: (47 × 115)/292 = 18.5
Expected number of mirex-exposed shrimp dead: (245 × 115)/292 = 96.5

A χ^2 statistic is calculated as described in Chapter 6 (e.g., Equation 6.125).

$$\chi^2 = \frac{(42-28.5)^2}{28.5} + \frac{(135-148.5)^2}{148.5} + \frac{(5-18.5)^2}{18.5} + \frac{(110-96.5)^2}{96.5} = 19.36$$

The calculated χ^2 statistic (19.36) is compared to a tabulated value with the appropriate α and degrees of freedom. The *df* is (number of rows – 1)(number of columns – 1) = (2 – 1)(2 – 1) = 1. The χ^2 for $df = 1$ and $\alpha = 0.05$ (3.841) is taken from a table such as Table 14 of Rohlf and Sokal (1981) or a p value is calculated with Excel CHIDIST(19.36,1). The calculated χ^2 is larger than this critical χ^2 and the p value from Excel is <0.0001. Consequently, the null hypothesis of independence (number of deaths from predation independent of treatment) is rejected: exposure to mirex did enhance predation.

This example is quite straightforward. Often several treatments are assessed as in the original work of Tagatz (1976). Further information on more detailed analyses is available in most statistical textbooks, e.g., Chapter 13 of Dixon and Massey (1969), Chapter 23 of Zar (1999), Chapter 9 of Levine et al. (2001), and Chapter 10 in Rosner (2006). As discussed in Chapter 4, Fisher's exact test might be used instead if an expected value in a table cell is less than 5. Finally, the difference between pairwise and experimentwise type I error rates should be kept in mind during such tests. (See Chapter 5 for a discussion of these error rates.) Adjustment of the critical χ^2 values might be necessary for multiple comparisons, as discussed in Chapter 5. Rohlf and Sokal (1981) provide a table (Table 15) of critical χ^2 values based on Šidàk's adjustment for multiple comparisons; however, generation of (one-tailed) p values with Excel CHIDIST() eliminates the need for such tables.

Analyses of predator–prey interactions framed with formal ecological models have also been applied effectively (e.g., Axelsen et al. 1997), but not frequently enough, in ecotoxicology (Clements 1999; Hammers and Krogh 1997; Preston 2002; Relyea and Hoverman 2006; Sih et al. 2004). Use of ecological models tends to be overshadowed by designs involving simple statistical hypothesis tests (e.g., Relyea 2003). Analyses involving ecological models have the advantage of direct linkage to underlying principles from which explicit predictions can be made. Basic predator–prey models are described below in an ecotoxicological context to foster more interest.

Predator–prey population dynamics have traditionally been explained with the Lotka–Volterra model. This model can be described with the clear notation of Emmel (1976) and Smith (1980) and the enriching details in Bonsall and Hassell (2007).

$$\frac{dN_1}{dt} = r_1 N_1 - P N_1 N_2 \quad \text{or} \quad N_{1t}(r_1 - P N_{2t}) \tag{7.1}$$

$$\frac{dN_2}{dt} = P_2 N_1 N_2 - d_2 N_2 \ or \ N_{2t}(\alpha c N_{1t} - d_2) \tag{7.2}$$

where N_1 = prey density, N_2 = predator density, r_1 = instantaneous rate of increase for the prey species (predator absent), P = predation coefficient, P_2 = predation effectiveness coefficient, d_2 = density-dependent death rate of the predator, α = predator attack rate, and c = positive effect of prey on the predator.

The prey population is assumed to increase exponentially in the absence of the predator ($dN_1/dt = r_1N_1$), and its growth is adversely affected by an increasing probability of predator–prey interaction (N_1N_2). In contrast, the growth of the predator is enhanced by any increase in the probability of predator–prey interaction. The PN_1N_2 in Equation (7.1) is the functional response term (May 1976). As reflected in this term, one possible effect of prey densities on the predator is a functional change in predatory behavior such as a shift in number of prey taken per predator, shift in size distribution of prey taken, or reallocation of time spent in various aspects of foraging for prey. The numerical response term ($P_2N_1N_2$) in Equation (7.2) reflects a change in the number of predators in response to change of the prey density. Note that the predator population is not density dependent. According to this model, the predator and prey populations will oscillate about each other through time.

Lags can be incorporated into these equations, as done in Chapter 6 with single species populations (Bonsall and Hassell 2007).

$$\frac{dN_1}{dt} = N_{1t-\tau_{N1}}(r_1 - P N_{2t}) \tag{7.3}$$

$$\frac{dN_2}{dt} = N_{2t-\tau_{N2}}(\alpha c N_{1t-\tau_{N2}} - d_2) \tag{7.4}$$

where τ_{N1} and τ_{N2} = time delays relative to new prey entry into the predated population and delay in the predator becoming effective.

The simple Lotka–Volterra model (Equations 7.1 and 7.2) has generally failed to accurately describe predator–prey dynamics, yet as done here, it is still widely used to introduce the oscillatory nature of predator and prey populations, e.g., Gause (1934) and Huffaker (1958). May (1976) points out mathematical difficulties associated with this model and outlines several modifications of Equations (7.1) and (7.2) resulting in more realistic models.

$$\frac{dN_1}{dt} = rN_1\left(1 - \frac{N_1}{K}\right) - N_2F(N_1, N_2) \tag{7.5}$$

$$\frac{dN_2}{dt} = N_2G(N_1, N_2) \tag{7.6}$$

where K = the prey carrying capacity, $F()$ = a specified function for the functional response, and $G()$ = a specified function for the numerical response.

The prey population is made density dependent with the term $1 - N_1/K$. Simple functional and numerical response terms are replaced by more realistic functions of N_1 and N_2 that account for such processes as predator saturation/satiation, search image acquisition, or switching to alternate prey. These functions, $F()$ and $G()$, take a variety of forms. Table 4.1 in May (1976) lists a series of candidate forms including $F() = PN_1$ and $G() = P_2N_1 - d_2$ for the simple Lotka–Volterra model (Equations 7.1 and 7.2). May (1974) provides details for each of these functions. For example, the predator can become less efficient at taking prey as prey densities increase, e.g., Equation (7.7) ("Holling Type II, Invertebrate" in May 1976) or Equation (7.8) ("Holling Type III, Vertebrate" in May 1976). The $G()$ function can be made logistic with N_1 controlling the predator carrying capacity (Equation 7.9). Alternatively, $G()$ can be linearly related to $F()$ (May 1976), e.g., Equation (7.10).

$$F() = \frac{kN_1}{N_1 + D} \tag{7.7}$$

$$F() = \frac{kN_1^2}{N_1^2 + D^2} \tag{7.8}$$

where D = prey density at which attack capability begins to saturate for the predator (May 1974), and k = the constant attack (or capture) rate (prey per predator).

$$G() = r_2\left(1 - \frac{N_2}{\gamma N_1}\right) \tag{7.9}$$

where the carrying capacity for the predator (γN_1) is proportional (γ) to the prey population density (N_1) (May 1974). The r_2 is the intrinsic rate of increase for the predator population.

$$G() = P_2 + d_2F() \tag{7.10}$$

From empirical data, Holling (1959) reasons that functional and numerical responses can take several general forms. He describes four types of functional responses (Figure 7.1). The number of prey consumed per predator is directly proportional to prey density in the simplest or type I functional response. It is a positive, linear relationship until saturation. Prey consumption is constant beyond a maximum prey density. Poole (1974) suggests that filter-feeding predators can display a type I response. The most commonly used response (type II) is often associated with predation by invertebrates (May 1976; Smith 1980). Poole (1974) suggests that it can also describe responses of some fish species. This functional response increases as prey density increases, but at a gradually decreasing rate. One model describing such a response is provided in Equation (7.7). It accommodates more aspects of predation than the type I response model, e.g., prey acquisition and handling

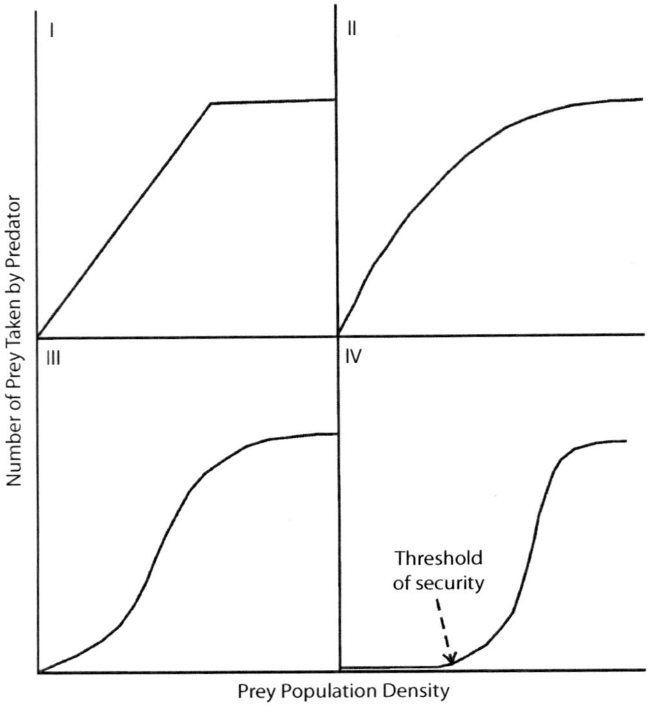

Figure 7.1 Four types of predator functional responses to prey density. See text for details. (Based on Holling, C.S., *Can. Entomol.*, 91, 293–303, 1959, Figure 8.)

times. Such details of prey acquisition and handling are illustrated with Holling's disc model (Equation 7.11). (See Poole 1974 for detailed discussion.) Again, the clear notation of Smith (1980) is borrowed here.

$$\frac{N_a}{P} = \frac{aNT}{1 + aT_hN} \tag{7.11}$$

where N_a = number of prey taken or attacked, N = total number of prey, P = number of predators, a = attack rate constant, T = total amount of time available for predator–prey interactions ($T = T_s + T_hN_a$), T_h = handling time spent for a prey item (pursue, capture, consumption, and digestion), and T_s = time spent searching for a prey item.

If learning is involved in the predator's actions, as is the case with many vertebrate predators, the type III response might be more appropriate (Poole 1974) than the type II model. Holling's study (Holling 1959) of mouse (*Peromyscus maniculatus bairsii*) predation on European pine sawfly (*Neodiprion sertifer*) cocoons is the classic example of a type III response. Type III functional responses can also involve several prey species with the predator choosing to allocate effort among them. For example, predation or other factors can result in the density of one prey species dropping below a certain level. The predator can then choose to spend an increasingly disproportionate amount of its attention on another, more abundant prey species as the density of the first prey species drops (prey switching). The first prey species population then has an opportunity to rebound. The type IV functional response mentioned in Holling's classic work (1959) is often discussed in recent texts as a form of type III response. With this relationship, there is a critical prey density (threshold of security) below which there is insufficient prey stimulus to elicit any material predator response.

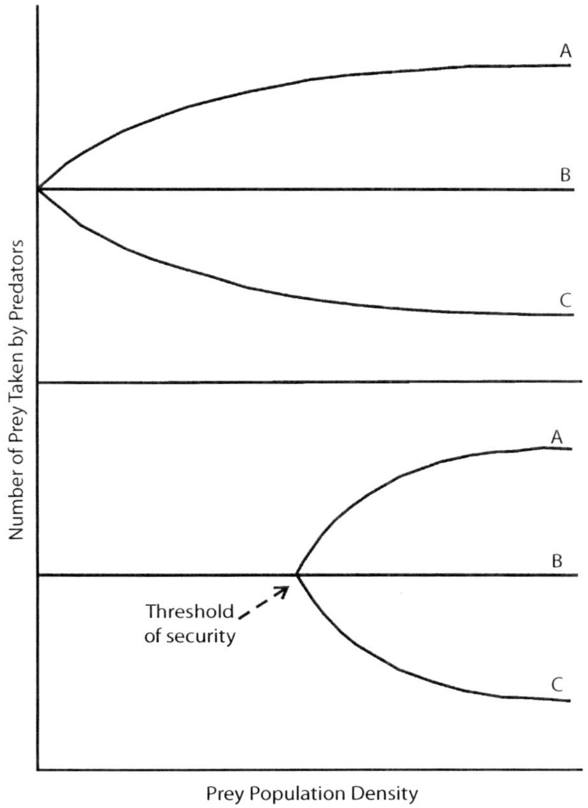

Figure 7.2 Three types of predator numerical responses to prey density. Positive (A), neutral (B), and negative (C) responses with (bottom panel) or without (top panel) a threshold of security are shown. See text for details. (Based on Holling, C.S., *Can. Entomol.*, 91, 293–303, 1959, Figure 8.)

Numerical responses can involve enhanced predator fecundity or immigration with increasing prey densities. Holling (1959) describes three numerical response models (Figure 7.2). Type A responses are direct (positive) responses to increasing prey density. Type B is simply no obvious numerical response to prey density, and Type C is an inverse response to prey density.

Taken together, functional and numerical responses produce a wide range of possible predator responses to prey density. Figure 7.3 shows only four of these responses. Underlying each is a complex of behaviors affecting the outcome of the predator–prey relationship over time.

The change in predator response to prey densities under the influence of toxicants has too infrequently drawn the attention of ecotoxicologists. A study of ammonia effects on largemouth bass consuming mosquitofish (*Gambusia affinis*) by Woltering et al. (1978) is an interesting early exception. They find that the shape of the functional response curve changes with increasing ammonia concentration. From control to 0.34 mg/L to 0.63 mg/L, there is a lowering of the rate at which consumption of mosquitofish increases with increasing prey densities. Indeed, the mosquitofish, being less sensitive to the effect of ammonia than the bass, harass the bass in treatment tanks with the highest concentrations (0.86 mg/L) and mosquitofish densities! This precipitates a drop in predation rate and loss of weight by the predator.

As should be obvious from Equation (7.11) and the weight loss noted for bass in the study just described, foraging behavior must be energetically efficient for a predator or grazer to optimize its fitness. Equation (7.11) indicates that a predator must expend time and energy to find, pursue,

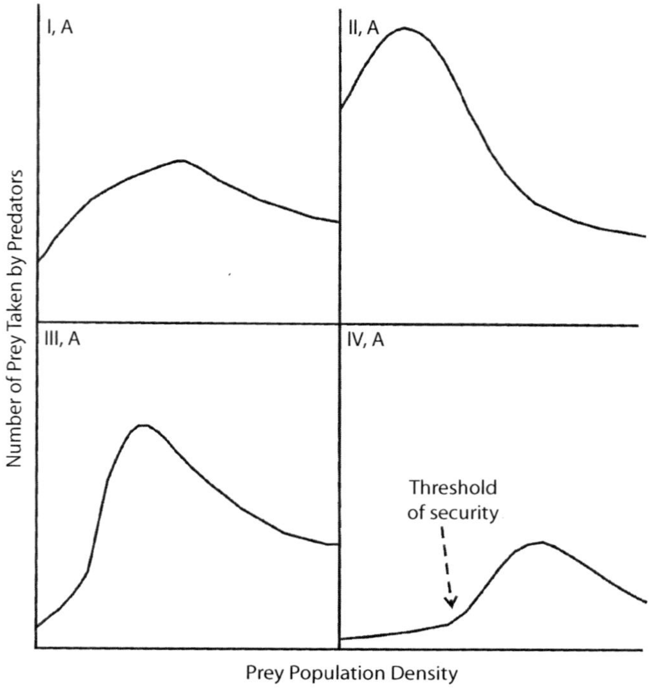

Figure 7.3 Four examples of the net response to prey density. Only a positive numerical response combined with Types I, II, III, and IV functional responses are shown. (Based on Holling, C.S., *Can. Entomol.*, 91, 293–303, 1959, Figure 8.)

capture, consume, and digest prey. Effective selection of prey items, e.g., large versus small prey, fast versus slow prey, widely dispersed versus aggregated prey, is crucial to optimum foraging. Decisions regarding time spent within a patch of prey versus time expended searching for new patches to exploit are also important. An enormous, theory-rich literature is available for addressing questions of optimum foraging behavior, e.g., Stephens and Krebs (1986), yet ecotoxicologists have only begun to exploit the associated theory and methodology. (See Su et al. 2008 for exceptions.)

Atchison and coworkers (Atchison et al. 1987; Henry and Atchison 1991; Sandheinrich and Atchison 1990) advocate the use of optimum foraging theory to assess toxicant effects. They observe that most studies of predator–prey behavior under toxicant stress draw on empirical data lacking any theoretical foundation. This severely restricts the generality of conclusions and the ability to formulate hypotheses for further elucidation of underlying mechanisms. They use a widely accepted model (Equation 7.12) as an example of one optimum foraging model that could be used by ecotoxicologists.

$$\frac{E_n}{T} = \frac{\sum_{i=1}^{z} B_i E_i}{1 + \sum_{i=1}^{z} B_i H_i} - C_s \tag{7.12}$$

where E_n = net energy (J), T = time (s), $E_i = e_i - C_h H_i$, e_i = net energy obtained from prey size i (J), B_i = prey encounter rate with size i (per s), H_i = prey handling time for size i (s), C_h = energetic cost of handling (J/s), and C_s = energetic cost of searching (J/s).

Given a range of prey sizes (z), the ideal predator selects a subset of prey sizes that maximizes net energy intake (E_n/T). Sandheinrich and Atchison (1990) provide numerous references suggesting that key components of Equation (7.12) are influenced for fish predators exposed to a wide range of toxicants. Application of optimum foraging theory to toxicant effects on predator (and grazer) foraging behavior promises to provide very useful information based on explanatory principles.

7.2.2 Interspecies Competition

7.2.2.1 *Two Species*

Competition between two species was described in the 1920s using the Lotka–Volterra (Volterra, Gause-Volterra) equations.

$$\frac{dN_1}{dt} = r_1 N_1 \left[1 - \frac{N_1}{K_1} - \frac{\propto_{12} N_2}{K_1}\right] \tag{7.13}$$

$$\frac{dN_2}{dt} = r_2 N_2 \left[1 - \frac{N_2}{K_2} - \frac{\propto_{21} N_1}{K_2}\right] \tag{7.14}$$

These equations can be made more general.

$$\frac{dN_i}{dt} = r_i N_i \left[1 - \frac{N_i}{K_i} - \frac{\propto_{ij} N_j}{K_i}\right] \tag{7.15}$$

where N_i, N_j = the densities of species i and j, K_i = the carrying capacity of species i, r_i = intrinsic rate of increase for species i, and α_{ij} = the competition coefficient or the linear reduction of growth rate of species i relative to its carrying capacity K_i as a result of competition with species j (Gilpin and Ayala 1973).

Gause and Witt (1935) developed graphical means for determining the outcome of two species' competition (Figure 7.4) based on the intersection of the two straight lines defined by (x-intercepts, y-intercepts) K_1/α_{12}, K_1 and K_2, K_2/α_{21} in a plot of species 1 density versus species 2 density (Figure 7.4). Assuming linear relationships, competitive coexistence can occur in the presence of slight mutual depression by the competitors (Figure 7.4A). If the two competitors had identical niches (Figure 7.4B), one would eventually exclude the other, i.e., the principle of competitive exclusion. If there is a strong mutual depression during competition, one or the other competitor would exclude the other; however, which species is the successful competitor would be determined by the initial N_1 and N_2 (Figure 7.4C). (See Gause and Witt 1935 or Vandermeer 1970 for a detailed discussion of this topic.)

Gilpin and Ayala (1973) add an additional parameter to Equation (7.15) based on results from competition experiments with *Drosophila* species. This parameter (θ_i) quantifies the nonlinearity of intraspecific population growth regulation as described previously for the modified logistic model (Equation 6.74 in Chapter 6).

$$\frac{dN_i}{dt} = r_i N_i \left[1 - \left[\frac{N_i}{K_i}\right]^{\theta_i} - \frac{\propto_{ij} N_j}{K_i}\right] \tag{7.16}$$

Let's examine Gilpin and Ayala's experiments used to select Equation (7.16) as they contain the means for extending the simple linear approach described above for determining equilibrium

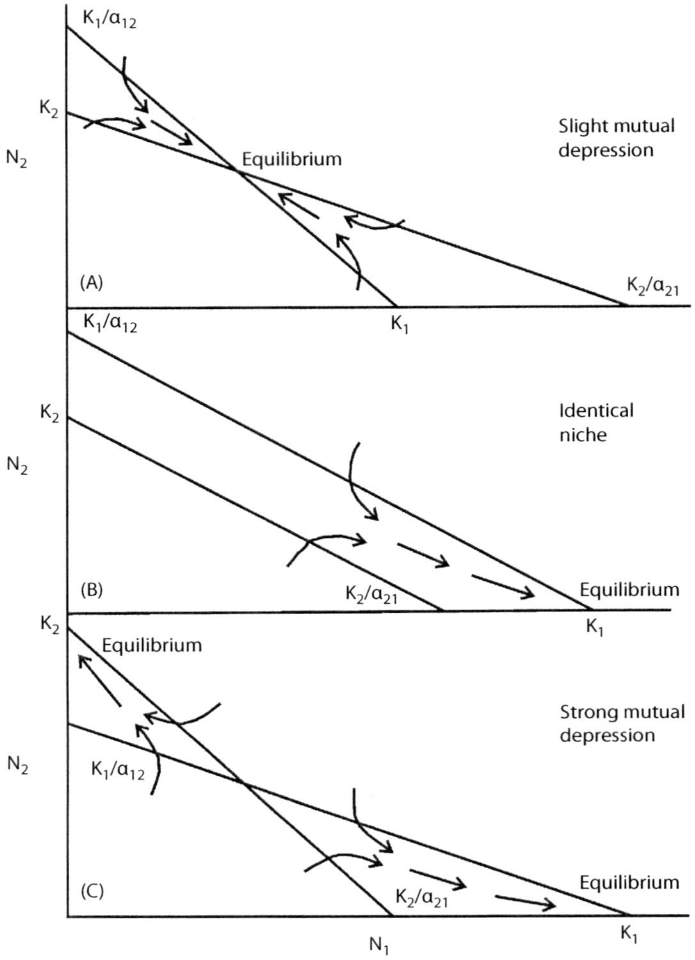

Figure 7.4 The possible outcomes of interspecies competition based on the linear model of Gause and Witt (1935). Coexistence (A) will result from slight mutual depression (at the noted point of equilibrium). With identical niches, an equilibrium with only one species will occur (competitive exclusion principle). In panel B, species 1 completely excludes species 2 at equilibrium. However, species 2 would prevail if line K_1/α_{12}, K_1 fell below line K_2/α_{21}, K_2. In the region of strong mutual depression, one species or the other remains at final equilibrium depending on the initial densities of species 1 and 2. (Modified from Gause, G.F., and A.A. Witt, *Am. Nat.*, 69, 596–609, 1935, Figures 1, 2, and 3.)

conditions. Various initial population densities of *D. willistoni* and *D. pseudoobscura* were combined, and after 1 week, the changes in densities for both species (dN_1/dt, dN_2/dt) were measured. The changes for each species in various initial combinations with the other species were fit by regression methods to Equation (7.16) to generate estimates of r_i, K_i, α_{ij}, and θ_i for each species. Furthermore, 19 vectors of change in species densities were generated and plotted on a graph of N_1 versus N_2 (Figure 7.5). The vectors for both species were used to visually fit lines for $dN_1/dt = 0$ and $dN_2/dt = 0$. The intersection of these two lines (N_1^e, N_2^e) is the two-species equilibrium point ($dN_1/dt = dN_2/dt = 0$). Note that the lines derived were not straight as assumed in the work by Gause and Witt (1935) (Figure 7.4) discussed above.

With traditional discussion of Lotka–Volterra equations, such an equilibrium point would be said to be stable (competitive coexistence) if intraspecific competition was stronger than interspecific

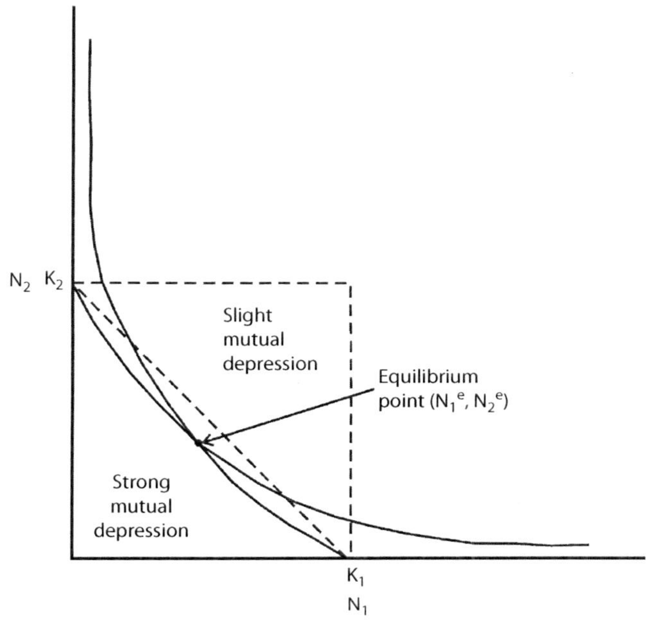

Figure 7.5 Model for interspecific competition demonstrating species coexistence in the region of strong mutual depression. (Modified from Gilpin, M.E., and F.J. Ayla, *Proc. Natl. Acad. Sci. U.S.A.*, 70, 3590–3593, 1973, Figure 1; and Gilpin, M.E., and K.E. Justice, *Nature*, 236, 273–301, 1972, Figure 1.)

competition (Ayala 1969; May 1976). In Figure 7.4 or 7.5, this would be in the region of slight mutual depression above the line connecting the carrying capacities for the two species (K_1, K_2). However, Gilpin and coworkers (Gilpin 1975; Gilpin and Ayala 1973; Gilpin and Justice 1972) indicate that a stable equilibrium (species coexistence) can also occur in the region of strong mutual depression if the conditions $K_1 < K_2/\alpha_{21}$ and $K_2 < K_1/\alpha_{12}$ were met. (As clearly seen with Figure 7.4, these conditions for species coexistence hold for the original Lotka–Volterra model as well as Gilpin's modification described by Equation 7.16.)

Example 7.2

Two *Daphnia* species are grown together for 2 weeks beginning with initial densities N_1 and N_2. Treatments consist of copper-contaminated or uncontaminated diets. Regression analyses of the changes in species densities are then used to generate the following estimates for Equation (7.16). (This example was produced by modifying Table 1 of Gilpin and Ayala 1973.)

	Species 1	Species 2
	Control	
r	1.496 ± 0.167	4.513 ± 0.259
K	1,332 ± 128	791 ± 43
α	0.713 ± 0.077	0.087 ± 0.006
θ	0.35 ± 0.04	0.12 ± 0.02
	Copper Diet	
r	1.210 ± 0.215	4.014 ± 0.318
K	450 ± 200	780 ± 53

	Species 1	Species 2
α	0.903 ± 0.103	0.090 ± 0.006
θ	0.34 ± 0.10	0.10 ± 0.08

Can the two species coexist indefinitely under the test conditions? The models above suggest that competitive coexistence is possible if $K_1 < K_2/\alpha_{21}$ and $K_2 < K_1/\alpha_{12}$.

With the control diet:

$$K_1 = 1{,}332 \qquad K_2/\alpha_{21} = 791/0.087 = 9{,}092$$
$$K_2 = 791 \qquad K_1/\alpha_{12} = 1{,}332/0.713 = 1{,}868$$

The inequalities are satisfied for the control populations.

With the elevated copper diet:

$$K_1 = 450 \qquad K_2/\alpha_{21} = 780/0.090 = 8{,}667$$
$$K_2 = 780 \qquad K_1/\alpha_{12} = 450/0.903 = 498$$

The second inequality ($K_2 < K_1/\alpha_{12}$) is not satisfied.

The decrease in carrying capacity of species 1 and the reduction in its growth due to competition increases with copper exposure. As a consequence, the necessary inequalities are not met when the diet contains elevated levels of copper, i.e., the ability to compete and coexist is lost for this species pair.

7.2.2.2 Several Species

Vandermeer (1969) and Gilpin and Ayala (1973) suggest that higher-order interactions, e.g., the effect of species 3 on the competition between species 1 and 2, can often be ignored. Since the linear effects model reflected by simple α_{ij} values for species competition is a reasonable caricature of species competition in communities involving several species, the interactions between all competitors can be summarized as a simple matrix of α_{ij} values. Such a competition matrix (Vandermeer 1970) is a specific type of community matrix in which all α_{ij} values have a positive sign, i.e., the effect of species j on i (α_{ij}) is inhibitory, not stimulatory. (The presence of mutualism and other beneficial species interactions results in negative α_{ij} values if interactions other than competition are incorporated in a more general matrix of interactions or community matrix.)

To generate a competition matrix, a series of competition experiments as described above is completed for all possible species pairs in the defined community of competitors (Poole 1974). The resulting α_{ij} values are then organized into a matrix. The diagonal series of 1's are the α values for all species effects on themselves.

$$\begin{matrix} 1 & \alpha_{12} & \alpha_{13} & \alpha_{14} & \cdot & \cdot & \alpha_{1m} \\ \alpha_{21} & 1 & \alpha_{23} & \alpha_{24} & \cdot & \cdot & \alpha_{2m} \\ \alpha_{31} & \alpha_{32} & 1 & \alpha_{34} & \cdot & \cdot & \alpha_{3m} \\ \cdot & \cdot & \cdot & 1 & \cdot & \cdot & \cdot \\ \cdot & \cdot & \cdot & \cdot & 1 & \cdot & \cdot \\ \cdot & \cdot & \cdot & \cdot & \cdot & 1 & \cdot \\ \alpha_{m1} & \alpha_{m2} & \alpha_{m3} & \alpha_{m4} & \cdot & \cdot & 1 \end{matrix}$$

In its simplest application, this matrix allows visualization of the distribution of intensities of competitive interactions within a community. More involved analyses are possible. For example, with the vector of carrying capacities generated in the experiments described above for all competitors, the matrix can be used to predict equilibrium species densities of each competing species. (See the appendix in Vandermeer (1970) and Chapter 15 in Poole (1974) for details.) Vandermeer (1970) uses the competition matrix approach to estimate the number of species that can be maintained in a community at equilibrium. May (1974) uses this approach to detail conditions for community stability.

Many species interactions can be incorporated into the generalized Lotka–Volterra model (Equation 7.15). Equation (7.17) is a modification of the multiple species model described in Vandermeer (1970).

$$\frac{dN_i}{dt} = r_i N_i \left[1 - \frac{N_i}{K_i} - \frac{\sum_{j \neq i}^{m} \alpha_{ij} N_j}{K_i} \right] \tag{7.17}$$

It can be modified to include Gilpin's θ_i (Equation 7.16).

$$\frac{dN_i}{dt} = r_i N_i \left[1 - \left[\frac{N_i}{K_i} \right]^{\theta_i} - \frac{\sum_{j \neq i}^{m} \alpha_{ij} N_j}{K_i} \right] \tag{7.18}$$

Such multiple species competition models can have relatively complex dynamics. Gilpin (1975) provides a means of assessing the dynamics within communities containing odd and even numbers of species.

To this point, multiple species competition has focused on contrived experimental designs that, for logistical reasons, involve relatively few species. It could be argued that, as a consequence, results from such studies have little value in environmental decision making because of their lack of realism. However, a strong counterpoint could be made that information from such studies is no less valuable or realistic than that used routinely and pervasively to make regulatory decisions, e.g., LC50 or EC50 based on death, or NOEC based on inhibition of reproduction or growth. Information generated by this means would be a valuable addition to that collected at other levels of ecological organization and that derived from field surveys. The ecological realism of such studies can be enhanced if the design involves field enclosures or mesocosms. Regardless, several important points emerge from these models. First, shifts in competition can eliminate species as surely as outright physiological death or reproductive failure. Second, by their very nature, population densities in systems involving species competition will exhibit cycles (Gilpin 1975). The dynamics of these cycles are complex functions of the vector of the competing species' carrying capacities and the matrix of competition coefficients (Equations 7.17 and 7.18). Both of these points should be kept in mind during assessment of toxicant impact.

Competition in natural communities is extremely difficult to quantify, and interpretation of the associated results is much more ambiguous than quantification of data from laboratory or enclosure studies. Regardless, it is possible (Connell 1961; MacArthur 1958) and highly desirable because of the associated enhanced ecological realism. Poole (1974) describes Levins's (1968) approximation of α_{ij} values for field populations sharing a common resource (Equation 7.17).

$$\alpha_{12} = \frac{\sum_{j=1}^{r} P_{1j}P_{2j}}{\sum_{j=1}^{r} P_{1j}^2} \tag{7.19}$$

where P_{1j} = proportion that resource j is of all resources used by species 1, P_{2j} = proportion that resource j is of all resources used by species 2, and r = number of resources used, e.g., habitat types, foods, and times available to feed.

Use of this measure of niche overlap in a community matrix is considered inappropriate by most ecologists (Krebs 1989). One difficulty is defining the qualities of natural habitats that influence the competition (Krebs 1989). Also, resource abundance or availability is ignored (Hurlbert 1978). If a resource is not limiting, the overlap of two species cannot be used to imply competition (May 1976). Hurlbert (1978) describes several similar indices linked to competition coefficients along with their application. He concludes that such use is valid only under specific, controlled conditions.

7.2.3 Symbiosis

Symbiotic relationships, including host-parasite, host-pathogen, or mutualistic associations, have received scant attention, but the literature suggests that these relationships are not trivial, e.g., Sures (2007, 2008). Sarokin and Schulkin (1992) review population level effects of pollutants and, in most of their examples, suggest that pollutant exposure plays a key role in the adverse outcome of infections. In their whole lake acidification studies, Schindler et al. (1985) indicate that the demise of a major species (*Orconectes virilis*) was a consequence of enhanced microsporozoan and fungal infections following lake acidification. In the few studies of contaminants and infectious diseases, significant effects are noted for diseases caused by bacteria (Snarski 1982), protozoans (Ewing et al. 1982), and metazoans (Evans 1982a, 1982b).

Other symbiotic relationships are important in the fate and effects of toxicants. An intriguing example involves the giant clam–algal zooxanthellae symbiotic relationship (Benson and Summons 1981). From the phosphorus-deficient waters of the Great Barrier Reef, arsenate is taken up, converted to organic form, and concentrated by zooxanthellae within clam tissues. As a consequence, arsenic is accumulated to extraordinarily high concentrations in giant clams, especially in the kidneys (500 to 1,000 μg/g).

Although ecological models exist for examination of symbiotic relationships (e.g., May 1976; Pacala et al. 1990; Grenfell and Keeling 2007), none have been applied to ecotoxicology, the exception being the occasional application of the epidemiological metrics described in the last chapter. Hopefully, this will change soon as ecotoxicology advances.

7.3 COMMUNITY STRUCTURE AND FUNCTION

7.3.1 General

Three general types of experimental units are used to study structural or functional shifts in communities exposed to toxicants: microcosms, mesocosms, and natural ecosystems. The term *microcosm* as used in ecotoxicology does not have precisely the same meaning as in general usage, i.e., a system acting as a representation of a larger system. Instead, microcosms are "laboratory study systems designed to simulate some component of an ecosystem such as multiple species assemblages" (Newman 2010). Mesocosms are defined by Newman (2010) as "relatively large experimental systems designed to simulate some component of an ecosystem. Mesocosms are delimited and enclosed to a lesser extent than are microcosms. They are normally used outdoors, or in some manner, incorporated intimately with the ecosystem that they are designed to reflect." Here, the combining form, *meso-*, is strictly consistent with general usage meaning middle, i.e., between the true ecosystem and a low-dimension laboratory system such as a microcosm. The distinction between microcosms and mesocosms appears to be a matter of location and size. Natural ecosystem studies involve manipulation of an entire system such as a small lake or watershed, or a large portion of a natural system.

The clear advantage of microcosms is the ability to examine community or ecosystem processes under strictly controlled conditions, to easily manipulate conditions, and to incorporate true replicates (Cairns and Niederlehner 1989; Taub 1989). Their obvious disadvantage is the loss of ecological realism relative to the other approaches. Mesocosms gain back some realism but become less yielding to control, manipulation, and replication (Caquet et al. 2000). They have been used very effectively to assess community and ecosystem level processes (e.g., Caquet et al. 2007; Crosland and La Point 1992; Hanson et al. 2007; Liber et al. 1992; Wilbur 1987). Whole or even partial ecosystem studies are expensive, difficult to manipulate, and often depend on pseudoreplication (Kimball and Levin 1985; Perry and Troelstrup 1988; Schindler et al. 1985). Regardless, their associated realism is invaluable. Intelligent melding of all three approaches has produced much valuable information.

Many methods are used to imply toxicant impact to communities or ecosystems. They range from semiquantitative indices, to ANOVA-based methods, to inverse regression methods. Three examples are provided below as applied to natural ecosystems, mesocosms, and microcosms.

Odum (1985) lists several changes expected in ecosystem qualities as a consequence of stress (Table 7.1). Most involve changes in biotic (community) rather than abiotic components of the ecosystem. Many (points 1, 2, 3, 6, 7, 8, 13, and 15 in Table 7.1) are similar to those outlined by Rapport et al. (1985). In addition to a list of consequences similar to Odum's (1985), Rapport et al. (1985) provide references to studies documenting such changes.

Cairns (1976, 1977) outlines the qualities contributing to natural ecosystem vulnerability to stress. (Sheenan (1984a, 1984b) details these and additional ecosystem qualities as influenced by toxicants.) Vulnerability is defined as the susceptibility of the ecosystem to irreversible damage. Implied in the phrase "irreversible damage" is a pragmatic timescale, i.e., irreversible within decades, not millennia. Vulnerability is a complex function of ecosystem elasticity (the ability to return to its original, prestress condition) and inertia (the ability to resist change in its function or structure). Resilience, a measure of the number of times that the ecosystem is able to recover to its normal state, is also identified as playing an important role. Cairns (1976, 1977) uses factors contributing to elasticity (Table 7.2) and inertia (Table 7.3) to develop an intentionally simplistic index for gauging ecosystem vulnerability. Although derived with considerable subjective opinion, these factors serve to focus attention on qualities underlying ecosystem vulnerability, therefore warranting discussion.

Six factors contributing to ecosystem elasticity are used to generate a simple recovery index (Equation 7.20).

$$Recovery\ Index = \prod_{i=a}^{f} Score_i \tag{7.20}$$

where a = rank for existence of epicenters, b = rank for transportability of dissemules, c = rank for habitat condition, d = rank for amount of residual toxicant, e = rank for physicochemical water quality, and f = rank for capability for regional management.

Table 7.1 Anticipated Changes in Ecosystems Experiencing Stress

Category	Trend
Energetics	1. Increase in community respiration
	2. Unbalanced production to respiration ratio ($P/R < 1$ or $P/R > 1$)
	3. Increase in maintenance to biomass, i.e., production/biomass and respiration/biomass
	4. Increase in importance of auxiliary energy (energy originating from outside the ecosystem)
	5. Increase in exported primary production
Nutrients	6. Increased turnover of nutrients
	7. Decreased cycling of nutrients
	8. Increased loss of nutrient as a consequence of 6 and 7
Community Structure	9. Increased proportion of species that are *r*-strategists
	10. Decreased size of organisms
	11. Decreased life spans
	12. Shortened food chains
	13. Decreased species diversity and increased species dominance (The reverse may occur if the original diversity was low)
Ecosystem	14. Internal cycling decreased and input/output from outside ecosystem becomes more important
	15. Regression to earlier successional condition
	16. Decreased efficiency of resource utilization
	17. Decreased positive (e.g., mutualism) and increased negative (e.g., parasitism) interactions
	18. Functional processes such as community metabolism tend to be more robust than species composition or other structural properties

Source: From Odum, E.P., *Bioscience,* 35, 419–422, 1985.

Table 7.2 Critical Factors and Qualitative Rankings for Ecosystem Elasticity

	Qualitative Rank of Importance		
Factor	1	2	3
a. Presence of nearby epicenters	Poor	Moderate	Good
b. Transportability of dissemules	Poor	Moderate	Good
c. Habitat condition	Poor	Moderate	Good
d. Presence of residual toxicants	Much	Intermediate	Low
e. Water quality	Very poor	Partially restored	Normal
f. Management capabilities	None	Some	Strong

Source: From Cairns, J., Jr. in eds. G.W. Esch and R. W. McFarlane, *Thermal Ecology II,* National Technical Information Center, Springfield, VA, 1976.

The first factor (*a*) is scored on the presence of refugia or other sources of species for reestablishment of the impacted area. The second (*b*) incorporates the ease with which eggs, young, or other forms of dissemules are transported into the area. The third (*c*) incorporates any change in the environment, such as a change in sediment composition due to siltation. The amount of toxicant remaining in the ecosystem (*d*) and extent to which the general water quality has been altered (*e*) are also included. The final factor (*f*) requires an arbitrary judgment regarding the potential for active remediation or some other action to lessen the impact. A calculated recovery index of 400 or more suggests rapid recovery. A score between 55 and 399 implies a fair to good chance of a rapid rate of recovery. Below 55, chances of a rapid recovery are judged to be poor.

Similarly, the subjective rankings from Table 7.3 can be used to generate an inertial index.

Table 7.3 Critical Factors and Qualitative Rankings for Ecosystem Inertia

Factor	Qualitative Rank of Importance		
	1	2	3
a. Biota adapted to significant variability in the environment	Poor	Moderate	Good
b. Much structural and functional redundancy	Poor	Moderate	Good
c. Mixing capability	Poor	Moderate	Good
d. Chemical characteristics	Poor	Moderate	Good
e. Proximity to ecological threshold	Close	Some margin of safety	Large margin
f. Management capabilities	Poor	Moderate	Good

Source: From Cairns, J., Jr. in eds. G.W. Esch and R. W. McFarlane, *Thermal Ecology II*, National Technical Information Center, Springfield, VA, 1976.

$$Inertial\ Index = \prod_{i=a}^{f} Score_i \tag{7.21}$$

where a = rank for the extent to which species are accustomed to wide variability in environmental conditions, b = rank for structural and functional redundancy, c = rank for ability to dilute and dissipate the toxicant, d = rank associated with the influence of water quality on the toxicant effect, e = rank associated with a nearby area of ecological threshold or transition such as an estuary, and f = rank for capability of regional management. Three general ranges of inertial stability (high, fair to good, poor) are defined by scores of 400 or more, 55 to 399, and less than 55.

Kersting (1984) develops a formal method of estimating ecosystem strain (distance moved in a state space relative to a reference state). Although it is applicable to many of the measures described in the remainder of this chapter, the approach is illustrated with zooplankton and algal densities (states) associated with a laboratory microcosm. *Daphnia magna* are grown in a microcosm with fluctuating algal densities, and a plot of *Daphnia* densities versus algal densities is constructed. As there is a lag before zooplankton abundance responds to changes in algal density (numerical response), the zooplankton density 1 week after an observed algal density is paired with that particular algal density. A 95% tolerance ellipse is then drawn about the points on this plot (state plane) from observations made prior to the introduction of the toxicant (Figure 7.6). (See Sokal and Rohlf 1981, pp. 594–601, for details on construction of a tolerance ellipse.) Data are next taken for the microcosm after toxicant introduction, placed onto the plot, and compared to the ellipse. For example, a 95% tolerance ellipse is constructed as shown in Figure 7.6. The point (X) associated with the microcosm after herbicide (Dichlobenil) application is placed on the plot and the distance from the center of the 95% tolerance ellipse (C) to point X in the state space (e.g., distance A) is divided by the distance from the center of the ellipse along that same line to a point on the 95% tolerance ellipse (e.g., distance B). The index A/B is called the normalized ecosystem strain (S). If $S \leq 1$, the system is behaving within its normal operating range. It is judged to be outside of its normal operating range if $S > 1$. Kersting (1984) suggests that this approach can also be taken for more complex systems provided interactions between fluctuating state variables are taken into consideration.

Liber et al. (1992) apply ANOVA (but see Chapter 5) and inverse regression methods to assess the effect of 2,3,4,6-tetrachlorophenol on zooplankton in mesocosms. Their inverse regression approach incorporates several enclosures spiked with varying amounts of toxicant. Inverse regression involves prediction of an X (not Y) value from the regression line of Y on X. (For more detail see Draper and Smith 1998, pp. 47–51; Sokal and Rohlf 1981, pp. 496–498; Neter et al. 1990, pp. 173–176.)

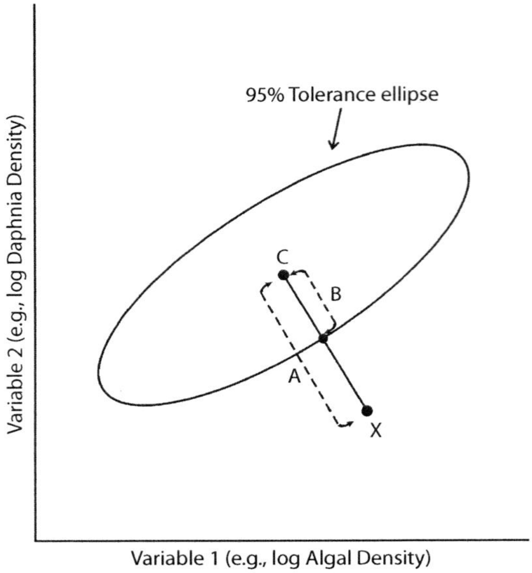

Figure 7.6 Estimation of normalized ecosystem strain (S = A/B). See text for details. (Based on Kersting, K., *Ecol. Bull.*, 36, 150–153, 1984, Figure 4.)

Initially, the log of zooplankton abundance is estimated in seven nonspiked mesocosms. The 95% confidence interval is estimated for the mean of the mesocosms (day 0 in Figure 7.7) prior to spiking.

$$Confidence\ Interval\ for\ \mu_0 = \bar{x} \pm \frac{t_{\alpha,2\text{-sided}} s}{\sqrt{n}} \tag{7.22}$$

where $t_{\alpha,2\text{-sided}} = t$ for two-sided test with $\alpha = 0.05$ (Appendix 14 with $p = 1$), and s = standard deviation for the n observations.

The mesocosms are then spiked and, after a certain time (day N in Figure 7.7), the resulting zooplankton abundances are determined. The observations are used to produce a regression line for the log of 2,3,4,6-tetrachlorophenol concentration versus the log of total zooplankton abundance. The concentration having an observed effect was estimated using the 95% confidence interval for this regression line.

The mean square deviation of the actual data points from the regression (mean square of the error (MSE); Equation 7.23) is used to generate the 95% confidence interval for the predicted mean values for log zooplankton abundance (Equation 7.24).

$$MSE = \frac{n-1}{n-2}(s_y^2 - b^2 s_x^2) \tag{7.23}$$

where n = the number of data pairs (x,y), b = the regression slope, $s_y^{\,2}$ = variance of the y values, and $s_x^{\,2}$ = variance of the x values.

$$95\%\ C.I. = Y_p \pm t_{\alpha,2\text{-sided}} MSE\sqrt{\frac{1}{n} + \frac{(x-\bar{x})^2}{(n-1)s_x^2}} \tag{7.24}$$

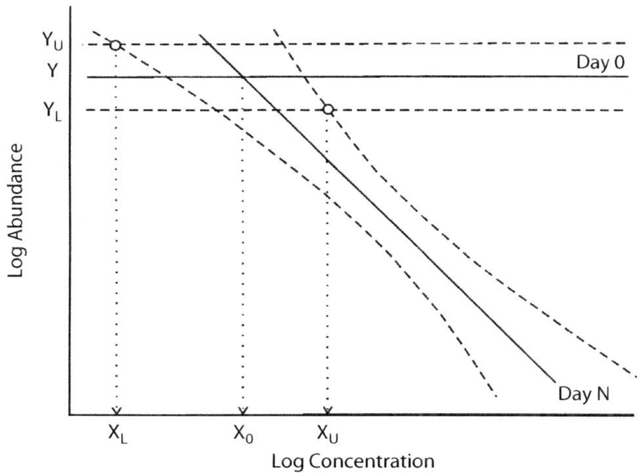

Figure 7.7 Inverse regression method for estimation of the concentration eliciting a significant effect on zooplankton abundance. See text including Example 3 for details. (Reprinted from Liber, K., N.K. Kaushik, K.R. Solomon, and J.H. Carey, *Environ. Toxicol. Chem.*, 11, 61–77, 1992, Figure 7. With kind permission from John Wiley and Sons. Copyright 1992.)

where Y_p = the predicted value of y for a given x, $t_{\alpha,2\text{-sided}} = t$ for a given α, $df = n - 2$ (e.g., Appendix 11 with $p = 1$), $\bar{x}$ = mean of the x values, and x = the specified x value at which the interval is estimated.

A statistical vantage is taken to define NOEC instead of determining the concentration having an unacceptable effect size (see Chapter 5). Liber et al. (1992) use the point at which the upper limit for the day N regression line intersects the lower limit of the day 0 log abundances (point X_U,Y_L in Figure 7.7) as the no observed effect concentration (NOEC). (See Chapter 5 for discussion of NOEC/NOEL.) The X_U is obtained by inverse regression methods with Y_L (the 95% lower limit for day 0 log abundances). Equation (23) from Draper and Smith (1998) predicts this point by making $Y_0 = Y_L$. The ± sign becomes positive in the numerator in this case. If the toxicant effect were to increase with concentration (i.e., the regression line has a positive slope), the Y_0 would have been set to Y_U to predict the log concentration at which the effect reaches the upper limit (X_L) defined with the day 0 observations.

$$X_U = \bar{X} + \frac{b(Y_0 - \bar{Y}) \pm ts\sqrt{\left(\frac{(Y_0 - \bar{Y})^2}{S_{xx}}\right) + \frac{b^2}{n} - \frac{t^2 s^2}{nS_{xx}}}}{b^2 - \frac{t^2 s^2}{S_{xx}}} \tag{7.25}$$

where $S_{xx} = \Sigma(x_i - \bar{x})^2$, $t = t$ for a given α (two-sided), $df = n - 2$ (e.g., Appendix 11 with $p = 1$), Y_0 = the lower 95% limit of log abundance, and s = MSE.

Example 7.3

Data of Liber et al. (1992) are used here to illustrate the inverse regression method described above. With these data, the NOEL (NOEC) is estimated for the 2,3,4,6-tetrachlorophenol effect on log zooplankton abundance in the mesocosms. The following SAS code contains the data supplied by K. Liber and generates the necessary summary statistics and regression analysis:

```
OPTIONS PS=58;
DATA LIBER;
   INPUT STATE $ ABUND CONC @@;
   DATALINES;
B 1944.0 0 B 2082.0 0 B 1619.0 0 B 1807.0 0 B 1575.0 0
B 1630.0 0 B 2016.0 0 A 2396.0 .25 A 2754.0 .25 A 3411.0 .25
A 692.0 .50 A 614.8 .50 A 793.3 .50 A 220.3 1.0 1 203.8 1.0
A 230.3 1.0 A 69.4 2.0 A 51.8 2.0 A 60.1 2.0 A 27.5 4.0
A 30.3 4.0 A 38.6 4.0 A 20.9 7.3 A 22.6 7.3 A 17.6 7.3
;
RUN;
DATA BEFORE;
   SET LIBER;
   IF STATE="B";
   LABUND=LOG10(ABUND);
RUN;
PROC UNIVARIATE DATA=BEFORE;
   VAR LABUND;
RUN;
DATA EXPOSED;
   SET LIBER;
   IF STATE="A";
   IF CONC<4.0; /* ONLY CONC'S BETWEEN 0.25 & 2.0 MG/L WERE USED */
LABUND=LOG10(ABUND);
LCONC=LOG10(CONC);
RUN;
PROC UNIVARIATE DATA=EXPOSED;
   VAR LABUND LCONC;
RUN;
PROC GLM DATA=EXPOSED;
   MODEL LABUND=LCONC;
RUN;
```

The 95% confidence interval is estimated with Equation (7.25) for the log of zooplankton abundances in the "prespiked" mesocosms.

$$n = 7 \qquad \bar{x} = 3.2553 \qquad s = 0.0498$$

The $t_{2\text{-sided}}$ for $\alpha = 0.05$ with $df = n - 1 = 6$ is 2.45 (Table 11 in the appendix with $p = 1$).

$$C.I. = 3.2553 \pm \frac{2.45 \cdot 0.0498}{\sqrt{7}}$$

$$\text{CI} = 3.2553 \pm 0.0461$$

$$3.2092 < \mu < 3.3014$$

The lower limit (3.2092) is used as Y_L.

Next, the MSE for the spiked mesocosm data is calculated with Equation (7.23) or taken from the SAS output resulting from the above code.

$$n = 12 \qquad s_y^2 = 0.41887 \qquad s_x^2 = 0.12357 \qquad b = -1.83449$$

$$MSE = \frac{12-1}{12-2}[0.41887 - ((-1.83449)^2(0.12357))]$$

$$MSE = 0.00331$$

The MSE is used as s in Equation (7.25).

$$Y_0 = Y_L = 3.2092 \qquad S_{xx} = 1.35929 \qquad b = -1.83449$$

$$t\ (\text{two-sided}, \alpha = 0.05, df = n - 2 = 10) = 2.2281$$

$$s = \text{MSE} = 0.00331$$

$$\bar{X} = -0.15051$$

$$\bar{Y} = 2.60247$$

$$X_U = -0.15051 + \frac{-1.11304 + 0.00738\sqrt{0.27082 + 0.28044 - 0.00000}}{3.36534 - \frac{0.00005}{1.35929}}$$

$$X_U = -0.47962$$

The antilog of X_U is 0.33 mg/L.

The inverse regression estimate of the NOEL for 2,3,4,6-tetrachlorophenol would be 0.33 mg/L. Note that this value is not the same as that calculated by Liber et al. (1992) (0.42 mg/L) because all 12 points were used here instead of the average for each concentration (n = 4 concentration averages).

The last two techniques discussed above and those in other chapters (e.g., ANOVA methods in Chapter 5) can be applied to microcosm, mesocosm, or natural ecosystem data to assess toxicant effects on community or ecosystem qualities. In the next several sections of this chapter, candidate qualities of study for communities and metacommunities will be discussed. The community qualities are divided into two broad categories: community structure and community function.

7.3.2 Community Structure

7.3.2.1 General

Although they are invaluable for summarizing data and assessing general shifts in community structure, no quantitative index or graph depicting community structure can replace a sound understanding of species natural histories combined with an accurate and thorough survey of species abundances. Often the roles of species such as keystone species* or the extreme sensitivity of indicator species must be understood in order to adequately predict or document the consequences of toxicant exposure (Preston 2002; Relyea and Hovermam 2006). For example, experimental removal of a starfish (Paine 1966) or sea urchin (Paine and Vadas 1969) from the intertidal zone significantly alters community structure. Removal of keystone species through the action of a toxicant would have a similar effect (e.g., Salminen et al. 2001). Recent examples of the importance of keystone species removal include the potential influence of ocean acidification on coral species viability and predicted collapse of coral reef systems (Hoegh-Guldberg et al. 2007), and the widespread decline of honeybees and resulting shifts in plant communities (Lettis and Delaplane 2010). Valuable clues such as the loss of a toxicant-sensitive indicator species could also easily be missed if one relied

* Species that influence the community by their activities or roles, not their numerical dominance.

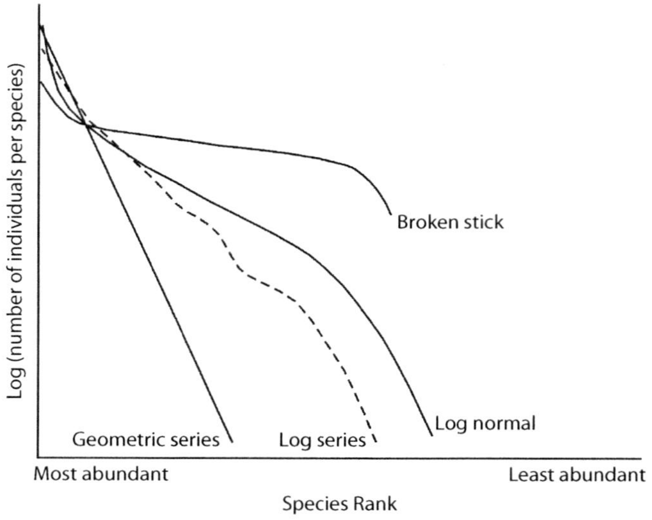

Figure 7.8 Four common models of species abundance. See text for a detailed description of these models.

solely on summary indices or graphs. Combined effects of toxicant sensitivity and species function within the community can produce results only interpretable with a detailed understanding of the associated species. For example, loss of one species of a mutualistic pair (e.g., Osman and Haugsness 1981; Lawrence et al. 2011) due to extreme toxicant sensitivity could result in the loss of the second species regardless of its tolerance to the toxicant.

7.3.2.2 Species Abundance

7.3.2.2.1 General

Often the relative abundances of species within an ecological community are used to summarize structure and imply processes occurring in the community. In one approach, all species are ranked, from those with highest number of individuals to those with the lowest number of individuals (Figure 7.8). Alternatively, relative cover or biomass may be used instead of numbers of individuals. Four types of abundance-species rank curves (geometric series, lognormal, log series, and broken stick) are used to suggest underlying mechanisms determining community structure. They are based on the assumption of resource competition. Although these explanations provide insight, they might or might not be helpful depending on how closely the operationally defined community ["all the organisms in a chosen area that belong to the taxonomic group the ecologist is studying" (Pielou 1969)] reflects the true ecological community or guild of species competing for resources. For example, interpretation of curve shape based on competition theory is questionable if only one genus of aquatic insects is enumerated but important competing species in other genera are ignored. However, such interpretation might provide very useful insight for a survey of a particular ecological guild ["a group of functionally similar species whose members interact strongly with one another but weakly with the remainder of the community" (Smith 1986)].

7.3.2.2.2 Geometric Series Model

The geometric series curve is interpreted with the niche preemption hypothesis. Species abundance (number of individuals, biomass, or percent cover) is thought to reflect the amount of resource that a particular species draws from the ecosystem. Magurran (1988, 2004) illustrates this hypothesis using the example of a vacant niche volume into which species arrive at regular intervals. When the most dominant species arrives, it takes a proportion (k) of the available resources. When the next most dominant species arrives, it takes the same proportion of the resources that remain. The repetition of this process for all competing species results in a geometric series curve for species abundance. There are a few dominant species, with most species being rare. Such curves are uncommon but can be associated with species-poor communities (Magurran 1988) with competition for one or a few crucial resources (Ludwig and Reynolds 1988), or with severe environments (Smith 1980) such as polluted environments (May 1976).

Magurran (1988, 2004) provides the following description of the geometric series. The S species are ranked from the most abundant ($i = 1$) to the least ($i = S$) abundant.

$$n_i = NC_k k(1-k)^{i-1} \tag{7.26}$$

where n_i = the number of individuals in species i, N = the total number of individuals in all species, and $C_k = [1 - (1 - k)^S]^{-1}$. A straight line should result from a plot of Ln abundance* versus species rank (Figure 7.8).

7.3.2.2.3 Lognormal Model

The lognormal is the most common species abundance curve. Indeed, May (1976) suggests that it rarely provides poorer description than the other abundance curves. The associated hypothesis for this curve is based on the assumption of a large community in which many factors influence species interactions during resource partitioning. These factors are assumed to be relatively independent (Magurran 1988, 2004; Smith 1980). The lognormal distribution results from the complex, multiplicative effects of these factors on the partitioning of resources among species (May 1976). As a consequence, there are many more species of intermediate abundances in the associated community than in the one just described for the geometric series.

The lognormal model is commonly analyzed using methods developed by Preston (1948, 1962) and illustrated in Figure 7.9. The $\log_2$-transformed abundances (number of individuals counted for a species) are plotted against the number of species in a particular abundance class, e.g., five species are represented by two individuals in the sample. The $\log_2$ is used most often, but any logarithmic transformation can be used. These $\log_2$ classes or octaves represent doublings in abundance (i.e., 1, 2, 4, 8, 16, 32, etc., individuals). In Figure 7.9, there is a point (veil line) below which the species are so rare that they are not found in the sample. The position of the veil line will vary depending on the size of the sample and the qualities of the community being used as a measure of resource use, e.g., number of individuals, biomass, or percent cover. With counts of individuals, the limit is less than one individual counted per species. The veil line indicates a point of truncation similar to the point of censoring, as discussed in Chapter 2 regarding observations below a limit of detection.

Above the veil line, the numbers of species per octave are plotted against the octave number. Units of the x-axis can be numbers of individuals, octaves, or R. The R transformation is $\log_2 [N/N_0]$, where N_0 is the number of individuals in the modal octave (R_0). It simply expresses octaves as deviations from the modal octave. See, for example, Figure 7.9, in which $N_0 = 8$. The R will be 0 for R_0 (octave 5) and increasingly positive or negative with each octave to the right or left of R_0.

The number of species in the Rth octave on either side of the modal octave (S_R) can be estimated (Magurran 1988, 2004).

* Although natural logarithm (Ln) is used here, Ln, $\log_2$, or $\log_{10}$ can be, and are, used in these models and indices.

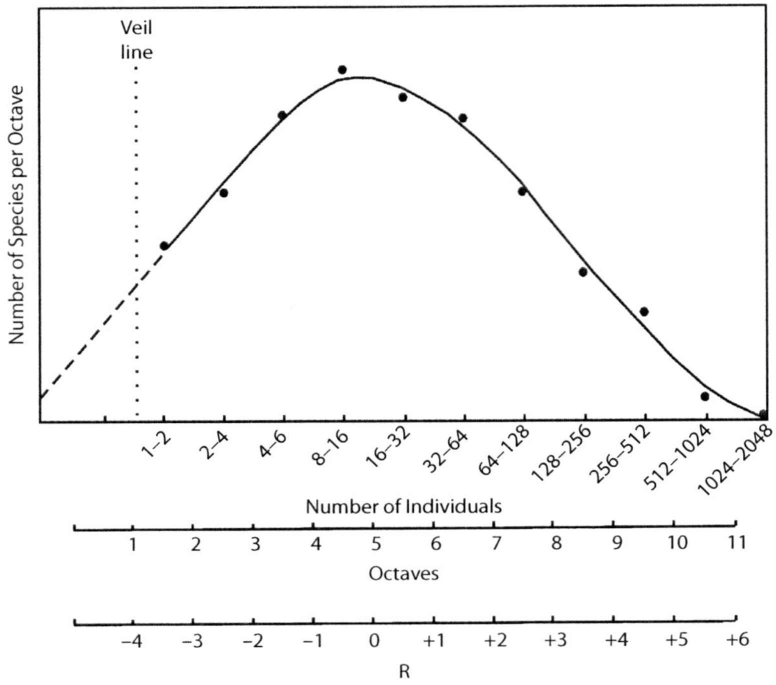

Figure 7.9 Typical plot used to analyze species abundance data under the assumption of a lognormal model. See text including Example 7.4 for details.

$$S_R = S_0 e^{-a^2 R^2} \tag{7.27}$$

where S_0 = the number of species in the modal octave, and a = the inverse width of the lognormal distribution. The a is approximately 0.2 in most cases when the calculations are done using $\log_2$. Under the assumption of a known mode, the a may be estimated with the relationship (Ludwig and Reynolds 1988)

$$a \approx \sqrt{\frac{Ln \dfrac{S_0}{S_{R_{MAX}}}}{R_{MAX}^2}} \tag{7.28}$$

where S_0 = the number of species in the modal octave, R_{MAX} = the octave with the maximum species abundance, and $S_{R_{MAX}}$ = the number of species in R_{MAX}.

Alternatively, Preston (1962) and Krebs (1989) provide a more involved estimate of a that does not require that the mode be fixed at a known value.

$$a = \frac{1}{\sqrt{2\sigma^2}} \tag{7.29}$$

where σ^2 = distribution variance estimated by s^2.

The presence of the veil line complicates estimation of s^2 because points below this line are not available for estimation with standard formulae. However, Cohen (1961) provides methods for such calculations. Estimation of the mean and standard deviation for the truncated lognormal distribution is very similar to estimation of these summary statistics for a censored lognormal distribution, as described in Chapter 2. However, because censoring and truncation are different, the reader should be aware that they require slightly different methods. Briefly, a data set is censored if a known number of values are missing below a known point for a data set, e.g., the limit of detection. For truncation, the specific point (e.g., veil line) is known but the number of observations below it is unknown. Traditionally, Cohen's (1961) maximum likelihood methods are applied to truncated species abundance curves, although some other methods in Chapter 2 are also appropriate. (See Equations 2.9 and 2.10 in Chapter 2, and associated text for details on Cohen's 1961 method for censored data.) The data are first $\log_2$ transformed and estimates of the mean ($\bar{x}_t$) and variance (s_t^2) for the data above the veil line are determined. Next, these statistics are used in the two equations below to derive estimates of the mean and variance for the complete data set, including values falling below the veil line.

$$\bar{x} = \bar{x}_t - \theta(\bar{x}_t - x_0) \tag{7.30}$$

$$s^2 = s_t^2 + \theta(\bar{x}_t - x_0)^2 \tag{7.31}$$

where x_0 = the $\log_2$ of the veil line midpoint position, and Θ = a value from Table 1 in Cohen (1961).

Cohen's table of Θ values is provided in Appendix 28. To extract a Θ from this table, γ must be calculated first.

$$\gamma = \frac{s_t^2}{(\bar{x}_t - x_0)^2} \tag{7.32}$$

Using linear interpolation, this γ is used to extract a Θ for use in estimating the mean and variance for the untruncated data. The variance estimate can then be used in Equation (7.29) to calculate a.

An additional (γ) parameter (not the same as that just described for Cohen's technique) is often estimated for lognormal species abundance data. It relates the species abundance curve discussed to this point with the analogous individuals curve. The individuals curve is a plot of the total number of individuals per octave (Y-axis) versus the abundance octave (X-axis). As such, it shares a common x-axis with and can be plotted with the species abundance curve, e.g., Figure 2.8 in Magurran (1988). (Preston 1962 also provides a detailed explanation of the individuals curve.) The estimated value of γ is normally in the range of 1.

$$\gamma = \frac{R_0}{R_{MAX}} \tag{7.33}$$

where R_{MAX} = the octave with the most abundant species from the species abundance curve, and R_N = the modal octave for the individuals curve.

It is important to note from the above equations that fitting a lognormal model to any data set lacking a clear mode, i.e., a mode assumed to be below the veil line, is unjustifiable (Krebs 1989). Such a curve cannot be distinguished from a log series, as discussed below.

With the information generated above, several parameters can be estimated, including a better estimate of S_0 (number of species in the modal octave), the predicted total number of species (S_T), and the predicted number of species in each octave (S_R). Ludwig and Reynolds (1988) provide the following estimator of S_0:

$$S_0 = e^{(Mean\ of\ Ln\ S_R\ values\ + a^2 \bar{R}^2)} \tag{7.34}$$

where mean of Ln S_R values = mean of the logarithms of the observed number of species in each octave, and $\bar{R}^2$ = mean of the R^2 values.

Krebs (1989) provided the following approach for predicting the total number of species. The $\log_2$ of the midpoint for the veil line (e.g., $\log_2 0.5$), and mean and standard deviation derived with Equations (7.30) and (7.31) are used to generate a z (standard normal deviate) for the veil point.

$$z = \frac{x_0 - \bar{x}}{s} \tag{7.35}$$

This z is used in a table of areas under the unit normal curve (or used as the argument for Excel NORMSDIST(z)) to estimate the area to the left of this point of truncation (p_0). This point of truncation expressed as an area under the normal curve (p_0) and the observed number of species (S_{ob}) are used in Equation (7.36) to estimate the total number of species (S_T).

$$S_T = \frac{S_{ob}}{1 - p_0} \tag{7.36}$$

A χ^2 goodness-of-fit test can be used to assess data fit to the lognormal model. Expected values from Equation (7.27) and observed values are used to calculate a χ^2 statistic with the associated *df* dependent on the method used. The χ^2 is used to measure goodness-of-fit. It can be recalculated after repeated substitutions of a and S_0 values until the pair producing the lowest χ^2 is obtained. Ludwig and Reynolds (1988) and Krebs (1999) describe several ways of performing these calculations and provide programs for fitting species abundance data to a lognormal model.* Ludwig and Reynolds (1988) use Equations (7.28) and (7.34) under the assumption of a known mode. Consequently, the *df* associated with the hypothesis test is the number of octaves minus 2. Krebs (1999) describes an approach (Equations 7.29 to 7.32) that does not make the assumption of a known mode and uses an estimate of μ (Equation 7.30). Consequently, the *df* must be reduced to the number of octaves minus 3 for the χ^2 hypothesis test. The approach described by Krebs has the advantage of avoiding the assumption of a known mode.

7.3.2.2.4 Fisher's Log Series Model

Magurran (1988) explains the log series as she did the geometric series, but the species arrivals to the vacant niche volume are at random, not regular intervals. Like the geometric series curve, the log series curve describes a situation with few dominants and many uncommon species (Fisher et al. 1943). It, along with the lognormal curve, reflects intermediate conditions between the geometric series and broken stick curves. The log series curve is used less often than the lognormal curve but has its advocates (e.g., Magurran 1988). It is appropriate for data sets of numbers of individuals but not those quantifying biomass or any continuous (noninteger) measurement (Magurran 2004).

* Excellent shareware are available from several sources. Estimates can be downloaded from Professor Robert K. Colwell of the University of Connecticut (http://viceroy.eeb.uconn.edu), and Biodiversity Professional can be downloaded from various webpages, including NASA's (http://gcmd.nasa.gov/records/NHML_Biopro.html). This second piece of shareware (BioDiversity Pro) was developed by the Natural History Museum in London and the Scottish Association for Marine Science, Oban, Scotland.

The total number of species (S) for a log series is defined by Equation (7.37).

$$S = \alpha[-Ln(1-x)] = \alpha Ln\left(1+\frac{N}{\alpha}\right) \tag{7.37}$$

where S = the total number of species, N = the total number of individuals, and x,α = parameters estimated as shown below.

The value of x is usually between 0.9 and 1.0 (Magurran 2004). The x will be >0.99 as N/S exceeds 20. It is estimated by an iterative process by substituting values of x into Equation (7.38) until the equality is close enough for the intended use (Krebs 1999).

$$\frac{S}{N} = \left(\frac{1-x}{x}\right)(-Ln(1-x)) \tag{7.38}$$

The α and its large sample variance are estimated using x (Krebs 1999).

$$\alpha = \frac{N(1-x)}{x} \tag{7.39}$$

$$Variance\ of\ \alpha = \frac{Ln\ 2\cdot\alpha}{\left[Ln\left(\frac{x}{1-x}\right)-1\right]^2} \tag{7.40}$$

This α is often used as a diversity index with large values indicating many species and small values indicating few species. Magurran (2004) indicates that this formula should not be used in the rare instance in which either $N/S \leq 1.44$ or $x \leq 0.50$.

Expected numbers of species in the various abundance classes can be predicted with these estimates of x and α: one species expected, αx; two species expected, $\alpha x^2/2$; three species expected, $\alpha x^3/3$; four species expected, $\alpha x^4/4$ … $\alpha x^n/n$. As described for the lognormal model, a χ^2 statistic may be used to assess data fit to the log series model (df = number of classes – 1).

7.3.2.2.5 Broken Stick Model

MacArthur's (1957) broken stick curve is usually interpreted with the random niche boundary hypothesis (e.g., Magurran 2004; Smith 1980). The niche volume is randomly divided into S portions for the available species to occupy, much the same way that a stick might be snapped simultaneously at random intervals along its length. No assumption of preemptive use of resources by species is made. The species utilization of the resource(s) is assumed to be nonoverlapping (Ludwig and Reynolds 1988) and much more equitable than in any of the other models described. There are many species with intermediate abundances in the community. The species are assumed to be influenced only by one shared resource (Magurran 1988, 2004). Despite the historical importance of the broken stick model, Smith (1980) and Ludwig and Reynolds (1988) suggest that it is seldom used, as it describes an uncommon situation. Magurran (1988) indicates that the broken stick curve can emerge during surveys of taxonomically very similar species.

To analyze data using the broken stick model, species are rank ordered by abundance. The number of individuals in the ith most abundant species (n_i) is estimated with Equation (7.41) (Magurran 2004).

$$n_i = \frac{N}{S}\sum_{n=i}^{S}\frac{1}{n} \tag{7.41}$$

where N = the total number of individuals in the sample from the community, and S = the total number of species in the sample.

The expected number of species in an abundance class (S_n) is predicted with Equation (7.42) (Magurran 2004).

$$S_n = \left(\frac{S(S-1)}{N}\right)\left(1-\frac{n}{N}\right)^{S-2} \tag{7.42}$$

where n = the number of individuals in the abundance class.

The expected values, the observed values, and *df* (number of abundance classes – 1) can be used in a χ^2 goodness-of-fit test.

7.3.2.2.6 Pollution Effects on Abundance Curves

Patrick (1973) indicated that the shapes of species abundance curves for communities in polluted environments exhibit characteristic shifts. With pollution, the species abundance curve moves away from the typical lognormal curve with good species richness toward an aberrant lognormal or geometric series curve (Gray 1979). The aberrant lognormal curve has a lowered mode and, perhaps, a broadening to cover more high abundance octaves. May (1976) hypothesizes that these shifts in communities inhabiting contaminated environments represent successional reversions. In early community succession, processes associated with the geometric series/niche preemption hypothesis are thought to be important, but with time, those processes are replaced by those associated with the lognormal hypothesis. This is consistent with the general theme that community structure is shaped by the rates of competitive exclusion among species with different life history strategies during fluctuations in environmental conditions (Huston 1979). Community structure results from a dynamic equilibrium between r- and K-strategy species under shifting intensities of perturbation to their populations due to environmental factors. Pollutants can also influence this dynamic equilibrium through the direct removal of individuals from populations and adversely impacting vital rates (Gray 1979).* The different combinations of these factors (killing individuals and lowering vital rates) tip the advantage of r- and K-selection strategy species in the community.

Warwick and coworkers (Clarke and Warwick 1999, 2001; McManus and Pauly 1990; Warwick 1986; Warwick and Clarke 1995, 1998) have explored a range of approaches to showing pollution effects on abundance curves. Warwick (1986) introduced an especially helpful graphical method based on a k-dominance abundance curve for species number and also biomass (Figure 7.10). A k-dominance curve is simply one in which the cumulative proportion or percentage dominance and logarithm of rank are plotted (Lambshead et al. 1983). Numerous software packages such as Biodiversity Pro produce k-dominance curves. Species abundance biomass comparison (ABC) curves are constructed with the logarithm of each species rank on the x-axis and the cumulative dominance expressed in terms of numbers of individuals and also biomass on the y-axis. For example, the first, most abundant species might comprise 25% of the total number of individuals surveyed and 50% of the total biomass of the surveyed species. The second species might comprise an additional

* Gray (1979) uses the terms *disturbance* and *stress* instead of the phrases "direct removal of individuals" and "adversely impacting some vital rate" used here.

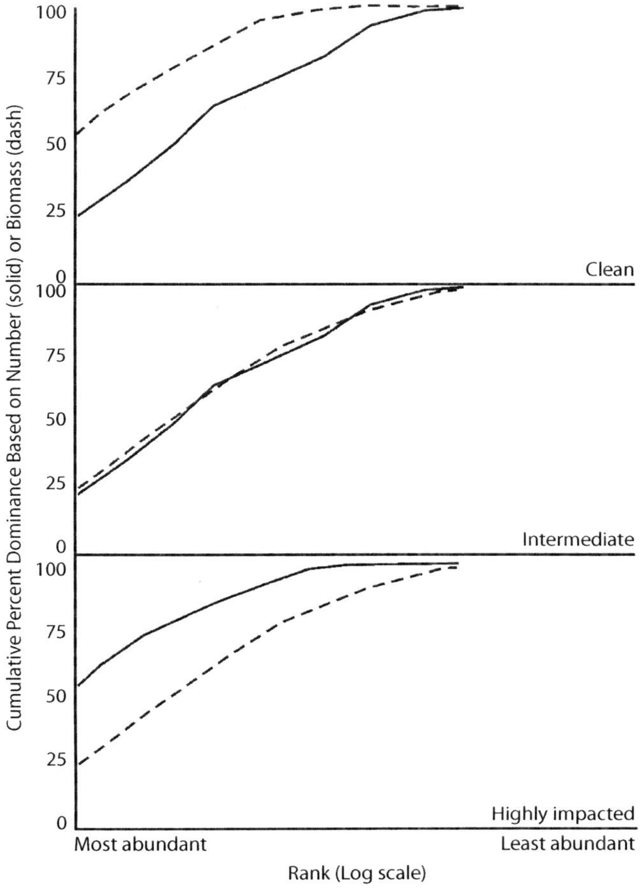

Figure 7.10 Illustrative species Abundance/Biomass Comparison (ABC) k-dominance curves based on the work of Warwick (1986) and Warwick et al. (1987). The curves shift in a predictable manner with increasing pollution effect. (Based on panels in Warwick, R.M., *Mar. Biol.*, 92, 557–562, 1986, Figure 1.)

10% of individuals and another 25% of biomas, so the cumulative percent dominance value for this second species would be 35% of species and 75% of biomass. The plotting continues to the last (least abundant) species that brings the cumulative number and biomass dominance to 100%.

Warwick (1986) interprets ABC curves based on several community characteristics. First, it is assumed that competitive displacement of species occurs under stable conditions and the K-selection strategy species will emerge as dominants whose populations will remain relatively stable. Such species were dominants in terms of biomass in the marine benthic communities that Warwick studied. The smaller, r-selection strategy species tend to be the numerical dominants in such stable systems. This situation is reflected in the top panel of Figure 7.10. Second, these benthic communities will not recover quickly enough between pollution impact episodes so the K-selection strategy species (biomass dominants) will fair less well than the r-selection strategy species (numerical dominants). The r-selection strategists can become the biomass dominants also in such a situation (middle panel of Figure 7.10). Finally, more intense pollution pressure on the community can begin impacting the numbers of some of the r-selection strategy species also, so that the number-based diversity begins to decrease (bottom panel of Figure 7.10).

Example 7.4

Diatom species sampled from a stream above (clean) and below (polluted) an outfall are counted and the number of species per octave tabulated. As done by Krebs (1989), intervals of 2–3, 4–7, etc., were used in calculations to account for species numbers at octave junctures, e.g., 4 or 8. Alternatively, if several species had abundances at an octave juncture, half can be assigned to the lower octave and half to the higher octave (Ludwig and Reynolds 1988). Log_2 is used as per the original method descriptions of Preston (1948, 1962), although $\log_{10}$ or Ln could be used. (Note: $\log_2 x = (\log_{10} x)/(\log_{10} 2)$.)

Octave (Interval)	Interval Midpoint (Log_2 Midpoint)	Number of Species	R
	Clean Site		
1 (1)	1 (0)	80	−3
2 (2–3)	2.5 (1.3219)	110	−2
3 (4–7)	5.5 (2.4594)	129	−1
4 (8–15)	11.5 (3.5236)	138	(R_0) 0
5 (16–31)	23.5 (4.5546)	131	+1
6 (32–63)	47.5 (5.5699)	109	+2
7 (64–127)	99.5 (6.6366)	78	+3
8 (128–255)	191.5 (7.5812)	53	+4
9 (256–511)	383.5 (8.5831)	29	+5
10 (512–1,023)	767.5 (9.5840)	14	+6
11 (1,024–2,047)	1,535.5 (10.5845)	5	+7
12 (2,048–4,095)	3,071.5 (11.5847)	3	+8
13 (4,096–8,191)	6,143.5 (12.5848)	2	(R_{MAX}) +9
	Polluted Site		
1 (1)	1 (0)	34	−4
2 (2–3)	2.5 (1.3219)	42	−3
3 (4–7)	5.5 (2.4594)	48	−2
4 (8–15)	1.5 (3.5236)	54	−1
5 (16–31)	23.5 (4.5546)	56	(R_0) 0
6 (32–63)	47.5 (5.5699)	52	+1
7 (64–127)	99.5 (6.6366)	50	+2
8 (128–255)	191.5 (7.5812)	42	+3
9 (256–511)	383.5 (8.5831)	34	+4
10 (512–1,023)	767.5 (9.5840)	23	+5
11 (1,024–2,047)	1,535.5 (10.5845)	17	+6
12 (2,048–4,095)	3,071.5 (11.5847)	10	+7
13 (4,096–8,191)	6,143.5 (12.5848)	7	+8
14 (8,192–16383)	12,287.5 (13.5849)	4	+9
15 (16,384–32,767)	24,575.5 (14.5849)	2	+10
16 (32,768–65,535)	49,151.5 (15.5849)	1	(R_{MAX}) +11

Estimates of a for these two collections are made with Equation (7.28): clean = 0.228, polluted = 0.182. The other, more involved methods described in Krebs (1989) (i.e., Equations 7.29 to 7.32) are also used. The mean and variance for the truncated data (data above the veil line) are estimated using the $\log_2$(interval midpoint) and the associated number of species in each octave.

$$\text{Clean: } \bar{x}_T = 4.0500 \qquad \text{Polluted: } \bar{x}_T = 5.3656$$
$$s_T^2 = 6.2699 \qquad s_T^2 = 10.7383$$

Cohen's (1961) methods are used to correct for the effect of truncation on the means and variances. First, γ values are calculated with Equation (7.32). The midpoint between 0 and the smallest abundance value (one individual per species) is used as an estimate of x_0. The $x_0 = \log_2((0 + 1)/2) = \log_2(0.5) = -1$.

$$\text{Clean: } \gamma = 6.2699/(4.0500 - (-1))^2 = 0.246$$

$$\text{Polluted: } \gamma = 10.7383/(5.3656 - (-1))^2 = 0.265$$

The Θ values are then extracted from Appendix 28 with these γ values.

$$\text{Clean: } \Theta = 0.04845 \qquad \text{Polluted: } \Theta = 0.06332$$

Equations (7.30) and (7.31) are used to estimate the mean and variance for the two data sets with correction for truncation at the veil line.

$$\text{Clean: } \bar{x} = 4.0500 - 0.04845(4.0500 - (-1)) = 3.8053$$
$$s^2 = 6.2699 + 0.04845(4.0500 - (-1))^2 = 7.5055$$

$$\text{Polluted: } \bar{x} = 5.3656 - 0.06332(5.3656 - (-1)) = 4.9625$$
$$s^2 = 10.7383 + 0.06332(5.3656 - (-1))^2 = 13.3041$$

Another set of estimates for a is generated with Equation (7.29) and the estimates of σ^2.

$$\text{Clean: } a = 0.2581 \qquad \text{Polluted: } a = 0.1939$$

The number of species in the modal octave can be calculated with Equation (7.34).

$$\text{Clean: Mean of Ln } S_R \text{ values} = 3.5516$$
$$\bar{R}^2 = 23$$
$$S_0 = 161.4$$

$$\text{Polluted: Mean of Ln } S_R \text{ values} = 2.9108$$
$$\bar{R}^2 = 33.5$$
$$S_0 = 64.7$$

The predicted total number of species is then calculated. First, Equation (7.35) is used to calculate z.

$$\text{Clean: } z = (-1 - 3.8053)/2.7396 = -1.7540$$

$$\text{Polluted: } z = (-1 - 4.9625)/3.6475 = -1.6347$$

From Appendix 5, the N.E.D. values corresponding with these z values are used to extract p_0 values for Equation (7.36). Alternatively, Table 11 in Rohlf and Sokal (1981) or a similar table could be used to estimate the area to the left of the veil line or the N.E.D. could be generated with the Excel NORMDIST(z,0,1,TRUE) function.

Clean: p_0 . 0.040 Polluted: p_0 . 0.050

The predicted total number of diatom species in these two samples was estimated with Equation (7.36).

$$\text{Clean: } S_T = 881/(1 - 0.040) = 917$$

$$\text{Polluted: } S_T = 476/(1 - 0.050) = 501$$

Next, the values of a and S_0 can be used to predict S_R values. However, as there were two estimates for each of these parameters, the question arises as to which estimates are best. Initially, all four possible combinations of a (from Equation 7.28 or Equation 7.29) and S_0 (observed or from Equation 7.34) are used to predict S_R values, and χ^2 statistics are calculated from the observed and expected species numbers in each R. The combination with the minimum χ^2 is judged to have the best fit. (Several of the programs mentioned previously can be used to do these tedious calculations.)

a	S_0	χ^2
	Clean	
0.228	138	10.62
0.228	161.4	47.36
0.258	138	5.27
0.258	161.4	17.59
	Polluted	
0.182	56	0.41
0.182	64.7	10.77
0.194	56	3.85
0.194	64.7	7.51

The best combinations of estimates for the samples from the clean and polluted sites are $a = 0.258$ and $S_0 = 138$, and $a = 0.182$ and $S_0 = 56$, respectively. Ludwig and Reynolds (1988) suggested that iterative substitution of a and S_0 values may be used at this point with the goal of minimizing the associated χ^2 to improve these estimates. When this is done, the final estimates remain the same for the sample from the polluted site, but the best estimates for the sample from the clean site are $a = 0.247$ and $S_0 = 138$ (final $\chi^2 = 1.87$). Comparing the two samples, the total and modal numbers of species drops and the R_{MAX} increases in the sample from below the effluent relative to the sample from above the point of discharge.

7.3.2.3 Species Richness

Species richness is the number of species in a community. Although the species richness concept is easily grasped, the difficulty in clearly delineating the community can make it difficult to interpret the associated measures of richness. The community is often some taxonomic group such as fish or insects in a region of interest. Pielou (1974) refers to such an operationally defined community as a taxocene. It is extremely important that the distinction between the abstract concept of a community and the actual sample from which an index is derived be kept in mind during data interpretation.

The number of species counted in an area will be influenced by the number of samples taken. It will be some increasing function of the logarithm of the area collected (Gleason 1922). With increased numbers of samples or quadrats, the cumulative number of species counted increases at a decreasing rate toward an asymptotic value. One could laboriously sample more and more quadrats at each site until the curve approaches its asymptotic value for the number of species. Alternatively, species richness can be normalized to a set sample size, e.g., James and Rathbun (1981). Hurlbert (1971) and Krebs (1999) provide a rarefaction method for determining species richness this way.

$$E(\hat{S}_n) = \sum_{i=1}^{S} \left[1 - \frac{\begin{pmatrix} N - N_i \\ n \end{pmatrix}}{\begin{pmatrix} N \\ n \end{pmatrix}} \right] \tag{7.43}$$

where $E(\hat{S}_n)$ = the expected number of species in a sample of size n, N_i = the number of individuals of species i, N = the total number of individuals in the sample, and n = the sample size (number of individuals) to which normalization is to be done.

For readers unfamiliar with the binomial coefficient notation in this equation, an explanation with solution is provided in Chapter 4 (Equation 4.42). Tables, explanation, and simple properties of the binomial coefficient are also provided on pages 362–366 of Beyer (1991).

The large sample variance for $\hat{S}_n$ can be estimated with a rather involved equation (Heck et al. 1975; Krebs 1989) that can be made clearer by simplifying some terms first. Let start with the following simplification of terms:

$$a = \begin{pmatrix} N \\ n \end{pmatrix}$$

$$b = \begin{pmatrix} N - N_i \\ n \end{pmatrix}$$

$$c = \begin{pmatrix} N - N_j \\ n \end{pmatrix}$$

$$d = \begin{pmatrix} N - N_i - N_j \\ n \end{pmatrix}$$

The variance is estimated with these terms.

$$\textit{Variance of } \hat{S}_n = a^{-1} \left[\sum_{i=1}^{S} b\left(1 - \frac{b}{a}\right) + 2 \sum_{i=1}^{S-1} \sum_{j=i+1}^{S} \left(d - \frac{bc}{a}\right) \right] \tag{7.44}$$

This rarefaction method is preferred over simple enumeration methods when samples (number of individuals) vary in size. Krebs (1999) suggests that it should be used only for comparing taxonomically similar groups from similar habitats sampled by identical techniques. The assumption is made that the underlying species-individual relationship is the same among units being compared (Peet 1974). The methods are also based on the assumption that individuals are randomly distributed, and with very large samples, the bias associated with any clumping tends to be minimized. Hurlbert (1971), Ludwig and Reynolds (1988), Magurran (1988), and Krebs (1989, 1999) provide further details regarding this index. The Biodiversity Pro software and that associated with Krebs (1999) produce such estimates and associated plots.

Heltshe and Forrester (1983) develop a jackknife procedure for estimating species richness from quadrat data. Only the presence or absence of the species is noted for each quadrat. Calculations focus on the number of unique species (those occurring in only one quadrat regardless of their numbers within that quadrat).

$$\hat{S} = S + \left(\frac{n-1}{n}\right)k \tag{7.45}$$

where n = the number of quadrats sampled, S = the total number of species in the n quadrats, and k = the number of unique species in the n quadrats.

The variance and approximate 95% confidence interval for this index of species richness can be estimated. (The software associated with Krebs (1999) does these calculations.)

$$Variance\ of\ \hat{S} = \frac{n-1}{n}\left[\sum_{j=0}^{S} j^2 f_j - \frac{k^2}{n}\right] \tag{7.46}$$

where f_j = the number of quadrats containing j unique species.

$$C.I. = \hat{S} \pm t_{2\text{-sided},\alpha}\sqrt{Variance\ of\ \hat{S}} \tag{7.47}$$

where $t_{2\text{-sided},\alpha} = t$ from Appendix 11 for $p = 1$, $df = n - 1$, and $\alpha = 0.05$.

This estimator is biased (overestimates the number of species) (Heltshe and Forrester 1983), but the bias is usually smaller than that for the actual observed number of species (underestimates the number of species). Heltshe and Forrester (1983) and Krebs (1999) suggest that this jackknife method not be used if there is a large number of unique species or S is low.

If a large number of quadrats are sampled, Krebs (1989) recommends a bootstrap method described by Smith and van Belle (1984) instead of the jackknife method described above. Both the bootstrap and jackknife methods described by Smith and van Belle (1984) are biased. Smith and van Belle (1984) recommend the jackknife method when the number of quadrats sampled is small and the bootstrap method when the number of quadrats is large. Mingoti and Meeden (1992) argue that both the jackknife and bootstrap methods are unacceptably biased and suggest an empirical Bayes estimator instead.

Example 7.5

Species in a particular guild are sampled at eight sites around an outfall and the number of individuals per species (N_i) tabulated. Site locations are designated as kilometers upstream (–) or downstream (+) of the outfall. As the total number of individuals per sample (N) differs, the rarefaction method is used to estimate species richness at a sample size of 40 individuals ($n = 40$).

	N_i at Sites (site = kilometers from the outfall)							
Species	**–1.0**	**–0.5**	**0**	**+0.5**	**+1.0**	**+1.7**	**+2.7**	**+5.3**
1	12	12	58	18	11	8	12	10
2	11	12	21	16	11	12	10	11
3	10	10	3	15	11	13	8	8

	N_i at Sites (site = kilometers from the outfall)							
Species	**−1.0**	**−0.5**	**0**	**+0.5**	**+1.0**	**+1.7**	**+2.7**	**+5.3**
4	8	8	2	5	8	3	11	6
5	8	8	1	4	7	3	8	6
6	5	4			3	3	4	5
7	2	3			2	1	1	4
8	2	2			1	2	1	2
9	1	2			1		2	1
10		1						1
N	59	62	85	58	55	45	57	54

Using Equations (7.43) and (7.44), an estimate of species richnesses and the associated variances for each sample is made at $n = 40$.

	Sites (site = kilometers from the outfall)							
Statistic	**−1.0**	**−0.5**	**0**	**+0.5**	**+1.0**	**+1.7**	**+2.7**	**+5.3**
$\hat{S}_n$	8.476	9.347	4.050	4.991	8.367	7.877	8.312	9.414
Variance	0.386	0.472	0.556	0.009	0.461	0.110	0.485	0.432

Above the outfall, the species richness is estimated to be in the range of 8.476 to 9.347 (expected species in a sample of $n = 40$). The species richness drops at the outfall but gradually increases again with distance downstream. Figure 7.11 is derived from a BioDiversity Pro software graphic output, and BioDiversity Pro was also used to produce is a k-dominance plot (Figure 7.12).

Calculations for the sample taken from the +0.5 km site are detailed here. Equation (7.43) is used to estimate $\hat{S}_n$.

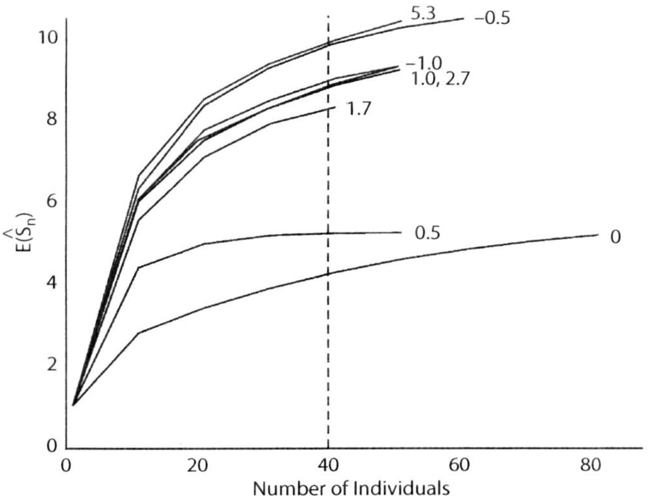

Figure 7.11 Rarefaction indices for Example 7.5 sampling sites. An estimated species richness was estimated using a sample size of 40 in that example. Notice the drop in richness at and just below the discharge and the gradual increase to reference site (−1.0 and −0.5 kilometers) species richness.

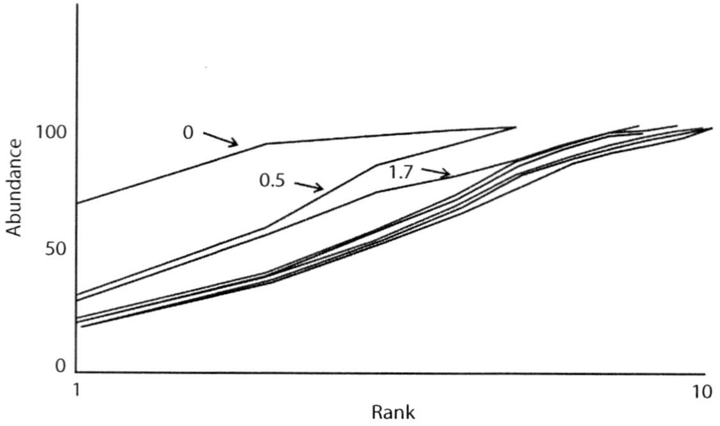

Figure 7.12 k-dominance plot for Example 7.5 sampling sites. Notice that the two curves for the most impacted sites (0 and 0.5 kilometers) were higher in the plot than those for the other sites. This was also the prediction for the general case of pollution impact as depicted for individual-based abundance in Figure 7.10. The more elevated a line on this type of plot, the less diverse the community or sampled taxocene.

$$\hat{S}_n = \left[1 - \frac{\binom{40}{40}}{\binom{58}{40}}\right] + \left[1 - \frac{\binom{42}{40}}{\binom{58}{40}}\right] + \left[1 - \frac{\binom{43}{40}}{\binom{58}{40}}\right]$$

$$+ \left[1 - \frac{\binom{53}{40}}{\binom{58}{40}}\right] + \left[1 - \frac{\binom{54}{40}}{\binom{58}{40}}\right]$$

$$\hat{S}_n \approx [1-0] + [1-0] + [1-0] + [1-0.002] + [1-0.007] \approx 4.991$$

The variance calculation (Equation 7.44) is tedious and best done with a computer to minimize error.

A software of Krebs (1999) was used here to do the calculations. Version 7.0 of this software can be obtained from Exeter Software (www.exetersoftware.com).

A wide range of designs can be used to analyze toxicant effect on species richness. Perhaps, ANOVA could be applied to the design described above. Alternative methods outlined in other chapters could be applied with appropriate care. For example, Niederlehner et al. (1986) use an index of percentage of taxa affected (Probit([1 – (species richness with exposure/species richness without exposure)] · 100) and Ln of concentration to perform a traditional probit analysis of protozoan community change with exposure to cadmium.

7.3.2.4 Species Diversity ("Heterogeneity")

In addition to species richness, the concept of species diversity includes the evenness or equitability with which individuals in a community are allocated among species. For example, if all species in a community have the same number of individuals, this community's diversity is thought to be higher than that of another community having the same number of species but most of the individuals associated with only a few species. Many species' diversity indices (also called heterogeneity or recently even biodiversity indices) account simultaneously for both species richness and evenness. Three will be discussed, although under the assumption of a log series species abundance model and sufficient number of species to construct the curve, the α described already may be used as a diversity measure (Pielou 1969). Because these indices measure the combined influence of species richness and evenness, interpretation of any diversity index is most effective with estimation of both species richness and evenness.

7.3.2.4.1 Scalar Aspects of Diversity

The qualities of communities vary with the spatial scale at which they are studied, yet much of the discussion so far has focused on spatially discrete assemblages. The diversity metrics for such units of study quantify what is called α diversity. However, the larger-scale landscape or region could contain many such communities, and the diversity for that region might be the issue of interest. Consequently, some additional spatial terminology is needed so that such issues can be addressed. A microsite is a spatial area that can hold only one individual, whereas a patch is an area of habitat that contains the many microsites in which the community exists (Leibold et al. 2004). A patch might also be called a locality. For example, Leibold et al. (2004) defines a locality as being a discrete area of habitat. A region is a large habitat area "containing multiple localities and capable of supporting a metacommunity" (Leibold et al. 2004). The scale of a region is often referred to as mesoscale.

The metrics that quantify changes within a locality (e.g., Simpson, Shannon's, and Brillouin's indices below) quantify α diversity, and those that quantify changes among localities that make up a region quantify β diversity. The community similarity indices described below can be seen as metrics of β diversity. Together, α and β diversity define the total (γ) diversity of a region (Whittaker 1960; Sarkar 2006). Measures of β diversity might be generated from the scale of survey transects, to much larger landscapes, to collections of discrete physical islands.

7.3.2.4.2 Metrics of Diversity

The Simpson (Yule) index (Simpson 1949), a species dominance type of diversity index, is a weighted mean of the species proportional abundances. It assumes no particular distribution for species abundance. According to Poole (1974), this index quantifies the dominance of one or a few of the commonest species in the community. It is relatively insensitive to species richness because it weights the most abundant species most heavily. Depending on its form, it estimates the probability that two individuals randomly taken from an infinitely large (indefinitely large) population will be of the same (λ or D) or different ($1 - \lambda$ or $1 - D$) species (Hurlbert 1971; Peet 1974; Pielou 1969, 1974).*

$$\lambda \text{ (or } D) = \sum_{i=1}^{S} p_i^2 \tag{7.48}$$

where S = the total number of species, and p_i = the proportion of the total number of individuals in the community that are members of the ith species.

Practical application of Simpson's index entails substitution of n_i (the number of individuals of species i in the total sample) divided by N (the total number of individuals in the sample) for p_i. Further, because the entire population is not sampled, a bias appears that might be corrected by

* The term *population* refers to the statistical population, i.e., all of the individuals in the community.

adding the terms (n_i – 1) and (N – 1) to the index. Thus, substitution of n_i/N into Equation (7.48) provides an estimate of λ for an (effectively) infinite population and Equation (7.49) provides an estimate for a finite population (Pielou 1969).

$$\hat{\lambda} = \sum_{i=1}^{S} \frac{n_i(n_i - 1)}{N(N-1)} \quad (7.49)$$

Simpson's index can range from 0 to 1. Obviously, species diversity is low when $\hat{\lambda}$ (probability of encounter between two individuals of the same species) is high. A simple modification of this index is often made so that its value increases as diversity increases, i.e., $1-\hat{\lambda}$. With this form of Simpson's index, it can be interpreted as the probability of sampling in sequence two individuals that are different species. The reciprocal, $1/\hat{\lambda}$ or $1/D$, is also used in many cases. Simpson (1949) provides methods for estimating the standard error of Simpson's index. Heltshe and Forrester (1985) describe jackknife methods for estimating the confidence interval for this index. The Krebs (1999), Biodiversity Pro, and EstimateS (see Colwell 2009) software do these computations (and those for the following two diversity metrics). EstimateS provides standard deviations for estimates and the Krebs (1999) software estimates bootstrap confidence intervals.

Two information theory-based indices (Shannon's and Brillouin's indices) are used routinely for measuring α diversity. Both attempt to quantify the amount of information contained in each individual in the community. However, Shannon's index produces a measure for the effectively infinite population from which the sample was taken, but Brillouin's index provides a measure for the specific collection of individuals in the sample. The Shannon (also known as Shannon-Wiener or incorrectly Shannon-Weaver) index (Equation 7.50), being based on the often dubious assumption that all species are represented in the sample from a population of infinite (effectively infinite) size, is ideally used with truly random samples from a very large population and a known number of individuals (Krebs 1989). Brillouin's index (Equation 7.51) is preferable if the sample is taken in a nonrandom fashion, as is often the case with field collections (Magurran 1988). Poole (1974) recommends it also if all the individuals in a population (community) are counted and assigned to species.

$$H' = -\sum_{i=1}^{S} p_i Ln\ p_i \quad (7.50)$$

$$H = \left(\frac{1}{N}\right) Ln \left(\frac{N!}{\prod_{i=1}^{S} n_i!} \right) \quad (7.51)$$

Bases for the logarithms in Equations (7.50) and (7.51) vary among applications. Units can be bits/individual ($\log_2$: binary units), decits/individual ($\log_{10}$: bel or decimal units), or nits/individual or nats/individual (Ln: natural bel) (Krebs 1999; Magurran 1988; Poole 1974). However, as Krebs (1999) points out, any H' or H can be interconverted easily because H' (base 2) = $H'/\log_{10}(2)$ = 3.321928H' (base 10) and H'(base e) = $H'/\log_{10}(e)$ = 2.302585H'(base 10).

As with Simpson's index, n_i/N estimates are used instead of known p_i values to estimate Shannon's index from a sample. Use of n_i/N values instead of p_i values (i.e., the maximum likelihood estimator of H') introduces a bias that worsens as the sample size decreases. Although Peet (1974) and Magurran (1988) suggest this bias is rarely of a magnitude that seriously impacts results,

Kaesler and Herricks (1976) put this bias forward as one reason for avoiding Shannon's index and using Brillouin's index instead. The variance for this estimate of H' can be derived as detailed in Poole (1974) and Magurran (1988). (Heltshe and Forrester (1985) provide jackknife procedures for estimating a confidence interval about Brillouin's index.) The EstimateS and Krebs (1999) software produce standard errors or confidence intervals for estimates.

Unlike Shannon's index, Brillouin's index does not treat the sample as a random sample from an infinitely large population. Instead, it treats the sample as a collection of individuals and estimates the diversity of that collection (Kaesler and Herricks 1976; Magurran 1988). Consequently, Shannon's index provides a diversity measure for a community, but Brillouin's index provides a diversity measure for a sample. Although Kaesler and Herricks (1976) recommend avoiding Shannon's index in applied ecology due to its unrealistic assumptions and bias, Krebs (1989) argues that the use of Shannon's or Brillouin's index usually does not produce very different results for large samples. (Brillouin's index will be slightly lower than Shannon's index, as seen in Example 7.6.) Importantly, Brillouin's index can be used for individual counts but not for measures such as biomass or coverage because the factorial operations require integers (Krebs 1999).

Example 7.6

These indices are estimated for the fictitious data used in Example 7.5. Krebs' (1999) software produces index estimates for all sites and, as shown in the table below, indicates a distinct drop in diversity regardless of the index. Only calculations for the +0.5 km site are detailed here. The $\log_2$ is used in the example so $\log_2$ would be substituted for Ln in Equations (7.50) and (7.51). The units are bits/individual for H' and H. (Biodiversity Pro and EstimateS also will do these calculations.)

Index	N_i at Sites (site = kilometers from the outfall)							
	–1.0	**–0.5**	**0**	**+0.5**	**+1.0**	**+1.7**	**+2.7**	**+5.3**
$1-\hat{\lambda}$	0.863	0.871	0.477	0.762	0.853	0.816	0.857	0.878
H'	2.867	2.986	1.248	2.112	2.789	2.572	2.810	3.016
H	2.557	2.658	1.145	1.928	2.471	2.241	2.499	2.653

Estimation of $1-\hat{\lambda}$ for the +0.5 km site (Equation 7.49 for a finite population):

$$\hat{\lambda} = \frac{18 \cdot 17}{58 \cdot 57} + \frac{16 \cdot 15}{58 \cdot 57} + \frac{15 \cdot 14}{58 \cdot 57} + \frac{5 \cdot 4}{58 \cdot 57} + \frac{4 \cdot 3}{58 \cdot 57} = 0.2384$$

$$1-\hat{\lambda} = 0.7616$$

Estimation of $1-\hat{\lambda}$ for the +0.5 km site (Equation 7.48 for an infinite population):

$$\hat{\lambda} = \left(\frac{18}{58}\right)^2 + \left(\frac{16}{58}\right)^2 + \left(\frac{15}{58}\right)^2 + \left(\frac{5}{58}\right)^2 + \left(\frac{4}{58}\right)^2 = 0.2515$$

$$1-\hat{\lambda} = 0.749$$

Estimation of H' for the +0.5 km site:

$$H' = -\left[\frac{18}{58} Log_2\left(\frac{18}{58}\right) + \frac{16}{58} Log_2\left(\frac{16}{58}\right) + \ldots + \frac{4}{58} Log_2\left(\frac{4}{58}\right)\right]$$

$$= -(-0.5239 - 0.5125 - 0.5046 - 0.3048 - 0.2661) = 2.112 \text{ bits/individual}$$

Estimation of H for the +0.5 km site:

$$H = \frac{1}{58} Log_2 \left(\frac{58!}{18! \cdot 16! \cdot 15! \cdot 5! \cdot 4!} \right) = 1.929 \text{ bits/individual}$$

Other useful approaches exist in addition to those based on presence and relative abundance of species. For example, Warwick and Clarke (1998) and Clarke and Warwick (1999, 2001) generate a taxonomic index, Δ^+, that successfully allowed them to define the impact of human activities on the *biodiversity* of marine benthic assemblages. This index of biodiversity includes taxonomic distance between pairs of species as quantified with path lengths. In so doing, it broadens the context of the above indices to incorporate the second component of biodiversity, taxonomic diversity.*

7.3.2.5 Species Evenness

Species evenness is the equitability of abundances among species in an assemblage, taxocene, or community (Alatalo 1981). Many evenness indices have been developed, although only three will be described here. The first two are the quotients of the observed H' or H, and their estimated maxima. As such, the index J' based on H' estimates the evenness for the (effectively infinite) population from which the sample was taken, and the index J based on H estimates the evenness for the collection of individuals in the sample (Pielou 1969). The J' will overestimate the true evenness of the community because it is impossible to census all species in the infinite population (Pielou 1969).

The maximum of Shannon's index is Ln S. The associated estimate of evenness (Pielou's J') is calculated with the following equation:

$$J' = \frac{H'}{Ln\ S} \tag{7.52}$$

where S = the number of species in the sample.

An evenness index (Pielou's J) can also be estimated with the quotient of the Brillouin index (H) to its maximum,

$$J = \frac{H}{H_{MAX}} \tag{7.53}$$

The maximum value for H is generated with Equation (7.54) (Pielou 1969, 1974; Poole 1974).

$$H_{MAX} = \frac{1}{N} Ln \left[\frac{N!}{\left(\left[\frac{N}{S} \right] ! \right)^{S-r} \left(\left(\left[\frac{N}{S} \right] + 1 \right) ! \right)^{r}} \right] \tag{7.54}$$

* Biodiversity has several definitions, but that given by Magurran (2004) is sufficient here: "the variety and abundance of species in a defined unit of study." The context for variety can range from genetic, to taxonomic, to species richness, to spatial scale. Diversity and biodiversity are increasingly being used interchangeably.

where N = total number of individuals summed over all species, $[N/S]$ = the integer part of the quotient N/S, and $r = N - S[N/S]$.

Both J' and J are familiar tools to ecologists, yet the first is the most widely used. Unfortunately, they are dependent on species richness as well as species evenness (Peet 1974). Alatalo (1981) recommends a modified Hill ratio estimator of evenness as an index insensitive to species richness. Ludwig and Reynolds (1988) also point out that this modified Hill ratio estimator was not as sensitive to variation in the number of species among samples. It is the ratio of $e^{H'}$ to $1/\hat{\lambda}$ after correction of each for their minima (i.e., subtraction of 1). Simpson's index ($\hat{\lambda}$) is estimated with Equation (7.49).

$$E = \frac{\frac{1}{\hat{\lambda}} - 1}{e^{H'} - 1} \tag{7.55}$$

The $e^{H'}$ (Hill's N_1) reflects the number of "abundant species," and $1/\hat{\lambda}$ (Hill's N_2) reflects the number of "very abundant species" (Alatalo 1981) (for more information see Hill 1973). This index decreases as evenness decreases and approaches 0 as one species approaches complete dominance (Luwig and Reynolds 1988). Krebs (1999) also describes a Simpson's index of evenness also based on $\hat{\lambda}$:

$$E_{1/\hat{\lambda}} = \frac{\frac{1}{\hat{\lambda}}}{S} = \frac{Hill's\ N_2}{S} = \frac{\frac{1}{\left(\sum_{i=1}^{S} p_i^2\right)}}{S} \tag{7.56}$$

Example 7.7

Evenness indices are estimated for the fictitious data used in Examples 7.5 and 7.6. Calculations are shown only for the sample from the +0.5 km site. For illustrative purposes, the natural logarithm is used instead of the $\log_2$ used in the previous examples. Consequently, units change from bits/individual to nits/individual. Calculations are done with Krebs' (1999) software.

	N_i at Sites (site = kilometers from the outfall)							
Index	**−1.0**	**−0.5**	**0**	**+0.5**	**+1.0**	**+1.7**	**+2.7**	**+5.3**
S	9	10	5	5	9	8	9	10
N	59	62	85	58	55	45	57	54
a	2.960	3.373	1.162	1.313	3.056	2.828	3.005	3.609
$\hat{\lambda}$ (Eq. 7.49)	0.137	0.129	0.523	0.238	0.147	0.184	0.143	0.122
$1-\hat{\lambda}$	0.863	0.871	0.477	0.772	0.853	0.816	0.857	0.878
H'	1.987	2.070	0.865	1.464	1.933	1.783	1.948	2.091
H	1.772	1.842	0.793	1.337	1.712	1.554	1.732	1.839
J'	0.905	0.899	0.537	0.910	0.880	0.857	0.887	0.908
J	0.905	0.898	0.526	0.907	0.879	0.854	0.887	0.909
E (Simpson)	1.002	0.975	0.664	0.962	0.983	0.898	0.993	1.012

All indices (J', J, and E) show that species evenness decreases near the outfall. This is consistent with the species list (Example 7.5) that indicates species 1 becomes more dominant and the abundances of the other species decrease at the outfall.

Estimation of J' for the +0.5 km site:

Before J' can be calculated with Equation (7.52), H' must be calculated with Equation (7.50).

$$H' = -\left[\frac{18}{58} Ln\left(\frac{18}{58}\right) + \frac{16}{58} Ln\left(\frac{16}{58}\right) + \ldots + \frac{4}{58} Ln\left(\frac{4}{58}\right)\right]$$

$$= -[-0.3631 - 0.3553 - 0.3498 - 0.2113 - 0.1844]$$

$$= 1.464 \text{ nits/individual}$$

$$J' = H' / Ln\ S = 1.464 / Ln(5) = 0.910$$

Estimation of J for the +0.5 km site:

Before J can be calculated with Equation (7.53), H and H_{MAX} must be derived with Equations (7.51) and (7.54), respectively.

$$H = \frac{1}{58} Ln\left(\frac{58!}{18! \cdot 16! \cdot 15! \cdot 5! \cdot 4!}\right)$$

$$= 1.337 \text{ nits/individual}$$

For Equation (7.54), $[N/S]$ is estimated from 58/5 to be 11. The r is 58 – 5(11), or 3.

$$H_{MAX} = \frac{1}{58} Ln\left[\frac{58!}{(11!)^2((11+1)!)^3)}\right] = 1.474$$

$$J = \frac{H}{H_{MAX}} = \frac{1.337}{1.474} = 0.907$$

Estimation of E for the +0.5 km site:

Before E can be calculated with Equation (7.55), $\hat{\lambda}$ was determined with Equation (7.49). The $\hat{\lambda}$ was found in Example 7.6 to be 0.2384.

$$E = \frac{\frac{1}{\hat{\lambda}} - 1}{e^{H'} - 1} = \frac{\frac{1}{0.2384} - 1}{e^{1.4639} - 1} \approx 0.961$$

7.3.2.6 SHE Analysis

With diversity, richness, and evenness metrics described, SHE analysis (Buzas and Hayek 1996) can now be introduced. The SHE acronym stands for richness (S being number of species), diversity (Shannon-Wiener diversity, **H**), and **E**venness. The method resolves longstanding issues that (1) the influences of E (evenness) and S (the number of species) on any estimated H are difficult to define, (2) diversity and its components with different abundance models change in different ways with increasing number of sampled individuals (N), and (3) there is a correlation between diversity measures and N (Hayek and Buzas 1997).

Buzas and Hayek (1996) introduced the SHE approach by providing a straightforward decomposition of diversity (H) into its richness and evenness components. (H was defined in their treatment in the form of Equation 7.50.) The goal was to estimate how much of the value of H was due to richness and how much due to evenness. They begin by restating that the maximum value for $H = \text{Ln}\ S$. This equation can be rearranged to $e^H = S$ or $e^H/S = 1$. The quotient e^H/S is 1 with maximum evenness but drops below 1 with departure from this maximum. Relabeling e^H/S as E, this can be expressed explicitly as $0 < E \leq 1$. It follows that Ln E will be negative if E is less than 1. So the quotient (e^H/S) is a straightforward measure of evenness (E). Rearranging this relationship to $e^H = SE$ and taking the logarithms of both sides, Buzas and Hayak establish their central SHE model,

$$H = Ln\ S + Ln\ E \tag{7.57}$$

The richness and evenness components of diversity are now distinct in this model so the first of the three issues noted above is resolved. "The equation shows that the value of H always will be equal to its maximum possible value (Ln S) in the assemblage, minus the amount of evenness (Ln E)" (Buzas and Hayek 1996). This can be expressed succinctly in terms of either $H_{Maximum}$ or Pielou's J (Hayek and Buzas 1997),

$$H = H_{Maximum} + Ln\ E \tag{7.58}$$

$$H = J \cdot Ln\ S \tag{7.59}$$

Next, Buzas and Hayek (1996) examined the influence of sample size (N, number of individuals in the sample) for three species abundance models: broken stick, log series, and lognormal. The Ln E is constant with N for the broken stick model, whereas H is constant for the log series. The Ln E/Ln S is constant for the lognormal model. So examining the behavior of Ln E, H, and Ln E/Ln S with different N can help decide the appropriate abundance distribution model to use for a sample. For example, one might calculate these quantities for an increasing number of quadrats or amount of sampling effort. Hayek and Buzas (1997) also suggest that the three abundance models might be distinguished if small N and large N samples are taken from an assemblage and the following metrics obtained.

Broken stick abundance model (Ln E constant):

$$H_1 \neq H_2,\ S_1 \neq S_2, \text{ and } E_1 = E_2$$

Log series abundance model (H constant with E and S changes, balancing each other):

$$H_1 = H_2,\ S_1 \neq S_2, \text{ and } E_1 \neq E_2$$

Lognormal abundance model (only Ln S/Ln E constant):

$$H_1 \neq H_2,\ S_1 \neq S_2, \text{ and } E_1 \neq E_2$$

In summary, the SHE approach allows diversity and its components to be examined over a wide range of situations. It also allows comparison of candidate abundance models (Hayek and Buzas 1997). Hayek and Buzas (2006) used SHE analysis to understand the impact of humans on meiofaunal assemblages of Indian River Lagoon (Florida). Leponce et al. (2004) examined the influence of scale on ant assemblage diversity estimation and Kindt et al. (2006) explored tree communities in Kenya with SHE analyses. For the interested reader, the Biodiversity Pro shareware can be downloaded and used to do SHE analysis.

7.3.2.7 Community Similarity

Indices of species abundance, evenness, and diversity are most effectively used together in assessing changes in community structure. Additionally, an index of similarity between communities can be used, as in the studies of Pontasch et al. (1989) and Snoeijs (1991). Indices of similarity compare two communities based on species presence/absence data or species abundance data. From a certain viewpoint, such indices begin to assess what we earlier identified as β diversity issues. From the vantage of a conservation biologist attempting to identify the most valuable areas to set aside, Magurran (2004) envisions them as complementarity measures for locations or assemblages.* From the vantage of the ecotoxicologist, they can be applied as metrics for suggesting differences between assemblages or sites possibly linked to toxicant impact. The complementarity context might also be relevant to the ecotoxicologist attempting to inform resource management decisions about the best site to set aside as part of a natural resource damage compensation (NRDC) plan. Similarity indices based on presence/absence data are very popular, e.g., Snoeijs (1991). Two such indices are described here using the modified notations of Magurran (1988, 2004) and Krebs (1999).

The most common similarity indices based on presence/absence data are the Jaccard and Sorenson indices. Both range from 0 (the communities are completely dissimilar) to 1 (the communities are identical). Magurran (1988, 2004) suggests that the Sorenson index (apparently called classic Sorensen or Bray–Curtis index in Estimates) performed best of those similarity indices based on presence/absence data.

$$Jaccard\ Index\ (JI) = \frac{j}{a+b-j} \tag{7.60}$$

$$Sorenson\ Index\ (SI) = \frac{2j}{a+b} \tag{7.61}$$

where j = the number of species occurring in both samples (sites), a = the number of species occurring in the first sample, and b = the number of species occurring in the second sample.

There are numerous indices based on species abundance data. Only four (Sorenson, Euclidean distance, Bray–Curtis, and Morisita–Horn) are described here. Magurran's (1988) review of the literature suggests the Morisita–Horn index is best. She notes that the Morisita–Horn index is not as sensitive to species richness and sample size as the other abundance-based indices.

The Sorenson quantitative index (apparently called a quantitative Bray-Curtis index on the EstimationS software) is based on species abundances (number of individuals, biomass, percent cover) instead of presence/absence information (Magurran 1988, 2004).

$$Soreson\ Index\ (Abundance\ or\ Quantitative)(SIA) = \frac{2N_i}{N_a + N_b} \tag{7.62}$$

where N_i = the sum of the smallest of the two abundances for the i species collected from the sites being compared, N_a = the abundance (total number of individuals) for the first site (a), and N_b = the abundance for the second site (b).

* Complementarity (Kirkpatrick 1983; Vane-Wright et al. 1991) is the "difference between sites in terms of the species they support" (Magurran 2004).

The Euclidean distance index for samples for sites a and b is defined in Equation (7.63) (Krebs 1989).

$$\Delta = \sqrt{\sum_{i=1}^{n} (N_{ia} - N_{ib})^2} \quad (7.63)$$

where N_{ia} = abundance of species i in the sample from site a, N_{ib} = abundance of species i in the sample from site b, and n = the total number of species. Krebs (1989) notes that this index increases as the number of species in the samples increases. Consequently, he suggests that the following average distance measure be used.

$$d = \sqrt{\frac{\Delta^2}{n}} \quad (7.64)$$

Values of d range from 0 (sites are completely similar) to infinity (Krebs 1989; Ludwig and Reynolds 1988). Another abundance-based index, the Bray–Curtis index ranges from 0 (complete similarity) to 1. According to Krebs (1989), it weights most heavily the most abundant species and, as a consequence, is insensitive to rare species.

$$B = \frac{\sum_{i=1}^{n} |N_{ia} - N_{ib}|}{\sum_{i=1}^{n} (N_{ia} + N_{ib})} \quad (7.65)$$

Magurran (1988) favors the final abundance-based index (Morisita–Horn index). Wolda (1981) suggests that this index is the best abundance-based index. Krebs (1989) refers to this index as the simplified Morisita index and also notes that it is nearly independent of sample size.

$$MH = \frac{2\sum_{i=1}^{n} N_{ia} \cdot N_{ib}}{(da + db)N_a N_b} \quad (7.66)$$

where N_{ia},N_{ib} = number of individuals of species i in the samples from sites a and b, respectively.

As previously defined, N_a and N_b are the number of individuals in the samples from sites a and b. The total number of species is n. The da and db in Equation (7.66) are calculated as shown.

$$da = \frac{\sum_{i=1}^{n} N_{ia}^2}{N_a^2}$$

$$db = \frac{\sum_{i=1}^{n} N_{ib}^2}{N_b^2}$$

Example 7.8

The data for the −0.5 and +0.5 km sites in Example 7.5 are used to illustrate the calculation of these indices. The Krebs (1999) software was used to calculate these indices with the exception of the Morisita–Horn index recommended by Magurran (1988).

$$JI = \frac{5}{10 + 5 - 5} = 0.500$$

$$SI = \frac{2 \cdot 5}{10 + 5} = 0.667$$

$$SIA = \frac{2 \cdot (12 + 12 + 10 + \ldots + 0)}{62 + 58} = 0.7167$$

$$\Delta = \sqrt{(12 - 18)^2 + (12 - 16)^2 + (10 - 15)^2 + \ldots + (1 - 0)^2} = 11.662$$

$$d = \sqrt{\frac{11.662^2}{10}} = 3.688$$

$$B = \frac{|12 - 18| + |12 - 16| + \ldots + |1 - 0|}{(12 + 18) + (12 + 16) + \ldots + (1 + 0)} = 0.283$$

To calculate the Morisita–Horn index, *da* and *db* are calculated first.

$$da = \frac{12^2 + 12^2 + 10^2 + \ldots + 1^2}{62^2} = 0.143$$

$$db = \frac{18^2 + 16^2 + \ldots + 4^2}{58^2} = 0.251$$

$$MH = \frac{2(12 \cdot 18 + 12 \cdot 16 + \ldots + 0 \cdot 1)}{(0.143 + 0.251) \cdot 62 \cdot 58} = 0.889$$

7.3.3 Community Function

7.3.3.1 General

Although a wide range of functional responses exists, there is a tendency for ecotoxicologists to favor qualities of community structure (Matthews et al. 1982). Matthews et al. (1982) break down functions into two general groups. The first includes biological, chemical, and physical rate constants such as photosynthesis, respiration, sedimentation, bioturbation, nitrogen fixation, sediment oxygen consumption, zooplankton filtration rates, and chelation. The second includes taxonomic functions such as rates of recovery after exposure, rates of colonization, or rates of succession. They also note that enumeration of functional groups, e.g., scrapers, shredders, collectors, specialized feeders, and predators, can be used to assess functional changes in benthic macroinvertebrate communities. Four examples reflecting community functions will be discussed here. Sheenan (1984a, 1984b) and more recently Clements (Newman and Clements 2008) provide more detail and specific examples of studies documenting toxicant-induced functional changes in ecosystems.

It is critical to keep in mind the natural variation and trends in these functions over which the effects of a toxicant or effluent might be imposed. Failure to do so can result in falsely attributing an effect to a contaminant. Examples of such natural trends will be elucidated here using the river continuum theory. This theory predicts that the community functions associated with different reaches of a river will differ due to physical changes in the river from headwaters to mouth (Vannote et al. 1980). For example, in the headwaters (stream orders 1 to 3) where there is considerable shading and input of allochthonous material, the productivity to respiration ratio (*P*/*R*) can be less than 1. However, the *P*/*R* can increase above 1 in the middle reaches of the river (stream orders 4 to 6) where shading is reduced and ample nutrients are present. In the lower reaches, turbidity and depth can limit productivity and the *P*/*R* ratio can drop below 1 again. Similarly, the macroinvertebrate functional groups (detritus shredders, collectors, scrapers, and predators) shift with the availability of associated resources. Shredders will be prominent in headwaters where there is an abundance of coarse organic particulate matter such as leaves. They will decrease with distance downriver. Grazers will be most abundant in the middle regions where productivity is high. Collectors will be most abundant in the lower reaches. These very general patterns along the gradient from first-order to higher-order channels are modified in areas of confluence with lower-order tributaries. It is critical to keep such natural gradients as these in mind when interpreting community functional shifts at contaminated sites.

7.3.3.2 Productivity and Respiration

A wide range of studies, e.g., Cairns (1977), Kondratieff et al. (1984), Blanck (1985), Goldsborough and Robinson (1986), Amblard et al. (1990), and Day (1993), use measures of community photosynthesis and respiration in combination with structural changes to imply toxicant impact. For example, Amblard et al. (1990) measured chlorophyll a, ATP/ADP, photosynthetic activity (C^{14} method), and photoheterotrophic incorporation of amino acids in periphyton communities. They found an increase in heterotrophic biomass and activity near a pulp and paper mill effluent. Kondratieff et al. (1984) also measured chlorophyll a and ATP. They calculated an autotrophic index with these indicators of community function.

$$AI = \frac{\hat{B}_{ATP(\text{mg/L})}}{Chlorophyll\ a\ (\text{mg/L})} \tag{7.67}$$

The biomass used in the autotrophic index was estimated with the ATP measurements.

$$\hat{B}_{ATP} = \frac{ATP\left(\frac{ng}{L}\right)}{2400\ (mg\ biomass\ per\ ng\ ATP)} \tag{7.68}$$

This index (AI) increases with a proportional increase in heterotrophic activity. The autotrophic index suggested a shift to heterotrophy at sites closest to the outfall (sewage and electroplating waste). Kondratieff et al. (1984) also used ANOVA to detect a shift in invertebrate functional groups. Near the effluent, differences in means of invertebrate densities and ash-free dry weights suggested that collector-gatherers and filter feeders dominated.

7.3.3.3 Detritus Processing

A wide range of stressors influence detritus processing (Webster and Benfield 1986; Caquet et al. 2007; Bundschuh et al. 2011). For a quantitative example, Giesy (1978) found that cadmium slows leaf decomposition and leaf colonization by fungi and bacteria. Leaf (Mulholland et al. 1987), macrophyte, and algal (Schoenberg et al. 1990) detritus processing can also be modified by shifts in pH and acidification-related phenomena.

Allred and Giesy (1988) examined leaf litter processing in streams using open-mesh bags of leaves. The following model is fit to the weight loss in the bags over time (211 days).

$$W_t = W_0 e^{-kt} \tag{7.69}$$

where W_t = weight (total or ash-free dry weight) at time t, W_0 = initial weight (t = 0), and k = the rate constant for weight loss.

The form of this model is the same as that used in Chapter 3 (Equation 3.2) for toxicant elimination kinetics. All elaborations (e.g., half-life or mean lifetime estimates), alternatives (e.g., Equation 3.12), and examples (e.g., Examples 3.1 to 3.3) in Chapter 3 are potentially pertinent to the analysis of such data. For example, Mulholland et al. (1987) found that a linear model (Equation 3.1 in Chapter 3) can fit leaf decomposition data as well as Equation (7.69). Multiple compartment formulations can also be appropriate (Minderman 1968) in describing detrital decomposition (e.g., Equation 3.8 in Chapter 3).

7.3.3.4 Nutrient Spiraling

Cairns (1977) argued that it is important to consider nutrient cycling during assessment of biological integrity ("maintenance of the community structure and function characteristic of a particular locale or deemed satisfactory to society"). Later, Cairns and Pratt (1986) added that such basic functional attributes as nutrient spiraling could be used for this purpose. Unfortunately, studies of such attributes are uncommon. The basic premise and approach to nutrient spiraling is sketched out here to stimulate more interest in this and similar attributes of nutrient cycling.

Nutrient cycles in lotic (flowing) systems are open. As nutrients move downstream with the hydraulic flow, they are taken up by, and become associated with, benthic communities and abiotic solid phases. Elwood et al. (1983) suggest that biotic processes dominate in healthy streams. The nutrients are eventually released to continue their movement downstream. This process of uptake and release during net movement downstream is called nutrient spiraling. Because decomposition rates and algal biomass accumulation in lotic systems can be nutrient limited (Elwood et al. 1981, 1983), this attribute of nutrient movement is important in flowing systems. Productivity tends to be high with tight spiraling because nutrients are retained efficiently (Newbold et al. 1982).

Spiraling length (S) is used as a measure of nutrient retention in lotic systems (Elwood et al. 1983; Newbold et al. 1982). It is defined as the average distance downstream to complete one cycle for a nutrient. It is estimated with the following equation (Newbold et al. 1982):

$$S = \frac{F_T}{Uw} \tag{7.70}$$

where F_T = the total downstream flux of the nutrient (g/s), U = the uptake rate of the nutrient per unit of benthic surface (g/m^2/s), and w = the stream width (m).

Although toxicants likely affect spiraling length, this aspect of nutrient cycling in flowing systems has been neglected.

7.3.3.5 Colonization and Succession

The potential impacts of toxicants on successional processes are touched upon very briefly in earlier discussions of species abundance models. Cairns and coworkers' studies of toxicant effects

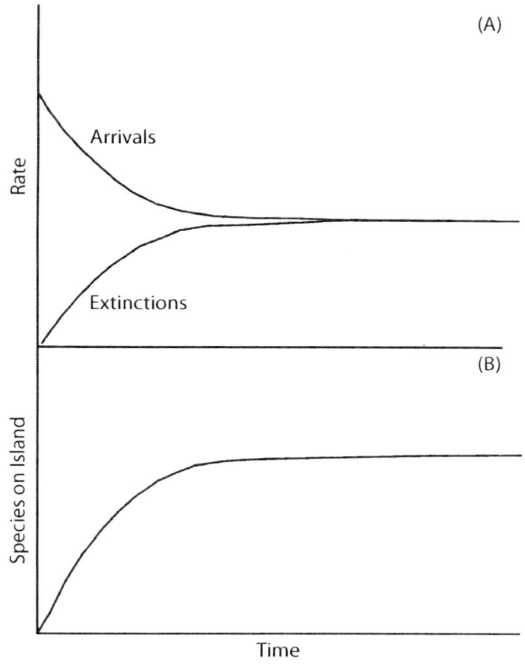

Figure 7.13 Time course of island colonization based on the MacArthur–Wilson model. (From MacArthur, R.H., and E.O. Wilson, *The Theory of Island Biogeography*, Princeton University Press, Princeton, NJ, 1967. Reprinted by permission of Princeton University Press.)

on protozoan colonization (increase to species abundance during primary succession) are examined here. They (Cairns et al. 1986; Cairns and Pratt 1985; Niederlehner et al. 1985; Pratt and Cairns 1985) use the MacArthur-Wilson theory of island colonization to model protozoan colonization of polyurethane foam substrates (islands). (For details see MacArthur and Wilson 1967; Simberloff and Wilson 1969.)

The MacArthur-Wilson model is identical in form to that described in Chapter 3 for bioaccumulation. Indeed, substituting C_s in Equation (3.43) from Chapter 3 into Equation (3.42) will produce the same model formulation except the process being modeled is toxicant accumulation in an organism rather than species accumulation on an island. Therefore, many of the comments regarding the fit of models to bioaccumulation data in Chapter 3 are pertinent here. The reader is encouraged to review Chapter 3 prior to application of these methods.

The model considers the arrival (invasion or immigration) and extinction rates of colonizing species from a source community, e.g., a mainland (Figure 7.13). The arrival rate is a function of time and the distance from the source community to the island. Once on the island, species can remain extant or become extinct. The extinction rate is a function of time and the size of the island. Extinction rates are generally lower for larger islands. For a given island with a fixed size and distance from species source, the rate of species arrivals decreases with time and the rate of species extinctions increases with time until an equilibrium is reached, i.e., the rate of species additions equals that of species loss. Equations (7.71) and (7.72) from MacArthur and Wilson (1967) describe this approach to an equilibrium number of species.

$$G = \frac{d\mu}{dS} - \frac{d\lambda}{dS} \tag{7.71}$$

where S = the number of species, μ = the rate of species extinctions when S species are present, and λ = the rate of species arrivals when S species are present.

The constant G is analogous to k_e in Chapter 3. It is a summary statistic incorporating individual species constants and island qualities such as size and distance from the source community.

$$S_t = S_{EQ}(1 - e^{-Gt}) \tag{7.72}$$

where S_{EQ} = the equilibrium number of species, S_t = the number of species at time t, and t = the time (duration) over which colonization has occurred.

MacArthur and Wilson (1967) estimate the time to approach 90% of the equilibrium number of species in a manner analogous to that used for bioaccumulation kinetics in Chapter 3 (Equations 3.3 to 3.7).

$$t_{0.90} = \frac{2.303}{G} \tag{7.73}$$

This is the time to reach all but 10% (0.10) of the equilibrium number of species. More generally, the time to reach any proportion of the S_{EQ} can be calculated,

$$t_{0.90} = \frac{Ln\dfrac{1}{0.10}}{G} \tag{7.74}$$

Cairns et al. (1986) use this model to assess the effect of cadmium and an organic pesticide used against larval lamprey (3-trifluoro-methyl-4-nitrophenyl (TFM)) on protozoan colonization. A naturally colonized polyurethane foam substrate is placed in the center of a microcosm containing noncolonized foam substrates. The noncolonized substrates form a ring of islands equidistant from the naturally colonized epicenter. The microcosms then receive various treatments, and the increase in species abundance in the islands is quantified over a period of time. These data are fit to the MacArthur-Wilson model using nonlinear least-squares regression. For TFM, S_{EQ} and G decrease abruptly between the 1.0 and 10.0 mg/L TFM treatments. The S_{EQ} decreases linearly with the Ln of cadmium concentration in the microcosm water. Inverse regression (Cairns and Pratt 1985) methods are used to estimate the concentration of cadmium corresponding to an EC20, the concentration at which the S_{EQ} is reduced by 20% of the control S_{EQ}.

Example 7.9

As described above, Niederlehner et al. (1985) measured protozoan colonization of foam substrates in the presence of a series of cadmium concentrations. Data from the control and highest cadmium concentration (9.5 μg/L) treatments were visually extracted from their Figure 2 and used here to illustrate the associated data analysis. These analyses are very similar to those shown in Example 3.8 in Chapter 3 for bioaccumulation. Although not shown here, the actual numbers of species and those predicted from the regression results should be plotted with time to assess the fit of the nonlinear model. Also, the regression residuals should be plotted against time to detect any trends in the residuals. The codes to perform these plots for assessing regression model fit to these data are given in Example 3.2 in Chapter 3.

```
DATA COLONIZ;
   INPUT CD DAY SPECIES @@;
   DATALINES;
   0.0  0  0 0.0  1  1 0.0  1  5 0.0  3  8
```

```
   0.0  7 29 0.0  7 27 0.0 14 38 0.0 14 36
   0.0 21 37 0.0 21 32 0.0 28 32 0.0 28 28
   9.5  0  0 9.5  1  2 9.5  1  3 9.5  3  9
   9.5  3  7 9.5  7 14 9.5  7 16 9.5 14 13
   9.5 14 18 9.5 21  9 9.5 21 12 9.5 28 10
   9.5 28 14
   ;
PROC SORT;
   BY CD DAY;
RUN;
DATA ZERO;
   SET COLONIZ;
   IF CD=0.0;
RUN;
PROC NLIN;
   PARMS SEQ=36 G=0.5;
   BOUNDS 0<SEQ<50 0<G<1;
   MODEL SPECIES = SEQ*(1-EXP(-G*DAY));
   OUTPUT OUT=SPEC1 P=PSPEC1 R=RSPEC1;
RUN;
DATA NINE;
SET COLONIZ;
   IF CD=9.5;
RUN;
PROC NLIN;
   PARMS SEQ=12 G=0.5;
   BOUNDS 0<SEQ<50 0<G<1;
   MODEL SPECIES = SEQ*(1-EXP(-G*DAY));
   OUTPUT OUT=SPEC2 P=PSPEC2 R=RSPEC2;
RUN;
```

Control treatment (0 μg/L):
- $S_{EQ} = 34.69$
- Asymptotic standard error for $S_{EQ} = 2.31$
- Asymptotic 95% confidence intervals for $S_{EQ} = 29.54$ to 39.84
- $G = 0.19$
- Asymptotic standard error for $G = 0.05$
- Asymptotic 95% confidence intervals for $G = 0.08$ to 0.30

Cadmium treatment (9.5 μg/L):
- $S_{EQ} = 13.25$
- Asymptotic standard error for $S_{EQ} = 1.08$
- Asymptotic 95% confidence intervals for $S_{EQ} = 10.88$ to 15.62
- $G = 0.36$
- Asymptotic standard error for $G = 0.12$
- Asymptotic 95% confidence intervals for $G = 0.09$ to 0.63

The S_{EQ} in the 9.5 μg/L treatment (approximately 13 species) was less than half of that for the control treatment (approximately 35 species). Although the confidence intervals for the G estimates for the two treatments overlapped, the predicted G values can be used to generate approximate times to 90% equilibrium (Equation 7.73). The $t_{0.90}$ was shortened by cadmium exposure from 12 to 6 days.

7.4 COMPOSITE INDICES

Species richness, diversity, and evenness, and community similarity indices as described above, are among the most commonly used indices in studies of community level effects of toxicants. Sheenan (1984a, 1984b) and more recently Clements (Newman and Clements 2008, Chapter 24) describe and give specific examples of the successful application of these indices. They also describe more specialized indices for human disturbance and pollution effects. The purpose of these other indices is to include additional information about the community so as to gain further insight into shifts due to the presence of toxicant. These multimetric indices range in complexity from simple to complex indices requiring considerable expertise for accurate application. Some have the distinct disadvantage of no clear causal linkage between observed difference and the intensity of toxicant exposure (Weiss and Reice 2005). For illustrative purposes, several of these composite indices are discussed below.

Karr (1991) defines the biological integrity of an ecological system as the ability to support and maintain "a balanced, integrated, adaptive community of organisms having a species composition, diversity, and functional organization comparable to that of natural habitat of the region." Multimetric indices for measuring biological integrity or some similar property have been developed that combine various community qualities. Most are calibrated to qualities of undisturbed similar systems in the same region, e.g., a low-gradient, warm-water stream in the midwestern United States.

Composite indices can be as simple as those described by Gammon (1989) and Maurer et al. (1991). Gammon (1989) bases his index for lotic systems on the assumption of high abundance and species diversity in healthy fish communities. His index of well-being (I_{wb}) is calculated with fish species and abundance data generated by electrofishing streams. High values of I_{wb} indicate a relatively healthy state.

$$I_{wb} = 0.5\ Ln\ N + 0.5\ Ln\ W + D_N + D_W \tag{7.75}$$

where N = the number of fish taken/km of stream, W = weight (kg) of fish taken/km of stream, D_N = Shannon's index based on number of individuals, and D_W = Shannon's index based on weight of fish.

Maurer et al. (1991) generate a coefficient of pollution (p) applicable to coastal marine systems. The number of benthic species and total number of individuals expected in an area are estimated using empirical relationships between these metrics, and sediment type and water depth. The expected number of species (g') and individuals (i_0) if the site were undisturbed are then combined with the observed number of species (g) and individuals (i) to generate a coefficient of pollution (p).

$$p = \frac{g'}{g\sqrt{\frac{i}{i_0}}} \tag{7.76}$$

Both of these simple indices draw on measures of total abundance and species diversity. They are relative indices interpreted by comparison of values from suspect sites to those expected for unimpacted sites.

More complex indices are also common. The index of biological integrity (IBI) is a useful index that, in its original form, assesses impact using 12 attributes (metrics) of fish communities from warm-water, low-gradient, lotic systems (Karr 1991; Karr et al. 1986). These community metrics are grouped into three types: species richness and composition, trophic composition, and abundance and condition metrics. Species richness and composition metrics include the total number of species, number of darter species, number of sunfish species, number of sucker species, number of intolerant species, and proportion of the total number of individuals that are green sunfish (a species indicative of a disturbed system). Trophic composition metrics include the proportion of all individuals that are omnivores, insectivorous cyprinids, or piscivores. The abundance and condition metrics include the total number of individuals in the sample, proportion of the total number of fish that are hybrids, and proportion of the total number of individuals with obvious signs of disease or other anomalies. Each metric is scored based on expected values for systems with various intensities of stress. The total of the scores for all 12 metrics is used to categorize the integrity of the system in question. These metrics were selected based on the assumptions listed below (summarized from Table 2 of Fausch et al. 1990).

1. Native species richness declines with increased degradation.
2. Fish abundance decreases with increased degradation.
3. As degradation increases, the number of sensitive species drops and the number of tolerant species increases.
4. As degradation increases, species that are specialized feeders and top predators decline but the number of trophic generalists increases.
5. Siltation associated with stream degradation will decrease the availability of optimum spawning substrate with a consequent increase in species hybridization.
6. Incidences of infectious disease and abnormalities increase as degradation worsens.
7. Introduced species increase with degradation.

The overlap in metrics used in the IBI imparts a degree of robustness to the index (Fausch et al. 1990). As discussed, the IBI includes functional in addition to structural community qualities. It also combines qualities of individuals (e.g., disease state), populations (e.g., incidence of hybridization), and the community (e.g., species abundance). These qualities enhance its value in reflecting overall community response (Fausch et al. 1990). Yoder (1989) suggests from the relatively narrow coefficient of variation generated during its application that the IBI is as reliable as the bioassay and chemical analyses commonly used for environmental assessments. The enhanced realism associated with community level metrics offsets the necessity for considerable taxonomic and field expertise to generate the associated metrics and, if needed, to intelligently modify the index.

The IBI is modified prior to application to a particular ecological region (e.g., Belpaire et al. 2000; Fausch et al. 1990; Harris and Silveira 1999; Uzarski et al. 2005). Leonard and Orth (1986) apply a modified IBI to a fish community of cool, high-gradient streams. Karr and Dionne (1991) develop an IBI for lakes and reservoirs. Karr and Kerans (1992) and Burton et al. (1999) discuss several invertebrate-based IBI. Maxted (1989) develops a similar type of index based on marine invertebrates. Hill et al. (2000) and Wang et al. (2005) developed IBI based on diatoms. Moving on to dry land, O'Connell et al. (1998) describe a bird community IBI.

7.5 METACOMMUNITIES

> Community ecology as a field is concerned with explaining the patterns of distribution, abundance and interaction of species. Such patterns occur at different spatial scales and can vary with the scale of observation … much of formal community theory is focused on a single scale, assuming local communities are closed and isolated.
>
> **—Leibold et al. (2004)**

Our exploration of natural multispecies collections can now be broadened to include the metacommunity context. A metacommunity—analogous to a metapopulation—is a collection of local communities among which species can potentially interdisperse (Leibold et al. 2004). This broadened spatial scale allows study of additional, fundamental features relevant to ecotoxicology. The increase in species richness with increasing spatial scale readily comes to mind as one such feature. Indeed, the slope of a curve of species richness versus sampled area is one measure of β diversity (Caswell and Cohen 1993).

The fundamental assumption is that species distributions among localities reflect responses to abiotic and biotic gradients or differences. Metacommunity studies explore dispersal and co-occurrence of species collections that are defined taxonomically (e.g., species assemblages such as macrobenthic molluskan fauna), functionally (e.g., guilds such as filter feeders), or perhaps both (e.g., filter-feeding macrobenthic mollusks). Basing studies on such qualities establishes a certain level of metacommunity coherence, although it might not be obvious for species-poor metacommunities (Preslet et al. 2010). From the species' vantage, species respond generally to abiotic and biotic factors in a coherent manner without gaps in responses (presence/absence) along gradients of these factors. From a broader community context, the species making up any studied community might respond to the same environmental conditions and gradients in a coherent and nonrandom manner. Leibold and Mikkelson (2002) suggest that such coherence is one of three key factors that shape metacommunities. The second is the tendency of species to replace each other in patches/localities (species turnover) with complete competitive exclusion being an extreme instance of this tendency. The third factor is boundary clumping, "the degree to which the boundaries of the different species' ranges are clustered together" (Leibold and Mikkelson 2002).

Though other approaches exist (Williamson 1978; Caswell and Cohen 1993; Holt 1993), many metacommunity publications utilize site-by-species incidence matrices (Maron et al. 2004; Presley et al. 2010). This approach indicates presence or absence for species (rows) in a series of patches or localities (columns). It was developed by conservation biologists attempting to best manage reintroductions of species into sets of localities comprising a metacommunity (Maron et al. 2004). Toward this end, the presence/absence information for the species most widespread among the sampled sites was placed in the matrix top row, that of the next most widespread species was placed in the second row, and so forth, until the data for the most restricted species were entered in the bottommost row. Similarly, information for the localities was ordered from that with the most species in the leftmost column to that with the least species in the rightmost column. This arrangement of species-by-site ordering attempts to find the maximum packing for the metacommunity. A maximum packed incidence matrix contains the lowest number of unexpected absences in the upper lefthand corner. (Two pieces of shareware described by Guimaraes and Guimaraes (2006) produce maximum packed incidence matrices and incidence matrix information.)

Six types of spatial patterns of species presence/absence in metapopulations have been defined: checkerboard, Clementsian, Gleasonian, evenly spaced, nested distributions, and random (Hu et al. 2009; Leibold et al. 2002; Presley et al. 2010). These patterns are described by expanding details of Table 1 in Leibold and Mikkelson (2002).

1. Checkerboard: Species pairs are mutually exclusive, as in the case of competitive exclusion. If one appears in a patch/locality, the other one will not. The coherence in this situation is said to be negative. As an example, this pattern has been examined by Stone and Roberts (1990) for islands in which one species could exclude another from becoming established.
2. Clementsian: Communities as units replace each other along a gradient. In contrast to the Gleasonian pattern, this pattern involves equilibrium communities stabilized by niche partitioning among species. The species in the community interact based on evolved mechanisms of interspecies interactions to produce an equilibrium community (Presley et al. 2010). These communities respond as a whole to gradients among localities. Coherence, turnover, and boundary clumping are all thought to be positive.
3. Gleasonian: There are gradients in species turnover, but the actual arrangement of species ranges is random along any gradient. The communities contain individual species with characteristic tolerance ranges that respond as individual species, not as interacting parts of a community. The species associations result simply from similarities of individual species' tolerance and requirements (Presley et al. 2010). Coherence and turnover will be positive, but in contrast to the Clementsian pattern above, boundary clumping will be random.
4. Evenly spaced: Species ranges vary along a gradient but communities do not change as a unit. However, the species ranges are distributed more evenly than expected by chance alone. Strong interspecific competition fosters this type of pattern (Presley et al. 2010). Coherence and species turnover will be positive, but boundary clumping will be negative.

5. Nested: Species-poor communities exist with nested groups or subsets of species-rich communities. Coherence will be positive but turnover will tend to be negative.
6. Random: There is no apparent pattern or gradient, and the species do not interact. Any measure of coherence will simply reflect a random pattern.

These spatial patterns can be illustrated with incidence matrices in terms of coherence, species turnover, and boundary clumping (Figure 7.14). The level of coherence (Figure 7.14, top) can be estimated by the number of presences in a species row that are not interrupted by an absence. The perfectly coherent incidence matrix in this figure has no interruptions. The 10 species rows in the imperfectly coherent incidence matrix have 0 + 3 + 1 + 3 + 3 + 2 + 3 + 3 + 1 + 2 + 1, or 22, absences embedded between presences. Embedded absences that occur at random are indicative of the random pattern described above. A checkerboard pattern has negative coherence and the remaining four patterns (ignoring random) have positive coherence. Species turnover rates, the rate at which one species replaces another, differ among these four remaining coherent patterns. The species turnover rate is the inverse of the degree of nesting, so a highly nested pattern reflects low species turnover. The highly nested pattern in Figure 7.14 (middle left) has minimal species turnover and contrasts with the incidence matrix to its right. Several metrics are applied as estimators of nestedness or

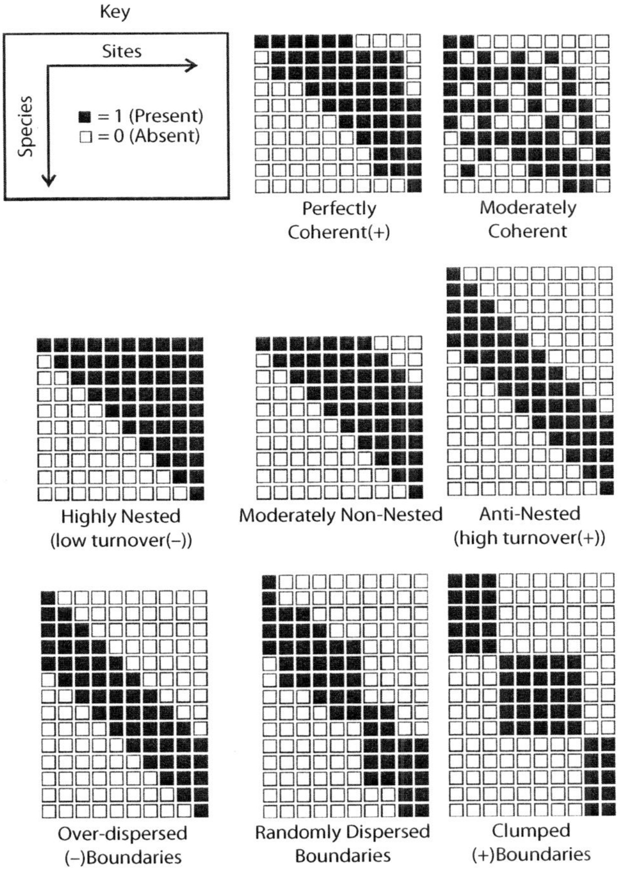

Figure 7.14 Illustrative examples of site-by-species incidence matrices illustrating coherence (top two matrices), species turnover (center three matrices), and boundary clumping (bottom three matrices). (Combination and modification from Leibold, M.A., and G.M. Mikkelson, *Oikos*, 97, 237–250, 2002, Figures 1, 2, and 3.).)

species turnover. Leibold and Mikkelson (2002) use the number of times a species replaces another between localities. The nested spatial pattern described above is the only pattern with extremely low (–) turnover. The three remaining patterns (Clementsian, Gleasonian, and evenly spaced) can be differentiated based on boundary clumping, that is, the tendency of the boundaries of the different species to be clustered (Figure 7.13, bottom). A Clementsian pattern is indicated by high clumping of boundaries (rightmost incidence matrix): the community is acting as a unit with many feedback loops. Such a situation reflects less species sharing among patches/localities. Overdispersed boundaries (leftmost incidence matrix) suggest a Gleasonian pattern with species acting more as independent units responding to gradients than as a single community unit. Randomly dispersed boundaries (evenly spaced pattern) are depicted as boundary clumping intermediate between the Clementsian and Gleasonian patterns. There are several metrics used for boundary clumping, including Morisita 1, which was used by Leibold and Mikkelson (2002) to describe the matrices shown at the bottom of Figure 7.14.

7.6 TROPHIC EXCHANGE

7.6.1 General

In the 1960s, the conditional nature of the dilution paradigm ("the solution to pollution is dilution") became painfully clear. Fallout from atmospheric testing of nuclear weapons introduced low but detectable levels of radionuclides over vast regions of the globe. Some, like strontium-90 (^{90}Sr), accumulated to discomfortingly high concentrations in foodstuffs. Epidemics of metal poisoning (Minamata and Itai-itai diseases) were also linked to accumulation of toxicants into food. After dissipation into the environment, DDT (dichloro-diphenyl-trichloroethane) reappeared at extremely high concentrations in species occupying upper trophic levels. *Silent Spring* (Carson 1962) and articles published in broadly read journals (e.g., Woodwell 1967a, 1967b) presented this trophic "boomerang effect" (Ramade 1987) to the public as one of the first signs of a failing paradigm. The dilution paradigm was replaced as a central theme in environmental management, and the specter of toxicant transfer through trophic levels has become part of our collective knowledge.

Unfortunately, as an important but favored child of our environmental conscience, the hypothesis of toxicant transfer to dangerous levels tends to be invoked dogmatically despite equally viable, alternative hypotheses (Beyer 1986; Moriarty 1983). (See ruling theories and precipitate explanation in Chapter 1.)

As toxicants can enter aquatic organisms from both food and water, two terms were needed to initially distinguish between the associated processes. Bioconcentration is the increase in toxicant concentration in an organism resulting from direct uptake from water alone. As discussed in Chapter 3, bioconcentration is often quantified with the bioconcentration factor (BCF). In contrast, bioaccumulation is the increase in toxicant concentration in an organism resulting from the combined uptake from food and water. The term *bioaccumulation* is often applied in field surveys or other studies in which sources are poorly understood or vaguely defined. In our present discussion of community level phenomena, an additional term and associated methods must be described. Biomagnification (bioamplification) is the increase in toxicant concentration of organisms in successive trophic levels (Moriarty 1983). With biomagnification, the concentration of a toxicant increases with passage upward through the trophic chain or web. Often the implied consequence is damage to species at the highest trophic levels. Hunt and Bischoff (1960) describe such a scenario involving DDD (dichloro-diphenyl-dichloroethane) spraying with fatal consequences for the western grebe (*Aechmophorus occidentalis*).

In theory, the degree to which a toxicant is transferred between successive trophic levels can be measured as the ratio of the toxicant concentrations in the two trophic levels of interest (the bioaccumulation factor or BAF) (Thomann 1989; Walker 1987). Implied in this term is the potential for significant accumulation from water as well as food. Other, very similar terms used in studies focusing only on food sources are biomagnification factor (Laskowski 1991), enrichment ratio (Paasivirta 1991), and transfer factor (Ramade 1987). Although normally presented in the simple form below, formulation of the BAF can be more involved if many individuals with different weights are sampled from each trophic level (Laskowski 1991).

$$BAF = BAF_{n,n+1} = \frac{C_{n+1}}{C_n} \tag{7.77}$$

where C_{n+1} = toxicant concentration in members of trophic level $n + 1$, and C_n = toxicant concentration in members of the next lowest level, n.

Biomagnification exists if BAF > 1. Simple transfer occurs if BAF = 1. Obviously, concentrations decrease (trophic dilution) if BAF < 1. A BAF can be estimated across several trophic levels as illustrated below with Ramade's (1987) formulations. Here, accumulation from water is ignored. The concentration of a toxicant at trophic level $n + 1$ is estimated with the BAF and the concentration at trophic level n by rearranging Equation (7.77).

$$C_{n+1} = BAF_{n,n+1}C_n \tag{7.78}$$

If uptake from water is assumed to be insignificant, the BAF can be expressed in terms of the weight (g) of organisms in level n (b_n), weight (g) of level n individuals consumed (a_n), fraction of the toxicant absorbed after ingestion (f_n), and daily fraction of the pollutant excreted at rate k_n.

$$BAF_{n,n+1} = \frac{a_n f_n}{b_n k_n} \tag{7.79}$$

The following model predicting C_{n+1} from C_n results from combining the two equations above.

$$C_{n+1} = \frac{a_n f_n}{b_n k_n} C_n \tag{7.80}$$

The model can be expanded to predict C_{n+2}.

$$C_{n+2} = \left[\frac{a_{n+1} f_{n+1}}{b_{n+1} k_{n+1}}\right]\left[\frac{a_n f_n}{b_n k_n}\right] C_n \tag{7.81}$$

Expression of the concentration of toxicant at any trophic level based on the concentration in the lowest level (0) can be generalized.

$$C_r = \left[\prod_{i=1}^{r} \frac{a_i f_i}{b_i k_i}\right] C_0 \tag{7.82}$$

Thomann (1981) integrated organism growth into this general model using the following function for the food chain transfer constant:

$$f = \frac{\alpha C}{k} + G \tag{7.83}$$

where α = the chemical absorption efficiency analogous to f_i in the above BAF equations, C = the specific consumption (units of mass of prey/mass of predator per day), k = excretion rate (k_i in the above BAF model), and G = net organism growth rate.

Other detailed models of such trophic transfer for terrestrial (e.g., Holleman et al. 1971) and aquatic (e.g., Conover and Francis 1973; Patrick and Loutit 1978) systems exist. Equation (3.55) in Chapter 3 is one such simple model. Models are also common for toxicant transfers from both water and food sources (e.g., Spacie and Hamelink 1985; Thomann 1989).

$$C_{n+1} = \frac{k_{u(n+1)}}{k_{n+1}} C_w + \frac{\alpha_{n+1,n} R_{n+1,n}}{k_{n+1}} C_n \tag{7.84}$$

where C_w = concentration in the water, C_n = concentration in the nth trophic level (prey), C_{n+1} = concentration in the trophic level n + 1 (predator), $\alpha_{n+1,n}$ = assimilation efficiency (amount absorbed/amount ingested), $R_{n+1,n}$ = weight-specific ration or ingestion (amount ingested/g of total weight), $k_{n+1} = k_{e(n+1)} + g_{n+1}$, $k_{e(n+1)}$ = elimination rate constant for the predator, and g_{n+1} = growth rate constant for the predator.

Bioaccumulation to high concentrations of metals (Newman et al. 1983, 1985), metalloids (Newman et al. 1985), radionuclides (Neal et al. 1967), and organic compounds (Connell 1990) can occur at the lowest trophic levels; therefore, the potential still exists for adverse effects on consumers even in the absence of biomagnification. However, although there is reason for concern, consumption of contaminated food does not necessarily imply significant exposure because the toxicant must also be in a chemical form available for assimilation. For example, material accumulating on submerged surfaces and often defined as periphyton (aufwuchs) accumulates high concentrations of metals. But the bioavailability of the metals can be ambiguous because the metals can be associated primarily with hydrous oxides, not the periphytic microflora. Another example involves the significant incorporation of zinc into intracellular phosphate granules as a detoxification mechanism of mollusks. A significant portion of the zinc in these granules is not available for assimilation when the common periwinkle (*Littorina littorea*) is consumed by another snail (*Nassarius reticulatus*) (Nott and Nicolaidou 1993).

Reichle and Van Hook (1970) describe biomagnification of essential elements along a terrestrial food chain and find that potassium, sodium, phosphorus, and nitrogen can exhibit biomagnification in a terrestrial food chain, but calcium is diluted at each transfer. After finding no biomagnification of cadmium, lead, or zinc in a terrestrial (arthropod) food chain, Van Straalen and Van Wensem (1986) indicate that many other factors correlated with trophic level, including animal size and longevity, can render questionable the interpretation of biomagnification results. Predators tend to be larger and live longer than their prey. Consequently, they likely have different (size-dependent) rate constants and have longer periods of time to accumulate metals than their prey. The complex mechanisms determining body concentration in each species can be more important than trophic status. Beyer (1986) also suggests that the favored status given to biomagnification in many studies of terrestrial food chains results in inadequate consideration of alternative mechanisms such as size-related kinetics, longevity, and individual species variation in accumulation. Beyer (1986) notes that the essential element zinc can display biomagnification in a zinc-deficient, terrestrial ecosystem but not in one with adequate zinc.

Davis and Foster (1958) indicate that biomagnification in aquatic food chains occurs for some radionuclides of essential elements such as ^{32}P. But cadmium, copper, lead, and zinc show no biomagnification in aquatic food chains studied by Gächter and Geiger (1979), Cushing et al. (1981), or Soto-Jimenez et al. (2011). Bryan and Gibbs (1991) indicate that tributyltin (TBT) shows no biomagnification during marine trophic exchanges. Wren et al. (1983) examined a series of elements (Al, B, Ba, Be, Ca, Cd, Co, Cu, Fe, Hg, Mg, Mn, Mo, Ni, P, Pb, S, Sr, Ti, Zn) along an aquatic food chain and found that only mercury displays biomagnification. Terhaar et al. (1977) detect no biomagnification of silver in an artificial aquatic system, although biomagnification of mercury was measured. Mercury biomagnification has also been documented or suggested by Potter et al. (1975), Boudou et al. (1979), Wren and MacCrimmon (1986), Tom et al. (2010), and Newman et al. (2011).

Moriarty (1983) suggests that studies documenting biomagnification of organic compounds have also tended toward precipitate explanation and, as a consequence, do not represent critical evaluations of toxicant distribution among trophic levels. He indicates that species variation within a trophic level can be of sufficient magnitude to make casual documentation of biomagnification across trophic levels questionable. Connell (1990) also suggests that the relative roles of bioconcentration (accumulation from water) and biomagnification are often unclear in studies of aquatic food chains.

Which organic contaminants are prone to biomagnify is a function of their tendency to accumulate in organisms, and their resistance to breakdown in the environment and tissues of organisms, that is to say, the persistent organic pollutants (POPs) are prone to biomagnification. Many excellent examples of POP biomagnfication modeling in diverse trophic webs have been published, including Armitage and Gobas (2007), Arnot and Gobas (2004), Kelly et al. (2007), and Mackay and Fraser (2000). Generally, the extent of biomagnification for an organic compound will depend on the amount of material consumed, fraction of the ingested compound that is absorbed, the excretion rate, uptake from water, and the duration of exposure. The duration of exposure is correlated with longevity, and therefore often trophic status. Many constants can be animal size dependent and animal size is correlated with trophic level. Generally, biomagnification is enhanced by increased species longevity and near proximity to the top of the food chain (Connell 1990). Finally, many of these qualities can be predicted with the K_{ow} of an organic compound. Connolly and Pederson (1988) and Thomann (1989) indicate that biomagnification will be significant if the log K_{ow} for a compound is greater than approximately 4 to 5.

7.6.2 Models Based on Light Isotopes

A pragmatic approach to quantifying chemical movement through trophic webs takes advantage of isotopic discrimination, specifically that occurring between the naturally occurring nitrogen isotopes, ^{14}N and ^{15}N (Dawson and Siegwolf 2007; Fry 2006; West et al. 2010). There is a tendency for the lighter isotope (^{14}N) to be favored kinetically as atoms of nitrogen enter, react within, and eventually leave organisms. Of the ingested nitrogen in food, ^{14}N is retained less readily than ^{15}N within the organisms comprising a trophic web. Consequently, ^{15}N tends to become enriched relative to ^{14}N with movement upward through trophic webs: the ratio of ^{15}N to ^{14}N in a consumer will be higher than that in its food. Because nitrogen isotopic ratio reflects trophic position as a consequence, a quantitative tool emerges for relating trophic position to chemical concentration for members of the trophic web. The ratio most often used for this purpose is normalized to the N isotopic ratio of the ultimate N source, the atmosphere,

$$\delta\ {}^{15}N = 1{,}000\left[\frac{\dfrac{{}^{15}N_{sample}}{{}^{14}N_{sample}}}{\dfrac{{}^{15}N_{air}}{{}^{14}N_{air}}} - 1\right] \tag{7.85}$$

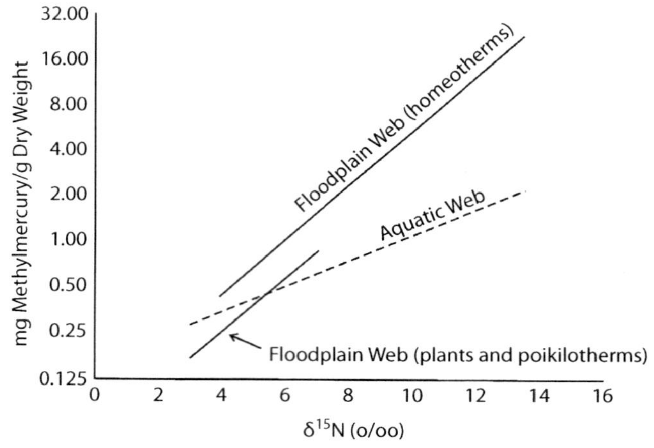

Figure 7.15 Methylmercury biomagnifications in a contaminated river food web (South River, Virginia, USA) (Tom et al. 2010) and the contiguous floodplain (Newman et al. 2011). Dashed line = aquatic trophic web with bass as the apex predator; Solid lines = floodplain trophic web with Eastern screech owl and Carolina wren as the top predators. The floodplain model has separate lines showing the influence of homeothermy (small mammals and birds) on biomagnification within the trophic web.

The units are per mill (‰). Other light isotopes such as those of carbon and sulfur are also applied in studies of trophic structure and interactions, although $\delta\ ^{15}N$ is generally the best metric of trophic position. The ratios (R) of $^{13}C/^{12}C$ or $^{34}S/^{32}S$ are inserted in the following equation to generate $\delta\ ^{13}C$ and $\delta\ ^{34}S$.

$$\delta\ X = 1{,}000\left[\frac{R_{Sample}}{R_{Reference}} - 1\right] \tag{7.86}$$

Different reference standards might be used; however, that for carbon is usually South Carolina Pee Dee limestone formation and that for sulfur is usually a meteorite standard, Canyon Diablo Troilite (Peterson and Fry 1987).

Discrimination or fractionation of isotopes is generally estimated as the difference in δX between initial and final states (Peterson and Fry 1987). In the case of trophic discrimination,

$$D = \delta\ ^{15}N_{Prey} - \delta\ ^{15}N_{Predator}$$

As shown below, trophic discrimination is often expressed with a discrimination factor, $\Delta\ ^{15}N$.

Often a plot of Ln or $\log_{10}$ of concentration versus $\delta\ ^{15}N$ for sampled organisms within a trophic web results in a straight line, as illustrated in Figure 7.15 for methylmercury concentration in a river (South River, Virginia) and the adjacent floodplain trophic webs (Tom et al. 2010; Newman et al. 2011).

$$Log_{10}\ Concentration_i = a + b\ \delta\ ^{15}N_i + \varepsilon \tag{7.87}$$

Often this linearized model is fit to data and then backtransformed to the following model that, as discussed in Newman et al. (2011), requires correction for a backtransformation bias ($10^{\varepsilon} \approx 10^{MSE/2}$).

$$Concentration_i = 10^{a}10^{\varepsilon}10^{b(\delta\ ^{15}N_i)} \tag{7.88}$$

The slope b estimates the change in $\log_{10}$ concentration per change of one mill in the $\delta\ ^{15}N$. The change in $\delta\ ^{15}N$ per trophic level ($\Delta\ ^{15}N$, discrimination factor) is generally in the range of 3.4‰ per trophic level (TL).* [The average $\Delta\ ^{13}C$ is lower, ranging from 0.4 to 0.8‰ (Vander Zanden and Rasmussen 2001; Post 2002).] Using this general estimate or a derived estimate of $\Delta\ ^{15}N$, the models above can be converted to change in $\log_{10}$ concentration per TL. The difference in $\delta\ ^{15}N$ between the consumer of interest and primary producers might be used.

$$TL_{Consumer} = 1 + \frac{\delta\ ^{15}N_{Consumer} - \delta\ ^{15}N_{Primary\ Producers}}{\Delta\ ^{15}N} \tag{7.89}$$

Wide variation in primary producer $\delta\ ^{15}N$ values might lead to choosing a long-lived primary consumer instead in these calculations (Jardine et al. 2006).

$$TL_{Consumer} = 2 + \frac{\delta\ ^{15}N_{Consumer} - \delta\ ^{15}N_{Primary\ Consumer}}{\Delta\ ^{15}N} \tag{7.90}$$

The model (Equation 7.87) can be expressed in terms of TL.

$$Log_{10}\ Concentration_i = a + b\ TL + \varepsilon \tag{7.91}$$

The slope of this TL-based model can be used as an estimate of a biomagnification factor, that is, the food web magnification factor (FWMF).

$$FWMF = 10^b \tag{7.92}$$

For example, a FWMF of 2 indicates that, averaged over all TL included in the model, the chemical concentration will double per TL increase.

This general framework for analyzing trophic web movement of contaminants has been applied to describing movement of numerous contaminants in a wide range of trophic webs. More recently it has been used to explain contaminant concentration shifts when trophic web structure was changed due to a species removal (Lepak et al. 2009) or introduction (Swanson et al. 2006) from an ecological community. Given the global trend in trophic downgrading (Estes et al. 2011), it would also be useful for predicting global transitions in contaminant movement in the world's trophic webs that might be occurring now.

Example 7.10

Between 1929 and 1950, the South River (Virginia) received elemental and ionic mercury from a DuPont acetate fiber production facility. A 23-mile river reach remains contaminated. We (Tom et al. 2010; Newman et al. 2011) modeled the movement of mercury through the aquatic and also floodplain trophic webs (Figure 7.15). The data for the aquatic web will be used here to illustrate model selection and assessment of predictive capability for supporting management decision making. Replicate samples taken of periphyton, insect larvae, and fish species were analyzed for methylmercury concentration and $\delta\ ^{15}N$. Candidate models ($\log_{10}$ methylmercury concentration = f(explanatory variables)) with combinations of several explanatory variables were built and the best model selected. Model selection involved AIC (see Equation 3.9 in Chapter 3;

* The 3.4 estimate of $\Delta\ ^{15}N$ is a common one, although the $\Delta\ ^{15}N$ can vary among trophic webs (Caut et al. 2009; Jardine et al. 2006; Vander Zanden and Rasmussen 1996).

also Bozodogan 2000; Wada and Kashiwagi 1990), graphical exploration of regression residuals, and subject knowledge. Candidate explanatory variables included $\delta\ ^{15}N$, $\delta\ ^{13}C$, and RM (river miles downriver of the historic source). The best model incorporated $\delta\ ^{15}N$ and RM.

Prior to beginning the study, a criterion for acceptable prediction was established for the final selected model based on the $r^2_{prediction}$. The $r^2_{prediction}$ is calculated with the model total sum of squares (SS_T) and PRESS. The PRESS is similar to, but not the same as, the error sum of squares (SS_e) that is used to estimate the conventional regression r^2 (see Example 3.1 in Chapter 3 for details). The PRESS is calculated in several steps. First, an observation is removed from the data set with n observations and a model then built from the remaining $n - 1$ observations. Second, a prediction is made for the omitted observation by placing the explanatory variable value for that point into the model. Third, a prediction residual is calculated for that point and then squared, $(y_i - \hat{y}_i)^2$. Fourth, the omitted point is placed back into the data set and the next observation is omitted and its $(y_i - \hat{y}_i)^2$ calculated. Finally, the process is repeated for all observations and the prediction residuals sum of squares (PRESS) produced.

$$PRESS = \sum_{i=1}^{n} (y_i - \hat{y}_i)^2 \tag{7.93}$$

The prediction r^2 can now be calculated for the model.

$$r^2_{prediction} = 1 - \frac{PRESS}{SS_T} \tag{7.94}$$

Whereas the correlation coefficient, r^2, quantifies the amount of variation in the response variable that can be explained by the model, the $r^2_{prediction}$ quantifies the amount of variation expected in prediction from a model built without the point for which prediction is being made. It indicates how good predictions might be expected to be, not how much of the variation in the response variable can be explained by the model. Prior to beginning the study, an $r^2_{prediction}$ in the range of 0.80 was judged to be needed to support sound river manager decision making.

The SAS code below was used to fit the selected model to these data. The selection of the CLPARM SOLUTION CLM P options in the PROC GLM Model statement resulted in 95% confidence intervals being generated for the estimated model parameters and also the generation of the PRESS.

```
DATA MERC;
INPUT SAMPDATE $ 1-8 SITE $ 10-14 ORGANISM $ 16-29 REP $ 31 DRYWT 33-37
WETWT 39-43 THG 45-51 MHG 53-59 DELN15 61-65 DELC13 67-72 perOC 74-77 RM
79-82;
MHG=MHG/1000; /* CONVERTING TO UG/G DRY WGT */
LMHG=LOG(MHG);
DATALINES;
05232007 AFC Baetidae A 0.02 0.17 700.4 8.75 -23.02 51.2 11.6
05232007 AFC Baetidae C 0.03 0.18 756.6 8.24 -24.44 48.7 11.6
...
RUN;
PROC SORT;
  BY SITE;
RUN;
PROC GLM;
  MODEL LMHG=DELN15 RM/CLPARM SOLUTION CLM P;
  OUTPUT OUT=TROPHIC PREDICTED=PLNMHG RESIDUAL=RLNMHG;
RUN;
PROC GPLOT;
  PLOT LMHG*DELN15 PLNMHG*DELN15 RLNMHG*DELN15;
RUN;
```

Relevant portions of the resulting output are pasted below.

```
                          Dependent Variable: LMHG

                                     Sum of
Source                 DF           Squares    Mean Square    F Value    Pr > F
Model                   2       118.7848943     59.3924471     112.40    <.0001
Error                  63        33.2907157      0.5284241
Corrected Total        65       152.0756100

            R-Square    Coeff Var    Root MSE    LMHG Mean
            0.781091    -133.3643    0.726928    -0.545069

                              Standard
Parameter        Estimate        Error  t Value  Pr > |t|     95% Confidence Limits
Intercept    -5.216059790   0.33095947   -15.76    <.0001  -5.877429511  -4.554690069
DELN15        0.448036500   0.03411241    13.13    <.0001   0.379868292   0.516204707
RM            0.054059975   0.01104129     4.90    <.0001   0.031995710   0.076124239
PRESS Statistic                              36.53221520
```

The correlation coefficient, r^2, was 0.78. The prediction r^2 can be estimated as described above, using the total sum of squares (152.0756) and PRESS (36.5322):

$$r^2_{prediction} = 1 - \frac{PRESS}{SS_T} = 1 - \frac{36.5322}{152.076} \approx 0.76$$

The prediction r^2 is slightly lower than the correlation coefficient but still in the general range judged to be required for the purposes of the study. The estimated model parameters are provided in addition to their 95% confidence intervals. The model is the following:

$$Log_{10}\ Methylmercury\ Concentration = -5.216 + 0.448\delta\ ^{15}N + 0.054RM + \varepsilon$$

If this model is backtransformed, the associated bias can be corrected with the model MSE of 0.528.

Newman et al. (2011) also fit the model in terms of *TL* instead of $\delta\ ^{15}$N for these same data.

$$Log_{10}\ Methylmercury\ Concentration = -1.09 + 0.66TL + 0.02RM + \varepsilon$$

The FWMF could then be estimated for the aquatic trophic web of the contaminated reach of the South River.

$$FWMF = 10^b = 10^{0.66} \approx 4.6$$

Averaged over the entire trophic web, the methylmercury concentration will increase 4.6-fold every trophic level.

There are instances in which omnivory might be integrated into trophic transfer models (Vander Zanden and Rasmussen 1996; Thompson et al. 2007). Consideration of omnivory might be prompted by a $\Delta\ ^{15}$N lower than the generally assumed 3.4‰ per TL. Addressing omnivory that involves only the two TL below a particular TL, Cabana and Rasmussen (1994) provided the following matrix formulation and then linked it to the trophic transfer model of Thomann (1981). In their approach, the total ration for a trophic level (TL_i) was initially defined as the weighted sum of the contributions coming from the lower trophic levels, C_i.

$$C_i = \sum^{j} \rho_{ij} C_i \tag{7.95}$$

where ρ_{ij} = the fraction of level i's diet comprised of j level items, and $\Sigma^i \rho_{ij} = 1$. The trophic web is then defined* with a vector of C_i (**C**) and a matrix of r_{ij}, (**r**). Cabana and Rasmussen state that **C** for the entire food web is **rC**. The matrix of dietary fractions is the following:

$$\rho = \begin{bmatrix} 0 & 0 & 0 & 0 \\ \rho_{21} & 0 & 0 & 0 \\ \rho_{31} & \rho_{32} & 0 & 0 \\ 0 & \rho_{42} & \rho_{43} & 0 \end{bmatrix} \tag{7.96}$$

All of the elements in rows other than $i = 1$ will sum to 1, e.g., $r_{21} = 1$, $r_{31} + r_{32} = 1$, and $r_{42} + r_{43} = 1$. Notice again that omnivory for a particular trophic level only involves the two levels below it. If there is no omnivory (a linear trophic chain), this matrix becomes one with all $r_{i,i-1}$ elements equal to 1 and all other r_{ij} elements being 0.

$$\rho = \begin{bmatrix} 0 & 0 & 0 & 0 \\ 1 & 0 & 0 & 0 \\ 0 & 1 & 0 & 0 \\ 0 & 0 & 1 & 0 \end{bmatrix} \tag{7.97}$$

In terms of $\delta\,^{15}N$ for an entire trophic web without omnivory, a vector of $\delta\,^{15}N$ values ($\delta\,^{15}\mathbf{N}$) could be multiplied by the matrix **r** to predict change in $\delta\,^{15}N$ in the trophic web: $\delta\,^{15}\mathbf{N} - 3.4 = \mathbf{r}\,\delta\,^{15}\mathbf{N}$. The $\Delta\,^{15}N$ would be less than 3.4 in this formula if omnivory were present.

Next this model was linked with that of Thomann (1981). Letting **B** = vector of biomagnifications ratios (concentration in consumer/concentration in consumed), **a** = vector of assimilation efficiencies, **C** = vector of rations, **K** = vector of contaminant excretion rates, **G** = vector of growth rates, and **I** = identity matrix:

$$\boldsymbol{B} = \alpha \boldsymbol{C}[(\boldsymbol{K} + \boldsymbol{G})\boldsymbol{I}]^{-1} \tag{7.98}$$

Letting **V** = vector of concentrations in the i levels, **P** = the omnivory matrix, and **PV** = the weighted mean concentration for each level with omnivory,

$$\boldsymbol{V}[\boldsymbol{PVI}]^{-1} = \alpha \boldsymbol{C}[(\boldsymbol{K} + \boldsymbol{G})\boldsymbol{I}]^{-1} \tag{7.99}$$

Isotopic ratios have also been applied by community ecologists to estimate the relative contributions of various food sources to a consumer's diet (e.g., Akamatsu et al. 2004; Anderson and

* Notations used here are those of Cabana and Rasmussen (1994) that, in some instances, do not adhere strictly to conventional vector and matrix notation described in Appendix 27.

Cabana 2005; Briand and Cohen 1987; Caut et al. 2009; Chase 2000; Finlay et al. 2002; Hobson et al. 1994; McNaughton et al. 1989; Peterson and Fry 1987; Post 2002; Rounick and Winterbourn 1986; Thompson et al. 2007; Vander Zanden and Rasmussen 2001). The approach can be illustrated graphically using stable isotope data from the mercury-contaminated South River that is modeled in Example 7.11. Replicate samples of possible food sources (leaf litter, sediment, seston, periphyton, and submerged aquatic vegetation (SAV)) were sampled on June 2010. Replicate samples of two consumers, caddisfly larvae and the Asian clam (*Corbicula fluminea*), were also taken. Stable nitrogen and carbon isotopic ratios were determined in these samples and then plotted (Figure 7.16, top panel). A polygon was created by connecting the data points for the candidate food sources. The data for the two consumers were adjusted (black dot to open dot) for $\Delta\,^{15}N$ and $\Delta\,^{13}C$ using the average values reported by Vander Zanden and Rasmussen (2001). The adjusted consumer points are close to those for the seston and sediments and distant from the SAV and leaf litter points. This

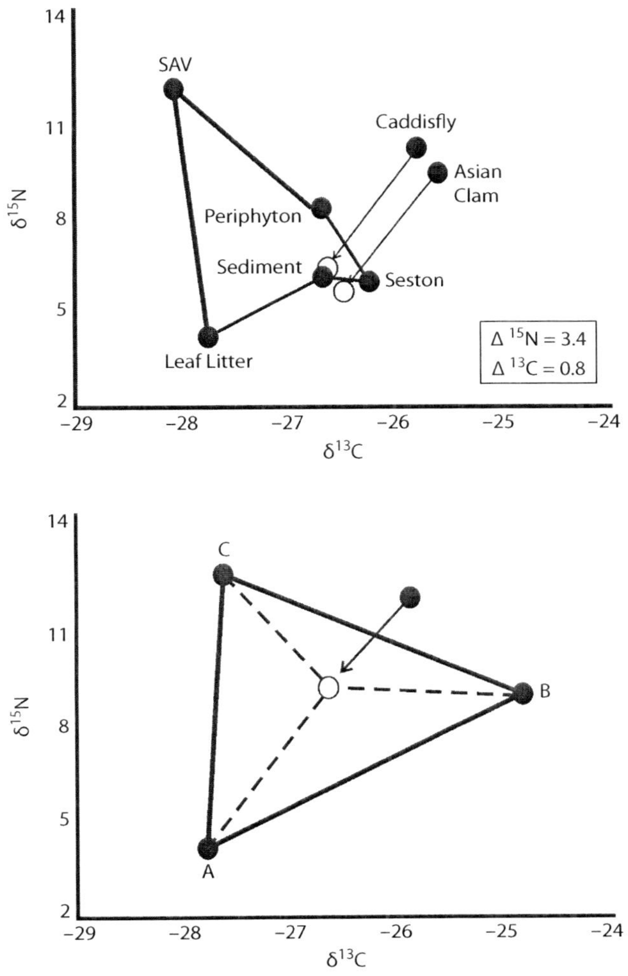

Figure 7.16 Top panel depicts a mixing polygon for several potential food sources for caddisfly larvae and Asian clam (*Corbicula fluminea*) taken from RM 23.5 of the mercury contaminated reach of the South River (Virginia, USA). The bottom panel is a mixture triangle employed to estimate fractional contributions from three sources (A, B, and C). Two isotopic ratios are required for calculation involving three sources.

suggests that the sediment and seston were much more prominent sources than the other sources to the diets of these consumers.

The mixing polygon approach just described might be the only avenue for assessing contributions to diets; however, a more quantitative approach is possible under certain conditions (Benstead et al. 2006; Phillips 2001; Phillips and Gregg 2003; Phillips and Koch 2002; Phillips et al. 2005). Phillips and Koch (2002) stipulate that these methods can be applied for situations involving up to three sources. They work best if the isotopic signatures of the candidate sources are quite different and have low variance. The mixing triangle shown in the bottom panel of Figure 7.16 can be used to visualize the approach with three potential sources. The distances from the points for each potential source (*A*, *B*, and *C*) to the adjusted consumer point reflect the contribution of each source to the consumer's diet. The further the consumer's point is from that of a source, the less that source contributes to the consumer's diet.

The mathematic rendering of this framework can be provided for two- and three-candidate source situations. Only one isotopic ratio is required for a situation involving two potential sources. To illustrate the two source situation, let $\delta\ ^{13}C_{Mixture}$ = the $\delta\ ^{13}C$ measured for the consumer that reflects two food sources, *A* and *B*, $\Delta\ ^{13}C_{Tissue\text{-}A}$ = the trophic fractionation between food source *A* and the consumer's tissues, $\Delta\ ^{13}C_{Tissue\text{-}B}$ = the trophic fractionation between food source *B* and the consumer's tissues, f_A = fractional contribution of carbon to the consumer's diet from *A*, and f_B = fractional contribution of carbon to the consumer's diet from *B*. Two equations are required to define this situation:

$$\delta\ ^{13}C_{Mixture} = f_A(\delta\ ^{13}C_A + \Delta\ ^{13}C_{Tissue-A}) + f_B(\delta\ ^{13}C_B + \Delta^{13}C_{Tissue-B}) \tag{7.100}$$

$$1 = f_A + f_B \tag{7.101}$$

These two equations can be solved simultaneously to estimate f_A and f_B. The approach is not as convenient for three sources that require two sets of isotopic ratios. Three equations now define this situation.

$$\begin{aligned}\delta\ ^{13}C_{Mixture} &= f_A(\delta\ ^{13}C_A + \Delta\ ^{13}C_{Tissue-A}) + f_B(\delta\ ^{13}C_B + \Delta^{13}C_{Tissue-B}) \\ &+ f_C(\delta\ ^{13}C_C + \Delta^{13}C_{Tissue-C})\end{aligned} \tag{7.102}$$

$$\begin{aligned}\delta\ ^{15}N_{Mixture} &= f_A(\delta\ ^{15}N_A + \Delta\ ^{15}N_{Tissue-A}) + f_B(\delta\ ^{15}N_B + \Delta^{15}N_{Tissue-B}) \\ &+ f_C(\delta\ ^{15}N_C + \Delta^{15}N_{Tissue-C})\end{aligned} \tag{7.103}$$

$$1 = f_A + f_B + f_C \tag{7.104}$$

The tedium of doing the associated calculations, including estimates of uncertainty, can be eliminated by using the shareware IsoConc, IsoSource, and IsoError, provided by the EPA Western Ecology Division (http://www.epa.gov/wed/pages/models/stableIsotopes/isotopes.htm).

7.7 SUMMARY

In this overview of community level effects of toxicants, commonly used and potentially useful techniques are described. Emphasis is provided in many sections on underlying mechanisms often overlooked by ecotoxicologists. Basic models of predator–prey interactions and foraging theory are presented along with examples of their application. Models of interspecies competition and general discussion of symbiotic relationships are also presented to reinforce the importance of considering all aspects of a species' niche integrity. Traditional and potentially useful measures of community structure and function are outlined along with details of their applications in ecotoxicology. Metacommunity features are also examined and related to toxicant effects. Finally, the transfer of toxicants through trophic levels is explored using theoretical models and a practical light isotope-based approach.

REFERENCES

Akamatsu, F., H. Toda, and T. Okino. Food source of riparian spiders analyzed by using stable isotope ratios. *Ecol. Res.* 19:655–662 (2004).

Alatalo, R.V. Problems in the measurement of evenness in ecology. *Oikos* 37:199–204 (1981).

Allred, P.M., and J.P. Giesy. Use of *in situ* microcosms to study mass loss and chemical composition of leaf litter being processed in a blackwater stream. *Arch. Hydrobiol.* 114:231–250 (1988).

Amblard, C., P. Couture, and G. Bourdier. Effects of a pulp and paper mill on the structure and metabolism of periphytic algae in experimental streams. *Aquat. Toxicol.* 18:137–162 (1990).

Anderson, C., and G. Cabana. $\delta\,^{15}N$ in riverine food webs: Effects of N inputs from agricultural watersheds. *Can. J. Fish. Aquat. Sci.* 62:333–340 (2005).

Armitage, J.M., and F.A.P.C. Gobas. A terrestrial food-chain bioaccumulation model for POPs. *Environ. Sci. Technol.* 41:4019–4025 (2007).

Arnot, J.A., and F.A.P.C. Gobas. A food web bioaccumulation model for organic chemicals in aquatic ecosystems. *Environ. Toxicol. Chem.* 23:2343–2355 (2004).

Atchison, G.J., M.G. Henry, and M.B. Sanheinrich. Effects of metals on fish behavior: A review. *Environ. Biol. Fish.* 18:11–25 (1987).

Axelsen, J.A., N. Holst, T. Hamers, and P.H. Krogh. Simulations of the predator-prey interactions in a two species ecotoxicological test system. *Ecol. Model.* 101:15–25 (1997).

Ayala, F.J. Experimental invalidation of the principle of competitive exclusion. *Nature* 224:1076–1079 (1969).

Belpaire, C., R. Smolders, I. Vanden Auweele, D. Ercken, J. Breine, G. Van Thuyne, and F. Ollevier. An index of biotic integrity characterizing fish populations and the ecological quality of Flandrian water bodies. *Hydrobiologia* 434:17–33 (2000).

Benson, A.A., and R.E. Summons. Arsenic accumulation in Great Barrier Reef invertebrates. *Science* 211:482–483 (1981).

Benstead, J.P., J.G. March, B. Fry, K.C. Ewel, and C.M. Pringle. Testing ISOSOURCE: Stable isotope analysis of a tropical fishery with diverse organic matter sources. *Ecology* 87:326–333 (2006).

Beyer, W.H. *CRC standard probability and statistics tables and formulae.* Boca Raton, FL: CRC Press, 1991.

Beyer, W.N. A reexamination of biomagnification of metals in terrestrial food chains. *Environ. Toxicol. Chem.* 5:863–864 (1986).

Blanck, H. A simple, community level, ecotoxicological test using samples of periphyton. *Hydrobiologia* 124:251–261 (1985).

Bonsall, M.B., and M.P. Hassell. Predator-prey interactions. In *Theoretical ecology. Principles and applications*, ed. R.M. May and A.R. McLean. Oxford: Oxford University Press, 2007, pp. 46–61.

Boudou, A., A. Delarche, F. Ribeyre, and R. Larty. Bioaccumulation and bioamplification of mercury compounds in a second level consumer, *Gambusia affinis*—Temperature effects. *Bull. Environ. Contam. Toxicol.* 22:813–818 (1979).

Bozodogan, H. Akaike's information criterion and recent developments in information complexity. *J. Math. Psychol.* 44:62–91 (2000).

Briand, F., and J.E. Cohen. Environmental correlates of food chain length. *Science* 238:956–960 (1987).

Bryan, G.W., and P.E. Gibbs. Impact of low concentrations of tributyltin (TBT) on marine organisms: A review. In *Metal ecotoxicology. Concepts and applications,* ed. M.C. Newman and A.W. McIntosh. Chelsea, MI: Lewis Publishers, 1991, pp. 323–361.

Bundschuh, M., J.P. Zubrod, F. Seitz, M.C. Newman, and R. Schulz. Mercury-contaminated sediments affect amphipod feeding. *Arch. Environ. Contam. Toxicol.* 60:437–443 (2011).

Burton, T.M., D.G. Uzarski, J.P. Gathman, J.A. Genet, B.E. Keas, and C.A. Stricker. Development of a preliminary invertebrate index of biotic integrity for Lake Huron coastal wetlands. *Wetlands* 19:869–882 (1999).

Buzas, M.A., and L.-A.C. Hayek. Biodiversity resolution: An integrated approach. *Biodiversity Lett.* 3:40–43 (1996).

Cabana, G., and J.B. Rasmussen. Modeling food chain structure and contaminant bioaccumulation using stable nitrogen isotopes. *Nature* 372:255–257 (1994).

Cairns, J., Jr. Heated waste-water effects on aquatic ecosystems. In *Thermal ecology II*, ed. G.W. Esch and R.W. McFarlane. Springfield, VA: National Technical Information Center, 1976.

Cairns, J., Jr. Quantification of biological integrity. In *The Integrity of water*, ed. R.K. Ballentine and L.J. Guarraia. Washington, DC: U.S. EPA, Office of Water and Hazardous Materials, 1977, pp. 171–187.

Cairns, J., Jr., and B.R. Niederlehner. Adaptation and resistance of ecosystems to stress: A major knowledge gap in understanding anthropogenic perturbations. *Speculations Sci. Technol.* 12:23–30 (1989).

Cairns, J., Jr., and J.R. Pratt. Multispecies toxicity testing using indigenous organisms—A new cost-effective approach to ecosystem protection. In *1985 Environmental Conference, TAPPI Proceedings*. Atlanta, GA: TAPPI Press, 1985, pp. 149–159.

Cairns, J., Jr., and J.R. Pratt. Developing a sampling strategy. In *Rationale for sampling and interpretation of ecological data in the assessment of freshwater ecosystems*, ed. B.G. Isom. ASTM STP 894. Philadelphia: American Society for Testing and Materials, 1986.

Cairns, J., Jr., J.R. Pratt, B.R. Niederlehner, and P.V. McCormick. A simple cost-effective multispecies toxicity test using organisms with a cosmopolitan distribution. *Environ. Monit. Assess.* 6:207–220 (1986).

Caquet, T., M.L. Hanson, M. Roucaute, D.W. Graham, and L. Lagadic. Influence of isolation on the recovery of pond mesocosms from the application of an insecticide. II. Benthic macroinvertebrate responses. *Environ. Toxicol. Chem.* 26:1280–1290 (2007).

Caquet, T., L. Lagadic, and S.R. Sheffield. Mesocosms in ecotoxicology. *Rev. Environ. Contam. Toxicol.* 165:1–38 (2000).

Carson, R. *Silent spring*. Boston: Houghton Mifflin, 1962.

Caswell, H., and J.E. Cohen. Local and regional regulation of species-area relations: A patch-occupancy model. In *Species diversity in ecological communities*, ed. R.E. Ricklefs and D. Schluter. Chicago: University of Chicago, 1993, pp. 99–107.

Caut, S., E. Angulo, and F. Courchamp. Variation in discrimination factors ($\Delta\ ^{15}N$ and $\Delta\ ^{13}C$): The effect of diet isotopic values and applications for diet reconstruction. *J. Appl. Ecol.* 46:443–453 (2009).

Chase, J.M. Are there real differences among aquatic and terrestrial food webs? *TREE* 15:408–412 (2000).

Clarke, K.R., and R.M. Warwick. The taxonomic distinctness measure of biodiversity weighting of step lengths between hierarchical levels. *Mar. Ecol.-Prog. Ser.* 184:21–29 (1999).

Clarke, K.R., and R.M. Warwick. A further biodiversity index applicable to species lists variation in taxonomic distinctness. *Mar. Ecol.-Prog. Ser.* 216:265–278 (2001).

Clements, W.H. Metal tolerance and predator-prey interactions in benthic macroinvertebrate stream communities. *Ecol. Appl.* 9:1073–1084 (1999).

Clements, W.H., J.H. Van Hassel, D.S. Cherry, and J. Cairns Jr. Colonization, variability, and the application of substratum-filled trays for biomonitoring benthic communities. *Hydrobiologia* 173:45–53 (1989).

Cohen, A.C., Jr. Tables for maximum likelihood estimates: Singly truncated and singly censored samples. *Technometrics* 3:535–541 (1961).

Colwell, R.K. 2009. *EstimateS: Statistical estimation of species richness and shared species from samples*. Version 8.2. User's guide and application published at http://purl.oclc.org/estimates.

Connell, D.W. *Bioaccumulation of xenobiotic compounds*. Boca Raton, FL: CRC Press, 1990.

Connell, J.H. The influence of interspecific competition and other factors on the distribution of the barnacle *Chthamalus stellatus*. *Ecology* 42:710–723 (1961).

Connolly, J.P., and C.J. Pedersen. A thermodynamic-based evaluation of organic chemical accumulation in aquatic organisms. *Environ. Sci. Technol.* 22:99–103 (1988).

Conover, R.J., and V. Francis. The use of radioactive isotopes to measure the transfer of materials in aquatic food chains. *Mar. Biol. (Berl.)* 18:272–283 (1973).

Crossland, R. O., and T. W. La Point. The design of the mesocosm experiments. *Environ. Toxicol. Chem.* 11:1–4 (1992).

Cushing, C.E., D.G. Watson, A.J. Scott, and J.M. Gurtisen. Decrease of radionuclides in Columbia River biota following closure of the Hanford reactors. *Health Phys.* 41:59–67 (1981).
Davis, J.J., and R.F. Foster. Bioaccumulation of radioisotopes through aquatic food chains. *Ecology* 39:530–535 (1958).
Dawson, T.E., and R.T.W. Siegwolf. *Stable isotopes as indicators of ecological change*. Boston: Elsevier, 2007.
Day, K.E. Short-term effects of herbicides on primary productivity of periphyton in lotic environments. *Ecotoxicology* 2:123–138 (1993).
Dixon, W.J., and F.J. Massey Jr. *Introduction to statistics*. New York: McGraw-Hill Book Company, 1969.
Draper, N.R., and H. Smith. *Applied regression analysis*. New York: John Wiley & Sons, 1998.
Elwood, J.W., J.D. Newbold, R.V. O'Neill, and W. Van Winkle. Resource spiraling: An operational paradigm for analyzing lotic ecosystems. In *Dynamics of lotic ecosystems*, ed. T.D. Fontaine and S.M. Bartell. Ann Arbor, MI: Ann Arbor Science, 1983, pp. 3–27.
Elwood, J.W., J.D. Newbold, A.F. Trimble, and R.W. Stark. The limiting role of phosphorus in a woodland stream ecosystem: Effects of P enrichment on leaf decomposition and primary producers. *Ecology* 62:146–158 (1981).
Emmel, T.C. *Population biology*. New York: Harper & Row, 1976.
Estes, J.A., J. Terborgh, J.S. Brashares, M.E. Power, J. Berger, W.J. Bond, S.R. Carpenter, T.E. Essington, R.D. Holt, J.B.C. Jackson, R.J. Marquis, L. Oksanen, T. Oksanen, R.T. Paine, E.K. Pikitch, W.J. Ripple, S.A. Sandin, M. Scheffer, T.W. Schoener, J.B. Shurin, A.R.E. Sinclair, M.E. Soule, R. Virtanen, and D.A. Wardle. Trophic downgrading of planet earth. *Science* 333:301–306 (2011).
Evans, N.A. Effect of copper and zinc upon the survival and infectivity of *Echinoparyphium recurvatum* cercariae. *Parasitology* 85:295–303 (1982a).
Evans, N.A. Effects of copper and zinc on the life cycle of *Notocotylus attenuatus* (Digenea: Notocotylidae). *Int. J. Parasitol.* 12:363–369 (1982b).
Ewing, M.S., S.A. Ewing, and M.A. Zimmer. Sublethal copper stress and susceptibility of channel catfish to experimental infections with *Ichthyophthirius multifilis*. *Bull. Environ. Contam. Toxicol.* 28:674–681 (1982).
Fausch, K.D., J. Lyons, J.R. Karr, and P.L. Angermeier. Fish communities as indicators of environmental degradation. *Am. Fish. Soc. Symp.* 8:123–144 (1990).
Finlay, J.C., S. Khandwala, and M.E. Power. Spatial scales of carbon flow in a river food web. *Ecology* 83:1845–1859 (2002).
Fisher, R.A., A.S. Corbet, and C.B. Williams. The relation between the number of species and the number of individuals in a random sample of an animal population. *J. Anim. Ecol.* 12:42–58 (1943).
Fry, B. *Stable isotope ecology*. New York: Springer Science+Business Media, LLC, 2006.
Gächter, R., and W. Geiger. MELIMEX, an experimental heavy metal pollution study: Behavior of heavy metals in an aquatic food chain. *Schweiz. Z. Hydrol.* 41:277–290 (1979).
Gammon, J.R. Biological monitoring in the Wabash River and its tributaries. In *Water quality standards for the 21st century*. Washington, DC: U.S. EPA Criteria and Standards Division, 1989, pp. 105–111.
Gause, G.F. *The struggle for existence*. Baltimore: William and Wilkins, 1934.
Gause, G.F., and A.A. Witt. Behavior of mixed populations and the problem of natural selection. *Am. Nat.* 69:596–609 (1935).
Giesy, J.P., Jr. Cadmium inhibition of leaf decomposition in an aquatic microcosm. *Chemosphere* 6:467–475 (1978).
Gilpin, M.E. Limit cycles in competition communities. *Am. Nat.* 109:51–60 (1975).
Gilpin, M.E., and F.J. Ayala. Global models of growth and competition. *Proc. Natl. Acad. Sci. U.S.A.* 70:3590–3593 (1973).
Gilpin, M.E., and K.E. Justice. Reinterpretation of the invalidation of the principle of competitive exclusion. *Nature* 236:273–301 (1972).
Gleason, H.A. On the relation between species and area. *Ecology* 3:158–162 (1922).
Goldsborough, L.G., and G.G.C. Robinson. Changes in periphytic algal community structure as a consequence of short herbicide exposures. *Hydrobiologia* 139:177–192 (1986).
Goodyear, C.P. A simple technique for detecting effects of toxicants or other stresses on a predator-prey interaction. *Trans. Am. Fish. Soc.* 101:367–370 (1972).
Gray, J.S. Pollution-induced changes in populations. *Phil. Trans. R. Soc. Lond. B* 286:545–561 (1979).
Green, R.H. A multivariate statistical approach to the Hutchinsonian niche: Bivalve molluscs of central Canada. *Ecology* 52:543–556 (1971).

Grenfell, B., and M. Keeling. Dynamics of infectious disease. In *Theoretical ecology. Principles and applications*, ed. R.M. May and A.R. McLean. Oxford: Oxford University Press, 2007.

Guimaraes, Jr., P.R., and P. Guimaraes. Improving the analyses of nestedness for large sets of matrices. *Environ. Modell. Softw.* 21:1512–1513 (2006).

Hamer, T., and P.H. Krogh. Predator-prey relationships in a two-species toxicity test system. *Ecotoxicol. Environ. Safe.* 37:203–212 (1997).

Hanson, M.L., D.W. Graham, E. Babin, D. Azam, M.-A. Coutellec, C.W. Knapp, L. Lagadic, and T. Caquet. Influence of isolation on the recovery of pond mesocosms from the application of an insecticide. 1. Study design and plankton community responses. *Environ. Toxicol. Chem.* 26:1265–1279 (2007).

Harris, J.H., and R. Silveira. Large-scale assessments of river health using an index of biotic integrity with low-diversity fish communities. *Freshw. Biol.* 41:235–252 (1999).

Hayek, L.-A.C., and M.A. Buzas. *Surveying natural populations*. New York: Columbia University Press, 1997.

Hayek, L.-A.C., and M.A. Buzas. The martyrdom of St. Lucie: Decimation of a meiofauna. *B. Mar. Sci.* 79:341–352 (2006).

Heck, K.L.J., G. van Belle, and D. Simberloff. Explicit calculation of the rarefaction diversity measurement and the determination of sufficient sample size. *Ecology* 56:1459–1461 (1975).

Heltshe, J.F., and N.E. Forrester. Estimating species richness using the jackknife procedure. *Biometrics* 39:1–11 (1983).

Heltshe, J.F., and N.E. Forrester. Statistical evaluation of the jackknife estimate of diversity when using quadrat samples. *Ecology* 66:107–111 (1985).

Henry, M.G., and G.J. Atchison. Metal effects on fish behavior—Advances in determining the ecological significance of responses. In *Metal ecotoxicology: Concepts and applications*, ed. M.C. Newman and A.W. McIntosh. Chelsea, MI: Lewis Publishers, 1991.

Hill, B.A., A.T. Herlihy, P.R. Kaufmann, R.J. Stevenson, F.H. McCormick, and C. Burch Johnson. Use of periphyton assemblage data as an index of biotic integrity. *J. N. Am. Benthol. Soc.* 19:50–67 (2000).

Hill, M.O. Diversity and evenness: A unifying notation and its consequences. *Ecology* 54:427–432 (1973).

Hobson, K.A., J.F. Piatt, and J. Pitocchelli. Using stable isotopes to determine seabird trophic relationships. *J. Anim. Ecol.* 63:786–798 (1994).

Hoegh-Guldberg, O., P.J. Mumby, A.J. Hooten, R.S. Steneck, P. Greenfield, E. Gomez, C.D. Harvell, P.F. Sale, A.J. Edwards, K. Caldeira, N. Knowlton, C.M. Eakin, R. Iglesias-Prieto, N. Muthiga, R.H. Bradbury, A. Dubi, and M.E. Hatziolos. Coral reefs under rapid climate change and ocean acidification. *Science* 318:1737–1742 (2007).

Holleman, D.F., J.R. Luick, and F.W. Whicker. Transfer of radiocesium from lichen to reindeer. *Health Phys.* 21:657–666 (1971).

Holling, C.S. The components of predation as revealed by a study of small-mammal predation of the European pine sawfly. *Can. Entomol.* 91:293–320 (1959).

Holt, R.D. Ecology at the mesoscale: The influence of regional processes on local communities. In *Species diversity in ecological communities*, ed. R.E. Ricklefs and D. Schluter. Chicago: University of Chicago, 1993, pp. 77–88.

Hu, X.-S., F. He, and S.P. Hubbell. Community differentiation on landscapes: Drift, migration and speciation. *Oikos* 118:1515–1523 (2009).

Huffaker, C.B. Experimental studies on predation: Dispersion factors and predator-prey oscillations. *Hilgardia* 27:343–383 (1958).

Hunt, E.G., and A.I. Bischoff. Inimical effects on wildlife of periodic DDD applications to Clear Lake. *Calif. Fish Game* 46:91–106 (1960).

Hurlbert, S.H. The non-concept of species diversity: A critique and alternative parameters. *Ecology* 52:577–586 (1971).

Hurlbert, S.H. The measurement of niche overlap and some relatives. *Ecology* 59:67–77 (1978).

Huston, M. A general hypothesis of species diversity. *Am. Nat.* 113:81–101 (1979).

Hutchinson, G.E. Concluding remarks. *Cold Spring Harbor Symp. Quant. Biol.* 22:415–427 (1957).

James, F.C., and S. Rathbun. Rarefaction, relative abundance, and diversity of avian communities. *Auk* 98:785–800 (1981).

Jardine, T.D., K.A. Kidd, and A.T. Fisk. Application, considerations, and sources of uncertainty when using stable isotope analysis in ecotoxicology. *Environ. Sci. Technol.* 40:7501–7511 (2006).

Kaesler, R.J., and E.E. Herricks. Analysis of data from biological surveys of streams: Diversity and sample size. *Water Resour. Bull.* 12:125–135 (1976).

Kania, H.J., and J. O'Hara. Behavioral alterations in a simple predator-prey system due to sublethal exposure to mercury. *Trans. Am. Fish. Soc.* 103:134–136 (1974).

Karr, J.R. Biological integrity: A long-neglected aspect of water resource management. *Ecol. Appl.* 1:66–84 (1991).

Karr, J.R., and M. Dionne. Designing surveys to assess biological integrity in lakes and reservoirs. In *U.S. EPA biological criteria: Research and regulation*. EPA-440/5-91-005. Washington, DC: Office of Water, U.S. EPA, 1991.

Karr, J.R., K.D. Fausch, P.L. Angermeier, P.R. Yant, and I.J. Schlosser. *Assessing biological integrity in running waters. A method and its rationale*. Illinois Natural History Survey Special Publication 5. Champagne, IL: Illinois Natural History Survey, 1986.

Karr, J.R., and B.L. Kerans. Components of biological integrity: Their definition and use in development of an invertebrate IBI. In *Proceedings of the 1991 Midwest Pollution Control Biologists Meeting*, ed. W.S. Davis and T.P. Simon. EPA-905/R-92/003. Chicago: U.S. EPA, Region V, Instream Biological Criteria and Ecological Assessment Committee, 1992, pp. 1–16.

Kelly, B.C., M.G. Ikonomou, J.D. Blair, A.E. Morin, and F.A.P.C. Gobas. Food web-specific biomagnifications of persistent organic pollutants. *Science* 317:236–239 (2007).

Kersting, K. Normalizing ecosystem strain: A system parameter for analysis of toxic stress in (micro-) ecosystems. *Ecol. Bull.* 36:150–153 (1984).

Kettlewell, H.B.D. Selection experiments on industrial melanism in the Lepidoptera. *Heredity* 9:323–342 (1955).

Kimball, K.D., and S.A. Levin. Limitations of laboratory bioassays: The need for ecosystem-level testing. *Bioscience* 35:165–171 (1985).

Kindt, R., P. van Damme, and A.J. Simmons. Tree diversity in western Kenya: Using profiles to characterize richness and evenness. *Biodivers. Conserv.* 15:1253–1270 (2006).

Kirkpatrick, J.B. An iterative method for establishing priorities for the selection of nature reserves: An example from Tasmania. *Biol. Conserv.* 25:127–134 (1983).

Kondratieff, P.F., R.A. Matthews, and A.L. Buikema Jr. A stressed stream ecosystem: Macroinvertebrate community integrity and microbial trophic response. *Hydrobiologia* 111:81–91 (1984).

Krebs, C.J. *Ecological methodology.* New York: Harper Collins Publishers, 1989.

Krebs, C.J. *Ecological methodology*, 2nd ed. Menlo Park, CA: Addison-Wesley Publishers, 1999.

Lambshead, P.J.D., H.M. Platt, and K.M. Shaw. The detection of differences among assemblages of marine benthic species based on an assessment of dominance and diversity. *J. Nat. Hist.* 17:859–874 (1983).

Laskowski, R. Are the top predators endangered by heavy metal biomagnification? *Oikos* 60:387–390 (1991).

Lawrence, J.M., M.J. Samways, J. Henwood, and J. Kelly. Effect of an invasive ant and its chemical control on a threatened endemic Seychelles millipede. *Ecotoxicology* 20:731–738 (2011).

Leibold, M.A., M. Holyoak, N. Mouquet, P. Amarasekare, J.M. Chase, M.F. Hoopes, R.D. Holt, J.B. Shurin, R. Law, D. Tilman, M. Loreau, and A. Gonzalez. The metacommunity concept: A framework for multi-scale community ecology. *Ecol. Lett.* 7:601–613 (2004).

Leibold, M.A., and G.M. Mikkelson. Coherence, species turnover, and boundary clumping: Elements of metacommunity structure. *Oikos* 97:237–250 (2002).

Leonard, P.M., and D.J. Orth. Application and testing of an index of biotic integrity in small, coolwater streams. *Trans. Am. Fish. Soc.* 115:401–414 (1986).

Lepak, J.M., J.M. Robinson, C.E. Kraft, and D.C. Josephson. Changes in mercury bioaccumulation in an apex predator in response to removal of an introduced competitor. *Ecotoxicology* 18:488–498 (2009).

Leponce, M., L. Theunis, J.H.C. Delabie, and Y. Roisin. Scale dependence of diversity measures in a leaf-litter ant assemblage. *Ecography* 27:253–267 (2004).

Lettis, J.S., and K.S. Delaplane. Coordinated responses to honey bee decline in the USA. *Apidolgie* 41:256–263 (2010).

Levine, D.M., P.P. Ramsey, and R.K. Smidt. *Applied statistics for engineers and scientists*. Upper Saddle River, NJ: Prentice Hall, 2001.

Levins, R. Evolution in changing environments: Some theoretical explorations. *Monogr. Popul. Biol.* 2 (1968).

Liber, K., N.K. Kaushik, K.R. Solomon, and J.H. Carey. Experimental designs for aquatic mesocosm studies: Comparison of the "ANOVA" and "regression" design for assessing the impact of tetrachlorophenol on zooplankton populations in limnocorrals. *Environ. Toxicol. Chem.* 11:61–77 (1992).

Ludwig, J.A., and J.F. Reynolds. *Statistical ecology. A primer on methods and computing.* New York: John Wiley & Sons, 1988.

MacArthur, R.H. On the relative abundance of bird species. *Proc. Natl. Acad. Sci. U.S.A.* 43:293–295 (1957).

MacArthur, R.H. Population ecology of some warblers of northeastern coniferous forests. *Ecology* 39:599–619 (1958).

MacArthur, R.H., and E.O. Wilson. *The theory of island biogeography.* Princeton, NJ: Princeton University Press, 1967.

Magurran, A.E. *Ecological diversity and its measurement.* Princeton, NJ: Princeton University Press, 1988.

Magurran, A.E. *Measuring biological diversity.* Malden, MA: Blackwell Science, Ltd., 2004.

Mackay, D., and A. Fraser. Bioaccumulation of persistent organic chemicals: Mechanisms and models. *Environ. Pollut.* 110:375–391 (2000).

Maron, M., R. Mac Nally, D.M. Watson, and A. Lill. Can the biotic nestedness matrix be used predictively? *Oikos* 106:433–444 (2004).

Matthews, R.A., A.L. Buikema Jr., J. Cairns Jr., and J.H. Rodgers Jr. Biological monitoring. Part IIA. Receiving system functional methods. Relationships and indices. *Water Res.* 16:129–139 (1982).

Maurer, D., G. Robertson, and I. Haydock. Coefficient of pollution (p): The Southern California Shelf and some ocean outfalls. *Marine. Pollut. Bull.* 22:141–148 (1991).

Maxted, J.R. The development of biocriteria in marine and estuarine waters in Delaware. In *Water quality standards for the 21st century.* Washington, DC: U.S. EPA Criteria and Standards Division, 1989, pp. 169–175.

May, R.M. *Stability and complexity in model ecosystems.* Princeton: Princeton University Press, 1974.

May, R.M. *Theoretical ecology. Principles and applications.* Philadelphia: W.B. Saunders Company, 1976.

McManus, J., and D. Pauly. Measuring ecological stress: Variations on a theme by R.M. Warwick. *Mar. Biol.* 106:305–308 (1990).

McNaughton, S.J., M. Oesterheld, D.A. Frank, and K.J. Williams. Ecosystem-level patterns of primary productivity and herbivory in terrestrial habitats. *Nature* 341:142–144 (1989).

Minderman, G. Addition, decomposition and accumulation of organic matter in forests. *J. Ecol.* 56:355–362 (1968).

Mingoti, S.A., and G. Meeden. Estimating the total number of distinct species using presence and absence data. *Biometrics* 48:863–875 (1992).

Moriarty, F. *Ecotoxicology: The study of pollutants in ecosystems.* New York: Academic Press, 1983.

Mulholland, P.J., A.V. Palumbo, J.W. Elwood, and A.D. Rosemond. Effects of acidification on leaf decomposition in streams. *J. N. Am. Benthol. Soc.* 6:147–158 (1987).

Neal, E.C., B.C. Patten, and C.E. DePoe. Periphyton growth on artificial substrates in a radioactively contaminated lake. *Ecology* 48:918–924 (1967).

Neter, J., W. Wasserman, and M.H. Kutner. *Applied linear statistical models. Regression, analysis of variance, and experimental design.* Homeland, IL: Irwin, 1990.

Newbold, J.D., R.V. O'Neill, J.W. Elwood, and W. Van Winkle. Nutrient spiraling in streams: Implications for nutrient limitation and invertebrate activity. *Am. Nat.* 120:628–652 (1982).

Newman, M.C. *Fundamentals of ecotoxicology*. 3rd ed. Boca Raton, FL: CRC Press, 2010.

Newman, M.C., J.J. Alberts, and V.A. Greenhut. Geochemical factors complicating the use of aufwuchs to monitor accumulation of arsenic, cadmium, chromium, copper, and zinc. *Water Res.* 19:1157–1165 (1985).

Newman, M.C., and W.H. Clements. *Ecotoxicology: A comprehensive treatment.* Boca Raton, FL: CRC Press, 2008.

Newman, M.C., A.W. McIntosh, and V.A. Greenhut. Geochemical factors complicating the use of aufwuchs as a biomonitor for lead levels in two New Jersey reservoirs. *Water Res.* 17:625–630 (1983).

Newman, M.C., X. Xu, A. Condon, and L. Liang. Floodplain methylmercury biomagnifications factor higher than that of the contiguous river (South River, Virginia USA). *Environ. Pollut.* 159:2840–2844 (2011).

Niederlehner, B.R., J.R. Pratt, A.L. Buikema Jr., and J. Cairns Jr. Laboratory tests evaluating the effects of cadmium on freshwater protozoan communities. *Environ. Toxicol. Chem.* 4:155–165 (1985).

Niederlehner, B.R., J.R. Pratt, A.L. Buikema Jr., and J. Cairns Jr. Comparison of estimates of hazard derived at three levels of complexity. In *Community toxicity testing*, ed. J. Cairns Jr. ASTM STP 920. Philadelphia: American Society for Testing and Materials, 1986, pp. 30–48.

Nott, J.A., and A. Nicolaidou. Bioreduction of zinc and manganese along a molluscan food chain. *Comp. Biochem. Physiol.* 104A:235–238 (1993).

O'Connell, T.J., L.E., Jackson, and R.P. Brooks. A bird community index of biotic integrity for the mid-Atlantic highlands. *Environ. Monit. Assess.* 51:145–156 (1998).

Odum, E.P. Trends expected in stressed ecosystems. *Bioscience* 35:419–422 (1985).

Osman, R.W., and J.A. Haugsness. Mutualism among sessile invertebrates: A mediator of competition and predation. *Science* 211:846–848 (1981).

Paasivirta, J. *Chemical ecotoxicology.* Chelsea, MI: Lewis Publishers, 1991.

Pacala, S.W., M.P. Hassell, and R.M. May. Host-parasite associations in patchy environments. *Nature* 344:150–153 (1990).

Paine, R.T. Food web complexity and species diversity. *Am. Nat.* 100:65–75 (1966).

Paine, R.T., and R.L. Vadas. The effects of grazing by sea urchins, *Strongylocentrotus* spp., on benthic algal populations. *Limnol. Oceanogr.* 14:710–719 (1969).

Patrick, F.M., and M.W. Loutit. Passage of metals to freshwater fish from their food. *Water Res.* 12:395–398 (1978).

Patrick, R. Use of algae, especially diatoms, in the assessment of water quality. *ASTM Spec. Tech. Bull.* 528:76–95 (1973).

Peet, R.K. The measurement of species diversity. *Annu. Rev. Ecol. Syst.* 5:285–307 (1974).

Perry, J.A., and N.H. Troelstrup Jr. Whole ecosystem manipulation: A productive avenue for test system research? *Environ. Toxicol. Chem.* 7:941–951 (1988).

Peterson, B.J., and B. Fry. Stable isotopes in ecosystem studies. *Am. Rev. Ecol. Syst.* 18:293–320 (1987).

Phillips, D.L. Mixing models in analyses of diet using multiple stable isotopes: A critique. *Oecologia* 127:166–170 (2001).

Phillips, D.L., and J.W. Gregg. Source partitioning using stable isotopes: Coping with too many sources. *Oecologia* 136:261–269 (2003).

Phillips, D.L., and P.L. Koch. Incorporating concentration dependence in stable isotope mixing models. *Oecologia* 130:114–125 (2002).

Phillips, D.L., S.D. Newsome, and J.W. Gregg. Combining sources in stable isotope mixing models: Alternative methods. *Oecologia* 144:520–527 (2005).

Pielou, E.C. *An introduction to mathematical ecology.* New York: John Wiley & Sons, 1969.

Pielou, E.C. *Population and community ecology. Principles and methods.* New York: Gordon and Breach Science Publishers, 1974.

Pontasch, K.W., E.P. Smith, and J. Cairns Jr. Diversity indices, community comparison indices and canonical discriminant analysis: Interpreting the results of multispecies toxicity tests. *Wat. Res.* 23:1229–1238 (1989).

Poole, R.W. *An introduction to quantitative ecology.* New York: McGraw-Hill Book Company, 1974.

Post, D.M. Using stable isotopes to estimate trophic position: Models, methods, and assumptions. *Ecology* 83:703–718 (2002).

Potter, L., D. Kidd, and D. Standiford. Mercury levels in Lake Powell. Bioamplification of mercury in man-made desert reservoir. *Environ. Sci. Technol.* 9:41–46 (1975).

Pratt, J.R., and J. Cairns Jr. Long-term patterns of protozoan colonization in Douglas Lake, Michigan. *J. Protozool.* 32:95–99 (1985).

Presley, S.J., C.L. Higgins, and M.R. Willig. A comprehensive framework for the evaluation of a metacommunity structure. *Oikos* 119:908–917 (2010).

Preston, B.L. Indirect effects in aquatic ecotoxicology: Implications for ecological risk assessment. *Environ. Manage.* 29:311–323 (2002).

Preston, F.W. The commonness, and rarity, of species. *Ecology* 29:254–283 (1948).

Preston, F.W. The canonical distribution of commonness and rarity: Part 1. *Ecology* 43:185–215 (1962).

Ramade, F. *Ecotoxicology.* New York: John Wiley & Sons, 1987.

Rapport, D.J., H.A. Regier, and T.C. Hutchinson. Ecosystem behavior under stress. *Am. Nat.* 125:617–640 (1985).

Reichle, D.E., and R.I. Van Hook Jr. Radionuclide dynamics in insect food chains. *Manit. Entomol.* 4:22–32 (1970).

Relyea. R.A. Predator cues and pesticides: A double dose of danger from amphibians. *Ecol. Appl.* 13:1515–1521 (2003).

Relyea, R., and J. Hoverman. Assessing the ecology in ecotoxicology: A review and synthesis in freshwater systems. *Ecol. Lett.* 9:1157–1171 (2006).

Rohlf, F.J., and R.R. Sokal. *Statistical tables.* New York: W.H. Freeman and Company, 1981.

Rosner, B. *Fundamentals of biostatistics.* 6th ed. Belmont, CA: Thomson Higher Education, 2006.

Rounick, J.S., and M.J. Winterbourn. Stable carbon isotopes and carbon flow in ecosystems. *Bioscience* 36:171–177 (1986).

Salminen, J., B.T. Anh, and C.A.M. Van Gestel. Indirect effects of zinc on soil microbes via a keystone enchytraeid species. *Environ. Toxicol. Chem.* 20:1167–1174 (2001).

Sandheinrich, M.B., and G.J. Atchison. Sublethal toxicant effects on fish foraging behavior: Empirical vs. mechanistic approaches. *Environ. Toxicol. Chem.* 9:107–119 (1990).

Sarkar, S. Ecological diversity and biodiversity as concepts for conservation planning: Comments on Ricotta. *Acta Biotheor.* 54:133–140 (2006).

Sarokin, D., and J. Schulkin. The role of pollution in large-scale population disturbances. Part 1. Aquatic populations. *Environ. Sci. Technol.* 26:1476–1484 (1992).

Schindler, D.W., K.H. Mills, D.F. Malley, D.L. Findlay, J.A. Shearer, I.J. Davies, M.A. Turner, G.A. Linsey, and D.R. Cruikshank. Long-term ecosystem stress: The effects of years of experimental acidification on a small lake. *Science* 228:1395–1401 (1985).

Schneider, M.J., S.A. Barraclough, R.G. Genoway, and M.L. Wolford. Effects of phenol on predation of juvenile rainbow trout *Salmo gairdneri. Environ. Pollut. A Ecol. Biol.* 23:121–130 (1980).

Schoenberg, S.A., R. Benner, A. Armstrong, P. Sobecky, and R.E. Hodson. Effects of acid stress on aerobic decomposition of algal and aquatic macrophyte detritus: Direct comparison in a radiocarbon assay. *Appl. Environ. Microbiol.* 56:237–244 (1990).

Sheenan, P.J. Effect on community and ecosystem structure and dynamics. In *Effects of pollutants at the ecosystem level*, ed. P.J. Sheenan, D.R. Miller, G.C. Butler, and P. Bourdeau. New York: John Wiley & Sons, Ltd., 1984a, pp. 51–99.

Sheenan, P.J. Functional changes in the ecosystem. In *Effects of pollutants at the ecosystem level*, ed. P.J. Sheenan, D.R. Miller, G.C. Butler, and P. Bourdeau. New York: John Wiley & Sons, Ltd., 1984b.

Sih, A., A.M. Bell, and J.L. Kerby. Two stressors are far deadlier than one. *Trends Ecol. Evol.* 19:274–275 (2004).

Simberloff, D.S., and E.O. Wilson. Experimental zoogeography of islands: The colonization of empty islands. *Ecology* 50:278–296 (1969).

Simpson, E.H. Measurement of diversity. *Nature* 163:688 (1949).

Smith, E.P., and G. van Belle. Nonparametric estimation of species richness. *Biometrics* 40:119–129 (1984).

Smith, R.L. *Ecology and field biology.* New York: Harper & Row Publishers, 1980.

Smith, R.L. *Elements of ecology.* New York: Harper & Row Publishers, 1986.

Snarski, V.M. The response of rainbow trout *Salmo gairdneri* to *Aeromonas hydrophila* after sublethal exposures to PCB and copper. *Environ. Pollut.* 28:219–232 (1982).

Snoeijs, P.J.M. Monitoring pollution effects by diatom community composition. A comparison of sampling methods. *Arch. Hydrobiol.* 121:497–510 (1991).

Sokal, R.R., and F.J. Rohlf. *Biometry. The principles and practice of statistics in biological research.* New York: W.H. Freeman and Company, 1981.

Soto-Jimenez, M.F., C. Arellano, R. Rocha-Velarde, M.E. Jara-Marini, J. Ruelas-Inzunza, and F. Paez-Osuna. Trophic transfer of lead through a model marine four-level food chain: *Tetraselmis suecica*, *Artemia franciscana*, *Litopenaeus vannamei*, and *Haemulon scudderi*. *Arch. Environ. Contam. Toxicol.* 61:280–291 (2011).

Spacie, A., and J.L. Hamelink. Bioaccumulation. In *Fundamentals of aquatic toxicology*, ed. G.M. Rand and S.R. Petrocelli. New York: Hemisphere Publishing Corp., 1985, pp. 495–525.

Stephens, D.W., and J.R. Krebs. *Foraging theory.* Princeton, NJ: Princeton University Press, 1986.

Stone, L., and A. Roberts. The checkerboard score and species distributions. *Oecologia* 85:74–79 (1990).

Su, H., B. Dai, Y. Chen, and K. Li. Dynamic complexities of a predator-prey model with generalized Holling type III functional response and impulsive effects. *Comput. Math. Appl.* 56:1715–1725 (2008).

Sullivan, J.F., G.J. Atchison, D.J. Kolar, and A.W. McIntosh. Changes in the predator-prey behavior of fathead minnows (*Pimephales promelas*) and largemouth bass (*Micropterus salmoides*) caused by cadmium. *J. Fish. Res. Board Can.* 35:446–451 (1978).

Sures, B. Host-parasite interactions from an ecotoxicological perspective. *Passitologia* 49:173–176 (2007).

Sures, B. Environmental parasitology. Interactions between parasites and pollutants in the aquatic environment. *Parasite* 15:434–438 (2008).

Swanson, H.K., T.A. Johnston, D.W. Schindler, R.A. Bodaly, and D.M. Whittle. Mercury bioaccumulation in forage fish communities invaded by rainbow smelt (*Osmerus mordax*). *Environ. Sci. Technol.* 40:1439–1446 (2006).

Tagatz, M.E. Effect of mirex on predator-prey interaction in an experimental estuarine ecosystem. *Trans. Am. Fish. Soc.* 105:546–549 (1976).

Taub, F.B. Standardized aquatic microcosm. *Environ. Sci. Technol.* 23:1064–1066 (1989).

Terhaar, C.J., W.S. Ewell, S.P. Dziuba, W.W. White, and P.J. Murphy. A laboratory model for evaluating the behavior of heavy metals in an aquatic environment. *Water Res.* 11:101–110 (1977).

Thomann, R.V. Equilibrium model of fate of microcontaminants in diverse aquatic food chains. *Can. J. Fish. Aquat. Sci.* 38:280–296 (1981).

Thomann, R.V. Bioaccumulation model of organic chemical distribution in aquatic food chains. *Environ. Sci. Technol.* 23:699–707 (1989).

Thompson, R.M., M. Hemberg, B.M. Starzomski, and J.B. Shurin. Trophic level and trophic tangles: The prevalence of omnivory in real world food webs. *Ecology* 88:612–617 (2007).

Tom, K.R., M.C. Newman, and J. Schmerfeld. Modeling mercury biomagnification (South River, Virginia USA) to inform river management decision making. *Environ. Toxicol. Chem.* 29:1013–1020 (2010).

Uzarski, D.G., T.M. Burton, M.J. Cooper, J.W. Ingram, and S.T.A. Timmermans. Fish habitat use within and across wetland classes in coastal wetlands of the five Great Lakes: Development of a fish-based index of biotic integrity. *J. Great Lakes Res.* 31(Suppl.):171–187 (2005).

Vandermeer, J.H. The competitive structure of communities: An experimental approach with protozoa. *Ecology* 50:362–371 (1969).

Vandermeer, J.H. The community matrix and the number of species in a community. *Am. Nat.* 104:73–83 (1970).

Vander Zanden, M.J., and J.B. Rasmussen. A trophic position model of pelagic food webs: Impact on contaminant bioaccumulation in lake trout. *Ecol. Monogr.* 66:451–477 (1996).

Vander Zanden, M.J., and J.B. Rasmussen. Variation in $\delta\,^{15}N$ and $\delta\,^{13}C$ trophic fractionation: Implications for aquatic food web studies. *Limnol. Oceanogr.* 46:2061–2066 (2001).

Vane-Wright, R.I., C.J. Humphres, and P.H. Williams. What to protect? Systematics and the agony of choice. *Biol. Conserv.* 55:235–254 (1991).

Vannote, R.L., G.W. Minshall, K.W. Cummins, J.R. Sedell, and C.E. Cushing. The river continuum concept. *Can. J. Fish. Aquat. Sci.* 37:130–137 (1980).

Van Straalen, N.M., and J. Van Wensem. Heavy metal content of forest litter arthropods as related to body-size and trophic level. *Environ. Pollut.* 42:209–221 (1986).

Wada, Y., and N. Kashiwagi. Selecting statistical models with information statistics. *J. Dairy Sci.* 73:3575–3582 (1990).

Walker, C.H. Kinetic models for predicting bioaccumulation of pollutants in ecosystems. *Environ. Pollut.* 44:227–240 (1987).

Wang, Y.-K., R.J. Stevenson, and L. Metzmeier. Development and evaluation of a diatom-based index of biotic integrity for the Interior Plateau Ecoregion, USA. *J. N. Am. Benthol. Soc.*24:990–1008 (2005).

Warwick, R.M. A new method for detecting pollution effects on marine macrobenthic communities. *Mar. Biol.* 92:557–562 (1986).

Warwick, R.M., and K.R. Clarke. New "biodiversity" measures reveal a decrease in taxonomic distinctness with increasing stress. *Mar. Ecol.-Prog. Ser.* 129:301–305 (1995).

Warwick, R.M., and K.R. Clarke. Taxonomic distinctness and environmental assessment. *J. Appl. Ecol.* 35:532–543 (1998).

Warwick, R.M., T.H. Pearson, and Ruswahyuni. Detection of pollution effects on marine macrobenthos: Further evaluation of the species abundance/biomass method. *Mar. Biol.* 95:193–200 (1987).

Webster, J.R., and E.F. Benfield. Vascular plant breakdown in freshwater ecosystems. *Ann. Rev. Ecol. Syst.* 17:567–594 (1986).

Weiss, J.M., and S.R. Reice. The aggregation of impacts: Using species-specific effects to infer community-level disturbances. *Ecol. Appl.* 15:599–617 (2005).

West, J.B., G.J. Bowen, T.E. Dawson, and K.P. Tu. *Isoscapes. Understanding movement, pattern, and process on earth through isotope mapping*. New York: Springer Science+Business Media, 2010.

Wetzel, R.G. *Limnology.* Philadelphia: Saunders College Publishing, 1983.

Whittaker, R.H. Vegetation of the Siskiyou Mountains, Oregon and California. *Ecol. Monogr.* 30:279–338 (1960).

Wilbur, H.M. Regulation of structure in complex systems: Experimental temporary pond communities. *Ecology* 68:1437–1452 (1987).

Williamson, M.H. The ordination of incidence data. *J. Ecol.* 66:911–920 (1978).

Wolda, H. Similarity indices, sample size and diversity. *Oecologia* 50:296–302 (1981).

Woltering, D.M., J.L. Hedtke, and L.J. Weber. Predator-prey interactions of fishes under the influence of ammonia. *Trans. Am. Fish. Soc.* 107:500–504 (1978).

Woodwell, G.M. Toxic substances and ecological cycle. *Sci. Am.* 216:24–31 (1967a).

Woodwell, G.M. DDT residues in an East Coast estuary: A case of biological concentration of a persistent insecticide. *Science* 156:821–823 (1967b).

Wren, C.D., and H.R. MacCrimmon. Comparative bioaccumulation of mercury in two adjacent freshwater ecosystems. *Water Res.* 20:763–769 (1986).
Wren, C.D., H.R. MacCrimmon, and B.R. Loescher. Examination of bioaccumulation and biomagnification of metals in a Precambrian shield lake. *Water Air Soil Pollut.* 19:277–291 (1983).
Yoder, C.O. Answering some concerns about biological criteria based on experiences in Ohio. In *Water quality standards for the 21st century*. Washington, DC: U.S. EPA Criteria and Standards Division, 1989.
Zar, J.H. *Biostatistical analysis*, 4th ed. Upper Saddle River, NJ: Simon & Schuster, 1999.

CHAPTER 8

Summary

> But if I abstain from giving my judgement on any thing when I do not perceive it with sufficient clearness and distinction, it is plain that I act rightly and am not deceived. But if I determine to deny or affirm, I no longer make use as I should of my free will, ... understanding should always precede the determination of the will. And it is in the misuse of the free will that the privation which constitutes the characteristic nature of error is met with.
>
> **—Descartes (1637)**

> Science without utility is intellectual vanity.
>
> **—Johnson (1750)**

8.1 APPLICATION

How might the concepts and methods described in this book be bent to ecological effects assessment or to natural resource damage determination? Immediately upon posing this question, the too familiar dilemma exemplified by the above quotes presents itself. Socially mandated, but logically untenable, judgments must be made without complete understanding. Consequently, the best possible decisions incorporate a clear understanding of uncertainties associated with the underlying assumptions and models, and the likelihood of alternative conclusions being true.* Fortunately, considerable effort is now being expended to meet this need. Several publications provide details for addressing pollutant effects, such as Bartell et al. (1992), Bruins and Heberling (2005), EPA (1991, 1998), Landis (2004), Norton et al. (1988), Suter (1993, 2000), and Warren-Hicks et al. (1989). Several frame risk assessment in watershed and larger spatial contexts, e.g., Bruins and Hemberling (2005) and Landis (2004). Consequently, only general comment is needed about connecting the material in this book to that coalescing for ecological risk and natural resource damage assessments.

Despite many advances, most of the quantitative methods applied in these activities still assume that effect metrics for lower levels of biological organization adequately predict effects at higher levels; i.e., effects to individuals predict population consequences, and population effects predict community change (Kooijman 1987; Van Straalen and Denneman 1989; Wagner and Løkke 1991). For instance, the core premise of tiered risk assessment for pesticides (Dialogue Group Members 1994) is that effects are unlikely at higher levels if one is not detected at a lower level of biological organization. Cairns (1983) argued aggressively that there should be simultaneous testing of several levels of ecological organization. He argued that effects at lower levels, although grossly indicative of the probability of effects at higher levels (e.g., Slooff et al. 1986), lack sufficient predictive potential to

* Although remarkable progress has occurred for assessing causality and making decisions in the presence of uncertainty (e.g., Ayyub 2001; Cooke 1991; Friedman and Sandow 2011; Josephson and Josephson 1996; O'Hagan et al. 2006; Pearl 2000), the tools created in other fields are rarely drawn upon in applied ecotoxicology.

warrant their exclusive use in the first tiers of testing. The materials described in this book support such an argument. Given the abundant contrary evidence in the ecological literature, why does this issue remain unresolved in ecotoxicology 30 years after Cairns voiced his objections?*

The explanation might best be introduced using Otto Neurath's ship at sea metaphor for the conduct of science. "Neurath has likened science to a boat which, if we are to rebuild it, we must rebuild plank by plank while staying afloat in it.... Our boat stays afloat because at each alteration we keep the bulk of it intact as a going concern" (Quine 1960). Known as the Neurathian bootstrap, it seems to give context to the difficulty in our discipline that displays such slow change. Understanding can be broadened further by acknowledging that ideas flow more freely within any group than among groups. This innate group behavior fosters conceptual isolation in our field. But, according to the strength of weak ties concept (Granovetter 1973, 1978, 1983), the solution to a novel problem for any group—ecotoxicologists, for example—is most likely to come from outside that group.

An obvious instance of a Neurathian bootstrap in ecotoxicology is the species sensitivity distribution technique (e.g., Posthuma et al. 2002). The extensive compilation of LC50, EC50, NOEC, and LOEC data for effects to individuals is mined with the intent of producing a distribution of sensitivities for many species. A concentration is then found on this curve at which all but a certain percentage of species will be "protected" in some vague sense. That concentration is the basis for decisions about acceptable concentrations in natural ecosystems. The technique now enjoys widespread use despite published and well-reasoned criticisms.† A quick review of Chapters 6 and 7 will reveal more realistic metrics that might be used if available. At the population level, these metrics might include rate of population increase or probability of local extinction at the predicted exposure duration-concentration combination. Abandoning the species sensitivity distribution approach for inferring impact on communities in ecosystems in favor of several outlined in Chapter 7 would be even better. Clearly, we are forced to work with the planks available onboard while requiring some major refitting at a shipyard.

Should understanding the Neurathian bootstrap justify reluctant acceptance of the slow advancement and the confined scale of currently applied ecotoxicology? The answer depends on whether we judge the problems facing applied ecotoxicologists to be adequately addressed with the existing, albeit suboptimal, tools. Are the consequences of poor decisions about today's ecotoxicologcal problems not serious enough to argue for accelerated progress?

The trends in pollution problems through recent history can be gathered using a tool developed in the new field of culturomics. N-Gram shareware (ngrams.googlelabs.com) was just developed to extract sociological insight from the millions of books digitized by Google (Hand 2011). It searches back as far as two centuries for usage frequency of phrases such as "acid precipitation" in books written in various languages. Figure 8.1 shows the results of doing so for various English phrases (bigrams here) since 1900. The top panel shows the frequency of appearance of bigrams representing four root causes of environment contamination. Predictably, the "Industrial Revolution" bigram increased steadily and remains highly used to this day. The onset of what was dubbed the "Chemical Revolution" is clearly depicted in the middle of the twentieth century. The bigrams "Population Explosion" and "Green Revolution" entered the Anglosphere in the mid-1950s and early 1960s.

The bottom panel depicts the changes in several representative pollution-related bigrams during the twentieth century. "Sewage Pollution" rose to prominence early in the century and decreased to a steady frequency thereafter. The spatial scale of sewage pollution was that of point sources or subwatershed. Several issue bigrams increased in frequency by roughly the mid-1960s, the time of awakening to environmental issues in many countries. Mercury and PCB pollution resulting from point sources became concerns. DDT use in agricultural and residential regions came under

* Several instances exist in which higher-level effects were addressed directly, and they have been pointed out in the previous chapters. However, there are many more instances where this is not the case.

† These criticisms are summarized on pages 205–208 of Newman and Clements (2008).

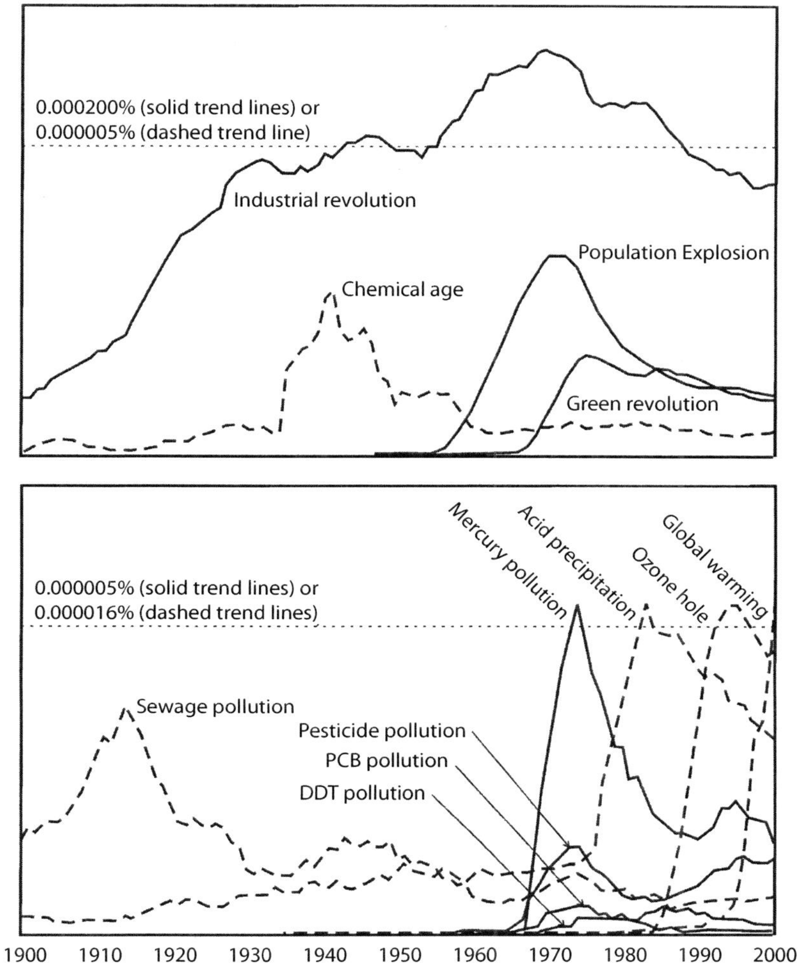

Figure 8.1 Results of N-Gram analysis of several bigrams related to pollution generating transitions in society (top panel) and types of pollution (bottom panel). The origin for both panels is 0 percentage. The horizontal dashed line on both panels reflects a frequency of occurrence in Google-scanned books. For example, the horizontal dashed line in the top panel reflects 0.000005% occurrence of the bigram in scanned books for a publication year.

scrutiny as raptor and piscivorous bird populations began to decline over wide geographical regions. "Acid Precipitation" that involved wide geographical scales also became a concern in the 1970s. Pollution issues framed in a global scale ("Ozone Hole" and "Global Warming") emerged as prominent more recently. Clearly, the combined influence of the four factors depicted in the top panel produced pollution issues of broadening spatial scale through the century. The current small spatial scale, individual effect-based approach of ecotoxicology is not expanding fast enough to address today's broad-scale ecotoxicological concerns. The unprecedented information explosion associated with the global Internet provides an opportunity to accelerate progress by the next generation of ecotoxicologists. If not taken, the ecotoxicological problems requiring solution at population and higher levels will be taken up by scientists and technologists of other disciplines. Ecotoxicologists would benefit at this time by bringing their ship for refitting into a shipyard patronized by ecologists, conservation biologists, or even atmospheric scientists.

8.2 FACILITATING GROWTH OF THE SCIENCE

The underlying intent in the preceding chapters is to detail inferential and quantitative ways for accelerating the advancement of ecotoxicology and, consequently, the effectiveness of applied ecotoxicology. A scientific context for ecotoxicology was developed and the concept of stress qualified. The commonly held belief that progress in the field is slow because of the complexity of ecological systems was rejected. Not only is this belief inconsistent with rapid progress noted in equally complex areas of study, but it imbues a hobbling passivity to our movement toward a solid knowledge base. Why strive for more rigor if populations, communities, metacommunities, and ecosystems can never be understood sufficiently? Instead, a lingering tradition of weak inference is put forward as a major, yet resolvable, impediment to progress in the field. Evidence to support this proposition is provided by John Cairns's observations that many ecotoxicological approaches are used today more as a consequence of the history of the field than their scientific merit (Cairns 1992), and that a more vigorous inferential mode is needed in the field. To this end, guidelines for the growth of a strongest possible inferential approach in the field are presented in Chapter 1. Bayesian inference is also introduced for this reason in several chapters and in the companion book *Ecotoxicology: A Comprehensive Treatment* (Newman and Clements 2008).

In Chapter 2, the necessary steps to control and define the quality of measurements are presented because inaccurate or imprecise data render results from any subsequent tests useless at best. In Chapters 3 to 7, the topics are arranged in order of increasing ecological organization with both familiar and unfamiliar quantitative methods being illustrated with numerical examples. Details for the application of these quantitative methods dominate most of this book because quantitative methods are most amenable to rigorous inference (Popper 1968). The intent of each chapter is to enhance the use of presently accepted quantitative methods and introduce useful methods from other disciplines. Particularly in the last chapters, these methods include many that have been firmly established for decades in other disciplines yet are inexplicably underexploited or deemed too esoteric by ecotoxicologists. Indeed, many of the descriptions of such techniques in this volume are unelaborated composites of those outlined in basic toxicological (e.g., survival time models in Chapter 4), epidemiological (e.g., logistic regression in Chapter 6), ecological (e.g., demographic analysis via matrices in Chapter 6 and community level metrics in Chapter 7), and statistical (e.g., power analysis in Appendix 27) methods textbooks.

REFERENCES

Ayyub, B.M. *Elicitation of expert opinions for uncertainty and risks.* Boca Raton, FL: CRC Press, 2001.

Bartell, S.M., R.H. Gardner, and R.V. O'Neill. *Ecological risk estimation.* Chelsea, MI: Lewis Publishers, 1992.

Bruins, R.J.F., and M.T. Heberling. *Economics and ecological risk assessment. Applications to watershed management.* Boca Raton, FL: CRC Press, 2005.

Cairns, J., Jr. The case of simultaneous toxicity testing at different levels of biological organization. In *Aquatic toxicology and hazard assessment: Sixth symposium,* ed. W.E. Bishop, R.D. Cardwell, and B.B. Heidolph. ASTM STP 802. Philadelphia: American Society for Testing and Materials, 1983.

Cairns, J., Jr. Paradigms flossed: The coming of age of environmental toxicology. *Environ. Toxicol. Chem.* 11:285–287 (1992).

Cooke, R.M. *Experts in uncertainty. Opinion and subjective probability in science.* New York: Oxford University Press, 1991.

Descartes, R. *A discourse on method and other works,* trans. E.S. Haldane and G.R.T. Ross. New York: Washington Square Press, 1637 (trans. 1965), p. 287.

Dialogue Group Members/Editors. *Aquatic dialogue group: Pesticide risk assessment and mitigation.* Pensacola, FL: SETAC Press, 1994.

Friedman, C., and S. Sandow. *Utility-based learning from data.* Boca Raton, FL: Chapman & Hall/CRC Press, 2011.

Granovetter, M.S. The strength of weak ties. *Am. J. Sociol.* 78:1360–1380 (1973).

Granovetter, M.S. Threshold models of collective behavior. *Am. J. Sociol.* 83:1420–1443 (1978).

Granovetter, M.S. The strength of weak ties: A network theory revisited. *Sociol. Theor.* 1:201–233 (1983).

Hand, E. Word play. *Nature* 474:436–440 (2011).

Johnson, S. *The complete works of Samuel Johnson.* Troy, NY: Pafraets Book Company, 1750; reprinted, 1903.

Josephson, J.R., and S.G. Josephson. *Abductive inference. Computation, philosophy, technology.* Cambridge: Cambridge University Press, 1996.

Kooijman, S.A.L.M. A safety factor for LC_{50} values allowing for differences in sensitivity among species. *Water Res.* 21:269–276 (1987).

Landis, W.G. *Regional scale ecological risk assessment: Using the relative risk method.* Boca Raton, FL: CRC Press, 2004.

Newman, M.C., and W.H. Clements. *Ecotoxicology: A comprehensive treatment.* Boca Raton, FL: Taylor & Francis/CRC Press, 2008.

Norton, S., M. McVey, J. Colt, J. Durda, and R. Hegner. *Review of ecological risk assessment methods.* EPA/230-10-88-041. Washington, DC: U.S. EPA, 1988.

O'Hagan, A., C.E. Buck, A. Daneshkhah, J.R. Eiser, P.H. Garthwaite, D.J. Jenkinson, J.E. Oakley, and T. Rakow. *Uncertain judgements. Eliciting experts' probabilities.* Chichester, UK: John Wiley & Sons Ltd., 2006.

Pearl, J. *Causality. Models, reasoning, and inference.* Cambridge: Cambridge University Press, 2000.

Popper, K.R. *The logic of scientific discovery.* London: Hutchinson and Company, 1968.

Posthuma, L., G.W. Suter II, and T.P. Traas, eds. *Species sensitivity distributions in ecotoxicology.* Boca Raton, FL: Lewis Publishers/CRC Press, 2002.

Quine, W. van O. *Word and object.* Cambridge, MA: MIT Press, 1960.

Slooff, W., J.A.M. van Oers, and D. De Zwart. Margins of uncertainty in ecotoxicological hazard assessment. *Environ. Toxicol. Chem.* 5:841–852 (1986).

Suter, II, G.W. *Ecological risk assessment.* Chelsea, MI: Lewis Publishers, 1993.

Suter, II, G.W. *Ecological risk assessment for contaminated sites.* Boca Raton, FL: CRC/Lewis Publishers, 2000.

U.S. EPA. *Summary report on issues in ecological risk assessment.* EPA/625/3-91/018. Washington, DC: U.S. EPA, 1991.

U.S. EPA. Guidelines for ecological risk assessment. *Federal Register* 63(93):26846–26924 (1998).

Van Straalen, N.M., and C.A.J. Denneman. Ecotoxicological evaluation of soil quality criteria. *Ecotoxicol. Environ. Saf.* 18:241–251 (1989).

Wagner, C., and H. Løkke. Estimation of ecotoxicological protection levels from NOEC toxicity data. *Water Res.* 25:1237–1242 (1991).

Warren-Hicks, W., B.R. Parkhurst, and S.S. Baker Jr. *Ecological assessments of hazardous waste sites: A field and laboratory reference document.* EPA/600/3-89/013. Washington, DC: U.S. EPA, 1989.

Appendix 1: Factors for Estimating Standard Deviation and Control Limits for Range

Number of Observations in the Subgroup	d_2	D_3	D_4
2	1.128	0	3.267
3	1.693	0	2.575
4	2.059	0	2.282
5	2.326	0	2.115
6	2.534	0	2.004
7	2.704	0.076	1.924
8	2.847	0.136	1.864
9	2.970	0.184	1.816
10	3.078	0.223	1.777
11	3.173	0.256	1.744
12	3.258	0.284	1.716
13	3.336	0.308	1.692
14	3.407	0.329	1.671
15	3.472	0.348	1.652
16	3.532	0.364	1.636
17	3.588	0.379	1.621
18	3.640	0.392	1.608
19	3.689	0.404	1.596
20	3.735	0.414	1.586

Source: The values in this table were taken from Tables C (d_2) and D (D_3 and D_4) of Grant, E.L., and R.S. Leavenworth, *Statistical Quality Control*, 7th ed., McGraw-Hill Book Company, New York, 1996. With permission from McGraw-Hill Companies (Boston).

Appendix 2: One-Sample Tolerance Probability Comparisons between n_m^* and n_m

		n_m^*				
n_m	$(1-\gamma')$	$(1-\gamma) = 0.70$	$(1-\gamma) = 0.80$	$(1-\gamma) = 0.90$	$(1-\gamma) = 0.95$	$(1-\gamma) = 0.99$
			$\alpha = 0.10$			
5	0.33	8	9	11	12	13
10	0.39	14	15	17	19	21
15	0.41	20	21	23	25	28
20	0.42	25	27	29	32	35
25	0.43	30	33	35	38	42
30	0.44	36	38	41	44	49
35	0.44	41	44	47	50	55
40	0.45	46	49	53	56	61
45	0.45	52	55	58	62	68
50	0.45	57	60	64	67	74
55	0.45	62	65	70	73	80
60	0.46	67	71	75	79	86
65	0.46	73	76	81	85	92
70	0.46	78	81	86	90	98
75	0.46	83	87	92	96	104
80	0.46	88	92	97	102	110
85	0.46	93	97	103	107	116
90	0.46	99	103	108	113	122
95	0.46	104	108	114	119	127
100	0.47	109	113	119	124	133
			$\alpha = 0.05$			
5	0.26	9	10	11	12	14
10	0.34	15	16	18	19	22
15	0.37	20	22	24	26	29
20	0.39	26	27	30	32	36
25	0.40	31	33	36	38	43
30	0.41	36	39	42	44	49
35	0.42	42	44	48	50	55
40	0.42	47	50	53	56	62
45	0.43	52	55	59	62	68
50	0.43	57	60	65	68	74
55	0.43	63	66	70	74	80
60	0.44	68	71	76	80	86
65	0.44	73	77	81	85	92
70	0.44	78	82	87	91	98
75	0.44	84	87	92	97	104
80	0.44	89	93	98	102	110
85	0.45	94	98	103	108	116
90	0.45	99	103	109	114	112
95	0.45	104	109	114	119	128
100	0.45	110	114	120	125	134

		n_m^*				
n_m	$(1-\gamma')$	$(1-\gamma) = 0.70$	$(1-\gamma) = 0.80$	$(1-\gamma) = 0.90$	$(1-\gamma) = 0.95$	$(1-\gamma) = 0.99$
			$\alpha = 0.01$			
5	0.13	10	11	12	13	15
10	0.23	16	17	19	20	23
15	0.27	21	23	25	27	30
20	0.30	27	29	31	33	37
25	0.32	32	34	37	39	44
30	0.34	38	40	43	46	50
35	0.35	43	45	49	52	57
40	0.36	48	51	55	58	63
45	0.37	53	56	60	63	69
50	0.38	59	62	66	69	75
55	0.38	64	67	72	75	82
60	0.39	69	73	77	81	88
65	0.39	74	78	83	87	94
70	0.39	80	83	88	92	100
75	0.40	85	89	94	98	106
80	0.40	90	94	99	104	112
85	0.40	95	99	105	109	117
90	0.41	101	105	110	115	123
95	0.41	106	110	116	120	129
100	0.41	111	115	121	126	135

Source: Reprinted from Kupper, L.L., and K.B. Hafner, *Am. Stat.,* 43(2), 101–105, 1989, Table 1. With permission from the *American Statistician*.

Note: Used first in Chapter 2. $(1-\gamma)$ is the tolerance probability using n_m^* and $(1-\gamma')$ is the tolerance probability using n_m.

Appendix 3: Critical Values of *T* Used to Test for Single Outliers (One-Sided Test)

Number of Observations *n*	5% Significance Level	2.5% Significance Level	1% Significance Level
3	1.153	1.154	1.154
4	1.462	1.481	1.492
5	1.671	1.715	1.749
6	1.822	1.887	1.944
7	1.938	2.020	2.097
8	2.032	2.127	2.221
9	2.110	2.215	2.323
10	2.176	2.290	2.410
11	2.234	2.355	2.484
12	2.285	2.411	2.549
13	2.330	2.462	2.607
14	2.372	2.507	2.658
15	2.409	2.550	2.705
16	2.443	2.586	2.747
17	2.475	2.620	2.785
18	2.504	2.652	2.821
19	2.531	2.681	2.853
20	2.557	2.708	2.884
21	2.580	2.734	2.912
22	2.603	2.758	2.938
23	2.624	2.780	2.963
24	2.644	2.801	2.987
25	2.662	2.822	3.009
30	2.745	2.908	3.103
35	2.812	2.978	3.178
40	2.867	3.036	3.239
45	2.915	3.085	3.292
50	2.957	3.128	3.337
60	3.027	3.200	3.411
70	3.084	3.258	3.341
80	3.132	3.306	3.521
90	3.173	3.348	3.563
100	3.209	3.384	3.600

Note: Used first in Chapter 2. See Grubbs (1969) in Chapter 2.

Grubbs test critical T values placed into this table were calculated using the equation below that was described in the online Engineering Statistics Handbook (http://www.itl.nist.gov/div898/handbook/eda/section3/eda35h1.htm). See also more details in Barnett and Lewis (1994), cited in Chapter 2. The *T* calculated from the data is compared to the result of the right side of the equations below, and the hull hypothesis is rejected if greater than the value on the right side: an outlier is detected. The *t* statistics required in these computations can be generated with the Excel function TINV(probability, df). Because the TINV() function is for a two-sided test, the α should be multiplied by 2 for a one-sided test before being placed as an argument into this function. *(Continued)*

$$T_{Calculated} > T_{Critical} = \frac{n-1}{\sqrt{n}}\sqrt{\frac{[t_{(\alpha/n,n-2)}]^2}{n-2+[t_{(\alpha/n,n-2)}]^2}}$$

Although described in this book only in the context of a one-sided test, critical values for two-sided can be estimated.

$$T_{Calculated} > T_{Critical} = \frac{n-1}{\sqrt{n}}\sqrt{\frac{[t_{(\alpha/(2n),n-2)}]^2}{n-2+[t_{(\alpha/(2n),n-2)}]^2}}$$

Appendix 4: Critical Values for λ Used to Test for Multiple Outliers ($\alpha = 0.05$)

n	$\ell + 1$	λ
25	1	2.82
	2	2.80
	3	2.78
	4	2.76
	5	2.73
	10	2.59
26	1	2.84
	2	2.82
	3	2.80
	4	2.78
	5	2.76
	10	2.62
27	1	2.86
	2	2.84
	3	2.82
	4	2.80
	5	2.78
	10	2.65
28	1	2.88
	2	2.86
	3	2.84
	4	2.82
	5	2.80
	10	2.68
29	1	2.89
	2	2.88
	3	2.86
	4	2.84
	5	2.82
	10	2.71
30	1	2.91
	2	2.89
	3	2.88
	4	2.86
	5	2.84
	10	2.73
31	1	2.92
	2	2.91
	3	2.89
	4	2.88
	5	2.86
	10	2.76

n	ℓ + 1	λ
32	1	2.94
	2	2.92
	3	2.91
	4	2.89
	5	2.88
	10	2.78
33	1	2.95
	2	2.94
	3	2.92
	4	2.91
	5	2.89
	10	2.80
34	1	2.97
	2	2.95
	3	2.94
	4	2.92
	5	2.91
	10	2.82
35	1	2.98
	2	2.97
	3	2.95
	4	2.94
	5	2.92
	10	2.84
36	1	2.99
	2	2.98
	3	2.97
	4	2.95
	5	2.94
	10	2.86
37	1	3.00
	2	2.99
	3	2.98
	4	2.97
	5	2.95
	10	2.88
38	1	3.01
	2	3.00
	3	2.99
	4	2.98
	5	2.97
	10	2.89
39	1	3.03
	2	3.01
	3	3.00
	4	2.99
	5	2.98
	10	2.91

n	$\ell + 1$	λ
40	1	3.04
	2	3.03
	3	3.01
	4	3.00
	5	2.99
	10	2.92
41	1	3.05
	2	3.04
	3	3.03
	4	3.01
	5	3.00
	10	2.94
42	1	3.06
	2	3.05
	3	3.04
	4	3.03
	5	3.01
	10	2.95
43	1	3.07
	2	3.06
	3	3.05
	4	3.04
	5	3.03
	10	2.97
44	1	3.08
	2	3.07
	3	3.06
	4	3.05
	5	3.04
	10	2.98
45	1	3.09
	2	3.08
	3	3.07
	4	3.06
	5	3.05
	10	2.99
46	1	3.09
	2	3.09
	3	3.08
	4	3.07
	5	3.06
	10	3.00
47	1	3.10
	2	3.09
	3	3.09
	4	3.08
	5	3.07
	10	3.01

n	ℓ + 1	λ
48	1	3.11
	2	3.10
	3	3.09
	4	3.09
	5	3.08
	10	3.03
49	1	3.12
	2	3.11
	3	3.10
	4	3.09
	5	3.09
	10	3.04
50	1	3.13
	2	3.12
	3	3.11
	4	3.10
	5	3.09
	10	3.05
60	1	3.20
	2	3.19
	3	3.19
	4	3.18
	5	3.17
	10	3.14
70	1	3.26
	2	3.25
	3	3.25
	4	3.24
	5	3.24
	10	3.21
80	1	3.31
	2	3.30
	3	3.30
	4	3.29
	5	3.29
	10	3.26
90	1	3.35
	2	3.34
	3	3.34
	4	3.34
	5	3.33
	10	3.31
100	1	3.38
	2	3.38
	3	3.38
	4	3.37
	5	3.37
	10	3.35

n	ℓ + 1	λ
150	1	3.52
	2	3.51
	3	3.51
	4	3.51
	5	3.51
	10	3.50
200	1	3.61
	2	3.60
	3	3.60
	4	3.60
	5	3.60
	10	3.59
250	1	3.67
	5	3.67
	10	3.66
300	1	3.72
	5	3.72
	10	3.71
350	1	3.77
	5	3.76
	10	3.76
400	1	3.80
	5	3.80
	10	3.80
450	1	3.84
	5	3.83
	10	3.83
500	1	3.86
	5	3.86
	10	3.86

Note: Used first in Chapter 2 and structured to conform to the more extensive table in Rosner (1983) (see Chapter 2 references). Produced from the following equation from Rosner (1983):

$$\lambda_{\ell+1} = \frac{t_{p,n-\ell-2}(n-\ell-1)}{\sqrt{[n-\ell-2+(t_{p,n-\ell-2})^2][n-\ell]}}$$

With p in this equation being defined as

$$p = 1-[(\alpha/2)/(n-\ell)]$$

the author found that if the Excel TINV() function is used, p should be estimated in the above equation with α instead of $\alpha/2$.

Appendix 5: Response Metameters for Proportion Affected

Proportion (no. affected/ total no.)	Arcsin $\sqrt{P}$ (radians)	Weibull	N.E.D.	Probit	Logit	Transformed Logit	Percent (%)
0.000	0.00000						0.0
0.005	0.07077	−5.29581	−2.57583	2.42417	−5.29330	2.35335	0.5
0.010	0.10017	−4.60015	−2.32635	2.67365	−4.59512	2.70244	1.0
0.015	0.12278	−4.19216	−2.17009	2.82991	−4.18459	2.90770	1.5
0.020	0.14190	−3.90194	−2.05375	2.94625	−3.89182	3.05409	2.0
0.025	0.15878	−3.67625	−1.95996	3.04004	−3.66356	3.16822	2.5
0.030	0.17408	−3.49137	−1.88079	3.11921	−3.47610	3.26195	3.0
0.035	0.18819	−3.33465	−1.81191	3.18809	−3.31678	3.34161	3.5
0.040	0.20136	−3.19853	−1.75069	3.24931	−3.17805	3.41097	4.0
0.045	0.21376	−3.07816	−1.69540	3.30460	−3.05505	3.47248	4.5
0.050	0.22551	−2.97020	−1.64485	3.35515	−2.94444	3.52778	5.0
0.055	0.23673	−2.87227	−1.59819	3.40181	−2.84385	3.57807	5.5
0.060	0.24747	−2.78263	−1.55477	3.44523	−2.75154	3.62423	6.0
0.065	0.25780	−2.69995	−1.51410	3.48590	−2.66616	3.66692	6.5
0.070	0.26776	−2.62319	−1.47579	3.52421	−2.58669	3.70666	7.0
0.075	0.27741	−2.55154	−1.43953	3.56047	−2.51231	3.74385	7.5
0.080	0.28676	−2.48433	−1.40507	3.59493	−2.44235	3.77883	8.0
0.085	0.29584	−2.42102	−1.37220	3.62780	−2.37627	3.81186	8.5
0.090	0.30469	−2.36116	−1.34076	3.65924	−2.31363	3.84318	9.0
0.095	0.31332	−2.30438	−1.31058	3.68942	−2.25406	3.87297	9.5
0.100	0.32175	−2.25037	−1.28155	3.71845	−2.19722	3.90139	10.0
0.105	0.32999	−2.19884	−1.25357	3.74643	−2.14286	3.92857	10.5
0.110	0.33807	−2.14957	−1.22653	3.77347	−2.09074	3.95463	11.0
0.115	0.34598	−2.10236	−1.20036	3.79964	−2.04066	3.97967	11.5
0.120	0.35374	−2.05703	−1.17499	3.82501	−1.99243	4.00378	12.0
0.125	0.36137	−2.01342	−1.15035	3.84965	−1.94591	4.02704	12.5
0.130	0.36886	−1.97140	−1.12639	3.87361	−1.90096	4.04952	13.0
0.135	0.37624	−1.93084	−1.10306	3.89694	−1.85745	4.07127	13.5
0.140	0.38350	−1.89165	−1.08032	3.91968	−1.81529	4.09236	14.0
0.145	0.39065	−1.85372	−1.05812	3.94188	−1.77437	4.11282	14.5
0.150	0.39770	−1.81696	−1.03643	3.96357	−1.73460	4.13270	15.0
0.155	0.40465	−1.78130	−1.01522	3.98478	−1.69591	4.15204	15.5
0.160	0.41152	−1.74667	−0.99446	4.00554	−1.65823	4.17089	16.0
0.165	0.41829	−1.71300	−0.97411	4.02589	−1.62149	4.18926	16.5
0.170	0.42499	−1.68024	−0.95417	4.04583	−1.58563	4.20719	17.0
0.175	0.43161	−1.64832	−0.93459	4.06541	−1.55060	4.22470	17.5
0.180	0.43815	−1.61721	−0.91537	4.08463	−1.51635	4.24183	18.0
0.185	0.44462	−1.58686	−0.89647	4.10353	−1.48283	4.25858	18.5
0.190	0.45103	−1.55722	−0.87790	4.12210	−1.45001	4.27499	19.0
0.195	0.45737	−1.52826	−0.85962	4.14038	−1.41784	4.29108	19.5
0.200	0.46365	−1.49994	−0.84162	4.15838	−1.38629	4.30685	20.0

Proportion (no. affected/ total no.)	Arcsin $\sqrt{P}$ (radians)	Weibull	N.E.D.	Probit	Logit	Transformed Logit	Percent (%)
0.205	0.46987	−1.47223	−0.82389	4.17611	−1.35533	4.32233	20.5
0.210	0.47603	−1.44510	−0.80642	4.19358	−1.32493	4.33754	21.0
0.215	0.48215	−1.41852	−0.78919	4.21081	−1.29505	4.35248	21.5
0.220	0.48821	−1.39247	−0.77219	4.22781	−1.26567	4.36717	22.0
0.225	0.49422	−1.36691	−0.75542	4.24458	−1.23676	4.38162	22.5
0.230	0.50018	−1.34184	−0.73885	4.26115	−1.20831	4.39584	23.0
0.235	0.50610	−1.31722	−0.72248	4.27752	−1.18029	4.40985	23.5
0.240	0.51197	−1.29303	−0.70630	4.29370	−1.15268	4.42366	24.0
0.245	0.51781	−1.26927	−0.69031	4.30969	−1.12546	4.43727	24.5
0.250	0.52360	−1.24590	−0.67449	4.32551	−1.09861	4.45069	25.0
0.255	0.52935	−1.22291	−0.65884	4.34116	−1.07212	4.46394	25.5
0.260	0.53507	−1.20030	−0.64335	4.35665	−1.04597	4.47702	26.0
0.265	0.54075	−1.17803	−0.62801	4.37199	−1.02014	4.48993	26.5
0.270	0.54640	−1.15610	−0.61281	4.38719	−0.99462	4.50269	27.0
0.275	0.55202	−1.13450	−0.59776	4.40224	−0.96940	4.51530	27.5
0.280	0.55760	−1.11321	−0.58284	4.41716	−0.94446	4.52777	28.0
0.285	0.56315	−1.09221	−0.56805	4.43195	−0.91979	4.54010	28.5
0.290	0.56868	−1.07151	−0.55338	4.44662	−0.89538	4.55231	29.0
0.295	0.57417	−1.05109	−0.53884	4.46116	−0.87122	4.56439	29.5
0.300	0.57964	−1.03093	−0.52440	4.47560	−0.84730	4.57635	30.0
0.305	0.58508	−1.01103	−0.51007	4.48993	−0.82360	4.58820	30.5
0.310	0.59050	−0.99138	−0.49585	4.50415	−0.80012	4.59994	31.0
0.315	0.59589	−0.97197	−0.48173	4.51827	−0.77685	4.61158	31.5
0.320	0.60126	−0.95279	−0.46770	4.53230	−0.75377	4.62311	32.0
0.325	0.60661	−0.93384	−0.45376	4.54624	−0.73089	4.63456	32.5
0.330	0.61194	−0.91510	−0.43991	4.56009	−0.70819	4.64591	33.0
0.335	0.61725	−0.89657	−0.42615	4.57385	−0.68566	4.65717	33.5
0.340	0.62253	−0.87824	−0.41246	4.58754	−0.66329	4.66835	34.0
0.345	0.62780	−0.86010	−0.39886	4.60114	−0.64109	4.67945	34.5
0.350	0.63305	−0.84215	−0.38532	4.61468	−0.61904	4.69048	35.0
0.355	0.63828	−0.82438	−0.37186	4.62814	−0.59713	4.70143	35.5
0.360	0.64350	−0.80679	−0.35846	4.64154	−0.57536	4.71232	36.0
0.365	0.64870	−0.78937	−0.34513	4.65487	−0.55373	4.72314	36.5
0.370	0.65389	−0.77211	−0.33185	4.66815	−0.53222	4.73389	37.0
0.375	0.65906	−0.75501	−0.31864	4.68136	−0.51083	4.74459	37.5
0.380	0.66422	−0.73807	−0.30548	4.69452	−0.48955	4.75523	38.0
0.385	0.66936	−0.72127	−0.29237	4.70763	−0.46838	4.76581	38.5
0.390	0.67449	−0.70462	−0.27932	4.72068	−0.44731	4.77634	39.0
0.395	0.67961	−0.68811	−0.26631	4.73369	−0.42634	4.78683	39.5
0.400	0.68472	−0.67173	−0.25335	4.74665	−0.40547	4.79727	40.0
0.405	0.68982	−0.65548	−0.24043	4.75957	−0.38467	4.80766	40.5
0.410	0.69490	−0.63935	−0.22754	4.77246	−0.36397	4.81802	41.0
0.415	0.69998	−0.62335	−0.21470	4.78530	−0.34333	4.82833	41.5
0.420	0.70505	−0.60747	−0.20189	4.79811	−0.32277	4.83861	42.0

Proportion (no. affected/ total no.)	Arcsin $\sqrt{P}$ (radians)	Weibull	N.E.D.	Probit	Logit	Transformed Logit	Percent (%)
0.425	0.71011	–0.59170	–0.18912	4.81088	–0.30228	4.84886	42.5
0.430	0.71517	–0.57604	–0.17637	4.82363	–0.28185	4.85907	43.0
0.435	0.72021	–0.56049	–0.16366	4.83634	–0.26148	4.86926	43.5
0.440	0.72525	–0.54504	–0.15097	4.84903	–0.24116	4.87942	44.0
0.445	0.73029	–0.52969	–0.13830	4.86170	–0.22089	4.88955	44.5
0.450	0.73531	–0.51444	–0.12566	4.87434	–0.20067	4.89966	45.0
0.455	0.74034	–0.49928	–0.11304	4.88696	–0.18049	4.90976	45.5
0.460	0.74536	–0.48421	–0.10043	4.89957	–0.16034	4.91983	46.0
0.465	0.75037	–0.46922	–0.08784	4.91216	–0.14023	4.92989	46.5
0.470	0.75538	–0.45432	–0.07527	4.92473	–0.12014	4.93993	47.0
0.475	0.76039	–0.43950	–0.06271	4.93729	–0.10008	4.94996	47.5
0.480	0.76539	–0.42476	–0.05015	4.94985	–0.08004	4.95998	48.0
0.485	0.77040	–0.41009	–0.03761	4.96239	–0.06002	4.96999	48.5
0.490	0.77540	–0.39550	–0.02507	4.97493	–0.04001	4.98000	49.0
0.495	0.78040	–0.38097	–0.01253	4.98747	–0.02000	4.99000	49.5
0.500	0.78540	–0.36651	0.00000	5.00000	0.00000	5.00000	50.0
0.505	0.79040	–0.35212	0.01253	5.01253	0.02000	5.01000	50.5
0.510	0.79540	–0.33778	0.02507	5.02507	0.04001	5.02000	51.0
0.515	0.80040	–0.32351	0.03761	5.03761	0.06002	5.03001	51.5
0.520	0.80540	–0.30929	0.05015	5.05015	0.08004	5.04002	52.0
0.525	0.81041	–0.29512	0.06271	5.06271	0.10008	5.05004	52.5
0.530	0.81542	–0.28101	0.07527	5.07527	0.12014	5.06007	53.0
0.535	0.82043	–0.26694	0.08784	5.08784	0.14023	5.07011	53.5
0.540	0.82544	–0.25292	0.10043	5.10043	0.16034	5.08017	54.0
0.545	0.83046	–0.23895	0.11304	5.11304	0.18049	5.09024	54.5
0.550	0.83548	–0.22501	0.12566	5.12566	0.20067	5.10034	55.0
0.555	0.84051	–0.21111	0.13830	5.13830	0.22089	5.11045	55.5
0.560	0.84554	–0.19726	0.15097	5.15097	0.24116	5.12058	56.0
0.565	0.85058	–0.18343	0.16366	5.16366	0.26148	5.13074	56.5
0.570	0.85563	–0.16964	0.17637	5.17637	0.28185	5.14093	57.0
0.575	0.86068	–0.15588	0.18912	5.18912	0.30228	5.15114	57.5
0.580	0.86574	–0.14214	0.20189	5.20189	0.32277	5.16139	58.0
0.585	0.87081	–0.12843	0.21470	5.21470	0.34333	5.17167	58.5
0.590	0.87589	–0.11474	0.22754	5.22754	0.36397	5.18198	59.0
0.595	0.88098	–0.10107	0.24043	5.24043	0.38467	5.19234	59.5
0.600	0.88608	–0.08742	0.25335	5.25335	0.40547	5.20273	60.0
0.605	0.89119	–0.07379	0.26631	5.26631	0.42634	5.21317	60.5
0.610	0.89631	–0.06017	0.27932	5.27932	0.44731	5.22366	61.0
0.615	0.90144	–0.04656	0.29237	5.29237	0.46838	5.23419	61.5
0.620	0.90658	0.03295	0.30548	5.30548	0.48955	5.24477	62.0
0.625	0.91174	0.01936	0.31864	5.31864	0.51083	5.25541	62.5
0.630	0.91691	0.00576	0.33185	5.33185	0.53222	5.26611	63.0
0.635	0.92209	0.00783	0.34513	5.34513	0.55373	5.27686	63.5
0.640	0.92730	0.02142	0.35846	5.35846	0.57536	5.28768	64.0

Proportion (no. affected/ total no.)	Arcsin $\sqrt{P}$ (radians)	Weibull	N.E.D.	Probit	Logit	Transformed Logit	Percent (%)
0.645	0.93251	0.03502	0.37186	5.37186	0.59713	5.29857	64.5
0.650	0.93774	0.04862	0.38532	5.38532	0.61904	5.30952	65.0
0.655	0.94299	0.06223	0.39886	5.39886	0.64109	5.32055	65.5
0.660	0.94826	0.07586	0.41246	5.41246	0.66329	5.33165	66.0
0.665	0.95355	0.08950	0.42615	5.42615	0.68566	5.34283	66.5
0.670	0.95886	0.10315	0.43991	5.43991	0.70819	5.35409	67.0
0.675	0.96418	0.11683	0.45376	5.45376	0.73089	5.36544	67.5
0.680	0.96953	0.13053	0.46770	5.46770	0.75377	5.37689	68.0
0.685	0.97490	0.14426	0.48173	5.48173	0.77685	5.38842	68.5
0.690	0.98030	0.15801	0.49585	5.49585	0.80012	5.40006	69.0
0.695	0.98571	0.17180	0.51007	5.51007	0.82360	5.41180	69.5
0.700	0.99116	0.18563	0.52440	5.52440	0.84730	5.42365	70.0
0.705	0.99663	0.19949	0.53884	5.53884	0.87122	5.43561	70.5
0.710	1.00212	0.21340	0.55338	5.55338	0.89538	5.44769	71.0
0.715	1.00764	0.22735	0.56805	5.56805	0.91979	5.45990	71.5
0.720	1.01320	0.24135	0.58284	5.58284	0.94446	5.47223	72.0
0.725	1.01878	0.25540	0.59776	5.59776	0.96940	5.48470	72.5
0.730	1.02440	0.26952	0.61281	5.61281	0.99462	5.49731	73.0
0.735	1.03004	0.28369	0.62801	5.62801	1.02014	5.51007	73.5
0.740	1.03573	0.29793	0.64335	5.64335	1.04597	5.52298	74.0
0.745	1.04144	0.31225	0.65884	5.65884	1.07212	5.53606	74.5
0.750	1.04720	0.32663	0.67449	5.67449	1.09861	5.54931	75.0
0.755	1.05299	0.34110	0.69031	5.69031	1.12546	5.56273	75.5
0.760	1.05882	0.35566	0.70630	5.70630	1.15268	5.57634	76.0
0.765	1.06470	0.37030	0.72248	5.72248	1.18029	5.59015	76.5
0.770	1.07062	0.38504	0.73885	5.73885	1.20831	5.60416	77.0
0.775	1.07658	0.39989	0.75542	5.75542	1.23676	5.61838	77.5
0.780	1.08259	0.41484	0.77219	5.77219	1.26567	5.63283	78.0
0.785	1.08865	0.42991	0.78919	5.78919	1.29505	5.64752	78.5
0.790	1.09476	0.44510	0.80642	5.80642	1.32493	5.66246	79.0
0.795	1.10093	0.46042	0.82389	5.82389	1.35533	5.67767	79.5
0.800	1.10715	0.47588	0.84162	5.84162	1.38629	5.69315	80.0
0.805	1.11343	0.49149	0.85962	5.85962	1.41784	5.70892	80.5
0.810	1.11977	0.50726	0.87790	5.87790	1.45001	5.72501	81.0
0.815	1.12617	0.52319	0.89647	5.89647	1.48283	5.74142	81.5
0.820	1.13265	0.53930	0.91537	5.91537	1.51635	5.75817	82.0
0.825	1.13919	0.55559	0.93459	5.93459	1.55060	5.77530	82.5
0.830	1.14581	0.57208	0.95417	5.95417	1.58563	5.79281	83.0
0.835	1.15250	0.58879	0.97411	5.97411	1.62149	5.81074	83.5
0.840	1.15928	0.60573	0.99446	5.99446	1.65823	5.82911	84.0
0.845	1.16614	0.62290	1.01522	6.01522	1.69591	5.84796	84.5
0.850	1.17310	0.64034	1.03643	6.03643	1.73460	5.86730	85.0
0.855	1.18015	0.65805	1.05812	6.05812	1.77437	5.88718	85.5
0.860	1.18730	0.67606	1.08032	6.08032	1.81529	5.90764	86.0

Proportion (no. affected/ total no.)	Arcsin $\sqrt{P}$ (radians)	Weibull	N.E.D.	Probit	Logit	Transformed Logit	Percent (%)
0.865	1.19456	0.69439	1.10306	6.10306	1.85745	5.92873	86.5
0.870	1.20193	0.71306	1.12639	6.12639	1.90096	5.95048	87.0
0.875	1.20943	0.73210	1.15035	6.15035	1.94591	5.97296	87.5
0.880	1.21705	0.75154	1.17499	6.17499	1.99243	5.99622	88.0
0.885	1.22482	0.77141	1.20036	6.20036	2.04066	6.02033	88.5
0.890	1.23273	0.79176	1.22653	6.22653	2.09074	6.04537	89.0
0.895	1.24080	0.81262	1.25357	6.25357	2.14286	6.07143	89.5
0.900	1.24905	0.83403	1.28155	6.28155	2.19722	6.09861	90.0
0.905	1.25747	0.85606	1.31058	6.31058	2.25406	6.12703	90.5
0.910	1.26610	0.87877	1.34076	6.34076	2.31363	6.15682	91.0
0.915	1.27495	0.90223	1.37220	6.37220	2.37627	6.18814	91.5
0.920	1.28404	0.92653	1.40507	6.40507	2.44235	6.22117	92.0
0.925	1.29339	0.95176	1.43953	6.43953	2.51231	6.25615	92.5
0.930	1.30303	0.97805	1.47579	6.47579	2.58669	6.29334	93.0
0.935	1.31300	1.00553	1.51410	6.51410	2.66616	6.33308	93.5
0.940	1.32333	1.03440	1.55477	6.55477	2.75154	6.37577	94.0
0.945	1.33407	1.06486	1.59819	6.59819	2.84385	6.42193	94.5
0.950	1.34528	1.09719	1.64485	6.64485	2.94444	6.47222	95.0
0.955	1.35704	1.13175	1.69540	6.69540	3.05505	6.52752	95.5
0.960	1.36944	1.16903	1.75069	6.75069	3.17805	6.58903	96.0
0.965	1.38260	1.20968	1.81191	6.81191	3.31678	6.65839	96.5
0.970	1.39671	1.25463	1.88079	6.88079	3.47610	6.73805	97.0
0.975	1.41202	1.30532	1.95996	6.95996	3.66356	6.83178	97.5
0.980	1.42890	1.36405	2.05375	7.05375	3.89182	6.94591	98.0
0.985	1.44801	1.43501	2.17009	7.17009	4.18459	7.09230	98.5
0.990	1.47063	1.52718	2.32635	7.32635	4.59512	7.29756	99.0
0.995	1.50003	1.66739	2.57583	7.57583	5.29330	7.64665	99.5
1.000	1.57080						100.0

Note: Used first in Chapter 4. The tabulated values can be generated with most spreadsheet software. The following Excel® functions produce the metameters: Arcsin $\sqrt{P}$ = ASIN(SQRT(*P*)), Weibull = *LN*(–*LN*(*P*)), N.E.D. = NORMINV(*P*,0,1) or NORMSINV(*P*), Probit = NORMINV(*P*,0,1) + 5, Logit = *LN*(*P*/(1 – *P*)), and Transformed Logit = *LN*(*P*/(1 – *P*)) + 5.

Appendix 6: Maximum Likelihood Values for Dixon's Up-and-Down Method

***k* Values for Series Beginning with Initial Survivals**

N	2nd Part of Series	Series That Begin With →	1st Part of Series O	OO	OOO	OOOO
2	X		−0.500	−0.388	−0.378	−0.377
3	XO		0.842	0.890	0.894	0.894
3	XX		−0.178	0.000	0.026	0.028
4	XOO		0.299	0.314	0.315	0.315
4	XOX		−0.500	−0.439	−0.432	−0.432
4	XXO		1.000	1.122	1.139	1.140
4	XXX		0.194	0.449	0.500	0.506
5	XOOO		−0.157	−0.154	−0.154	−0.154
5	XOOX		−0.878	−0.861	−0.860	−0.860
5	XOXO		0.701	0.737	0.741	0.741
5	XOXX		0.084	0.169	0.181	0.182
5	XXOO		0.305	0.372	0.380	0.381
5	XXOX		−0.305	−0.169	−0.144	−0.142
5	XXXO		1.288	1.500	1.544	1.549
5	XXXX		0.555	0.897	0.985	**1.000**
6	XOOOO		−0.547	−0.547	−0.547	−0.547
6	XOOOX		−1.250	−1.247	−1.246	−1.246
6	XOOXO		0.372	0.380	0.381	0.381
6	XOOXX		−0.169	−0.144	−0.142	−0.142
6	XOXOO		0.022	0.039	0.040	0.040
6	XOXOX		−0.500	−0.458	−0.453	−0.453
6	XOXXO		1.169	1.237	1.247	1.248
6	XOXXX		0.611	0.732	0.756	0.758
6	XXOOO		−0.296	−0.266	−0.263	−0.263
6	XXOOX		−0.831	−0.763	−0.753	−0.752
6	XXOXO		0.831	0.935	0.952	0.954
6	XXOXX		0.296	0.463	0.500	**0.504**
6	XXXOO		0.500	0.648	0.678	0.681
6	XXXOX		−0.043	0.187	0.244	**0.252**
6	XXXXO		1.603	1.917	2.000	**2.014**
6	XXXXX		0.893	1.329	1.465	**1.496**

***k* Values for Series Beginning with Initial Deaths**

			1st Part of Series			
N	**2nd Part of Series**	**Series That Begin With →**	**X**	**XX**	**XXX**	**XXXX**
2	O		0.500	0.388	0.378	0.377
3	OX		–0.842	–0.890	–0.894	–0.894
3	OO		0.178	0.000	–0.026	–0.028
4	OXX		–0.299	–0.314	–0.315	–0.315
4	OXO		0.500	0.439	0.432	0.432
4	OOX		–1.000	–1.122	–1.139	–1.140
4	OOO		–0.194	–0.449	–0.500	–0.506
5	OXXX		0.157	0.154	0.154	0.154
5	OXXO		0.878	0.861	0.860	0.860
5	OXOX		–0.701	–0.737	–0.741	–0.741
5	OXOO		–0.084	–0.169	–0.181	–0.182
5	OOXX		–0.305	–0.372	–0.380	–0.381
5	OOXO		0.305	0.169	0.144	0.142
5	OOOX		–1.288	–1.500	–1.544	–1.549
5	OOOO		–0.555	–0.897	–0.985	**–1.000**
6	OXXXX		0.547	0.547	0.547	0.547
6	OXXXO		1.250	1.247	1.246	1.246
6	OXXOX		–0.372	–0.380	–0.381	–0.381
6	OXXOO		0.169	0.144	0.142	0.142
6	OXOXX		–0.022	–0.039	–0.040	–0.040
6	OXOXO		0.500	0.458	0.453	0.453
6	OXOOX		–1.169	–1.237	–1.247	–1.248
6	OXOOO		–0.611	–0.732	–0.756	–0.758
6	OOXXX		0.296	0.266	0.263	0.263
6	OOXXO		0.831	0.763	0.753	0.752
6	OOXOX		–0.831	–0.935	–0.952	–0.954
6	OOXOO		–0.296	–0.463	–0.500	**–0.504**
6	OOOXX		–0.500	–0.648	–0.678	–0.681
6	OOOXO		0.043	–0.187	–0.244	**–0.252**
6	OOOOX		–1.603	–1.917	–2.000	**–2.014**
6	OOOOO		–0.893	–1.329	–1.465	**–1.496**

Note: Derived from Table 1 in Dixon (1965) and Table 19.3 in Dixon and Massey (1969) with the omission of the standard error column (which are tabulated in Chapter 4), the separation of series starting with survivals (first table) and deaths (second table), and other reformattings. (First used in Chapter 4.)

The five bolded entries in both tables carry an explanation in the original publications that is helpful if more than four like responses make up the first part of a series. Dixon (1965) states, "[The tables list] all solutions for all N' and for $N \leq 6$. If the series begins with more than four like responses, i.e., $N' - N > 3$, the entry in the final column [of the tables] may be used (except for the five [**bolded**] tabular entries where an additional increment in the third decimal place is indicated)." This comment appears to suggest that the values have 0.001 ("survival") or –0.001 ("dead") added if they are being used for series beginning with more than four like responses. For example, *k* for OOOOOXXXX would be 1.001, and that for XXXXXOOOO would be –1.001.

Appendix 7: E Values Used to Estimate 95% Confidence Intervals for LT50 with the Litchfield Method

x'	E
–2.5	1.00056
–2.4	1.00078
–2.3	1.00107
–2.2	1.00147
–2.1	1.00200
–2.0	1.00270
–1.9	1.00363
–1.8	1.00485
–1.7	1.00645
–1.6	1.00852
–1.5	1.01120
–1.4	1.01467
–1.3	1.01914
–1.2	1.02488
–1.1	1.03224
–1.0	1.04168
–0.9	1.05376
–0.8	1.06923
–0.7	1.08904
–0.6	1.11442
–0.5	1.14696
–0.4	1.18876
–0.3	1.24252
–0.2	1.31180
–0.1	1.40127
0.0	1.51709
0.1	1.66743
0.2	1.86310
0.3	2.11857
0.4	2.45318
0.5	2.89293
0.6	3.47293
0.7	4.24075
0.8	5.26121
0.9	6.62291
1.0	8.44766

x'	E
1.1	10.90365
1.2	14.22420
1.3	18.73495
1.4	24.89192
1.5	33.33860
1.6	44.98586
1.7	61.13204
1.8	83.63826
1.9	115.18668
2.0	159.66335

Source: Adapted from Bliss, C.I., *Ann. Appl. Biol.,* 24, 815–852, 1936, Table III. With permission from the Company of Biologists Ltd. (Cambridge, UK).

Note: Used first in Chapter 4.

Appendix 8: Coefficients (a_{n-i+1}) for Shapiro–Wilk's Test for Normality

$i \backslash n$	2	3	4	5	6	7	8	9	10
1	0.7071	0.7071	0.6872	0.6646	0.6431	0.6233	0.6052	0.5888	0.5739
2		0.0000	0.1677	0.2413	0.2806	0.3031	0.3164	0.3244	0.3291
3				0.0000	0.0875	0.1401	0.1743	0.1976	0.2141
4						0.0000	0.0561	0.0947	0.1224
5								0.0000	0.0399

$i \backslash n$	11	12	13	14	15	16	17	18	19	20
1	0.5601	0.5475	0.5359	0.5251	0.5150	0.5056	0.4968	0.4886	0.4808	0.4734
2	0.3315	0.3325	0.3325	0.3318	0.3306	0.3290	0.3273	0.3253	0.3232	0.3211
3	0.2260	0.2347	0.2412	0.2460	0.2495	0.2521	0.2540	0.2553	0.2561	0.2565
4	0.1429	0.1586	0.1707	0.1802	0.1878	0.1939	0.1988	0.2027	0.2059	0.2085
5	0.0695	0.0922	0.1099	0.1240	0.1353	0.1447	0.1524	0.1587	0.1641	0.1686
6	0.0000	0.0303	0.0539	0.0727	0.0880	0.1005	0.1109	0.1197	0.1271	0.1334
7			0.0000	0.0240	0.0433	0.0593	0.0725	0.0837	0.0932	0.1013
8					0.0000	0.0196	0.0359	0.0496	0.0612	0.0711
9							0.0000	0.0163	0.0303	0.0422
10									0.0000	0.0140

$i \backslash n$	21	22	23	24	25	26	27	28	29	30
1	0.4643	0.4590	0.4542	0.4493	0.4450	0.4407	0.4366	0.4328	0.4291	0.4254
2	0.3185	0.3156	0.3126	0.3098	0.3069	0.3043	0.3018	0.2992	0.2968	0.2944
3	0.2578	0.2571	0.2563	0.2554	0.2543	0.2533	0.2522	0.2510	0.2499	0.2487
4	0.2119	0.2131	0.2139	0.2145	0.2148	0.2151	0.2152	0.2151	0.2150	0.2148
5	0.1736	0.1764	0.1787	0.1807	0.1822	0.1836	0.1848	0.1857	0.1864	0.1870
6	0.1399	0.1443	0.1480	0.1512	0.1539	0.1563	0.1584	0.1601	0.1616	0.1630
7	0.1092	0.1150	0.1201	0.1245	0.1283	0.1316	0.1346	0.1372	0.1395	0.1415
8	0.0804	0.0878	0.0941	0.0997	0.1046	0.1089	0.1128	0.1162	0.1192	0.1219
9	0.0530	0.0618	0.0696	0.0764	0.0823	0.0876	0.0923	0.0965	0.1002	0.1036
10	0.0263	0.0368	0.0459	0.0539	0.0610	0.0672	0.0728	0.0778	0.0822	0.0862
11	0.0000	0.0122	0.0228	0.0321	0.0403	0.0476	0.0540	0.0598	0.0650	0.0697
12			0.0000	0.0107	0.0200	0.0284	0.0358	0.0424	0.0483	0.0537
13					0.0000	0.0094	0.0178	0.0253	0.0320	0.0381
14							0.0000	0.0084	0.0159	0.0227
15									0.0000	0.0076

$i \backslash n$	31	32	33	34	35	36	37	38	39	40
1	0.4220	0.4188	0.4156	0.4127	0.4096	0.4068	0.4040	0.4015	0.3989	0.3964
2	0.2921	0.2898	0.2876	0.2854	0.2834	0.2813	0.2794	0.2774	0.2755	0.2737
3	0.2475	0.2463	0.2451	0.2439	0.2427	0.2415	0.2403	0.2391	0.2380	0.2368
4	0.2145	0.2141	0.2137	0.2132	0.2127	0.2121	0.2116	0.2110	0.2104	0.2098
5	0.1874	0.1878	0.1880	0.1882	0.1883	0.1883	0.1883	0.1881	0.1880	0.1878
6	0.1641	0.1651	0.1660	0.1667	0.1673	0.1678	0.1683	0.1686	0.1689	0.1691
7	0.1433	0.1449	0.1463	0.1475	0.1487	0.1496	0.1505	0.1513	0.1520	0.1526
8	0.1243	0.1265	0.1284	0.1301	0.1317	0.1331	0.1344	0.1356	0.1366	0.1376
9	0.1066	0.1093	0.1118	0.1140	0.1160	0.1179	0.1196	0.1211	0.1225	0.1237
10	0.0899	0.0931	0.0961	0.0988	0.1013	0.1036	0.1056	0.1075	0.1092	0.1108
11	0.0739	0.0777	0.0812	0.0844	0.0873	0.0900	0.0924	0.0947	0.0967	0.0986
12	0.0585	0.0629	0.0669	0.0706	0.0739	0.0770	0.0798	0.0824	0.0848	0.0870
13	0.0435	0.0485	0.0530	0.0572	0.0610	0.0645	0.0677	0.0706	0.0733	0.0759
14	0.0289	0.0344	0.0395	0.0441	0.0484	0.0523	0.0559	0.0592	0.0622	0.0651
15	0.0144	0.0206	0.0262	0.0314	0.0361	0.0404	0.0444	0.0481	0.0515	0.0546

$i \backslash n$	31	32	33	34	35	36	37	38	39	40
16	0.0000	0.0068	0.0131	0.0187	0.0239	0.0287	0.0331	0.0372	0.0409	0.0444
17			0.0000	0.0062	0.0119	0.0172	0.0220	0.0264	0.0305	0.0343
18					0.0000	0.0057	0.0110	0.0158	0.0203	0.0244
19							0.0000	0.0053	0.0101	0.0146
20									0.0000	0.0049

$i \backslash n$	41	42	43	44	45	46	47	48	49	50
1	0.3940	0.3917	0.3894	0.3872	0.3850	0.3830	0.3808	0.3789	0.3770	0.3751
2	0.2719	0.2701	0.2684	0.2667	0.2651	0.2635	0.2620	0.2604	0.2589	0.2574
3	0.2357	0.2345	0.2334	0.2323	0.2313	0.2302	0.2291	0.2281	0.2271	0.2260
4	0.2091	0.2085	0.2078	0.2072	0.2065	0.2058	0.2052	0.2045	0.2038	0.2032
5	0.1876	0.1874	0.1871	0.1868	0.1865	0.1862	0.1859	0.1855	0.1851	0.1847
6	0.1693	0.1694	0.1695	0.1695	0.1695	0.1695	0.1695	0.1693	0.1692	0.1691
7	0.1531	0.1535	0.1539	0.1542	0.1545	0.1548	0.1550	0.1551	0.1553	0.1554
8	0.1384	0.1392	0.1398	0.1405	0.1410	0.1415	0.1420	0.1423	0.1427	0.1430
9	0.1249	0.1259	0.1269	0.1278	0.1286	0.1293	0.1300	0.1306	0.1312	0.1317
10	0.1123	0.1136	0.1149	0.1160	0.1170	0.1180	0.1189	0.1197	0.1205	0.1212
11	0.1004	0.1020	0.1035	0.1049	0.1062	0.1073	0.1085	0.1095	0.1105	0.1113
12	0.0891	0.0909	0.0927	0.0943	0.0959	0.0972	0.0986	0.0998	0.1010	0.1020
13	0.0782	0.0804	0.0824	0.0842	0.0860	0.0876	0.0892	0.0906	0.0919	0.0932
14	0.0677	0.0701	0.0724	0.0745	0.0765	0.0783	0.0801	0.0817	0.0832	0.0846
15	0.0575	0.0602	0.0628	0.0651	0.0673	0.0694	0.0713	0.0731	0.0748	0.0764

$i \backslash n$	41	42	43	44	45	46	47	48	49	50
16	0.0476	0.0506	0.0534	0.0560	0.0584	0.0607	0.0628	0.0648	0.0667	0.0685
17	0.0379	0.0411	0.0442	0.0471	0.0497	0.0522	0.0546	0.0568	0.0588	0.0698
18	0.0283	0.0318	0.0352	0.0383	0.0412	0.0439	0.0465	0.0489	0.0511	0.0532
19	0.0188	0.0227	0.0263	0.0296	0.0328	0.0357	0.0385	0.0411	0.0436	0.0459
20	0.0094	0.0136	0.0175	0.0211	0.0245	0.0277	0.0307	0.0335	0.0361	0.0386
21	0.0000	0.0045	0.0087	0.0126	0.0163	0.0197	0.0229	0.0259	0.0288	0.0314
22			0.0000	0.0042	0.0081	0.0118	0.0153	0.0185	0.0215	0.0244
23					0.0000	0.0039	0.0076	0.0111	0.0143	0.0174
24							0.0000	0.0037	0.0071	0.0104
25									0.0000	0.0035

Source: Reproduced from Shapiro, S.S., and Wilk, M.B., *Biometrika,* 52, 591–611, 1965. With permission from the Biometrika Trust (London).

Note: Used first in Chapter 5.

Appendix 9: Percentage Points of Shapiro–Wilk's *W* Test for Normality

$n \setminus \alpha$	0.01	0.02	0.05	0.10	0.50	0.90	0.95	0.98	0.99
3	0.753	0.756	0.767	0.789	0.959	0.998	0.999	1.000	1.000
4	0.687	0.707	0.748	0.792	0.935	0.987	0.992	0.996	0.997
5	0.686	0.715	0.762	0.806	0.927	0.979	0.986	0.991	0.993
6	0.713	0.743	0.788	0.826	0.927	0.974	0.981	0.986	0.989
7	0.730	0.760	0.803	0.838	0.928	0.972	0.979	0.985	0.988
8	0.749	0.778	0.818	0.851	0.932	0.972	0.978	0.984	0.987
9	0.764	0.791	0.829	0.859	0.935	0.972	0.978	0.984	0.986
10	0.781	0.806	0.842	0.869	0.938	0.972	0.978	0.983	0.986
11	0.792	0.817	0.850	0.876	0.940	0.973	0.979	0.984	0.986
12	0.805	0.828	0.859	0.883	0.943	0.973	0.979	0.984	0.986
13	0.814	0.837	0.866	0.889	0.945	0.974	0.979	0.984	0.986
14	0.825	0.846	0.874	0.895	0.947	0.975	0.980	0.984	0.986
15	0.835	0.855	0.881	0.901	0.950	0.975	0.980	0.984	0.987
16	0.844	0.863	0.887	0.906	0.952	0.976	0.981	0.985	0.987
17	0.851	0.869	0.892	0.910	0.954	0.977	0.981	0.985	0.987
18	0.858	0.874	0.897	0.914	0.956	0.978	0.982	0.986	0.988
19	0.863	0.879	0.901	0.917	0.957	0.978	0.982	0.986	0.988
20	0.868	0.884	0.905	0.920	0.959	0.979	0.983	0.986	0.988
21	0.873	0.888	0.908	0.923	0.960	0.980	0.983	0.987	0.989
22	0.878	0.892	0.911	0.926	0.961	0.980	0.984	0.987	0.989
23	0.881	0.895	0.914	0.928	0.962	0.981	0.984	0.987	0.989
24	0.884	0.898	0.916	0.930	0.963	0.981	0.984	0.987	0.989
25	0.888	0.901	0.918	0.931	0.964	0.981	0.985	0.988	0.989
26	0.891	0.904	0.920	0.933	0.965	0.982	0.985	0.988	0.989
27	0.894	0.906	0.923	0.935	0.965	0.982	0.985	0.988	0.990
28	0.896	0.908	0.924	0.936	0.966	0.982	0.985	0.988	0.990
29	0.898	0.910	0.926	0.937	0.966	0.982	0.985	0.988	0.990
30	0.900	0.912	0.927	0.939	0.967	0.983	0.985	0.988	0.990
31	0.902	0.914	0.929	0.940	0.967	0.983	0.986	0.988	0.990
32	0.904	0.915	0.930	0.941	0.968	0.983	0.986	0.988	0.990
33	0.906	0.917	0.931	0.942	0.968	0.983	0.986	0.989	0.990
34	0.908	0.919	0.933	0.943	0.969	0.983	0.986	0.989	0.990
35	0.910	0.920	0.934	0.944	0.969	0.984	0.986	0.989	0.990
36	0.912	0.922	0.935	0.945	0.970	0.984	0.986	0.989	0.990
37	0.914	0.924	0.936	0.946	0.970	0.984	0.987	0.989	0.990
38	0.916	0.925	0.938	0.947	0.971	0.984	0.987	0.989	0.990
39	0.917	0.927	0.939	0.948	0.971	0.984	0.987	0.989	0.991
40	0.919	0.928	0.940	0.949	0.972	0.985	0.987	0.989	0.991
41	0.920	0.929	0.941	0.950	0.972	0.985	0.987	0.989	0.991
42	0.922	0.930	0.942	0.951	0.972	0.985	0.987	0.989	0.991
43	0.923	0.932	0.943	0.951	0.973	0.985	0.987	0.990	0.991
44	0.924	0.933	0.944	0.952	0.973	0.985	0.987	0.990	0.991
45	0.926	0.934	0.945	0.953	0.973	0.985	0.988	0.990	0.991
46	0.927	0.935	0.945	0.953	0.974	0.985	0.988	0.990	0.991
47	0.928	0.936	0.946	0.954	0.974	0.985	0.988	0.990	0.991
48	0.929	0.937	0.947	0.954	0.974	0.985	0.988	0.990	0.991
49	0.929	0.937	0.947	0.955	0.974	0.985	0.988	0.990	0.991
50	0.930	0.938	0.947	0.955	0.974	0.985	0.988	0.990	0.991

Source: Reproduced from Shapiro, S.S., and M.B. Wilk, *Biometrika,* 52, 591–611, 1965, Table 6. With permission from Biometrika Trust (London).

Note: Used first in Chapter 5.

Appendix 10: Dunnett's t for One-Sided Comparisons between p Treatment Means and a Control for $\alpha = 0.05$

$df \setminus p$	1	2	3	4	5	6	7	8	9
5	2.02	2.44	2.68	2.85	2.98	3.08	3.16	3.24	3.30
6	1.94	2.34	2.56	2.71	2.83	2.92	3.00	3.07	3.12
7	1.89	2.27	2.48	2.62	2.73	2.82	2.89	2.95	3.01
8	1.86	2.22	2.42	2.55	2.66	2.74	2.81	2.87	2.92
9	1.83	2.18	2.37	2.50	2.60	2.68	2.75	2.81	2.86
10	1.81	2.15	2.34	2.47	2.56	2.64	2.70	2.76	2.81
11	1.80	2.13	2.31	2.44	2.53	2.60	2.67	2.72	2.77
12	1.78	2.11	2.29	2.41	2.50	2.58	2.64	2.69	2.74
13	1.77	2.09	2.27	2.39	2.48	2.55	2.61	2.66	2.71
14	1.76	2.08	2.25	2.37	2.46	2.53	2.59	2.64	2.69
15	1.75	2.07	2.24	2.36	2.44	2.51	2.57	2.62	2.67
16	1.75	2.06	2.23	2.34	2.43	2.50	2.56	2.61	2.65
17	1.74	2.05	2.22	2.33	2.42	2.49	2.54	2.59	2.64
18	1.73	2.04	2.21	2.32	2.41	2.48	2.53	2.58	2.62
19	1.73	2.03	2.20	2.31	2.40	2.47	2.52	2.57	2.61
20	1.72	2.03	2.19	2.30	2.39	2.46	2.51	2.56	2.60
24	1.71	2.01	2.17	2.28	2.36	2.43	2.48	2.53	2.57
30	1.70	1.99	2.15	2.25	2.33	2.40	2.45	2.50	2.54
40	1.68	1.97	2.13	2.23	2.31	2.37	2.42	2.47	2.51
60	1.67	1.95	2.10	2.21	2.28	2.35	2.39	2.44	2.48
120	1.66	1.93	2.08	2.18	2.26	2.32	2.37	2.41	2.45
∞	1.64	1.92	2.06	2.16	2.23	2.29	2.34	2.38	2.42

Source: Reprinted from Dunnett, C.W., *J. Am. Stat. Assoc.*, 50, 1096–1121, 1955, Table 1a. With permission from the *Journal of the American Statistical Association*.

Note: Used first in Chapter 5.

Appendix 11: Dunnett's *t* for Two-Sided Comparisons between *p* Treatment Means and a Control for $\alpha = 0.05$

df \ *p*	1	2	3	4	5	6	7	8	9	10	11	12	15	20
5	2.57	$3.03^{2.3}$	$3.29^{3.6}$	$3.48^{4.6}$	$3.62^{5.4}$	$3.73^{5.9}$	$3.82^{6.4}$	$3.90^{6.8}$	$3.97^{7.2}$	$4.03^{7.5}$	$4.09^{7.8}$	$4.14^{8.0}$	$4.26^{8.7}$	$4.42^{9.4}$
6	2.45	$2.86^{2.1}$	$3.10^{3.4}$	$3.26^{4.3}$	$3.39^{5.0}$	$3.49^{5.6}$	$3.57^{6.0}$	$3.64^{6.4}$	$3.71^{6.8}$	$3.76^{7.1}$	$3.81^{7.4}$	$3.86^{7.6}$	$3.97^{8.2}$	$4.11^{9.0}$
7	2.36	$2.75^{2.0}$	$2.97^{3.2}$	$3.12^{4.1}$	$3.24^{4.8}$	$3.33^{5.3}$	$3.41^{5.7}$	$3.47^{6.1}$	$3.53^{6.5}$	$3.58^{6.7}$	$3.63^{7.0}$	$3.67^{7.2}$	$3.78^{7.8}$	$3.91^{8.6}$
8	2.31	$2.67^{2.0}$	$2.88^{3.1}$	$3.02^{3.9}$	$3.13^{4.5}$	$3.22^{5.1}$	$3.29^{5.5}$	$3.35^{5.9}$	$3.41^{6.2}$	$3.46^{6.5}$	$3.50^{6.7}$	$3.54^{6.9}$	$3.64^{7.5}$	$3.76^{8.2}$
9	2.26	$2.61^{1.9}$	$2.81^{3.0}$	$2.95^{3.8}$	$3.05^{4.4}$	$3.14^{4.9}$	$3.20^{5.3}$	$3.26^{5.6}$	$3.32^{5.9}$	$3.36^{6.2}$	$3.40^{6.5}$	$3.44^{6.7}$	$3.53^{7.2}$	$3.65^{7.9}$
10	2.23	$2.57^{1.8}$	$2.76^{2.9}$	$2.89^{3.6}$	$2.99^{4.2}$	$3.07^{4.7}$	$3.14^{5.1}$	$3.19^{5.4}$	$3.24^{5.7}$	$3.29^{6.0}$	$3.33^{6.2}$	$3.36^{6.5}$	$3.45^{7.0}$	$3.57^{7.7}$
11	2.20	$2.53^{1.8}$	$2.72^{2.8}$	$2.84^{3.5}$	$2.94^{4.1}$	$3.02^{4.6}$	$3.08^{4.9}$	$3.14^{5.3}$	$3.19^{5.6}$	$3.23^{5.8}$	$3.27^{6.1}$	$3.30^{6.3}$	$3.39^{6.8}$	$3.50^{7.5}$
12	2.18	$2.50^{1.7}$	$2.68^{2.7}$	$2.81^{3.4}$	$2.90^{4.0}$	$2.98^{4.4}$	$3.04^{4.8}$	$3.09^{5.1}$	$3.14^{5.4}$	$3.18^{5.7}$	$3.22^{5.9}$	$3.25^{6.1}$	$3.34^{6.6}$	$3.45^{7.3}$
13	2.16	$2.48^{1.7}$	$2.65^{2.7}$	$2.78^{3.4}$	$2.87^{3.9}$	$2.94^{4.3}$	$3.00^{4.7}$	$3.06^{5.0}$	$3.10^{5.3}$	$3.14^{5.5}$	$3.18^{5.8}$	$3.21^{6.0}$	$3.29^{6.5}$	$3.40^{7.1}$
14	2.14	$2.46^{1.7}$	$2.63^{2.6}$	$2.75^{3.3}$	$2.84^{3.8}$	$2.91^{4.2}$	$2.97^{4.6}$	$3.02^{4.9}$	$3.07^{5.2}$	$3.11^{5.4}$	$3.14^{5.6}$	$3.18^{5.8}$	$3.26^{6.3}$	$3.36^{7.0}$
15	2.13	$2.44^{1.7}$	$2.61^{2.6}$	$2.73^{3.2}$	$2.82^{3.8}$	$2.89^{4.2}$	$2.95^{4.5}$	$3.00^{4.8}$	$3.04^{5.1}$	$3.08^{5.3}$	$3.12^{5.5}$	$3.15^{5.7}$	$3.23^{6.2}$	$3.33^{6.8}$
16	2.12	$2.42^{1.6}$	$2.59^{2.5}$	$2.71^{3.2}$	$2.80^{3.7}$	$2.87^{4.1}$	$2.92^{4.4}$	$2.97^{4.7}$	$3.02^{5.0}$	$3.06^{5.2}$	$3.09^{5.4}$	$3.12^{5.6}$	$3.20^{6.1}$	$3.30^{6.7}$
17	2.11	$2.41^{1.6}$	$2.58^{2.5}$	$2.69^{3.1}$	$2.78^{3.6}$	$2.85^{4.0}$	$2.90^{4.4}$	$2.95^{4.7}$	$3.00^{4.9}$	$3.03^{5.1}$	$3.07^{5.3}$	$3.10^{5.5}$	$3.18^{6.0}$	$3.27^{6.6}$
18	2.10	$2.40^{1.6}$	$2.56^{2.5}$	$2.68^{3.1}$	$2.76^{3.6}$	$2.83^{4.0}$	$2.89^{4.3}$	$2.94^{4.6}$	$2.98^{4.8}$	$3.01^{5.1}$	$3.05^{5.3}$	$3.08^{5.4}$	$3.16^{5.9}$	$3.25^{6.5}$
19	2.09	$2.39^{1.6}$	$2.55^{2.5}$	$2.66^{3.1}$	$2.75^{3.5}$	$2.81^{3.9}$	$2.87^{4.2}$	$2.92^{4.5}$	$2.96^{4.8}$	$3.00^{5.0}$	$3.03^{5.2}$	$3.06^{5.4}$	$3.14^{5.8}$	$3.23^{6.4}$
20	2.09	$2.38^{1.6}$	$2.54^{2.4}$	$2.65^{3.0}$	$2.73^{3.5}$	$2.80^{3.9}$	$2.86^{4.2}$	$2.90^{4.5}$	$2.95^{4.7}$	$2.98^{4.9}$	$3.02^{5.1}$	$3.05^{5.3}$	$3.12^{5.7}$	$3.22^{6.3}$
24	2.06	$2.35^{1.5}$	$2.51^{2.3}$	$2.61^{2.9}$	$2.70^{3.4}$	$2.76^{3.7}$	$2.81^{4.0}$	$2.86^{4.3}$	$2.90^{4.5}$	$2.94^{4.7}$	$2.97^{4.9}$	$3.00^{5.1}$	$3.07^{5.5}$	$3.16^{6.0}$
30	2.04	$2.32^{1.5}$	$2.47^{2.3}$	$2.58^{2.8}$	$2.66^{3.2}$	$2.72^{3.6}$	$2.77^{3.9}$	$2.82^{4.1}$	$2.86^{4.3}$	$2.89^{4.5}$	$2.92^{4.7}$	$2.95^{4.8}$	$3.02^{5.2}$	$3.11^{5.8}$
40	2.02	$2.29^{1.4}$	$2.44^{2.2}$	$2.54^{2.7}$	$2.62^{3.1}$	$2.68^{3.4}$	$2.73^{3.7}$	$2.77^{3.9}$	$2.81^{4.1}$	$2.85^{4.3}$	$2.87^{4.5}$	$2.90^{4.6}$	$2.97^{5.0}$	$3.06^{5.5}$
60	2.00	$2.27^{1.4}$	$2.41^{2.1}$	$2.51^{2.6}$	$2.58^{3.0}$	$2.64^{3.3}$	$2.69^{3.5}$	$2.73^{3.7}$	$2.77^{3.9}$	$2.80^{4.1}$	$2.83^{4.2}$	$2.86^{4.4}$	$2.92^{4.7}$	$3.00^{5.1}$
120	1.98	$2.24^{1.3}$	$2.38^{2.0}$	$2.47^{2.5}$	$2.55^{2.8}$	$2.60^{3.1}$	$2.65^{3.3}$	$2.69^{3.5}$	$2.73^{3.7}$	$2.76^{3.8}$	$2.79^{4.0}$	$2.81^{4.1}$	$2.87^{4.4}$	$2.95^{4.8}$
∞	1.96	$2.21^{1.3}$	$2.35^{1.9}$	$2.44^{2.3}$	$2.51^{2.7}$	$2.57^{2.9}$	$2.61^{3.1}$	$2.65^{3.3}$	$2.69^{3.5}$	$2.72^{3.6}$	$2.74^{3.7}$	$2.77^{3.8}$	$2.83^{4.1}$	$2.91^{4.5}$

Source: Reproduced from Dunnett, C.W., *Biometrics,* 20, 482–491, 1964, Table II. With permission from the Biometric Society.
Note: Used first in Chapter 5.

Appendix 12: Bonferroni's Adjusted *t* Values for One-Sided Test and $\alpha = 0.01$

df \ *p*	1	2	3	4	5	6	7	8	9	10
2	6.9646	9.9248	12.1861	14.0890	15.7639	17.2772	18.6682	19.9625	21.1778	22.3271
3	4.5407	5.8909	6.7411	7.4533	8.0526	8.5752	9.0416	9.4649	9.8538	10.2145
4	3.7470	4.6041	5.1668	5.5976	5.9514	6.2541	6.5201	6.7583	6.9746	7.1732
5	3.3649	4.0321	4.4558	4.7733	5.0302	5.2474	5.4364	5.6042	5.7554	5.8934
6	3.1427	3.7074	4.0579	4.3168	4.5241	4.6979	4.8481	4.9807	5.0996	5.2076
7	2.9980	3.4995	3.8055	4.0293	4.2071	4.3553	4.4827	4.5946	4.6947	4.7853
8	2.8965	3.3554	3.6319	3.8325	3.9910	4.1224	4.2349	4.3335	4.4214	4.5008
9	2.8214	3.2498	3.5054	3.6897	3.8345	3.9542	4.0564	4.1458	4.2252	4.2968
10	2.7638	3.1693	3.4093	3.5814	3.7162	3.8273	3.9220	4.0045	4.0778	4.1437
11	2.7181	3.1058	3.3338	3.4966	3.6238	3.7283	3.8172	3.8945	3.9631	4.0247
12	2.6810	3.0545	3.2729	3.4284	3.5495	3.6489	3.7332	3.8065	3.8714	3.9296
13	2.6503	3.0123	3.2229	3.3725	3.4887	3.5838	3.6645	3.7345	3.7965	3.8520
14	2.6245	2.9768	3.1811	3.3257	3.4379	3.5296	3.6072	3.6746	3.7341	3.7874
15	2.6025	2.9467	3.1456	3.2860	3.3948	3.4837	3.5588	3.6239	3.6814	3.7328
16	2.5835	2.9208	3.1150	3.2520	3.3579	3.4443	3.5173	3.5805	3.6363	3.6862
17	2.5669	2.8982	3.0885	3.2224	3.3259	3.4102	3.4814	3.5429	3.5972	3.6458
18	2.5524	2.8784	3.0653	3.1966	3.2979	3.3804	3.4499	3.5101	3.5631	3.6105
19	2.5395	2.8609	3.0447	3.1737	3.2731	3.3540	3.4222	3.4812	3.5330	3.5794
20	2.5280	2.8453	3.0264	3.1534	3.2512	3.3306	3.3976	3.4554	3.5063	3.5518
21	2.5177	2.8314	3.0101	3.1352	3.2315	3.3097	3.3756	3.4325	3.4825	3.5272
22	2.5083	2.8188	2.9953	3.1188	3.2138	3.2909	3.3558	3.4118	3.4610	3.5050
23	2.4999	2.8073	2.9820	3.1040	3.1978	3.2739	3.3379	3.3931	3.4416	3.4850
24	2.4922	2.7969	2.9698	3.0905	3.1832	3.2584	3.3216	3.3761	3.4240	3.4668
25	2.4851	2.7874	2.9587	3.0782	3.1699	3.2443	3.3067	3.3606	3.4080	3.4502
26	2.4786	2.7787	2.9485	3.0669	3.1577	3.2313	3.2931	3.3464	3.3933	3.4350
27	2.4727	2.7707	2.9391	3.0565	3.1465	3.2194	3.2806	3.3334	3.3797	3.4210
28	2.4671	2.7633	2.9305	3.0469	3.1362	3.2084	3.2691	3.3214	3.3673	3.4082
29	2.4620	2.7564	2.9225	3.0380	3.1266	3.1982	3.2584	3.3102	3.3557	3.3962
30	2.4573	2.7500	2.9150	3.0298	3.1177	3.1888	3.2485	3.2999	3.3450	3.3852
35	2.4377	2.7238	2.8845	2.9960	3.0813	3.1502	3.2080	3.2577	3.3013	3.3400
40	2.4233	2.7045	2.8620	2.9712	3.0545	3.1218	3.1782	3.2266	3.2691	3.3069
45	2.4121	2.6896	2.8447	2.9521	3.0340	3.1,000	3.1553	3.2028	3.2445	3.2815
50	2.4033	2.6778	2.8310	2.9370	3.0177	3.0828	3.1373	3.1840	3.2250	3.2614
55	2.3961	2.6682	2.8199	2.9247	3.0045	3.0688	3.1226	3.1688	3.2092	3.2451
60	2.3901	2.6603	2.8107	2.9146	2.9936	3.0573	3.1105	3.1562	3.1962	3.2317
70	2.3808	2.6479	2.7964	2.8987	2.9766	3.0393	3.0916	3.1366	3.1759	3.2108
80	2.3739	2.6387	2.7857	2.8870	2.9640	3.0259	3.0776	3.1220	3.1608	3.1953
90	2.3685	2.6316	2.7774	2.8779	3.9542	3.0156	3.0668	3.1108	3.1492	3.1833
100	2.3642	2.6259	2.7709	2.8707	2.9464	3.0073	3.0582	3.1018	3.1399	3.1737
110	2.3607	2.6213	2.7655	2.8648	2.9401	3.0007	3.0512	3.0945	3.1324	3.1660
120	2.3578	2.6174	2.7611	2.8599	2.9348	2.9951	3.0454	3.0885	3.1261	3.1595
250	2.3414	2.5956	2.7359	2.8322	2.9051	2.9637	3.0125	3.0543	3.0908	3.1232
500	2.3338	2.5857	2.7244	2.8195	2.8916	2.9494	2.9975	3.0387	3.0747	3.1066
1,000	2.3301	2.5808	2.7187	2.8133	2.8849	2.9423	2.9901	3.0310	3.0667	3.0984

$df \setminus p$	11	12	13	14	15	16	17	18	19
2	23.4201	24.4643	25.4657	26.4292	27.3587	28.2577	29.1290	29.9750	30.7977
3	10.5517	10.8688	11.1685	11.4532	11.7244	11.9838	12.2324	12.4715	12.7017
4	7.3571	7.5287	7.6896	7.8414	7.9851	8.1216	8.2518	8.3763	8.4957
5	6.0205	6.1384	6.2484	6.3518	6.4492	6.5414	6.6291	6.7126	6.7924
6	5.3067	5.3982	5.4834	5.5632	5.6382	5.7090	5.7761	5.8399	5.9007
7	4.8681	4.9445	5.0154	5.0815	5.1437	5.2022	5.2576	5.3101	5.3601
8	4.5732	4.6398	4.7015	4.7590	4.8129	4.8636	4.9116	4.9570	5.0001
9	4.3620	4.4219	4.4773	4.5288	4.5771	4.6224	4.6652	4.7058	4.7443
10	4.2036	4.2586	4.3094	4.3567	4.4008	4.4423	4.4814	4.5184	4.5535
11	4.0807	4.1319	4.1793	4.2232	4.2643	4.3028	4.3391	4.3735	4.4060
12	3.9825	4.0308	4.0755	4.1169	4.1555	4.1918	4.2259	4.2582	4.2887
13	3.9023	3.9484	3.9908	4.0302	4.0669	4.1013	4.1337	4.1643	4.1933
14	3.8357	3.8798	3.9205	3.9582	3.9933	4.0263	4.0572	4.0865	4.1142
15	3.7794	3.8220	3.8612	3.8975	3.9313	3.9630	3.9928	4.0209	4.0475
16	3.7313	3.7725	3.8104	3.8456	3.8783	3.9089	3.9377	3.9649	3.9906
17	3.6897	3.7297	3.7666	3.8007	3.8325	3.8623	3.8902	3.9165	3.9415
18	3.6533	3.6924	3.7283	3.7616	3.7926	3.8215	3.8487	3.8744	3.8986
19	3.6213	3.6595	3.6946	3.7271	3.7574	3.7857	3.8122	3.8373	3.8609
20	3.5929	3.6303	3.6647	3.6966	3.7262	3.7539	3.7799	3.8044	3.8275
21	3.5675	3.6043	3.6380	3.6693	3.6983	3.7255	3.7510	3.7750	3.7977
22	3.5447	3.5808	3.6141	3.6448	3.6733	3.7000	3.7251	3.7487	3.7710
23	3.5241	3.5597	3.5924	3.6226	3.6507	3.6770	3.7017	3.7249	3.7468
24	3.5053	3.5405	3.5727	3.6025	3.6302	3.6561	3.6804	3.7033	3.7249
25	3.4883	3.5230	3.5548	3.5842	3.6116	3.6371	3.6611	3.6836	3.7049
26	3.4726	3.5069	3.5384	3.5674	3.5945	3.6197	3.6433	3.6656	3.6867
27	3.4583	3.4922	3.5233	3.5520	3.5787	3.6037	3.6271	3.6491	3.6699
28	3.4450	3.4786	3.5094	3.5378	3.5643	3.5889	3.6121	3.6338	3.6544
29	3.4328	3.4660	3.4965	3.5247	3.5509	3.5753	3.5982	3.6198	3.6401
30	3.4214	3.4544	3.4846	3.5125	3.5384	3.5626	3.5853	3.6067	3.6269
35	3.3750	3.4068	3.4359	3.4628	3.4877	3.5110	3.5329	3.5534	3.5728
40	3.3409	3.3718	3.4001	3.4263	3.4506	3.4732	3.4944	3.5143	3.5331
45	3.3148	3.3451	3.3728	3.3984	3.4221	3.4442	3.4650	3.4845	3.5028
50	3.2942	3.3239	3.3512	3.3763	3.3996	3.4214	3.4417	3.4609	3.4789
55	3.2775	3.3068	3.3337	3.3585	3.3814	3.4029	3.4229	3.4418	3.4596
60	3.2637	3.2927	3.3192	3.3437	3.3664	3.3876	3.4074	3.4260	3.4436
70	3.2422	3.2707	3.2967	3.3208	3.3430	3.3638	3.3832	3.4015	3.4187
80	3.2262	3.2543	3.2800	3.3037	3.3257	3.3462	3.3653	3.3833	3.4003
90	3.2139	3.2417	3.2672	3.2906	3.3123	3.3326	3.3515	3.3693	3.3861
100	3.2041	3.2317	3.2569	3.2802	3.3017	3.3218	3.3405	3.3582	3.3748
110	3.1962	3.2235	3.2486	3.2717	3.2930	3.3130	3.3316	3.3491	3.3656
120	3.1896	3.2168	3.2417	3.2646	3.2859	3.3057	3.3242	3.3416	3.3580
250	3.1522	3.1785	3.2026	3.2248	3.2453	3.2644	3.2823	3.2991	3.3149
500	3.1352	3.1612	3.1849	3.2067	3.2269	3.2457	3.2633	3.2798	3.2954
1,000	3.1268	3.1526	3.1761	3.1977	3.2178	3.2365	3.2539	3.2703	3.2857

df \ *p*	20	21	28	36	45	55	66	78
2	31.5991	32.3806	37.3965	42.4087	47.4184	52.4261	57.4326	62.4380
3	12.9240	13.1389	14.4787	15.7577	16.9858	18.1703	19.3169	20.4300
4	8.6103	8.7207	9.3983	10.0298	10.6237	11.1860	11.7211	12.2329
5	6.8688	6.9422	7.3884	7.7981	8.1783	8.5341	8.8692	9.1865
6	5.9588	6.0145	6.3510	6.6568	6.9379	7.1988	7.4428	7.6724
7	5.4079	5.4536	5.7282	5.9757	6.2019	6.4105	6.6045	6.7862
8	5.0413	5.0806	5.3162	5.5274	5.7192	5.8954	6.0586	6.2107
9	4.7809	4.8159	5.0249	5.2114	5.3801	5.5345	5.6770	5.8095
10	4.5869	4.6188	4.8087	4.9774	5.1296	5.2684	5.3962	5.5148
11	4.4370	4.4665	4.6420	4.7975	4.9374	5.0646	5.1815	5.2898
12	4.3178	4.3455	4.5099	4.6551	4.7854	4.9037	5.0123	5.1125
13	4.2208	4.2471	4.4026	4.5396	4.6624	4.7737	4.8755	4.9695
14	4.1405	4.1655	6.3138	4.4442	4.5608	4.6663	4.7628	4.8518
15	4.0728	4.0968	4.2391	3.3640	4.4756	4.5764	4.6684	4.7532
16	4.0150	4.0382	4.1754	4.2958	4.4030	4.4999	4.5882	4.6695
17	3.9651	3.9876	4.1205	4.2369	4.3406	4.4340	4.5192	4.5975
18	3.9216	3.9435	4.0727	4.1857	4.2862	4.3768	4.4593	4.5350
19	3.8834	3.9048	4.0307	4.1408	4.2386	4.3266	4.4068	4.4803
20	3.8495	3.8704	3.9935	4.1010	4.1964	4.2823	4.3603	4.4319
21	3.8193	3.8398	3.9603	4.0655	4.1588	4.2427	4.3190	4.3888
22	3.7921	3.8122	3.9306	4.0337	4.1252	4.2073	4.2819	4.3503
23	3.7676	3.7874	3.9037	4.0050	4.0948	4.1754	4.2486	4.3156
24	3.7454	3.7649	3.8794	3.9790	4.0673	4.1465	4.2184	4.2842
25	3.7251	3.7443	3.8572	3.9554	4.0423	4.1202	4.1909	4.2556
26	3.7066	3.7256	3.8369	3.9337	4.0194	4.0962	4.1658	4.2295
27	3.6896	3.7083	3.8183	3.9139	3.9984	4.0742	4.1428	4.2056
28	3.6739	3.6924	3.8012	3.8956	3.9791	4.0539	4.1216	4.1835
29	3.6594	3.6777	3.7853	3.8787	3.9612	4.0351	4.1021	4.1632
30	3.6460	3.6641	3.7706	3.8631	3.9447	4.0178	4.0839	4.1444
35	3.5911	3.6086	3.7108	3.7993	3.8774	3.9472	4.0103	4.0679
40	3.5510	3.5679	3.6670	3.7527	3.8282	3.8956	3.9565	4.0121
45	3.5203	3.5368	3.6335	3.7171	3.7907	3.8563	3.9156	3.9696
50	3.4960	3.5122	3.6071	3.6890	3.7611	3.8253	3.8833	3.9361
55	3.4764	3.4924	3.5858	3.6664	3.7372	3.8003	3.8573	3.9091
60	3.4602	3.4760	3.5682	3.6477	3.7175	3.7797	3.8358	3.8868
70	3.4350	3.4505	3.5408	3.6186	3.6869	3.7477	3.8024	3.8523
80	3.4163	3.4316	3.5205	3.5970	3.6642	3.7240	3.7778	3.8267
90	3.4019	3.4170	3.5048	3.5804	3.6467	3.7057	3.7588	3.8070
100	3.3905	3.4054	3.4924	3.5673	3.6329	3.6912	3.7437	3.7914
110	3.3812	3.3960	3.4823	3.5565	3.6216	3.6794	3.7314	3.7787
120	3.3735	3.3882	3.4739	3.5477	3.6122	3.6697	3.7213	3.7682
250	3.3299	3.3440	3.4267	3.4976	3.5596	3.6146	3.6641	3.7090
500	3.3101	3.3240	3.4052	3.4749	3.5357	3.5897	3.6382	3.6822
1,000	3.3003	3.3141	3.3946	3.4636	3.5239	3.5774	3.6254	3.6689

df* \ *p	**91**	**105**	**120**	**136**	**153**	**171**	**190**
2	67.4426	72.4465	77.4500	82.4530	87.4557	92.4581	97.4602
3	21.5132	22.5696	23.6016	24.6113	25.6006	26.5711	27.5240
4	12.7239	13.1967	13.6531	14.0947	14.5229	14.9387	15.3433
5	9.4884	9.7768	10.0530	10.3185	10.5741	10.8210	11.0597
6	7.8894	8.0956	8.2921	8.4799	8.6601	8.8333	9.0001
7	6.9572	7.1189	7.2725	7.4188	7.5586	7.6926	7.8213
8	6.3535	6.4881	6.6155	6.7366	6.8520	6.9624	7.0681
9	5.9335	6.0500	6.1602	6.2646	6.3639	6.4586	6.5492
10	5.6255	5.7294	5.8273	5.9199	6.0079	6.0918	6.1718
11	5.3906	5.4851	5.5740	5.6580	5.7377	5.8135	5.8858
12	5.2058	5.2931	5.3751	5.4525	5.5259	5.5956	5.6620
13	5.0569	5.1385	5.2151	5.2873	5.3556	5.4205	5.4823
14	4.9343	5.0113	5.0836	5.1516	5.2160	5.2770	5.3351
15	4.8318	4.9050	4.9737	5.0383	5.0993	5.1572	5.2122
16	4.7447	4.8148	4.8805	4.9423	5.0006	5.0558	5.1083
17	4.6700	4.7374	4.8005	4.8599	4.9159	4.9689	5.0192
18	4.6051	4.6703	4.7312	4.7885	4.8424	4.8935	4.9421
19	4.5482	4.6114	4.6705	4.7259	4.7782	4.8277	4.8746
20	4.4980	4.5595	4.6169	4.6708	4.7216	4.7696	4.8151
21	4.4534	4.5133	4.5693	4.6218	4.6712	4.7180	4.7623
22	4.4134	4.4720	4.5266	4.5779	4.6262	4.6718	4.7151
23	4.3774	4.4348	4.4883	4.5385	4.5857	4.6303	4.6726
24	4.3448	4.4011	4.4536	4.5028	4.5491	4.5928	4.6342
25	4.3152	4.3705	4.4221	4.4704	4.5158	4.5587	4.5994
26	4.2882	4.3426	4.3933	4.4408	4.4855	4.5276	4.5676
27	4.2634	4.3170	4.3669	4.4137	4.4576	4.4991	4.5384
28	4.2406	4.2934	4.3426	4.3887	4.4321	4.4730	4.5116
29	4.2195	4.2717	4.3203	4.3657	4.4085	4.4488	4.4869
30	4.2000	4.2515	4.2995	3.3444	4.3866	4.4264	4.4641
35	4.1208	4.1698	4.2154	4.2581	4.2981	4.3358	4.3715
40	4.0631	4.1103	4.1541	4.1951	4.2336	4.2698	4.3041
45	4.0191	4.0649	4.1075	4.1473	4.1846	4.2197	4.2528
50	3.9845	4.0293	4.0708	4.1097	4.1460	4.1803	4.2126
55	3.9566	4.0005	4.0413	4.0793	4.1150	4.1485	4.1802
60	3.9336	3.9768	4.0169	4.0543	4.0894	4.1223	4.1535
70	3.8979	3.9400	3.9791	4.0155	4.0497	4.0818	4.1121
80	3.8715	3.9128	3.9512	3.9869	4.0204	4.0518	4.0815
90	3.8512	3.8919	3.9297	3.9649	3.9978	4.0288	4.0580
100	3.8350	3.8753	3.9126	3.9474	3.9799	4.0105	4.0394
110	3.8219	3.8618	3.8988	3.9332	3.9654	3.9957	4.0242
120	3.8111	3.8506	3.8873	3.9214	3.9534	3.9834	4.0117
250	3.7500	3.7878	3.8227	3.8553	3.8857	3.9143	3.9412
500	3.7224	3.7594	3.7936	3.8254	3.8552	3.8831	3.9095
1,000	3.7087	3.7453	3.7792	3.8107	3.8401	3.8677	3.8937

Note: Values generated as described in Chapter 5.

Appendix 13: Bonferroni's Adjusted *t* Values for One-Sided Test and $\alpha = 0.05$

df \ *p*	1	2	3	4	5	6	7	8	9	10
2	2.9200	4.3027	5.3393	6.2054	6.9646	7.6488	8.2767	8.8602	9.4076	9.9248
3	2.3534	3.1825	3.7405	4.1765	4.5407	4.8567	5.1377	5.3920	5.6251	5.8409
4	2.1319	2.7765	3.1863	3.4954	3.7470	3.9608	4.1478	4.3147	4.4657	4.6041
5	2.0151	2.5706	2.9117	3.1634	3.3649	3.5341	3.6805	3.8100	3.9263	4.0321
6	1.9432	2.4469	2.7491	2.9687	3.1427	3.2875	3.4119	3.5212	3.6190	3.7074
7	1.8946	2.3646	2.6419	2.8412	2.9980	3.1276	3.2384	3.3353	3.4216	3.4995
8	1.8596	2.3060	2.5660	2.7515	2.8965	3.0158	3.1174	3.2060	3.2846	3.3554
9	1.8331	2.2622	2.5096	2.6850	2.8214	2.9333	3.0283	3.1109	3.1841	3.2498
10	1.8125	2.2281	2.4660	2.6338	2.7638	2.8701	2.9601	3.0382	3.1073	3.1693
11	1.7959	2.2010	2.4313	2.5931	2.7181	2.8200	2.9062	2.9809	3.0468	3.1058
12	1.7823	2.1788	2.4030	2.5600	2.6810	2.7795	2.8626	2.9345	2.9978	3.0545
13	1.7709	2.1604	2.3796	2.5326	2.6503	2.7459	2.8265	2.8962	2.9575	3.0123
14	1.7613	2.1448	2.3598	2.5096	3.6245	2.7178	2.7962	2.8640	2.9236	2.9768
15	1.7531	2.1315	2.3429	2.4899	3.6025	2.6937	2.7705	2.8366	2.8948	2.9467
16	1.7459	2.1199	2.3283	2.4729	2.5835	2.6730	2.7482	2.8131	2.8700	2.9208
17	1.7396	2.1098	2.3156	2.4581	2.5669	2.6550	2.7289	2.7925	2.8484	2.8982
18	1.7341	2.1009	2.3044	2.4450	2.5524	2.6391	2.7119	2.7745	2.8295	2.8784
19	1.7291	2.0930	2.2944	2.4334	2.5395	2.6251	2.6969	2.7586	2.8127	2.8609
20	1.7247	2.0860	2.2855	2.4231	2.5280	2.6126	2.6835	2.7444	2.7978	2.8453
21	1.7207	2.0796	2.2775	2.4139	2.5177	2.6014	2.6714	2.7316	2.7844	2.8314
22	1.7171	2.0739	2.2703	2.4055	2.5083	2.5912	2.6606	2.7201	2.7723	2.8188
23	1.7139	2.0687	2.2637	2.3979	2.4999	2.5820	2.6507	2.7097	2.7614	2.8073
24	1.7109	2.0639	2.2578	2.3910	2.4922	2.5736	2.6418	2.7002	2.7514	2.7969
25	1.7081	2.0595	2.2523	2.3846	2.4851	2.5660	2.6336	2.6916	2.7423	2.7874
26	1.7056	2.0555	2.2472	2.3788	2.4786	2.5589	2.6260	2.6836	2.7340	2.7787
27	1.7033	2.0518	2.2426	2.3734	2.4727	2.5525	2.6191	2.6763	2.7263	2.7707
28	1.7011	2.0484	2.2383	2.3685	2.4671	2.5465	2.6127	2.6695	2.7192	2.7633
29	1.6991	2.0452	2.2343	2.3639	2.4620	2.5409	2.6068	2.6632	2.7126	2.7564
30	1.6973	2.0423	2.2306	2.3596	2.4573	2.5357	2.6012	2.6574	2.7064	2.7500
35	1.6896	2.0301	2.2154	2.3420	2.4377	2.5145	2.5786	2.6334	2.6813	2.7238
40	1.6839	2.0211	2.2041	2.3289	2.4233	2.4989	2.5618	2.6157	2.6627	2.7045
45	1.6794	2.0141	2.1954	2.3189	2.4121	2.4868	2.5489	2.6021	2.6485	2.6896
50	1.6759	2.0086	2.1885	2.3109	2.4033	2.4772	2.5387	2.5913	2.6372	2.6778
55	1.6730	2.0040	2.1829	2.3044	2.3961	2.4694	2.5304	2.5825	2.6280	2.6682
60	1.6707	2.0003	2.1782	2.2991	2.3901	2.4630	2.5235	2.5752	2.6203	2.6603
70	1.6669	1.9944	2.1709	2.2906	2.3808	2.4529	2.5128	2.5639	2.6085	2.6479
80	1.6641	1.9901	2.1654	2.2844	2.3739	2.4454	2.5047	2.5554	2.5996	2.6387
90	1.6620	1.9867	2.1612	2.2795	2.3685	2.4396	2.4986	2.5489	2.5928	2.6316
100	1.6602	1.9840	2.1579	2.2757	2.3642	2.4349	2.4936	2.5437	2.5873	2.6259
110	1.6588	1.9818	2.1551	2.2725	2.3607	2.4311	2.4896	2.5394	2.5829	2.6213
120	1.6577	1.9799	2.1528	2.2699	2.3578	2.4280	2.4862	2.5359	2.5792	2.6174
250	1.6510	1.9650	2.1399	2.2550	2.3414	2.4102	2.4673	2.5159	2.5582	2.5956
500	1.6479	1.9647	2.1339	2.2482	2.3338	2.4021	2.4586	2.5068	2.5487	2.5857
1,000	1.6464	1.9623	2.1310	2.2448	2.3301	2.3980	2.4543	2.5022	2.5439	2.5808

df*\ *p	**11**	**12**	**13**	**14**	**15**	**16**	**17**	**18**	**19**
2	10.4164	10.8859	11.3359	11.7687	12.1861	12.5897	12.9808	13.3604	13.7296
3	6.0423	6.2315	6.4102	6.5797	6.7411	6.8952	7.0430	7.1849	7.3215
4	4.7320	4.8510	4.9625	5.0675	5.1668	5.2611	5.3508	5.4366	5.5187
5	4.1293	4.2193	4.3032	4.3818	4.4558	4.5257	4.5921	4.6553	4.7156
6	3.7884	3.8630	3.9323	3.9971	4.0579	4.1152	4.1694	4.2209	4.2700
7	3.5705	3.6358	3.6963	3.7527	3.8055	3.8552	3.9022	3.9467	3.9890
8	3.4198	3.4789	3.5335	3.5844	3.6319	3.6766	3.7187	3.7586	3.7965
9	3.3095	3.3642	3.4147	3.4616	3.5054	2.5465	3.5852	3.6219	3.6566
10	3.2254	3.2768	3.3242	3.3682	3.4093	3.4477	3.4839	3.5182	3.5506
11	3.1593	3.2081	3.2531	3.2949	3.3338	3.3702	3.4045	3.4368	3.4675
12	3.1058	3.1527	3.1958	3.2357	3.2729	3.3078	3.3405	3.3714	3.4007
13	3.0618	3.1070	3.1486	3.1871	3.2229	3.2565	3.2880	3.3177	3.3458
14	3.0249	3.0688	3.1091	3.1464	3.1811	3.2135	3.2440	3.2727	3.2999
15	2.9936	3.0363	3.0755	3.1118	3.1456	3.1771	3.2067	3.2346	3.2610
16	2.9666	3.0083	3.0467	3.0821	3.1150	3.1458	3.1747	3.2019	3.2276
17	2.9431	2.9840	3.0216	3.0563	3.0885	3.1186	3.1469	3.1735	3.1986
18	2.9226	2.9627	2.9996	3.0336	3.0653	3.0948	3.1225	3.1486	3.1732
19	2.9044	2.9439	2.9801	3.0136	3.0447	3.0738	3.1010	3.1266	3.1508
20	2.8882	2.9271	2.9628	2.9958	3.0264	3.0550	3.0818	3.1070	3.1308
21	2.8736	2.9121	2.9473	2.9799	3.0101	3.0382	3.0647	3.0895	3.1130
22	2.8605	2.8985	2.9334	2.9655	2.9953	3.0231	3.0492	3.0737	3.0969
23	2.8487	2.8863	2.9207	2.9525	2.9820	3.0095	3.0352	3.0595	3.0823
24	2.8379	2.8751	2.9092	2.9406	2.9698	2.9970	3.0225	3.0465	3.0691
25	2.8280	2.8649	2.8987	2.9298	2.9587	2.9856	3.0109	3.0346	3.0570
26	2.8190	2.8555	2.8890	2.9199	2.9485	2.9752	3.0002	3.0237	3.0459
27	2.8106	2.8469	2.8801	2.9107	2.9391	2.9656	2.9904	3.0137	3.0357
28	2.8029	2.8389	2.8719	2.9023	2.9305	2.9567	2.9813	3.0045	3.0263
29	2.7958	2.8316	2.8643	2.8945	2.9225	2.9485	2.9730	2.9959	3.0176
30	2.7892	2.8247	2.8572	2.8872	2.9150	2.9409	2.9652	2.9880	3.0095
35	2.7620	2.7966	2.8283	2.8575	2.8845	2.9097	2.9333	2.9554	2.9763
40	2.7419	2.7759	2.8069	2.8355	2.8620	2.8867	2.9097	2.9314	2.9519
45	2.7265	2.7599	2.7905	2.8187	2.8447	2.8690	2.8917	2.9130	2.9331
50	2.7143	2.7473	2.7775	2.8053	2.8310	2.8550	2.8774	2.8984	2.9182
55	2.7043	2.7370	2.7669	2.7944	2.8199	2.8436	2.8658	2.8866	2.9062
60	2.6961	2.7286	2.7582	2.7855	2.8107	2.8342	2.8562	2.8768	2.8962
70	2.6833	2.7153	2.7446	2.7715	2.7964	2.8195	2.8412	2.8615	2.8807
80	2.6737	2.7054	2.7344	2.7610	2.7857	2.8086	2.8301	2.8502	2.8691
90	2.6663	2.6978	2.7266	2.7530	2.7774	2.8002	2.8214	2.8414	2.8602
100	2.6605	2.6918	2.7203	2.7466	2.7709	2.7935	2.8146	2.8344	2.8530
110	2.6557	2.6868	2.7152	2.7414	2.7655	2.7880	2.8090	2.8287	2.8472
120	2.6517	2.6827	2.7110	2.7370	2.7611	2.7835	2.8044	2.8240	2.8424
250	2.6291	2.6594	2.6870	2.7124	2.7359	2.7577	2.7781	2.7972	2.8152
500	2.6188	2.6488	2.6761	2.7012	2.7244	2.7460	2.7661	2.7850	2.8028
1,000	2.6137	2.6435	2.6707	2.6957	2.7187	2.7402	2.7602	2.7790	2.7966

df \ *p*	20	21	28	36	45	55	66	78
2	14.0890	14.4396	16.6883	18.9341	21.1778	23.4201	25.6613	27.9016
3	7.4533	7.5807	8.3738	9.1294	9.8538	10.5517	11.2266	11.8814
4	5.5976	5.6734	6.1380	6.5697	6.9746	7.3571	7.7207	8.0678
5	4.7733	4.8287	5.1644	5.4715	5.7554	6.0205	6.2696	6.5051
6	4.3168	4.3617	4.6317	4.8759	5.0996	5.3067	5.4998	5.6811
7	4.0293	4.0679	4.2989	4.5062	4.6947	4.8681	5.0289	5.1792
8	3.8325	3.8669	4.0724	4.2556	4.4214	4.5732	4.7133	4.8437
9	3.6897	3.7212	3.9088	4.0752	4.2252	4.3620	4.4879	4.6046
10	3.5814	3.6108	3.7852	3.9394	4.0778	4.2036	3.3191	4.4260
11	3.4966	3.5243	3.6887	3.8335	3.9631	4.0807	4.1883	4.2877
12	3.4284	3.4549	3.6112	3.7487	3.8714	3.9825	4.0840	4.1776
13	3.3725	3.3979	3.5478	3.6793	3.7965	3.9023	3.9989	4.0878
14	3.3257	3.3502	3.4949	3.6214	3.7341	3.8357	3.9283	4.0134
15	3.2860	3.3098	3.4501	3.5725	3.6814	3.7794	3.8686	3.9506
16	3.2520	3.2752	3.4116	3.5306	3.6363	3.7313	3.8177	3.8969
17	3.2224	3.2451	3.3783	3.4944	3.5972	3.6897	3.7736	3.8506
18	3.1966	3.2188	3.3492	3.4626	3.5631	3.6533	3.7352	3.8102
19	3.1737	3.1955	3.3235	3.4347	3.5330	3.6213	3.7013	3.7746
20	3.1534	3.1748	3.3006	3.4098	3.5063	3.5929	3.6713	3.7430
21	3.1352	3.1563	3.2802	3.3876	3.4825	3.5675	3.6445	3.7149
22	3.1188	3.1396	3.2618	3.3676	3.4610	3.5447	3.6204	3.6896
23	3.1040	3.1246	3.2451	3.3495	3.4416	3.5241	3.5986	3.6667
24	3.0905	3.1108	3.2300	3.3331	3.4240	3.5053	3.5789	3.6460
25	3.0782	3.0983	3.2162	3.3181	3.4080	3.4883	3.5609	3.6271
26	3.0669	3.0868	3.2035	3.3044	3.3933	3.4726	3.5444	3.6098
27	3.0565	3.0763	3.1919	3.2918	3.3797	3.4583	3.5292	3.5939
28	3.0469	3.0665	3.1811	3.2801	3.3673	3.4450	3.5152	3.5793
29	3.0680	3.0575	3.1712	3.2694	3.3557	3.4328	3.5023	3.5657
30	3.0298	3.0491	3.1620	3.2594	3.3450	3.4214	3.4903	3.5531
35	2.9960	3.0148	3.1242	3.2185	3.3013	3.3750	3.4414	3.5019
40	2.9712	2.9895	3.0964	3.1884	3.2691	3.3409	3.4055	3.4643
45	2.9521	2.9701	3.0751	3.1654	3.2445	3.3148	3.3781	3.4356
50	2.9370	2.9547	3.0582	3.1472	3.2250	3.2942	3.3564	3.4129
55	2.9247	2.9423	3.0446	3.1324	3.2092	3.2775	3.3388	3.3945
60	2.9146	2.9319	3.0333	3.1202	3.1962	3.2637	3.3243	3.3793
70	2.8987	2.9159	3.0156	3.1012	3.1759	3.2422	3.3017	3.3557
80	2.8870	2.9039	3.0026	3.0870	3.1608	3.2262	3.2849	3.3382
90	2.8779	2.8947	2.9924	3.0761	3.1492	3.2139	3.2720	3.3246
100	2.8707	2.8873	2.9844	3.0674	3.1399	3.2041	3.2617	3.3139
110	2.8648	2.8813	2.9778	3.0604	3.1324	3.1962	2.2534	3.3052
120	2.8599	2.8764	2.9724	3.0545	3.1261	3.1896	3.2464	3.2979
250	2.8322	2.8482	2.9416	3.0213	3.0908	3.1522	3.2072	3.2570
500	2.9195	2.8354	2.9276	3.0063	3.0747	3.1352	3.1894	3.2384
1,000	2.8133	2.8291	2.9207	2.9988	3.0667	3.1268	3.1806	3.2291

df \ *p*	91	105	120	136	153	171	190
2	30.1413	32.3806	34.6194	36.8578	39.0960	41.3340	43.5718
3	12.5182	13.1389	13.7450	14.3379	14.9186	15.4880	16.0471
4	8.4006	8.7207	9.0294	9.3278	9.6171	9.8978	10.1708
5	6.7288	6.9422	7.1464	7.3424	7.5310	7.7129	7.8888
6	5.8522	6.0145	6.1690	6.3165	6.4578	6.5934	6.7240
7	5.3203	5.4536	5.5799	5.7001	5.8148	5.9246	6.0300
8	4.9658	5.0806	5.1892	5.2922	5.3902	5.4838	5.5735
9	4.7136	4.8159	4.9124	5.0037	5.0904	5.1730	5.2520
10	4.5256	4.6188	4.7065	4.7894	4.8680	4.9427	5.0141
11	4.3801	4.4665	4.5477	4.6242	4.6967	4.7656	4.8312
12	4.2644	4.3455	4.4215	4.4932	4.5610	4.6253	4.6865
13	4.1702	4.2471	4.3191	4.3868	4.4509	4.5116	4.5693
14	4.0921	4.1655	4.2342	4.2988	4.3597	4.4175	4.4724
15	4.0263	4.0968	4.1628	4.2247	4.2831	4.3385	4.3910
16	3.9701	4.0382	4.1018	4.1616	4.2179	4.2712	4.3217
17	3.9216	3.9876	4.0493	4.1071	4.1616	4.2131	4.2620
18	3.8793	3.9435	4.0035	4.0597	4.1126	4.1626	4.2101
19	3.8421	3.9048	3.9632	4.0180	4.0696	4.1183	4.1645
20	3.8091	3.8704	3.9276	3.9811	4.0315	4.0790	4.1241
21	3.7797	3.8398	3.8958	3.9482	3.9975	4.0441	4.0881
22	3.7532	3.8122	3.8672	3.9187	3.9670	4.0127	4.0559
23	3.7294	3.7874	3.8414	3.8920	3.9395	3.9844	4.0268
24	3.7077	3.7649	3.8181	3.8679	3.9146	3.9587	4.0004
25	3.6880	3.7443	3.7968	3.8459	3.8919	3.9354	3.9765
26	3.6699	3.7256	3.7773	3.8257	3.8712	3.9140	3.9545
27	3.6533	3.7083	3.7595	3.8073	3.8521	3.8944	3.9344
28	3.6381	3.6924	3.7430	3.7902	3.8346	3.8764	3.9159
29	3.6239	3.6777	3.7278	3.7745	3.8184	3.8597	3.8987
30	3.6108	3.6641	3.7136	3.7599	3.8033	3.8442	3.8829
35	3.5574	3.6086	3.6561	3.7005	3.7421	3.7813	3.8183
40	3.5182	3.5679	3.6140	3.6570	3.6973	3.7353	3.7710
45	3.4882	3.5368	3.5818	3.6238	3.6631	3.7001	3.7350
50	3.4646	3.5122	3.5564	3.5976	3.6362	3.6724	3.7066
55	3.4454	3.4924	3.5359	3.5764	3.6143	3.6500	3.6836
60	3.4296	3.4760	3.5189	3.5589	3.5963	3.6315	3.6646
70	3.4050	3.4505	3.4926	3.5317	3.5684	3.6028	3.6352
80	3.3868	3.4316	3.4730	3.5116	3.5476	3.5815	3.6134
90	3.3727	3.4170	3.4579	3.4960	3.5317	3.5651	3.5966
100	3.3616	3.4054	3.4460	3.4837	3.5190	3.5521	3.5832
110	3.3525	3.3960	3.4362	3.4737	3.5087	3.5415	3.5724
120	3.3449	3.3882	3.4281	3.4653	3.5001	3.5327	3.5634
250	3.3023	3.3440	3.3826	3.4184	3.4518	3.4832	3.5127
500	3.2830	3.3240	3.3619	3.3971	3.4300	3.4607	3.4897
1,000	3.2734	3.3141	3.3517	3.3866	3.4191	3.4496	3.4783

Note: Values generated as described in Chapter 5.

Appendix 14: Bonferroni's Adjusted *t* Values for Two-Sided Test and α = 0.01

df \ *p*	1	2	3	4	5	6	7	8	9	10
2	9.9248	14.0890	17.2772	19.9625	22.3271	24.4643	26.4292	28.2577	29.9750	31.5991
3	5.8409	7.4533	8.5752	9.4649	10.2145	10.8688	11.4532	11.9838	12.4715	12.9240
4	4.6041	5.5976	6.2541	6.7583	7.1732	7.5287	7.8414	8.1216	8.3763	8.6103
5	4.0321	4.7733	5.2474	5.6042	5.8934	6.1384	6.3518	6.5414	6.7126	6.8688
6	3.7074	4.3168	4.6979	4.9807	5.2076	5.3982	5.5632	5.7090	5.8399	5.9588
7	3.4995	4.0293	4.3553	4.5946	4.7853	4.9445	5.0815	5.2022	5.3101	5.4079
8	3.3554	3.8325	4.1224	4.3335	4.5008	4.6398	4.7590	4.8636	4.9570	5.0413
9	3.2498	3.6897	3.9542	4.1458	4.2968	4.4219	4.5288	4.6224	4.7058	4.7809
10	3.1693	3.5814	3.8273	4.0045	4.1437	4.2586	4.3567	4.4423	4.5184	4.5869
11	3.1058	3.4966	3.7283	3.8945	4.0247	4.1319	4.2232	4.3028	4.3735	4.4370
12	3.0545	3.4284	3.6489	3.8065	3.9296	4.0308	4.1169	4.1918	4.2582	4.3178
13	3.0123	3.3725	3.5838	3.7345	3.8520	3.9484	4.0302	4.1013	4.1643	4.2208
14	2.9768	3.3257	3.5296	3.6746	3.7874	3.8798	3.9582	4.0263	4.0865	4.1405
15	2.9467	3.2860	3.4837	3.6239	3.7328	3.8220	3.8975	3.9630	4.0209	4.0728
16	2.9208	3.2520	3.4443	3.5805	3.6862	3.7725	3.8456	3.9089	3.9649	4.0150
17	2.8982	3.2224	3.4102	3.5429	3.6458	3.7297	3.8007	3.8623	3.9165	3.9651
18	2.8784	3.1966	3.3804	3.5101	3.6105	3.6924	3.7616	3.8215	3.8744	3.9216
19	2.8609	3.1737	3.3540	3.4812	3.5794	3.6595	3.7271	3.7857	3.8373	3.8834
20	2.8453	3.1534	3.3306	3.4554	3.5518	3.6303	3.6966	3.7539	3.8044	3.8495
21	2.8314	3.1352	3.3097	3.4325	3.5272	3.6043	3.6693	3.7255	3.7750	3.8193
22	2.8188	3.1188	3.2909	3.4118	3.5050	3.5808	3.6448	3.7000	3.7487	3.7921
23	2.8073	3.1040	3.2739	3.3931	3.4850	3.5597	3.6226	3.6700	3.7249	3.7676
24	2.7969	3.0905	3.2584	3.3761	3.4668	3.5405	3.6025	3.6561	3.7033	3.7454
25	2.7874	3.0782	3.2443	3.3606	3.4502	3.5230	3.5842	3.6371	3.6836	3.7251
26	2.7787	3.0669	3.2313	3.3464	3.4350	3.5069	3.5674	3.6197	3.6656	3.7066
27	2.7707	3.0565	3.2194	3.3334	3.4210	3.4922	3.5520	3.6037	3.6491	3.6896
28	2.7633	3.0469	3.2084	3.3214	3.4082	3.4786	3.5378	3.5889	3.6338	3.6739
29	2.7564	3.0380	3.1982	3.3102	3.3962	3.4660	3.5247	3.5753	3.6198	3.6594
30	2.7500	3.0298	3.1888	3.2999	3.3852	3.4544	3.5125	3.5626	3.6067	3.6460
35	2.7238	2.9960	3.1502	3.2577	3.3400	3.4068	3.4628	3.5110	3.5534	3.5911
40	2.7045	2.9712	3.1218	3.2266	3.3069	3.3718	3.4263	3.4732	3.5143	3.5510
45	2.6896	2.9521	3.1,000	3.2028	3.2815	3.3451	3.3984	3.4442	3.4845	3.5203
50	3.6778	2.9370	3.0828	3.1840	3.2614	3.3239	3.3763	3.4214	3.4609	3.4960
55	2.6682	2.9247	3.0688	3.1688	3.2451	3.3068	3.3585	3.4029	3.4418	3.4764
60	2.6603	2.9146	3.0573	3.1562	3.2317	3.2927	3.3437	3.3876	3.4260	3.4602
70	2.6479	2.8987	3.0393	3.1366	3.2108	3.2707	3.3208	3.3638	3.4015	3.4350
80	2.6387	2.8870	3.0259	3.1220	3.1953	3.2543	3.3037	3.3462	3.3833	3.4163
90	2.6316	2.8779	3.0156	3.1108	3.1833	3.2417	3.2906	3.3326	3.3693	3.4019
100	2.6259	2.8707	3.0073	3.1018	3.1737	3.2317	3.2802	3.3218	3.3582	3.3905
110	2.6213	2.8648	3.0007	3.0945	3.1660	3.2235	3.2717	3.3130	3.3491	3.3812
120	2.6174	2.8599	2.9951	3.0885	3.1595	3.2168	3.2646	3.3057	3.3416	3.3735
250	2.5956	2.8322	2.9637	3.0543	3.1232	3.1785	3.2248	3.2644	3.2991	3.3299
500	2.5857	2.8195	2.9494	3.0387	3.1066	3.1612	3.2067	3.2457	3.2798	3.3101
1,000	2.5808	2.8133	2.9423	3.0310	3.0984	3.1526	3.1977	3.2365	3.2703	3.3003
∞	2.5758	2.8070	2.9352	3.0233	3.0902	3.1440	3.1888	3.2272	3.2608	3.2905

df \ *p*	11	12	13	14	15	16	17	18	19
2	33.1436	34.6194	36.0347	37.3965	38.7105	39.9812	41.2129	42.4087	43.5718
3	13.3471	13.7450	14.1214	14.4787	14.8194	15.1451	15.4575	15.7577	16.0471
4	8.8271	9.0294	9.2192	9.3983	9.5679	9.7291	9.8828	10.0298	10.1708
5	7.0128	7.1464	7.2712	7.3884	7.4990	7.6037	7.7032	7.7981	7.8888
6	6.0680	6.1690	6.2630	6.3510	6.4338	6.5121	6.5862	6.6568	6.7240
7	5.4973	5.5799	5.6565	5.7282	5.7954	5.8588	5.9188	5.9757	6.0300
8	5.1183	5.1892	5.2549	5.3162	5.3737	5.4278	5.4789	5.5274	5.5735
9	4.8494	4.9124	4.9706	5.0249	5.0757	5.1235	5.1686	5.2114	5.2520
10	4.6492	4.7065	4.7594	4.8087	4.8547	4.8980	4.9388	4.9774	5.0141
11	4.4947	4.5477	4.5966	4.6420	4.6845	4.7244	4.7620	4.7975	4.8312
12	4.3719	4.4215	4.4673	4.5099	4.5496	4.5868	4.6219	4.6551	4.6865
13	4.2721	4.3191	4.3624	4.4026	4.4401	4.4752	4.5083	4.5396	4.5693
14	4.1894	4.2342	4.2755	4.3138	4.3495	4.3829	4.4144	4.4442	4.4724
15	4.1198	4.1628	4.2024	4.2391	4.2733	4.3054	4.3355	4.3640	4.3910
16	4.0604	4.1018	4.1400	4.1754	4.2084	4.2393	4.2683	4.2958	4.3217
17	4.0091	4.0493	4.0863	4.1205	4.1525	4.1823	4.2104	4.2369	4.2620
18	3.9644	4.0035	4.0394	4.0727	4.1037	4.1327	4.1600	4.1857	4.2101
19	3.9251	3.9632	3.9983	4.0307	4.0609	4.0892	4.1157	4.1408	4.1645
20	3.8903	3.9276	3.9618	3.9935	4.0230	4.0506	4.0765	4.1010	4.1241
21	3.8593	3.8958	3.9293	3.9603	3.9892	4.0162	4.0416	4.0655	4.0881
22	3.8314	3.8672	3.9001	3.9306	3.9589	3.9854	4.0103	4.0337	4.0559
23	3.8062	3.8414	3.8738	3.9037	3.9316	3.9576	3.9820	4.0050	4.0268
24	3.7834	3.8181	3.8499	3.8794	3.9068	3.9324	3.9564	3.9790	4.0004
25	3.7626	3.7968	3.8282	3.8572	3.8842	3.9094	3.9331	3.9554	3.9765
26	3.7436	3.7773	3.8083	3.8369	3.8635	3.8884	3.9118	3.9337	3.9545
27	3.7261	3.7595	3.7900	3.8183	3.8446	3.8692	3.8922	3.9139	3.9344
28	3.7101	3.7430	3.7732	3.8012	3.8271	3.8514	3.8742	3.8956	3.9159
29	3.6952	3.7278	3.7577	3.7853	3.8110	3.8350	3.8575	3.8787	3.8987
30	3.6814	3.7136	3.7433	3.7706	3.7961	3.8198	3.8421	3.8631	3.8829
35	3.6252	3.6561	3.6845	3.7108	3.7352	3.7579	3.7792	3.7993	3.8183
40	3.5840	3.6140	3.6415	3.6670	3.6906	3.7126	3.7333	3.7527	3.7710
45	3.5525	3.5818	3.6087	3.6335	3.6565	3.6780	3.6982	3.7171	3.7350
50	3.5277	3.5564	3.5828	3.6071	3.6297	3.6508	3.6705	3.6890	3.7066
55	3.5076	3.5359	3.5618	3.5858	3.6080	3.6287	3.6481	3.6664	3.6836
60	3.4910	3.5189	3.5445	3.5682	3.5901	3.6105	3.6297	3.6477	3.6646
70	3.4652	3.4926	3.5176	3.5408	3.5622	3.5822	3.6010	3.6186	3.6352
80	3.4460	3.4730	3.4977	3.5205	3.5416	3.5613	3.5797	3.5970	3.6134
90	3.4313	3.4579	3.4823	3.5048	3.5257	3.5451	3.5633	3.5804	3.5966
100	3.4196	3.4460	3.4701	3.4924	3.5131	3.5323	3.5503	3.5673	3.5832
110	3.4100	3.4362	3.4602	3.4823	3.5028	3.5219	3.5398	3.5565	3.5724
120	3.4021	3.4281	3.4520	3.4739	3.4943	3.5132	3.5310	3.5477	3.5634
250	3.3575	3.3826	3.4055	3.4267	3.4462	3.4645	3.4815	3.4976	3.5127
500	3.3373	3.3619	3.3845	3.4052	3.4245	3.4424	3.4591	3.4749	3.4897
1,000	3.3272	3.3517	3.3740	3.3946	3.4137	3.4314	3.4480	3.4636	3.4783
∞	3.3172	3.3415	3.3636	3.3840	3.4029	3.4205	3.4370	3.4524	3.4670

df \ *p*	20	21	28	36	45	55	66	78
2	44.7046	45.8094	52.9009	59.9875	67.0709	74.1519	81.2312	88.3091
3	16.3263	16.5964	18.2806	19.8889	21.4337	22.9239	24.3667	25.7675
4	10.3063	10.4367	11.2378	11.9851	12.6881	13.3540	13.9882	14.5946
5	7.9757	8.0591	8.5667	9.0332	9.4665	9.8722	10.2546	10.6168
6	6.7883	6.8500	7.2226	7.5617	7.8737	8.1636	8.4348	8.6901
7	6.0818	6.1313	6.4295	6.6987	6.9448	7.1721	7.3837	7.5819
8	5.6174	5.6594	5.9114	6.1375	6.3432	6.5323	6.7076	6.8712
9	5.2907	5.3276	5.5484	5.7458	5.9245	6.0883	6.2395	6.3803
10	5.0490	5.0823	5.2810	5.4578	3.6175	5.7634	5.8978	6.0225
11	4.8633	4.8939	5.0761	5.2378	5.3833	5.5160	5.6379	5.7509
12	4.7165	4.7450	4.9144	5.0644	5.1991	5.3216	5.4340	5.5380
13	4.5975	4.6243	4.7837	4.9244	5.0506	5.1651	5.2700	5.3670
14	4.4992	4.5247	4.6759	4.8091	4.9284	5.0364	5.1354	5.2266
15	4.4166	4.4410	4.5854	4.7125	4.8261	4.9289	5.0229	5.1094
16	4.3463	4.3698	4.5086	4.6305	4.7393	4.8377	4.9275	5.0102
17	4.2858	4.3085	4.4425	4.5600	4.6648	4.7594	4.8457	4.9251
18	4.2332	4.2551	4.3850	4.4987	4.6001	4.6915	4.7748	4.8514
19	4.1869	4.2083	4.3345	4.4450	4.5434	4.6320	4.7127	4.7868
20	4.1460	4.1669	4.2900	4.3976	4.4933	4.5795	4.6579	4.7299
21	4.1096	4.1300	4.2503	4.3554	4.4487	4.5328	4.6092	4.6794
22	4.0769	4.0969	4.2147	4.3175	4.4089	4.4910	4.5657	4.6342
23	4.0474	4.0671	4.1826	4.2835	4.3730	4.4534	4.5265	4.5935
24	4.0207	4.0400	4.1536	4.2527	4.3405	4.4194	4.4911	4.5567
25	3.9964	4.0154	4.1272	4.2246	4.3109	4.3885	4.4589	4.5233
26	3.9742	3.9929	4.1031	4.1990	4.2840	4.3602	4.4295	4.4928
27	3.9538	3.9723	4.0809	4.1755	4.2592	4.3344	4.4025	4.4649
28	3.9351	3.9533	4.0606	4.1539	4.2365	4.3106	4.3778	4.4392
29	3.9177	3.9357	4.0418	4.1339	4.2155	4.2886	4.3549	4.4155
30	3.9016	3.9195	4.0243	4.1154	4.1960	4.2683	4.3337	4.3936
35	3.8362	3.8533	3.9534	4.0403	4.1170	4.1857	4.2479	4.3047
40	3.7884	3.8049	3.9017	3.9855	4.0594	4.1256	4.1854	4.2399
45	3.7519	3.7680	3.8622	3.9437	4.0156	4.0798	4.1378	4.1907
50	3.7231	3.7389	3.8311	3.9108	3.9811	4.0438	4.1004	4.1520
55	3.6999	3.7154	3.8060	3.8843	3.9532	4.0147	4.0702	4.1208
60	3.6807	3.6960	3.7853	3.8624	3.9303	3.9908	4.0454	4.0951
70	3.6509	3.6658	3.7531	3.8284	3.8946	3.9537	4.0069	4.0553
80	3.6288	3.6435	3.7293	3.8033	3.8683	3.9262	3.9784	4.0259
90	3.6118	3.6263	3.7110	3.7839	3.8480	3.9051	3.9565	4.0032
100	3.5983	3.6127	3.6964	3.7686	3.8319	3.8883	3.9391	3.9853
110	3.5874	3.6016	3.6846	3.7561	3.8189	3.8747	3.9250	3.9707
120	3.5783	3.5924	3.6748	3.7458	3.8080	3.8634	3.9133	3.9586
250	3.5270	3.5405	3.6196	3.6875	3.7471	3.8000	3.8475	3.8907
500	3.5037	3.5170	3.5946	3.6612	3.7195	3.7713	3.8179	3.8601
1,000	4.4922	3.5054	3.5822	3.6481	3.7059	3.7571	3.8032	3.8449
∞	3.4808	3.4938	3.5699	3.6352	3.6923	3.7430	3.7886	3.8299

$df \setminus p$	91	105	120	136	153	171	190
2	95.3861	102.4622	109.5377	116.6126	123.6871	130.7612	137.8350
3	27.1309	28.4606	29.7598	31.0310	32.2766	33.4985	34.6984
4	15.1768	15.7375	16.2788	16.8026	17.3105	17.8040	18.2841
5	10.9616	11.2910	11.6067	11.9102	12.2025	12.4848	12.7578
6	8.9317	9.1612	9.3800	9.5893	9.7901	9.9831	10.1692
7	7.7685	7.9452	8.1130	8.2729	8.4258	8.5724	8.7132
8	7.0248	7.1696	7.3069	7.4373	7.5617	7.6806	7.7947
9	6.5121	6.6361	6.7533	6.8645	6.9073	7.0713	7.1679
10	6.1391	6.2485	6.3517	6.4495	6.5423	6.6308	6.7154
11	5.8562	5.9550	6.0480	6.1359	6.2193	6.2987	6.3745
12	5.6348	5.7254	5.8107	5.8911	5.9674	6.0399	6.1091
13	5.4571	5.5413	5.6204	5.6951	5.7658	5.8329	5.8969
14	5.3113	5.3904	5.4647	5.5347	5.6009	5.6637	5.7235
15	5.1897	5.2647	5.3350	5.4011	5.4637	5.5230	5.5794
16	5.0868	5.1583	5.2252	5.2882	5.3478	5.4042	5.4578
17	4.9986	5.0671	5.1313	5.1916	5.2486	5.3025	5.3538
18	4.9222	4.9882	5.0500	5.1080	5.1628	5.2146	5.2639
19	4.8554	4.9192	4.9789	5.0350	5.0879	5.1379	5.1854
20	4.7965	4.8584	4.9163	4.9707	5.0219	5.0704	5.1163
21	4.7442	4.8044	4.8607	4.9136	4.9633	5.0104	5.0551
22	4.6974	4.7562	4.8111	4.8625	4.9111	4.9569	5.0004
23	4.6553	4.7128	4.7664	4.8167	4.8641	4.9089	4.9513
24	4.6173	4.6736	4.7261	4.7753	4.8217	4.8654	4.9070
25	4.5828	4.6380	4.6894	4.7377	4.7831	4.8261	4.8667
26	4.5513	4.6055	4.6560	4.7034	4.7480	4.7901	4.8300
27	4.5224	4.5757	4.6255	4.6721	4.7159	4.7573	4.7965
28	4.4959	4.5484	4.5974	4.6432	4.6864	4.7271	4.7657
29	4.4714	4.5232	4.5714	4.6166	4.6591	4.6993	4.7372
30	4.4487	4.4998	4.5475	4.5921	4.6340	4.6735	4.7110
35	4.3569	4.4053	4.4503	4.4924	4.5320	4.5694	4.6047
40	4.2901	4.3365	4.3797	4.4201	4.4580	4.4937	4.5275
45	4.2393	4.2843	4.3261	4.3651	4.4018	4.4363	4.4689
50	4.1994	4.2432	4.2840	4.3220	4.3577	4.3913	4.4230
55	4.1672	4.2101	4.2500	4.2872	4.3222	4.3550	4.3860
60	4.1408	4.1829	4.2221	4.2586	4.2929	4.3252	4.3556
70	4.0997	4.1407	4.1788	4.2143	4.2476	4.2790	4.3085
80	4.0694	4.1096	4.1469	4.1816	4.2142	4.2449	4.2738
90	4.0461	4.0856	4.1223	4.1565	4.1886	4.2187	4.2471
100	4.0276	4.0666	4.1028	4.1366	4.1682	4.1979	4.2260
110	4.0126	4.0512	4.0870	4.1204	4.1517	4.1811	4.2088
120	4.0001	4.0384	4.0739	4.1070	4.1380	4.1671	4.1946
250	3.9303	3.9667	4.0004	4.0318	4.0612	4.0889	4.1149
500	3.8987	3.9343	3.9673	3.9980	4.0266	4.0536	4.0790
1,000	3.8831	3.9183	3.9509	3.9812	4.0095	4.0362	4.0612
∞	3.8676	3.9024	3.9346	3.9646	3.9926	4.0189	4.0436

Source: Reprinted from Bailey, B.J.R., *J. Am. Stat. Assoc.*, 72, 469–478, 1977, Table 2. With permission from the *Journal of the American Statistical Association*.

Note: Used first in Chapter 5.

Appendix 15: Bonferroni's Adjusted *t* Values for Two-Sided Test and $\alpha = 0.05$

df \ *p*	1	2	3	4	5	6	7	8	9	10
2	4.3027	6.2053	7.6488	8.8602	9.9248	10.8859	11.7687	12.5897	13.3604	14.0890
3	3.1824	4.1765	4.8567	5.3919	5.8409	6.2315	6.5797	6.8952	7.1849	7.4533
4	2.7764	3.4954	3.9608	4.3147	4.6041	4.8510	5.0675	5.2611	5.4366	5.5976
5	2.5706	3.1634	3.5341	3.8100	4.0321	4.2193	4.3818	4.5257	4.6553	4.7733
6	2.4469	2.9687	3.2875	3.5212	3.7074	3.8630	3.9971	4.1152	4.2209	4.3168
7	2.3646	2.8412	3.1276	3.3353	3.4995	3.6358	3.7527	3.8552	3.9467	4.0293
8	2.3060	2.7515	3.0158	3.2060	3.3554	3.4789	3.5844	3.6766	3.7586	3.8325
9	2.2622	2.6850	2.9333	3.1109	3.2498	3.3642	3.4616	3.5465	3.6219	3.6897
10	2.2281	2.6338	2.8701	3.0382	3.1693	3.2768	3.3682	3.4477	3.5182	3.5814
11	2.2010	2.5931	2.8200	2.9809	3.1058	3.2081	3.2949	3.3702	3.4368	3.4966
12	2.1788	2.5600	2.7795	2.9345	3.0545	3.1527	3.2357	3.3078	3.3714	3.4284
13	2.1604	2.5326	2.7459	2.8961	3.0123	3.1070	3.1871	3.2565	3.3177	3.3725
14	2.1448	2.5096	2.7178	2.8640	2.9768	3.0688	3.1464	3.2135	3.2727	3.3257
15	2.1314	2.4899	2.6937	2.8366	2.9467	3.0363	3.1118	3.1771	3.2346	3.2860
16	2.1199	2.4729	2.6730	2.8131	2.9208	3.0083	3.0821	3.1458	3.2019	3.2520
17	2.1098	2.4581	2.6550	2.7925	2.8982	2.9840	3.0563	3.1186	3.1735	3.2224
18	2.1009	2.4450	2.6391	2.7745	2.8784	2.9627	3.0336	3.0948	3.1486	3.1966
19	2.0930	2.4334	2.6251	2.7586	2.8609	2.9439	3.0136	3.0738	3.1266	3.1737
20	2.0860	2.4231	2.6126	2.7444	2.8453	2.9271	2.9958	3.0550	3.1070	3.1534
21	2.0796	2.4138	2.6013	2.7316	2.8314	2.9121	2.9799	3.0382	3.0895	3.1352
22	2.0739	2.4055	2.5912	2.7201	2.8188	2.8985	2.9655	3.0231	3.0737	3.1188
23	2.0687	2.3979	2.5820	2.7097	2.8073	2.8863	2.9525	3.0095	3.0595	3.1040
24	2.0639	2.3909	2.5736	2.7002	2.7969	2.8751	2.9406	2.9970	3.0465	3.0905
25	2.0595	2.3846	2.5660	2.6916	2.7874	2.8649	2.9298	2.9856	3.0346	3.0782
26	2.0555	2.3788	2.5589	2.6836	2.7787	2.8555	2.9199	2.9752	3.0237	3.0669
27	2.0518	2.3734	2.5525	2.6763	2.7707	2.8469	2.9107	2.9656	3.0137	3.0565
28	2.0484	2.3685	2.5465	2.6695	2.7633	2.8389	2.9023	2.9567	3.0045	3.0469
29	2.0452	2.3638	2.5409	2.6632	2.7564	2.8316	2.8945	2.9485	2.9959	3.0380
30	2.0423	2.3596	2.5357	2.6574	2.7500	2.8247	2.8872	2.9409	2.9880	3.0298
35	2.0301	2.3420	2.5145	2.6334	2.7238	2.7966	2.8575	2.9097	2.9554	2.9960
40	2.0211	2.3289	2.4989	2.6157	2.7045	2.7759	2.8355	2.8867	2.9314	2.9712
45	2.0141	2.3189	2.4868	2.6021	2.6896	2.7599	2.8187	2.8690	2.9130	2.9521
50	2.0086	2.3109	2.4772	2.5913	2.6778	2.7473	2.8053	2.8550	2.8984	2.9370
55	2.0040	2.3044	2.4694	2.5825	2.6682	2.7370	2.7944	2.8436	2.8866	2.9247
60	2.0003	2.2990	2.4630	2.5752	2.6603	2.7286	2.7855	2.8342	2.8768	2.9146
70	1.9944	2.2906	2.4529	2.5639	2.6479	2.7153	2.7715	2.8195	2.8615	2.8987
80	1.9901	2.2844	2.4454	2.5554	2.6387	2.7054	2.7610	2.8086	2.8502	2.8870
90	1.9867	2.2795	2.4395	2.5489	2.6316	2.6978	2.7530	2.8002	2.8414	2.8779
100	1.9840	2.2757	2.4349	2.5437	2.6259	2.6918	2.7466	2.7935	2.8344	2.8707
110	1.9818	2.2725	2.4311	2.5394	2.6213	2.6868	2.7414	2.7880	2.8287	2.8648
120	1.9799	2.2699	2.4280	2.5359	2.6174	2.6827	2.7370	2.7835	2.8240	2.8599
250	1.9695	2.2550	2.4102	2.5159	2.5956	2.6594	2.7124	2.7577	2.7972	2.8322
500	1.9647	2.2482	2.4021	2.5068	2.5857	2.6488	2.7012	2.7460	2.7850	2.8195
1,000	1.9623	2.2448	2.3980	2.5022	2.5808	2.6435	2.6957	2.7402	2.7790	2.8133
∞	1.9600	2.2414	2.3940	2.4977	2.5758	2.6383	2.6901	2.7344	2.7729	2.8070

df \ *p*	11	12	13	14	15	16	17	18	19
2	14.7818	15.4435	16.0780	16.6883	17.2772	17.8466	18.3984	18.9341	19.4551
3	7.7041	7.9398	8.1625	8.3738	8.5752	8.7676	8.9521	9.1294	9.3001
4	5.7465	5.8853	6.0154	6.1380	6.2541	6.3643	6.4693	6.5697	6.6659
5	4.8819	4.9825	5.0764	5.1644	5.2474	5.3259	5.4005	5.4715	5.5393
6	4.4047	4.4858	4.5612	4.6317	4.6979	4.7604	4.8196	4.8759	4.9295
7	4.1048	4.1743	4.2388	4.2989	4.3553	4.4084	4.4586	4.5062	4.5514
8	3.8999	3.9618	4.0191	4.0724	4.1224	4.1693	4.2137	4.2556	4.2955
9	3.7513	3.8079	3.8602	3.9088	3.9542	3.9969	4.0371	4.0752	4.1114
10	3.6388	3.6915	3.7401	3.7852	3.8273	3.8669	3.9041	3.9394	3.9728
11	3.5508	3.6004	3.6462	3.6887	3.7283	3.7654	3.8004	3.8335	3.8648
12	3.4801	3.5274	3.5709	3.6112	3.6489	3.6842	3.7173	3.7487	3.7783
13	3.4221	3.4674	3.5091	3.5478	3.5838	3.6176	3.6493	3.6793	3.7076
14	3.3736	3.4173	3.4576	3.4949	3.5296	3.5621	3.5926	3.6214	3.6487
15	3.3325	3.3749	3.4139	3.4501	3.4837	3.5151	3.5447	3.5725	3.5989
16	3.2973	3.3386	3.3765	3.4116	3.4443	3.4749	3.5036	3.5306	3.5562
17	3.2667	3.3070	3.3440	3.3783	3.4102	3.4400	3.4680	3.4944	3.5193
18	3.2399	3.2794	3.3156	3.3492	3.3804	3.4095	3.4369	3.4626	3.4870
19	3.2163	3.2550	3.2906	3.3235	3.3540	3.3826	3.4094	3.4347	3.4585
20	3.1952	3.2333	3.2683	3.3006	3.3306	3.3587	3.3850	3.4098	3.4332
21	3.1764	3.2139	3.2483	3.2802	3.3097	3.3373	3.3632	3.3876	3.4106
22	3.1595	3.1965	3.2304	3.2618	3.2909	3.3181	3.3436	3.3676	3.3903
23	3.1441	3.1807	3.2142	3.2451	3.2739	3.3007	3.3259	3.3495	3.3719
24	3.1302	3.1663	3.1994	3.2300	3.2584	3.2849	3.3097	3.3331	3.3552
25	3.1175	3.1532	3.1859	3.2162	3.2443	3.2705	3.2950	3.3181	3.3400
26	3.1058	3.1412	3.1736	3.2035	3.2313	3.2572	3.2815	3.3044	3.3260
27	3.0951	3.1301	3.1622	3.1919	3.2194	3.2451	3.2691	3.2918	3.3132
28	3.0852	3.1199	3.1517	3.1811	3.2084	3.2339	3.2577	3.2801	3.3013
29	3.0760	3.1105	3.1420	3.1712	3.1982	3.2235	3.2471	3.2694	3.2904
30	3.0675	3.1017	3.1330	3.1620	3.1888	3.2138	3.2373	3.2594	3.2802
35	3.0326	3.0658	3.0962	3.1242	3.1502	3.1744	3.1971	3.2185	3.2386
40	3.0069	3.0393	3.0690	3.0964	3.1218	3.1455	3.1676	3.1884	3.2081
45	2.9872	3.0191	3.0482	3.0751	3.1,000	3.1232	3.1450	3.1654	3.1846
50	2.9716	3.0030	3.0318	3.0582	3.0828	3.1057	3.1271	3.1472	3.1661
55	2.9589	2.9900	3.0184	3.0446	3.0688	3.0914	3.1125	3.1324	3.1511
60	2.9485	2.9792	3.0074	3.0333	3.0573	3.0796	3.1005	3.1202	3.1387
70	2.9321	2.9624	2.9901	3.0156	3.0393	3.0613	3.0818	3.1012	3.1194
80	2.9200	2.9500	2.9773	3.0026	3.0259	3.0476	3.0679	3.0870	3.1050
90	2.9106	2.9403	2.9675	2.9924	3.0156	3.0371	3.0572	3.0761	3.0939
100	2.9032	2.9327	2.9596	2.9844	3.0073	3.0287	3.0487	3.0674	3.0851
110	2.8971	2.9264	2.9532	2.9778	3.0007	3.0219	3.0417	3.0604	3.0779
120	2.8921	2.9212	2.9479	2.9724	2.9951	3.0162	3.0360	3.0545	3.0720
250	2.8635	2.8919	2.9178	2.9416	2.9637	2.9842	3.0034	3.0213	3.0383
500	2.8505	2.8785	2.9041	2.9276	2.9494	2.9696	2.9885	3.0063	3.0230
1,000	2.8440	2.8719	2.8973	2.9207	2.9423	2.9624	2.9812	2.9988	3.0154
∞	2.8376	2.8653	2.8905	2.9137	2.9352	2.9552	2.9738	2.9913	3.0078

df \ *p*	20	21	28	36	45	55	66	78
2	19.9625	20.4573	23.6326	26.8049	29.9750	33.1436	36.3112	39.4778
3	9.4649	9.6242	10.6166	11.5632	12.4715	13.3471	14.1943	15.0165
4	6.7583	6.8471	7.3924	7.8998	8.3763	8.8271	9.2558	9.6655
5	5.6042	5.6665	6.0447	6.3914	6.7126	7.0128	7.2952	7.5625
6	4.9807	5.0297	5.3255	5.5937	5.8399	6.0680	6.2810	6.4813
7	4.5946	4.6359	4.8839	5.1068	5.3101	5.4973	5.6712	5.8339
8	4.3335	4.3699	4.5869	4.7810	4.9570	5.1183	5.2675	5.4065
9	4.1458	4.1786	4.3744	4.5485	4.7058	4.8494	4.9818	5.1048
10	4.0045	4.0348	4.2150	4.3747	4.5184	4.6492	4.7695	4.8810
11	3.8945	3.9229	4.0913	4.2400	4.3735	4.4947	4.6059	4.7087
12	3.8065	3.8334	3.9925	4.1327	4.2582	4.3719	4.4761	4.5722
13	3.7345	3.7602	3.9118	4.0452	4.1643	4.2721	4.3706	4.4614
14	3.6746	3.6992	3.8448	3.9725	4.0865	4.1894	4.2833	4.3698
15	3.6239	3.6477	3.7882	3.9113	4.0209	4.1198	4.2099	4.2928
16	3.5805	3.6036	3.7398	3.8589	3.9649	4.0604	4.1473	4.2272
17	3.5429	3.5654	3.6980	3.8137	3.9165	4.0091	4.0933	4.1706
18	3.5101	3.5321	3.6614	3.7742	3.8744	3.9644	4.0463	4.1214
19	3.4812	3.5027	3.6292	3.7395	3.8373	3.9251	4.0050	4.0781
20	3.4554	3.4765	3.6006	3.7087	3.8044	3.8903	3.9683	4.0398
21	3.4325	3.4532	3.5751	3.6812	3.7750	3.8593	3.9357	4.0056
22	3.4118	3.4322	3.5522	3.6564	3.7487	3.8314	3.9064	3.9750
23	3.3931	3.4132	3.5314	3.6341	3.7249	3.8062	3.8800	3.9474
24	3.3761	3.3960	3.5126	3.6139	3.7033	3.7834	3.8560	3.9223
25	3.3606	3.3803	3.4955	3.5954	3.6836	3.7626	3.8342	3.8995
26	3.3464	3.3659	3.4797	3.5785	3.6656	3.7436	3.8142	3.8787
27	3.3334	3.3526	3.4653	3.5629	3.6491	3.7261	3.7959	3.8595
28	3.3214	3.3404	3.4520	3.5486	3.6338	3.7101	3.7790	3.8419
29	3.3102	3.3291	3.4397	3.5354	3.6198	3.6952	3.7634	3.8256
30	3.2999	3.3186	3.4282	3.5231	3.6067	3.6814	3.7489	3.8105
35	3.2577	3.2758	3.3816	3.4730	3.5534	3.6252	3.6900	3.7490
40	3.2266	3.2443	3.3473	3.4362	3.5143	3.5840	3.6468	3.7040
45	3.2028	3.2201	3.3211	3.4081	3.4845	3.5525	3.6138	3.6696
50	3.1840	3.2010	3.3003	3.3858	3.4609	3.5277	3.5878	3.6425
55	3.1688	3.1856	3.2836	3.3679	3.4418	3.5076	3.5668	3.6206
60	3.1562	3.1728	3.2697	3.3530	3.4260	3.4910	3.5494	3.6025
70	3.1366	3.1529	3.2481	3.3299	3.4015	3.4652	3.5224	6.5744
80	3.1220	3.1381	3.2321	3.3127	3.3833	3.4460	3.5024	3.5536
90	3.1108	3.1267	3.2197	3.2995	3.3693	3.4313	3.4870	3.5375
100	3.1018	3.1176	3.2099	3.2890	3.3582	3.4196	3.4747	3.5248
110	3.0945	3.1102	3.2018	3.2804	3.3491	3.4100	3.4648	3.5144
120	3.0885	3.1041	3.1952	3.2733	3.3416	3.4021	3.4565	3.5058
250	3.0543	3.0694	3.1577	3.2332	3.2991	3.3575	3.4099	3.4573
500	3.0387	3.0537	3.1406	3.2150	3.2798	3.3373	3.3887	3.4354
1,000	3.0310	3.0459	3.1322	3.2059	3.2703	3.3272	3.3783	3.4245
∞	3.0233	3.0381	3.1237	3.1970	3.2608	3.3172	3.3678	3.4136

df \ p	**91**	**105**	**120**	**136**	**153**	**171**	**190**
2	42.6439	45.8094	48.9745	52.1392	55.3037	58.4679	61.6320
3	15.8165	16.5964	17.3582	18.1035	18.8336	19.5497	20.2528
4	10.0585	10.4367	10.8016	11.1545	11.4966	11.8288	12.1519
5	7.8166	8.0591	8.2913	8.5143	8.7290	8.9362	9.1365
6	6.6705	6.8500	7.0210	7.1844	7.3410	7.4914	7.6363
7	5.9868	6.1313	6.2684	6.3990	6.5236	6.6430	6.7577
8	5.5368	5.6594	5.7755	5.8857	5.9906	6.0909	6.1869
9	5.2197	5.3276	5.4295	5.5260	5.6177	5.7051	5.7888
10	4.9849	5.0823	5.1740	5.2608	5.3431	5.4215	5.4963
11	4.8044	4.8939	4.9781	5.0576	5.1330	5.2046	5.2729
12	4.6615	4.7450	4.8233	4.8972	4.9672	5.0336	5.0969
13	4.5457	4.6243	4.6981	4.7675	4.8332	4.8956	4.9549
14	4.4500	4.5247	4.5947	4.6606	4.7228	4.7818	4.8379
15	4.3695	4.4410	4.5079	4.5708	4.6302	4.6865	4.7400
16	4.3011	4.3698	4.4341	4.4946	4.5516	4.6056	4.6568
17	4.2421	4.3085	4.3706	4.4289	4.4839	4.5360	4.5853
18	4.1907	4.2551	4.3154	4.3719	4.4251	4.4755	4.5232
19	4.1456	4.2083	4.2669	4.3218	4.3736	4.4225	4.4688
20	4.1057	4.1669	4.2240	4.2776	4.3280	4.3756	4.4208
21	4.0701	4.1300	4.1858	4.2381	4.2874	4.3339	4.3780
22	4.0382	4.0969	4.1516	4.2028	4.2510	4.2966	4.3397
23	4.0095	4.0671	4.1207	4.1710	4.2183	4.2629	4.3052
24	3.9834	4.0400	4.0928	4.1422	4.1886	4.2325	4.2739
25	3.9597	4.0154	4.0674	4.1160	4.1616	4.2047	4.2455
26	3.9380	3.9929	4.0441	4.0920	4.1370	4.1794	4.2196
27	3.9181	3.9723	4.0228	4.0700	4.1144	4.1562	4.1958
28	3.8997	3.9533	4.0032	4.0498	4.0936	4.1349	4.1739
29	3.8828	3.9357	3.9850	4.0311	4.0744	4.1151	4.1537
30	3.8671	3.9195	3.9682	4.0138	4.0566	4.0969	4.1350
35	3.8032	3.8533	3.8999	3.9434	3.9842	4.0226	4.0590
40	3.7564	3.8049	3.8499	3.8919	3.9314	3.9684	4.0035
45	3.7208	3.7680	3.8118	3.8527	3.8911	3.9271	3.9612
50	3.6926	3.7389	3.7818	3.8218	3.8594	3.8946	3.9279
55	3.6699	3.7154	3.7576	3.7969	3.8337	3.8684	3.9010
60	3.6511	3.6960	3.7376	3.7763	3.8126	3.8467	3.8789
70	3.6220	3.6658	3.7065	3.7444	3.7798	3.8131	3.8445
80	3.6004	3.6435	3.6835	3.7207	3.7555	3.7883	3.8191
90	3.5837	3.6263	3.6658	3.7025	3.7369	3.7691	3.7995
100	3.5705	3.6127	3.6517	3.6880	3.7220	3.7539	3.7840
110	3.5598	3.6016	3.6403	3.6763	3.7100	3.7416	3.7714
120	3.5509	3.5924	3.6308	3.6665	3.7000	3.7313	3.7609
250	3.5007	3.5405	3.5774	3.6117	3.6437	3.6737	3.7020
500	3.4779	3.5170	3.5532	3.5868	3.6182	3.6477	3.6754
1,000	3.4666	3.5054	3.5412	3.5745	3.6056	3.6348	3.6622
∞	3.4554	3.4938	3.5293	3.5623	3.5931	3.6219	3.6491

Source: Reprinted from Bailey, B.J.R., *J. Am. Stat. Assoc.*, 72, 469–478, 1977, Table 1. With permission from the *Journal of the American Statistical Association*.

Note: Used first in Chapter 5.

Appendix 16: Dunn–Šidák's *t* for Comparisons between *p* Treatment Means and a Control for $\alpha = 0.01, 0.05, 0.10$, and 0.20 (One-Sided Test)

df \ *p*	α	2	3	4	5	6	7	8	9	10	15	20	25	30	35	40	45	50
2	0.01	9.912	12.166	14.062	15.732	17.241	18.628	19.919	21.130	22.277	27.295	31.524	35.249	38.617	41.713	44.595	47.302	49.862
	0.05	4.273	5.292	6.144	6.892	7.566	8.185	8.760	9.300	9.810	12.040	13.918	15.571	17.064	18.437	19.715	20.914	22.049
	0.10	2.876	3.606	4.213	4.742	5.219	5.655	6.060	6.439	6.798	8.363	9.678	10.835	11.879	12.839	13.732	14.571	15.363
	0.20	1.815	2.348	2.783	3.159	3.495	3.802	4.085	4.351	4.601	5.689	6.601	7.401	8.123	8.785	9.401	9.979	10.526
3	0.01	5.836	6.733	7.444	8.041	8.563	9.028	9.451	9.839	10.199	11.706	12.903	13.913	14.795	15.583	16.299	16.958	17.569
	0.05	3.166	3.716	4.146	4.506	4.819	5.097	5.349	5.580	5.793	6.685	7.391	7.985	8.503	8.965	9.385	9.770	10.128
	0.10	2.325	2.779	3.132	3.425	3.678	3.902	4.105	4.290	4.462	5.176	5.738	6.211	6.623	6.990	7.323	7.628	7.911
	0.20	1.585	1.971	2.264	2.505	2.712	2.895	3.059	3.209	3.348	3.920	4.368	4.743	5.069	5.360	5.622	5.863	6.087
4	0.01	4.601	5.162	5.592	5.945	6.247	6.513	6.750	6.966	7.165	7.975	8.600	9.114	9.556	9.945	10.293	10.610	10.902
	0.05	2.764	3.169	3.474	3.723	3.935	4.121	4.286	4.436	4.574	5.132	5.560	5.912	6.212	6.477	6.713	6.928	7.126
	0.10	2.109	2.469	2.738	2.956	3.141	3.302	3.445	3.575	3.693	4.173	4.539	4.838	5.094	5.319	5.519	5.702	5.869
	0.20	1.487	1.817	2.059	2.254	2.417	2.558	2.684	2.797	2.900	3.314	3.628	3.885	4.103	4.294	4.464	4.618	4.760
5	0.01	4.030	4.452	4.769	5.026	5.242	5.431	5.599	5.750	5.887	6.443	6.862	7.202	7.491	7.743	7.967	8.170	8.355
	0.05	2.560	2.897	3.146	3.346	3.514	3.660	3.788	3.904	4.009	4.430	4.746	5.002	5.218	5.406	5.573	5.723	5.860
	0.10	1.995	2.309	2.538	2.721	2.874	3.006	3.123	3.227	3.322	3.700	3.982	4.210	4.402	4.568	4.716	4.849	4.970
	0.20	1.434	1.734	1.951	2.122	2.264	2.385	2.492	2.588	2.674	3.017	3.270	3.474	3.645	3.794	3.925	4.043	4.150
6	0.01	3.705	4.055	4.313	4.520	4.694	4.844	4.976	5.095	5.203	5.633	5.953	6.211	6.428	6.616	6.782	6.932	7.068
	0.05	2.438	2.736	2.954	3.127	3.270	3.394	3.503	3.600	3.688	4.037	4.295	4.501	4.674	4.824	4.956	5.074	5.182
	0.10	1.924	2.211	2.418	2.581	2.716	2.832	2.934	3.024	3.106	3.428	3.666	3.855	4.013	4.150	4.270	4.378	4.476
	0.20	1.400	1.683	1.884	2.041	2.170	2.281	2.377	2.462	2.539	2.841	3.061	3.236	3.382	3.507	3.618	3.717	3.806
7	0.01	3.498	3.803	4.026	4.204	4.352	4.479	4.591	4.691	4.781	5.139	5.403	5.614	5.791	5.943	6.077	6.197	6.306
	0.05	2.356	2.630	2.828	2.984	3.112	3.223	3.319	3.405	3.482	3.787	4.010	4.187	4.335	4.462	4.574	4.673	4.764
	0.10	1.877	2.146	2.338	2.488	2.612	2.717	2.809	2.891	2.965	3.253	3.463	3.629	3.767	3.885	3.989	4.082	4.166
	0.20	1.376	1.648	1.839	1.987	2.108	2.211	2.300	2.379	2.450	2.725	2.925	3.082	3.212	3.323	3.421	3.507	3.586
8	0.01	3.354	3.630	3.830	3.988	4.119	4.232	4.330	4.418	4.497	4.809	5.037	5.219	5.370	5.499	5.613	5.715	5.807
	0.05	2.298	2.555	2.739	2.883	3.002	3.103	3.191	3.269	3.340	3.615	3.815	3.973	4.104	4.217	4.315	4.403	4.482
	0.10	1.843	2.099	2.281	2.422	2.538	2.636	2.722	2.798	2.866	3.131	3.322	3.473	3.597	3.704	3.797	3.880	3.954
	0.20	1.359	1.622	1.806	1.948	2.063	2.160	2.245	2.319	2.386	2.644	2.829	2.974	3.093	3.195	3.284	3.363	3.434
9	0.01	3.248	3.503	3.687	3.832	3.951	4.054	4.143	4.222	4.294	4.574	4.777	4.939	5.072	5.187	5.287	5.376	5.457
	0.05	2.254	2.499	2.673	2.809	2.920	3.015	3.097	3.170	3.235	3.490	3.674	3.819	3.938	4.040	4.129	4.208	4.280
	0.10	1.817	2.064	2.238	2.373	2.483	2.576	2.657	2.729	2.793	3.041	3.219	3.358	3.473	3.571	3.657	3.733	3.801
	0.20	1.346	1.603	1.782	1.918	2.029	2.123	2.204	2.275	2.338	2.583	2.758	2.894	3.006	3.101	3.183	3.257	3.322
10	0.01	3.168	3.407	3.579	3.714	3.825	3.919	4.002	4.075	4.141	4.398	4.584	4.730	4.851	4.955	5.046	5.126	5.199
	0.05	2.221	2.456	2.623	2.752	2.858	2.947	3.025	3.094	3.156	3.395	3.567	3.701	3.812	3.907	3.989	4.062	4.128
	0.10	1.797	2.037	2.205	2.335	2.441	2.530	2.607	2.675	2.736	2.972	3.140	3.271	3.379	3.471	3.551	3.622	3.685
	0.20	1.336	1.588	1.762	1.895	2.003	2.094	2.172	2.240	2.302	2.537	2.703	2.833	2.939	3.028	3.107	3.176	3.238

11	0.01	3.104	3.332	3.494	3.621	3.726	3.815	3.892	3.960	4.022	4.262	4.434	4.570	4.682	4.777	4.860	4.934	5.001
	0.05	2.194	2.422	2.582	2.707	2.808	2.894	2.968	3.034	3.093	3.320	3.483	3.610	3.714	3.803	3.880	3.949	4.010
	0.10	1.780	2.015	2.179	2.304	2.407	2.493	2.567	2.633	2.692	2.918	3.078	3.203	3.305	3.392	3.468	3.535	3.595
	0.20	1.328	1.576	1.747	1.877	1.982	2.070	2.146	2.213	2.272	2.500	2.660	2.784	2.886	2.972	3.046	3.112	3.171
12	0.01	3.053	3.271	3.426	3.547	3.647	3.731	3.804	3.869	3.927	4.153	4.315	4.442	4.547	4.636	4.714	4.783	4.844
	0.05	2.172	2.394	2.550	2.670	2.768	2.851	2.922	2.986	3.042	3.260	3.415	3.536	3.635	3.720	3.793	3.858	3.916
	0.10	1.767	1.997	2.157	2.280	2.379	2.463	2.535	2.599	2.656	2.873	3.028	3.148	3.246	3.329	3.401	3.465	3.522
	0.20	1.321	1.566	1.734	1.862	1.965	2.051	2.125	2.190	2.248	2.469	2.625	2.745	2.843	2.925	2.997	3.060	3.117
13	0.01	3.011	3.221	3.371	3.487	3.582	3.662	3.732	3.794	3.850	4.064	4.218	4.339	4.437	4.522	4.595	4.660	4.718
	0.05	2.153	2.370	2.523	2.640	2.735	2.815	2.885	2.946	3.000	3.211	3.360	3.476	3.571	3.652	3.721	3.783	3.839
	0.10	1.756	1.982	2.139	2.259	2.356	2.438	2.508	2.570	2.626	2.837	2.986	3.102	3.197	3.277	3.346	3.407	3.462
	0.20	1.315	1.557	1.723	1.849	1.950	2.035	2.108	2.172	2.228	2.444	2.596	2.712	2.807	2.887	2.957	3.018	3.073
14	0.01	2.976	3.179	3.324	3.436	3.527	3.605	3.672	3.732	3.785	3.991	4.138	4.253	4.347	4.427	4.497	4.558	4.614
	0.05	2.138	2.351	2.500	2.614	2.707	2.785	2.853	2.912	2.965	3.169	3.314	3.426	3.517	3.595	3.662	3.721	3.775
	0.10	1.746	1.969	2.124	2.242	2.337	2.417	2.486	2.546	2.600	2.807	2.952	3.064	3.156	3.233	3.300	3.359	3.412
	0.20	1.310	1.550	1.714	1.839	1.938	2.022	2.093	2.156	2.212	2.423	2.571	2.685	2.778	2.856	2.923	2.982	3.035
15	0.01	2.945	3.144	3.284	3.393	3.482	3.557	3.622	3.679	3.731	3.929	4.070	4.181	4.271	4.348	4.414	4.473	4.526
	0.05	2.125	2.334	2.480	2.592	2.683	2.760	2.826	2.884	2.935	3.134	3.274	3.383	3.472	3.547	3.612	3.669	3.721
	0.10	1.738	1.959	2.111	2.227	2.321	2.399	2.467	2.526	2.579	2.781	2.922	3.032	3.121	3.196	3.261	3.319	3.370
	0.20	1.306	1.544	1.707	1.829	1.928	2.010	2.081	2.142	2.197	2.405	2.550	2.662	2.752	2.828	2.894	2.952	3.004
16	0.01	2.920	3.113	3.250	3.356	3.442	3.515	3.578	3.634	3.684	3.876	4.013	4.119	4.206	4.280	4.344	4.401	4.451
	0.05	2.113	2.320	2.463	2.573	2.663	2.738	2.802	2.859	2.910	3.104	3.240	3.346	3.433	3.506	3.569	3.624	3.674
	0.10	1.731	1.949	2.099	2.214	2.306	2.384	2.450	2.508	2.560	2.758	2.897	3.004	3.091	3.164	3.228	3.284	3.334
	0.20	1.303	1.539	1.700	1.821	1.919	2.000	2.070	2.131	2.185	2.389	2.532	2.642	2.730	2.805	2.869	2.926	2.976
17	0.01	2.897	3.087	3.221	3.324	3.408	3.479	3.541	3.595	3.644	3.830	3.963	4.066	4.150	4.222	4.284	4.338	4.387
	0.05	2.103	2.307	2.449	2.557	2.645	2.718	2.782	2.838	2.887	3.077	3.211	3.315	3.399	3.470	3.531	3.586	3.634
	0.10	1.725	1.941	2.090	2.203	2.294	2.370	2.436	2.493	2.544	2.739	2.875	2.980	3.065	3.137	3.199	3.253	3.302
	0.20	1.299	1.534	1.694	1.814	1.911	1.991	2.060	2.120	2.174	2.376	2.516	2.624	2.711	2.785	2.848	2.903	2.953
18	0.01	2.877	3.064	3.195	3.296	3.378	3.448	3.508	3.561	3.608	3.790	3.920	4.020	4.102	4.171	4.231	4.284	4.332
	0.05	2.094	2.296	2.436	2.543	2.629	2.702	2.764	2.819	2.868	3.054	3.186	3.287	3.369	3.439	3.499	3.552	3.599
	0.10	1.720	1.934	2.081	2.193	2.283	2.358	2.423	2.480	2.530	2.721	2.855	2.958	3.042	3.113	3.173	3.227	3.275
	0.20	1.297	1.530	1.689	1.808	1.904	1.984	2.052	2.111	2.164	2.364	2.503	2.609	2.695	2.767	2.829	2.883	2.932
19	0.01	2.860	3.043	3.172	3.271	3.352	3.420	3.479	3.531	3.577	3.755	3.881	3.979	4.059	4.126	4.185	4.236	4.283
	0.05	2.087	2.286	2.424	2.530	2.615	2.687	2.748	2.802	2.850	3.034	3.163	3.262	3.343	3.411	3.470	3.522	3.568
	0.10	1.715	1.927	2.073	2.184	2.273	2.348	2.412	2.468	2.517	2.706	2.838	2.940	3.022	3.091	3.151	3.204	3.250
	0.20	1.294	1.526	1.684	1.803	1.898	1.977	2.044	2.103	2.156	2.353	2.490	2.595	2.680	2.751	2.812	2.866	2.914

Continued

df \ *p*	α	2	3	4	5	6	7	8	9	10	15	20	25	30	35	40	45	50
20	0.01	2.844	3.025	3.152	3.249	3.329	3.396	3.454	3.504	3.550	3.724	3.847	3.943	4.021	4.087	4.144	4.194	4.239
	0.05	2.080	2.277	2.414	2.518	2.603	2.673	2.734	2.788	2.835	3.016	3.143	3.240	3.320	3.387	3.445	3.496	3.541
	0.10	1.711	1.922	2.066	2.176	2.264	2.338	2.401	2.457	2.506	2.693	2.823	2.923	3.004	3.072	3.131	3.183	3.229
	0.20	1.292	1.523	1.680	1.798	1.892	1.971	2.038	2.096	2.148	2.344	2.479	2.583	2.667	2.737	2.797	2.850	2.897
21	0.01	2.830	3.009	3.134	3.230	3.308	3.374	3.431	3.481	3.525	3.696	3.817	3.911	3.987	4.052	4.108	4.157	4.201
	0.05	2.073	2.269	2.405	2.508	2.592	2.662	2.722	2.774	2.821	3.000	3.125	3.221	3.299	3.365	3.422	3.472	3.517
	0.10	1.707	1.916	2.060	2.169	2.257	2.330	2.392	2.447	2.496	2.681	2.809	2.908	2.988	3.055	3.113	3.164	3.209
	0.20	1.290	1.520	1.676	1.794	1.887	1.965	2.032	2.090	2.141	3.335	2.470	2.572	2.655	2.724	2.784	2.836	2.883
22	0.01	2.818	2.994	3.117	3.212	3.289	3.354	3.410	3.459	3.503	3.671	3.790	3.882	3.957	4.020	4.075	4.123	4.166
	0.05	2.068	2.262	2.397	2.499	2.582	2.651	2.710	2.762	2.809	2.985	3.108	3.203	3.280	3.345	3.401	3.451	3.495
	0.10	1.703	1.912	2.055	2.163	2.250	2.322	2.384	2.439	2.487	2.670	2.797	2.894	2.974	3.040	3.097	3.147	3.192
	0.20	1.288	1.518	1.673	1.790	1.883	1.960	2.026	2.084	2.135	2.328	2.461	2.562	2.644	2.713	2.772	2.823	2.870
23	0.01	2.806	2.981	3.102	3.196	3.272	3.336	3.391	3.440	3.483	3.649	3.766	3.856	3.930	3.992	4.045	4.093	4.135
	0.05	2.062	2.256	2.389	2.491	2.572	2.641	2.700	2.751	2.797	2.972	3.094	3.187	3.264	3.328	3.383	3.431	3.475
	0.10	1.700	1.907	2.049	2.157	2.243	2.315	2.377	2.431	2.479	2.660	2.786	2.882	2.960	3.026	3.083	3.132	3.176
	0.20	1.286	1.515	1.670	1.786	1.879	1.956	2.021	2.079	2.130	2.321	2.453	2.553	2.634	2.702	2.761	2.812	2.858
24	0.01	2.796	2.968	3.089	3.182	3.257	3.320	3.374	3.422	3.465	3.628	3.743	3.832	3.905	3.966	4.019	4.065	4.107
	0.05	2.058	2.250	2.382	2.483	2.564	2.632	2.690	2.742	2.787	2.960	3.080	3.173	3.248	3.311	3.366	3.414	3.457
	0.10	1.697	1.904	2.045	2.152	2.237	2.309	2.370	2.424	2.471	2.651	2.776	2.871	2.948	3.013	3.069	3.118	3.162
	0.20	1.285	1.513	1.667	1.783	1.875	1.952	2.017	2.074	2.124	2.314	2.445	2.545	2.626	2.693	2.751	2.802	2.847
25	0.01	2.786	2.957	3.077	3.168	3.243	3.305	3.359	3.406	3.448	3.610	3.723	3.811	3.882	3.942	3.995	4.040	4.081
	0.05	2.053	2.244	2.376	2.476	2.557	2.624	2.682	2.733	2.778	2.949	3.068	3.160	3.234	3.297	3.351	3.398	3.440
	0.10	1.694	1.900	2.041	2.147	2.232	2.303	2.364	2.417	2.464	2.642	2.766	2.861	2.938	3.002	3.057	3.106	3.149
	0.20	1.283	1.511	1.664	1.780	1.872	1.948	2.013	2.070	2.120	2.308	2.439	2.538	2.617	2.684	2.742	2.792	2.837
26	0.01	2.778	2.947	3.065	3.156	3.230	3.291	3.345	3.391	3.433	3.593	3.705	3.791	3.862	3.921	3.972	4.018	4.058
	0.05	2.049	2.239	2.370	2.470	2.550	2.617	2.674	2.724	2.769	2.939	3.057	3.148	3.221	3.283	3.337	3.383	3.425
	0.10	1.692	1.897	2.037	2.142	2.227	2.298	2.358	2.411	2.458	2.635	2.758	2.852	2.928	2.991	3.046	3.094	3.137
	0.20	1.282	1.509	1.662	1.777	1.868	1.944	2.009	2.065	2.115	2.303	2.432	2.531	2.610	2.676	2.733	2.783	2.828
27	0.01	2.770	2.938	3.055	3.145	3.218	3.279	3.332	3.378	3.419	3.577	3.688	3.773	3.843	3.901	3.952	3.997	4.036
	0.05	2.046	2.235	2.365	2.464	2.543	2.610	2.667	2.717	2.761	2.929	3.047	3.137	3.210	3.271	3.324	3.370	3.411
	0.10	1.690	1.894	2.033	2.138	2.223	2.293	2.353	2.405	2.452	2.628	2.750	2.843	2.918	2.982	3.036	3.084	3.126
	0.20	1.281	1.507	1.660	1.774	1.865	1.941	2.006	2.062	2.111	2.298	2.427	2.524	2.603	2.669	2.726	2.775	2.819

28	0.01	2.762	2.929	3.045	3.135	3.207	3.267	3.320	3.366	3.406	3.562	3.672	3.757	3.825	3.883	3.933	3.977	4.017
	0.05	2.042	2.231	2.360	2.458	2.537	2.603	2.660	2.710	2.754	2.921	3.037	3.126	3.199	3.259	3.312	3.358	3.398
	0.10	1.687	1.891	2.030	2.134	2.218	2.288	2.348	2.400	2.447	2.621	2.743	2.835	2.910	2.973	3.026	3.074	3.116
	0.20	1.280	1.506	1.658	1.772	1.863	1.938	2.002	2.058	2.108	2.294	2.421	2.519	2.597	2.662	2.719	2.768	2.811
29	0.01	2.755	2.921	3.037	3.125	3.197	3.257	3.309	3.354	3.395	3.549	3.658	3.741	3.809	3.866	3.916	3.959	3.998
	0.05	2.039	2.227	2.355	2.453	2.532	2.597	2.654	2.703	2.747	2.913	3.028	3.117	3.189	3.249	3.301	3.346	3.387
	0.10	1.685	1.888	2.027	2.131	2.214	2.284	2.344	2.395	2.442	2.615	2.736	2.828	2.902	2.964	3.018	3.065	3.106
	0.20	1.279	1.504	1.656	1.770	1.860	1.935	1.999	2.055	2.104	2.289	2.417	2.513	2.591	2.656	2.712	2.761	2.804
30	0.01	2.749	2.914	3.028	3.116	3.187	3.247	3.298	3.343	3.383	3.537	3.644	3.727	3.794	3.851	3.900	3.943	3.981
	0.05	2.036	2.223	2.351	2.448	2.527	2.592	2.648	2.697	2.741	2.905	3.020	3.108	3.179	3.239	3.290	3.335	3.376
	0.10	1.684	1.886	2.024	2.128	2.211	2.280	2.339	2.391	2.437	2.610	2.730	2.821	2.895	2.956	3.010	3.056	3.098
	0.20	1.278	1.503	1.654	1.768	1.858	1.933	1.997	2.052	2.101	2.285	2.412	2.508	2.586	2.650	2.706	2.754	2.798
40	0.01	2.703	2.861	2.970	3.053	3.120	3.177	3.225	3.267	3.305	3.449	3.549	3.626	3.689	3.741	3.787	3.827	3.862
	0.05	2.015	2.197	2.321	2.415	2.490	2.553	2.607	2.654	2.695	2.853	2.962	3.045	3.113	3.169	3.218	3.260	3.298
	0.10	1.670	1.869	2.003	2.104	2.185	2.252	2.309	2.359	2.404	2.570	2.685	2.772	2.843	2.901	2.952	2.996	3.036
	0.20	1.271	1.493	1.642	1.753	1.841	1.914	1.977	2.031	2.078	2.257	2.380	2.472	2.547	2.609	2.662	2.709	2.750
60	0.01	2.659	2.809	2.913	2.992	3.056	3.109	3.155	3.195	3.230	3.365	3.459	3.530	3.589	3.637	3.679	3.716	3.749
	0.05	1.995	2.171	2.291	2.382	2.455	2.515	2.567	2.612	2.652	2.802	2.906	2.985	3.049	3.102	3.148	3.188	3.223
	0.10	1.658	1.851	1.983	2.081	2.160	2.225	2.280	2.328	2.371	2.531	2.642	2.725	2.792	2.848	2.896	2.938	2.976
	0.20	1.264	1.483	1.629	1.738	1.825	1.896	1.957	2.010	2.056	2.230	2.348	2.437	2.509	2.569	2.620	2.664	2.704
120	0.01	2.617	2.760	2.859	2.934	2.994	3.044	3.087	3.125	3.158	3.284	3.372	3.439	3.493	3.538	3.577	3.611	3.641
	0.05	1.974	2.146	2.262	2.350	2.420	2.478	2.528	2.571	2.609	2.753	2.852	2.927	2.987	3.037	3.080	3.118	3.152
	0.10	1.645	1.835	1.963	2.059	2.135	2.198	2.252	2.298	2.340	2.494	2.599	2.679	2.743	2.797	2.843	2.882	2.918
	0.20	1.257	1.473	1.617	1.724	1.809	1.878	1.938	1.989	2.034	2.203	2.317	2.403	2.472	2.530	2.579	2.621	2.659

Note: Values generated as described in Chapter 5.

Appendix 17: Dunn–Šidák's *t* for Comparisons between *p* Treatment Means and a Control for $\alpha = 0.01, 0.05, 0.10$, and 0.20 (Two-Sided Test)

df \ *p*	α	2	3	4	5	6	7	8	9	10	15	20	25	30	35	40	45	50
2	0.01	14.071	17.248	19.925	22.282	24.413	26.372	28.196	29.908	31.528	38.620	44.598	49.865	54.626	59.004	63.079	66.906	70.526
	0.05	6.164	7.582	8.774	9.823	10.769	11.639	12.449	13.208	13.927	17.072	19.721	22.054	24.163	26.103	27.908	29.603	31.206
	0.10	4.243	5.243	6.081	6.816	7.480	8.090	8.656	9.188	9.691	11.890	13.741	15.371	16.845	18.199	19.459	20.642	21.761
	0.20	2.828	3.531	4.116	4.428	5.089	5.512	5.904	6.272	6.620	8.138	9.414	10.537	11.552	12.484	13.351	14.166	14.936
3	0.01	7.447	8.565	9.453	10.201	10.853	11.436	11.966	12.453	12.904	14.796	16.300	17.569	18.678	19.670	20.570	21.398	22.167
	0.05	4.156	4.826	5.355	5.799	6.185	6.529	6.842	7.128	7.394	8.505	9.387	10.129	10.778	11.357	11.883	12.366	12.815
	0.10	3.149	3.690	4.115	4.471	4.780	5.055	5.304	5.532	5.744	6.627	7.326	7.914	8.427	8.886	9.301	9.683	10.038
	0.20	2.294	2.734	3.077	3.363	3.610	3.829	4.028	4.209	4.377	5.076	5.628	6.091	6.495	6.855	7.181	7.481	7.759
4	0.01	5.594	6.248	6.751	7.166	7.520	7.832	8.112	8.367	8.600	9.556	10.294	10.902	11.424	11.884	12.297	12.672	13.017
	0.05	3.481	3.941	4.290	4.577	4.822	5.036	5.228	5.402	5.562	6.214	6.714	7.127	7.480	7.790	8.069	8.322	8.554
	0.10	2.751	3.150	3.452	3.699	3.909	4.093	4.257	4.406	4.542	5.097	5.521	5.870	6.169	6.432	6.667	6.880	7.076
	0.20	2.084	2.434	2.697	2.911	3.092	3.250	3.391	3.518	3.635	4.107	4.468	4.763	5.015	5.237	5.435	5.614	5.779
5	0.01	4.771	5.243	5.599	5.888	6.133	6.346	6.535	6.706	6.862	7.491	7.968	8.355	8.684	8.971	9.226	9.457	9.668
	0.05	3.152	3.518	3.791	4.012	4.197	4.358	4.501	4.630	4.747	5.219	5.573	5.861	6.105	6.317	6.506	6.676	6.831
	0.10	2.549	2.882	3.129	3.327	3.493	3.638	3.765	3.880	3.985	4.403	4.718	4.972	5.187	5.374	5.540	5.689	5.826
	0.20	1.973	2.278	2.503	2.683	2.834	2.964	3.079	3.182	3.275	3.649	3.928	4.153	4.343	4.508	4.654	4.786	4.906
6	0.01	4.315	4.695	4.977	5.203	5.394	5.559	5.704	5.835	5.954	6.428	6.782	7.068	7.308	7.516	7.701	7.867	8.018
	0.05	2.959	3.274	3.505	3.690	3.845	3.978	4.095	4.200	4.296	4.675	4.956	5.182	5.372	5.536	5.682	5.812	5.930
	0.10	2.428	2.723	2.939	3.110	3.253	3.376	3.484	3.580	3.668	4.015	4.272	4.477	4.649	4.798	4.930	5.048	5.155
	0.20	1.904	2.184	2.387	2.547	2.681	2.795	2.895	2.985	3.066	3.385	3.620	3.808	3.965	4.100	4.220	4.327	4.424
7	0.01	4.027	4.353	4.591	4.782	4.941	5.078	5.198	5.306	5.404	5.791	6.077	6.306	6.497	6.663	6.809	6.936	7.058
	0.05	2.832	3.115	3.321	3.484	3.620	3.736	3.838	3.929	4.011	4.336	4.574	4.764	4.923	5.059	5.180	5.287	5.385
	0.10	2.347	2.618	2.814	2.969	3.097	3.206	3.302	3.388	3.465	3.768	3.990	4.167	4.314	4.441	4.552	4.651	4.741
	0.20	1.858	2.120	2.309	2.457	2.579	2.684	2.775	2.856	2.929	3.214	3.423	3.588	3.725	3.842	3.946	4.038	4.121
8	0.01	3.831	4.120	4.331	4.498	4.637	4.756	4.860	4.953	5.038	5.370	5.613	5.807	5.969	6.107	6.230	6.339	6.437
	0.05	2.743	3.005	3.193	3.342	3.464	3.569	3.661	3.743	3.816	4.105	4.316	4.482	4.621	4.740	4.844	4.937	5.021
	0.10	2.289	2.544	2.726	2.869	2.987	3.088	3.176	3.254	3.324	3.598	3.798	3.955	4.086	4.198	4.296	4.383	4.462
	0.20	1.824	2.075	2.254	2.393	2.508	2.605	2.690	2.765	2.832	3.095	3.286	3.435	3.559	3.665	3.758	3.840	3.914
9	0.01	3.688	3.952	4.143	4.294	4.419	4.526	4.619	4.703	4.778	5.072	5.287	5.457	5.598	5.720	5.826	5.921	6.006
	0.05	2.677	2.923	3.099	3.237	3.351	3.448	3.532	3.607	3.675	3.938	4.129	4.280	4.405	4.512	4.605	4.688	4.763
	0.10	2.246	2.488	2.661	2.796	2.907	3.001	3.083	3.155	3.221	3.474	3.658	3.802	3.921	4.023	4.112	4.191	4.262
	0.20	1.799	2.041	2.212	2.345	2.454	2.546	2.627	2.698	2.761	3.008	3.185	3.324	3.438	3.536	3.621	3.697	3.765
10	0.01	3.580	3.825	4.002	4.141	4.256	4.354	4.439	4.515	4.584	4.852	5.046	5.199	5.326	5.434	5.529	5.614	5.690
	0.05	2.626	2.860	3.027	3.157	3.264	3.355	3.434	3.505	3.568	3.813	3.989	4.128	4.243	4.341	4.426	4.502	4.571
	0.10	2.213	2.446	2.611	2.739	2.845	2.934	3.012	3.080	3.142	3.380	3.552	3.686	3.796	3.891	3.973	4.046	4.112
	0.20	1.779	2.014	2.180	2.308	2.413	2.501	2.578	2.646	2.706	2.941	3.108	3.239	3.346	3.438	3.517	3.588	3.651

11	0.01	3.495	3.726	3.892	4.022	4.129	4.221	4.300	4.371	4.434	4.682	4.860	5.001	5.117	5.216	5.303	5.380	5.450
	0.05	2.586	2.811	2.970	3.094	3.196	3.283	3.358	3.424	3.484	3.715	3.880	4.010	4.117	4.208	4.288	4.358	4.422
	0.10	2.186	2.412	2.571	2.695	2.796	2.881	2.955	3.021	3.079	3.306	3.468	3.595	3.699	3.788	3.865	3.933	3.995
	0.20	1.763	1.993	2.154	2.279	2.380	2.465	2.539	2.605	2.663	2.888	3.048	3.172	3.274	3.361	3.436	3.503	3.563
12	0.01	3.427	3.647	3.804	3.927	4.029	4.114	4.189	4.256	4.315	4.547	4.714	4.845	4.953	5.045	5.125	5.196	5.260
	0.05	2.553	2.770	2.924	3.044	3.141	3.224	3.296	3.359	3.416	3.636	3.793	3.916	4.017	4.103	4.178	4.244	4.304
	0.10	2.164	2.384	2.539	2.658	2.756	2.838	2.910	2.973	3.029	3.247	3.402	3.522	3.621	3.705	3.779	3.843	3.901
	0.20	1.750	1.975	2.133	2.254	2.353	2.436	2.508	2.571	2.628	2.845	2.999	3.118	3.216	3.299	3.371	3.434	3.491
13	0.01	3.371	3.582	3.733	3.850	3.946	4.028	4.099	4.162	4.218	4.438	4.595	4.718	4.819	4.906	4.981	5.048	5.108
	0.05	2.526	2.737	2.886	3.002	3.096	3.176	3.245	3.306	3.361	3.571	3.722	3.839	3.935	4.017	4.088	4.151	4.207
	0.10	2.146	2.361	2.512	2.628	2.723	2.803	2.872	2.933	2.988	3.198	3.347	3.463	3.557	3.638	3.708	3.770	3.825
	0.20	1.739	1.961	2.116	2.234	2.331	2.412	2.482	2.544	2.599	2.809	2.958	3.074	3.168	3.248	3.317	3.378	3.433
14	0.01	3.324	3.528	3.673	3.785	3.878	3.956	4.024	4.084	4.138	4.347	4.497	4.614	4.710	4.792	4.863	4.926	4.982
	0.05	2.503	2.709	2.854	2.967	3.058	3.135	3.202	3.261	3.314	3.518	3.662	3.775	3.867	3.946	4.014	4.074	4.128
	0.10	2.131	2.342	2.489	2.603	2.696	2.774	2.841	2.900	2.953	3.157	3.301	3.413	3.504	3.582	3.649	3.708	3.761
	0.20	1.730	1.949	2.101	2.217	2.312	2.392	2.460	2.520	2.574	2.779	2.924	3.036	3.128	3.205	3.272	3.331	3.384
15	0.01	3.285	3.482	3.622	3.731	3.820	3.895	3.961	4.019	4.071	4.271	4.414	4.526	4.618	4.696	4.764	4.824	4.877
	0.05	2.483	2.685	2.827	2.937	3.026	3.101	3.166	3.224	3.275	3.472	3.612	3.721	3.810	3.885	3.951	4.009	4.060
	0.10	2.118	2.325	2.470	2.582	2.672	2.748	2.814	2.872	2.924	3.122	3.262	3.370	3.459	3.534	3.599	3.656	3.708
	0.20	1.722	1.938	2.088	2.203	2.296	2.374	2.441	2.500	2.553	2.754	2.896	3.005	3.094	3.169	3.234	3.291	3.343
16	0.01	3.251	3.443	3.579	3.684	3.771	3.844	3.907	3.963	4.013	4.206	4.344	4.451	4.540	4.614	4.679	4.737	4.788
	0.05	2.467	2.665	2.804	2.911	2.998	3.072	3.135	3.191	3.241	3.433	3.569	3.675	3.761	3.834	3.897	3.953	4.003
	0.10	2.106	2.311	2.453	2.563	2.652	2.726	2.791	2.848	2.898	3.092	3.228	3.334	3.420	3.493	3.556	3.612	3.662
	0.20	1.715	1.929	2.077	2.190	2.282	2.359	2.425	2.483	2.535	2.732	2.871	2.978	3.064	3.138	3.201	3.257	3.307
17	0.01	3.221	3.409	3.541	3.644	3.728	3.799	3.860	3.914	3.963	4.150	4.284	4.387	4.472	4.544	4.607	4.662	4.712
	0.05	2.452	2.647	2.783	2.889	2.974	3.046	3.108	3.163	3.212	3.399	3.532	3.634	3.718	3.789	3.851	3.905	3.954
	0.10	2.096	2.298	2.439	2.547	2.634	2.708	2.771	2.826	2.876	3.066	3.199	3.303	3.387	3.458	3.519	3.574	3.622
	0.20	1.709	1.921	2.068	2.179	2.270	2.346	2.411	2.468	2.519	2.713	2.849	2.954	3.039	3.111	3.173	3.227	3.276
18	0.01	3.195	3.379	3.508	3.609	3.691	3.760	3.820	3.872	3.920	4.102	4.231	4.332	4.414	4.484	4.544	4.598	4.646
	0.05	2.439	2.631	2.766	2.869	2.953	3.024	3.085	3.138	3.186	3.370	3.499	3.599	3.681	3.750	3.810	3.863	3.910
	0.10	2.088	2.287	2.426	2.532	2.619	2.691	2.753	2.808	2.857	3.043	3.174	3.275	3.358	3.427	3.487	3.540	3.587
	0.20	1.704	1.914	2.059	2.170	2.259	2.334	2.399	2.455	2.505	2.696	2.830	2.933	3.017	3.087	3.148	3.201	3.249
19	0.01	3.173	3.353	3.479	3.578	3.658	3.725	3.784	3.835	3.881	4.059	4.185	4.283	4.363	4.430	4.489	4.541	4.588
	0.05	2.427	2.617	2.750	2.852	2.934	3.004	3.064	3.116	3.163	3.343	3.470	3.569	3.649	3.716	3.775	3.826	3.872
	0.10	2.080	2.277	2.415	2.520	2.605	2.676	2.738	2.791	2.839	3.023	3.152	3.251	3.332	3.400	3.459	3.511	3.557
	0.20	1.699	1.908	2.052	2.161	2.250	2.324	2.388	2.443	2.493	2.682	2.813	2.915	2.997	3.066	3.126	3.179	3.225

Continued

df \ *p*	α	2	3	4	5	6	7	8	9	10	15	20	25	30	35	40	45	50
20	0.01	3.152	3.329	3.454	3.550	3.629	3.695	3.752	3.802	3.848	4.021	4.144	4.239	4.317	4.383	4.441	4.491	4.536
	0.05	2.417	2.605	2.736	2.836	2.918	2.986	3.045	3.097	3.143	3.320	3.445	3.541	3.620	3.686	3.743	3.794	3.839
	0.10	2.073	2.269	2.405	2.508	2.593	2.663	2.724	2.777	2.824	3.005	3.132	3.229	3.309	3.376	3.433	3.484	3.530
	0.20	1.695	1.902	2.045	2.154	2.241	2.315	2.378	2.433	2.482	2.668	2.798	2.898	2.979	3.048	3.106	3.158	3.204
21	0.01	3.134	3.308	3.431	3.525	3.602	3.667	3.724	3.773	3.817	3.987	4.108	4.201	4.277	4.342	4.397	4.447	4.491
	0.05	2.408	2.594	2.723	2.822	2.903	2.970	3.028	3.080	3.125	3.300	3.422	3.517	3.594	3.659	3.715	3.765	3.809
	0.10	2.067	2.261	2.396	2.498	2.581	2.651	2.711	2.764	2.810	2.989	3.114	3.210	3.288	3.354	3.411	3.461	3.505
	0.20	1.691	1.897	2.039	2.147	2.234	2.306	2.369	2.424	2.472	2.656	2.785	2.884	2.964	3.031	3.089	3.140	3.185
22	0.01	3.118	3.289	3.410	3.503	3.579	3.643	3.698	3.747	3.790	3.957	4.075	4.166	4.241	4.304	4.359	4.407	4.450
	0.05	2.400	2.584	2.712	2.810	2.889	2.956	3.014	3.064	3.109	3.281	3.402	3.495	3.571	3.634	3.690	3.738	3.782
	0.10	2.061	2.254	2.387	2.489	2.572	2.641	2.700	2.752	2.798	2.974	3.098	3.193	3.270	3.334	3.390	3.440	3.484
	0.20	1.688	1.892	2.033	2.141	2.227	2.299	2.361	2.415	2.463	2.646	2.773	2.871	2.950	3.016	3.073	3.123	3.168
23	0.01	3.103	3.272	3.392	3.483	3.558	3.621	3.675	3.723	3.766	3.930	4.046	4.135	4.208	4.270	4.324	4.371	4.413
	0.05	2.392	2.574	2.701	2.798	2.877	2.943	3.000	3.050	3.094	3.264	3.383	3.475	3.550	3.613	3.667	3.715	3.757
	0.10	2.056	2.247	2.380	2.481	2.563	2.631	2.690	2.741	2.787	2.961	3.083	3.177	3.253	3.317	3.372	3.421	3.464
	0.20	1.685	1.888	2.028	2.135	2.221	2.292	2.354	2.407	2.455	2.636	2.762	2.859	2.937	3.002	3.059	3.109	3.153
24	0.01	3.089	3.257	3.375	3.465	3.539	3.601	3.654	3.702	3.744	3.905	4.019	4.107	4.179	4.240	4.292	4.339	4.380
	0.05	2.385	2.566	2.692	2.788	2.866	2.931	2.988	3.037	3.081	3.249	3.366	3.457	3.531	3.593	3.646	3.693	3.735
	0.10	2.051	2.241	2.373	2.473	2.554	2.622	2.680	2.731	2.777	2.949	3.070	3.162	3.238	3.301	3.355	3.403	3.446
	0.20	1.682	1.884	2.024	2.130	2.215	2.286	2.347	2.400	2.448	2.627	2.752	2.848	2.925	2.990	3.046	3.095	3.139
25	0.01	3.077	3.243	3.359	3.449	3.521	3.583	3.635	3.682	3.723	3.882	3.995	4.081	4.152	4.212	4.263	4.309	4.350
	0.05	2.379	2.558	2.683	2.779	2.856	2.921	2.976	3.025	3.069	3.235	3.351	3.440	3.513	3.574	3.627	3.674	3.715
	0.10	2.047	2.236	2.367	2.466	2.547	2.614	2.672	2.722	2.767	2.938	3.058	3.149	3.224	3.286	3.340	3.387	3.430
	0.20	1.679	1.881	2.020	2.125	2.210	2.280	2.341	2.394	2.441	2.619	2.743	2.838	2.914	2.979	3.034	3.083	3.126
26	0.01	3.066	3.230	3.345	3.433	3.505	3.566	3.618	3.664	3.705	3.862	3.972	4.058	4.128	4.186	4.237	4.282	4.322
	0.05	2.373	2.551	2.675	2.770	2.847	2.911	2.966	3.014	3.058	3.222	3.337	3.425	3.497	3.558	3.610	3.656	3.697
	0.10	2.043	2.231	2.361	2.460	2.540	2.607	2.664	2.714	2.759	2.928	3.047	3.137	3.211	3.273	3.326	3.373	3.415
	0.20	1.677	1.878	2.016	2.121	2.205	2.275	2.335	2.388	2.435	2.612	2.735	2.829	2.905	2.968	3.023	3.071	3.114
27	0.01	3.056	3.218	3.332	3.419	3.491	3.550	3.602	3.647	3.688	3.843	3.952	4.036	4.105	4.163	4.213	4.257	4.297
	0.05	2.368	2.545	2.668	2.762	2.838	2.902	2.956	3.004	3.047	3.210	3.324	3.411	3.483	3.542	3.594	3.639	3.680
	0.10	2.039	2.227	2.356	2.454	2.534	2.600	2.657	2.707	2.751	2.919	3.036	3.126	3.199	3.261	3.313	3.360	3.401
	0.20	1.675	1.875	2.012	2.117	2.201	2.270	2.330	2.383	2.429	2.605	2.727	2.820	2.896	2.959	3.013	3.061	3.103

28	0.01	3.046	3.207	3.320	3.407	3.477	3.536	3.587	3.632	3.672	3.825	3.933	4.017	4.084	4.142	4.191	4.235	4.274
	0.05	2.363	2.539	2.661	2.755	2.830	2.893	2.948	2.995	3.038	3.199	3.312	3.399	3.469	3.528	3.579	3.624	3.664
	0.10	2.036	2.222	2.351	2.449	2.528	2.594	2.650	2.700	2.744	2.911	3.027	3.116	3.188	3.249	3.301	3.347	3.388
	0.20	1.672	1.872	2.009	2.113	2.196	2.266	2.326	2.378	2.424	2.599	2.720	2.812	2.887	2.950	3.004	3.051	3.093
29	0.01	3.037	3.197	3.309	3.395	3.464	3.523	3.574	3.618	3.658	3.809	3.916	3.998	4.065	4.122	4.171	4.214	4.252
	0.05	2.358	2.534	2.655	2.748	2.823	2.886	2.940	2.987	3.029	3.189	3.301	3.387	3.457	3.515	3.566	3.610	3.650
	0.10	2.033	2.218	2.346	2.444	2.522	2.588	2.644	2.693	2.737	2.903	3.018	3.107	3.178	3.239	3.291	3.336	3.377
	0.20	1.671	1.869	2.006	2.110	2.193	2.262	2.321	2.373	2.419	2.593	2.713	2.805	2.880	2.942	2.995	3.042	3.084
30	0.01	3.029	3.188	3.298	3.384	3.453	3.511	3.561	3.605	3.644	3.794	3.900	3.981	4.048	4.103	4.152	4.194	4.232
	0.05	2.354	2.528	2.649	2.742	2.816	2.878	2.932	2.979	3.021	3.180	3.291	3.376	3.445	3.503	3.553	3.597	3.637
	0.10	2.030	2.215	2.342	2.439	2.517	2.582	2.638	2.687	2.731	2.895	3.010	3.098	3.169	3.229	3.280	3.325	3.366
	0.20	1.669	1.867	2.003	2.106	2.189	2.258	2.317	2.369	2.414	2.587	2.707	2.798	2.872	2.934	2.987	3.034	3.076
40	0.01	2.970	3.121	3.225	3.305	3.370	3.425	3.472	3.513	3.549	3.689	3.787	3.862	3.923	3.975	4.019	4.058	4.093
	0.05	2.323	2.492	2.608	2.696	2.768	2.827	2.878	2.923	2.963	3.113	3.218	3.298	3.363	3.418	3.464	3.506	3.542
	0.10	2.009	2.189	2.312	2.406	2.481	2.544	2.597	2.644	2.686	2.843	2.952	3.036	3.103	3.160	3.208	3.251	3.289
	0.20	1.656	1.850	1.983	2.083	2.164	2.231	2.288	2.338	2.382	2.548	2.663	2.751	2.821	2.880	2.931	2.975	3.015
60	0.01	2.914	3.056	3.155	3.230	3.291	3.342	3.386	3.425	3.459	3.589	3.679	3.749	3.805	3.852	3.893	3.929	3.961
	0.05	2.294	2.456	2.568	2.653	2.721	2.777	2.826	2.869	2.906	3.049	3.148	3.223	3.284	3.336	3.379	3.418	3.452
	0.10	1.989	2.163	2.283	2.373	2.446	2.506	2.558	2.603	2.643	2.793	2.897	2.976	3.040	3.093	3.139	3.179	3.214
	0.20	1.643	1.834	1.963	2.061	2.139	2.204	2.259	2.308	2.350	2.511	2.621	2.705	2.772	2.828	2.876	2.918	2.956
120	0.01	2.859	2.994	3.087	3.158	3.215	3.263	3.304	3.340	3.372	3.493	3.577	3.641	3.693	3.736	3.774	3.807	3.836
	0.05	2.265	2.422	2.529	2.610	2.675	2.729	2.776	2.816	2.852	2.987	3.081	3.152	3.209	3.257	3.298	3.334	3.366
	0.10	1.968	2.138	2.254	2.342	2.411	2.469	2.519	2.562	2.600	2.744	2.843	2.918	2.978	3.029	3.072	3.110	3.143
	0.20	1.631	1.817	1.944	2.039	2.115	2.178	2.231	2.278	2.319	2.474	2.580	2.660	2.724	2.778	2.824	2.864	2.899
∞	0.01	2.806	2.934	3.022	3.089	3.143	3.188	3.226	3.260	3.289	3.402	3.480	3.539	3.587	3.627	3.661	3.691	3.718
	0.05	2.237	2.388	2.491	2.569	2.631	2.683	2.727	2.766	2.800	2.928	3.016	3.083	3.137	3.182	3.220	3.254	3.284
	0.10	1.949	2.114	2.226	2.311	2.378	2.434	2.482	2.523	2.560	2.697	2.791	2.862	2.920	2.967	3.008	3.044	3.075
	0.20	1.618	1.801	1.925	2.018	2.091	2.152	2.204	2.249	2.289	2.438	2.540	2.617	2.678	2.729	2.773	2.811	2.844

Source: Reprinted from Games, P.A., *J. Am. Stat. Assoc.*, 72, 531–534, 1977, Table 1. With permission from the *Journal of the American Statistical Association*.

Note: Used first in Chapter 5.

Appendix 18: Williams's $\bar{t}_{i,\alpha}$ for $w = 1$ and Extrapolation β_t (Superscript) for a One-Sided Test and $\alpha = 0.01$

df \ *p*	2	3	4	5	6	8	10
5	3.501^{4}	3.548^{6}	3.572^{7}	3.586^{8}	3.595^{9}	3.607^{9}	3.614^{10}
6	3.256^{4}	3.294^{6}	3.313^{7}	3.324^{8}	3.332^{8}	3.341^{9}	3.346^{9}
7	3.097^{4}	3.130^{6}	3.146^{7}	3.155^{7}	3.161^{8}	3.169^{8}	3.173^{9}
8	2.985^{4}	3.015^{6}	3.029^{6}	3.037^{7}	3.042^{7}	3.049^{8}	3.053^{9}
9	2.903^{4}	2.930^{5}	2.943^{6}	2.950^{7}	2.955^{7}	2.961^{8}	2.964^{8}
10	2.840^{3}	2.865^{5}	2.877^{6}	2.883^{7}	2.888^{7}	2.893^{8}	2.896^{8}
11	2.791^{3}	2.814^{5}	2.824^{6}	2.831^{7}	2.835^{7}	2.840^{7}	2.843^{8}
12	2.750^{3}	2.772^{5}	2.782^{6}	2.788^{6}	2.792^{7}	2.797^{7}	2.799^{7}
13	2.717^{3}	2.738^{5}	2.747^{6}	2.753^{6}	2.757^{7}	2.761^{7}	2.764^{7}
14	2.689^{3}	2.709^{5}	2.718^{6}	2.723^{6}	2.727^{7}	2.731^{7}	2.733^{7}
15	2.665^{3}	2.684^{5}	2.693^{6}	2.698^{6}	2.701^{7}	2.705^{7}	2.708^{7}
16	2.644^{3}	2.663^{5}	2.671^{6}	2.676^{6}	2.680^{7}	2.683^{7}	2.686^{7}
17	2.626^{3}	2.644^{5}	2.653^{6}	2.658^{6}	2.661^{6}	2.664^{7}	2.666^{7}
18	2.610^{3}	2.628^{5}	2.636^{5}	2.641^{6}	2.644^{6}	2.647^{7}	2.650^{7}
19	2.596^{3}	2.614^{5}	2.622^{5}	2.626^{6}	2.629^{6}	2.633^{7}	2.635^{7}
20	2.584^{3}	2.601^{5}	2.609^{5}	2.613^{6}	2.616^{6}	2.619^{7}	2.621^{7}
22	2.563^{3}	2.579^{5}	2.586^{5}	2.591^{6}	2.593^{6}	2.597^{7}	2.599^{7}
24	2.545^{3}	2.561^{5}	2.568^{5}	2.572^{6}	2.575^{6}	2.578^{6}	2.580^{7}
26	2.531^{3}	2.546^{4}	2.553^{5}	2.557^{6}	2.559^{6}	2.562^{6}	2.564^{6}
28	2.518^{3}	2.533^{4}	2.540^{5}	2.544^{6}	2.546^{6}	2.549^{6}	2.551^{6}
30	2.507^{3}	2.522^{4}	2.529^{5}	2.533^{6}	2.535^{6}	2.538^{6}	2.540^{6}
35	2.486^{3}	2.501^{4}	2.507^{5}	2.511^{6}	2.513^{6}	2.516^{6}	2.517^{6}
40	2.471^{3}	2.484^{4}	2.491^{5}	2.494^{5}	2.496^{6}	2.499^{6}	2.500^{6}
60	2.453^{3}	2.448^{4}	2.435^{5}	2.457^{5}	2.459^{5}	2.461^{6}	2.462^{6}
120	2.400^{3}	2.412^{4}	2.417^{5}	2.420^{5}	2.422^{5}	2.424^{5}	2.425^{6}
∞	2.366^{3}	2.377^{4}	2.382^{5}	2.385^{5}	2.386^{5}	2.388^{5}	2.389^{6}

Source: Reproduced from Williams, D.A., *Bicmetrics,* 28, 519–531, 1972, Table 3. With permission from the Biometric Society.

Note: Used first in Chapter 5.

Appendix 19: Williams's $\bar{t}_{i,\alpha}$ for $w = 1$ and Extrapolation β_t (Superscript) for a One-Sided Test and $\alpha = 0.05$

df \ *p*	2	3	4	5	6	8	10
5	2.142^2	2.186^4	2.209^5	2.223^5	2.232^6	2.243^6	2.250^6
6	2.058^2	2.098^4	2.119^5	2.131^5	2.139^6	2.149^6	2.154^6
7	2.002^2	2.039^4	2.058^5	2.069^5	2.076^6	2.085^6	2.091^7
8	1.962^2	1.997^4	2.014^5	2.024^5	2.031^6	2.040^6	2.045^7
9	1.931^2	1.965^4	1.981^5	1.991^5	1.998^6	2.006^6	2.010^7
10	1.908^3	1.940^4	1.956^5	1.965^5	1.971^6	1.979^6	1.983^7
11	1.889^3	1.920^4	1.935^5	1.944^5	1.950^6	1.958^6	1.962^7
12	1.873^3	1.903^4	1.918^5	1.927^5	1.933^6	1.940^6	1.944^7
13	1.860^3	1.890^4	1.904^5	1.913^5	1.919^6	1.926^6	1.930^7
14	1.849^3	1.878^4	1.892^5	1.901^5	1.906^6	1.913^6	1.917^7
15	1.840^3	1.868^4	1.882^5	1.891^5	1.896^6	1.903^6	1.907^7
16	1.831^3	1.860^4	1.873^5	1.882^5	1.887^6	1.893^6	1.897^7
17	1.824^3	1.852^4	1.866^5	1.874^5	1.879^6	1.885^6	1.889^7
18	1.818^3	1.845^4	1.859^5	1.867^5	1.872^6	1.878^6	1.882^7
19	1.812^3	1.840^4	1.853^5	1.861^5	1.866^6	1.872^6	1.876^7
20	1.807^3	1.834^4	1.847^5	1.855^5	1.860^6	1.866^6	1.870^7
22	1.798^3	1.825^4	1.838^5	1.846^5	1.851^6	1.857^6	1.860^7
24	1.791^3	1.818^4	1.830^5	1.838^5	1.843^6	1.849^6	1.852^7
26	1.785^3	1.811^4	1.824^5	1.831^5	1.836^6	1.842^6	1.846^7
28	1.780^3	1.806^4	1.819^5	1.826^5	1.831^6	1.836^6	1.840^7
30	1.776^3	1.801^4	1.814^5	1.821^5	1.826^6	1.832^6	1.835^7
35	1.767^3	1.792^4	1.804^5	1.811^5	1.816^6	1.822^6	1.825^7
40	1.761^3	1.785^4	1.797^5	1.804^5	1.809^6	1.814^6	1.818^7
60	1.746^3	1.770^4	1.781^5	1.788^5	1.792^6	1.798^6	1.801^7
120	1.731^3	1.754^4	1.765^5	1.772^5	1.776^6	1.781^6	1.784^7
∞	1.716^3	1.739^4	1.750^5	1.756^5	1.760^6	1.765^6	1.768^7

Source: Reproduced from Williams, D.A., *Biometrics,* 28, 519–531, 1972, Table 1. With permission from the Biometric Society.

Note: Used first in Chapter 5.

Appendix 20: Williams's $\bar{t}_{i,\alpha}$ for $w = 1$ and Extrapolation β_t (Superscript) for a Two-Sided Test and $\alpha = 0.01$

$df \setminus p$	2	3	4	5	6	8	10
5	4.179^{5}	4.229^{7}	4.255^{9}	4.270^{10}	4.279^{10}	4.292^{11}	4.299^{11}
6	3.825^{5}	3.864^{7}	3.883^{8}	3.895^{9}	3.902^{9}	3.912^{10}	3.917^{10}
7	3.599^{4}	3.631^{6}	3.647^{7}	3.657^{8}	3.663^{9}	3.670^{9}	3.674^{10}
8	3.443^{4}	3.471^{6}	3.484^{7}	3.492^{8}	3.497^{8}	3.504^{9}	3.507^{9}
9	3.329^{4}	3.354^{6}	3.366^{7}	3.373^{7}	3.377^{8}	3.383^{8}	3.886^{9}
10	3.242^{4}	3.265^{6}	3.275^{6}	3.281^{7}	3.286^{7}	3.290^{8}	3.293^{8}
11	3.173^{4}	3.194^{5}	3.204^{6}	3.210^{7}	3.214^{7}	3.218^{8}	3.221^{8}
12	3.118^{4}	3.138^{5}	3.147^{6}	3.152^{7}	3.156^{7}	3.160^{7}	3.162^{8}
13	3.073^{4}	3.091^{5}	3.100^{6}	3.105^{6}	3.108^{7}	3.112^{7}	3.114^{7}
14	3.035^{4}	3.052^{5}	3.060^{6}	3.065^{6}	3.068^{6}	3.072^{7}	3.074^{7}
15	3.003^{3}	3.019^{5}	3.027^{6}	3.031^{6}	3.034^{6}	3.037^{7}	3.039^{7}
16	2.957^{3}	2.991^{5}	2.998^{5}	3.002^{6}	3.005^{6}	3.008^{7}	3.010^{7}
17	2.951^{3}	2.966^{5}	2.973^{5}	2.977^{6}	2.980^{6}	2.938^{7}	2.984^{7}
18	2.929^{3}	3.944^{5}	2.951^{5}	2.955^{6}	2.958^{6}	2.960^{6}	2.962^{7}
19	2.911^{3}	2.925^{4}	2.932^{5}	2.936^{6}	2.938^{6}	2.941^{6}	2.942^{7}
20	2.894^{3}	2.908^{4}	2.915^{5}	2.918^{6}	2.920^{6}	2.923^{6}	2.925^{7}
22	2.866^{3}	2.879^{4}	2.885^{5}	2.889^{6}	2.891^{6}	2.893^{6}	2.895^{6}
24	2.842^{3}	2.855^{4}	2.861^{5}	2.864^{5}	2.866^{6}	2.869^{6}	2.870^{6}
26	2.823^{3}	2.835^{4}	2.841^{5}	2.844^{5}	2.846^{6}	2.848^{6}	2.850^{6}
28	2.806^{3}	2.819^{4}	2.824^{5}	2.827^{5}	2.829^{5}	2.831^{6}	2.832^{6}
30	2.792^{3}	2.804^{4}	2.809^{5}	2.812^{5}	2.814^{5}	2.816^{6}	2.817^{6}
35	2.764^{3}	2.776^{4}	2.781^{5}	2.783^{5}	2.785^{5}	2.787^{5}	2.788^{6}
40	2.744^{3}	2.755^{4}	2.759^{5}	2.762^{5}	2.764^{5}	2.765^{5}	2.766^{5}
60	2.697^{3}	2.707^{4}	2.711^{4}	2.713^{5}	2.715^{5}	2.716^{5}	2.717^{5}
120	2.651^{3}	2.660^{4}	2.664^{4}	2.666^{4}	2.667^{5}	2.669^{5}	2.669^{5}
∞	2.607^{3}	2.615^{4}	2.618^{4}	2.620^{4}	2.621^{4}	2.623^{5}	2.623^{5}

Source: Reproduced from Williams, D.A., *Biometrics,* 28, 519–531, 1972, Table 4. With permission from the Biometric Society.

Note: Used first in Chapter 5.

Appendix 21: Williams's $\bar{t}_{i,\alpha}$ for $w = 1$ and Extrapolation β_t (Superscript) for a Two-Sided Test and $\alpha = 0.05$

$df \setminus p$	2	3	4	5	6	8	10
5	2.699^3	2.743^5	2.766^6	2.779^6	2.788^7	2.799^7	2.806^8
6	2.559^3	2.597^5	2.617^6	2.628^6	2.635^7	2.645^7	2.650^8
7	2.466^3	2.501^5	2.518^6	2.528^6	2.535^7	2.543^7	2.548^8
8	2.400^3	2.432^5	2.448^6	2.457^6	2.436^7	2.470^7	2.475^8
9	2.351^3	2.381^5	2.395^6	2.404^6	2.410^7	2.416^7	2.421^8
10	2.313^3	2.341^5	2.355^6	2.363^6	2.368^6	2.375^7	2.379^7
11	2.283^3	2.310^5	2.323^6	2.330^6	2.335^6	2.342^7	2.345^7
12	2.258^3	2.284^5	2.297^6	2.304^6	2.309^6	2.315^7	2.318^7
13	2.238^3	2.263^5	2.275^5	2.282^6	2.286^6	2.292^7	2.295^7
14	2.220^3	2.245^5	2.256^5	2.263^6	2.268^6	2.273^7	2.276^7
15	2.205^3	2.229^5	2.241^5	2.247^6	2.252^6	2.257^7	2.260^7
16	2.193^3	2.216^5	2.227^5	2.234^6	2.238^6	2.243^7	2.246^7
17	2.181^3	2.204^4	2.215^5	2.222^6	2.226^6	2.231^7	2.234^7
18	2.171^3	2.194^4	2.205^5	2.211^6	2.215^6	2.220^7	2.223^7
19	2.163^3	2.185^4	2.195^5	2.202^6	2.205^6	2.210^7	2.213^7
20	2.155^3	2.177^4	2.187^5	2.193^6	2.197^6	2.202^7	2.205^7
22	2.141^3	2.163^4	2.173^5	2.179^6	2.183^6	2.187^7	2.190^7
24	2.130^3	2.151^4	2.161^5	2.167^6	2.171^6	2.175^7	2.178^7
26	2.121^3	2.142^4	2.151^5	2.157^6	2.161^6	2.165^7	2.168^7
28	2.113^3	2.133^4	2.143^5	2.149^6	2.152^6	2.156^7	2.159^7
30	2.106^3	2.126^4	2.136^5	2.141^6	2.145^6	2.149^7	2.151^7
35	2.093^3	2.112^4	2.122^5	2.127^6	2.130^6	2.134^7	2.137^7
40	2.083^3	2.102^4	2.111^5	2.116^6	2.119^6	2.123^6	2.126^7
60	2.060^3	2.078^4	2.087^5	2.092^6	2.095^6	2.099^6	2.101^7
120	2.037^3	2.055^4	2.063^5	2.068^6	2.071^6	2.074^6	2.076^7
∞	2.015^3	2.032^4	2.040^5	2.044^6	2.047^6	2.050^6	2.052^6

Source: Reproduced from Williams, D.A., *Biometrics,* 28, 519–531, 1972, Table 2. With permission from the Biometric Society.

Note: Used first in Chapter 5.

Appendix 22: Significant Values of Steel's Rank Sums for a One-Sided Test with $\alpha = 0.05$ or 0.01

$n \backslash p$	α	2	3	4	5	6	7	8	9
4	0.05	11	10	10	10	10			
	0.01								
5	0.05	18	17	17	16	16	16	16	15
	0.01	15							
6	0.05	27	26	25	25	24	24	24	23
	0.01	23	22	21	21				
7	0.05	37	36	35	35	34	34	33	33
	0.01	32	31	30	30	29	29	29	29
8	0.05	49	48	47	46	46	45	45	44
	0.01	43	42	41	40	40	40	39	39
9	0.05	63	62	61	60	59	59	58	58
	0.01	56	55	54	53	52	52	51	51
10	0.05	79	77	76	75	74	74	73	72
	0.01	71	69	68	67	66	66	65	65
11	0.05	97	95	93	92	91	90	90	89
	0.01	87	85	84	83	82	81	81	80
12	0.05	116	114	112	111	110	109	108	108
	0.01	105	103	102	100	99	99	98	98
13	0.05	138	135	133	132	130	129	129	128
	0.01	125	123	121	120	119	118	117	117
14	0.05	161	158	155	154	153	152	151	150
	0.01	147	144	142	141	140	139	138	137
15	0.05	186	182	180	178	177	176	175	174
	0.01	170	167	165	164	162	161	160	160
16	0.05	213	209	206	204	203	201	200	199
	0.01	196	192	190	188	187	186	185	184
17	0.05	241	237	234	232	231	229	228	227
	0.01	223	219	217	215	213	212	211	210
18	0.05	272	267	264	262	260	259	257	256
	0.01	252	248	245	243	241	240	239	238
19	0.05	304	299	296	294	292	290	288	287
	0.01	282	278	275	273	271	270	268	267

Source: Reproduced from Steel, R.G.D., *Biometrics,* 15, 560–572, 1959, Table 2. With permission from the Biometric Society.

Note: Used first in Chapter 5.

Appendix 23: Significant Values of Steel's Rank Sums for a Two-Sided Test with α = 0.05 or 0.01

$n \setminus p$	α	2	3	4	5	6	7	8	9
4	0.05	10							
	0.01								
5	0.05	16	16	15	15				
	0.01								
6	0.05	25	24	23	23	22	22	22	21
	0.01	21							
7	0.05	35	33	33	32	32	31	31	30
	0.01	30	29	28	28				
8	0.05	46	45	44	43	43	42	42	41
	0.01	41	40	39	38	38	37	37	37
9	0.05	60	58	57	56	55	55	54	54
	0.01	53	52	51	50	49	49	49	48
10	0.05	75	73	72	71	70	69	69	68
	0.01	68	66	65	64	63	62	62	62
11	0.05	92	90	88	87	86	85	85	84
	0.01	84	82	80	79	78	78	77	77
12	0.05	111	108	107	105	104	103	103	102
	0.01	101	99	97	96	95	94	94	93
13	0.05	132	129	127	125	124	123	122	121
	0.01	121	118	116	115	114	113	112	112
14	0.05	154	151	149	147	145	144	144	143
	0.01	142	139	137	135	134	133	132	132
15	0.05	179	175	172	171	169	168	167	166
	0.01	165	162	159	158	156	155	154	154
16	0.05	205	201	196	196	194	193	192	191
	0.01	189	186	184	182	180	179	178	177
17	0.05	233	228	225	223	221	219	218	217
	0.01	216	212	210	208	206	205	204	203
18	0.05	263	258	254	252	250	248	247	246
	0.01	244	240	237	235	233	232	231	230
19	0.05	294	289	285	283	280	279	277	276
	0.01	274	270	267	265	262	261	260	259
20	0.05	328	322	318	315	313	311	309	308
	0.01	306	302	298	296	293	292	290	289

Source: Reproduced from Steel, R.G.D., *Biometrics,* 15, 560–572, 1959, Table 3. With permission from the Biometric Society.

Note: Used first in Chapter 5.

Appendix 24: Wilcoxon (Mann–Whitney) Rank-Sum Test Critical Values with Bonferroni's Adjustments: One-Sided Test and α = 0.05

P	Observations in Control (*m*)	Observations in Experimental Treatment (*n*)							
		3	4	5	6	7	8	9	10
1 (α′ = 0.05)	3	6	10	16	23	30	39	49	59
	4	6	11	17	24	32	41	51	62
	5	7	12	19	26	34	44	54	66
	6	8	13	20	28	36	46	57	69
	7	8	14	21	29	39	49	60	72
	8	9	15	23	31	41	51	63	72
	9	10	16	24	33	43	54	66	79
	10	10	17	26	35	45	56	69	82
2 (α′ = 0.025)	3			15	22	29	38	47	58
	4		10	16	23	31	40	49	60
	5	6	11	17	24	33	42	52	63
	6	7	12	18	26	34	44	55	66
	7	7	13	20	27	36	46	57	69
	8	8	14	21	29	38	49	60	72
	9	8	14	22	31	40	51	62	75
	10	9	15	23	32	42	43	65	78
3 (α′ = 0.016)	3				21	28	37	46	57
	4		10	16	22	30	39	48	59
	5		11	17	24	32	41	51	62
	6	6	11	18	25	33	42	53	65
	7	6	12	19	26	35	45	56	68
	8	7	13	20	28	37	47	58	70
	9	7	13	21	29	39	49	61	73
	10	8	14	22	31	41	51	63	76
4 (α′ = 0.0125)	3				21	28	37	46	56
	4			15	22	30	38	48	59
	5		10	16	23	31	40	50	61
	6	6	11	17	24	33	42	52	64
	7	6	12	18	26	34	44	55	67
	8	7	12	19	27	36	46	57	69
	9	7	13	20	28	38	48	60	72
	10	7	14	21	30	40	50	62	75
5 (α′ = 0.0100)	3					28	36	46	56
	4			15	22	29	38	48	58
	5		10	16	23	31	40	50	61
	6		11	17	24	32	42	52	63
	7	6	11	18	25	34	44	54	66
	8	6	12	19	27	36	46	56	68
	9	7	13	20	28	37	47	59	71
	10	7	13	21	29	39	49	61	74
6 (α′ = 0.0083)	3					28	36	45	56
	4			15	21	29	38	47	58
	5		10	16	22	30	39	49	60
	6		10	16	24	32	41	51	63
	7	6	11	17	25	33	43	54	65
	8	6	12	18	26	35	45	56	68
	9	6	12	19	27	37	47	58	70
	10	7	13	20	29	38	49	60	73

		Observations in Experimental Treatment (n)							
P	**Observations in Control (m)**	**3**	**4**	**5**	**6**	**7**	**8**	**9**	**10**
7 ($\alpha' = 0.0071$)	3						36	45	56
	4				21	29	37	47	58
	5			15	22	30	39	49	60
	6		10	16	23	32	41	51	62
	7		11	17	25	33	43	53	65
	8	6	11	18	26	35	45	55	67
	9	6	12	19	27	36	46	58	70
	10	7	13	20	28	38	48	60	72
8 ($\alpha' = 0.00625$)	3						36	45	55
	4				21	29	37	47	57
	5			15	22	30	39	49	59
	6		10	16	23	31	41	51	62
	7		11	17	24	33	42	53	64
	8	6	11	18	26	34	44	55	67
	9	6	12	19	27	36	46	57	69
	10	6	12	19	28	37	48	59	72
9 ($\alpha' = 0.0055$)	3							45	55
	4				21	28	37	46	57
	5			15	22	30	39	48	59
	6		10	16	23	31	40	50	62
	7		10	17	24	32	42	52	64
	8		11	18	25	34	44	55	66
	9	6	11	18	26	35	46	57	69
	10	6	12	19	28	37	47	59	71
10 ($\alpha' = 0.0050$)	3							45	55
	4				21	28	37	46	57
	5			15	22	30	39	48	59
	6		10	16	23	31	40	50	61
	7		10	17	24	32	42	52	64
	8		11	18	25	34	44	54	66
	9	6	11	18	26	35	45	56	68
	10	6	12	19	27	37	47	58	71

Note: First used in Chapter 5.

Appendix 25: Wilcoxon Rank-Sum Test Critical Values with Bonferroni's Adjustments: Two-Sided Test and $\alpha = 0.05$

P	Observations in Control (*m*)	Observations in Experimental Treatment (*n*)							
		3	4	5	6	7	8	9	10
1 ($\alpha' = 0.0250$)	3			15	22	29	38	47	58
	4		10	16	23	31	40	50	60
	5	6	11	17	24	33	42	52	63
	6	7	12	18	26	34	44	55	66
	7	7	13	20	27	36	46	57	69
	8	8	14	21	29	38	49	60	72
	9	8	15	22	31	40	51	63	75
	10	9	15	23	32	42	53	65	78
2 ($\alpha' = 0.0125$)	3				21	28	37	46	56
	4			15	22	30	38	48	59
	5		10	16	23	31	40	50	61
	6	6	11	17	24	33	42	52	64
	7	6	12	18	26	34	44	55	67
	8	7	12	19	27	36	46	57	69
	9	7	13	20	28	38	48	60	72
	10	7	14	21	30	40	50	62	75
3 ($\alpha' = 0.0083$)	3					28	36	45	56
	4			15	21	29	38	47	58
	5		10	16	22	30	39	49	60
	6		10	16	24	32	41	51	62
	7	6	11	17	25	33	43	54	65
	8	6	12	18	26	35	45	56	68
	9	6	12	19	27	37	47	58	70
	10	7	13	20	28	38	49	60	73
4 ($\alpha' = 0.0625$)	3						36	45	55
	4				21	29	37	47	57
	5			15	22	30	39	49	60
	6		10	16	23	31	41	51	62
	7		11	17	24	33	42	53	64
	8	6	11	18	26	34	44	55	67
	9	6	12	19	27	36	46	57	69
	10	6	12	20	28	37	48	59	72
5 ($\alpha' = 0.005$)	3							45	55
	4				21	28	37	46	57
	5			15	22	30	39	48	59
	6		10	16	23	31	40	50	62
	7		10	17	24	32	42	52	64
	8		11	18	25	34	44	54	66
	9	6	11	18	26	35	45	57	69
	10	6	12	19	28	37	47	59	71
6 ($\alpha' = 0.0042$)	3								55
	4					28	37	46	57
	5			15	22	29	38	48	59
	6			16	23	31	40	50	61
	7		10	16	24	32	41	52	63
	8		11	17	25	33	43	54	66
	9		11	18	26	35	45	56	68
	10	6	12	19	27	36	47	58	71

P	**Observations in Control (m)**	**Observations in Experimental Treatment (n)**							
		3	**4**	**5**	**6**	**7**	**8**	**9**	**10**
7 (α′ = 0.0036)	3								55
	4					28	36	46	56
	5				21	29	38	48	58
	6			15	22	30	39	49	60
	7		10	16	23	32	41	53	63
	8		10	17	24	33	43	53	65
	9		11	18	25	36	44	55	67
	10	6	11	18	26	36	46	57	70
8 (α′ = 0.0031)	3								55
	4					28	36	46	56
	5				21	29	38	48	58
	6			15	22	30	39	49	60
	7		10	16	23	32	41	51	63
	8		10	17	24	33	43	53	65
	9		11	18	25	34	44	55	67
	10	6	11	18	26	36	46	57	70
9 (α′ = 0.0028)	3								
	4						36	45	56
	5				21	28	37	47	58
	6			15	22	30	39	49	60
	7			15	23	31	40	50	62
	8		10	16	24	32	42	52	64
	9		10	17	25	33	43	54	66
	10		11	18	26	35	45	56	68
10 (α′ = 0.0025)	3								
	4						36	45	56
	5				21	28	37	47	58
	6			15	22	30	39	49	60
	7			15	23	31	40	50	62
	8		10	16	24	32	42	52	64
	9		10	17	25	33	43	54	66
	10		11	18	26	35	45	56	68

Note: First used in Chapter 5.

Appendix 26: SAS Code for Implementing the Jonckheere–Terpstra Test

```
/* THIS CODE WAS WRITTEN AND PROVIDED BY DR. JOHN GREEN */
/* DUPONT CORP, WILMINGTON, DE; AUGUST 10 2006 */
/* MODIFIED SLIGHTLY BY M. NEWMAN 2011 */
  DATA FATHEAD;
     INPUT ARC CONC @@;
     CARDS;
     1.4120   0 1.4120   0 1.2490   0 1.2490   0
     1.1071  32 1.1071  32 1.4120  32 1.1071  32
     1.2490  64 1.4120  64 1.4120  64 1.4120  64
     1.2490 128 1.2490 128 1.1071 128 1.4120 128
     0.9912 256 1.2490 256 1.4120 256 0.7854 256
     0.6847 512 0.5796 512 0.6847 512 0.4636 512
     ;
PROC GLM DATA=FATHEAD;
CLASS CONC;
MODEL ARC=CONC;
LSMEANS CONC/PDIFF=CONTROL('0') ADJUST=DUNNETT;
TITLE 'FATHEAD DATA';
RUN;
************************************************************;
* BEGIN JONCKHEERE TESTING                                 *;
************************************************************;
%macro JTTEST(DATAIN=,DOSEVAR=,RESPONSE=,DIRECTION=);
PROC DATASETS LIBRARY=WORK;
DELETE JT JTRESULTS JTRES2 JTRES3 DOSES_ TEMP REPMNS CHKJT CHKJT1;
QUIT;
PROC SORT DATA=&DATAIN OUT=TEMP;
BY &DOSEVAR;
RUN;
DATA TEMP;
SET TEMP;
BY &DOSEVAR;
RETAIN DOSENUM 0 ;
IF (FIRST.&DOSEVAR) THEN DO;
     DOSENUM=DOSENUM+1;
     CALL SYMPUT('DOSES',DOSENUM);
END;
RUN;
%PUT DOSES=&DOSES;
DATA DOSES_;
DOSES_=&DOSES;
RUN;
************************************************************************;
* DO JT TEST WITH ALL DOSES PRESENT TO DETERMINE DIRECTION OF TREND, IF
ANY *;
************************************************************************;
     data REPMNS;
     set temp;
```

```
IF DOSENUM = &DOSES THEN DO;
  CALL SYMPUT('HIDOSE',&DOSEVAR);
  CALL SYMPUT('HINUM',DOSENUM);
END;
RUN;
ODS LISTING CLOSE;
ODS OUTPUT JTTEST=JT;
PROC FREQ DATA =REPMNS;
TABLES &RESPONSE * DOSENUM/JT;
RUN;
ODS LISTING;
DATA _NULL_;
X=1;
RUN;
DATA CHKJT;
SET JT;
RUN;
DATA JTRESULTS;
SET JT;
IF UPCASE(NAME1) IN ("PL_JT","PR_JT") ;
DOSENUM=&HINUM;
&DOSEVAR=&HIDOSE;
P_JT=NVALUE1;
DIR_TESTED="&DIRECTION";
KEEP &DOSEVAR DOSENUM NAME1 P_JT DIR_TESTED;
RUN;
DATA CHKJT1;
SET JTRESULTS;
CALL SYMPUT('DIRECTN',NAME1);
RUN;
%PUT DIRECTN=&DIRECTN AND DIRECTION=&DIRECTION;
*********************************************************;
* CHECK DIRECTION OF OBSERVED TREND AGAINST SPECIFICATION ;
*********************************************************;
%LET PCRIT=0.05;
DATA JTRESULTS;
SET JTRESULTS;
%IF %UPCASE(&DIRECTION)=INCREASE %THEN %DO;
  IF NAME1='PL_JT' THEN DO;
    JT_NOEC=&DOSEVAR;
    P_JTDIR=1-P_JT;
    CALL SYMPUT('DEST','FINAL');
  END;
  IF NAME1='PR_JT' AND P_JT>=0.05 THEN DO;
    JT_NOEC=&DOSEVAR;
    P_JTDIR=P_JT;
    CALL SYMPUT('DEST','FINAL');
  END;
  IF NAME1='PR_JT' AND P_JT<0.05 THEN DO;
    JT_NOEC=.;
    P_JTDIR=P_JT;
    CALL SYMPUT('DEST','NEXT');
  END;
%END;
```

```
    %IF %UPCASE(&DIRECTION)=DECREASE %THEN %DO;
      IF NAME1='PR_JT' THEN DO;
        JT_NOEC=&DOSEVAR;
        P_JTDIR=1-P_JT;
        CALL SYMPUT('DEST','FINAL');
      END;
      IF NAME1='PL_JT' AND P_JT>=0.05 THEN DO;
        JT_NOEC=&DOSEVAR;
        P_JTDIR=P_JT;
        CALL SYMPUT('DEST','FINAL');
      END;
      IF NAME1='PL_JT' AND P_JT<0.05 THEN DO;
        JT_NOEC=.;
        P_JTDIR=P_JT;
        CALL SYMPUT('DEST','NEXT');
      END;
    %END;
    %IF %UPCASE(&DIRECTION)=BOTH %THEN %DO;
      IF P_JT>=0.025 THEN DO;
        JT_NOEC=&DOSEVAR;
        P_JTDIR=P_JT;
        %LET PCRIT=0.025;
        CALL SYMPUT('DEST','FINAL');
      END;
      IF P_JT<0.025 THEN DO;
        JT_NOEC=.;
        P_JTDIR=P_JT;
        CALL SYMPUT('DEST','NEXT');
        %LET PCRIT=0.025;
      END;
    %END;
    RUN;
    DATA CHKRESULTS1;
    SET JTRESULTS;
    RUN;
    %PUT AT STEP 1 DIRECTN=&DIRECTN AND PCRIT=&PCRIT AND
DESTINATION=&DEST;
    %GOTO &DEST;
%NEXT:
*************************************************************************;
* IF THERE IS A SIGNIFICANT TREND IN THE FULL DATASET CONSISTENT WITH
DIRECTION SPECIFIED;
* THEN CONTINUE STEPPING DOWN THROUGH THE DOSES TO FIND THE NOEC. ;
*************************************************************************;
%LET DOSESM1=%EVAL(&DOSES -1);
%DO J=2 %TO &DOSESM1;
    %LET HINUM=%EVAL(&DOSES-&J+1);
    DATA REPMNS;
    SET TEMP;
    IF DOSENUM <=&HINUM;
    IF DOSENUM = &HINUM THEN DO;
      CALL SYMPUT('HIDOSE',&DOSEVAR);
    END;
    RUN;
```

```
ODS LISTING CLOSE;
ODS OUTPUT JTTEST=JT;
PROC FREQ DATA =REPMNS;
TABLES &RESPONSE * DOSENUM/JT;
RUN;
ODS LISTING;
DATA _NULL_;
X=1;
RUN;
DATA CHKJT;
SET JT;
RUN;
DATA JT;
SET JT;
IF UPCASE(NAME1) IN ("PL_JT","PR_JT") ;
DOSENUM=&HINUM;
&DOSEVAR=&HIDOSE;
P_JT=NVALUE1;
DIR_TESTED="&DIRECTION";
KEEP &DOSEVAR DOSENUM NAME1 P_JT DIR_TESTED;
RUN;
*********************************************************;
* CHECK DIRECTION OF OBSERVED TREND AGAINST SPECIFICATION ;
*********************************************************;
DATA JTRES2B;
SET JT;
  IF NAME1 NE "&DIRECTN" THEN DO;
    JT_NOEC=&DOSEVAR;
    P_JTDIR=1-P_JT;
    CALL SYMPUT('DEST','FINAL');
  END;
  IF NAME1="&DIRECTN" AND P_JT>=&PCRIT THEN DO;
    JT_NOEC=&DOSEVAR;
    P_JTDIR=P_JT;
    CALL SYMPUT('DEST','FINAL');
  END;
  IF NAME1="&DIRECTN" AND P_JT<&PCRIT THEN DO;
    P_JTDIR=P_JT;
    JT_NOEC=.;
    CALL SYMPUT('DEST','NEXT');
  END;
RUN;
%PUT AT STEP &J DESTINATION=&DEST;
DATA CHK&J;
SET JTRES2B;
RUN;
DATA JTRES2B;
SET JTRES2B;
IF DOSENUM=2 AND JT_NOEC=. THEN JT_NOEC=0;
RUN;
PROC DATASETS;
APPEND BASE=JTRESULTS DATA=JTRES2B;
QUIT;
%IF &DEST=FINAL %THEN %DO;
```

```
       %GOTO FINAL;
     %END;
%end;
%GOTO FINAL;
%FINAL:
    ****************************************;
    * DROP ALL RESULTS EXCEPT FOR NOEC     *;
    ****************************************;
    PROC SORT DATA=JTRESULTS ;
    BY DESCENDING DOSENUM;
    RUN;
    PROC PRINT DATA=JTRESULTS NOOBS;
    TITLE1 'P_JTDIR IS THE P-VALUE FOR THE JT TEST IN THE DIRECTION
SPECIFIED';
    TITLE2 "DATASET=&DATAIN, RESPONSE=&RESPONSE,DOSEVAR=&DOSEVAR";
    RUN;
    TITLE;
%MEND;
%JTTEST(DATAIN=FATHEAD,DOSEVAR=CONC,RESPONSE=ARC,DIRECTION=DECREASE);
%JTTEST(DATAIN=FATHEAD,DOSEVAR=CONC,RESPONSE=ARC,DIRECTION=INCREASE);
%JTTEST(DATAIN=FATHEAD,DOSEVAR=CONC,RESPONSE=ARC,DIRECTION=BOTH    );
```

Appendix 27: Balancing α, β, and Effect Size (ES)

The investigator who insists that he has absolutely no way of knowing how large an ES to posit fails to appreciate that this necessarily means that he has no rational basis for deciding whether he needs to make ten observations or ten thousand.

—Cohen (1988)

As emphasized in Chapter 5 and elsewhere, *a priori* specification of meaningful effect size (ES) and desired error rates (α and β) is essential in most applications to making valid inferences during statistical testing. Some discussion was already provided in Chapter 2 of the related task of sample size estimation to characterize statistical populations with confidence intervals. This appendix lays out basic sample size estimation techniques for confidence limit precision, and also power and sample size estimation methods for many of the statistical tests mentioned throughout this book. The reader can find a much more thorough and authoritative description of power analysis in Cohen's classic 1988 book. As another general source, Dattalo (2008) provides a concise guide to methods with frequent reference back to Cohen (1988) and linkage to current software that facilitates these computations. Chow et al. (2008) and also Chapter 8 in Woodward (2005) give details for tests often done with epidemiological or clinical data (e.g., Williams's test), including various tests associated with proportions, relative risk, and case-control data.

A few overarching comments should be made before discussing details. (1) The most important is a reiteration of the Chapter 4 discussion about inferential difficulties from conventional statistical tests that do not set α, β, and effect size *a priori*. The pervasive inattention to β and effect size in the ecotoxicology literature should not be considered sound evidence that this point is merely a statistical nicety or *bête noire* of a few fussy professionals. Calls for more attention to β and ES are becoming increasingly common in ecotoxicology, e.g., Munkittrick et al. (2009). It is very difficult to make valid inferences or to estimate positive predictive power from experiments lacking them. (2) That having been said, the tempering point must also be made that α and β should be chosen astutely, not by rigid dogma or regulatory convention. As an important instance, many valid explanations could be prematurely dismissed if demanding test limits are imposed at the early exploratory stage of a research program. Often initial explorations involve insufficient information and experience with which to formulate definitive tests of the mettle of competing explanations. Loehle (1987) refers to the unthinking insistence on rigid testing standards at an early stage of research as premature or dogmatic falsification. Normal science* will dominate initially until theory has matured to a stage at which effective, rigorous testing is possible. An adaptive inference scheme along this vein was proposed by Holling and Allen (2002) to appropriately balance α and β during the progression of a research program from the initial exploratory phase (minimize type II error so useful hypotheses are not rejected prematurely) to the final mature phase (minimize type I error so hypotheses can be culled away effectively). From another vantage discussed in Chapter 5, the seriousness of consequences of making a type I or type II error needs to be considered carefully in an applied science such as ecotoxicology.† (3) Another point to be made before discussing details of power and sample size estimation is that effective inference can occur without conventional statistical testing.

* Two kinds of scientific activity exist, normal and innovative. Normal science is conducted to gather information and produce better measurements. Innovative scientists use the results of normal science to test causal hypotheses and to formulate novel hypotheses. The explanatory hypotheses of innovative science are then used as the framework for further normal science. Both scientific activities are essential to a healthy science. (See Newman 2010 for more discussion of this topic.)

† On page 56 of his classic book on power, Cohen (1988) repeats the recommendation made in Chapter 4 that the quotient α/β be used as a measure of the relative seriousness of the consequences of making these two kinds of errors.

Darwin formulated evolution through natural selection before formal statistical tests existed,* yet few scientists would dismiss his work as unsound. The use of Bayes' factors mentioned in Chapter 6 is one effective contemporary method that does not depend on conventional significance testing. Information-theoretic methods that emphasize model construction and comparison (see AIC discussed in Chapters 3, 4, 6, and 7) are also emerging as valuable inferential tools (Anderson and Burnham 2002; Burnham and Anderson 2001; Richards 2005). (4) Finally, as described in Chapter 5, confidence intervals often are effective alternatives to formal statistical testing. Indeed, they have recently been advocated as an alternative to Dunnett's testing to generate ecotoxicity test metrics (Delignette-Muller et al. 2011). Sample number estimation to achieve a desired confidence interval precision is included below for that reason.

A range of software packages exist for the calculation of the sample size needed to obtain a desired precision or power. Dattalo (2008, 2009) provides synthesis of recent software, including descriptions of the most common methods and web links to applicable software. Some specialized software packages to perform sample size/power analysis also generate maps of spatial placement of survey samples. These include the Pacific Northwest National Laboratory's shareware, Visual Sample Plan (VSP) (VSP Development Team 2011, http://vsl.pnl.gov), which was described recently in the published literature by Matzke et al. (2010). The VSP software has the added advantage of facilitating adherence to required EPA sampling design specifications (U.S. EPA 2002), as does the EPA Deft shareware (http://www.epa.gov/esd/databases/deft/install.htm). Other general shareware choices are the exceptional G*Power 3 shareware (http://www.psycho.uni-duesseldorf.de/abteilungen/aap/gpower3; Faul and Erdfelder 1992; Faul et al. 2007, 2009) developed principally for psychological and biomedical studies, and the Vanderbilt University PS shareware (http://biostat.mc.vanderbilt.edu/PowerSampleSize; DuPont and Plummer 1990, 1998) developed to support clinical studies. Notably, the PS shareware will do power calculations for certain specialty epidemiological tests. One such test is the Mantel–Haenszel test (Mantel and Haenzel 1959; Wittes and Wallenstein 1987) for odds ratios from case-control studies such as those described in Chapter 6. Power or precision estimation for some of the more specialized ecological methods such as the Poisson and negative binomial curve fit testing, and population estimation metrics described in Chapter 6 can be implemented with the 2009 version of the software supporting Krebs (1999). At the other extreme is the PASS 11 package (http://www.ncss.com/pass_procedures.html) that is arguably the most comprehensive tool performing a wide range of power analyses and sample size calculations. Although other popular packages, such as Minitab, are capable of many of these computations, the SAS statistical package (SAS Institute 2008) and a few exemplary software packages will be used here to illustrate methods mentioned throughout this book. The most useful SAS procedures are PROC POWER and PROC GLMPOWER (SAS Institute 2008). The interested reader might want to also explore the SAS macro UnifyPow written by R. O'Brien (http://www.bio.ri.ccf.org/power.html).

A27.1 CONFIDENCE INTERVAL PRECISION FOR A MEAN

Whereas the goal of most techniques explored in this appendix is to ascertain the combination of effect size (ES), α and β needed for doing a useful hypothesis test, the goal in this section will be to obtain enough precision for a confidence interval to generate meaningful results.

* Darwin proposed evolution by natural selection in 1838, prior to R.A. Fisher's formulation of significance testing circa 1925 to 1935. Pierre-Simon Laplace's elaboration of (Bayesian) inverse probability theory (Laplace 1925) did exist by then but was uncommon knowledge outside of a few fields such as astronomy and mathematics.

$$Two\text{-}sided: \bar{x} \pm t_{1-\propto/2,df=n-1} \frac{s}{\sqrt{n}} \quad (A27.1)$$

$$One\text{-}sided: \bar{x} - t_{1-\propto,df=n-1} \frac{s}{\sqrt{n}} \quad (A27.2)$$

or

$$\bar{x} + t_{1-\propto,df=n-1} \frac{s}{\sqrt{n}}$$

Precision estimation might draw on an initial estimate of the population standard deviation from a pilot study or from the literature. Then a desired half width (h) might be set for the specified (perhaps two-sided 95%) confidence interval (SAS Institute 2008).

$$Two\text{-}sided\ Half\ Width\ (h): t_{1-\propto/2,df=n-1} \frac{s}{\sqrt{n}} \quad (A27.3)$$

$$One\text{-}sided\ Half\ Width\ (h): t_{1-\propto,df=n-1} \frac{s}{\sqrt{n}} \quad (A27.4)$$

As discussed in Chapter 2,* estimates can be made for the number of samples needed to achieve a desired confidence interval half width (h) or less. The relationship between sample size (n), standard deviation (s), and desired h might be calculated with a t statistic (Graybill 1958). For example, sample number for a two-sided confidence interval might be estimated with Equation (A27.5).

$$n \geq \left(\frac{t_{\propto/2,df=n-1} s}{h} \right)^2 \quad (A27.5)$$

Although the variance is usually unknown, sample size approximation is sometimes done by substituting $z_{1-\alpha/2}$ for the $t_{n-1,1-\alpha/2}$ (Rosner 2006). Often that approach is taken if $n \geq 30$. With the SAS PROC POWER and other software packages, the probability of obtaining a particular half width (*Prob*(*Width*)) is calculated for a certain sample size and a pilot sample standard deviation. The *Prob*(*Width*) is treated like the targeted power in the other methods described below.

$$Two\text{-}sided\ Interval: Prob(Half\ Width \leq h) = P\left(\chi^2_{n-1} \leq \frac{h^2 n(n-1)}{\sigma^2 (t^2_{1-\alpha/2,df=n-1})} \right) \quad (A27.6)$$

$$One\text{-}sided\ Interval: Prob(Half\ Width \leq h) = P\left(\chi^2_{n-1} \leq \frac{h^2 n(n-1)}{\sigma^2 (t^2_{1-\alpha,df=n-1})} \right) \quad (A27.7)$$

* See discussions of Equations (2.38) to (2.41).

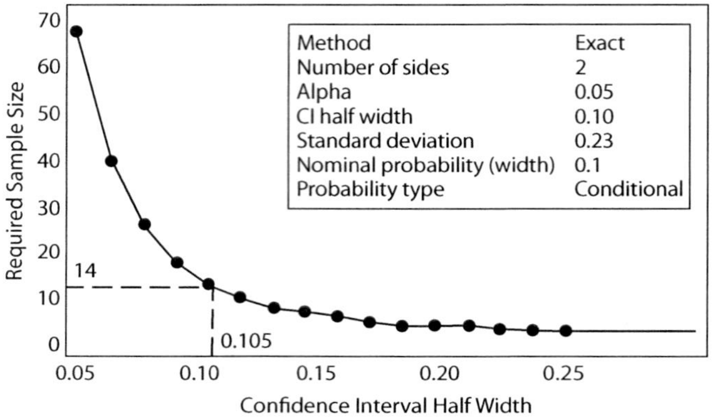

Figure A27.1 Confidence interval precision estimation using the SAS package PROC POWER. The inset table contains the details used in the SAS procedure.

The SAS code below produces the results shown in Figure A27.1. With 14 samples, the nominal confidence interval half width from $\bar{x}$ to the lower or upper confidence limit will be 0.105.

```
/* ONESAMPLEMEANS CI=T STARTS THE SAMPLE SIZE DETERMINATION PROCEDURE.  */
/* NTOTAL=. INDICATES THAT N BE DETERMINED BY THE PROCEDURE, NOT HALF   */
/* WIDTH. THE N's NEEDED FOR TARGET INTERVAL WIDTHS FROM 0.05 TO 0.30   */
/* ARE REQUESTED. OUTPUT IS A PLOT OF SAMPLE SIZE VS HALF WIDTH, AND N  */
/* FOR A SPECIFIED ALPHA OF 0.05. CI=T ASKS THAT A T STATISTIC BE       */
/* USED TO ESTIMATE THE PROBABILITY FOR THE CONFIDENCE INTERVAL.        */
PROC POWER;
ONESAMPLEMEANS CI=T ALPHA= 0.05 HALFWIDTH = 0.1 STDDEV = 0.23 SIDES=2
     PROBWIDTH = 0.1 NTOTAL=.;
PLOT X=EFFECT MIN = 0.05 MAX=0.30;
RUN;
```

A computer-intensive bootstrap approach might draw from a pilot data set to simulate the influence of sample size on confidence interval width. Many programs do bootstrapping; however, the Resampling Stat software (Simon 1995; http://www.resample.com/) will be used here. In the Resampling Stat code below, 20 measurements from a pilot survey are placed into an array labeled *A*. A bootstrap resampling with replacement is done 1,000 times from this array. During each of the 1,000 resamplings, a specified number of observations (10 in the code below) is taken randomly with replacement from *A* and placed into a temporary array called *B*. The mean of the 10 numbers in *B* is calculated and placed into another array called *BB*. This resampling is repeated 1,000 times to produce 1,000 means for bootstrap samples from array *A*. The 1,000 means in array *BB* are ranked from the smallest to largest, and the means corresponding to 2.5 and 97.5% are used as the bootstrap lower and upper 95% confidence limits. This is done for different sample sizes by changing the 10 in the line SAMPLE 10 A B to whatever sample size is desired.

```
'HERE WE HAVE 20 MEASUREMENTS FROM A PILOT SAMPLING AND WANT TO ESTIMATE
'HOW MANY WOULD BE NEEDED TO REASONABLY MINIMIZE THE VARIABILITY IN
'ESTIMATES OF THE MEAN.
DATA (.27 .22 .21 .31 .23 .23 .25 .24 .25 .23 .27 .28 .2 .25 .22 .26 .22
.23 .24 .26) A
REPEAT 1000
  SAMPLE 10 A B
```

```
  MEAN B BB
  SCORE BB Z
END
HISTOGRAM Z
PERCENTILE Z (2.5 50 97.5) K
PRINT K
'GRADUALLY INCREASE SAMPLE NUMBER, NOTING THE WIDTH OF THE CI WITH
'EACH INCREASE
'IN SAMPLE NUMBER. NOTE WHAT SAMPLE NUMBER IS NEEDED TO APPROACH A POINT
'OF 'MINIMUM IMPROVEMENT.
```

For illustration, the 2.5, 50, and 97.5% bootstrap values were generated for sample sizes from 2 to 20 observations.

n	2.5%	50%	97.5%
2	0.215	0.245	0.282
4	0.220	0.243	0.273
6	0.223	0.243	0.265
8	0.228	0.244	0.263
10	0.229	0.243	0.260
12	0.230	0.243	0.259
14	0.230	0.242	0.259
16	0.231	0.244	0.258
18	0.232	0.243	0.255
20	0.232	0.243	0.256

There is a rapid narrowing of confidence intervals as bootstrap sample size increases, with a point of diminishing returns occurring at approximately 10 or 12 observations. Calculated half widths for the bootstrapped confidence intervals could be used to judge acceptability instead of visual inspection of results. The lower and upper confidence limit half widths for $n = 10$ are 0.014 and 0.017, respectively. Notice that the distribution might not be symmetrical. This increasingly commonplace resampling approach was applied by Daskalakis (1996) to determine the number of oysters to sample in order to obtain narrow confidence intervals for estimated mean metal concentrations in their soft tissues.

A27.2 COMPARING ONE SAMPLE MEAN TO A HYPOTHETICAL MEAN WITH A *t*-TEST

A t-test might be used to compare a mean from a set of observations to some hypothetical mean. Recollect that the following statistic (T with $df = n - 1$) has a characteristic t distribution so it can be compared to a t value from a table or a software function to find the associated probability.

$$T = \frac{\bar{x} - \mu}{\frac{s}{\sqrt{n}}} \tag{A27.8}$$

where n = number of observations, $\bar{x}$ = estimated mean, s = estimated standard deviation, and μ = the true mean. A t-test for an estimated mean against a hypothetical mean (μ_0) can be done,* H_0: $\mu = \mu_0$. The alternative hypothesis, H_A, for a two-sided test would be $\mu \neq \mu_0$. It would be either $\mu < \mu_0$ or $\mu_0 > \mu$ for a one-sided test depending on whether the hypothesis was that the mean is smaller or larger than the hypothetical mean. The noncentrality parameter here is $\delta = \sqrt{n}(\mu - \mu_0 / \sigma)$ and the test t statistic given δ is $\sqrt{n}(\bar{x} - \mu_0 / s)$. SAS (2008) indicates that this t is approximately $F_{1,n-1}$ for a given δ^2 so power can be estimated with the following:

$$\textit{Two-sided Test: Power} = \textit{Prob}[(F_{1,n-1}, \delta^2) \geq F_{1-\propto, df=1,n-1}] \quad \text{(A27.9)}$$

$$\textit{One-sided Test: Power} = \textit{Prob}[(t_{n-1}, \delta) \geq t_{1-\propto, df=n-1}] \quad \text{(A27.10)}$$

or

$$\textit{Power} = \textit{Prob}[(t_{n-1}, \delta) \leq t_{1-\propto, df=n-1}]$$

The example below uses the SAS PROC POWER to do sample size estimation for a one-sided t-test comparing a mean to a hypothesized null mean.

```
/* THE MEAN COMES FROM A PREVIOUS PILOT STUDY OR ANOTHER SOURCE.   */
/* THE STDDEV ALSO IS FROM THAT PILOT STUDY. THE NULLMEAN IS THE   */
/* HYPOTHESIZED MEAN THAT YOU ARE TESTING THE DATA AGAINST. AN     */
/* ALPHA OF 0.05 IS ASSUMED BECAUSE IT WAS NOT SPECIFIED. A ONE    */
/* SIDED TEST WAS SELECTED. */
PROC POWER;
   ONESAMPLEMEANS
      NULLMEAN=0.1
      SIDES=1
      MEAN = .15
      NTOTAL = 3 TO 10
      STDDEV = .02
      POWER = .;
RUN;
```

The following output indicates that four observations would be enough to get the desired power. (See also Figure A27.2.)

```
              The POWER Procedure
           One-sample t Test for Mean

            Fixed Scenario Elements
           Distribution           Normal
           Method                 Exact
           Number of Sides        1
           Null Mean              0.1
           Mean                   0.15
           Standard Deviation     0.02
           Alpha                  0.05
```

* As discussed in Chapter 2, the initial issue in this approach is that z requires that the standard deviation for the sample be known before you draw the sample. Extracting a t value requires that you know the degrees of freedom that is produced for a specified n. This issue can be resolved by using the estimated standard deviation for the population and a z (standard normal) score initially or as an approximation for t. Recollect that that the z score for a value (X_i) is $(X_i - \mu)/\sigma$.

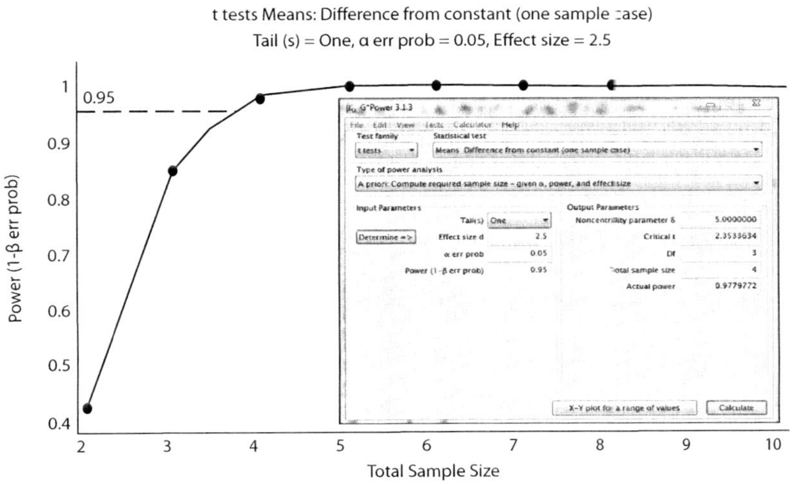

Figure A27.2 Estimation of minimal sample size needed to obtain a specified α and β for a test of a sample mean versus a hypothetical mean. The graph depicts results from the SAS package PROC POWER procedure. The inset is a screen capture from the G*Power 3 shareware package.

Computed Power

Index	N Total	Power
1	3	0.848
2	4	0.978
3	5	0.998
4	6	>.999
5	7	>.999
6	8	>.999
7	9	>.999
8	10	>.999

Similarly, the G*Power 3 software produced the Figure A27.2 inset using the same input, except the input included the difference (effect size) between the sampled population mean, $\bar{x}_A$, and the hypothetical mean, μ_0, as estimated by $d = (\bar{x}_A - \mu_0)/s = |0.10 - 0.15/0.2| = 2.5$. The noncentrality parameter was $\delta = \sqrt{n}(\bar{x}_A - \mu_0/s)$.

A27.3 COMPARING TWO SAMPLE MEANS (ASSUMING EQUAL VARIANCES) WITH A *t*-TEST

A *t*-test for comparing sample means was discussed in several places in this book. An associated power analysis can be done with many software packages. Similar to the description above for a *t*-test involving only one estimated mean, sample size or power estimation begins with the effect size for the difference between the means of *A* and *B*,

$$d = \frac{\mu_A - \mu_B}{\sigma} \approx \frac{\bar{x}_A - \bar{x}_B}{s} \tag{A27.11}$$

The standard deviation in Equation (A27.11) is the root mean square of σ_A^2 and σ_A^2

$$\sigma = \sqrt{\frac{\sigma_A^2 + \sigma_B^2}{2}}$$

For planning purposes in situations where only a general effect size can be specified, Cohen (1988) categorizes effect sizes for differences between means as small, medium, or large when d values are in the ranges of 0.20, 0.50, and 0.80, respectively. If one wished to be able to identify a small effect, a d of 0.20 might be used in *a priori* sample size calculations. A noncentrality parameter is estimated with N being the total number of observations if the same number of observations is being drawn from both, i.e., $N = n_A + n_B$ with $n_A = n_B$.

$$\delta = d\sqrt{\frac{N}{4}} \tag{A27.12}$$

The sample size might be estimated based on this relationship. If the number of observations will be different for the two independent samples, the harmonic mean $(2n_An_B)/(n_A + n_B)$ is used instead of $n_A + n_B$ (Cohen 1988; Dattalo 2008).

$$\delta = d\sqrt{\frac{(2n_An_B)/(n_A + n_B)}{2}} \tag{A27.13}$$

The SAS software applies F and t statistics as described above for calculating the power (SAS Institute 2008).

$$\textit{Two-sided Test}: \textit{Power} = \textit{Prob}[(F_{1,n-2}, \delta^2) \geq F_{1-\infty,df=1,n-2}] \tag{A27.14}$$

$$\textit{One-sided Test}: \textit{Power} = \textit{Prob}[(t_{n-2}, \delta) \geq t_{1-\infty,df=n-2}] \tag{A27.15}$$

or

$$\textit{Power} = \textit{Prob}[(t_{n-2}, \delta) \leq t_{\infty,df=n-2}]$$

The SAS code below implements power analysis for a t-test comparing means of two samples. It does the calculations for differences in means of 0.2, 0.3, and 0.4, and samples sizes of 2 to 25 per group.

```
/* THE NULL DIFFERENCE BY DEFAULT IS A TRUE DIFFERENCE OF ZERO      */
/* BUT ADDING A NULLDIFF=0.2, FOR EXAMPLE, TO THE SPECIFICATIONS    */
/* BELOW WILL ALLOW ANOTHER NULL DIFFERENCE TO BE SPECIFIED.        */
PROC POWER;
  TWOSAMPLEMEANS TEST=DIFF
  MEANDIFF= .2 .3 .4
  STDDEV = 0.17
  ALPHA=0.01
  POWER = .
  SIDES=1
  NPERGROUP = 2 TO 25;
  PLOT;
RUN;
```

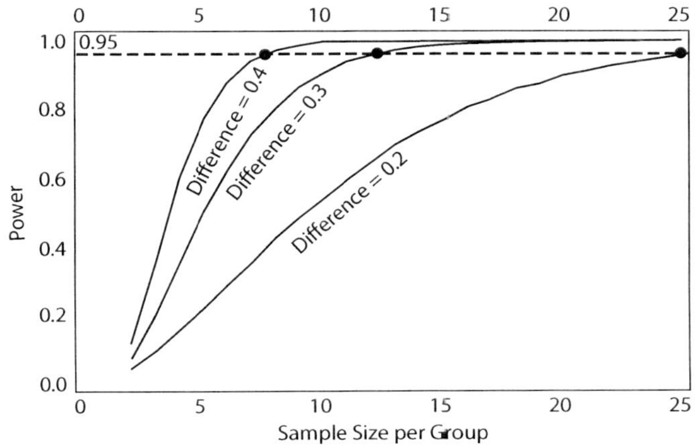

Figure A27.3 SAS PROC POWER estimation of sample size needed to obtain a specified α and β for a test of difference between two sample means. Estimates are done with three hypothetical differences and a one-sided test.

The SAS results (Figure A27.3) indicate that samples sizes per group of 25, 12, or 8 are needed for differences of 0.2, 0.3, and 0.4 to obtain power of 0.95 with an α of 0.01 (one-sided test).

A27.4 ANOVA *F* TEST

A different effect metric might be applied for the overall test of several means (Cohen 1988). With more than two means being compared, Cohen's f (= standard deviation of the treatment means/common standard deviation) might be applied instead of the standardized difference in means (d). Assuming all treatments have the same number of observations, let σ = the standard deviation within the treatment populations, which is assumed to be the same for all treatments, σ_m = the standard deviation of the means for the G treatments, μ_i = the mean of the ith treatment, and μ = the mean of the G means.

$$f = \frac{\sigma_m}{\sigma} = \frac{\sum_{i=1}^{G}(\bar{x}_i - \mu)^2}{\sigma} \quad \text{(A27.16)}$$

The f is the standard deviation for the means standardized to the within-treatment standard deviation. (Compare it to d in Equation (A27.11), which is the range of two means standardized to their common standard deviation.) The f is 0 if all treatments are the same but becomes larger as they become increasingly different. The f is related to sample size (n). The alternative metric, ϕ, is used in some power calculations (Cohen 1988).

$$f = \frac{\phi}{\sqrt{n}} \quad \text{(A27.17)}$$

Like f, ϕ is a standardization, but the standardization is done to the standard error instead of the standard deviation.

The f can be related to the standardized range of the means (d) described for t-tests. The f would be equal to half of the d for two means being compared (Equation A27.11). The range from the smallest to the largest mean is used for situations with more than two means. Cohen (1988) estimates the relationship between d and f for three possible patterns for the $G-2$ treatment means between the smallest and largest means. For a given d, Cohen states that Equation (A27.18) is appropriate if the other $G-2$ means are clustered at the mean of the treatment means.

$$f_1 = d\sqrt{\frac{1}{2G}} \tag{A27.18}$$

Equation (A27.19) defines f given d if the $G-2$ treatment means are evenly distributed between the two extreme treatment means.

$$f_2 = \frac{d}{2}\sqrt{(G+1)/(3(G-1))} \tag{A27.19}$$

In the case where the $G-2$ means are distributed toward the range extremes, the f for a given d will be one of two values depending on whether G is even or odd. Equation (A27.20) is appropriate if G is an even number. Otherwise, Equation (A27.21) is the appropriate equation.

$$f_3 = \frac{1}{2}d \tag{A27.20}$$

$$f_3 = d\frac{\sqrt{G^2-1}}{2G} \tag{A27.21}$$

Effect size calculations can involve d or f for ANOVA F ratio-based power calculations. For purposes of *a priori* calculation where one only specifies a general effect size that needs to be detected, Cohen (1988) separates f values into those reflecting large ($f = 0.4$), medium ($f = 0.25$), and small ($f = 0.10$) effect sizes. These general categories can be used to establish the desired design given a particular number of treatments (G) and some estimate of expected variation. The standard deviation for the treatments is only 10% of the within-treatment standard deviation if f is 0.10 and is 40% if f is 0.40. Dattalo (2008) describes a less precise approach to power analyses if it is difficult to choose among f_1, f_2, and f_3. It is based on a noncentrality parameter (Equation A27.22), δ.

$$\delta = N\left(\frac{\frac{\mu_{Largest}-\mu_{Smallest}}{\sigma}}{2}\right) \tag{A27.22}$$

where $\mu_{Largest}$ and $\mu_{Smallest}$ = largest and smallest means for the G treatments, respectively. This approach carries the assumptions of equal numbers of samples in all treatments and any differences being associated with the two extreme means.

Power analyses can also be done in which effect size is expressed with a correlation ratio (η), that is, "f is described … in terms of correlation between [treatment] population membership and the dependent variable" (Cohen 1988). The η^2 defined in Equation (A27.23) reflects the dispersion of treatment means with η^2 being 0 if there is no treatment effect.

$$\eta^2 = \frac{f^2}{1+f^2} \tag{A27.23}$$

Power estimation for an ANOVA F ratio test for differences among treatment means will be illustrated with SAS PROC POWER, although PROC GLMPOWER could have been used. Power for specified contrasts could also be done, but the specified TEST=OVERALL option focuses on the F ratio-based test of H_0: $\mu_1 = \mu_2 = \ldots = \mu_G$ for G treatments. The power is expressed as a probability (*Prob*) given the number of treatments (G), total number of observations in all groups (N), initial estimates of treatment means ($\bar{x}_i$ as estimates of μ_i), and perhaps any allocation weights (w_i), although weightings are not relevant in the example here because the same number of observations will be used for each treatment. The $F(v_1,v_2)$ indicates the F distribution with v_1 and v_2 degrees of freedom.

$$Power = Prob[F_{G-1,N-G},\lambda \geq F_{1-\alpha,G-1,N-G}] \tag{A27.24}$$

where the noncentrality parameter, λ, is estimated with Equation (A27.25).

$$\lambda = N\frac{\sum_{i=1}^{G} w_i(\mu_i - \bar{\mu})^2}{\sigma^2} \tag{A27.25}$$

For Equation (A27.25),

$$\bar{\mu} = \frac{\left(\sum_{i=1}^{G} w_i\mu_i\right)}{G}$$

and σ^2 = the common variance.

The following SAS example uses group means and the estimated MSE from a pilot study with six treatments ($G = 6$) and a specified type I error rate of 0.05 (α). The range for number of observations per treatment in the definitive study (NPERGROUP) is set to 2 to 8 in the hopes that a useful power of 0.95 would be obtained within that range.

```
PROC POWER;
  ONEWAYANOVA
  TEST=OVERALL
  GROUPMEANS = 1.33| 1.183 | 1.371 | 1.254 | 1.109 | 0.604
  STDDEV = 0.17
  ALPHA = 0.05
  NPERGROUP = 2 TO 8
  POWER=.;
RUN;
/* THE "OVERALL" SPECIFIES THAT THE POWER CALCULATIONS WILL BE              */
/* FOCUSED ON THE OVERALL F TEST FOR EQUALITY OF THE TREATMENT MEANS.       */
/* TEST=CONTRAST WOULD FOCUS ON A 1 DF CONTRAST OF MEANS. THE SIX           */
```

```
/* MEANS ARE THOSE FROM A PILOT STUDY. THE STDDEV (ERROR STANDARD     */
/* DEVIATION) IS ALSO FROM A PILOT. POWER (MISSING VALUE ".") WILL    */
/* BE ESTIMATED HERE FOR 2 TO 8 OBSERVATIONS PER TREATMENT.           */
```

The output copied below was generated, suggesting that triplicates for each treatment (group) would be adequate for the definitive study if both α and β values were set at 0.05.

```
                    The POWER Procedure
              Overall F Test for One-Way ANOVA
                  Fixed Scenario Elements

Method                  Exact
Alpha                   0.05
Group Means             1.33 1.183 1.371 1.254 1.109 0.604
Standard Deviation      0.17

                     Computed Power

                          N Per
                  Index   Group   Power
                  1       2       0.753
                  2       3       0.987
                  3       4       >.999
                  4       5       >.999
                  5       6       >.999
                  6       7       >.999
                  7       8       >.999
```

The shareware package G*Power 3 was also used to estimate sample size for this example (Figure A27.4). Notice the effect size expressed as *f* was quite large (1.505). Again, only triplicates were required to obtain α and β of 0.05.

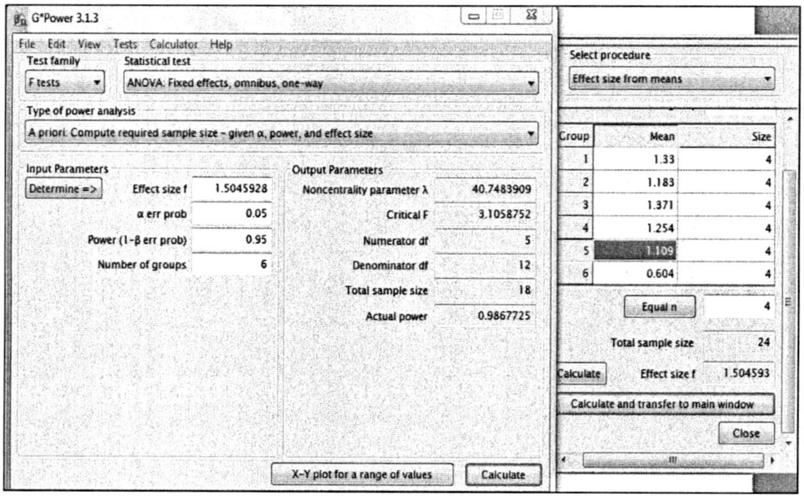

Figure A27.4 Screen capture of G*Power 3 sample size estimation for a one-way ANOVA *F* ratio test of equal means.

A27.5 POST (ONE-WAY)-ANOVA COMPARISONS

Post-ANOVA treatment comparisons of means were discussed in Chapter 5 and elsewhere. They could be done in several ways, but that of comparing a reference mean to the other treatment means was the focus of most of our discussions. Power can be calculated for these *t*-based tests as illustrated below with SAS PROC POWER. Unlike in Section A27.4, with the ANOVA *F* test in which TEST=OVERALL was specified, the contrasts of interest can be examined by specifying TEST=CONTRAST instead. For this illustration, the fathead minnow pentachlorophenol data used throughout Chapter 5 were now considered the pilot study data. The treatment means are taken from that study and the estimated standard deviation estimated as the square root of the MSE. Reflecting discussions in Chapter 5 about the advantage of having more replicates in the reference treatment, the GROUPWEIGHTS = (2 1 1 1 1 1) specifies that twice as many replicates be used in the reference than the other treatments, and that all nonreference treatments have the same number of replicates. In this example, the code below was executed and this execution repeated after changing to GROUPWEIGHTS = (1 1 1 1 1 1) to produce results for a test with equal numbers of replicates in all treatments. The CONTRASTS = (5 -1 -1 -1 -1 -1) specifies that the contrast will be between the reference and the other five treatments. (See SAS Institute 2008 for more details on specifying contrasts.)

```
/* THE FATHEAD MINNOW SURVIVAL OF PENTACHLOROPHENOL EXPOSURE DATA      */
/* USED IN CHAPTER 5 WILL BE USED HERE TO ILLUSTRATE POWER             */
/* ESTIMATION FOR POST ANOVA CONTRASTS.                                */
/* THE SIX CONTRAST VALUES COMPRISE A VECTOR (L) FOR TESTING THE       */
/* HYPOTHESIS THAT SOME B*L IS ZERO. FOR THIS PARTICULAR DESIGN 5 IS   */
/* CHOSEN FOR THE REFERENCE TREATMENT AND -1 FOR THE OTHERS WHICH      */
/* MAKES THEIR SUM EQUAL TO -5.0. THE NONREFERENCE TREATMENTS ARE      */
/* TREATED THE SAME IN THE TESTING.                                    */
PROC POWER PLOTONLY;
  ONEWAYANOVA
  TEST=CONTRAST
  GROUPMEANS = 1.33| 1.183 | 1.371 | 1.254 | 1.109 | 0.604
  STDDEV = 0.155 /* SQUARE ROOT OF THE MSE USED HERE FOR STDDEV.       */
  GROUPWEIGHTS=(2 1 1 1 1 1 )
  NTOTAL= 14
  ALPHA = 0.05
  POWER= .
  CONTRAST=(5 -1 -1 -1 -1 -1);
  PLOT X=N MIN=14 MAX=49;
RUN;
```

Figure A27.5 is a plot of total sample size (total number of replicates in all treatments) versus power for a design with twice as many replicates in the reference treatment than in any nonreference treatment (solid line, *n* = replicates in reference, replicates in each nonreference treatment) and the same number of replicates in treatments (dashed line, *n* = number of replicates in a treatment). The results reinforce the point in Chapter 5 that power can be enhanced by using more replicates in the reference treatment than in the nonreference treatments.

Two specialized tests were also discussed in Chapter 5, Dunnett's and Williams's tests. Recollect that Dunnett's test carries the assumptions of normality and equal variance among treatments. It compares the means of the control/reference treatment to those of the other treatments. The equations specifying power or required sample size are too involved to warrant their detailing in this brief appendix. The interested reader is directed to the original publication

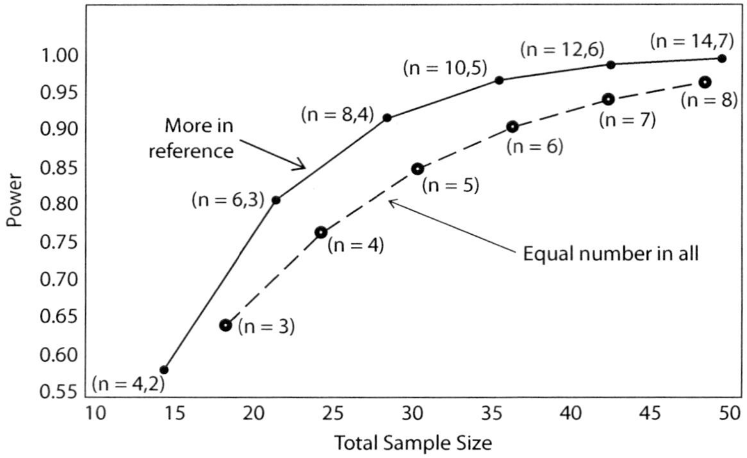

Figure A27.5 SAS PROC POWER results for post-ANOVA contrasting a reference treatment to the other nonreference treatments. Two scenarios are plotted, including one with equal numbers of observations in all treatments (dashed line) and twice as many observations from the reference treatment as in any of the other nonreference treatments (solid line).

(Dunnett 1955) and also NCSS (2011) for details. In very general terms, the difference between means (range), within-group variance, and sample size are used to estimate power for this test. Williams's method tests for monotonic trend over changing intensity of some treatment such as toxicant concentration. Power estimation for Williams's test is detailed by Chow et al. (2008). Generally, the meaningful difference one wishes to detect (Δ), the number of groups (*G*), within-treatment group standard deviation (σ), observation number (*n*), and Williams's critical *t* ($t_{G,}\alpha$) are used to estimate power of sample size (NCSS 2011). (The Φ() in these equations is the standard normal distribution function and *z* is the conventional *z* statistic associated with a particular probability such as β.)

$$Power\ (as\ a\ Probability) = 1 - \Phi\left[t_{G,\propto} - \frac{\Delta}{\sigma\sqrt{2/n}}\right] \quad \text{(A27.26)}$$

$$n = \frac{2\sigma(t_{G,\alpha} + z_\beta)^2}{\Delta^2} \quad \text{(A27.27)}$$

The PASS 11 software package does power analyses for both of these tests, so it will be used with the pentachlorophenol data set again. The desired detectable difference between means will be set at 0.5. Estimates were performed for equal numbers of observations in all treatment groups and also twice as many observations in the reference group for Dunnett's test (Figure A27.6). For the unequal numbers of observation design, the necessary *n* values in the reference and nonreference groups were 12 and 6, respectively, that is, a total of 42 observations. The required number per group was 7 (i.e., 42 total) to get the same power assuming equal numbers of observations in all groups. If Williams's test is applied (Figure A27.7) with equal numbers of observations per group and the same 0.5 difference, the required number of observations per group was 2, resulting in a total number of observations of 12 to meet the specified conditions.

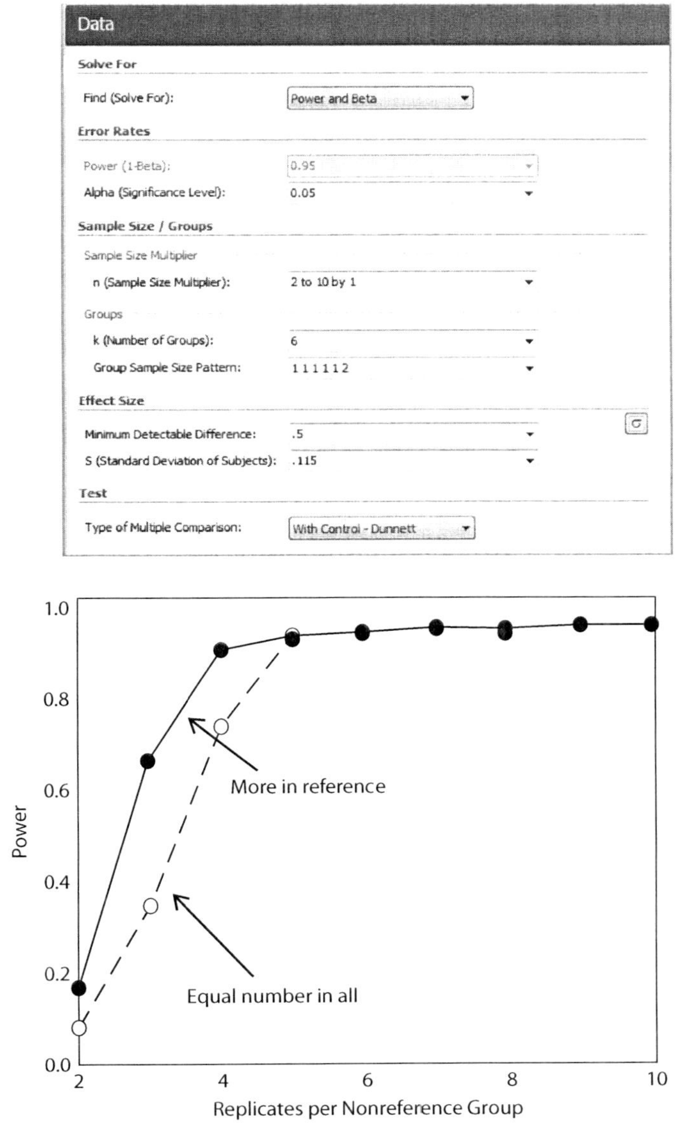

Figure A27.6 PASS 11 software analysis for Dunnett's test (reference versus nonreference groups) with the same number of observations per group (dashed line) and twice as many observations from the reference group (solid line). The top panel is a screen capture of the PASS 11 input for estimation with more reference than nonreference observations.

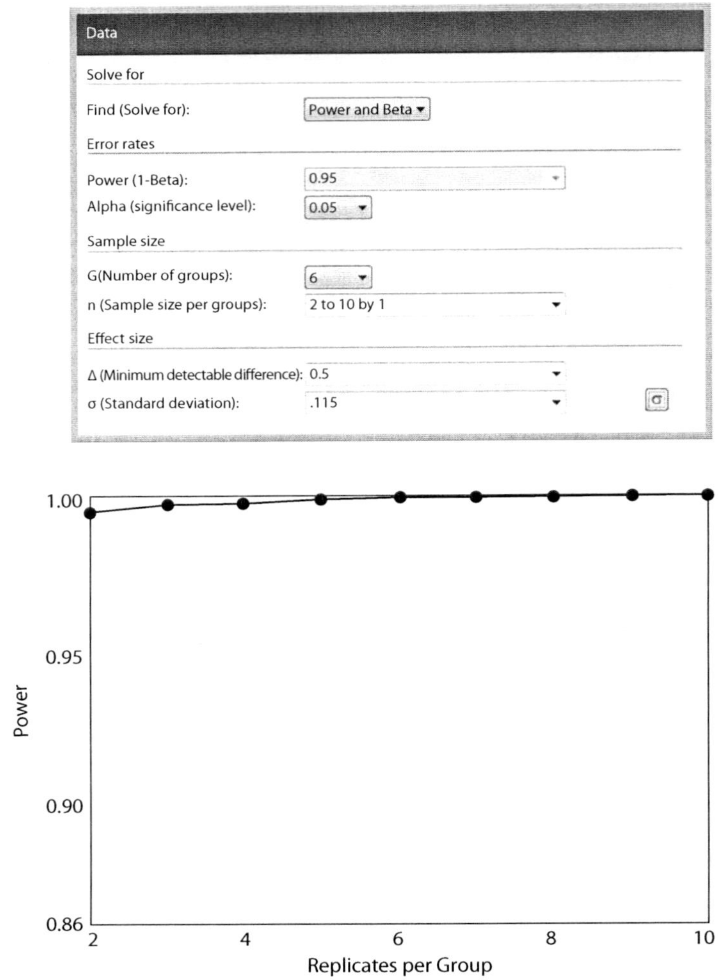

Figure A27.7 PASS 11 software analysis for Williams's test with the same number of observations per group. The top panel is a screen capture of the PASS input.

A27.6 TWO-WAY ANOVA

The SAS code below involves two-way ANOVA of survival time for minnows exposed to five different cyanide concentrations at three temperatures.* The PROC GLM code does two-way ANOVA with in-hand data from an experiment involving combinations of cyanide concentration and temperature treatments.

```
/* DATA FOR MINNOW TIME TO INTOXICATION AT 5 CYNIDE CONCENTRATIONS AT 3 */
/* TEMPERATURES. FROM SOKAL AND RHOLF. 1981. BIOMETRY PAGES 374-387.    */
DATA CYANIDE;
     INPUT TEMP $ CN $ OXYGEN $ TIME @@;
     IF OXYGEN=9.0; /* USING ONLY THE 9.0 OXYGEN CONCENTRATION          */
```

* The original data from Wuhrmann and Woker (1953) also included dissolved oxygen concentration, but that information was ignored here for the sake of simplicity. Only results for the 9.0 oxygen treatment were used.

```
    /* TREATMENTS.                                                    */
    LTIME=LOG(TIME);
    DATALINES;
    5 0.16 1.5 201 5 0.80 1.5 150 5 4.00 1.5 131 5 20.0 1.5 130 5 100
    1.5 97 5 0.16 3.0 246
    5 0.80 3.0 164 5 4.00 3.0 138 5 20.0 3.0 136 5 100 3.0 102 5 0.16
    9.0 271 5 0.80 9.0 170
    5 4.00 9.0 149 5 20.0 9.0 127 5 100 9.0 99 15 0.16 1.5 124 15 0.80
    1.5 104 15 4.00 1.5 86
    15 20.0 1.5 89 15 100 1.5 60 15 0.16 3.0 158 15 0.80 3.0 111 15 4.00
    3.0 99 15 20.0 3.0 91
    15 100 3.0 74 15 0.16 9.0 207 15 0.80 9.0 117 15 4.00 9.0 81 15 20.0
    9.0 87 15 100 9.0 72
    25 0.16 1.5 79 25 0.80 1.5 63 25 4.00 1.5 50 25 20.0 1.5 51 25 100
    1.5 32 25 0.16 3.0 129
    25 0.80 3.0 54 25 4.00 3.0 51 25 20.0 3.0 52 25 100 3.0 46 25 0.16
    9.0 142 25 0.80 9.0 93
    25 4.00 9.0 62 25 20.0 9.0 51 25 100 9.0 52
;
PROC GLM;
  CLASS CN TEMP;
  MODEL LTIME=CN TEMP;
RUN;
PROC GLMPOWER DATA=CYANIDE;
  CLASS CN TEMP;
  MODEL LTIME=CN|TEMP;
  POWER
  STDDEV=0.0935
  NTOTAL= 30 45 60 75 /* EXPLORING 2, 3, 4 OR 5 REPLICATES PER      */
  /* COMBINATION IN THIS FULL                                         */
  POWER=.; /* FACTORIAL DESIGN.                                       */
RUN;
```

The output for this portion of the code indicates that the p value for the F statistic associated with either cyanide concentration or temperature is less than 0.0001, and the root MSE was 0.0935. The goal was to estimate the number of replicates that would be needed to obtain a specified power. The decision is to use 0.05 for both α and β. Hypothetical samples sizes of two, three, four, and five replicates per treatment combination were judged to be workable. The error standard deviation from the PROC GLM is used in the PROC POWERGLM procedure to produce the power estimates shown below. Duplicates would be adequate to test for the main effects, but the power associated with the interaction term would not be high enough without four or more replicates. If understanding interactions is important in the definitive study, another design such as a fixed-ratio mixture ray design might be explored (Casey et al. 2004; Coffey et al. 2005).

```
                         The GLM Procedure
                     Dependent Variable: LTIME

                                  Sum of
   Source              DF        Squares   Mean Square   F Value   Pr > F
   Model                6     3.33883543    0.55647257     63.64   <.0001
   Error                8     0.06994836    0.00874355
   Corrected Total     14     3.40878379
```

```
R-Square    Coeff Var    Root MSE    LTIME Mean
0.979480    2.006250     0.093507    4.660781

Source   DF   Type I SS    Mean Square   F Value   Pr > F
CN        4   1.98117044   0.49529261     56.65    <.0001
TEMP      2   1.35766499   0.67883249     77.64    <.0001

                The GLMPOWER Procedure
               Fixed Scenario Elements

          Dependent Variable            LTIME
          Error Standard Deviation     0.0935
          Alpha                          0.05

                   Computed Power

                            N    Test   Error
Index   Source          Total      DF      DF   Power
    1   CN                 30       4      15   >.999
    2   CN                 45       4      30   >.999
    3   CN                 60       4      45   >.999
    4   CN                 75       4      60   >.999
    5   TEMP               30       2      15   >.999
    6   TEMP               45       2      30   >.999
    7   TEMP               60       2      45   >.999
    8   TEMP               75       2      60   >.999
    9   CN*TEMP            30       8      15   0.595
   10   CN*TEMP            45       8      30   0.894
   11   CN*TEMP            60       8      45   0.979
   12   CN*TEMP            75       8      60   0.997
```

A27.7 WILCOXON–MANN–WHITNEY NONPARAMETRIC TEST

The nonparametric Wilcoxon test was mentioned in terms of ranked data. Here an example of power estimation is shown using ordinal data.

```
/* WILCOXON RANK TEST WITH THE SPECIFICATION OF ORDINAL DATA.              */
/* IF THE DATA ARE CONTINUOUS, YOU CAN MAKE THEM DISCRETE AND THE          */
/* PROBABILITY ASSOCIATED WITH EACH DISCRETE CATEGORY MADE EQUAL, I.E.,    */
/* 1/NUMBER OF DISCRETE CATEGORIES. THE PROCEDURE CAN BE USED FOR DATA     */
/* ANALYZED BY THE WILCOXON OPTION OF PROC NPAR1WAY. HERE WE HAVE 4        */
/* CATEGORIES WITH 1 BEING "BEST" AND 4 BEING "WORST." IN TREATMENT        */
/* REF, 0.50 0.25, 0.20, AND 0.05 OF PILOT STUDY RESPONSES WERE IN         */
/* CATEGORIES 1, 2, 3, AND 4. 0.20, 0.20, 0.40 AND 0.20 OF PILOT STUDY    */
/* EXPOSED TREATMENT WERE 1,2, 3, AND 4. A ONE-SIDED TEST IS USED AS       */
/* THE EXPSOURE EFFECT IS JUDGED TO GO ONLY ONE WAY (U =UPPER. L WOULD     */
/* HAVE BEEN SELECTED TO GET A LOWER. SIDES=2 WOULD GET A TWO-SIDED        */
/* TEST.                                                                   */
  PROC POWER; /* WILCOXON-MANN-WHITNEY TEST FOR TWO DISTRIBUTIONS          */
  TWOSAMPLEWILCOXON TEST=WMW   /* SERIES OF RANKED/ORDINAL DATA SETS.      */
```

```
  VARDIST("REFERNC") = ORDINAL ((1 2 3 4) : (.5 .25 .20 .05))
  VARDIST("EXPOSED")= ORDINAL ((1 2 3 4) : (.20 .25 .35 .20))
VARIABLES = "REFERNC" | "EXPOSED"
  NPERGROUP=5 TO 50 BY 5
  ALPHA=0.05
  SIDES=U
  POWER=.;
  PLOT;
RUN;
```

The following output is produced in addition to a plot (Figure A27.8). Approximately 40 observations per group would be required to achieve a power of 0.95 with the specified α of 0.05. A one-sided test was done in which *U* (upper) reflects the alternate hypothesis being greater than the null hypothesis.

```
                    The POWER Procedure
                Wilcoxon-Mann-Whitney Test
                 Fixed Scenario Elements

Method                           O'Brien-Castelloe approximation
Number of Sides                                                U
Alpha                                                       0.05
Group 1 Variable                                         REFERNC
Group 2 Variable                                         EXPOSED
Pooled Number of Bins                                          4
NBins Per Group                                             1000

                      Computed Power

                          N Per
                  Index   Group   Power
                      1       5   0.339
                      2      10   0.520
```

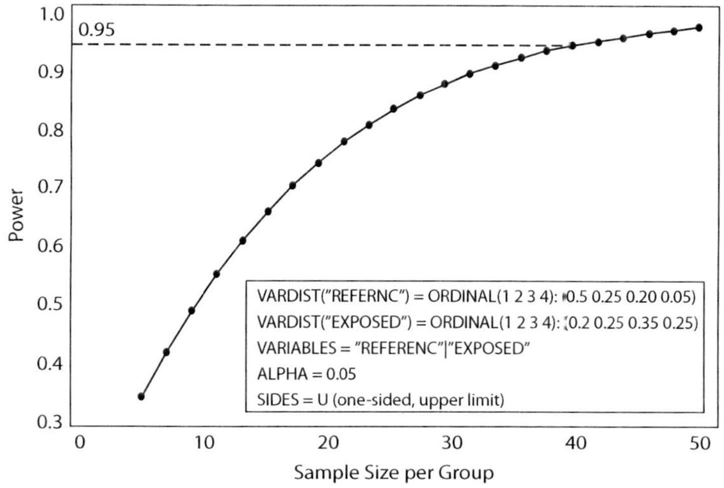

Figure A27.8 SAS PROC POWER results for a one-sided Wilcoxon–Mann–Whitney nonparametric test involving ordinal data.

Index	N Per Group	Power
3	15	0.659
4	20	0.761
5	25	0.835
6	30	0.888
7	35	0.924
8	40	0.950
9	45	0.967
10	50	0.978

A27.8 χ^2 AND FISHER'S EXACT TESTS

Cohen (1988) defines an effect size index, *w*, for the relative frequencies of a conventional table of a χ^2 test.

$$w = \sqrt{\sum_{i=1}^{m} \frac{(P_{1i} - P_{01})^2}{P_{0i}}} \tag{A27.28}$$

where m = the number of table cells, P_{oi} = the expected proportion in the ith cell assuming H_0, and P_{1i} = the proportion in the ith cell assuming the alternate hypothesis, H_1. Dattalo (2008) presents w in a slightly different manner, because in practice, one uses the sample proportions to do these computations. (The χ^2 is estimated to be equivalent to $w^2 \cdot N$ (Cohen 1988).) The following equations are relevant for 2 × 2 (Equation A27.29) and 2 × 3 or larger (Equation A27.30) tables.

$$w = \phi = \sqrt{\frac{\chi^2}{N}} \tag{A27.29}$$

$$w = \phi = \sqrt{\frac{\chi^2}{N(k-1)}} \tag{A27.30}$$

where N = the total number of observations and k = the smaller of the number of rows or number of columns in the table. The noncentrality parameter used in power and sample size calculations is defined with Equation (A27.29).

$$\delta = \left(\frac{2\phi}{\sqrt{1-\phi^2}} \right) \sqrt{N} \tag{A27.31}$$

The SAS PROC POWER code below does the power calculations for the Pearson χ^2 and also Fisher's exact one-sided tests. The hypothesized difference in proportions is set at 0.05, and the required α and β are both judged *a priori* to be 0.05. Study proportions of 0.10 and 0.30 are specified based on results of a pilot study. Practically, the maximum number of observations that could be gathered for each group was determined to be 100.

```
/* ALTHOUGH N PER GROUP FOR BOTH GROUPS IS EQUAL HERE, THIS CAN BE       */
/* CHANGED BY USING THE GROUPNS INSTEAD OF NPERGROUP LINE BELOW. E.G.,   */
/* GROUPNS = 10 15 | 20 40 WILL ESTIMATE POWER FOR FOUR COMBINATIONS:    */
/* 10,20; 10,40; 15,20 AND 15,40.                                        */
/* POWER FOR PEARSON'S CHISQUARE TEST                                    */
PROC POWER;
      TWOSAMPLEFREQ TEST=PCHI
      GROUPPROPORTIONS = (0.10 0.30)
      NULLPROPORTIONDIFF = 0.05
      NPERGROUP = 10 TO 100 BY 10
      ALPHA=0.05
      SIDES=1
      POWER = .;
      PLOT;
RUN;
/* POWER FOR FISHER'S EXACT TEST */
PROC POWER;
    TWOSAMPLEFREQ TEST=FISHER
    GROUPPROPORTIONS = (0.10 0.30)
    NPERGROUP = 10 TO 100 BY 10
    ALPHA=0.05
    SIDES=1
    POWER = .;
    PLOT;
RUN;
/* GROUPPROPORTIONS INDICATES THE PROPORTIONS OBTAINED FROM A PILOT      */
/* STUDY OR FROM SOME OTHER SOURCE. THE NPERGROUP IS THE SAME FOR        */
/* THE TWO GROUPS TESTED BUT THIS COULD BE CHANGED AS INDICATED IN       */
/* THE COMMENTS ABOVE OR USING A GROUPWEIGHTS SPECIFICATION AS           */
/* DESCRIBED IN THE SAS HELP.                                            */
```

The results for both tests are pasted below. Adequate power was not realistically obtainable with the χ^2 test but was obtained with Fisher's exact test with approximately 90 observations per group.

```
                    The POWER Procedure
         Pearson Chi-square Test for Two Proportions
                  Fixed Scenario Elements

        Distribution                         Asymptotic normal
        Method                            Normal approximation
        Number of Sides                                      1
        Null Proportion Difference                        0.05
        Alpha                                             0.05
        Group 1 Proportion                                 0.1
        Group 2 Proportion                                 0.3

                       Computed Power

                            N Per
                    Index   Group   Power
                        1      10   0.202
                        2      20   0.318
                        3      30   0.421
```

Index	N Per Group	Power
4	40	0.513
5	50	0.594
6	60	0.664
7	70	0.723
8	80	0.774
9	90	0.816
10	100	0.851

```
The POWER Procedure
Fisher's Exact Conditional Test for Two Proportions
Fixed Scenario Elements
```

Distribution	Exact conditional
Method	Walters normal approximation
Number of Sides	1
Alpha	0.05
Group 1 Proportion	0.1
Group 2 Proportion	0.3

```
Computed Power
```

Index	N Per Group	Power
1	10	0.140
2	20	0.330
3	30	0.501
4	40	0.640
5	50	0.746
6	60	0.824
7	70	0.880
8	80	0.919
9	90	0.946
10	100	0.965

A27.9 CORRELATION AND REGRESSION

The Pearson product-limit correlation coefficient, r, is an effect size measure for correlation analyses involving n pairs of observations and a specified significance level for testing the null hypothesis that $r = 0$. Such a test could be one- or two-sided. Cohen's (1988) description of effect size assigns $r = 0.10$, 0.30, and 0.50 as the cutoff values for small, medium, and large effects. The difference in correlation coefficient between the hypothesis-defined populations can be applied as the effect size for sample size or power calculations. For the familiar H_0 of no correlation ($r_0 = 0$) and the hypothesized alternate population correlation (r_1), d and δ might be calculated with the following equations (Cohen 1988; Dattalo 2008).*

* Cohen (1988) lists r values and equivalent d values in his Table 2.2.1.

$$d = r_1 - r_0 = r_1 - 0 = r_1 \tag{A27.32}$$

$$\delta = d\sqrt{n-1} \tag{A27.33}$$

The small, medium, and large effect sizes given in terms of the r above correspond to the previously discussed effect sizes for mean differences of $d = 0.20$, 0.50, and 0.80 (Cohen 1988).

The following SAS PROC POWER code calculates power for $r = 0.40$ and a number of observations ranging from 10 to 100.

```
/* ONE SET OF X,Y DATA USING A T STATISTIC TO ESTIMATE      */
/* POWER FOR A RANGE OF NUMBERS OF PAIRED SETS OF POINTS.   */
PROC POWER;
      ONECORR TEST=PEARSON DIST=T
      CORR=0.40
      NTOTAL=10 TO 100 BY 10
      SIDES=1
      POWER=.;
      PLOT MIN=10 MAX=100;
RUN;
```

```
                    The POWER Procedure
               t Test for Pearson Correlation
                  Fixed Scenario Elements

     Distribution                             t transformation of r
     Method                                                   Exact
     Number of Sides                                              1
     Correlation                                                0.4
     Model                                                 Random X
     Alpha                                                     0.05
     Number of Variables Partialled Out                           0

                       Computed Power

                                 N
                    Index    Total    Power
                        1       10    0.322
                        2       20    0.561
                        3       30    0.726
                        4       40    0.834
                        5       50    0.902
                        6       60    0.943
                        7       70    0.968
                        8       80    0.982
                        9       90    0.990
                       10      100    0.994
```

Approximately 65 observations would be needed for a one-sided test to obtain both, α and β of 0.05 with an r of 0.40.

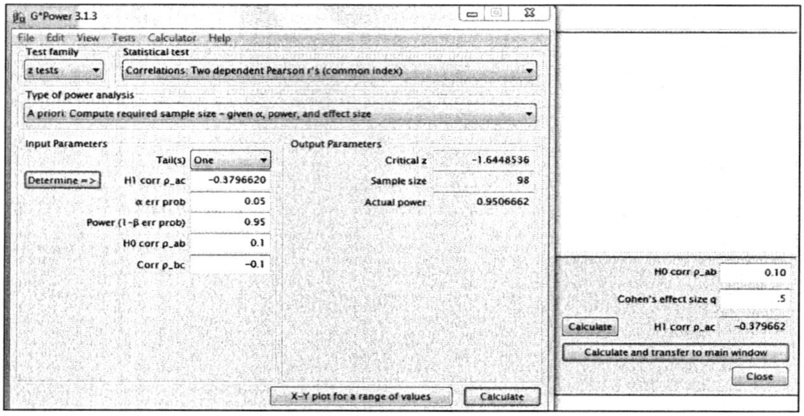

Figure A27.9 Screen capture of G*Power 3 sample size analysis for comparing two Pearson product-limit correlation coefficients.

The Fisher transformation of r to z can be applied during power calculations for tests of difference between two correlation coefficients (Cohen 1988; Sheskin 2004). A z is estimated for each of the two correlation coefficients with Equation (A27.34).

$$z = \frac{1}{2}\left(Ln \, \frac{1+r}{1-r} \right) \tag{A27.34}$$

The measure of effect size (q) is defined by Equation (A27.35) if the direction in the difference is specified. Equation (A27.36) is appropriate if the difference is nondirectional (Cohen 1988).

$$q = z_1 - z_2 \tag{A27.35}$$

$$q = \left| z_1 - z_2 \right| \tag{A27.36}$$

Figure A27.9 is a screen capture image for G*Power 3 showing an example of such a calculation.

Relative to regression analysis, the regression coefficient, r^2, is used to generate the effect size metric if the alternate hypothesis is that the regression coefficient is 0 (Cohen 1988; Dattalo 2008).

$$f^2 = \frac{r^2}{1-r^2} \tag{A27.37}$$

The noncentrality parameter is λ (Equation 9.3.3 of Cohen 1988).

$$\lambda = \frac{r^2}{1-r^2} N = f^2 N \tag{A27.38}$$

The N in Equation (A27.38) is inappropriate if there are several predictor variables in the regression. The number of predictor variables (p) and degrees of freedom ($v = N - p - 1$) must be considered (Equation 9.3.1 in Cohen 1988). Cohen (1988) provides guidance for initial design sampling in terms of small, medium, and large effect sizes in these terms also.

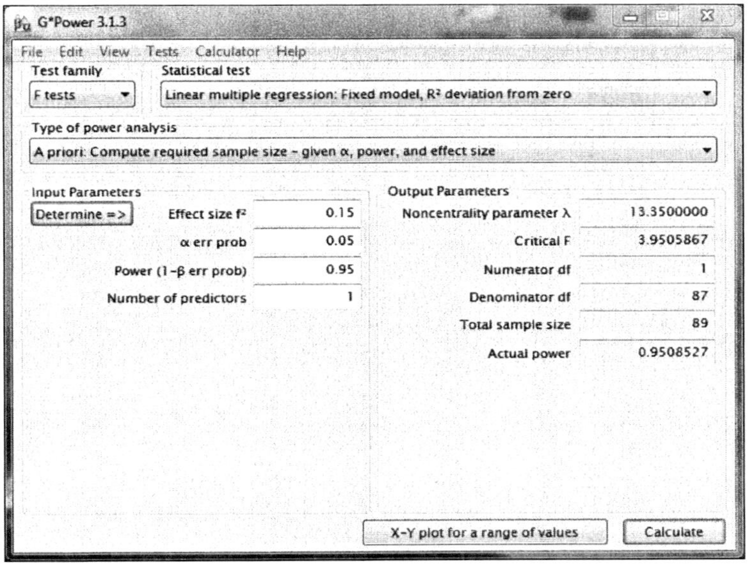

Figure A27.10 Screen capture of G*Power 3 sample size analysis for a test of regression r^2 with the H_0 of $r^2 = 0$.

Effect	r^2	f^2
Small	0.02	0.02
Medium	0.13	0.15
Large	0.26	0.35

The SAS PROC POWER (MULTREG option) or G*Power 3 can be used to do the power and sample size estimations for multiple regressions. A screen capture for the associated G*Power 3 regression option is shown in Figure A27.10 assuming a medium effect level with a model with only one predictor variable and $\alpha = \beta = 0.05$.

A27.10 COMPARING PROPORTIONS

Many epidemiological studies, such as that described in Example 6.2, focus on numbers of observations from sampled groups categorized in terms of whether or not they have a pathological condition, and also whether or not they were exposed to a suspected disease agent. In Example 6.2, fish from a polycyclic aromatic hydrocarbons (PAH)-contaminated site and a reference site were scored as having or not having liver lesions. In the more conventional context of a case-control study, the proportions of individuals exposed or not exposed who did or did not manifest the pathological condition might be estimated.

$$\pi_1 = p(\textit{Pathological Condition Present} \mid \textit{Exposed})$$

$$\pi_2 = p(\textit{Pathological Condition Present} \mid \textit{Not Exposed})$$

Woodward (2005) provides the following description of the testing of the related hypotheses and sample size estimation. The hypotheses below are often tested formally for these kinds of data.

$$H_0\colon \pi_1 = \pi_2 \quad \text{or} \quad \frac{\pi_1}{\pi_2} = 1$$

$$H_1\colon \frac{\pi_1}{\pi_2} = RR$$

where RR = relative risk. The H_1 would be expressed differently for one- and two-sided tests. The actual π_1 and π_2 are unknown to the researcher conducting a case-control study, so she must work with π_1^* and π_2^* instead, and the relevant hypotheses become those below.

$$\pi_1^* = p\left(Exposed \backslash Pathological\ Condition\ Present\right) = p(Exposed \mid Case)$$

$$\pi_2^* = p\left(Exposed \backslash Pathological\ Condition\ Absent\right) = p(Exposed \mid Control)$$

$$H_0^*\colon \pi_1^* = \pi_2^*$$

$$H_1^*\colon \frac{\pi_1^*}{\pi_2^*} = RR^*$$

Woodward uses these formulations, the probability of exposure for an individual in the population (P), and Bayes's theorem to generate the following.

$$\pi_1^* = \frac{RR \cdot P}{1 + (RR - 1)P} \tag{A27.39}$$

$$\pi_2^* \approx P \tag{A27.40}$$

This results in the following expression of RR^*:

$$RR^* \approx \frac{RR}{1 + (RR - 1)P} \tag{A27.41}$$

Woodward states that Equation (A27.40) is adequate for a rare pathological condition in the exposed population. He uses these relationships as the basis for hypothesis tests (H_0^* vs. H_1^*) and generation of required total sample size (Equation A27.42). Equation (A27.42) specifies z_α which is relevant to a one-sided test, but the equation can be used for a two-sided test if $z_{\alpha/2}$ is used instead. The r in this equation is the case:control ratio for clinical data, that is, the cases with the condition:individuals without the condition. The r would be 1 if there were equal numbers of cases and controls in the study. The RR, expressed as the odds ratio, is that which one judges to be necessary to detect if present.

$$n = \frac{(r+1)(1+(RR-1)P)^2}{rP^2(P-1)^2(RR-1)^2}\left[z_\alpha\sqrt{(r+1)P_c^*(1-P_c^*)} + z_\beta\sqrt{\frac{RR \cdot P(1-P)}{\left[1+(RR-1)P\right]^2} + rP(1-P)}\right]^2 \tag{A27.42}$$

where the P_c^* is defined by Equation (A27.43).

$$P_c^* = \frac{P}{r+1}\left[\frac{r \cdot RR}{1+(RR-1)P}+1\right] \tag{A27.43}$$

Although mentioned only briefly in Chapter 6, there might be confounding variables in these kinds of studies. In the PAH-induced liver lesion example, age might confound the ratios. Samples might be comprised of two groups of fish, juveniles (young of year) and adults (>2 years old). Rosner (2006) provides a very clear illustration of the Mantel and Haenszel (1959) approach to testing such confounded data sets. He also provides an equivalently clear illustration of the Wittes and Wallenstein (1987) approach to estimating power for hypothesis tests for such data sets. Generally the H_0 and H_A are expressed in terms of odds ratios or proportions. DuPont and Plumer (1990, 1998) developed the PS V3 (2009) shareware to do the sample size and power calculations for clinical or epidemiological studies. They include an option that does this for the Mantel and Haenszel (1959) testing approach. The α, power, and number of strata (one per confounding factor, T), and total number of cases in all strata (n) are needed for the calculations. The disease odds ratio for exposed versus control individuals (ψ) is estimated from some pilot study data or another source.

A27.11 SURVIVAL TESTING AND LOGISTIC REGRESSION

Survival analysis techniques explored in Chapter 4 included tests of significant differences among treatments or classes. For example, nonparametric survival time methods were illustrated in Example 4.9, including log-rank or Wilcoxon nonparametric tests. Several of the software packages can do the associated power calculations, including PS V3 and SAS. The following SAS PROC POWER code illustrates the general approach.

```
/* TWO GROUPS ARE DEFINED WITH A PILOT STUDY TO HAVE MEDIAN SURVIVAL    */
/* TIMES OF 19 AND 39 HOURS. A 48 HOUR SURVIVAL TEST IS TO BE DONE IN   */
/* WHICH IT WOULD BE PRACTICAL TO HAVE UP TO 70 INDIVIDUALS PER GROUP.  */
/* ALPHA AND BETA OF 0.05 FOR A ONE-SIDED TEST IS DEEMED APPROPRIATE.   */
/* ACCURALTIME OF 0 SIMPLY SPECIFIES THAT ALL INDIVIDUALS ENTERED THE   */
/* TEST AT TIME 0 WITH NO NEW INDIVIDUALS BEING ADDED DURING THE        */
/* EXPOSURE DURATION. A LOG RANK TEST IS SPECIFIED HERE BUT A GEHAN     */
/* TEST COULD ALSO BE DONE.                                             */
PROC POWER;
  TWOSAMPLESURVIVAL TEST=LOGRANK
  GROUPMEDSURVTIMES=(19 39)
  ACCRUALTIME=0
  TOTALTIME=48
  GROUPNS = 40 50 60 70 | 40 50 60 70
  SIDES=1
  ALPHA=0.05
  POWER=.;
RUN;
```

The following output is generated from this code. The conditions specified for an acceptable test would be met with 60 or more individuals in each group.

```
                    The POWER Procedure
            Log-Rank Test for Two Survival Curves
                  Fixed Scenario Elements

Method                                Lakatos normal approximation
Form of Survival Curve 1                               Exponential
Form of Survival Curve 2                               Exponential
Number of Sides                                                  1
Accrual Time                                                     0
Total Time                                                      48
Alpha                                                         0.05
Group 1 Median Survival Time                                    19
Group 2 Median Survival Time                                    39
Number of Time Sub-Intervals                                    12
Group 1 Loss Exponential Hazard                                  0
Group 2 Loss Exponential Hazard                                  0

                      Computed Power

                  Index   N1   N2   Power
                      1   40   40   0.850
                      2   40   50   0.887
                      3   40   60   0.911
                      4   40   70   0.928
                      5   50   40   0.876
                      6   50   50   0.912
                      7   50   60   0.935
                      8   50   70   0.950
                      9   60   40   0.893
                     10   60   50   0.928
                     11   60   60   0.950
                     12   60   70   0.963
                     13   70   40   0.905
                     14   70   50   0.939
                     15   70   60   0.959
                     16   70   70   0.971
```

Logistic regression was also mentioned in Chapter 4 as a worthwhile tool for dealing with effects to individuals. Sample size or power estimation for this approach also can be done using SAS (PROC POWER LOGISTIC), PS V3, and G*Power 3.

A27.12 SPECIALIZED MEASURES

The 2009 software supporting Krebs (1999) does sample size estimations for a series of special ecological applications relevant to Chapters 6 and 7. It estimates sample sizes for Poisson and negative binomial counts data. Sample size for populations size/density are provided based on mark-recapture and distance (e.g., nearest-neighbor) sampling techniques. The software also estimates required length of transects. Several other useful power or sample size estimation techniques relevant to the topic of this book are implemented by PASS 11. Methods associated with the early chapters are those for tests for normality and quality control. Later chapters discuss topics (bioequivalence, test sensitivity/specificity, ROC charts) that also are addressed in the PASS 11 program.

REFERENCES

Anderson, D.R., and K.P. Burnham. Avoiding pitfalls when using information-theoretic methods. *J. Wildl. Manag.* 66:912–918 (2002).

Burnham, K.P., and D.R. Anderson. Kullback-Leibler information as a basis for strong inference in ecological studies. *Wildl. Res.* 28:111–119 (2001).

Casey, M., C. Gennings, W.H. Carter Jr., V.C. Moser, and J.E. Simmons. Detecting interaction(s) and assessing the impact of component subsets in a chemical mixture using fixed-ratio mixture ray designs. *Agric. Biol. Environ. Stat.* 9(3):339–361 (2004).

Chow, S.-C., J. Shao, and H. Wang. *Sample size calculations in clinical research*. 2nd ed. Boca Raton, FL: Chapman & Hall/CRC, 2008.

Coffey, T., C. Gennings, J.E. Simmons, and D.W. Herr. D-optimal experimental designs to test for departure from additivity in a fixed-ratio mixture ray. *Toxicol. Sci.* 88(2):467–476 (2005).

Cohen, J. *Statistical power analysis for the behavioral sciences*. 2nd ed. Hillsdale, NJ: Lawrence Erlbaum Association, 1988.

Daskalakis, K.D. Variability of metal concentrations in oyster tissue and implications to biomonitoring. *Mar. Pollut. Bull.* 32:794–801 (1996).

Dattalo, P. *Determining sample size. Balancing power, precision, and practicality*. New York: Oxford University Press, 2008.

Dattalo, P. A review of software for sample size determination. *Eval. Health Prof.* 32:229–248 (2009).

Delignette-Muller, M.-L., C. Forfait, E. Billoir, and C. Charles. A new perspective on the Dunnett procedure: Filling the gap between NOEC/LOEC and ECx concepts. *Environ. Toxicol. Chem.* 30 (2011). DOI: 10.1002/etc.686.

Dupont, W.D., and W.D. Plummer. Power and sample size calculations: A review and computer program. *Control. Clin. Trials* 11:116–28 (1990).

Dupont, W.D., and W.D. Plummer. Power and sample size calculations for studies involving linear regression. *Control. Clin. Trials* 19:589–601 (1998).

Faul, F., and E. Erdfelder. GPOWER: A priori, post-hoc, and compromise power analyses for MS-DOS. Bonn, FRG: Bonn University, Department of Psychology, 1992.

Faul, F., E. Erdfelder, A. Buchner, and A.-G. Lang. Statistical power analyses using G*Power 3.1: Test for correlation and regression analyses. *Behav. Res. Methods* 41:1149–1160 (2009).

Faul, F., E. Erdfelder, A.-G. Lang, and A. Buchner. G*Power 3: A flexible statistical power analysis program for the social, behavioral, and biomedical sciences. *Behav. Res. Methods* 39:175–191 (2007).

Graybill, F.A. Determining sample size for a specified width confidence interval. *Ann. Math. Stat.* 29:282–287 (1958).

Holling, C.S., and C.R. Allen. Adaptive inference for distinguishing credible from incredible patterns in nature. *Ecosystems* 5:319–328 (2002).

Krebs, C.J. *Ecological methodology*. Menlo Park, CA: Addison-Wesley Educational Publishers, 1999.

Laplace, P.-S. *Essai Philosophique sur les Probabilités*. Paris: Libraire pour les Mathématiques, 1925.

Loehle, C. Hypothesis testing in ecology: Psychological aspects and the importance of theory maturation. *Q. Rev. Biol.* 31:145–164 (1987).

Mantel, N., and W. Haenszel. Statistical aspects of the analysis of data from retrospective studies of disease. *J. Natl. Cancer Inst.* 22:719–748 (1959).

Matzke, B.D., L.L. Nuffer, J.E. Hathaway, L.H. Sego, B.A. Pulsipher, S. McKenna, J.E. Wilson, S.T. Dowson, N.L. Hassig, C.J. Murray, and B. Roberts. *Visual Sample Plan Version 6.0 User's Guide*. PNNL-19915. Springfield, VA: Pacific Northwest National Laboratory, NIST, 2010.

Munkittrick, K.R., C.J. Arens, R.D. Lowell, and G.P. Kaminski. A review of potential methods of determining critical effect size for designing environmental monitoring programs. *Environ. Toxicol. Chem.* 28:1361–1371 (2009).

NCSS (J.L. Hintze). *PASS 11 power analysis and sample size user's guide*. Kaysville, UT: NCSS, 2011.

Newman, M.C. *Fundamentals of ecotoxicology*. 3rd ed. Boca Raton, FL: Taylor & Francis/CRC Press, 2010.

Richards, S.A. Testing ecological theory using the information-theoretic approach: Examples and cautionary results. *Ecology* 86:2805–2814 (2005).

Rosner, B. *Fundamentals of biostatistics*. 6th ed. Belmont, CA: Thomson Higher Education, 2006.

SAS Institute. *SAS/STAT® 9.2 user's guide*. Cary, NC: SAS Institute, 2008.

Sheskin, D.J. *Handbook of parametric and nonparametric statistical procedures*. 3rd ed. Boca Raton, FL: Chapman & Hall, 2004.

Simon, J.L. *Resampling: The new statistics*. Arlington, VA: Resampling Stats, 1995.

Sokal, R.R., and F.J. Rohlf. *Biometry. The principles and practice of statistics in biological research*. 2nd ed. New York: W.H. Freeman and Company, 1981.

U.S. Environmental Protection Agency. *A guidance on choosing a sampling design for environmental data collection*. EPA QA/G5S, EPA/240/R-02/005. Springfield, VA: 2002.

VSP Development Team. *Visual Sample Plan: A tool for design and analysis of environmental sampling*. Version 6.1b. Richland, WA: Pacific Northwest National Laboratory, 2011.

Wittes, J., and S. Wallenstein. The power of the Mantel-Haenszel test. *J. Am. Stat. Assoc.* 82:1104–1109 (1987).

Woodward, M. *Epidemiology. Study design and data analysis*. 2nd ed. Boca Raton, FL: Chapman & Hall/CRC, 2005.

Wuhrmann, K., and H. Woker. Uber die giftwirkungen von ammoniak und zyanidlosungen mit verschiedener sauerstoffspannung und temperatur auf fische. *Schweiz. Z. Hydrol.* 15:235–260 (1953).

Appendix 28: Basic Matrix Methods

This appendix is based on Chapter 15, Section 15.2.1 of Newman and Clements (2008). Discussion of eigenvectors and eigenvalues draws on the clear explanations set down in Appendix A in Caswell (2001) and also Chapter 6 in Neter et al. (1990). (Full references are provided in Chapter 6.) First used in Chapter 6 but also relevant to sections of Chapter 7. The reader should note that the MATLAB®* software (MathWorks, Natick, Massachusetts), including the Simulink® add-on, allows many of these computations to be done very quickly and accurately, including matrix-based modeling. Dixon (2012) implements matrix-based methods via MATLAB® to do many of the computations discussed in this book.

A matrix is simply a rectangular array of numbers or variables. Its size is usually designated by the number of rows (i) and columns (j). A square matrix is one with the same numbers of rows as columns, i.e., $i = j$. A vector is a matrix made up of only one row or one column, that is, either a column vector or a row vector. A 1×1 matrix is called a scalar.

By convention, matrices are designated with boldfaced, capital letters (e.g., **A** below) and vectors with boldface, small letters (e.g., **a** below). Scalars are written simply as small letters without boldfacing (e.g., a below). A matrix can be specified as $\mathbf{A} = \{a_{ij}\}$, where i is the row position and j is the column position of an element. For example, element a_{21} of **a** below is 5 and element a_{13} of **A** below is 17.

$$[12] = 1 \times 1 \text{ scalar} = a \tag{A28.1}$$

$$\begin{bmatrix} 2 \\ 5 \\ 6 \\ 3 \end{bmatrix} = 4 \times 1 \text{ matrix} = \mathbf{a} \tag{A28.2}$$

$$\begin{bmatrix} 12 & 5 & 17 & 3 \\ 23 & 0 & 12 & 10 \\ 5 & 5 & 5 & 3 \\ 12 & 7 & 13 & 5 \end{bmatrix} = 4 \times 4 \text{ matrix} = \mathbf{A} \tag{A28.3}$$

To multiply a scalar by a matrix, each element of the matrix is multiplied by the scalar. Let b be a scalar with value 12 and **A** be a 2×2 matrix.

$$12 \times \mathbf{A} = 12\begin{bmatrix} a_{11} & a_{12} \\ a_{21} & a_{22} \end{bmatrix} = \begin{bmatrix} 12a_{11} & 12a_{12} \\ 12a_{21} & 12a_{22} \end{bmatrix} \tag{A28.4}$$

Multiplication of a matrix **A** by another matrix **B** is more involved computationally and can only be done if the number of columns in **A** is equal to the number in **B**. The cross-products of the rows of **A** and columns of **B** are generated and then summed. The **A** matrix described above is multiplied here by another 2×2 matrix, **B**,

* Mention of software does not imply endorsement.

$$\mathbf{A}\times\mathbf{B}=\begin{bmatrix} a_{11} & a_{12} \\ a_{21} & a_{22} \end{bmatrix}\times\begin{bmatrix} b_{11} & b_{12} \\ b_{21} & b_{22} \end{bmatrix}=\begin{bmatrix} a_{11}b_{11}+a_{12}b_{21} & a_{11}b_{12}+a_{12}b_{22} \\ a_{21}b_{11}+a_{22}b_{21} & a_{21}b_{12}+a_{22}b_{22} \end{bmatrix} \quad \text{(A28.5)}$$

As a specific example,

$$\mathbf{A}\times\mathbf{B}=\begin{bmatrix} 1 & 3 \\ 5 & 2 \end{bmatrix}\times\begin{bmatrix} 4 & 1 \\ 5 & 6 \end{bmatrix}=\begin{bmatrix} 4+15 & 1+18 \\ 20+10 & 5+12 \end{bmatrix}=\begin{bmatrix} 19 & 19 \\ 30 & 17 \end{bmatrix}$$

It is important to note that **A** × **B** will not produce the same result as multiplying **B** × **A**. Multiplication of a matrix (**A**) and a vector (**b**) is done in the same way.

$$\mathbf{A}\times\mathbf{b}=\begin{bmatrix} a_{11} & a_{12} \\ a_{21} & a_{22} \end{bmatrix}\times\begin{bmatrix} b_{11} \\ b_{21} \end{bmatrix}=\begin{bmatrix} a_{11}b_{11}+a_{12}b_{21} \\ a_{21}b_{11}+a_{22}b_{21} \end{bmatrix} \quad \text{(A28.6)}$$

Multiplication of matrix and vector can be illustrated by modifying the specific example above.

$$\mathbf{A}\times\mathbf{b}=\begin{bmatrix} 1 & 3 \\ 5 & 2 \end{bmatrix}\times\begin{bmatrix} 4 \\ 5 \end{bmatrix}=\begin{bmatrix} 4+15 \\ 20+10 \end{bmatrix}=\begin{bmatrix} 19 \\ 30 \end{bmatrix}$$

As pointed out by Vandermeer and Goldberg (2003; full reference in Chapter 6), a quick glance back and forth between Equations (A28.5) and (A28.6) should make it clear that multiplication of the two square matrices only involves multiplying the first matrix by successive column vectors making up the second matrix. This matrix × vector multiplication will be useful because projections of the change in numbers of individuals at each age or stage will be made in Chapter 6 by multiplying a matrix of demographic information by a vector of the initial number of individuals at the different ages or stages.

Finally, multiplication (taking the cross-product) of two column vectors involves the following:

$$\mathbf{a}\times\mathbf{b}=\begin{bmatrix} a_{11} \\ a_{21} \\ a_{31} \end{bmatrix}\times\begin{bmatrix} b_{11} \\ b_{21} \\ b_{31} \end{bmatrix}=\begin{bmatrix} a_{21}b_{31}-a_{31}b_{21} \\ a_{31}b_{11}-a_{11}b_{31} \\ a_{11}b_{21}-a_{21}b_{11} \end{bmatrix} \quad \text{(A28.7)}$$

Addition and subtraction of matrices of the same dimensions can also be done.

$$\mathbf{A}+\mathbf{B}=\begin{bmatrix} a_{11} & a_{12} & a_{13} \\ a_{21} & a_{22} & a_{23} \\ a_{31} & a_{32} & a_{33} \end{bmatrix}+\begin{bmatrix} b_{11} & b_{12} & b_{13} \\ b_{21} & b_{22} & b_{23} \\ b_{31} & b_{32} & b_{33} \end{bmatrix} \quad \text{(A28.8)}$$

$$=\begin{bmatrix} a_{11}+b_{11} & a_{12}+b_{12} & a_{13}+b_{13} \\ a_{21}+b_{21} & a_{22}+b_{22} & a_{23}+b_{23} \\ a_{31}+b_{31} & a_{32}+b_{32} & a_{33}+b_{33} \end{bmatrix}$$

$$\mathbf{A}-\mathbf{B}=\begin{bmatrix} a_{11} & a_{12} & a_{13} \\ a_{21} & a_{22} & a_{23} \\ a_{31} & a_{32} & a_{33} \end{bmatrix}-\begin{bmatrix} b_{11} & b_{12} & b_{13} \\ b_{21} & b_{22} & b_{23} \\ b_{31} & b_{32} & b_{33} \end{bmatrix} \tag{A28.9}$$

$$=\begin{bmatrix} a_{11}-b_{11} & a_{12}-b_{12} & a_{13}-b_{13} \\ a_{21}-b_{21} & a_{22}-b_{22} & a_{23}-b_{23} \\ a_{31}-b_{31} & a_{32}-b_{32} & a_{33}-b_{33} \end{bmatrix}$$

Before moving on to other important matrix concepts, some additional definitions need to be provided. A diagonal matrix is one whose elements other than those on the diagonal are zero.

$$\begin{bmatrix} a_{11} & 0 & 0 \\ 0 & a_{22} & 0 \\ 0 & 0 & a_{33} \end{bmatrix}$$

The identity matrix (**I**) is a diagonal matrix with 1 as the value for all its nonzero elements.

$$\begin{bmatrix} 1 & 0 & 0 \\ 0 & 1 & 0 \\ 0 & 0 & 1 \end{bmatrix}$$

A scalar matrix is a diagonal matrix with a scalar (ρ) as the elements on the diagonal.

$$\mathbf{B}=\begin{bmatrix} \rho & 0 & 0 \\ 0 & \rho & 0 \\ 0 & 0 & \rho \end{bmatrix}$$

A matrix multiplied by **I** will be equal to itself. The result of multiplying a matrix **A** by the scalar matrix **B** (see Equation A28.4) would be analogous to multiplying **A** by the scalar ρ.

The inverse of a number x in mathematics is $1/x$, and the number multiplied by its inverse is 1. Although not all matrices have inverses, a square matrix (**A**) can have an inverse ($\mathbf{A}^{-1}$) that is unique for it. The $\mathbf{A}^{-1}$ multiplied by **A** will produce an identity matrix, **I**. The inverse for a diagonal matrix might be the following:

$$\mathbf{A}^{-1}=\begin{bmatrix} a_{11} & 0 & 0 \\ 0 & a_{22} & 0 \\ 0 & 0 & a_{33} \end{bmatrix}\begin{bmatrix} 1/a_{11} & 0 & 0 \\ 0 & 1/a_{22} & 0 \\ 0 & 0 & 1/a_{33} \end{bmatrix}=\begin{bmatrix} 1 & 0 & 0 \\ 0 & 1 & 0 \\ 0 & 0 & 1 \end{bmatrix}$$

Generating inverses for other matrices is more difficult and often involves software to be convenient.

Transposes of matrices will be involved in the modeling of age- or stage-structured populations. To transpose a matrix (**A**), rows of the original matrix are simply made into the columns of the transpose ($\mathbf{A}^{T}$), that is, $\{a_{ij}\}$ becomes $\{a_{ji}\}$,

$$\mathbf{A}^T = \begin{bmatrix} 1 & 3 \\ 5 & 2 \end{bmatrix}^T = \begin{bmatrix} 1 & 5 \\ 3 & 2 \end{bmatrix} \qquad \text{(A28.10)}$$

Obviously, the transpose of any diagonal matrix, including identity and scalar matrices, is equal to itself, e.g., $\mathbf{I} = \mathbf{I}^T$. This is true for any matrix symmetrical on both sides of its diagonal.

In Equation (A28.6), we noted that a matrix multiplied by a column vector results in a column vector such as the following:

$$\begin{bmatrix} 2 \\ 5 \\ 6 \\ 3 \end{bmatrix}$$

Multiplication of a matrix transpose by a column vector produces a row vector.

$$\begin{bmatrix} 2 & 5 & 6 & 3 \end{bmatrix}$$

Also, multiplication of a matrix transpose by a row vector produces a column vector. Other features of transposes include the following (a = a scalar): $(\mathbf{A}^T)^T = \mathbf{A}$, $(a\mathbf{A})^T = a\mathbf{A}^T$, $(\mathbf{A} + \mathbf{B})^T = \mathbf{A}^T + \mathbf{B}^T$, and $(\mathbf{AB})^T = \mathbf{B}^T\mathbf{A}^T$.

The next matrix qualities to understand are eigenvectors (= latent vectors) and eigenvalues (= latent roots). From Wikipedia, we can read that an eigenvector in the broadest sense is a (non-zero) vector that, when multiplied by a matrix, remains proportional to the original (square) matrix. Further, the eigenvalue (scalar λ) of an eigenvector (**e**) is the proportion by which the eigenvector changes when multiplied by the matrix (**A**).

$$\mathbf{Ae} = \lambda\mathbf{e} \qquad \text{(A28.11)}$$

So, the eigenvalue λ modifies the eigenvector **e**. For example, $0 \leftarrow \lambda \rightarrow 1$ reflects expanding and contracting of **e** during multiplication of **e** and **A.** A $\lambda < 0$ will flip the eigenvector **e**.

In Equation (A28.11), the λ is called the right eigenvector. A left eigenvector will be produced by multiplying **A** by the row vector, $\mathbf{e}^T$, instead of the column vector, **e**.

These abstract mathematical notions likely seem quite removed from population modeling; however, they are critical to generating some central demographic features in Chapter 6. The reader interested in more details is directed to Appendix M in Nisbet and Gurney (1982; full reference in Chapter 6) and Appendix A in Caswell (2001).

REFERENCES

Dixon, K.R. *Modeling and simulation in ecotoxicology with applications in MATLAB and Simulink.* Boca Raton, FL: Taylor & Francis/CRC Press, 2012.

Appendix 29: Values of θ Used for Maximum Likelihood Estimation of Mean and Standard Deviation of Truncated Data

γ	0.000	0.001	0.002	0.003	0.004	0.005	0.006	0.007	0.008	0.009
0.05	0.00000	0.00000	0.00000	0.00001	0.00001	0.00001	0.00001	0.00001	0.00002	0.00002
0.06	0.00002	0.00003	0.00003	0.00003	0.00004	0.00004	0.00005	0.00006	0.00007	0.00007
0.07	0.00008	0.00009	0.00010	0.00011	0.00013	0.00014	0.00016	0.00017	0.00019	0.00020
0.08	0.00022	0.00024	0.00026	0.00028	0.00031	0.00033	0.00036	0.00039	0.00042	0.00045
0.09	0.00048	0.00051	0.00055	0.00059	0.00063	0.00067	0.00071	0.00075	0.00080	0.00085
0.10	0.00090	0.00095	0.00101	0.00106	0.00112	0.00118	0.00125	0.00131	0.00138	0.00145
0.11	0.00153	0.00160	0.00168	0.00176	0.00184	0.00193	0.00202	0.00211	0.00220	0.00230
0.12	0.00240	0.00250	0.00261	0.00272	0.00283	0.00294	0.00305	0.00317	0.00330	0.00342
0.13	0.00355	0.00369	0.00382	0.00396	0.00410	0.00425	0.00440	0.00455	0.00470	0.00486
0.14	0.00503	0.00519	0.00536	0.00553	0.00571	0.00589	0.00608	0.00627	0.00646	0.00665
0.15	0.00685	0.00705	0.00726	0.00747	0.00769	0.00791	0.00813	0.00835	0.00858	0.00882
0.16	0.00906	0.00930	0.00955	0.00980	0.01006	0.01032	0.01058	0.01085	0.01112	0.01140
0.17	0.01168	0.01197	0.01226	0.01256	0.01286	0.01316	0.01347	0.01378	0.01410	0.01443
0.18	0.01476	0.01509	0.01543	0.01577	0.01611	0.01646	0.01682	0.01718	0.01755	0.01792
0.19	0.01830	0.01868	0.01907	0.01946	0.01986	0.02026	0.02067	0.02108	0.02150	0.02193
0.20	0.02236	0.02279	0.02323	0.02368	0.02413	0.02458	0.02504	0.02551	0.02599	0.02647
0.21	0.02695	0.02744	0.02794	0.02844	0.02895	0.02946	0.02998	0.03050	0.03103	0.03157
0.22	0.03211	0.03266	0.03322	0.03378	0.03435	0.03492	0.03550	0.03609	0.03668	0.03728
0.23	0.03788	0.03849	0.03911	0.03973	0.04036	0.04100	0.04165	0.04230	0.04296	0.04362
0.24	0.04429	0.04497	0.04565	0.04634	0.04704	0.04774	0.04845	0.04917	0.04989	0.05062
0.25	0.05136	0.05211	0.05286	0.05362	0.05439	0.05516	0.05594	0.05673	0.05753	0.05834
0.26	0.05915	0.05997	0.06080	0.06163	0.06247	0.06332	0.06418	0.06504	0.06591	0.06679
0.27	0.06768	0.06858	0.06948	0.07039	0.07131	0.07224	0.07317	0.07412	0.07507	0.07603
0.28	0.07700	0.07797	0.07896	0.07995	0.08095	0.08196	0.08298	0.08401	0.08504	0.08609
0.29	0.08714	0.08820	0.08927	0.09035	0.09144	0.09254	0.09364	0.09476	0.09588	0.09701
0.30	0.09815	0.09930	0.10046	0.10163	0.10281	0.10400	0.10520	0.10641	0.10762	0.10885
0.31	0.1101	0.1113	0.1126	0.1138	0.1151	0.1164	0.1177	0.1190	0.1203	0.1216
0.32	0.1230	0.1243	0.1257	0.1270	0.1284	0.1298	0.1312	0.1326	0.1340	0.1355
0.33	0.1369	0.1383	0.1398	0.1413	0.1428	0.1443	0.1458	0.1473	0.1448	0.1503
0.34	0.1519	0.1534	0.1550	0.1566	0.1582	0.1598	0.1614	0.1630	0.1647	0.1663
0.35	0.1680	0.1697	0.1714	0.1731	0.1748	0.1765	0.1782	0.1800	0.1817	0.1835
0.36	0.1853	0.1871	0.1889	0.1907	0.1926	0.1944	0.1963	0.1982	0.2001	0.2020
0.37	0.2039	0.2058	0.2077	0.2097	0.2117	0.2136	0.2156	0.2176	0.2197	0.2217
0.38	0.2238	0.2258	0.2279	0.2300	0.2321	0.2342	0.2364	0.2385	0.2407	0.2429
0.39	0.2451	0.2473	0.2495	0.2517	0.2540	0.2562	0.2585	0.2608	0.2631	0.2655
0.40	0.2678	0.2702	0.2726	0.2750	0.2774	0.2798	0.2822	0.2847	0.2871	0.2896
0.41	0.2921	0.2947	0.2972	0.2998	0.3023	0.3049	0.3075	0.3102	0.3128	0.3155
0.42	0.3181	0.3208	0.3235	0.3263	0.3290	0.3318	0.3346	0.3374	0.3402	0.3430
0.43	0.3459	0.3487	0.3516	0.3545	0.3575	0.3604	0.3634	0.3664	0.3694	0.3724
0.44	0.3755	0.3785	0.3816	0.3847	0.3878	0.3910	0.3941	0.3973	0.4005	0.4038
0.45	0.4070	0.4103	0.4136	0.4169	0.4202	0.4236	0.4269	0.4303	0.4338	0.4372
0.46	0.4407	0.4442	0.4477	0.4512	0.4547	0.4583	0.4619	0.4655	0.4692	0.4728
0.47	0.4765	0.4802	0.4840	0.4877	0.4915	0.4953	0.4992	0.5030	0.5069	0.5108
0.48	0.5148	0.5187	0.5227	0.5267	0.5307	0.5348	0.5389	0.5430	0.5471	0.5513
0.49	0.5555	0.5597	0.5639	0.5682	0.5725	0.5768	0.5812	0.5856	0.5900	0.5944
0.50	0.5989	0.6034	0.6079	0.6124	0.6170	0.6216	0.6263	0.6309	0.6356	0.6404

γ	**0.000**	**0.001**	**0.002**	**0.003**	**0.004**	**0.005**	**0.006**	**0.007**	**0.008**	**0.009**
0.51	0.6451	0.6499	0.6547	0.6596	0.6645	0.6694	0.6743	0.6793	0.6843	0.6893
0.52	0.6944	0.6995	0.7046	0.7098	0.7150	0.7202	0.7255	0.7308	0.7361	0.7415
0.53	0.7469	0.7524	0.7578	0.7633	0.7689	0.7745	0.7801	0.7857	0.7914	0.7972
0.54	0.8029	0.8087	0.8146	0.8204	0.8263	0.8323	0.8383	0.8443	0.8504	0.8565
0.55	0.8627	0.8689	0.8751	0.8813	0.8876	0.8940	0.9004	0.9068	0.9133	0.9198
0.56	0.9264	0.9330	0.9396	0.9463	0.9530	0.9598	0.9666	0.9735	0.9804	0.9874
0.57	0.9944	1.001	1.009	1.016	1.023	1.030	1.037	1.045	1.052	1.060
0.58	1.067	1.075	1.082	1.090	1.097	1.105	1.113	1.121	1.129	1.137
0.59	1.145	1.153	1.161	1.169	1.177	1.185	1.194	1.202	1.211	1.219
0.60	1.228	1.236	1.245	1.254	1.262	1.271	1.280	1.289	1.298	1.307
0.61	1.316	1.326	1.335	1.344	1.353	1.363	1.373	1.382	1.392	1.402
0.62	1.411	1.421	1.431	1.441	1.451	1.461	1.472	1.482	1.492	1.503
0.63	1.513	1.524	1.534	1.545	1.556	1.567	1.578	1.589	1.600	1.611
0.64	1.622	1.634	1.645	1.657	1.668	1.680	1.692	1.704	1.716	1.728
0.65	1.740	1.752	1.764	1.777	1.789	1.802	1.814	1.827	1.840	1.853
0.66	1.866	1.879	1.892	1.905	1.919	1.932	1.946	1.960	1.974	1.988
0.67	2.002	2.016	2.030	2.044	2.059	2.073	2.088	2.103	2.118	2.133
0.68	2.148	2.163	2.179	2.194	2.210	2.225	2.241	2.257	2.273	2.290
0.69	2.306	2.322	2.339	2.356	2.373	2.390	2.407	2.424	2.441	2.459
0.70	2.477	2.495	2.512	2.531	2.549	2.567	2.586	2.605	2.623	2.643
0.71	2.662	2.681	2.701	2.720	2.740	2.760	2.780	2.800	2.821	2.842
0.72	2.863	2.884	2.905	2.926	2.948	2.969	2.991	3.013	3.036	3.058
0.73	3.081	3.104	3.127	3.150	3.173	3.197	3.221	3.245	3.270	3.294
0.74	3.319	3.344	3.369	3.394	3.420	3.446	3.472	3.498	3.525	3.552
0.75	3.579	3.606	3.634	3.662	3.690	3.718	3.747	3.776	3.805	3.834
0.76	3.864	3.894	3.924	3.955	3.986	4.017	4.048	4.080	4.112	4.144
0.77	4.177	4.210	4.243	4.277	4.311	4.345	4.380	4.415	4.450	4.486
0.78	4.52	4.56	4.60	4.63	4.67	4.71	4.75	4.79	4.82	4.86
0.79	4.90	4.94	4.99	5.03	5.07	5.11	5.15	5.20	5.24	5.28
0.80	5.33	5.37	5.42	5.46	5.51	5.56	5.61	5.65	5.70	5.75
0.81	5.80	5.85	5.90	5.95	6.01	6.06	6.11	6.17	6.22	6.28
0.82	6.33	6.39	6.45	6.50	6.56	6.62	6.68	6.74	6.81	6.87
0.83	6.93	7.00	7.06	7.13	7.19	7.26	7.33	7.40	7.47	7.54
0.84	7.61	7.68	7.76	7.83	7.91	7.98	8.06	8.14	8.22	8.30
0.85	8.39	8.47	8.55	8.64	8.73	8.82	8.91	9.00	9.09	9.18

Source: Reprinted from Cohen, A.C., Jr., *Technometrics,* 3, 535–541, 1961. With permission from *Technometrics*.

Note: Used first in Chapter 7.

Index

C

D

E

F

I

J

K

L

M

N

Q

R

S

T

U

V

W

X

Y

Z